Understanding Drug Release and Absorption Mechanisms

A PHYSICAL and MATHEMATICAL APPROACH

Understanding Drug Release and Absorption Mechanisms

A PHYSICAL and MATHEMATICAL APPROACH

Mario Grassi • Gabriele Grassi
Romano Lapasin • Italo Colombo

CRC Press
Taylor & Francis Group
Boca Raton London New York

CRC Press is an imprint of the
Taylor & Francis Group, an informa business

CRC Press
Taylor & Francis Group
6000 Broken Sound Parkway NW, Suite 300
Boca Raton, FL 33487-2742

International Standard Book Number-10: 0-8493-3087-4 (Hardcover)
International Standard Book Number-13: 978-0-8493-3087-2 (Hardcover)

Library of Congress Cataloging-in-Publication Data

Understanding drug release and absorption mechanisms : a physical and mathematical approach / Mario Grassi ... [et al.].
    p. ; cm.
"A CRC title"
Includes bibliographical references and index.
ISBN-13: 978-0-8493-3087-2 (Hardcover : alk. paper)
ISBN-10: 0-8493-3087-4 (Hardcover : alk. paper)
    1. Pharmacokinetics--Mathematical models. 2. Drugs--Solubility--Mathematical models. 3. Drugs--Absorption and adsorption--Mathematical models. 4. Drugs--Controlled release--Mathematical models. I. Grassi, Mario, 1954-
    [DNLM: 1. Pharmacokinetics. 2. Models, Theoretical. 3. Pharmaceutical Preparations--metabolism. QV 38 U55 2007]

RM301.5.U53 2007
615'.7--dc22
                                         2006019649

Visit the Taylor & Francis Web site at
http://www.taylorandfrancis.com

and the CRC Press Web site at
http://www.crcpress.com

Dedication

Mario Grassi
dedicates this book to his parents
Elvira and Giancarlo, his wife Francesca, and
his sons Lucia and Giacomo

Gabriele Grassi
dedicates this book to his
parents Elvira and Giancarlo who taught him
how to live, to Chiara, his shining star of the morning,
and to Roberta, the constant support of his life

Preface

The aim of this book is to illustrate and discuss drug release and absorption mechanisms with the help of mathematical modeling approach. Nevertheless, in order to avoid mathematical formalism from hiding the physics underlying release/absorption mechanisms, the chapters, in which mathematical modeling is discussed, are organized in such a way that the initial part is devoted to explain and to comment on the physical frame and only afterward, mathematical model building is treated. Finally, a comparison between experimental tests and mathematical models is done. Due to this organization, the book can be read at different levels and a reader can stop at the physical description, skip the mathematical treatment, and pass directly onto the model/experiment comparison. This is the reason this book could be useful for both graduate and PhD students belonging to the pharmaceutical and engineering (in particular, chemical and biochemical engineering) fields. Nevertheless, we hope that this book is also useful to expert researchers who may find information and new ideas for new mathematical modeling approaches. Indeed, the continuous demand for designing more reliable delivery systems requires the use of the most updated theoretical tools. The old trial-and-error approach is accepted neither by academy nor by industry any more, whereas this new approach reduces product development time (and thus costs) and, what is of paramount importance, it minimizes the risk of unsuccessful investments. In addition, this strategy adapts well with industrial demand of simple and reliable tools aimed at improving product characteristics and at exploring new market regions. Although, traditionally, the adjective "simple" translates into the simplification of the theoretical analysis, this book tries to simplify the use of complex theoretical tools to avoid unrealistic simplification of experimental data analysis. Practically, this is realized by providing and illustrating proper user-friendly software.

The delivery systems studied are monolithic matrices (mainly polymeric matrices) and matrices in the form of an ensemble of polydisperse spheres (Chapter 7), microemulsions (emulsion) (Chapter 8), polydisperse spherical microcapsules, and systems using membranes (Chapter 9). Moreover, oil–water partitioning and drug dissolution (Chapter 5), mass transport principles and thermodynamics (Chapter 4), rheology (Chapter 3), and physiology (Chapter 2) constitute the necessary base that supports Chapter 7 through Chapter 9. Drug absorption is discussed in Chapter 2 (anatomy, physiology, mechanisms, both in humans and animals) and Chapter 9 (mathematical modeling) for what concerns gastrointestinal and skin absorption. Indeed, oral and transdermal (topical) administration routes represent two of the most common and widely used delivery strategies. Chapter 6 focuses attention on the intriguing topic of drug solubility dependence on drug crystal size from both an experimental and a theoretical viewpoint. Finally, Chapter 1 is an introductory section dealing with the concept of mathematical modeling and how this powerful tool can be used (data fitting, prediction, and model parameter determination). In the Appendix, the reader can find some comments and explanations about the provided software represented by Microsoft Excel files or executable files.

Mario Grassi would like to thank his students Alberta Dal Col, Alessia Valori, and Giuseppe Fasano for a careful reading of the book and his wife Francesca Peirano for her great help in writing book references. Gabriele Grassi would also like to thank Dr. Alessia Stocca for her excellent technical support. Finally, the authors would like to thank the Fondazione Cassa di Risparmio of Trieste, Fondazione per il Sostegno delle Strutture Cardiovascolari, Mirano, VE, Italy, Fondazione Casali of Trieste, Fondo Trieste 2006, and G. Grassi would like to acknowledge the program Rientro cervelli art. 1 DM n.13, MIUR (Ministero dell'Istruzione, dell'Università e della Ricerca Scientifica) for allowing the author's presence in Italy, which greatly simplified book writing.

Mario Grassi
Unversity of Trieste

Gabriele Grassi
University Hospital of Trieste

Romano Lapasin
University of Trieste

Italo Colombo
Eurand SpA

Authors

Mario Grassi was born in Genova, Italy on July 18, 1964 and lives in Trieste, Italy. He received his high school degree (scientific field) from the Liceo Scientifico Galileo Galilei in Trieste in 1983. He obtained his graduate degree in chemical engineering from the University of Trieste in 1991 defending a thesis entitled "Studio della diffusione in idrogeli di biopolimeri per la progettazione di sistemi farmaceutici a rilascio controllato" (Study on diffusion in biopolymeric hydrogels for the designing of pharmaceutical controlled release systems).

In 2005, he became an associate professor at the Faculty of Engineering at the University of Trieste. Now he is the professor of diffusive processes in the complex media and macromolecular materials engineering. The first course is focused on mass transport in complex systems (such as polymeric gel, matrices, fractal media, and living tissues) with particular attention to release processes from pharmaceutical controlled release systems.

Prof. Grassi has been guest editor of *Current Pharmaceutical Biotechnology* for the special issue on "Nucleic acid-based drugs as novel therapeutics in the treatment of human disease" in 2004. He has authored and coauthored more than 130 papers and congress communications.

At present, his research interests are concentrated in the field of mass transport and are mainly concerned with the study and modeling of the phenomena involved in release processes from pharmaceutical systems, and the study and modeling of the phenomena involved in drug absorption by living tissues.

Gabriele Grassi was born in Genova, Italy on July 18, 1964 and lives in Trieste, Italy. In 1990, he obtained a degree in medicine from the University of Trieste by defending a thesis entitled "Morphological and morphogenetical features of fetal polycystic kidney between the fourth and fifth month of development: morphogenetical evaluations." He passed the entrance exam at the International School for Advanced Studies (ISAS), Trieste in October 1990 and got the degree of Magister Philosophiae by defending a thesis entitled "Molecular genetics of chronic granulomatous disease" in November 1992.

On January 27, 1995, he earned his PhD from the ISAS defending a thesis entitled "Molecular characterization of the genetic defect and functional reconstitution of the NADPH oxidase activity in B-lymphoblasts from patients with X-linked chronic granulomatous disease."

Since 2003, Dr. Grassi has been consultant professor in the Department of Internal Medicine, University Hospital of Trieste, Italy. There, he continues his studies on the restenosis treatment with particular emphasis on the selection of optimal delivery systems for nucleic acid-based drugs (ribozymes, DNA enzymes, siRNAs). Additionally, he has started research aimed at the prevention of hepatocellular carcinoma cell growth by using ribozymes, DNA enzymes, and siRNAs. His working group, includes a PhD student in molecular medicine, a postdoctoral fellow (assegnista di ricerca), two students (Faculty of Biotechnology) in the laboratory of Trieste, and a PhD student (molecular biology) in the Department of Molecular Pathology, University of Tuebingen, Germany, directed by Prof. Dr. R. Kandolf.

Dr. Grassi is the author and coauthor of more than 60 papers and congress communications and his current research fields include the identification of active siRNA, able to prevent cell proliferation in tumor and nontumor pathological conditions together with the setup of novel approaches to deliver siRNA/ribozyme to cultured cells and living tissues.

Romano Lapasin was born in Aviano, Italy and received his degree in chemical engineering in 1971 from the University of Trieste. He is currently full professor of rheology in the Faculty of Engineering at the University of Trieste and lecturer of transport phenomena in the same faculty. He serves as president of the course in chemical engineering, and head of the Rheology and Polymer Laboratory of the Department of Chemical, Environmental and Raw Materials Engineering (DICAMP).

He was president of Rheotech and the Italian Society of Rheology, and served as an international delegate at the Italian Society of Rheology. He is member of the following: British Society of Rheology, The Groupe Française de Rheology, The Society of Rheology (USA), GRICU (Gruppo Ricercatori di Ingegneria Chimica dell'Università), and is also an honorary member of AITIVA. He is the author of more than 300 papers and congress communications, and of the book *Rheology of Industrial Polysaccharides: Theory and Applications* published by Chapman & Hall.

At present, his research interests mainly concern (a) rheology of polymeric and colloidal particle gels, and other structured fluids, such as concentrated emulsions and dispersions in polymeric matrices, as well as its application to the analysis and solution of industrial problems, and (b) study and modeling of the phenomena involved in release processes from pharmaceutical systems.

Italo Colombo was born in Casirate d'Adda, Italy on March 17, 1953 and he lives in Treviglio, Italy. He obtained a BSc in physical sciences from the State University of Milan, Italy in 1981.

In 1983, he secured a postgraduate certificate on a technologically important aspect of interface science from the Imperial College, London. He got a postgraduate certificate on progress in microemulsions at the

International School of Quantum Electronics, Centre for Scientific Culture E. Majorana in Erice, Italy in 1986.

He is the author or coauthor of more than 100 papers and congress communications.

Several patents in the fields of drug delivery and drug–polymer nanocomposites have been authored and coauthored by him. He has more than 25 years of experience in solid state physics and in physical chemistry of surfaces.

Since 2004, he is contract professor at the Department of Pharmaceutical Chemistry, the University of Pavia, and teaches physical chemistry of surfaces at the postgraduate master course on the pharmaceutical preformulation and development.

From 1998 to the present, he has been working with Eurand SpA (a drug delivery company with production and R&D sites in Italy, France, and USA), site of Pessano con Bornago, Milan, as manager of the Physical Pharmacy Department. His research interests are focused in the field of nanomaterials and are mainly concerned with.

- study of innovative technologies in order to improve the dissolution process of the poorly water-soluble drugs;
- study and development of advanced analytical tools for the physico-chemical characterization of nanocomposite systems for drug delivery systems;
- study of the influence of particles dimension on the melting and solubility characteristics of nanostructured drugs.

Table of Contents

Chapter 9

1 Mathematical Modeling

1.1 INTRODUCTION

According to dictionary definition, the word model means a miniature representation of something; a pattern of something to be made; an example for imitation or emulation; a description or analogy used to help visualize something (a molecule, for instance) that cannot be directly observed; a system of postulates, data, and inferences presented as a mathematical description of an entity or state of affairs [1]. Consequently, modeling can be viewed as a cognitive activity aimed at the description of how devices or objects behave. Obviously, modeling can be performed according to different strategies such as sketches, physical models, drawings, computer programs, or mathematical formulas. For its generality and potentiality, however, mathematics often represents the selection strategy for model building. In this light, the above-mentioned definition of a model can be particularized and the mathematical model can be defined as a mathematical metaphor of some aspects of reality (objects and devices behavior) [2]. This means that if A represents a determined phenomenological manifestation of reality, model building (modeling) means to find out, inside the mathematical world, one or more structures able to formally represent and interpret A allowing the simulation of internal A interactions and A interactions with the external environment [3]. Accordingly, the mathematical model represents the law linking the independent variable X (usually time) to the dependent one Y (drug concentration in the blood following oral administration, for example). Obviously, once the law is fixed, Y can assume different values in relation to the same X value by modifying model parameters. Consequently, model parameters, quantities that, by definition, neither depend on X nor on Y, allow the modification of the numerical relation between the independent and dependent variables, the law connecting Y to X remaining the same. Therefore, if, for example, a linear relation exists between Y and X ($Y = mX + q$), the modification of slope m or intercept q alters the numerical Y value in relation to the same X value, being the linear law still valid.

Basically, the mathematical model can be of two different kinds: empirical or theoretical. An empirical model is nothing more than a mathematical equation able to describe the experimental trend of a quantity of interest (e.g., drug concentration in the blood vs. time) by adopting proper values for its parameters. As this model is not a real mathematical metaphor of all the phenomena concurring to the experimental evidence, its parameters have no physical meaning. For this reason, an empirical model does not imply a theoretical knowledge improvement of the physical phenomenon under study, but it allows a more objective comparison between the different experimental data sets on the basis of model parameter variations. In this light, the mathematical model approach can improve the usual statistical analysis by making the detection of similarities or differences between different experimental data sets easier. On the contrary, a theoretical model represents a possible mathematical schematization of all the phenomena concurring to the experimental evidence to be studied. Accordingly, as its parameters do possess a physical meaning (e.g., drug diffusion coefficient in a membrane), by properly setting them, the model can be used to predict the experimental trend corresponding to the different conditions. Thus, the theoretical model is a tool superior to the empirical one. However, due to the complexity of reality, sometimes it is impossible to have a theoretical model and the constitution of mixed theoretical–empirical models is necessary. Basically, these models rely on a clear theoretical background but adopt empirical expressions for the description of some or all the submechanisms concurring to the behavior of the entire phenomenon. If mixed models also fail, empirical models represent the unique chance.

Although the reasons for undertaking the mathematical model approach implicitly emerge from the preceeding discussion on empirical and theoretical models, it is worth stressing the practical consequences of this approach. Indeed, in an industrial world that is more and more interested in increasing earnings, the possibility of cutting expenses and experimental trials is very attractive. In this context, mathematical modeling can be seen as a powerful tool allowing the reduction of the number of experiments and properly addressing them. In other words, in the place of many time-consuming experiments, it allows the selection of a small number of them to confirm or reject the correctness of the mechanisms believed on the basis of the experimental evidence [4]. In addition, for many practical applications such as skyscraper and bridges building, lack of a mathematical model approach would result in many bankruptcies making, *de facto*, engineers work just a pure bet. Moreover, a mathematical model represents an unavoidable tool when the object of interest cannot be directly seen and only its effects can be experimentally detected.

If it is well known that mathematical modeling is largely employed in engineering, physics, and chemistry, less known is its use in pharmacy, biology, medicine, and psychology. This probably depends on both cultural and practical reasons. Indeed, the human necessity of knowledge categorizing favored, in the past, the formation of impermeable barriers among knowledge fields that hindered the diffusion of transversal tools such as mathematical modeling. Accordingly, mathematical modeling remained in its historical nest, i.e., physical sciences field. In addition, the typical complexity of many medical, biological, and psychological problems did not surely encourage mathematical modelers to cross the above-mentioned barriers. Nowadays, fortunately, the diffusion of mathematical modeling in medical, biological, and psychological sciences is well established and a huge variety of examples can be cited. For example, Freguglia [3] edited a book on mathematical models in biological sciences dealing with the theoretical aspects of evolutionary biology, mathematical interpretation of Darwin's evolutionary theory, bacterial motion, and heartbeat. In addition, there exist many societies, such as the European Society for Mathematical and Theoretical Biology [5], the Society of Mathematical Biology [6], and the Japan Association for Mathematical Biology [7], that actively operate in this relatively new field promoting the interdisciplinary collaboration between mathematicians and bioscientists. Accordingly, mathematical modeling finds application, for example, in cardiovascular system, neuroscience, ecology, environment, evolution, immunology, infectious diseases, tumor growth, and therapy [8]. These aspects are nothing more than a modern translation of the Leonardo da Vinci idea (1651) according to which "... *niuna umana investigazione si può dimandare vera scienzia, s'essa non passa per le matematiche dimostrazioni* ..." (no human research can be defined as true science if it cannot be mathematically demonstrated). The aim of this chapter is to provide some information about model building and model fitting by means of some examples.

1.2 MODEL BUILDING

Despite of its basis on a clear scientific background, model building is an art more than a science as a rule for the realization of the optimal model in relation to a specific problem does not exist. Moreover, sometimes, it can also be difficult to individuate the best model among different models devoted to describe a particular phenomenon. However, fortunately, mathematical modelers can take advantage from some criteria of general validity. Indeed, all models share the same building procedure consisting in (1) delineating the phenomenon studied, (2) expressing this in mathematical terms, (3) fitting experimental data, and, finally, (4) predicting experimental behavior in conditions different from those considered for data fitting [9]. The first step (see Figure 1.1) consists in a physico-chemical-biological analysis of the whole phenomenon to enucleate the various mechanisms concurring to the determination of the experimental behavior. It is in this stage that many of the hypotheses on which the model will rely are formulated. Indeed, not only are the modelers called to individuate the truly important mechanisms among the great variety that could be reasonable, but they also need to establish the relations occurring among them. For example, focusing the attention on a stone sliding on an iced inclined plane, the fundamental mechanism ruling the entire phenomenon (stone movement) is represented by the action of the gravity field component along the plane maximum slope

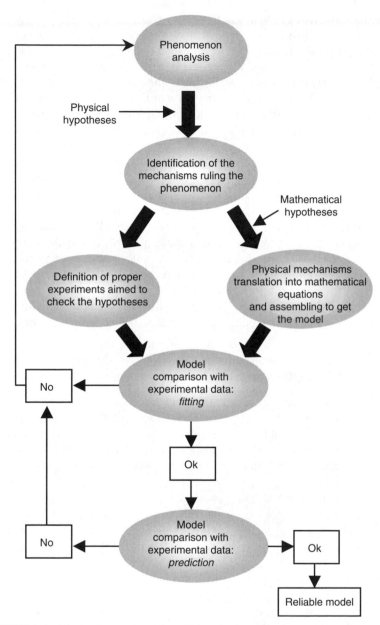

FIGURE 1.1 Schematic representation of the process needed to get a reliable mathematical model.

direction. Indeed, other concurring mechanisms such as the resistance exerted by air or the friction at the plane–stone surface can be neglected. If, on the contrary, our stone were sliding on a wrinkled plane, the effect of friction cannot be neglected and both friction and gravity field mechanisms should be simultaneously considered. In this particular case, the relation between the two mechanisms consists in being simultaneous and, thus, reciprocally influencing (friction effect depends on stone velocity modulus). Once the whole phenomenon has been decomposed in its fundamental mechanisms, the problem of translating them into mathematical terms and to assemble them altogether arises (Figure 1.1). This stage usually comports to formulate further hypotheses as mathematical translation of a mechanism, generally, is not unique. Indeed, it depends on how adherent to mechanism physics we would like to be and the mathematical complexity we can accept. Coming back to the example of the sliding stone on a wrinkled inclined plane, the simplest approximation about the friction effect is to assume that it is velocity independent. However, of course, a more precise description would require assuming that friction depends on velocity (e.g., the square of velocity modulus). The final form of the mathematical model results from the gravity field and friction equations assembling into a general balance equation that, in this case, is represented by Newton's law stating that the variation of stone acceleration depends on the sum of all the forces acting on its mass. Now, the third step (data fitting, see Figure 1.1) implies getting model solution, namely the way of converting model mathematical form into experimental variables such as stone position, velocity, and acceleration vs. time. This step may not be trivial especially when dealing with complex phenomena where sophisticated solution tools are required. In addition, the fitting procedure also *per se* requires the choice of the most appropriate mathematical technique. Indeed, basically, as it will be discussed in more detail later in this chapter, model fitting on experimental data means to select, according to precise criteria, the values of model parameter (also called model fitting parameters) allowing the best description of the experimental trend. If the best model description (or best fitting) of experimental data is satisfactory from both a statistical and a physical point of view (model parameters must assume physically reasonable values), the model can be accepted and considered for the last verification consisting of the prediction of experimental behavior. On the basis of fitting parameter values deduced from data fitting, the model should be able to predict experimental evolution in other situations. In the case of the sliding stone, for example, one fitting parameter could be the proportionality constant (β) connecting the frictional force (f_τ) to the squared velocity (v^2) ($f_\tau = \beta * v^2$). Once β is determined from data fitting (here experimental data are represented by stone position, velocity, and acceleration vs. time), the model should correctly predict stone position, velocity, and acceleration at time t, considering that stone initial position, velocity, and acceleration are different from those set in the data used for model fitting.

Model building first step (phenomenon analysis and decomposition in its fundamental mechanisms) also reveals of paramount importance for what concerns experiments designing (Figure 1.1). Indeed, only with a clear physical model of the real phenomenon in mind, it is possible to correctly decide the experiments necessary and the ways to perform them. It is well known that lacking a reliable set of experimental data, no reliable mathematical model can be built. Obviously, the reliable character of experimental data does not only refer to the technical accuracy with which they are collected, but also to their pertinence with hypotheses on which the model relies. For example, if we built the model on the assumption that the inclined plane surface is extremely smooth (friction is negligible), it would be absolutely meaningless to fit it on experimental data referring to a wrinkled surface or to a surface whose roughness is not known. Accordingly, it should turn out that this necessary intimate link between the mathematical model and experimental data can be ensured only if both are correctly designed. In particular, the model must accomplish some rules that can be condensed into the concepts of dimensional homogeneity and consistency (DOC), scaling (S), and balance or conservation principles (BCP) [1]. DOC simply states that all model equations must be dimensionally homogeneous or consistent, this meaning that all the addends constituting an equation must have the same dimensions (of energy, force, and so on). For this purpose, the Buckingham PI theorem [10] affirms that "*a dimensionally homogeneous equation involving* n *variables in* m *primary or fundamental dimensions can be reduced to a single relationship among* n−m *independent products*". This statement affirms that if our phenomenon is characterized by n variables and m of these are defined as primary or fundamental variables (as they contain all the dimensions involved in the definition of the n variables), we need n−m dimensionless groups to correlate, in a unique relation, all the n variables. For example, coming back to the sliding stone on a smooth surface, we can retain that stone initial position (x_0), velocity (v_0), acceleration (a_0), and plane slope (α) are the n (= 4) variables of interest. Assuming x_0 and v_0 as fundamental variables ($m = 2$), the following two (= $n - m$) dimensionless groups can be formed:

$$\pi_1 = (x_0)^{a_1}(v_0)^{b_1}a_0 \equiv (L)^{a_1}\left(\frac{L}{T}\right)^{b_1}\left(\frac{L}{T^2}\right), \qquad (1.1)$$

$$\pi_2 = (x_0)^{a_2}(v_0)^{b_2}\alpha \equiv (L)^{a_2}\left(\frac{L}{T}\right)^{b_2}, \qquad (1.2)$$

where, dimensionally, L is the length and T is the time. In order for π_1 and π_2 to be dimensionless, the net exponents of L and T must vanish. Accordingly, it follows

$$L: a_1 + b_1 + 1 = 0 \quad \text{from } \pi_1, \qquad (1.3a)$$

$$T: -b_1 - 2 = 0 \quad \text{from } \pi_1, \qquad (1.3b)$$

$$L: a_2 + b_2 = 0 \quad \text{from} \quad \pi_2, \tag{1.4a}$$

$$T: b_2 = 0 \quad \text{from} \quad \pi_2. \tag{1.4b}$$

Consequently,

$$a_1 = 1; \ b_1 = -2; \quad a_2 = 0; \ b_2 = 0; \tag{1.5}$$

$$\pi_1 = \frac{x_0 a_0}{(v_0)^2}, \tag{1.1a}$$

$$\pi_2 = \alpha. \tag{1.2a}$$

Thus, these two dimensionless groups should guide experiments for what concerns the simple stone sliding on the inclined smooth plane.

The concept of scaling (S) is strictly connected with the first phase of model building, that is, phenomenon decomposition into its submechanisms. Indeed, it is at this stage that we decide at which scale the model is supposed to work. This means, for example, that if we desire to model planets' trajectory in the solar system, each planet can be approximated by a point mass characterized by its own mass. In this situation, we can absolutely neglect the fact that each planet undergoes many other movements such as the rotation around its symmetry axis or the effect exerted by its satellites. If, on the contrary, the aim of our modeling goes further the simple estimation of trajectory, both the revolution around the sun and the rotation around the planet symmetry axis must be considered. Indeed, these two phenomena must be accounted for if we need to evaluate ground temperature in a precise planet surface area. In other words, ground temperature depends on the season (position of the whole planet with respect to the sun) and the hour (relative position of surface area under observation, and the apparent position of the sun in the sky, i.e., planet angular position with respect to its symmetry axis). The concept of scaling is also important in the scale-up process. Indeed, scale changes can result in unexpected problems. For example, let us focus the attention on the vortex shape that is formed when a fluid, contained in a cylindrical vessel, is posed in movement by a symmetric rotating impeller. In order to get the same vortex shape in a smaller cylindrical vessel, we are obliged to consider a less viscous fluid [11]. Accordingly, fluid properties (viscosity) must be accounted for in model building if scale changes are considered.

Finally, the use of BCP assumes a central role in model building. Indeed, in the majority of situations, we deal with the balance of something such as water mass in a vessel provided by inlet and outlet flows, the mass of a chemical element distributing among different phases at equilibrium, or the force acting on a moving body. This principle is so important that when it cannot be applied (as it sometimes happens in the financial world) the model building task can be very difficult if not impossible. Basically, the idea on which the balance approach relies is very simple and can be enunciated as follows: the variation of the physical properties $Z(t)$ vs. the independent variable t, in a fixed space region, is equal to the algebraic sum of the influx flow rate $(J_{in}(t))$, efflux flow rate $(J_{out}(t))$, the generative term $(G(t))$, and the consumption term $(C(t))$. In mathematical terms, it becomes

$$\frac{dZ(t)}{dt} = J_{in}(t) - J_{out}(t) + G(t) - C(t). \tag{1.6}$$

One of the simplest physical translations of Equation 1.6 is that occurring in a well-mixed reactor continuously fed by reagents and where products are continuously withdrawn. As conversion of reagents into products will probably not be complete, the temporal variation of reagent A mass in the reactor depends on the A amount entering the vessel, the A amount leaving it, and the A amount disappearing due to the chemical reaction (in this case the generative term is zero). Analogously, for a product B, its mass in the reactor depends on the B amount leaving the vessel and on the B amount produced due to the chemical reaction (in this case, B amount entering the reactor is zero as well as B consumption term). In the following chapters, the balance approach will be used extensively. In particular, a conspicuous part of Chapter 4 will be devoted to mass, momentum, and energy balances presentation and discussion.

1.2.1 MODEL BUILDING: AN EXAMPLE

The following chapters of this book will furnish many examples of mathematical models essentially devoted to the representation of drug release kinetics from different delivery systems and the subsequent absorption by living tissues; in this chapter, we would like to focus the attention on a particular example. Indeed, we are referring to a mathematical model of human memory. This example not only gives the opportunity of understanding in practice the above-discussed steps involved in model building, but it also allows the reader to see how topics far from the mathematical world can take advantage of this approach.

The activity on human memory modeling began in the early 1960s perhaps owing to the influence of the information-processing movement and the computer metaphor, which are heavily concerned with information storage and retrieval processes [12]. Although this research led to many powerful results, here we would like to focus the attention on the model that has been universally recognized to apply well to human memory [12,13], the Atkinson–Shiffrin model (ASM) [14–16]. According to this model, human memory can be subdivided into three different parts: the *sensory register* (SR), the *short-term store* (STS), and the *long-term store* (LTS) (see Figure 1.2). Incoming sensory information first enters the SR, where it resides for a very brief period of time, then decays, and is lost. The STS is the

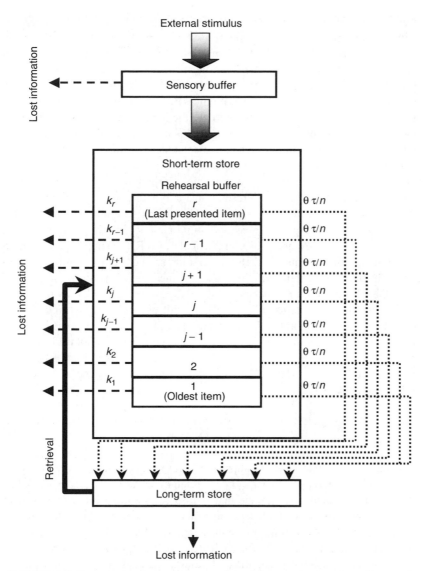

FIGURE 1.2 Human memory structure and working mechanism according to the Atkinson–Shiffrin model. (Adapted from Atkinson, R.C. and Shiffrin, R.M., in *Proceedings of the third conference on learning, remembering and forgetting*, New York Academy of Sciences, New York, 1965.)

subject's working memory as it receives selected inputs from the SR and also from the LTS. Information in the STS decays completely and is lost within a period of about 30 s, but a control process called *rehearsal* can maintain a limited amount of information in this store as long as the subject desires. The LTS is a fairly permanent repository for information transferred from the STS. It is important to underline the fact that this transfer means to copy the information in the destination store without canceling it from the supplying store. The amount of information that enters the SR is assumed to be a function of the exposure time of the stimulus (item) and that is why the presentation speed ν_p is a very important ASM parameter. Once the item is transferred from the SR to the STS, it enters the rehearsal buffer (RB) that is a particular STS structure aimed to decrease the item decaying probability from the STS adopting rehearsal strategies. In this manner, the first step of model building has been completed as the mechanisms ruling our phenomenon, human memory working, have been enucleated and defined. It is worth mentioning that this represents a possible interpretation of human memory structure but at this stage, it represents just a hypothesis to be verified. Now, we need to translate the above-mentioned memory structure into mathematical equations to complete the second step of model building. Accordingly, further hypotheses are necessary. In particular, (a) RB is supposed to contain a maximum of r items so that the first presented item occupies the rth position, the second item will be placed in the $(r-1)$th position, and so on until the RB saturation is reached (all the RB positions are filled by presented items); (b) the presentation of a new item will determine the elimination of one of the items lying in the RB according to a probabilistic process based on the probability δ that the item in the first position is dropped out (oldest item). In particular, defining k_j as the probability that the item in position j is dropped out from the RB, its mathematical evaluation can be given by the following expression:

$$k_j = \frac{\delta(1-\delta)^{j-1}}{1-(1-\delta)^r}. \tag{1.7}$$

It is easy to see that when δ approaches zero, k_j is equal to $1/r$ for all j. Conversely when δ approaches 1, $k_1 = 1$ and $k_2 = k_3 = \ldots; k_r = 0$; that is, the oldest item is always the one that is lost; (c) the probability that an item copy is transferred to the LTS is proportional to the item residence time in RS. Indeed, as the STS can send information to the LTS at a constant rate ϑ, the information sent to the LTS, for each item, will be given by the product of the item RB residence time (that, obviously, depends on k_j) and ϑ/n (the subject shares his attention among all the items present in the RB), where n represents the number of items contained in RS (after the RB saturation, n will be equal to r). Unfortunately, because of the interference effect exerted by the already presented items, the real amount of information reaching the LTS will be a fraction of the product $[(\vartheta/n) * \text{residence time}]$. The importance of this effect is accounted for by parameter τ: when τ is equal to 1 the interference effect vanishes, while it increases for τ approaching 0.

In conclusion, the subject will be able to remember one of the d presented items irrespective of whether at test time the item lies in RB or whether a successful retrieval process from the LTS takes place. The probability p_i that the ith item is correctly remembered at test time is given by the sum of the probability that the item is in the RB (p_{Ri}) and a successful retrieval from the LTS takes place (p_{Pi}) (of course, according to the ASM, when d is lower than r, the number of presented items is lower than the RS dimension, and the subject will be always able to correctly remember all the presented items). Accordingly, p_i mathematical expression reads

$$p_{Ri} = \pi_{j=1}^{r}(1 - k_{r+1-j})^{m_{ij}}, \tag{1.8}$$

$$p_{Pi} = 1 - e^{-\left(I_i^{\nabla} \tau^{d-1}\right)}, \tag{1.9}$$

$$p_i = p_{Pi} + p_{Ri}, \tag{1.10}$$

where m_{ij} and I_i^{∇} depend on $(r, \delta, \theta, \tau)$ and their mathematical expressions are omitted for the sake of simplicity. Thus, the ASM is characterized by four model parameters $(r, \delta, \theta, \tau)$. To verify the model (third model building step), Peirano [17] and Peirano et al. [18] tested it on memory performance of young subjects (19–28 years old) by means of a *free verbal recall* (FVR) *test*. This test consists in reading aloud to each subject eight lists containing 12 items (word) each, at a presentation speed v_p of one item every 2 s (0.5 items/s). At the end of each list presentation, the subject was requested to repeat all the items he could without paying attention to the items presentation order (FVR). The probability p_i of a correct remembrance is defined as the ratio between the number of remembered items in the ith position and the word's number remembered in position 12. Obviously, in the light of this definition, the word in position 12 is assigned probability $= 1$. Figure 1.3 shows good agreement between the experimental data (open circles) and the ASM best fitting (solid line). Fitting parameter values are: $r = 3$; $\vartheta = 1.75$; $\tau = 0.76$; $\delta = 0.27$. Final model reliability (fourth model building step) is ensured, for example, by the fact that the FVR curve predicted by the model assuming $r = 3$; $\vartheta = 1.75$; $\tau = 0.76$; $\delta = 0.27$ and a higher items presentation speed v_p (1 item/s) is in line with the experimental FVR curve relative to the same subjects.

1.3 MODEL FITTING

Once the phenomenon we are dealing with has been decomposed into its fundamental mechanisms and a mathematical expression accounting for all of them has been achieved (first two steps of model building), the first step of model building, model fitting to experimental data, is required. For this purpose, some definitions of statistical quantities are necessary. Accordingly, when an experimental data set shows a sufficiently strong tendency to centralize, that is, to group around a particular value, it can be useful to

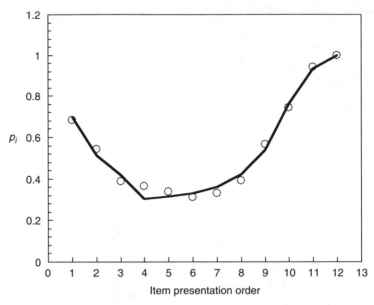

FIGURE 1.3 Comparison between ASM best fitting (solid line) and experimental data (open circles). p_i is the probability of remembering the ith presented item. (Adapted from Peirano, F., Rievocazoine libera e interpretazione della curva di posizione seriale negli anziani, Graduate Thesis, University of Padova (Italy), Department of General Psychology, 2000.)

characterize this data set by a few numbers called its *moments* [19]. In particular, we are interested in the mean $\bar{x}$ defined as

$$\bar{x} = \frac{1}{N} \sum_{i=1}^{N} x_i, \qquad (1.11)$$

where x_i represents the ith experimental datum value and N the number of data considered. A further characterization consists of measuring how these data are dispersed around the mean and the most common quantities considered at this purpose are the variance v_a or its square root, the standard deviation σ:

$$v_a = \frac{1}{N-1} \sum_{i=1}^{N} (x_i - \bar{x}),^2 \qquad (1.12)$$

$$\sigma = \sqrt{\frac{1}{N-1} \sum_{i=1}^{N} (x_i - \bar{x})^2}. \qquad (1.13)$$

Very often, it is necessary to establish whether two distributions are characterized by different means. This means that we are asked to decide whether the two means statistically differ from zero. The first step is to decide whether the two means are characterized by different variances or not. Indeed, on the basis of this result, a different strategy needs to be undertaken to decide whether the two means are statistically different. For this purpose, the F-test [19] is necessary. This test consists in evaluating the experimental F value defined as the ratio between the weighted higher v_{awh} and lower v_{awl} variances:

$$F = \frac{v_{awh}}{v_{awl}}, \quad v_{awh} = \frac{v_{ah}}{n_h - 1}, \quad v_{awl} = \frac{v_{al}}{n_l - 1}, \qquad (1.14)$$

where v_{ah} and v_{al} are, respectively, the higher and the lower variances (defined by Equation 1.12; $v_{ah} > v_{al}$) referring to two different data sets each one characterized by n_h and n_l elements, respectively. Fixing the value of probability $p = 1-\alpha$ of telling the truth (usually 0.95 or 0.99), if the F value calculated according to Equation 1.14 (with (n_h-1) degrees of freedom for numerator and (n_l-1) degrees of freedom for denominator) is bigger than that tabulated (see Table 1.1a and Table 1.1b), v_{ah} and v_{al} are statistically different. On the contrary, they are equal. If the two variances are equal, the following form of Student's t-test enables deciding whether the corresponding means ($\bar{x}_h$ and $\bar{x}_l$) are equal or not. At this purpose, the pooled variance needs to be estimated

$$s_D = \sqrt{\frac{\sum\limits_{i=1}^{n_h}(x_i - \bar{x}_h)^2 + \sum\limits_{i=1}^{n_l}(x_i - \bar{x}_l)^2}{n_h + n_l - 2}\left(\frac{1}{n_h} + \frac{1}{n_l}\right)}. \qquad (1.15)$$

Accordingly, the experimental t value reads

$$t = \frac{|\bar{x}_h - \bar{x}_l|}{s_D}, \qquad (1.16)$$

where $|\bar{x}_h - \bar{x}_l|$ indicates the absolute value of the difference $\bar{x}_h - \bar{x}_l$. Again, fixing a value of the probability $p = 1-\alpha$ of telling the truth (usually 0.95 or 0.99), if the calculated t value (with $(n_h + n_l - 2)$ degrees of freedom) is bigger than that tabulated (see Table 1.2), $\bar{x}_h$ and $\bar{x}_l$ are statistically different. On the contrary, they are equal.

If, conversely, v_{ah} and v_{al} are statistically different, the following form of Student's t-test enables deciding whether the corresponding means ($\bar{x}_h$ and $\bar{x}_l$) are equal or not:

$$t = \frac{|\bar{x}_h - \bar{x}_l|}{\sqrt{\dfrac{v_{ah}}{n_h} + \dfrac{v_{al}}{n_l}}}. \qquad (1.17)$$

TABLE 1.1a
F Distribution Values

Theoretical F Values with v_1 (Numerator) and v_2 (Denominator) Degrees of Freedom Fixing $p = 1 - \alpha = 0.95$ ($F[v_1, v_2, p]$)

$v_1 =$	1	2	3	4	5	6	7	8	9	10	12	15	20	24	30	40	60	120	∞
v_2																			
1	161.4	199.5	215.7	224.6	230.2	234.0	236.8	238.9	240.5	241.9	243.9	245.9	248.0	249.1	250.1	251.1	252.2	253.3	254.3
2	18.51	19.00	19.16	19.25	19.30	19.33	19.37	19.38	19.38	19.40	19.41	19.43	19.45	19.45	19.46	19.47	19.48	19.49	19.50
3	10.13	9.55	9.28	9.12	9.01	8.94	8.89	8.85	8.81	8.79	8.74	8.70	8.66	8.64	8.62	8.59	8.57	8.55	8.53
4	7.71	6.94	6.59	6.39	6.26	6.16	6.09	6.04	6.00	5.96	5.91	5.86	5.80	5.77	5.75	5.72	5.69	5.66	5.63
5	6.61	5.79	5.41	5.19	5.05	4.95	4.88	4.82	4.77	4.74	4.68	4.62	4.56	4.53	4.50	4.46	4.43	4.40	4.36
6	5.99	5.14	4.76	4.53	4.39	4.28	4.21	4.15	4.10	4.06	4.00	3.94	3.87	3.84	3.81	3.77	3.74	3.70	3.67
7	5.59	4.74	4.35	4.12	3.97	3.87	3.79	3.73	3.68	3.64	3.57	3.51	3.44	3.41	3.38	3.34	3.30	3.27	3.23
8	5.32	4.46	4.07	3.84	3.69	3.58	3.50	3.44	3.39	3.35	3.28	3.22	3.15	3.12	3.08	3.04	3.01	2.97	2.93
9	5.12	4.26	3.86	3.63	3.48	3.37	3.29	3.23	3.18	3.14	3.07	3.01	2.94	2.90	2.86	2.83	2.79	2.75	2.71
10	4.96	4.10	3.71	3.48	3.33	3.22	3.14	3.07	3.02	2.98	2.91	2.85	2.77	2.74	2.70	2.66	2.62	2.58	2.54
11	4.84	3.98	3.59	3.36	3.20	3.09	3.01	2.95	2.90	2.85	2.79	2.72	2.65	2.61	2.57	2.53	2.49	2.45	2.40
12	4.75	3.89	3.49	3.26	3.11	3.00	2.91	2.85	2.80	2.75	2.69	2.62	2.54	2.51	2.47	2.43	2.38	2.34	2.30
13	4.67	3.81	3.41	3.18	3.03	2.92	2.83	2.77	2.71	2.67	2.60	2.53	2.46	2.42	2.38	2.34	2.30	2.25	2.21
14	4.60	3.74	3.34	3.11	2.96	2.85	2.76	2.70	2.65	2.60	2.53	2.46	2.39	2.35	2.31	2.27	2.22	2.18	2.13

15	4.54	3.68	3.29	3.06	2.90	2.79	2.71	2.64	2.59	2.54	2.48	2.40	2.33	2.29	2.25	2.20	2.16	2.11	2.07
16	4.49	3.63	3.24	3.01	2.85	2.74	2.66	2.59	2.54	2.49	2.42	2.35	2.28	2.24	2.19	2.15	2.11	2.06	2.01
17	4.45	3.59	3.20	2.96	2.81	2.70	2.61	2.55	2.49	2.45	2.38	2.31	2.23	2.19	2.15	2.10	2.06	2.01	1.96
18	4.41	3.55	3.16	2.93	2.77	2.66	2.58	2.51	2.46	2.41	2.34	2.27	2.19	2.15	2.11	2.06	2.02	1.97	1.92
19	4.38	3.52	3.13	2.90	2.74	2.63	2.54	2.48	2.42	2.38	2.31	2.23	2.16	2.11	2.07	2.03	1.98	1.93	1.88
20	4.35	3.49	3.10	2.87	2.71	2.60	2.51	2.45	2.39	2.35	2.28	2.20	2.12	2.08	2.04	1.99	1.95	1.90	1.84
21	4.32	3.47	3.07	2.84	2.68	2.57	2.49	2.42	2.37	2.32	2.25	2.18	2.10	2.05	2.01	1.96	1.92	1.87	1.81
22	4.30	3.44	3.05	2.82	2.66	2.55	2.46	2.40	2.34	2.30	2.23	2.15	2.07	2.03	1.98	1.94	1.89	1.84	1.78
23	4.28	3.42	3.03	2.80	2.64	2.53	2.44	2.37	2.32	2.27	2.20	2.13	2.05	2.01	1.96	1.91	1.86	1.81	1.76
24	4.26	3.40	3.01	2.78	2.62	2.51	2.42	2.36	2.30	2.25	2.18	2.11	2.03	1.98	1.94	1.89	1.84	1.79	1.73
25	4.24	3.39	2.99	2.76	2.60	2.49	2.40	2.34	2.28	2.24	2.16	2.09	2.01	1.96	1.92	1.87	1.82	1.77	1.71
26	4.23	3.37	2.98	2.74	2.59	2.47	2.39	2.32	2.27	2.22	2.15	2.07	1.99	1.95	1.90	1.85	1.80	1.75	1.69
27	4.21	3.35	2.96	2.73	2.57	2.46	2.37	2.31	2.25	2.20	2.13	2.06	1.97	1.93	1.88	1.84	1.79	1.73	1.67
28	4.20	3.34	2.95	2.71	2.56	2.45	2.36	2.29	2.24	2.19	2.12	2.04	1.96	1.91	1.87	1.82	1.77	1.71	1.65
29	4.18	3.33	2.93	2.70	2.55	2.43	2.35	2.28	2.22	2.18	2.10	2.03	1.94	1.90	1.85	1.81	1.75	1.70	1.64
30	4.17	3.32	2.92	2.69	2.53	2.42	2.33	2.27	2.21	2.16	2.09	2.01	1.93	1.89	1.84	1.79	1.74	1.68	1.62
40	4.08	3.23	2.84	2.61	2.45	2.34	2.25	2.18	2.12	2.08	2.00	1.92	1.84	1.79	1.74	1.69	1.64	1.58	1.51
60	4.00	3.15	2.76	2.53	2.37	2.25	2.17	2.10	2.04	1.99	1.92	1.84	1.75	1.70	1.65	1.59	1.53	1.47	1.39
120	3.92	3.07	2.68	2.45	2.29	2.17	2.09	2.02	1.96	1.91	1.83	1.75	1.66	1.61	1.55	1.50	1.43	1.35	125.0
∞	3.84	3.00	2.60	2.37	2.21	2.10	2.01	1.94	1.88	1.83	1.75	1.67	1.57	1.52	1.46	1.39	1.32	1.22	1.00

TABLE 1.1b
F Distribution Values

Theoretical F Values with v_1 (Numerator) and v_2 (Denominator) Degrees of Freedom Fixing $p = 1 - \alpha = 0.99$ $(F[v_1,v_2,p])$

$v_1 =$	1	2	3	4	5	6	7	8	9	10	12	15	20	24	30	40	60	120	∞
v_2																			
1	4052	4999	5403	5625	5764	5859	5928	5982	6022	6056	6106	6157	6209	6235	6261	6287	6313	6339	6366
2	98.50	99.00	99.17	99.25	99.30	99.33	99.36	99.37	99.39	99.40	99.42	99.43	99.45	99.46	99.47	99.47	99.48	99.48	99.50
3	34.12	30.82	29.46	28.71	28.24	27.91	27.67	27.49	27.35	27.23	27.05	26.87	26.69	26.60	26.50	26.41	26.32	26.22	26.13
4	21.20	18.00	16.69	15.98	15.52	15.21	14.98	14.80	14.66	14.55	14.37	14.20	14.02	13.93	13.84	13.75	13.65	13.56	13.46
5	16.26	13.27	12.06	11.39	10.97	10.67	10.46	10.29	10.16	10.05	9.89	9.72	9.55	9.47	9.38	9.29	9.20	9.11	9.02
6	13.75	10.92	9.78	9.15	8.75	8.47	8.26	8.10	7.98	7.87	7.72	7.56	7.40	7.31	7.23	7.14	7.06	6.97	6.88
7	12.25	9.55	8.45	7.85	7.46	7.19	6.99	6.84	6.72	6.62	6.47	6.31	6.16	6.07	5.99	5.91	5.82	5.74	5.65
8	11.26	8.65	7.59	7.01	6.63	6.37	6.18	6.03	5.91	5.81	5.67	5.52	5.36	5.28	5.20	5.12	5.03	4.95	4.86
9	10.56	8.02	6.99	6.42	6.06	5.80	5.61	5.47	5.35	5.26	5.11	4.96	4.81	4.73	4.65	4.57	4.48	4.40	4.31
10	10.04	7.56	6.55	5.99	5.64	5.39	5.20	5.06	4.94	4.85	4.71	4.56	4.41	4.33	4.25	4.17	4.08	4.00	3.91
11	9.65	7.21	6.22	5.67	5.32	5.07	4.89	4.74	4.63	4.54	4.40	4.25	4.10	4.02	3.94	3.86	3.78	3.69	3.60
12	9.33	6.93	5.95	5.41	5.06	4.82	4.64	4.50	4.39	4.30	4.16	4.01	3.86	3.78	3.70	3.62	3.54	3.45	3.36
13	9.07	6.70	5.74	5.21	4.86	4.62	4.44	4.30	4.19	4.10	3.96	3.82	3.66	3.59	3.51	3.43	3.34	3.25	3.17
14	8.86	6.51	5.56	5.04	4.69	4.46	4.28	4.14	4.03	3.94	3.80	3.66	3.51	3.43	3.35	3.27	3.18	3.09	3.00

15	8.68	6.36	5.42	4.89	4.56	4.32	4.14	4.00	3.89	3.80	3.67	3.52	3.37	3.29	3.21	3.13	3.05	2.96	2.87
16	8.53	6.23	5.29	4.77	4.44	4.20	4.03	3.89	3.78	3.69	3.55	3.41	3.26	3.18	3.10	3.02	2.93	2.84	2.75
17	8.40	6.11	5.18	4.67	4.34	4.10	3.93	3.79	3.68	3.59	3.46	3.31	3.16	3.08	3.00	2.92	2.83	2.75	2.65
18	8.29	6.01	5.09	4.58	4.25	4.01	3.84	3.71	3.60	3.51	3.37	3.23	3.08	3.00	2.92	2.84	2.75	2.66	2.57
19	8.18	5.93	5.01	4.50	4.17	3.94	3.77	3.63	3.52	3.43	3.30	3.15	3.00	2.92	2.84	2.76	2.67	2.58	2.49
20	8.10	5.85	4.94	4.43	4.10	3.87	3.70	3.56	3.46	3.37	3.23	3.09	2.94	2.86	2.78	2.69	2.61	2.52	2.42
21	8.02	5.78	4.87	4.37	4.04	3.81	3.64	3.51	3.40	3.31	3.17	3.03	2.88	2.80	2.72	2.64	2.55	2.46	2.36
22	7.95	5.72	4.82	4.31	3.99	3.76	3.59	3.45	3.35	3.26	3.12	2.98	2.83	2.75	2.67	2.58	2.50	2.40	2.31
23	7.88	5.66	4.76	4.26	3.94	3.71	3.54	3.41	3.30	3.21	3.07	2.93	2.78	2.70	2.62	2.54	2.45	2.35	2.26
24	7.82	5.61	4.72	4.22	3.90	3.67	3.50	3.36	3.26	3.17	3.03	2.89	2.74	2.66	2.58	2.49	2.40	2.31	2.21
25	7.77	5.57	4.68	4.18	3.85	3.63	3.46	3.32	3.22	3.13	2.99	2.85	2.70	2.62	2.54	2.45	2.36	2.27	2.17
26	7.72	5.53	4.64	4.14	3.82	3.59	3.42	3.29	3.18	3.09	2.96	2.81	2.66	2.58	2.50	2.42	2.33	2.23	2.13
27	7.68	5.49	4.60	4.11	3.78	3.56	3.39	3.26	3.15	3.06	2.93	2.78	2.63	2.55	2.47	2.38	2.29	2.20	210
28	7.64	5.45	4.57	4.07	3.75	3.53	3.36	3.23	3.12	3.03	2.90	2.75	2.60	2.52	2.44	2.35	2.26	2.17	2.06
29	7.60	5.42	4.54	4.04	3.73	3.50	3.33	3.20	3.09	3.00	2.87	2.73	2.57	2.49	2.41	2.33	2.23	2.14	2.03
30	7.56	5.39	4.51	4.02	3.70	3.47	3.30	3.17	3.07	2.98	2.84	2.70	2.55	2.47	2.39	2.30	2.21	2.11	2.01
40	7.31	5.18	4.31	3.83	3.51	3.29	3.12	2.99	2.89	2.80	2.66	2.52	2.37	2.29	2.20	2.11	2.02	1.92	1.80
60	7.08	4.98	4.13	3.65	3.34	3.12	2.95	2.82	2.72	263	2.50	2.35	2.20	2.12	2.03	1.94	1.84	1.73	1.60
120	6.85	4.79	3.95	3.48	3.17	2.96	2.79	2.66	2.56	2.47	2.34	2.19	2.03	1.95	1.86	1.76	1.66	1.53	1.38
∞	6.63	4.61	3.78	3.32	3.02	2.80	2.64	2.51	2.41	2.32	2.18	2.04	1.88	1.79	1.70	1.59	1.47	1.32	1.00

TABLE 1.2
t Values Corresponding to Different Degrees of Freedom (dof)
and Probability $p = 1 - \alpha$

dof	$p = 0.999$	$p = 0.99$	$p = 0.98$	$p = 0.95$	$p = 0.90$	$p = 0.80$
1	636.619	63.657	31.821	12.706	6.314	3.078
2	31.598	9.925	6.965	4.303	2.920	1.886
3	12.924	5.841	4.541	3.182	2.353	1.638
4	8.610	4.604	3.747	2.776	2.132	1.533
5	6.869	4.032	3.365	2.571	2.015	1.476
6	5.959	3.707	3.143	2.447	1.943	1.440
7	5.408	3.499	2.998	2.365	1.895	1.415
8	5.041	3.355	2.896	2.306	1.860	1.397
9	4.781	3.250	2.821	2.262	1.833	1.383
10	4.587	3.169	2.764	2.228	1.812	1.372
11	4.437	3.106	2.718	2.201	1.796	1.363
12	4.318	3.055	2.681	2.179	1.782	1.356
13	4.221	3.012	2.650	2.160	1.771	1.350
14	3.140	2.977	2.624	2.145	1.761	1.345
15	4.073	2.947	2.602	2.131	1.753	1.341
16	4.015	2.921	2.583	2.120	1.746	1.337
17	3.965	2.898	2.567	2.110	1.740	1.333
18	3.922	2.878	2.552	2.101	1.734	1.330
19	3.883	2.861	2.539	2.093	1.729	1.328
20	3.850	2.845	2.528	2.086	1.725	1.325
21	3.819	2.831	2.518	2.080	1.721	1.323
22	3.792	2.819	2.508	2.074	1.717	1.321
23	3.767	2.807	2.500	2.069	1.714	1.319
24	3.745	2.797	2.492	2.064	1.711	1.318
25	3.725	2.787	2.485	2.060	1.708	1.316
26	3.707	2.779	2.479	2.056	1.706	1.315
27	3.690	2.771	2.473	2.052	1.703	1.314
28	3.674	2.763	2.467	2.048	1.701	1.313
29	3.659	2.756	2.462	2.045	1.699	1.311
30	3.646	2.750	2.457	2.042	1.697	1.310
40	3.551	2.704	2.423	2.021	1.684	1.303
60	3.460	2.660	2.390	2.000	1.671	1.296
120	3.373	2.617	2.358	1.980	1.658	1.289
∞	3.291	2.576	2.326	1.960	1.645	1.282

Fixing a value of the probability $p = 1-\alpha$ of telling the truth (usually 0.95 or 0.99), if the calculated t value characterised by the following degrees of freedom (dof)

$$dof = RTNI \left(\frac{\left(\frac{v_{ah}}{n_h} + \frac{v_{al}}{n_l} \right)^2}{\frac{(v_{ah}/n_h)^2}{n_h - 1} + \frac{(v_{al}/n_l)^2}{n_l - 1}} \right), \tag{1.18}$$

where *RTNI* means, "round to the nearest integer" is bigger than that tabulated (see Table 1.2), $\bar{x}_h$ and $\bar{x}_l$ are statistically different. On the contrary, they are equal.

1.3.1 Chi-Square Fitting: Straight Line

Let us suppose we have a mathematical model f, characterized by M adjustable parameters (a_j, $j = 1$ to M), to be fitted to N data points (x_i, y_i) ($i = 1$ to N) with x and y as the independent and the dependent variables, respectively. It can be seen [20] that the most probable set of model parameters is that minimizing chi-square χ^2:

$$\chi^2 \equiv \sum_{i=1}^{N} \left(\frac{y_i - f(x_i, a_1, \ldots, a_M)}{\sigma_i} \right)^2, \tag{1.19}$$

where σ_i is the ith datum standard error. Although Equation 1.19 strictly holds if measurement errors are normally distributed, it is useful also if this hypothesis does not hold.

It is now interesting to see how the fitting procedure develops according to Equation 1.19 in the simplest case where f is a straight line (linear regression):

$$f(x, a_1, \ldots, a_M) = mx + q, \tag{1.20}$$

where x is an independent variable while m and q are the two model parameters representing, respectively, straight line slope and intercept. In this case Equation 1.19 becomes

$$\chi^2(m, q) \equiv \sum_{i=1}^{N} \left(\frac{y_i - mx_i - q}{\sigma_i} \right)^2. \tag{1.21}$$

It is well known that the conditions required to render χ^2 minimum are

$$\frac{\partial \chi^2(m, q)}{\partial m} = -2 \sum_{i=1}^{N} \frac{(y_i - mx_i - q)x_i}{\sigma_i^2} = 0 \tag{1.22}$$

and

$$\frac{\partial \chi^2(m, q)}{\partial q} = -2 \sum_{i=1}^{N} \frac{y_i - mx_i - q}{\sigma_i^2} = 0. \tag{1.23}$$

The solution of these equations allows the calculation of the two unknowns m and q:

$$m = \frac{\sum_{i=1}^{N} \frac{1}{\sigma_i^2} \sum_{i=1}^{N} \frac{y_i x_i}{\sigma_i^2} - \sum_{i=1}^{N} \frac{x_i}{\sigma_i^2} \sum_{i=1}^{N} \frac{y_i}{\sigma_i^2}}{\sum_{i=1}^{N} \frac{1}{\sigma_i^2} \sum_{i=1}^{N} \frac{x_i^2}{\sigma_i^2} - \left(\sum_{i=1}^{N} \frac{x_i}{\sigma_i^2}\right)^2}, \quad q = \frac{\sum_{i=1}^{N} \frac{x_i^2}{\sigma_i^2} \sum_{i=1}^{N} \frac{y_i}{\sigma_i^2} - \sum_{i=1}^{N} \frac{x_i}{\sigma_i^2} \sum_{i=1}^{N} \frac{x_i y_i}{\sigma_i^2}}{\sum_{i=1}^{N} \frac{1}{\sigma_i^2} \sum_{i=1}^{N} \frac{x_i^2}{\sigma_i^2} - \left(\sum_{i=1}^{N} \frac{x_i}{\sigma_i^2}\right)^2}.$$

(1.24)

Remembering the error propagation law,

$$\sigma_g^2 = \sigma_x^2 \left(\left(\frac{\partial g(x, y, z, \ldots)}{\partial x}\right)\Big|_{y, z, \ldots}\right)^2 + \sigma_y^2 \left(\left(\frac{\partial g(x, y, z, \ldots)}{\partial y}\right)\Big|_{x, z, \ldots}\right)^2$$

$$+ \sigma_z^2 \left(\left(\frac{\partial g(x, y, z, \ldots)}{\partial z}\right)\Big|_{x, y, \ldots}\right)^2 + \cdots,$$

(1.25)

where g is a generic function of the independent variables x, y, z, and so on, and σ_x^2, σ_y^2, σ_z^2, and so on are the respective variances, it is possible to estimate the m and q variances σ_m^2 and σ_q^2 as follows:

$$\sigma_m^2 = \sum_{i=1}^{N} \sigma_i^2 \left(\frac{\partial m(x_i, y_i)}{\partial y_i}\Big|_{x_i}\right)^2 = \frac{\sum_{i=1}^{N} \frac{1}{\sigma_i^2}}{\sum_{i=1}^{N} \frac{1}{\sigma_i^2} \sum_{i=1}^{N} \frac{x_i^2}{\sigma_i^2} - \left(\sum_{i=1}^{N} \frac{x_i}{\sigma_i^2}\right)^2}, \quad (1.26)$$

$$\sigma_q^2 = \sum_{i=1}^{N} \sigma_i^2 \left(\frac{\partial q(x_i, y_i)}{\partial y_i}\Big|_{x_i}\right)^2 = \frac{\sum_{i=1}^{N} \frac{x_i}{\sigma_i^2}}{\sum_{i=1}^{N} \frac{1}{\sigma_i^2} \sum_{i=1}^{N} \frac{x_i^2}{\sigma_i^2} - \left(\sum_{i=1}^{N} \frac{x_i}{\sigma_i^2}\right)^2}. \quad (1.27)$$

It is worth mentioning that as we assume that x_i are error-free variables ($\sigma_{x_i}^2 = 0 \; \forall i$); in Equation 1.26 and Equation 1.27 only the partial derivative with respect to y_i appears. As the evaluation of the probable uncertainty associated to $m(\delta_m)$ and $q(\delta_q)$ (fitting parameters) is not an easy topic and it would imply further discussions [20] that are beyond the aim of this chapter, we limit the discussion to show an approximate solution adopted by a common software used for data fitting [21]. In particular, it is proposed:

$$\delta_m = \sqrt{\sum_{i=1}^{N} \left(\frac{y_i - mx_i - q}{\sigma_i}\right)^2 \Big/ (N - M) * \sqrt{\sigma_m^2}} \quad (1.28)$$

and

$$\delta_q = \sqrt{\sum_{i=1}^{N} \left(\frac{y_i - mx_i - q}{\sigma_i}\right)^2 \Big/ (N - M) * \sqrt{\sigma_q^2}}, \qquad (1.29)$$

where N and M are, respectively, the number of experimental points and model fitting parameters (for our straight line $M = 2$).

In order to be sure that parameters m and q are really meaningful, the goodness of the fit must be evaluated. For this purpose, we can recur to the F-test [21,22] to evaluate whether two particular variances, the mean square regression (MSR) and the mean square error (MSE), differ or not:

$$MSR = \frac{\sum_{i=1}^{N} ((y_i - \bar{y})/\sigma_i)^2 - \sum_{i=1}^{N} ((y_i - y_i^p)/\sigma_i)^2}{M - 1}, \quad \bar{y} = \frac{\sum_{i=1}^{N} (y_i/\sigma_i)^2}{\sum_{i=1}^{N} (1/\sigma_i)^2}, \qquad (1.30)$$

$$MSE = \frac{\sum_{i=1}^{N} ((y_i - y_i^p)/\sigma_i)^2}{N - M}, \qquad (1.31)$$

where $y_i^p = mx_i + q$ and the degrees of freedom associated to MSR are $\nu_1 = M-1$ while those associated to MSE are $\nu_2 = N-M$. Accordingly the calculated F value is

$$F = \frac{MSR}{MSE}. \qquad (1.32)$$

It is evident that in the presence of good data fitting, as MSE tends to zero but not MSR, F is very high. If the F value calculated according to Equation 1.32 (with ν_1 degrees of freedom for numerator and ν_2 degrees of freedom for denominator) is bigger than that tabulated corresponding to a fixed probability $p = 1-\alpha$ of being right (see Table 1.1), MSR is statistically bigger than MSE. Accordingly, data fitting is statistically acceptable.

1.3.2 Chi-Square Fitting: General Case

To generalize the previous discussion for the generic function $f(x, a_1, \ldots, a_M)$, a_j referring to its adjustable parameters, and x the independent variable, it is necessary to define the so-called design matrix A composed by N rows and M columns whose elements are [20]

$$A_{ij} = \frac{\partial f(x_i)}{\partial a_j} \frac{1}{\sigma_i}, \quad A = \begin{pmatrix} \dfrac{\partial f(x_i)}{\partial a_1} \dfrac{1}{\sigma_1} & \cdots & \dfrac{\partial f(x_1)}{\partial a_M} \dfrac{1}{\sigma_1} \\ \cdots & \cdots & \cdots \\ \dfrac{\partial f(x_N)}{\partial a_1} \dfrac{1}{\sigma_N} & \cdots & \dfrac{\partial f(x_N)}{\partial a_M} \dfrac{1}{\sigma_M} \end{pmatrix}, \quad (1.33)$$

where x_i and σ_i are, respectively, the mean value and the associated standard error corresponding to the ith experimental datum. Now, the condition of minimum χ^2

$$\chi^2(a_1, \ldots, a_M) \equiv \sum_{i=1}^{N} \left(\frac{y_i - f(x_i, a_1, \ldots, a_M)}{\sigma_i} \right)^2 \quad (1.21a)$$

requires that all the partial χ^2 derivatives with respect to the M parameters are equal to zero:

$$\frac{\partial \chi^2}{\partial a_j} = -2 \sum_{i=1}^{N} \frac{(y_i - f(x_i, a_1, \ldots, a_M))}{\sigma_i^2} \frac{\partial f(x_i, a_1, \ldots, a_M)}{\partial a_j} = 0. \quad (1.34)$$

Defining the matrix α as follows:

$$\alpha_{jk} = \sum_{i=1}^{N} \frac{1}{\sigma_i^2} \left(\frac{\partial f(x_i, a_1, \ldots, a_M)}{\partial a_j} \frac{\partial f(x_i, a_1, \ldots, a_M)}{\partial a_k} \right). \quad (1.35)$$

It is easy to demonstrate that $\alpha = A^T * A$, where A^T is the transpose of the design matrix descending from A by simply changing rows with columns. Finally, defining the matrix C as the inverse of α ($C = \alpha^{-1}$; this means $\alpha * C = I$, identity matrix), it is possible, in the same spirit leading to Equation 1.28 and Equation 1.29, to estimate the probable uncertainty (δ_{a_j}) associated to the fitting parameters a_j [21]:

$$\delta_{a_j} = \sqrt{\sum_{i=1}^{N} \left(\frac{y_i - f(x_i, a_1, \ldots a_M)}{\sigma_i} \right)^2 \Big/ (N - M)} * \sqrt{C_{jj}}, \quad (1.36)$$

where obviously, $\sigma_{a_j}^2 = C_{jj}$.

Also in this general case, the goodness of the fit can be determined according to the F-test proposed for the linear case (Equation 1.32).

1.3.3 ROBUST FITTING

Although it is not so usual, sometimes it can happen that both experimental coordinates x and y are affected by errors. This means that the generic experimental point (x_i, y_i) is associated to two standard errors σ_{xi} and σ_{yi}.

For the sake of clarity and simplicity, this theme will be matched in the simple case of straight line data fitting as its extension to the more general case [fitting function $f(x, a_1,..., a_M)$] implies additional difficulties that could deviate the reader from the main concept of robust fitting. Although different strategies can be adopted [20], a possible definition of the χ^2 function in the case of robust fitting is

$$\chi^2(m, q) \equiv \sum_{i=1}^{N} \frac{d_i^2}{\sigma_{x_i}^2 + \sigma_{y_i}^2}, \tag{1.37}$$

where d_i and $\sigma_i^2 = \sigma_{x_i}^2 + \sigma_{y_i}^2$ are, respectively, the distance between the experimental point (x_i, y_i) and the straight line $y = mx + q$ and the experimental point variance. Remembering that d_i is nothing more than the distance between (x_i, y_i) and the intersection of the straight line passing from (x_i, y_i) and perpendicular to $y = mx + q$ (accordingly its slope is $-1/m$), its expression reads

$$d_i^2 = \frac{(y_i - mx_i - q)^2}{m^2 + 1}. \tag{1.38}$$

The conditions required to minimize χ^2 are, obviously, those implying its partial derivatives zeroing

$$\frac{\partial \chi^2(m, q)}{\partial m} = -\frac{2}{(m^2 + 1)} \sum_{i=1}^{N} \frac{(y_i - mx_i - q)x_i(m^2 + 1) + 2m(y_i - mx_i - q)^2}{\sigma_i^2} = 0 \tag{1.39}$$

and

$$\frac{\partial \chi^2(m, q)}{\partial q} = -\frac{2}{m^2 + 1} \sum_{i=1}^{N} \frac{(y_i - mx_i - q)x_i}{\sigma_i^2} = 0. \tag{1.40}$$

The simultaneous solution of Equation 1.39 and Equation 1.40 leads to a quadratic equation in m whose solutions are

$$m_{1,2} = \frac{-B \pm \sqrt{B^2 + 4}}{2}, \qquad q_{1,2} = \frac{\sum_{i=1}^{N} \frac{y_i}{\sigma_i^2} - m_{1,2} \sum_{i=1}^{N} \frac{x_i}{\sigma_i^2}}{\sum_{i=1}^{N} \frac{1}{\sigma_i^2}} \tag{1.41}$$

$$B = \left(\left(\sum_{i=1}^{N} \left(\frac{y_i}{\sigma_i^2} \right)^2 - \sum_{i=1}^{N} \left(\frac{x_i}{\sigma_i^2} \right)^2 \right) \sum_{i=1}^{N} \frac{1}{\sigma_i^2} + \left(\sum_{i=1}^{N} \frac{x_i}{\sigma_i^2} \right)^2 - \left(\sum_{i=1}^{N} \frac{y_i}{\sigma_i^2} \right)^2 \right) \Big/$$

$$\left(\left(\sum_{i=1}^{N} \frac{x_i}{\sigma_i^2} \right) \left(\sum_{i=1}^{N} \frac{y_i}{\sigma_i^2} \right) - \left(\sum_{i=1}^{N} \frac{1}{\sigma_i^2} \right) \sum_{i=1}^{N} \frac{x_i y_i}{\sigma_i^2} \right), \tag{1.42}$$

It is clear that the solution we are interested in is that of minimizing χ^2. Accordingly, the solution will be m_1, q_1 if $\chi^2(m_1, q_1) \leq \chi^2(m_2, q_2)$, while it will be m_2, q_2 in the opposite case.

Remembering the error propagation law (Equation 1.25), m and q variances are

$$\sigma_m^2 = \sum_{i=1}^{N} \sigma_{x_i}^2 \left(\frac{\partial m(x_i, y_i)}{\partial x_i} \bigg|_{y_i} \right)^2 + \sigma_{y_i}^2 \left(\frac{\partial m(x_i, y_i)}{\partial y_i} \bigg|_{x_i} \right)^2 \qquad (1.43)$$

and

$$\sigma_q^2 = \sum_{i=1}^{N} \sigma_{x_i}^2 \left(\frac{\partial q(x_i, y_i)}{\partial x_i} \bigg|_{y_i} \right)^2 + \sigma_{y_i}^2 \left(\frac{\partial q(x_i, y_i)}{\partial y_i} \bigg|_{x_i} \right)^2. \qquad (1.44)$$

Consequently, the probable uncertainty associated to $m(\delta_m)$ and $q(\delta_q)$ (fitting parameters) can be evaluated in the light of Equation 1.28 and Equation 1.29 as

$$\delta_m = \sqrt{\sum_{i=1}^{N} \left(\frac{d_i}{\sigma_i} \right)^2 \bigg/ (N - M)} * \sqrt{\sigma_m^2} \qquad (1.45)$$

and

$$\delta_q = \sqrt{\sum_{i=1}^{N} \left(\frac{d_i}{\sigma_i} \right)^2 \bigg/ (N - M)} * \sqrt{\sigma_q^2}. \qquad (1.46)$$

The goodness of the fit can be estimated by means of the F-test (Equation 1.32) provided that the following expressions for the MSR and MSE are considered:

$$MSR = \frac{\sum_{i=1}^{N} ((x_i - \bar{x})/\sigma_{y_i})^2 + \sum_{i=1}^{N} ((y_i - \bar{y})/\sigma_{y_i})^2 - \sum_{i=1}^{N} (d_i/\sigma_i)^2}{M - 1}, \qquad (1.47)$$

$$\bar{x} = \frac{\sum_{i=1}^{N} (x_i/\sigma_{x_i})^2}{\sum_{i=1}^{N} (1/\sigma_{x_i})^2}, \qquad \bar{y} = \frac{\sum_{i=1}^{N} (y_i/\sigma_{y_i})^2}{\sum_{i=1}^{N} (1/\sigma_{y_i})^2}, \qquad (1.47a)$$

$$MSE = \frac{\sum\limits_{i=1}^{N} (d_i/\sigma_i)^2}{N - M}. \tag{1.48}$$

It is clear that this strategy (robust fitting) can also be applied to a general model $f(x, a_1, \ldots, a_M)$. Nevertheless, this task is not so straightforward because of the increased difficulty associated with the evaluation of d_i. Indeed, d_i can be evaluated remembering that the generic straight line passing from (x_i, y_i) must satisfy the condition $y_i = mx_i + q$ from which, we have, for example, $m = (y_i - q)/x_i$. Thus, the coordinates (x_i^∇, y_i^∇) of the intersection point between $f(x, a_1, \ldots, a_M)$ and the generic straight line passing from (x_i, y_i) are given by the solution of the following system of equations:

$$y = \frac{y_i - q}{x_i} x + q,$$
$$y = f(x, a_1, a_M). \tag{1.49}$$

The generic distance d_{ig} between (x_i, y_i) and (x_i^∇, y_i^∇) is, then, given by

$$d_{ig} = \sqrt{(x_i - x_i^\nabla)^2 + (y_i - y_i^\nabla)^2}. \tag{1.50}$$

Obviously, among the infinite distances Equation 1.50 gives the distance we are interested in, d_i represents the smallest one. Thus, q can be evaluated searching for d_{ig} minimum. If the system (1.49) does not yield an analytical solution, an iterative, numeric procedure is needed to obtain q. Once d_i is known for all the experimental points, χ^2 can be evaluated and a numerical technique is required for its minimization.

1.4   MODEL COMPARISON

In previous paragraphs, the attention was focused on how fitting a model to experimental data and on the fitting goodness. Basically, this task was developed referring to mathematical/statistical methods. Accordingly, if the F value is high enough and model parameters uncertainty is small, we can say that the model is suitable. Obviously, this is not enough to ensure model reliability as model fitting parameters must assume reasonable values (from the physics point of view) and the model must be able to yield reasonable, if not true, predictions for a different set of initial conditions. Another problem that can arise in mathematical modeling is to discern the best model among a group of models yielding a good fit.

Although different approaches can be followed at this purpose, for its generality and simplicity we would like to present Akaike's method [23] that is based on likelihood theory, information theory, and the concept of information entropy. Despite the complexity of its theoretical background, this method is very simple to use. Assuming the usual condition of a Gaussian distribution of points scattering around the model, the Akaike number AIC is defined as follows:

$$AIC = N \ln \left(\frac{\chi^2}{N} \right) + 2 \frac{(M+1)(N-2)}{N-M-2}. \tag{1.51}$$

The model showing the smallest AIC is the one preferred. In order to estimate how much the model showing the smallest AIC is more likely, it is sufficient to define the following probability p_{AIC}:

$$p_{AIC} = \frac{e^{-0.5\Delta}}{1 + e^{-0.5\Delta}}, \quad \Delta = AIC_{om} - AIC_{smallest}, \tag{1.52}$$

where $AIC_{smallest}$ is the Akaike number relative to the more likely model while AIC_{om} is the Akaike number relative to the other model ($AIC_{om} \geq AIC_{smallest}$). If $\Delta = 2$, for example, there is a probability $p_{AIC} = 0.73$ that the smallest AIC model is correct and a probability $p_{AIC} = 0.27$ that the other model is correct.

1.5 CONCLUSIONS

The aim of this chapter was to explain the concept of mathematical model and to furnish some tools for model building. In particular, the attention was focused on the theoretical process yielding model mathematical formulation, model fitting to experimental data, goodness of fit evaluation and, finally, on comparison between different models for the choice of the best one. Although all these topics would require an entire book to be deeply discussed and commented, we believe that these few pages should contain some clear and practical tools for data modeling, at least for what concerns the evaluation of the goodness of fit, of fitting parameter uncertainty, and model comparison. Conversely, as model building is an art more than a real science, only day-by-day work can confer the real ability to build models. Indeed, this activity, such as many others, consists in a trial and error process that takes a long time.

REFERENCES

1. Dym, C.L., What is mathematical modeling, in *Principles of Mathematical Modeling*, Elsevier, 2004, Chapter 1.
2. Israel, G., Balthazar van der pol e il primo modello matematico del battito cardiaco, in *Modelli Matematici nelle Scienze Biologiche*, Freguglia, P. Ed., QuattroVenti, 1998.

3. Freguglia, P., Prefazione, in *Modelli Matematici nelle Scienze Biologiche*, Freguglia, P. Ed., QuattroVenti, 1998.
4. Bourne, D.W.A., Why model the data, in *Mathematical Modeling of Pharmacokinetic Data*, Technomic Publishing Company, Lancaster, USA, Basel, 1995, Chapter 1.
5. European Society for Mathematical Biology, Grenoble, France, http://www.esmtb.org. 2006.
6. The Society for Mathematical Biology, Boulder, Colorado 80301 USA, http://www.smb.org/index.shtml. 2006.
7. Japan Association for Mathematical Biology, Kyushu, Japan, http://bio-math10.biology.kyushu-u.ac.jp/~jamb/. 2006.
8. Mathematical Modelling & Computing in Biology and Medicine. 5th ESMTB Conference 2002, Società Editrice Esculapio, Bologna (Italy).
9. Cartensen, J.T., *Modeling and Data Treatment in the Pharmaceutical Sciences*, Technomic Publishing Company, Lancaster, Pennsylvania, USA, 1996.
10. Dym, C.L., Dimensional analysis, in *Principles of Mathematical Modeling*, Elsevier, 2004, Chapter 2.
11. Bird, R.B., Stewart, W.E., and Lightfoot, E.N., *Transport Phenomena*, John Wiley & Sons, Inc., New York, London, 1960.
12. Raaijmakers, J.G.W. and Shiffrin, R.M., Models of memory, in *Stevens' Handbook of Experimental Psychology*, Pashler, H. and Medin D. Eds., John Wiley & Sons, Inc., Grenoble, France, 2002, Chapter 2.
13. Murdock, B., The buffer 30 years later: working memory in a theory of distributed associative memory (TODAM), in *On Human Memory: Evolution, Progress, and Reflections on the 30th Anniversary of the Atkinson–Shiffrin Buffer Model*, Izawa, I. Ed., Erlbaum, 1999.
14. Atkinson, R.C. and Shiffrin, R.M., Mathematical models for memory and learning, in *Proceedings of the third conference on learning, remembering and forgetting*, New York Academy of Sciences, New York, 1965.
15. Atkinson, R.C. and Shiffrin, R.M., Human memory: a proposed system and its control processes, in *The Psychology of Learning and Motivation: Advances in Research and Theory*, Spence, K.W. and Spence J.T. Eds., Academic Press, New York, 1968.
16. Atkinson, R.C. and Shiffrin, R.M., The control of short term memory. *Scientific American*, 224, 82, 1971.
17. Peirano, F., Rievocazione libera e interpretazione della curva di posizione seriale negli anziani, Graduate Thesis, University of Padova (Italy), Department of General Psychology, 2000.
18. Peirano, F., Grassi, M., and Bisiacchi, P.S., Mathematical interpretation of the serial position curve in elderly and young subjects. *Psychology and Aging*, 2006, in press.
19. Press, W.H. and others, Statistical description of data, in *Numerical Recipes in FORTRAN*, Cambridge University Press, 1992, Chapter 14.
20. Press, W.H. and others, Modelling of data, in *Numerical Recipes in FORTRAN*, Cambridge University Press, 1992, Chapter 15.
21. Table Curve TM 2D, SPSS Inc. (http://www.spss.com) 1997.
22. Draper, N. and Smith, H., *Applied Regression Analysis*, John Wiley & Sons, Inc., New York, 1966.
23. Burnham, K.P. and Anderson, D.R., *Model Selection and Multimodel Inference: A Practical Information-Theoretic Approach*, Springer, New York, 2002.

2 Part 1
Gastrointestinal Tract

2.1 MACRO- AND MICROSCOPIC ANATOMY OF THE GASTROINTESTINAL TRACT

2.1.1 THE STOMACH

Anatomically, the stomach can be divided into three major regions: fundus (the most proximal), corpus, and antrum. The regions close to the esophagus and duodenum are called cardias and pylorus, respectively. The main function of the stomach is to process and transport food. After feeding, the contractile activity of the stomach helps to mix, grind, and eventually evacuate small portions of chyme into the small bowel, while the rest of the chyme is mixed and ground. The stomach wall, like the wall of most other parts of the digestive canal, consists of three layers: the mucosal (the innermost), the muscularis, and the serosal (the outermost). The mucosal layer itself can be divided into three layers: the mucosa (the epithelial lining of the gastric cavity), the muscularis mucosae, and the submucosal layer (consisting of connective tissue interlaced with plexi of the enteric nervous system). The second gastric layer, the muscularis, can also be divided into three layers: the longitudinal (the most superficial), the circular, and the oblique.

Within the stomach there is an abrupt transition from stratified squamous epithelium extending from the esophagus to a columnar epithelium dedicated to secretion. In most species, this transition is very close to the esophageal orifice, but in some, particular horses and rodents, stratified squamous cells line much of the fundus and part of the body.

On the inner stomach surfaces are present numerous small holes which correspond to the openings of gastric pits which extend into the mucosa as straight and branched tubules, forming gastric glands. Four major types of secretory epithelial cells cover the surface of the stomach and extend down into gastric pits and glands: (1) mucous cells which secrete an alkaline mucus, (2) parietal cells which secrete hydrochloric acid, (3) chief cells which secrete pepsinogen, an inactive zymogen, and (4) G cells which secrete the hormone gastrin. The most abundant epithelial cells are mucous cells, which cover the entire lumenal surface and extend down into the glands as "mucous neck cells." These cells secrete a bicarbonate-rich mucus that coats and lubricates the gastric surface, and serves an important role in protecting the epithelium from acid and other chemical insults. Parietal cells secrete hydrochloric acid into the lumen where it establishes an extremely acidic environment. This acid is important for the conversion (activation) of pepsinogen to pepsin and inactivation of ingested microorganisms such as bacteria. Chief cells secrete pepsinogen, an inactive zymogen, which, once secreted, is activated by stomach acid into the active protease pepsin, which is largely responsible for the ability of the stomach to initiate digestion of proteins. In young animals, chief cells also secrete chymosin, a protease that coagulates milk protein allowing it to be retained more than briefly in the stomach. Finally, G cells secrete the principal gastric hormone, i.e., gastrin, a peptide that is important in the control of acid secretion and gastric motility.

The stomach absorbs very few substances, although small amounts of certain lipid-soluble compounds can be taken up, including aspirin, other nonsteroidal anti-infammatory drugs, and ethanol.

2.1.2 THE SMALL INTESTINE

The small intestine is made up of duodenum, jejunum, and ileum. In human beings, duodenum measures approximately 20 cm, while jejunum and ileum are approximately 2.5 m long. The small intestine is strategically located at the interface between the systemic circulation and the environment, and plays a major role as a selective permeability barrier. It permits the absorption of nutrients such as sugars, amino acids, peptides, lipids, and vitamins [1,2] and limits the absorption of xenobiotics, digestive enzymes, and bacteria.

The innermost part of the small intestine is made up of the submucosa and mucosa tunica characterized by macroscopic folding. The tunica mucosa is constituted, from the intestinal wall to the lumen, by the vascular endothelium, the lamina propria, and the epithelial cells. Every macroscopic folding shows, in turn, villi (diameter 0.1 mm, height 0.5–1 mm) that have proper artery, veins, and lymphatic capillaries (Figure 2.1).

The vascular endothelium belongs to the capillaries present within the villi. The lamina propria constitutes the connective tissue inside the villus and surrounds the crypt, the region present at the bottom of the villus. The lamina propria contains numerous defensive cells such as lymphocytes, plasma cells, and macrophages, which interact with foreign substances that enter this layer from the GI tract. The lamina propria also contains eosinophils, mast cells, blood and lymph vessels, and fibroblasts. In addition to carrying out several immunologic functions the lamina propria provides structural support to the epithelial cell layer. Finally, through its capillaries the lamina propria nourishes the epithelial cells and allows the transport of absorbed substances to the systemic circulation.

On the surface of the villi, there are the epithelial cells, mostly constituted by enterocytes, which represent 90% of all cells constituting villi. The apical

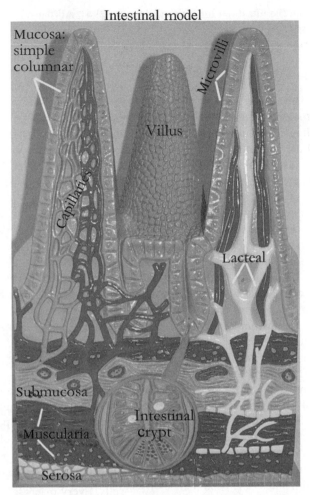

FIGURE 2.1 Villus anatomy. (From http://iws.ccccd.edu/mweis/Images/Models/
2402%20Models/digestive%20models/labeled%20gi%20models/gi_si_villi_labeled.
png.)

zone of enterocytes is characterized by the presence of a particular brushed
structure devoted to absorption (microvilli, diameter 0.1 μm, height 1 μm).
Microvilli are covered by a glycoproteic layer (glycocalix), 0.1 μm thick, continuously produced by the enterocytes. Enterocytes are produced (50 millions per
day) according to the mitoses of particular cells positioned among villi and called
Lieberkuhn crypt cells. As the average enterocyte life is 3 days, a continuous
stream of young enterocyte toward the upper part of the villi takes place.

Other cells present on the villi are represented by endocrine cells and
goblet cells, which secrete mucin. In certain parts of the intestine (Payer's
patches) the epithelial cell layer contains M (microfold) cells, which sample

antigens from the intestinal lumen to the lymph [3,4]. All these cells originate in the crypts from the undifferentiated Lieberkuhn crypt cells, which undergo differentiation as they migrate upward along the crypt–villus axis.

2.1.3 THE COLON

Also called the large intestine or large bowel, the colon is the part of the intestine from the cecum to the rectum. In mammals, it consists of the cecum, the ascending colon, the transverse colon, the descending colon, the sigmoid colon, and the rectum. The colon from cecum to the mid-transverse colon is also known as the right colon. The remainder is known as the left colon. Altogether, the colon measures approximately 1.5 m in length. The colon is constituted of four layers which, from the inner to the outer site are represented by the mucosa, the submucosa, the muscularis, and the adventitia. There are many similarities in the histological structure of the mucosa in the colon and the small intestine. The most obvious difference is that the mucosa of the large intestine is devoid of villi. It has numerous crypts which extend deeply and open onto a flat lumenal surface. The stem cells which support rapid and continuous renewal of the epithelium are located either at the bottom or midway down the crypts. These cells divide to populate the cryptal and surface epithelium. Mucus-secreting goblet cells are also much more abundant in the colonic epithelium than in the small gut.

Although there are differences in the large intestine between different organisms, the large intestine is mainly responsible for storing waste, reclaiming water, maintaining the water balance, and absorbing some vitamins, such as vitamin K. By the time the chyme has reached this tube, almost all nutrients and 90% of the water have been absorbed by the body. At this point some electrolytes like sodium, magnesium, and chloride are left as well as indigestible carbohydrates known as dietary fiber. As the chyme moves through the large intestine, most of the remaining water is removed, while the chyme is mixed with mucus and bacteria known as gut flora, and becomes feces. The bacteria break down some of the fiber for their own nourishment and create acetate, propionate, and butyrate as waste products, which in turn are used by the cells lining the colon for nourishment. This is an example of a symbiotic relationship and provides about 100 Cal a day to the body. The large intestine produces no digestive enzymes—chemical digestion is completed in the small intestine before the chyme reaches the large intestine. The pH in the colon varies between 5.5 and 7 (slightly acidic to neutral).

2.1.4 ABSORPTIVE CELLS

The epithelial cell layer of the villi consists mainly of enterocytes and undifferentiated crypt cells. Enterocytes, also termed absorptive cells, are highly polarized and columnar. The apical surface is characterized by closely packed microvilli about 0.5–1.5 mm in length and 0.1 mm in width, depending on the species [3]. The apical membrane has a high (1.7:1) protein-to-lipid molar ratio and an abundance of glycolipids and cholesterol [5]. The basolateral membrane of the absorption cells differs from the apical membrane in morphology, biochemical composition, and function [6]. For instance, it is about 7 nm wide and contains Na^+-K^+-ATPase.

Intestinal epithelial cells are joined at intercellular junctional complexes which connect the single enterocyte to the surrounding cells. These junctional complexes, which are approximately 0.5–2 nm wide, consist of tight junctions (TJs), intermediate junctions, and desmosomes [4]. The TJs, located at the apical end of the lateral membrane of adjacent cells, eliminate the intercellular space over a variable distance [4]. The TJs of ileal absorptive cells are deeper and denser than TJs of the jejunum. The depth and density of the TJs between absorptive cells are greater than those of undifferentiated crypt cells and goblet cells. Compared to the intermediate junctions and desmosomes, the TJs are the most significant diffusion barrier components of the junctional complexes.

2.1.5 Tight Junctions

TJs (for a recent and exhaustive review see [7]) provide barriers to prevent leakage of molecules across the epithelia through the gaps between the cells, thus regulating the so-called "paracellular transport" (see paragraph 2.2). Additional, TJs serve as fences between the apical and basolateral domains of the plasma membranes, preventing diffusion of integral proteins and lipids from one to the other [8–10]. TJs appear as continuous networks of parallel and interconnected strands that circumscribe the apex of lateral membranes of adjacent cells [11]. These sealing strands correspond to fusion sites of apposing plasma membranes and are composed of transmembrane cell adhesion proteins (CAMs). A variety of CAMs and adaptor proteins have been identified at TJs [12]. In particular, three CAMs have been identified at TJs: occludin [13], claudins [14], and junctional adhesion molecules (JAMs) [15], all connected to the actin cytoskeleton and cellular signaling elements. Occludin has four transmembrane domains and two extracellular loops, with the N- and C-termini facing the cytoplasm. Immunoelectron microscopic studies have revealed that occludin is associated with the TJ strand itself [16]. Occludin probably takes part in the regulation of paracellular permeability in epithelial cells. The second type of CAMs, i.e., claudins, has been identified as important structural and functional components of TJs involved in paracellular transport. Claudins are small proteins of 20–24 kDa, with two extracellular domains, four transmembrane domains, and cytoplasmic N- and C-termini. They generate a series of regulated channels within TJ membranes for the passage of ions and small molecules. The third transmembrane protein group, JAMs, are immunoglobulin (Ig)-like single-span transmembrane molecules which mediate Ca^{2+}-independent adhesion [15]. So far, four members of JAMs have been identified, JAM1–4. JAM-1 is colocalized with F-actin at the cell–cell contact and the membrane ruffles. JAM-1 facilitates cell adhesion through homophilic binding [17], and regulates TJ permeability. Of particular relevance to the small intestine is JAM-4, which interacts with MAGI-1 (membrane associated guanylate kinase interacting protein-like 1) a scaffolding protein at TJs, which may regulate the permeability of small-intestinal epithelial cells and kidney glomerulus [18].

In most membranes, with the exception of endothelial cells in the blood–brain barrier, TJs contain fenestrae, which can be regarded as pores filled with water. The dimensions of these pores have been estimated to be in the range of 3–10 Å. However, the radii are not fixed. Depending on the observed biomembrane and the presence of mechanistic control nutrients (e.g., Ca^{++}, glucose, and amino-acids), the pores size may vary. The number of TJs depends on the biomembrane type. For the small intestine, it amounts to about 0.01% of the whole surface. Thus, the surface area of the lipid part of a biological membrane is much greater than the aqueous part defined by the TJs.

2.1.6 CELLULAR MEMBRANE

Molecule absorption may occur via TJs or crossing enterocyte membranes (transcellular route). Cell membranes are bilayered, dynamic structures that function to form boundaries between cells and their environments, regulate movement of molecules into and out of cells and perform vital physiological roles. Lipids, proteins, and carbohydrates in various combinations make these tasks possible. The model used to show the composition of the membrane predicts proteins and carbohydrates floating in lipids and is defined as the "fluid mosaic model." The lipid portion of a cellular membrane provides structure and a barrier for water-soluble molecules. Membrane proteins are embedded in the lipid bilayer creating channels and transporting materials. Carbohydrates are attached to lipid or protein molecules on the membrane, generally on the outer surface and are involved in signaling and adhesion.

Most of the lipid molecules found in biological membranes are phospholipids. Each has a hydrophilic region, where the phosphate groups are located, and a hydrophobic region, the fatty acid "tails." The phospholipids organize themselves into a bilayer. The interior of the membrane is fluid, and allows some molecules to move laterally in the membrane. Although all biological membranes are structurally similar, some have quite different compositions of lipids and proteins. Phospholipids differ in length, degree of unsaturation, and degree of polarity. Cholesterol (a long-chained fat) and increased saturation of fatty acids decrease fluidity of the membrane. The membranes of cells that live at low temperatures tend to be high in unsaturated and short-chain fatty acids. The ratio of phospholipid to protein, the second component of cellular membranes, varies depending on membrane function. Many membrane proteins have both hydrophilic and hydrophobic regions. The association of protein molecules with lipid molecules is not covalent; both are free to move around laterally, according to the fluid mosaic model. Integral membrane proteins have hydrophobic regions of amino acids that penetrate or entirely cross the phospholipid bilayer. Transmembrane proteins have a specific orientation, showing different "faces" on the two sides of the membrane. Some of the proteins and lipids can move around in the membrane, others are anchored by the cytoskeleton. This causes an unequal distribution of proteins, allowing for specialization of certain regions of the cell membrane. Carbohydrates, the third component of cellular membranes, can be bound to lipid, forming the so-called glycolipid. Additionally, they can bind covalently with proteins, forming glycoproteins. Plasma membrane glycoproteins enable cells to be recognized by other cells and proteins.

2.2 INTESTINAL ABSORPTION

2.2.1 POSITIVE AND NEGATIVE FORCES

The positive forces regulating intestinal absorption consist of the concentration gradient, the electrical potential difference, and the hydrostatic pressure gradient. Opposing elements to absorption are related to the physical and biochemical barriers of the mucosa. The physical barrier is the result of the TJs and the lipid composition of the cell membrane. The biochemical barrier depends on the presence, on the mucosa, of enzymes able to metabolize drug and on the so-called "efflux transponders" which are able to expel the drug from the cells.

With regard to the physical barrier represented by TJs, it should be noted that the pores present in these cellular elements are approximately of 4–8 and 10–15 Å in humans and animals, respectively [19]. Thus, in humans, the absorption of molecules with diameters greater than ~8 Å through this route is unlikely to play an important role in the uptake of most pharmaceutical compounds which usually have bigger diameters. Additionally, the modest surface area represented by TJs further reduces the relevance of this absorption route. Nevertheless, the evidence of the possibility to increase TJs diameter by cellular regulatory process [20] suggests that it might be possible to increase absorption through TJs by coadministering the drug of interest with agents that can open TJs [21,22].

The biochemical barrier consists of drug metabolizing enzymes, present on the mucosa, and to the efflux transponder. The enzymes expressed by enterocytes include aminopeptidases N/P/W, dipeptidyl peptidase IV, isozymes of the cytochrome P450 (CYP) superfamily such as 1A1, 1A2, 2D6, 3A4, 2C9, and 2C19, esterases, phenol sulfotransferase (PST), and UDP glucuronyltransferase (UDPGT) [4,23,24]. CYP3A4, which accounts for approximately 60% of the total amount of CYP enzymes found in the intestine, metabolizes a large number of compounds and limits absorption [25,26].

With regard to the second component of the biochemical barrier, efflux transporters mediate the extrusion of compounds from the cell cytoplasm to the intestinal lumen through a process known as "apical efflux." Two families of transporters that belong to the ATP binding cassette (ABC) superfamily of transporters mediate apical drug efflux. The two families are represented by the multidrug resistance (MDR) and multidrug resistance-associated protein (MRP) families [27–30]. P-glycoprotein (Pgp), the most studied member of the apical efflux transporters, is the product of the MDR1 gene. Two isoforms have been described, the first being the product of MDR1a gene, the second of the MDR1b gene. Whereas MDR1b is expressed only in the liver and kidney, MDR1a is mostly expressed in brain capillaries, placenta, and intestinal tract. Pgp, the product of MDR1, has two subunits with six *trans*-membrane domains and two binding sites [31,32] for the ATP molecule, whose hydrolysis provides the energy required for the efflux process. The human form of Pgp transporter was initially identified because of its over-expression in cultured tumor cells and associated with an acquired cross-resistance to multiple cytotoxic anticancer agents. After this initial characterization of the Pgp transponder, it was found that it is expressed in many other human tissues [33]. In particular, it has been shown to be present on the inner surface of jejunum, duodenum, ileum, and colon, suggesting that it facilitates the excretion of substrates from the systemic circulation into the GI tract. Pgp has been shown to limit intestinal absorption of a large number of drugs [34–39] including paclitaxel [40].

Among the MRP transponder family, six MRP members have been discovered. Of these, three (MRP1, MRP2/cMOAT, and MRP3) have been found in human duodenum with MRP1 and MRP3 also detected in human colon together with MRP5 [41]. Their location on the apical membrane of enterocytes and colonocytes, together with their efflux function, make these transporters a potentially formidable barrier to GI drug absorption. It should be noted that the possible species-dependent differences in the expression or activity of many drug transporters may differently affect drug penetration. Thus, a drug that shows poor intestinal absorption in animal models may turn out to have far greater bioavailability in humans and vice versa.

2.2.2 DRUG ABSORPTION PATHWAYS

The intestinal mucosa is a selective permeability barrier which can be crossed by drugs through the transcellular and paracellular routes [42]. In the "transcellular transport process," the drug molecules have to go through the barrier cells (e.g., epithelial cells) to reach the systemic circulation. Transcellular transport is typically a two-step process, starting with drug uptake into the cells, and ending with drug efflux out of the cells. The second type is called "the paracellular transport process," where the drug molecules travel between the cells (or in the gaps) to reach the systemic circulation. It is possible that both routes contribute to the absorption of a particular drug.

The transport pathways can be divided into: (a) carrier-mediated transcellular transport, (b) vesicular transport, (c) passive paracellular transport, and (d) passive transcellular transport.

(a) Together with the efflux transporters above described, influx transporters also exist. The small-intestinal mucosa expresses large numbers of these biological units, which are responsible for the absorption of nutrients and vitamins. In addition to transporting vitamins and nutrients, these transporters have been shown also to mediate the absorption of some drugs [41,43–46]. Transporters for *di-/tri*peptides (PEPT1), large neutral amino acids, bile acids, nucleosides, and monocarboxylic acids have received a great deal of attention for their potential to deliver drugs across the intestinal mucosa [47–52]. Additionally, the presence of two peptide transporters, PTR3 and PHT1, in human and rat intestine [53] has been reported. Work carried out in a number of laboratories [30] has also demonstrated a family of uptake transporters known as organic anion transporting polypeptides (OATP, gene symbol *SLCO*), expressed in organs such as the intestine and liver, and blood–brain barrier, to be the key determinant in the cellular uptake of many endogenous and exogenous chemicals, including drugs in clinical use. The first human OATP, OATP1A2 (previously named OATP-A), was isolated from human liver. 22 OATP1A2 is truly a multispecific transporter and capable of transporting diverse compounds including bromosulfophthalein (BSP), bile acids, steroid sulfates [54], bulky organic cations [55,56], fexofenadine, thyroid hormones, and opioid peptides [57].

The full potential impact on drug absorption by these transporters, and others not yet known, remains to be established. Influx transporters can bind compounds that are dissolved in the intestinal fluid and translocate them across the apical membrane of enterocytes, thus facilitating the drug absorption process. Of these, PEPT1 mediates the transport of peptidomimetic drugs such as angiotensin converting enzyme (ACE) inhibitors, β-lactam antibiotics, and renin inhibitors [48–52,58]. Compounds that are substrates for these transporters exhibit intestinal absorption higher than expected from their diffusion across cell membranes. A list of the transponders involved in the transport of amino acids, oligopeptides, monosaccharides, monocarboxylic acids, phosphates, and several water-soluble vitamins across intestinal barrier are summarized in Reference [43].

(b) Enterocytes possess vesicular transport processes that can facilitate drug absorption. These include fluid-phase endocytosis (FPE), receptor-mediated endocytosis (RME), and transcytosis. FPE is a process by which solute molecules dissolved in the luminal fluid are incorporated by bulk transport into the fluid-phase of endocytic vesicles. This process starts when the plasma membrane forms invaginations that pinch off to form vesicles, which migrate inwardly. Molecules contained in the vesicle are transported to endosomes, which then fuse with lysosomes. Evidence exists for mucosal uptake of some peptides and proteins by means of FPE [59]. With regard to the RME, this process is generally considered to be of importance predominantly for the mucosal permeation of macromolecules. The mechanism regulating this process is based on the binding of the macromolecule to a receptor on the membrane followed by clustering of the receptor–ligand complex into clathrin-coated pits. After endocytosis, the fate of the elements of the receptor–ligand complex is determined through a process called sorting. Usually, the sorting process results in the destruction of the ligand in the lysosomes while the receptor can either undergo lysosomal destruction or recycling back to the cell membrane [60–62]. The ligand dissolved in the endocytic vesicle, following FPE or RME, can bypass the lysosomes and undergo release across the basolateral membrane [63]. This process, known as transcytosis, may result in the intestinal absorption of molecules unable to permeate the cell membrane by simple diffusion.

(c) The passive paracellular pathway occurs through an extracellular route across the epithelium. The forces regulating the passive paracellular diffusion are represented by the electrochemical potential gradients derived from differences in concentration, electrical potential, and hydrostatic pressure between the two sides of the epithelium. The TJs are the main barriers to passive paracellular diffusion. In general, hydrophilic compounds are mainly absorbed from the intestinal lumen by means of the paracellular route.

(d) The passive transcellular pathway involves the movement of solute molecules across the apical membrane, through the cell cytoplasm, and across the basolateral membrane. The surface area of the transcellular route (i.e., cell membrane) is much larger (99.9%) than the surface area of the paracellular route (0.01%, TJs) [64,65]. This implies that compounds whose permeability is restricted to the paracellular pathway have lower absorption rate compared to compounds that readily traverse the cell membrane [66]. However, the involvement of uptake or efflux transporters can affect this pattern. For example, some hydrophilic molecules such as peptides undergo carrier-mediated transport, and thus, exhibit absorption values higher than expected from their intrinsic membrane permeability characteristics.

For the description of molecules for which the oral bioavailability in man has been well documented, please refer to Reference [43].

2.2.3 ABSORPTION IN GI DISEASES

The above-described absorption mechanisms hold true for a healthy intestine wall. However, in many pathological conditions the integrity of the intestine wall can be altered to different degrees. Thus, abnormal absorption properties can occur in a diseased intestine. This aspect should be carefully considered by the clinicians when administering drugs through the oral route in intestine diseased patients. Additionally, this aspect is of particular interest also for researchers involved in the absorption study of novel drugs.

Without pretending to give an exhaustive spectrum of all pathological conditions potentially able to affect drug absorption, attention will be put on the description of some of the most commonly encountered GI diseases. In particular, Cronhn's disease, celiac disease (CS), food allergy, and pancreatitis will be considered.

Breeches in the integrity of the gut barrier have been associated with many inflammatory GI diseases, leading to the "leaky gut" hypothesis [67]. This is likely to be true in the case of Cronhn's disease. The pathogenesis of this pathological condition is unknown, but the currently accepted theory postulates several mechanisms. These include environmental factors such as antigens in the diet or the bacterial flora, increased gut permeability to environmental antigens and increased immune responsiveness to these environmental antigens in genetically susceptible individuals. Whatever the pathogenesis, many of the animal models for the disease suggest that breech of the intestinal barrier occurs early in the course of disease development. Additionally, knockout mice that are deficient in a major adhesion molecule of the epithelium, N-cadherin, develop intestinal inflammatory lesions similar to those seen in Cronhn's disease [68], pointing toward the implication of epithelium damage in the altered intestine permeability. Measurements of gut permeability with Cr–EDTA or differential urinary excretion of monosaccharides and disaccharides have shown increased intestinal permeability in many studies [69–73]. Also, other studies demonstrate increased pulmonary permeability in patients with Cronhn's disease, suggesting a systemic epithelial barrier dysfunction [74].

CS is a digestive disease that damages the small intestine. People who have CS cannot tolerate a protein called gluten found in wheat, rye, and barley. Gluten is found mainly not only in foods, but also found in products used every day, such as stamp, envelope adhesive, medicines, and vitamins. CS is a genetic disease which sometimes is triggered or becomes active for the first time after surgery, pregnancy, childbirth, viral infection, or severe emotional stress. When people affected by CS eat foods or use products containing gluten, their immune system responds by damaging the small intestine. In particular, the villi are damaged or destroyed with consequent influences on intestine absorption. Epithelial permeation tests in patients with CS demonstrate a reduced absorption of small markers of permeation, such as mannitol, and an increased absorption of relatively larger markers, such as lactulose. This finding fits well with the permeation hypothesis proposed by Hollander [75]. The increased absorption of large molecules may result from villous epithelial injury with greater exposure of these probes to the crypt TJs, increased cell shedding with less "mature" TJs, or changes in the TJs secondary to inflammatory cells and their mediators. The decreased absorption of small molecules may result from a reduction in the total number of paracellular junctions because of a reduced surface area [76,77].

Food allergy affects up to 6%–8% of children under the age of three and 2% of adults. Allergic reactions to food can cause serious illness and, in some cases, death. This pathological condition consists of an abnormal response to a food triggered by the body's immune system. The immune system mistakenly identifies a specific food or component of food as a harmful substance. The pathogenesis of the disease is based on the production of IgE, antibodies directed against the food or food component (the allergen). Disturbance in intestinal permeability is the result of the reaction to the allergen, possibly mediated by the release of inflammatory cytokines. This hypothesis is supported by findings of increased expression of inflammatory cytokines such as tumor necrosis factor and interferon gamma release after cow milk's challenges [78,79]. Additionally, *in vitro* experiments have indicated the upregulation of transport and processing of food antigens [79] and the increase in paracellular transport as a consequence of the release of these cytokines [79]. Finally, it has been shown that children with cow's milk allergy have increased [80] Cr–EDTA and lactulose–mannitol permeation [81–84].

Pancreatitis is an inflammation of the pancreas. The pancreas is a large gland behind the stomach and close to the duodenum. The pancreas secretes digestive enzymes into the small intestine through a tube called the pancreatic duct. These enzymes help to digest fats, proteins, and carbohydrates present in food. The pancreas also releases the hormones insulin and glucagon into the bloodstream. Normally, digestive enzymes do not become active until they reach the small intestine, where they begin digesting food. If these enzymes become active inside the pancreas, they start "digesting" the pancreas itself. Acute pancreatitis occurs suddenly and lasts for a short period of time and usually resolves. Chronic pancreatitis does not resolve itself and results in a slow destruction of the pancreas. Although increased intestinal permeation does not appear to play a role in the initiation of acute pancreatitis, it may contribute to an increase in the morbidity and mortality of this disease. The relationship between acute pancreatitis and changes in gut permeation has been studied in animal models. Using a model of cerulein-induced pancreatitis, it has been demonstrated that acute pancreatitis resulted in increased intestinal permeation to the macromolecular probe polyethylene glycol 3350. Additionally, the magnitude of this permeation defect correlated with the severity of the pancreatitis [85].

2.2.4 Models of Intestinal Absorption

To evaluate the feasibility of the oral administration of a given drug and study the mechanisms regulating the absorption, *in vivo* and *in vitro* approaches are used. Some *in vivo* models are reported below and some of the available *in vitro* models are also described here. Among the others which include the use of excised intestinal tissue, isolated primary enterocytes, the use of artificial membranes and in silico methods [86], intestinal cells lines are commonly used. Caco-2, HT-29, 2/4/A1, and MDCK cells have been previously used [87–91].

Caco-2 cell, first characterized as an intestinal permeability model in 1989 [92], is a human colon adenocarcinoma cell line, which undergoes spontaneous enterocytic differentiation in culture. When Caco-2 cells reach confluency on a semipermeable porous filter, the cell polarity, and TJs are well established. Additionally, they exhibit other morphological features of small-intestinal cells such as microvilli and the expression of intestinal enzymes such as aminopeptidases. Despite these similarities, relevant differences exist between CaCo2 cells and the *in vivo* conditions. It should be noted, in fact, that, although present, pharmaceutically relevant transporters are quantitatively less abundant compared to those present *in vivo*. For example, β-lactam antibiotics and ACE inhibitors, known substrates of dipeptide transporters, poorly cross the Caco-2 cell monolayer despite the fact that they are efficiently absorbed *in vivo* [93]. Additionally, low-molecular weight hydrophilic compounds (e.g., ranitidine, furosemide, hydrochlorothiazide) show poor permeability (i.e., equal or less than mannitol) in this cell model despite adequate absorption in humans. Thus, Caco-2 cell model is appropriate as a one-way screen meaning that highly permeable molecules in this model are typically well absorbed *in vivo*. However, compounds with low permeability cannot be ruled out as poorly absorbed compounds *in vivo*. Finally, the use of appreciable amount of organic cosolvent is limited as the integrity of TJs is easily compromised by commonly used organic solvents (e.g., methanol, ethanol, PEG) even at low concentration (more than $1 \pm 2\%$ v/v). Thus, a significant percentage of new drug candidates with poor aqueous solubility cannot be evaluated in this model.

Culturing wild-type HT29 cells in the presence of galactose instead of glucose leads to the selection of clones able to form monolayers of polarized cells which can be used to evaluate drug absorption. These cells are of particular interest as some clones can produce mucus. Clone HT29-H has this ability [94,95], and thus has been used to study the role of mucus on the permeability of drugs such as testosterone. The presence of a mucus layer in HT29-H monolayers resulted in lower permeability compared to Caco-2 monolayers. This indicates that the mucus accounts for most of the permeability resistance, at least with regard to the drug testosterone. This observation underlines the relevance of the barrier exerted by mucus, a variable—as yet—mostly neglected since most pharmaceutical scientists rely only on Caco-2 monolayers to predict intestinal permeability. In order to try to overcome this limitation, a coculture of HT29-H cells [96,97] and Caco-2 cells has been explored.

Recently, a cell model, 2/4/A1 that originates from fetal rat intestine, has been described [98]. This cell type mimics the permeability of the human small intestine rather well, especially with regard to passive transcellular and paracellular drug transport. This immortalized cell line forms viable differentiated monolayers characterized by the presence of TJs, brush border membrane enzymes, and transporter proteins. The transport rate of poorly permeable compounds (e.g., mannitol) in 2/4/A1 monolayers is comparable to that in the human jejunum, and is up to 300 times faster than that in the Caco-2 cell monolayers. This observation proposes 2/4/A1 cell line as a more predictive cell model for compounds that are absorbed via the paracellular route.

When cultured on semipermeable membranes, madin-darby canine kidney (MDCK) cells differentiate into columnar epithelial cells and like Caco-2 cells form TJs. The use of this dog-derived kidney cell line to study the intestinal transport was first proposed in 1989 [99]. More recently, it has been investigated for use as a tool for assessing the membrane permeability properties of drugs. Apparent permeability values of 55 compounds with known human absorption values were determined in MDCK cell system and compared with the values obtained using Caco-2 cells. Notably, not only the permeability was comparable between MDCK and Caco-2 cells, it also resembled that observed in humans. The similar behavior between MDCK and Caco-2 cells and the fact that, compared to Caco-2, MDCK cells require a shorter cultivation period (3 days vs. 3 weeks) make this cell type an attractive model for the *in vitro* evaluation of drug permeability. Despite these positive features, it should be considered that MDCK are dog-derived cells. This implies that they may not necessarily be always representative of the behavior of human-derived cells. For example, it is very likely that the expression level of various transporters may be grossly different from human-derived cells such as Caco-2. Thus, more extensive studies are required in MDCK cell line to confirm that the correlation of permeability to human absorption values would hold for actively transported/secreted drugs.

2.3 HUMAN AND ANIMAL GI TRACT: A COMPARISON

Whereas current data indicate that no single animal can perfectly mimic the GI tract characteristics of human, for a given purpose, selection of the more appropriate animal model is possible. Here it follows a series of anatomical/physio-physiological/biochemical consideration which may help in the selection of the most appropriate animal model for the specific investigation.

2.3.1 ANATOMY

Whereas different animal models are commonly used to study drug permeation in the GI tract, the differences existing among the various models and the human anatomy/physiology should be considered for a correct interpretation of the data and for the proper extrapolation to the human situation.

2.3.1.1 The Stomach

The stomachs of rodents consist of a glandular and nonglandular regions. The nonglandular stomach is generally thin walled and transparent and the glandular stomach is thick walled. The nonglandular part of the stomach contains a keratinized squamous epithelium and mainly serves as a food deposit. The glandular part contains tubular gastric glands where it is possible to distinguish mucus-secreting neck cells, pepsinogen-secreting chief cells, and HCl-secreting parietal cells [100]. In the human, pig, dog, and monkey, the stomach is of glandular type and with the internal mucosa divided into three regions defined as cardiac, gastric, and pyloric. It should be noted that, compared to the human stomach, the pig stomach is larger (from two up to three times) with a broader distribution of the cardiac mucosa.

The distribution and occurrence of the cells in the gastric glands differs considerably among the different animal models [101]. In mice, rats, hamsters, and gerbils, the lower one-third of the glandular lamina propria is occupied by a varying proportion of parietal and chief cells. In rabbits, the chief cells are mostly distributed into the lower three-quarters of the glands intermingling with parietal cells. Additionally, in guinea pigs the chief cells are not discernable. In hamsters, a gradual increase of chief cells is observed moving toward the pyloric region. With regard to the parietal cells, in all the species considered, they represent the dominant cell type in the upper one-third of the gastric glands. Moreover, parietal cells often extend to the neck of the gland between the mucus neck cells and occasionally chief cells.

2.3.1.2 The Small Intestine

Differences in the relative and absolute lengths of the different parts of the small intestine for different mammals, obtained from the post mortem examination, are reported in Table 2.1 and Table 2.2. The digestive tract of the dog is relatively short and simple as the one of the pig which, however, is longer (Table 2.1). The sheep has the longest small intestine among common laboratory animals while the mouse has a rather short small intestinal ranging from 35 to 45 cm [102]. With regard to the diameter of the small intestine, the human one is of approximately 5 cm (Table 2.2). In the pig, beagle dog, Rhesus monkey, and rat it corresponds to 2.5–3.5, 1.0 [103], 1.2–2, and 0.3–0.5 cm [104], respectively. As referred for human, also in animal models the small intestine represents the most relevant site for absorption of nutrients and drugs. The luminal surface is covered by villi with columnar absorptive cells (enterocytes) accounting for more than 90% of the cell population on the villi surface [105]. The luminal part of the enterocytes is equipped with fine extensions called microvilli which significantly increase the absorption surface. The number of villi/mm^2 in the freeze-dried jejunum and ileum of the rat, hamster, and dog corresponds to 23 and 38, 25 and 25, and 23 and 23, respectively [106]. The villi have characteristic shapes and patterns in different species with finger-shaped villi in the mouse, pig, and human and tongue-shaped villi in the rat. Notably, in stump-tailed monkeys (*Macaca speciosa*) the villi are tongue-shaped in the duodenum and finger shaped in the jejunum and ileum [107].

The density of microvilli on the villi corresponds to 65 microvilli/μm^2 of villi surface in the rat and 34 in the dog [108]. It should be noted, however, that after considering the size of the villi, the effective surface area per unit villus corresponds to a constant value of about 25 $\mu m^2/\mu m^2$.

2.3.1.3 The Colon

Whereas the small intestine is the most important site of nutrients and drug absorption, the colon is involved in the absorption of water and minerals. The colon is the major site of production and absorption of volatile fatty acid in

TABLE 2.1
Length of Parts of the Intestine at Autopsy

Animal	Part of Intestine	Average Absolute Length (m)
Horse	Small intestine	22.44
	Cecum	1.0
	Large colon	3.39
	Small colon	3.08
	Total	29.91
Ox	Small intestine	46.00
	Cecum	0.88
	Colon	10.18
	Total	57.06
Sheep and goat	Small intestine	26.20
	Cecum	0.36
	Colon	6.17
	Total	32.73
Pig	Small intestine	18.29
	Cecum	0.23
	Colon	4.99
	Total	23.51
Dog	Small intestine	4.14
	Cecum	0.08
	Colon	0.60
	Total	4.82
Cat	Small intestine	1.72
	Large intestine	0.35
	Total	2.07
Rabbit	Small intestine	3.56
	Cecum	0.61
	Colon	1.65
	Total	5.82

Source: Modified from Stevens, C.E., in *Dukes Physiology of Domestic Animals*, Swenson, M.J., Ed., Comstock, Ithaca, NY & London 1997.

human as well as in the sheep, pig, rabbit, and rat [109]. In contrast to the small intestine, the luminal surface of the colon, divided into geographical areas by transverse furrows, does not contain villi. The enterocyte differs slightly from that of the small intestinal as, for example, the microvilli are less closely packed [106,110].

In humans (Table 2.2), the colon, subdivided into four segments named ascending, transverse, descending, and sigmoid sections, ranges from 90 to

TABLE 2.2
Human Organ Size, Surface Area

	Average Length (cm)	Average Diameter (cm)	Absorbing Surface Area (m²)
Esophagus	25	2.5	~0.02
Stomach	20	15	~0.11
Duodenum	25	5	~0.09
Jejunum	300	5	~60
Ileum	300	5	~60
Cecum	10–30	7	~0.05
Colon	150	5	~0.25
Rectum	15–19	2,5	0.015

Source: Modified from Ritschel, W.A., *Exp. Clin. Pharmacol.*, 13, 313–336, 1991.

150 cm in length. Notably, as also reported for monkeys and pigs, all segments are sacculated. In contrast, the dog colon is not sacculated and is shorter, about 25 cm in length [103,111]. With regard to the cecal area, a small region which precedes the ascending colon, it should be noted that in humans this region is poorly defined and relatively small. In contrast, in the pig, the cecum is several orders of magnitude larger than that of the human. Among smaller animals such as guinea pigs, the cecum is three times as large as their stomach, is sacculated, capacious and never empty. In the rabbit, the cecum is large, sacculated and is characterized by a spirally arranged constriction related to the internal folding of the mucosa. The first part of the rabbit colon is structured like a cecum and constitutes the ampulla caecalis coli [112]. Finally, the rat cecum is large, with a rather short and nonsacculated colon.

2.3.2 PHYSIOLOGY

2.3.2.1 pH and Bile Fluid

The dissolution, solubilization, and absorption processes of ionizable drugs are largely dependent on the pH of the fluids present in the GI tract. Thus, the knowledge of this variable is of utmost relevance for the correct interpretation of the data obtained in animal models.

The parietal cells of the stomach are the main source of hydrogen ions in the GI tract. Neural (*n. vagus*) and hormonal (gastric) circuits regulate the secretion of hydrogen. Upon reaching the duodenum, the acidic content of the stomach induces the pancreas to secrete an alkaline fluid containing bicarbonate anions (pH 8). Together with the bile alkaline fluids, the pancreatic fluid neutralizes the acidic contents derived from the stomach substantially increasing its pH.

Table 2.3 through Table 2.4 and Reference [104] report different physiological parameters including the gastric acidity, secretion rates, gastric volume,

TABLE 2.3
pH Values of the Content of Different Parts of the Alimentary Tract in Different Species

pH Value (Median for Different Animals) of the Content of the Stomach and of the Small Intestine

Animals	A	P	(Average)	Cecum	Colon	Feces
Monkey	4.8	2.8	5.8	5.0	5.1	5.5
Dog	5.5	3.4	6.6	6.4	6.5	6.2
Cat	5.0	4.2	6.8	6.0	6.2	7.0
Ox	6.0	2.4	7.2	7.0	7.4	7.5
Sheep	6.4	3.0	7.0	7.3	7.8	8.0
Horse	5.4	3.3	7.2	7.0	7.4	7.5
Pig	4.3	2.2	6.6	6.3	6.8	7.1
Rabbit	1.9	1.9	7.0	6.6	7.2	7.2
Guinea Pig	4.5	4.1	7.8	7.0	6.7	6.7
Rat	5.0	3.3	6.8	6.8	6.6	6.9
Mouse	4.5	3.1		—	—	—
Hamster	6.9	2.9	6.6	7.1	—	—

A, Anterior portion of stomach; P, posterior portion; Feces, contents of posterior rectum.
Source: Modified from Smith, H.W., *J. Pathol. Bacterial*, 89, 95–122, 1965.

and pH values for humans, beagle dogs, pigs, and Rhesus monkeys. In humans, the stomach pH after food is initially higher due to the strong buffering action of food and it returns to a low value after about 1 h. Rather peculiar is the

TABLE 2.4
The pH Values of Human GI Fluids

Anotomical Site	Range
Stomach[a]	1.5–3.5
Duodenum[b]	5–7
Jejunum	6–7
Illeum	7.0
Colon[c]	5.5–7.0
Rectum	7.0

[a]Can be as high as 7. [b]pH of chyme entering duodenum can be as high as 6. [c]Depends on type of food ingested.
Source: From Gruber, P., Longer, M.A., and Robinson, J.R., *Adv. Drug Deliv. Rev.*, 1, 1987. With permission.

TABLE 2.5
Capacities of Parts of the Digestive Tract at Autopsy

Animal	Part of Canal	Average Absolute Capacity (L)
Horse	Stomach	17.96
	Small intestine	63.82
	Cecum	33.54
	Large colon	81.25
	Small colon and rectum	14.77
	Total	211.34
Ox	Stomach	252.5
	Small intestine	66.0
	Cecum	9.9
	Colon and rectum	28.0
	Total	356.4
Sheep and goat	Rumen	23.4
	Reticulum	2.0
	Omasum	0.9
	Abomasum	3.3
	Small intestine	9.0
	Cecum	1.0
	Colon and rectum	4.6
	Total	44.2
Pig	Stomach	8.0
	Small intestine	9.20
	Cecum	1.55
	Colon and rectum	8.70
	Total	27.45
Dog	Stomach	4.33
	Small intestine	1.62
	Cecum	0.09
	Colon and rectum	0.91
	Total	6.95
Cat	Stomach	0.341
	Small intestine	0.114
	Large intestine	0.124
	Total	0.579
Monkey[a]	Stomach	0.1
Human[a]	Stomach	1–1.6

[a]From Tsukita, S., Furuse, M., and Itoh, M., *Nat. Rev. Mol. Cell Biol.*, 2, 285, 2001.
Source: Modified from Stevens, C.E., *Dukes Physiology of Domestic Animals*, Swenson, M.J., Ed., Comstock, Ithaca, NY & London, 1977.

TABLE 2.6
Species Variations in Gastric Emptying Time of Fluid (F) (PEG) and Particulate (P) Markers (2 × 2 mm)

	Gastric Emptying Half-Time (h)		
Species	Diet	F	P
Dog	Chow	1.5	1.5
Pig (young)	High concentrate		
Pig (mature)	Horse diet	2	10
Rabbit	Rabbit diet	1.3	12
Pony	Horse diet	0.3	

Source: Modified from Stevens, C.E., *Dukes Physiology of Domestic Animals*, Swenson, M.J., Ed., Comstock, Ithaca, NY & London, 1977.

physiology of the dog. In this case, whereas the gastric acid secretion rate at the basal state is low [113], upon stimulation (food) gastric acid secretion increases sharply exceeding the levels found in human and pig.

The pH values of the different sections of the GI tract of different animal species and of human are reported in Table 2.3 and Table 2.4. In general, the cardiac region of the stomach was found to have a higher pH value than pyloric region. This is likely due to the fact that the parietal cells tend to be localized in the pyloric part of the stomach. An exception is represented by the rabbit where both the cardiac and pyloric regions display low pH values. It is thought that this phenomenon is due to the fact that the rabbit stomach is not separated into compartments, thus permitting the mix of its contents. Proceeding from the stomach to the small intestine, pH values became progressively more alkaline in the distal portions within the same animal. In the case of monkey, the pH value of the distal ileum is characterized by the lowest level among the animals studied. Further moving down, in the colon, the pH is more acidic than those observed in the small intestine, most likely due to fermentation process which takes place in the colon.

Bile plays a relevant role in the absorption of lipid-soluble vitamins, steroids and dietary lipids [114] acting by enhancing the dissolution rate and solubility. The action of bile is not only limited to this aspect, it extends to the lymphatic uptake of lipids and lipid-soluble drugs [115]. Bile is synthesized in the liver from where, through the hepatic duct, it is secreted into the gallbladder. Secretin and cholecystokinin stimulate the secretion of bile into the duodenum, upon food intake. From the anatomical point of view, the common bile duct enters the duodenum at the duodenal papilla in humans, dogs, and cats. Its orifice is surrounded by the Oddi sphincter usually together with the pancreatic duct, just before entering the duodenum. Anatomical variations are, however, possible. In contrast to humans, dogs, and cats, in rabbits and guinea pig the duodenal papilla is localized far from that of the pancreatic system [116]. In the rat, the strict interdependence among biliary and pancreatic systems is demonstrated by the observation that pancreatic ducts flow directly into the common bile duct [117].

Not only the anatomical differences among species have to be considered, also bile contents should be taken into account. Bile contains cholesterol, phospholipids, and bile salts, which, in human, are mostly represented by cholic acid. This is in contrast to the pig and dog where hyocholic acid [104] and four hydroxyl-group-containing bile acids are encountered [104], respectively.

To the peculiarities observed at the anatomical and biochemical levels among different animal models, the physiological differences should also be added. For example, it is known that the rate of bile fluid secretion within a given animal is regulated by the circadian rhythm [118]. Additionally, in animals such as rat and horse, which do not posses a gallbladder, bile fluid is continuously secreted in dilute form and large volumes. The composition of bile fluid together with the secretion rates in the different animal models is reported in Reference [114].

2.3.2.2 Microflora and Mucin

The microflora present in the GI tract has the capability to metabolize several chemicals, thus affecting intestinal permeation. This phenomenon differently affects different chemicals and the toxicological consequences of their metabolism depend on the chemical metabolized [119]. Among the chemical reactions occurring in the presence of the microflora, are included the hydrolysis of glucosides, glucuronides, sulfate esters, amides, esters, dehydroxylation, deamidation, decharbossilations, dealkylations, deamination, reduction of double bonds, nitro and diazo groups, acetylation, and esterification [119].

There are remarkable differences in the distribution and amount of the microflora in the GI tract among different species [120–122]. In human and rabbit the upper GI tract contains few organisms. More abundant are, in contrast, in the upper intestine of other animals [122]. Additionally, whereas bacteriodes and bifidobacteria are the most represented species in the upper intestine in humans and rabbit, other organisms are present in other animals. Despite the reported differences in the upper GI tract, all animals including humans have a comparable flora in the lower intestine [122]. Together, these observations indicate the relevance of the microflora variable in the interpretation of the drug permeation results obtained in animal models and further stress the caution in the extrapolation of the data to the human system.

In addition to the microflora which can differently affect absorption, another variable, i.e., the presence of mucin, should be considered. Mucin is a water-insoluble, free-flowing viscous gel which covers the epithelial surface of the GI. Mucin has a large protein core bound to oligosaccharide chains (1–39 units) attached on every other fourth amino acid in the backbone. The oligosaccharide chains are constituted by N-acetylgalactosamine, N-acetylglucosamine,

D-galactose, L-fucose, and sialic acid [123]. The presence of these anionic sugar groups confers to mucin an overall anionic character. In most animal species (human, pig, and rat), mucin is in tetramer form (MW, 2×106 Da) and required to protect the epithelium, in the stomach and duodenum, from luminal acid and pepsin [123,124]. The stomach mucin thickness varies among animal species [122] as well as its secretion rate, which, for example in the dog stomach, is significantly higher than that in the rat [125].

From the drug permeation point of view, mucin is important for its property to influence mucosal adhesion of drugs, which may lead to prolonged GI transit [123], thus affecting absorption.

2.3.2.3 Other Physiological Aspects

In addition to the aspects above described, other variables have to be considered for a comprehensive evaluation of the differences existing among humans and the other animal species with regard to the GI tract. The GI capacity for different animals is reported in Table 2.5, where volumes of the different sections of the GI tract are also indicated. The differences are obvious and almost proportional to the size of the animal considered. Additionally, it should be noted that animals such as the sheep and goat have a more complex stomach structure which includes the rumen, reticulum, omasum, and abomasums. More detailed information can be found in the specific scientific literature [103,104,126–128].

Not only the capacity, but also the motility of the GI tract can variably influence drug absorption. In this regard, gastric emptying is an important event, affecting the uptake of drug from the intestine. Gastric motility has two distinct modes: the fed and unfed states. The unfed state has several phases, which repeat every 2 h [125]. The intensity of the contractions is almost zero in phase 1, intermittent in phase 2, and high in phase 3, when nondigested stomach content is expulsed from the stomach to the duodenum. Notably, this cyclic motility pattern is found in commonly used laboratory animals [104]. In the fed state, the stomach contractions reduce compared to those of phase 3 and the mixing and grinding of introduced substances take place.

The movement of nondigestible particles from the stomach to the duodenum strongly depends on the sizes of the pyloric valve. In humans, particles up to 7 mm particles can easily leave the stomach. In contrast, larger particles can be retained for more than 12 h [129]. For comparison, in rabbit, 11 mm coated barium sulfate tablets were retained in the stomach for 24 h while 1 mm diameter granules were released, although not completely, during the same period [130].

Table 2.6 reports the GI transit of fluid and particulate markers in dogs, pigs, rabbits, and ponies. The fluid marker used to study the release from the stomach reveals that stomach–duodenum movement is faster in the pony compared to the other animals considered, with the pig and the rabbit showing the slowest release rate. With regard to the transit through the entire GI tract, dogs and pigs showed the highest percent of excretion in the first 24 h, thus reflecting a rapid transit in the GI tract. In contrast, the rabbit showed the lowest percent of excretion, reflecting the slowest GI transit. As expected, in all cases, particulate substances moved through the GI tract slower than fluid substances.

Concentrating the attention on the transit through the small intestine, the region of the GI tract where drug absorption is considered to be maximal, there appears to be a gradient of velocity in the small intestine. Indeed, for example, in the proximal portion of the small intestine of human and rat, the transit of substances is faster compared to the distal part [127,131]. With regard to the colon transit, in human it ranges from 8 to 72 h with the transit through the ascending colon around 5 h [132]. In general, the rate of passage of digestion markers through the large intestine of the different animal models such as the dog, pig, and pony seems to be related to the relative length and degree of sacculation of the colon [100].

Another physiological aspect to be considered deals with the presence of Peyer's patches (PPs). These are groups of lymphoid follicles distributed within the lamina propria and submucosa of the small intestine which have a central role of antigen uptake and induction of immune response. In humans [133], rodents [134], dogs [135], monkeys [107], and pigs [136], the epithelium covering PPs contains a combination of enterocytes and so-called M cells [137]. Given the implication of M cells in molecules absorption and the different absorption properties compared to enterocytes [137], the knowledge of their distribution may be useful for the specific application.

The number and size of the PPs are directly proportional to the age, body weight, and length of the small intestine. In humans the number of PPs varies from 59 before 30 weeks gestation to 239 at puberty [138]. In general, the ileum contains larger and more numerous patches than jejunum. The number of PPs in healthy adult animals varies from 2 to 10 with a mean of 4. The average size of the PPs is of 7×10 mm [139]. In many species, including ruminants, omnivora, and carnivore, there are several discrete patches in the jejunum and upper ileum. In pigs, there are a series of 20 long, disk-like discrete patches in the small intestine with an average size of 3–8 cm [140,141]. In sheep there are a total of 30–40 PPs. In the dog, there are a total of 26–39 PPs, which are randomly distributed with a large variation between the individual dogs [142]. Finally, in mice, 6 to 12 PPs are distributed along the entire small intestine and in the New Zeland white rabbit, the number of patches increases from the proximal small intestine to the ileum [139].

2.3.3 GI TRACT BIOCHEMISTRY

The positive forces regulating intestinal absorption consist of the concentration gradient, the electrical potential difference, and the hydrostatic pressure gradient, whereas opposing elements are related to the lipid composition of the cell membrane, to the presence of enzymes able to metabolize drug and to the protein complex called "transponder" which affects drug influx/efflux from the cells. Thus, differences at the lipid/protein levels in the membrane composition of the enterocyte may contribute to significantly alter the permeation properties of a given molecule. It is therefore important to take into account these variables when evaluating the drug absorption features of a given animal model.

The structure–function relationship in the intestinal brush border of various animal species has been reported [122,143]. The total lipid-to-protein ratios in the chick, pig, rat, and mouse microvilli membranes are similar (0.5–0.6) with phosphatidylcholine and phosphatidylethanolamine accounting for about 60%–70% of the lipids in all species. The cholesterol/phospholipid ratio appears to be the lowest in the pig and rabbit microvilli membranes [144, 145]. Whereas in the mouse, rat, and rabbit microvillus membranes, the major glycosphingolipid consists of monohexosylceramide [144,146–148], in the pig, digalactosylceramide and pentahesosylcereamide [144] are the major components. In rat microvillus membranes, the monohesosylceramide fraction contains glucose as a sugar constituent [148] whereas in pig membranes, both glucosylceramide and galactosylceramide are present [144].

Not only species variability in the lipid composition of enterocytes has been reported, also differences in relation to the age of the animal are known [149–151]. Microvillus membranes from the jejunum and ileum of suckling rats display a higher sphingomyelin/phosphatidylcholine ratio and are richer in total lipid, cholesterol, and phospholipid per milligram of protein compared to their counterparts from postweaning rats. Additionally, the molar ratio of saturated/*cis*-unsaturated fatty acids and the cholesterol/phospholipid ratio increased significantly from the natal to the suckling period [151,152]. Together these data suggest the relevance to carefully consider the age variable in the permeation test performed in animal model.

The protein component of the brush border membranes is as relevant as the above reported lipid composition. For example, proteins with enzymatic capacities in the brush border are important in the hydrolysis of oligopeptides and oligosaccharides. Furthermore, protein complexes with transport abilities are used for the influx/efflux of different molecules [122,153]. It follows that differences in the quality and location of the proteins in the microvillus membrane can contribute to affect the transport and metabolism of some molecules.

2 Part 2
Skin

2.4 ANATOMY

The human skin, representing the largest organ of the human body, is composed of three layers which differ with respect to the anatomical localization, structure, properties, and embryological origin (Figure 2.2) [154]. The outer layer, derived from the ectoderm, is represented by the epidermis. Epidermis is put in contact with the dermis by the so-called dermo-epidermic junction. The dermis is constituted by a connective tissue layer of mesenchymal origin. Finally, the inner layer is represented by the hypodermis, mainly constituted by connective tissue. The different layers of the skin are crossed by nerve endings and vessels. Additionally, hair follicles and glands, nonuniformly distributed throughout the skin, are present.

The total human skin weight is more than 3 kg and the total surface is of 1.5–2 m^2. The peripheral blood flow through the skin in the extremities and torsum is 0.3 mL/h/cm^3 [155]. The fluxes through the skin cheeks, front, fingers, foot–sole, or palms are somewhat higher. The total surface area of the intracutaneous blood vessels, available for the direct passage of drugs into the systemic circulation, amounts to 100%–200% of the skin area [156].

2.4.1 EPIDERMIS

The epidermis is constituted by a stratified epithelium with a thickness ranging from 50 μm to 1.5 mm. The epidermis can be divided into four layers which are represented by the stratum corneum (SC), granulosum, spinosum, and basale, from the outside to the inside. The SC represents the main barrier against loss of endogenous substances. This thin (1%–10% of the total) skin layer contributes over 80% to the skin permeability resistance, which is high enough to keep transcutaneous water loss at around 0.4 mg/cm^2/h under normal conditions, corresponding to approximately 150 g/day, transpiration excluded. This layer is consequently very dry (≤15% water [157]) and consists of a few dozen flat and partly overlapping, largely dead cells, so-called corneocytes. These cells represent the last step in the differentiation process which begins in the basal stratum with epidermis stem cells. Corneocytes are characterized by an irregular shape, a thick cellular membrane, reduced

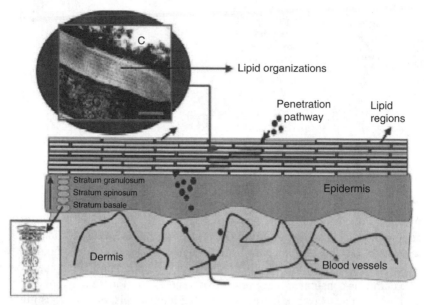

FIGURE 2.2 A schematic cartoon of a skin cross-section. In the basal layer of the epidermis cells proliferate. Upon leaving the basal layer cells start to differentiate and migrate in the direction of the skin surface. At the interface between stratum granulosum and stratum corneum (SC), the viable cells are transformed into dead keratin filled cells (corneocytes). The corneocytes are surrounded by a cell envelope composed of cross-linked proteins and a covalently bound lipid envelope. In SC, the corneocytes are embedded in lipid lamellar regions, which are oriented parallel to the corneocyte surface. Substances permeate mainly along the tortuous pathway in the intercellular lamellar regions. C: corneocyte filled with keratin. Bar 5100 nm. (From Browstra, J.A., and Honeywell-Nguyen, P.L., *Adv. Drug Deliv. Rev.*, 54 Suppl 1, S41, 2002. With permission.)

cytoplasmic water content, cytoplasmic accumulation of keratin and by degenerated nuclear and cellular organelle structures. Alterations are also present at the cell to cell contact structure, i.e., desmosomes, a fact which determines a looser adhesion among the most outer layer of the epidermis. This finally results in the detachment of corneocytes from the epidermis, a phenomenon which approximately involves one layer of corneocytes per day [158].

Corneocytes are organized in columnar clusters comprising groups of 3–10 corneocytes stacks [159]. Individual cell clusters are separated by gorges less than 5–6 μm wide [160]. The depth of gorges (≤3–5 m) is also fairly constant, but their length may differ considerably. The intraepidermal extension of intercluster gorges is 5–10 μm thick. The total estimated gorge area projected on the surface amounts to less than 20% of the entire projected intercellular area, or less than 1% of the total skin surface. Corneocytes in each cluster are very tightly packed and connected to each other through desmosomes [161], covering 15% of the intercellular space length [162]. The intercellular spaces contain specialized multilamellar lipid sheets with variable ultrastructures that are covalently attached to the corneocyte membranes [163]. Cells in each cluster stack tend to overlap at their edges with the cells in adjacent stacks. Such an organization contributes to the tortuosity of the intercellular space in the SC and improves the quality of the skin permeability barrier [164]. Each corneocyte layer (Figure 2.2), essentially coplanar, is approximately 0.3 μm thick and gives the impression of an irregular lattice [165], consisting of cellular hexagons or, more rarely, pentagons with an average cross-section of 25–30 μm [159].

Below the SC there is the stratum granulosum which is composed of 1 up to 4 layers of cells. It represents a transition region between the stratum spinosum which contains living cells and the upper SC containing the keratinized cells described above. The cells present in the stratum granulosum display some of the degeneration process clearly observed in the corneocytes, such as nuclear regression and the appearance of several basophil granulation within the cytoplasm. Below the stratum granulosum there is the stratum spinosum, typically composed of five or more cell layers. The cells present here (keratinocytes) are derived from those of the basal stratum from which they differ for a reduced mitothic capacity, a more condensed chromatin structure, and a flat shape. Keratinocytes are connected by desmosomes which represent a dynamic connection between cells. Indeed, they apparently can separate to allow the passage of cells such as melanocytes and leucocytes and can subsequently connect again. Among the intercellular space, nutrients, oxygen, and liquids can flow and support the cells of the outer layers. The basal stratum is considered to be the source of keratinocytes which subsequently migrate to the outer layers. These cells tend to have a prismatic shape and are connected to each other by desmosomes. In addition to keratinocytes, epidermis contains two other major cell types: melanocytes and langerhans cells. Melanocytes correspond to about one-fourth of the cells present in the basal stratum with a density of 1500 melanocytes/mm^2. These cells, typically surrounded by groups of about 36 keratinocytes, do not have desmoses. Their major function consists of the production of melanin. Langerhans cells, the third most represented cell type in the epidermis, are thought to be derived from macrophages coming from the dermis. They are characterized by the ability to perform phagocytosis.

In the epidermis, just before death, keratinocytes give birth to so-called lamellar, or "Odland" bodies [166,167], lysosomes filled by mostly nonpolar lipids [168]. The vast majority of the lipids in the horny layer are nonpolar and represented by ceramide species, representing ~50% of the total lipid mass, by free fatty acids (~15%), by cholesterol and cholesteryl-esters (~35%), with a total mass of a few mg/cm^2 of the SC. In deeper strata phospholipids are also found [169]. The lipidic component is required for normal barrier function [170]. Keratinocytes shrinkage and flattering, a phenomenon which transforms keratinocytes to corneocyte, provokes the expulsion of the lipid reach vesicles into the intercellular space. Here the vesicles stack together and fuse into ceramide-rich, extended multilamellae [171]. The resulting multilamellar domains are matched together by the domains of poorer organizations that comprise most of the nonceramide lipids.

2.4.1.1 Stratum Corneum Structure

Thus, in SC the corneocytes are embedded in lipid lamellar regions, which are oriented parallel to the corneocyte surface (Figure 2.2). This anatomical organization allows the diffusion of the exogenously applied molecules only through the lipid lamellae in the intercellular regions [172]. The lipid lamellae, which follow the contours of the cells, are arranged in a repeating pattern with electron translucent bands in a broad–narrow–broad sequence [162,171, 173–176]. Additionally, the lipids are organized in a crystalline sublattice and only a small proportion of them are in a liquid phase [177–179]. X-ray investigations revealed a lamellar phase with an unusual long periodicity of 13 nm in mouse (long periodicity) [180]. In human and pig, lipids are organized in two lamellar phases with periodicities of approximately 13 and 6, respectively, the latter is referred to as the short-periodicity phase.

The above-reported composition of the SC lipids strongly differs from that of cell membranes of living cells. The chain lengths of free fatty acids range between C22 and C24, and thus are longer than those of phospholipids present in plasma membranes. The most abundant lipid species, i.e., ceramides, is characterized by acyl chains between 16 and 33 carbons. Additionally, the CER head groups are very small and contain several functional groups that can form inter- and intramolecular hydrogen bonds. There are at least eight subclasses of ceramides (HCER) present in human SC [181,182]. These HCER, often referred to as HCER 1–8, differ from each other by the head-group architecture (sphingosine, phytosphingosine, and a 6-hydroxysphingosine base linked to a fatty acid or an α-hydroxy fatty acid) and the hydrocarbon chain length.

2.4.2 DERMIS

The dermis, an elastic membrane 0.3–4 mm thick, is mostly constituted by connective fibers and cells. The cells are represented by fibroblasts, macrophages, melanocytes, and leukocytes of blood origin. Within the derma there are also muscle fibers, both smooth and striated together with lymphatic and blood vessels, glands and pilosebaceous units. The pilosebaceous unit (Figure 2.3) is an integrated structure of the hair follicle, hair shaft, adjoining arrector pili muscle, and associated sebaceous gland(s) [183]. The hair follicle consists of a hair bulb and shaft enveloped in an inner root sheath, an outer root sheath, and an outermost acellular basement membrane termed the glassy membrane. The outer root sheath is a keratinized layer continuous with the

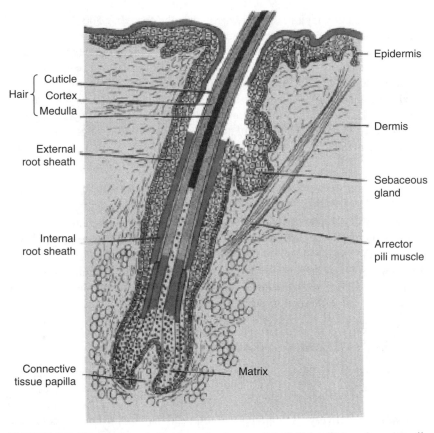

FIGURE 2.3 Cross-sectional diagram of a human hair follicle. (Taken from http://www.tgfolk.net/sites/gtg/hairroot.gif.)

epidermis while the inner root sheath ends about halfway up the follicle. Each hair follicle is associated with one or more sebaceous glands. Ducts join these multilobular holocrine glands to the upper part of the follicular canal. There are two types of human hairs—terminal hairs and vellus hairs. The terminal hairs are macroscopically long (>2 cm), thick (>0.03 mm) [183], and extend more than 3 mm into the hypodermis. The unpigmented vellus hairs are generally short (<2 cm), thin (<0.03 mm), and typically extend just 1 mm into the dermis. Among the other functions, the pilosebaceous unit synthesizes and releases sebum, a mixture of short-chain fatty acids with a fungistatic and bacteriostatic capacities. The secretion of sebum creates an environment rich in neutral, nonpolar lipids in this region of the follicle. Human sebum [184] is composed of 57% of triglycerides, 26% of wax esters, and 2% of squalene. Sebum excretion rate, regulated on the basis of circadian rhythm [185], is of 0.1 mg/cm^2 of skin/h which results in a skin surface content of 0.5 mg/cm^2. The skin density of pilosebaceous units varies greatly

according to body region. On the face and scalp, there are 500–1000 pilosebaceous units/cm^2 with each follicular opening exhibiting a diameter of
approximately 50–100 μm. The combined areas of these orifices may represent as much as 10% of the total surface area of the face and scalp. In other
parts of the body, the follicular openings constitute only about 0.1% of the
total skin area [186]. Whereas it has been calculated that follicular orifices
constitute 1.28% and 0.33% of the available surface on the forehead and back,
respectively, the potential area available for drug absorption is bigger. Indeed, it
includes also the internal follicular surfaces. Considering these areas, values of
13.7% and 1% have been calculated for the forehead and forearm, respectively.

2.4.3 HYPODERMIS

Hypodermis, the thickest layer of the skin, puts dermis in contact with the
underlying anatomical structures such as muscles. The thickness ranges from
0.5 to 2 cm, depending on the age, sex, and body region. Hypodermis is reach
of connective tissue which contains several elastic fibers. The adipose tissue
is present in variable amounts. Depending on the body region, striated
muscles as well as smooth muscles may be present.

2.5 SKIN DELIVERY STRATEGIES

Drug application to the skin can be devoted to treat dermatological disorders
(topical delivery), to treat deep tissues such as muscle and vein (regional
delivery), and to allow the penetration of drugs to the systemic circulation
(transdermal delivery). The goal of the topical delivery is to create a reservoir
of drug in the skin. Penetration of the active molecule to the deeper skin
layers or systemic circulation is not necessary and can even be detrimental.
Regional delivery requires deeper penetration than topical delivery as the
drug has to reach muscles and veins. Finally, transdermal delivery aims to
treat systemic conditions by penetration of drugs into the systemic circulation.

Unfortunately, the peculiar composition of the SC confers the skin remarkable barrier properties, negatively affecting the penetration of ionic and very
highly polar substances [187]. Penetration of lipophilic molecules is greatly
prevented by the hydrophilic epidermis and dermis. Small ampiphilic molecules with a low-melting point and solubility in both oil (SC) and water
(epidermis/dermis) have a better chance of crossing the skin. Large charged
molecules, such as peptides and proteins, are significantly more difficult to
deliver. That is why to date, the number of transdermal products in the market
is very limited.

Whereas the topical deliver substantially does not require to be improved
to achieve therapeutic dosage of the drug at the diseased site, regional and
even more transdermal delivery needs to be implemented by enhancement
methods able to interfere with the SC barrier. So far, four main approaches to

enhance the permeation of molecules through the skin have been investigated: electrically assisted enhancement, vesicular carriers, chemical enhancement, and microinvasive delivery. Of more recent development are the so-called velocity-based techniques [188].

2.5.1 ELECTRICALLY BASED ENHANCEMENT

Five main approaches have been developed in this field, namely ultrasound-facilitated strategies, electroporation, iontophoresis, photochemical waves, and laser ablation. Ultrasound-facilitated strategies (phonophoresis, sonophoresis) involve the use of sound waves (generated by low-energy frequency in the range of 20 kHz) to enhance the delivery of drugs through the skin [189]. It has been shown that sound waves act by increasing diffusion of the drug through the skin by determining either an ultrasound-induced temperature increase (thermal effects), or the formation of large-fluid velocities (acoustic streaming) and of gaseous cavities (cavitation). Cavitation was demonstrated by using nile read (NR) and calcein as lipophilic and hydrophilic model permeants, respectively [190,191]. This experiment showed that ultrasounds induced the formation of localized regions of increased skin permeability and not a uniform increase of skin permeation. This effect is consistent with a cavitation-based mechanism. Notably, the opening of essentially hydrophilic pathways across the skin in this way had a much greater impact on the transport of calcein as compared to nile red (NR).

Improved drug delivery by ultrasound-based approaches has been shown for corticosteroids and nonsteroidal anti-inflammatory drugs to tissues localized below the skin [192,193]. Additionally, low-frequency ultrasound was reported to increase the permeability of human cadaver skin to drugs such as insulin, interferon gamma, and erythropoietin [194]. Furthermore, diffusion coefficients of oestradiol, napthol, aldosterone, lidocaine, and testosterone were shown to be increased by up to 15 times [193,195,196]. With regard to the safety issue, several investigations suggest that low-frequency ultrasound is not detrimental for skin structure and does not cause skin damages. On the contrary, evidence exists for a low-frequency ultrasound-mediated dermal and muscular lesions in hairless rats. Whereas these effects are less pronounced on human skin, it is obvious that further safety studies need to be conducted [197].

Typically used to introduce genetic material into cultured cells for research purposes, electroporation has been subsequently investigated as a potential method for transdermal drug delivery. This method is based on the application of high voltages (<100 mV) for a very short time (ms), which most likely results in the formation of pores in the skin that render it more permeable to large molecules. The mechanisms allowing the drug penetration are thought to be due to a combination of diffusion, local electrophoresis, and electroosmosis [198–200]. So far, different molecules were reported to cross the skin *in vitro* using electroporation. Among these are small, hydrophilic molecules such as mannitol and large and highly charged molecules such as heparin [199]. The employment of this technique is not limited to conventional drugs such as the two mentioned above, it potentially extends to nucleic acid-based molecules [201,202] such as antisense oligodeoxynucleotides (ODN). In the latter case it has been reported that by this approach ODN can be delivered *in vitro* to human epidermis at a rate of >100 ng/cm^2 h [203] and that the transcellular route is followed [204]. Notably, these novel nucleic-acid based molecules attracted the attention of different pharmaceutical companies which are in the process to develop appropriate delivery systems such as electroporation and to test them into clinical trials. Finally, the delivery of therapeutic molecules by electroporation can be improved by combining the drug of interest to different macromolecules. For example, the delivery of mannitol in stripped human epidermis *in vitro* was increased by about 100 times by electroporation and inclusion of macromolecules such as heparin, dextran sulfate, neutral dextran increased delivery of mannitol by another five times. It is supposed that the macromolecules stabilize the pathways or pores that are induced by electroporation [205]. In conclusion, the acceptance of this approach for chronic transdermal administration will be determined by the efficiency/safety ratio assessed over long-term clinical investigations. So far, only few information are available with regard to the safety issue which report muscle contractions occurring with each electrical pulse and the occurrence of mild muscle fatigue after treatment [206].

Iontophoresis consists of the use of an electrical current to push charged molecules through the skin. Ionic molecules are directed into the body by being repulsed from an electrode with the same charge. Thus, electroosmosis drives cations into the skin and anions out of the skin. Additionally, iontophoresis promotes drug skin crossing by generating the electroosmosis effect where fluids move in the opposite direction of current flow. In this process, counter-ions and noncharged compounds are transported along with the fluids. This represents the most relevant mode of transport for neutral molecules, which are not affected by the electric flow. For these kinds of molecules, such as the lipophilic neutral compound NR, the iontophoretic delivery occurs via the lipid-filled intercellular regions of the SC. A peculiar aspect of iontophoresis is that current intensity is directly proportional to the flux of the drug. Thus, the proper modulation of current allows to adjust the amount of drug to be delivered [207–209]. In the case of charged molecules such as calcein, a hydrophilic dye, iontophoresis greatly enhances permeation, both via follicular and nonfollicular route. The fact that the follicular route is particularly enhanced suggests that this path is the predominant route of iontophoretic transport for charged species [210,211].

The iontophoretic approach has been used to deliver polypeptides such as the luteinizing hormone releasing hormone (LHRH), a molecule of 1.2 kDa [209,212]. Notably, as shown for electroporation, it is possible to improve

LHRH iontophoretic delivery by adding chemical enhancers [213,214]. Bigger molecules such as insulin (constituted by oligomer with a molecular weight of about 6 kDa) can be properly delivered by iontophoresis [215]. Other polypeptides successfully delivered by this approach are represented by thyrotropin releasing hormone [216], vasopressin [217], and calcitonin [218]. Additionally, local delivery of anesthetics or corticosteroids such as lidocaine and dexamethasone has been carried out. For a more complete list of drugs delivered by iontophoresis please refer to Reference [188]. Whereas different polypeptides are amenable for iontophoresis-based delivery, a strong inverse dependence between iontophoretic fluxes and molecular size has been shown [219]. Thus, this feature has to be carefully considered when planning an iontophoresis-based delivery approach. Iontophoresis has also the potential to deliver nucleic acid-based molecules such as oligonucleotides (ODN) [204,220] through the skin following the paracellular route. Notably, no skin penetration is observed for ODN by means of passive diffusion.

In conclusion, iontophoresis has the potential to significantly improve drug delivery through the skin. However, this technique is not free from side effects as skin irritation [221] has been reported. This effect seems to be due to [221,222] the microscopic cellular damage at sites of high-current density leading to cytokine and prostaglandin release with local vasodilatation [223]. Additionally, the release of substance-P and calcitonin gene-related peptide at nerve endings in the dermis is thought to contribute to the phenomenon. Notably, skin reactions depend on the site as they were noticed to be more evident at the chest than the abdomen or upper arm [221]. Together, these findings indicate that this aspect needs to be minimized before considering iontophoresis as a technique able to replace injections for the delivery of local anesthetics, analgesics, and biotechnological drugs.

Photomechanical waves (PWs) are generated by the pressure pulses produced by ablation of a material target (polystyrene) by Q-switched or mode-locked lasers [224,225]. It has been proposed that PWs render the SC more permeable to macromolecules via a possible transient permeabilization effect (duration of minutes) [226–230]. Permeabilization seems to depend on changes in the lacunar system of human SC [231] where the transient formation of channels is thought to take place [232]. Whereas PWs have not been extensively used so far, experimental works have been conducted with regard to the delivery of insulin through the skin of diabetic rats [228]. Further investigations are obviously required.

To enhance the topical drug delivery it has been also suggested to remove the SC barrier by controlled ablation [233]. Some success has been obtained in the delivery of [234] hydrocortisone and interferon in a pig model. Additionally, increased permeability of both lipophilic and hydrophilic drugs through nude mouse skin *in vitro* has been achieved [235]. However, the skin structural changes caused by this technique still need to be assessed for safety and reversibility, and thus the usefulness of this technique remains to be verified.

Most of the studies employing an electrically based enhancement of drug delivery did not recur to the combination of different electrically based techniques. Nonetheless, the works exploring the use of a combination of different techniques showed the possibility of achieving synergistic enhancement effects. For example, the combination of skin electroporation followed by iontophoresis has been shown to improve the flux of LHRH approximately fivefold over that of iontophoresis alone [207,236]. Electroporation combined with iontophoresis was also shown to increase the flux of salmon calcitonin through human epidermis compared to other technique used alone [237]. Finally, the combination of low-frequency ultrasound and iontophoresis also increased the flux of heparin across pig skin above that observed for each of the techniques alone [238].

2.5.2 VESICULAR CARRIERS

Ethosomes and transferosomes are phospholipid composed vesicular carriers developed to transfer therapeutic drugs through the skin [239,240]. The attractive feature of these types of vesicles is to be able to achieve a transdermal delivery. This is in contrast to classic liposomes, typically used to transfer molecules through the cytoplasmic membrane of cultured cells [241,242], which remain confined to the upper layers of the skin. This substantially limits the systemic effect [243–248] of the delivered molecule.

Transferosomes are composed mainly of phospholipid and sodium cholate as a biocompatible edge activator [240,249–251]. The possibility to tolerate high deformation stays at the base of their rapid penetration through the intercellular lipid of the SC. Indeed, the stress-induced shape transformation and the osmotic gradient within skin layers are thought to be the major driving forces for their skin penetration [240,251]. Transferosomes have been used to transfer corticosteroid drugs such as triamcinolone acetonide (TAC), used to treat psoriasis, as well as α-2b interferon for the treatment of cancer and viral infections [252].

Ethosomes are innovative vesicular systems [239] constituted by soft phospholipid vesicles together with high concentrations of ethanol. Ethanol fluidizes vesicle and skin lipids thus rendering the skin more penetrable and the vesicles less rigid. The size of ethosomes, vesicle structured in a unilamellar or multilamellar shape, ranges from 30 nm to a few microns. They can entrap molecules with various physico-chemical characteristics such as hydrophobic, hydrophilic, and amphiphilic molecules. Studies performed *in vitro*, *in vivo*, and in clinical trials have all verified the ethosomes' skin permeation-enhancing proprieties for dermal and transdermal delivery of drugs. Ethosomes have been shown effective in the transdermal delivery of trihexyphenidyl hydrochloride (THP), a drug used in the treatment of Parkinson's disease. Additionally, ethosome-mediated testosterone delivery has been shown to be markedly more effective compared to commercial preparation of testosterone. Finally, interesting results were obtained for the delivery of insulin and acyclovir. In the first case a hypoglycaemic effect was observed both for normal and diabetic rats up to 60% in blood glucose levels [253]. In the case of acyclovir, a drug used to treat recurrent herpes labialis, it has been documented that the ethosome-mediated delivery significantly reduced average time to crusting of the lesions and the time for loss of crust compared to available commercial preparation [254].

The studies reported above strongly suggest the higher efficacies of ethosome-mediated transdermal drug delivery compared to commonly used systems. This feature, together with the lack of significant skin irritant effects, makes ethosomes a very attractive delivery technology which may contribute to substantially improve the transdermal delivery of the drugs.

2.5.3 CHEMICAL ENHANCEMENT

Chemical penetration enhancers can influence a number of different parameters to facilitate the permeation of drugs through the skin. They can enhance diffusion through the SC, increase partitioning into the SC and extract skin lipids, essentially removing the barrier function. Azone (1 dodecylazacyclo-heptan-2-one), one of the most studied diffusion enhancers, facilitates the skin penetration of both lipophilic and hydrophilic drugs [255,256] by affecting the packing of SC lipids. This is thought to improve drug diffusivity through the skin [257]. Other related enhancers have been also studied [258,259], and some have been incorporated into products that were launched or are in advanced stage clinical trials. For example, enhancers have been used to improve the absorption of drugs involved in the treatment of erectile dysfunction, female sexual dysfunction, hairloss (minoxidil), pain (ibuprofen), oestrogen deficiencies (oestradiol), and antifungal infection (econazole) [260,261]. Fatty acids, esters, urea, terpene, and alcohols represent other commonly investigated enhancers used in commercial products. Among these, one of the most intensely studied is oleic acid. It has been suggested that it undergoes phase separation within the SC thus providing pathways of lesser resistance for the penetrating drug [179,198]. Oleic acid is used, for example, in the transdermal delivery of oestradiol [261]. Another frequently considered enhancer is represented by ethanol which enhances permeation of both hydrophobic and hydrophilic compounds [262,263]. It has been suggested that in concentrations <50% it functions as an agent fluidizing the lipids of the SC [264]. Interestingly, in combination with other enhancing agents, ethanol may further increase penetration [262–264]. Ethanol has been employed, for example, in enhancing the delivery of testosterone across nonscrotal skin. Additionally, it has been employed to deliver oestradiol and fentanyl, drugs used for the treatment of estrogen deficiencies and pain management, respectively. It should be reminded that the choice of the optimal enhancer for a given drug is not an easy task. However, databases constructed to help the selection of the most appropriate enhancer are available [265].

In conclusion, chemical enhancers are frequently used to improve drug skin penetration. They are especially indicated to augment permeation of drugs which needs a moderate degree of improvement. However, for molecules with extremely poor percutaneous absorption, such as large hydrophilic or ionized drugs, these enhancers will probably not be effective and more sophisticated technologies are needed.

2.5.4 MICROINVASIVE DELIVERY

The microinvasive delivery is based on the use of microfabricated needle. They represent a relatively new delivery technique, designed to create a physical pathway through the upper epidermis to increase skin permeability [266]. Silicon needles of approximately 150 μm in length and 80 μm base diameter are built onto arrays of approximately 3×3 mm. The sensation caused by microneedles is comparable to a smooth surface eliciting substantially no pain in comparison to injection by hypodermic needle.

With regard to the efficiency of skin delivery, it has been calculated that the permeability of the fluorescent dye calcein (molecular weight of 622 Da) delivered by microneedles increases [266] up to four orders of magnitude. In the therapeutic field, the delivery of anesthetics, the long-term administration of opioids, and the monitoring of blood chemistry have been proposed to be performed by microneedle. Other potential applications involve the delivery of insulin [267] as well as DNA and/or protein vaccines, with this last application probably representing the most attractive use of microneedle injection.

Apparently, no redness or swelling of the skin is reported [268] upon microneedle injection. However, the invasion of different anatomical components of the skin by the needle may destroy different skin structure possibly triggering side effects—as yet—unknown. Obviously, the possible detrimental effect on the skin is expected to depend on the microneedle length.

2.5.5 VELOCITY-BASED TECHNIQUE

The idea of skin-penetrating jet injectors dates back to over 50 years [269]. More recently, it has been evolved into the possibility to accelerate DNA- or RNA-coated gold or tungsten microparticles (1–3 μm in diameter) through the wall of target cells where the DNA or RNA can be expressed by the cell [270,271]. The transdermal application of this technique makes use of a device able to propel the drugs into the skin under helium gas pressure [188]. Particles speed and distance of administration are chosen on the basis of the physical properties of the delivery device, particle size, velocity of the carrier gas, and discharge pressure [272,273]. Notably, the targeting of the active agents to intracellular or intercellular sites has been achieved through manipulation of parameters such as particle size, density, and durability. For intracellular administration, particles are generally kept in the 1–3 μm size range whereas for intercellular targeting, agents are usually formulated into powder-like particles with sugar excipients to produce particles in the 20–70 μm size range [274].

Studies in the mid to late 1980s reported better absorption rates for insulin than traditional needle injection [275–277]. Encouraging results have also been obtained with the anaesthetic drug lidocaine [272]. Most interest is, however, devoted to the delivery of DNA and protein vaccines into the epidermis. The highly immune reactive nature of the epidermis makes it an ideal target for vaccine delivery compared to traditional needle delivery. Indeed, with this last technique, the vaccine agents are delivered below the epidermis into the dermis where a much lower level of immune stimulation is thought to be elicited [274]. Immunization studies performed by jet injectors in pigs (using DNA to elicit HbsAg-specific antibodies [278] or expressing influenza A hemagglutinin) [279] and monkeys [280] suggested the possibility to extend this technique to human being.

Together, these results suggest that jet injection administration of macromolecules is feasible although further refinement of the technique is still needed.

2.6 ANIMAL MODEL OF SKIN DELIVERY

2.6.1 PIG

Detailed information about skin absorption and percutaneous permeation is of utmost relevance for skin pharmacological and toxicological research. Advance in these fields depends largely on the availability of alternative methods to human experimentation. *In vitro* methods, using human or animal skin, can serve the purpose. However, without a reliable *in vivo* animal model, definitive conclusions are difficult to be obtained. Thus, animal models have been largely considered. Among these, the pig model is of particular interest. The anatomical and physiological characteristics, common between pig and man, indicate that the pig represents a reliable model to study percutaneous permeation in man. The pig, like man, is of median body size, large enough for collection of multiple samples (body fluids, biopsies) and at the same time is not too large to be conveniently handled in standard laboratory animal facilities. Additionally and importantly, the skin of pig and man are similar as both are only sparsely covered with hair. Moreover, it is possible to select pig strains with different amounts of hair [281–285].

Man and pig skins are both characterized by a sparse hair coat and a thick epidermis, which, when judged by parameters of tissue turnover time and the characterization of keratinous proteins [286], are comparable. Additionally, swine SC contains protein fractions grossly similar to human and presents a filament density and areas of cell overlapping [287] comparable as well. The epidermal–dermal junction [287] is also similar in the pig and man. Dermis is comparable between man and pig where it is characterized by a well-differentiated papillary body and a large content of elastic tissue [288,289]. Moreover, the architecture of collagen fibers and fiber bundles in the dermis of the pig generally corresponded to those observed in human skin, including the thickness of collagen fibrils [290,291]. Finally, biochemical similarities were also observed studying glycosphingolipids and ceramides in human and pig epidermis [292] as well as the enzyme patterns of the skin of the domestic pig and man [293].

The cutaneous vascular anatomy, and at least certain vascular reactions, is similar in man and the domestic pig [294]. However, some dissimilarities have been observed with regard to the amount of vessels which are more abundant in man than in pig [295]. Additionally, differences have been revealed in skin glands. Humans have mostly eccrine sweat glands over the body surface, whereas the pig has only apocrine glands [295].

Percutaneous absorption in man varies depending on the area of the body on which the chemical resides as well as [296,297] on the age and sex [298]. Although some information on regional variation of percutaneous absorption in animal is available, little is known about this issue in the pig. Most of the percutaneous permeation studies conducted in pigs were performed on their back [299–306], the ear [307–311], the flank [312,313], or the abdomen [314]. It is difficult to state whether the choice of a certain skin site is equivalent to another in the pig or if it can substantially influence the test. Therefore, this variable, i.e., skin site, should be carefully considered when doing *in vivo* test using the pig animal model.

Not only anatomical and biochemical analyses suggest that pig and man skin are comparable, also permeation tests substantially confirm the similarities. *In vitro* permeation of skin to water revealed that human and porcine skins were similar with respect to water permeation [299] as well as other compounds [300,301,314]. A study on the permeation of hexachlorophene through the skin indicated that concentrations of this compound in the blood in man and the pig were close, but different from those in rats and monkeys [302,315]. Permeation of paraquat, carbaryl, aldrin, and fluazifop-butyl was also very close in man and pig skin [307]. *In vivo* studies involving classified chemical warfare agents and agent simulants showed that the back skin of weanling pigs most closely approximated human forearm skin with regard to its resistance to permeation [305].

2.6.2 MOUSE

Given the reduced hair present on human skin, it is evident that a hairless mouse model is necessary for the investigations in the skin pharmacology field. With this model, shaving is unnecessary, and, even more importantly, the risk of damage is avoided. Despite these advantages, it is important to consider to what extent the skin of the hairless mouse resembles the human skin in terms of anatomy and molecule absorption. This evaluation can contribute to understanding whether or not the hairless mouse represents a reliable model for the study of the absorption of the molecule under consideration.

Whereas the observation of hairless mice has been first reported in the nineteenth century [316], a more precise characterization of this animal model has been performed in the twentieth century [317–324]. It should be pointed out that the hairless mice stem from a recessive character, whereas the "naked" animal is due to a partially dominant mutation [321]. Mice carrying the "naked gene" in a single copy resemble those homozygous for the recessive hairless character. Mice carrying the gene in a double copy are completely naked from birth.

As a result of the studies performed so far, it has been observed that, regarding the skin morphology, the SC of the hairless mouse is less than half as thick as that of the human tissue and thus characterized by lower barrier properties [325]. *In vivo* data, studying the percutaneous penetration of various compounds in several laboratory animals, including the nude mouse (Balb/c), corroborate the concept of the different absorption properties [301]. Statistically significant correlations were not obtained between the nude mouse and other species, including human. Additionally, another work suggested that the *in vivo* hairless mouse does not represent a useful predictive model for human skin *in vitro* permeability [326] as mouse skin displays a greater permeability than human skin. On the contrary, however, data have been reported in favor of a substantial similarity between nude mouse (BALB/cAnNCrj-nu/nu) and human skin absorption; at least for what concerns the absorption of benzene, toluene, and tetrachloroethylene vapor [327]. Together, these data suggest that the degree of suitability largely depends on the compound under investigation.

More experimental work has been conducted *in vitro* comparing the permeation properties of hairless mouse and human skin. A good correlation between the absorption properties has been revealed for the penetration of erythromycin esters [328], several glucocorticosteroids, and alcohols [329]. Additionally, a highly significant fit between human/mice skin (strain HRS-hr/hr) penetration of a series of phenols molecules was observed [330]. In SKH-HR-I hairless mice percutaneous penetration kinetics of nitroglycerin was comparable to that observed in man [331,332]. An intriguing experiment was performed to evaluate the skin barrier function after cold exposure. A hairless mice skin was used and the results indicated that the changes in transepidermal water loss parallelled that occurred in workers in the fish-processing industry (FPI), a phenomenon implicated in the well-known occupational dermatosis of FPI [333]. Different results were obtained in other experimental setups. A quantitatively different mass transfer was observed in SKH-hr-1 mice compared to human skin with regard to the percutaneous absorption of corticosteroid compounds. In particular, it was reported that human skin is at least one logarithmic order less permeable than the skin model used [334]. Another investigation regarding the permeation of hydrocortisone in human and hairless mouse skin (SKH-HR-1) reported that the peak permeability in human epidermis was an order of magnitude smaller than that observed for the mouse skin model considered [335]. Finally, differences were observed between the penetration of lidocaine through human and hairless mouse (SKH-HR-I) skin, suggesting that, under the experimental condition used, the hairless mouse skin was not an adequate model [336].

In conclusion, the suitability of the hairless mice skin model largely depends on the physicochemical properties of the compound under investigation as well as on the possible influences of enhancer molecules. The logical suggestion is to first establish a correlation between the results obtained in the hairless mouse and man for a given molecule. Only when the correlation is evident, further experiments in the mice skin model are fully authorized. It follows that the evaluation of the percutaneous penetration/skin absorption of new compounds using the mice model only is extremely questionable.

REFERENCES

1. Said, H.M., Redha, R., and Nylander,W., A carrier-mediated, Na^+ gradient-dependent transport for biotin in human intestinal brush-border membrane vesicles, *Am. J. Physiol.*, 253, G631, 1987.
2. Strum, W.B., Characteristics of the transport of pteroylglutamate and amethopterin in rat jejunum, *J. Pharmacol. Exp. Ther.*, 216, 329, 1981.
3. Farquhar, M.G. and Palade, G.E., Junctional complexes in various epithelia, *J. Cell Biol.*, 17, 375, 1963.
4. Trier, J.S. and Madara, J.L., Functional morphology of the mucosa of the small intestine, in *Physiology of the Gastrointestinal Tract*, Johnson, L.R., Ed., Raven Press, NY, 1981.
5. Forstner, G.G., Tanaka, K., and Isselbacher, K.J., Lipid composition of the isolated rat intestinal microvillus membrane, *Biochem. J.*, 109, 51, 1968.
6. Bouhours, J.F. and Glickman, R.M., Rat intestinal glycolipids. III. Fatty acids and long chain bases of glycolipids from villus and crypt cells, *Biochem. Biophys. Acta*, 487, 51, 1977.
7. Miyoshi, J. and Takai, Y., Molecular perspective on tight-junction assembly and epithelial polarity, *Adv. Drug Deliv. Rev.*, 57, 815, 2005.
8. Van Meer, G. and Simons, K., The function of tight junctions in maintaining differences in lipid composition between the apical and the basolateral cell surface domains of MDCK cells, *EMBO J.*, 5, 1455, 1986.
9. Tsukita, S., Furuse, M., and Itoh, M., Structural and signalling molecules come together at tight junctions, *Curr. Opin. Cell Biol.*, 11, 628, 1999.
10. Tsukita, S., Furuse, M., and Itoh, M., Multifunctional strands in tight junctions, *Nat. Rev. Mol. Cell Biol.*, 2, 285, 2001.
11. Claude, P., Morphological factors influencing transepithelial permeability: a model for the resistance of the zonula occludens, *J. Membr. Biol.*, 39, 219, 1978.

12. Takeichi, M., Cadherin cell adhesion receptors as a morphogenetic regulator, *Science*, 251, 1451, 1991.
13. Furuse, M., Hirase, T., Itoh, M., Nagafuchi, A., Yonemura, S., and Tsukita, S., Occludin: a novel integral membrane protein localizing at tight junctions, *J. Cell Biol.*, 123, 1777, 1993.
14. Furuse, M., Fujita, K., Hiiragi, T., Fujimoto, K., and Tsukita, S., Claudin-1 and -2: novel integral membrane proteins localizing at tight junctions with no sequence similarity to occludin, *J. Cell Biol.*, 141, 1539, 1998.
15. Martin-Padura, I., Lostaglio, S., Schneemann, M., Williams, L., Romano, M., Fruscella, P., Panzeri, C., Stoppacciaro, A., Ruco, L., Villa, A., Simmons, D., and Dejana, E., Junctional adhesion molecule, a novel member of the immunoglobulin superfamily that distributes at intercellular junctions and modulates monocyte transmigration, *J. Cell Biol.*, 142, 117, 1998.
16. Fujimoto, K., Freeze-fracture replica electron microscopy combined with SDS digestion for cytochemical labeling of integral membrane proteins. Application to the immunogold labeling of intercellular junctional complexes, *J. Cell Sci.*, 108 (Pt 11), 3443, 1995.
17. Naik, U.P., Naik, M.U., Eckfeld, K., Martin-DeLeon, P., and Spychala, J., Characterization and chromosomal localization of JAM-1, a platelet receptor for a stimulatory monoclonal antibody, *J. Cell Sci.*, 114, 539, 2001.
18. Hirabayashi, S., Tajima, M., Yao, I., Nishimura, W., Mori, H., and Hata, Y., JAM4, a junctional cell adhesion molecule interacting with a tight junction protein, MAGI-1, *Mol. Cell Biol.*, 23, 4267, 2003.
19. Smith, P.L., Wall, D.A., and Wilson, G., Drug carriers for the oral administration and transport of peptide drugs across the gastrointestinal epithelium, in *Multiparticulate Oral Drug delivery*, Ghebre-Sellassie, I., Ed., Marcel Dekker, NY, 1994.
20. Levitt, D.G., Hakim, A.A., and Lifson, N., Evaluation of components of transport of sugars by dog jejunum *in vivo*, *Am. J. Physiol.*, 217, 777, 1969.
21. Fix, J.A., Strategies for delivery of peptides utilizing absorption-enhancing agents, *J. Pharm. Sci.*, 85, 1282, 1996.
22. Brayden, D.J., Creed, E., Meehan, E., and O'Malley, K.E., Passive transepithelial diltiazem absorption across intestinal tissue leading to tight junction openings, *J. Control Release*, 38, 193, 1996.
23. McDonnell, W.M., Scheiman, J.M., and Traber, P.G., Induction of cytochrome P450IA genes (*CYP1A*) by omeprazole in the human alimentary tract, *Gastroenterology*, 103, 1509, 1992.
24. Kolars, J.C., Lown, K.S., Schmiedlin-Ren, P., Ghosh, M., Fang, C., Wrighton, S.A., Merion, R.M., and Watkins, P.B., *CYP3A* gene expression in human gut epithelium, *Pharmacogenetics*, 4, 247, 1994.
25. Watkins, P.B., Wrighton, S.A., Schuetz, E.G., Molowa, D.T., and Guzelian, P.S., Identification of glucocorticoid-inducible cytochromes P-450 in the intestinal mucosa of rats and man, *J. Clin. Invest.*, 80, 1029, 1987.
26. Hebert, M.F., Roberts, J.P., Prueksaritanont, T., and Benet, L.Z., Bioavailability of cyclosporine with concomitant rifampin administration is markedly less than predicted by hepatic enzyme induction, *Clin. Pharmacol. Ther.*, 52, 453, 1992.
27. Lorico, A., Rappa, G., Finch, R.A., Yang, D., Flavell, R.A., and Sartorelli, A.C., Disruption of the murine MRP (multidrug resistance protein) gene leads to

increased sensitivity to ctoposide (VP-16) and increased levels of glutathione, *Cancer Res.*, 57, 5238, 1997.

28. Gottesman, M.M. and Pastan, I., Biochemistry of multidrug resistance mediated by the multidrug transporter, *Annu. Rev. Biochem.*, 62, 385, 1993.

29. Keppler, D., Leier, I., and Jedlitschky, G., Transport of glutathione conjugates and glucuronides by the multidrug resistance proteins MRP1 and MRP2, *Biol. Chem.*, 378, 787, 1997.

30. Kim, R.B., Transporters and drug discovery: why, when, and how, *Mol. Pharm.*, 3, 26, 2006.

31. Ling, V., Multidrug resistance: molecular mechanisms and clinical relevance, *Cancer Chemother. Pharmacol.*, 40 Suppl, S3, 1997.

32. Patel, N.H. and Rothenberg, M.L., Multidrug resistance in cancer chemotherapy, *Invest. New Drugs*, 12, 1, 1994.

33. Thiebaut, F., Tsuruo, T., Hamada, H., Gottesman, M.M., Pastan, I., and Willingham, M.C., Cellular localization of the multidrug-resistance gene product P-glycoprotein in normal human tissues, *Proc. Natl. Acad. Sci. USA*, 84, 7735, 1987.

34. Stephens, R.H., O'Neill, C.A., Warhurst, A., Carlson, G.L., Rowland, M., and Warhurst, G., Kinetic profiling of P-glycoprotein-mediated drug efflux in rat and human intestinal epithelia, *J. Pharmacol. Exp. Ther.*, 296, 584, 2001.

35. Huisman, M.T., Smit, J.W., Wiltshire, H.R., Hoetelmans, R.M., Beijnen, J.H., and Schinkel, A.H., P-glycoprotein limits oral availability, brain, and fetal penetration of saquinavir even with high doses of ritonavir, *Mol. Pharmacol.*, 59, 806, 2001.

36. Hochman, J.H., Chiba, M., Yamazaki, M., Tang, C., and Lin, J.H., P-glycoprotein-mediated efflux of indinavir metabolites in Caco-2 cells expressing cytochrome P450 3A4, *J. Pharmacol. Exp. Ther.*, 298, 323, 2001.

37. Dautrey, S., Felice, K., Petiet, A., Lacour, B., Carbon, C., and Farinotti, R., Active intestinal elimination of ciprofloxacin in rats: modulation by different substrates, *Br. J. Pharmacol.*, 127, 1728, 1999.

38. Sparreboom, A., van Asperen, J., Mayer, U., Schinkel, A.H., Smit, J.W., Meijer, D.K., Borst, P., Nooijen, W.J., Beijnen, J.H., and van Tellingen, O., Limited oral bioavailability and active epithelial excretion of paclitaxel (Taxol) caused by P-glycoprotein in the intestine, *Proc. Natl. Acad. Sci. USA*, 94, 2031, 1997.

39. Wacher, V.J., Silverman, J.A., Zhang, Y., and Benet, L.Z., Role of P-glycoprotein and cytochrome P450 3A in limiting oral absorption of peptides and peptidomimetics, *J. Pharm. Sci.*, 87, 1322, 1998.

40. Marzolini, C., Paus, E., Buclin, T., and Kim, R.B., Polymorphisms in human MDR1 (P-glycoprotein): recent advances and clinical relevance, *Clin. Pharmacol. Ther.*, 75, 13, 2004.

41. Kool, M., de Haas, M., Scheffer, G.L., Scheper, R.J., van Eijk, M.J., Juijn, J.A., Baas, F., and Borst, P., Analysis of expression of cMOAT (MRP2), MRP3, MRP4, and MRP5, homologues of the multidrug resistance-associated protein gene (MRP1), in human cancer cell lines, *Cancer Res.*, 57, 3537, 1997.

42. Smith, P.L., Wall, D.A., Gochoco, C.H., and Wilson, G., Routes of delivery: case studies. Oral absorption of peptides and proteins, *Adv. Drug Deliv. Rev.*, 8, 253, 1992.

43. Tsuji, A. and Tamai, I., Carrier-mediated intestinal transport of drugs, *Pharm. Res.*, 13, 963, 1996.

44. Arimori, K. and Nakano, M., Drug exsorption from blood into the gastrointestinal tract, *Pharm. Res.*, 15, 371, 1998.
45. Lee, V.H., Membrane transporters, *Eur. J. Pharm. Sci.*, 11 Suppl 2, S41, 2000.
46. Hidalgo, I.J. and Li, J., Carrier-mediated transport and efflux mechanisms in Caco-2 cells, *Adv. Drug Deliv. Rev.*, 22, 53, 1996.
47. Fisher, R.B., Active transport of salicylate by rat jejunum, *Q.J. Exp. Physiol.*, 66, 91, 1981.
48. Bai, J.P., Hu, M., Subramanian, P., Mosberg, H.I., and Amidon, G.L., Utilization of peptide carrier system to improve intestinal absorption: targeting prolidase as a prodrug-converting enzyme, *J. Pharm. Sci.*, 81, 113, 1992.
49. Humphrey, M.J. and Ringrose, P.S., Peptides and related drugs: a review of their absorption, metabolism, and excretion, *Drug Metab. Rev.*, 17, 283, 1986.
50. Hu, M. and Amidon, G.L., Passive and carrier-mediated intestinal absorption components of captopril, *J. Pharm. Sci.*, 77, 1007, 1988.
51. Kramer, W., Girbig, F., Gutjahr, U., Kleemann, H.W., Leipe, I., Urbach, H., and Wagner, A., Interaction of renin inhibitors with the intestinal uptake system for oligopeptides and beta-lactam antibiotics, *Biochem. Biophys. Acta*, 1027, 25, 1990.
52. Yamashita, S., Yamazaki, Y., Masada, M., Nadai, T., Kimura, T., and Sazaki, H., Investigations of flux and accumulation processes of beta-lactam antibiotics and their role in the transmural transfer across rat jejunum, *J. Pharm. Dyn.*, 9, 369, 1986.
53. Herrera-Ruiz, D., Wang, Q., Gudmundsson, O.S., Cook, T., Smith, R.L., Faria, T.N., and Knipp, G.T., Spatial expression patterns of peptide transporters in the human and rat gastrointestinal tracts, Caco-2 *in vitro* cell culture model, and multiple human tissues, *AAPS Pharm. Sci.*, 3, 1, 2001.
54. Kullak-Ublick, G.A., Fisch, T., Oswald, M., Hagenbuch, B., Meier, P.J., Beuers, U., and Paumgartner, G., Dehydroepiandrosterone sulfate (DHEAS): identification of a carrier protein in human liver and brain, *FEBS Lett.*, 424, 173, 1998.
55. van Montfoort, J.E., Hagenbuch, B., Fattinger, K.E., Muller, M., Groothuis, G.M., Meijer, D.K., and Meier, P.J., Polyspecific organic anion transporting polypeptides mediate hepatic uptake of amphipathic type II organic cations, *J. Pharmacol. Exp. Ther.*, 291, 147, 1999.
56. van Montfoort, J.E., Muller, M., Groothuis, G.M., Meijer, D.K., Koepsell, H., and Meier, P.J., Comparison of "type I" and "type II" organic cation transport by organic cation transporters and organic anion-transporting polypeptides, *J. Pharmacol. Exp. Ther.*, 298, 110, 2001.
57. Lee, W., Glaeser, H., Smith, L.H., Roberts, R.L., Moeckel, G.W., Gervasini, G., Leake, B.F., and Kim, R.B., Polymorphisms in human organic anion-transporting polypeptide 1A2 (OATP1A2): implications for altered drug disposition and central nervous system drug entry, *J. Biol. Chem.*, 280, 9610, 2005.
58. Dyer, J., Beechey, R.B., Gorvel, J.P., Smith, R.T., Wootton, R., and Shirazi-Beechey, S.P., Glycyl-L-proline transport in rabbit enterocyte basolateral-membrane vesicles, *Biochem. J.*, 269, 565, 1990.
59. Clark, S.L., Jr., The ingestion of proteins and colloidal materials by columnar absorptive cells of the small intestine in suckling rats and mice, *J. Biophys. Biochem. Cytol.*, 5, 41, 1959.

60. Kraehenbuhl, J.P. and Campiche, M.A., Early stages of intestinal absorption of specific antibodies in the newborn. An ultrastructural, cytochemical, and immunological study in the pig, rat, and rabbit, *J. Cell Biol.*, 42, 345, 1969.
61. Rodewald, R., Selective antibody transport in the proximal small intestine of the neonatal rat, *J. Cell Biol.*, 45, 635, 1970.
62. Rodewald, R. and Kraehenbuhl, J.P., Receptor-mediated transport of IgG, *J. Cell Biol.*, 99, 159s, 1984.
63. Mostov, K.E. and Simister, N.E., Transcytosis, *Cell*, 43, 389, 1985.
64. Pappenheimer, J.R. and Reiss, K.Z., Contribution of solvent drag through intercellular junctions to absorption of nutrients by the small intestine of the rat, *J. Membr. Biol.*, 100, 123, 1987.
65. Madara, J.L. and Pappenheimer, J.R., Structural basis for physiological regulation of paracellular pathways in intestinal epithelia, *J. Membr. Biol.*, 100, 149, 1987.
66. Pade, V. and Stavchansky, S., Estimation of the relative contribution of the transcellular and paracellular pathway to the transport of passively absorbed drugs in the Caco-2 cell culture model, *Pharm. Res.*, 14, 1210, 1997.
67. DeMeo, M.T., Mutlu, E.A., Keshavarzian, A., and Tobin, M.C., Intestinal permeation and gastrointestinal disease, *J. Clin. Gastroenterol.*, 34, 385, 2002.
68. Hermiston, M.L. and Gordon, J.I., Inflammatory bowel disease and adenomas in mice expressing a dominant negative N-cadherin, *Science*, 270, 1203, 1995.
69. Ukabam, S.O., Clamp, J.R., and Cooper, B.T., Abnormal small intestinal permeability to sugars in patients with Crohn's disease of the terminal ileum and colon, *Digestion*, 27, 70, 1983.
70. Bjarnason, I., O'Morain, C., Levi, A.J., and Peters, T.J., Absorption of 51 chromium-labeled ethylenediaminetetraacetate in inflammatory bowel disease, *Gastroenterology*, 85, 318, 1983.
71. Andre, F., Andre, C., Emery, Y., Forichon, J., Descos, L., and Minaire, Y., Assessment of the lactulose–mannitol test in Crohn's disease, *Gut*, 29, 511, 1988.
72. Katz, K.D., Hollander, D., Vadheim, C.M., McElree, C., Delahunty, T., Dadufalza, V.D., Krugliak, P., and Rotter, J.I., Intestinal permeability in patients with Crohn's disease and their healthy relatives, *Gastroenterology*, 97, 927, 1989.
73. Wyatt, J., Vogelsang, H., Hubl, W., Waldhoer, T., and Lochs, H., Intestinal permeability and the prediction of relapse in Crohn's disease, *Lancet*, 341, 1437, 1993.
74. Adenis, A., Colombel, J.F., Lecouffe, P., Wallaert, B., Hecquet, B., Marchandise, X., and Cortot, A., Increased pulmonary and intestinal permeability in Crohn's disease, *Gut*, 33, 678, 1992.
75. Hollander, D., The intestinal permeability barrier. A hypothesis as to its regulation and involvement in Crohn's disease, *Scand. J. Gastroenterol.*, 27, 721, 1992.
76. Smecuol, E., Bai, J.C., Vazquez, H., Kogan, Z., Cabanne, A., Niveloni, S., Pedreira, S., Boerr, L., Maurino, E., and Meddings, J.B., Gastrointestinal permeability in celiac disease, *Gastroenterology*, 112, 1129, 1997.
77. Juby, L.D., Rothwell, J., and Axon, A.T., Cellobiose/mannitol sugar test—a sensitive tubeless test for celiac disease: results on 1010 unselected patients, *Gut*, 30, 476, 1989.
78. Heyman, M., Darmon, N., Dupont, C., Dugas, B., Hirribaren, A., Blaton, M.A., and Desjeux, J.F., Mononuclear cells from infants allergic to cow's milk secrete

tumor necrosis factor alpha, altering intestinal function, *Gastroenterology*, 106, 1514, 1994.

79. Heyman, M. and Desjeux, J.F., Cytokine-induced alteration of the epithelial barrier to food antigens in disease, *Ann. NY Acad. Sci.*, 915, 304, 2000.

80. Wyatt, J., Oberhuber, G., Pongratz, S., Puspok, A., Moser, G., Novacek, G., Lochs, H., and Vogelsang, H., Increased gastric and intestinal permeability in patients with Crohn's disease, *Am. J. Gastroenterol.*, 92, 1891, 1997.

81. van Elburg, R.M., Heymans, H.S., and De Monchy, J.G., Effect of disodiumcromoglycate on intestinal permeability changes and clinical response during cow's milk challenge, *Pediatr. Allergy Immunol.*, 4, 79, 1993.

82. Dupont, C., Barau, E., Molkhou, P., Raynaud, F., Barbet, J.P., and Dehennin, L., Food-induced alterations of intestinal permeability in children with cow's milk-sensitive enteropathy and atopic dermatitis, *J. Pediatr. Gastroenterol. Nutr.*, 8, 459, 1989.

83. Schrander, J.J., Unsalan-Hooyen, R.W., Forget, P.P., and Jansen, J., EDTA intestinal permeability in children with cow's milk intolerance, *J. Pediatr. Gastroenterol. Nutr.*, 10, 189, 1990.

84. Hamilton, I., Hill, A., Bose, B., Bouchier, I.A., and Forsyth, J.S., Small intestinal permeability in pediatric clinical practice, *J. Pediatr. Gastroenterol. Nutr.*, 6, 697, 1987.

85. Ryan, C.M., Schmidt, J., Lewandrowski, K., Compton, C.C., Rattner, D.W., Warshaw, A.L., and Tompkins, R.G., Gut macromolecular permeability in pancreatitis correlates with severity of disease in rats, *Gastroenterology*, 104, 890, 1993.

86. Hidalgo, I.J., Assessing the absorption of new pharmaceuticals, *Curr. Top. Med. Chem.*, 1, 385, 2001.

87. Grasset, E., Pinto, M., Dussaulx, E., Zweibaum, A., and Desjeux, J.F., Epithelial properties of human colonic carcinoma cell line Caco-2: electrical parameters, *Am. J. Physiol.*, 247, C260, 1984.

88. Hidalgo, I.J. and Borchardt, R.T., Transport of bile acids in a human intestinal epithelial cell line, Caco-2, *Biochem. Biophys. Acta*, 1035, 97, 1990.

89. Hidalgo, I.J. and Borchardt, R.T., Transport of a large neutral amino acid (phenylalanine) in a human intestinal epithelial cell line: Caco-2, *Biochem. Biophys. Acta*, 1028, 25, 1990.

90. Tavelin, S., Milovic, V., Ocklind, G., Olsson, S., and Artursson, P., A conditionally immortalized epithelial cell line for studies of intestinal drug transport, *J. Pharmacol. Exp. Ther.*, 290, 1212, 1999.

91. Pinto, M., Robine-Leon, S., Appay, M.D., Kedinger, M., Triadou, N., Dussaulx, E., Lacroix, B., Simon-Assman, P., Haffen, K., Fogh, J., and Zweibaum, A., Enterocytic-like differentiation and polarization of the human colon adenocarcinoma cell line Caco-2 in culture, *Cell*, 47, 323, 1983.

92. Hidalgo, I.J. and Borchardt, R.T., Transport of bile acids in a human intestinal epithelial cell line, Caco-2, *Biochem. Biophys. Acta*, 1035, 97, 1990.

93. Chong, S., Dando, S.A., Soucek, K.M., and Morrison, R.A., *In vitro* permeability through caco-2 cells is not quantitatively predictive of *in vivo* absorption for peptide-like drugs absorbed via the dipeptide transporter system, *Pharm. Res.*, 13, 120, 1996.

94. Carriere, V., Lesuffleur, T., Barbat, A., Rousset, M., Dussaulx, E., Costet, P., de, W., I Beaune, P., and Zweibaum, A., Expression of cytochrome P-450 3A in HT29-MTX cells and Caco-2 clone TC7, *FEBS Lett.*, 355, 247, 1994.

95. Hilgendorf, C., Spahn-Langguth, H., Regardh, C.G., Lipka, E., Amidon, G.L., and Langguth, P., Caco-2 versus Caco-2/HT29-MTX co-cultured cell lines: permeabilities via diffusion, inside- and outside-directed carrier-mediated transport, *J. Pharm. Sci.*, 89, 63, 2000.

96. Ong, S., Liu, H., and Pidgeon, C., Immobilized-artificial-membrane chromatography: measurements of membrane partition coefficient and predicting drug membrane permeability, *J. Chromatogr. A*, 728, 113, 1996.

97. Pidgeon, C., Ong, S., Liu, H., Qiu, X., Pidgeon, M., Dantzig, A.H., Munroe, J., Hornback, W.J., Kasher, J.S., and Glunz, L., IAM chromatography: an *in vitro* screen for predicting drug membrane permeability, *J. Med. Chem.*, 38, 590, 1995.

98. Tavelin, S., Milovic, V., Ocklind, G., Olsson, S., and Artursson, P., A conditionally immortalized epithelial cell line for studies of intestinal drug transport, *J. Pharmacol. Exp. Ther.*, 290, 1212, 1999.

99. Cho, M.J., Thompson, D.P., Cramer, C.T., Vidmar, T.J., and Scieszka, J.F., The madin darby canine kidney (MDCK) epithelial cell monolayer as a model cellular transport barrier, *Pharm. Res.*, 6, 71, 1989.

100. Stevens, C.E., Comparative physiology of the digestive system, in *Dukes Physiology of Domestic Animals*, Swenson, M.J., Ed., Comstock, Ithaca, NY & London, 1977.

101. Ghoshal, N.G. and Bal, H.S., Comparative morphology of the stomach of some laboratory mammals, *Lab. Anim.*, 23, 21, 1989.

102. Abe, K. and Ito, T., A qualitative and quantitative morphological study of Peyer's patches of the mouse, *Arch. Histol. Jpn.*, 40, 407, 1977.

103. Anderson, A.C., Digestive system, in *Beagle Dog as an Experimental Animal*, Anderson, A.C. and Good, L.S., Eds., Iowa state University Press, Ames, IA, 1970.

104. Dressman, J.B. and Yamada, K., Animal models for oral drug absorption, in *Pharmaceutical Bioequivalence*, Welting, P. and Tae, F.L., Eds., Dekker, NY, 1991.

105. Carr, K.E. and Toner, P.G., Morphology of the intestinal mucosa, in *Pharmacology of the Intestinal Permeation*, Csaky, T.Z., Ed., Springer, Berlin, 1984.

106. Taylor, A.B. and Anderson, J.H., Scanning electron microscope observations of mammalian intestinal villi, intervilli floor and crypt tubules, *Micron*, 3, 453, 1972.

107. Burke, J.A. and Holland, P., The epithelial surface of the monkey gastrointestinal tract. A scanning electron-microscopic study, *Digestion*, 14, 68, 1976.

108. Robinson, J.W., Menge, H., Sepulveda, F.V., and Mirkovitch, V., Functional and structural characteristics of the jejunum and ileum in the dog and the rat, *Digestion*, 15, 188, 1977.

109. Stevens, C.E., Physiological implications of microbial digestion in the large intestine of mammals: relation to dietary factors, *Am. J. Clin. Nutr.*, 31, S161, 1978.

110. Toner, P.G. and Carr, K.E., *Scientific Foundations of Gastroenterology*, Saunders, Philadelphia, 1980.

111. Evans, H.F., *Miller's Anatomy of the Dog*, Saunders, Philadelphia, PA, 1993.
112. Kozma, C., Macktin, W., Cummins, L.M., and Mauer, R., Anatomy, physiology and biochemistry of the rabbit, in *The Biology of Laboratory Rabbit*, Weisbroth, S.H., Flart, R.E., and Kraus, A.L., Eds., Academic, NY, 1974.
113. Yamada, I. and Haga, K., Measurement of gastric pH during digestion of a solid meal in dogs, *Chem. Pharm. Bull. (Tokyo)*, 38, 1755, 1990.
114. Kararli, T.T., Gastrointestinal absorption of drugs, *Crit. Rev. Ther. Drug Carrier Syst.*, 6, 39, 1989.
115. Charman, W.N., Lipid vehicle and formulations effects on intestinal lymphatic drug transport, in *Lymphatic Transport of Drugs*, Charman, W.N. and Stella, V.J., Eds., Chemical Rubber Company, Boca Raton, FL, 1992.
116. Farinon, A.M., Lampugnani, R., and Berri, T., Comparative anatomo-surgical study on the biliary and pancreatic excretory apparatus of the most common experimental animals, *Chir. Patol. Sper.*, 23, 363, 1975.
117. Hebel, R. and Stromberg, M.W., Digestive system, in *Anatomy of the Laboratory Rat*, Hebel, R. and Stromberg, M.W., Eds., Williams and Wilkins, Baltimore, 1976.
118. Vonk, R.J., van Doorn, A.B., and Strubbe, J.H., Bile secretion and bile composition in the freely moving, unanaesthetized rat with a permanent biliary drainage: influence of food intake on bile flow, *Clin. Sci. Mol. Med.*, 55, 253, 1978.
119. Calabrese, E.J., Intestinal microflora, in *Principles of Animal Extrapolation,* Calabrese, E.J., Ed., Wiley, NY, 1983.
120. Drasar, B.S., Hill, M.J., and Williams, R.E.O., The significance of the gut flora in safety testing of food additives, in *Metabolic Aspects of Food Safety,* Roe, F.J.C., Ed., Academic, NY, 1970.
121. Goldman, P., Biochemical pharmacology of the intestinal flora, *Annu. Rev. Pharmacol. Toxicol.*, 18, 523, 1978.
122. Kararli, T.T., Comparison of the gastrointestinal anatomy, physiology, and biochemistry of humans and commonly used laboratory animals, *Biopharm. Drug Dispos.*, 16, 351, 1995.
123. Park, K., Ching, H.S., and Robinson, J.R., Alternative approaches to oral controlled drug delivery, bioadhesives and *in situ* systems, in *Recent Advances in Drug Delivery,* Anderson, J.M. and Kim, S.W., Eds., Plenum, NY, 1984.
124. Allen, A. and Garner, A., Mucus and bicarbonate secretion in the stomach and their possible role in mucosal protection, *Gut*, 21, 249, 1980.
125. Gruber, P., Longer, M.A., and Robinson, J.R., Some biological issues in oral, controlled drug delivery, *Adv. Drug Deliv. Rev.*, 1, 1, 1987.
126. Fordtran, J.S. and Locklear, T.W., Ionic constituents and osmolality of gastric and small-intestinal fluids after eating, *Am. J. Dig. Dis.*, 11, 503, 1966.
127. Soenrgel, K.H., Flow measurements of test meals and fasting contents in human small intestine, in *Proceedings of the International Symposium on Motility of the Gastrointestinal Tract*, Demling, L., Ed., Thieme, Struttgart, 1971.
128. Harpur, R.P. and Popkin, J.S., Osmolality of blood and intestinal contents in the pig, guinea pig, and *Ascaris lumbricoides*, *Can. J. Biochem.*, 43, 1157, 1965.
129. Coupe, A.J., Davis, S.S., Evans, D.F., and Wilding, I.R., Correlation of the gastric emptying of nondisintegrating tablets with gastrointestinal motility, *Pharm. Res.*, 8, 1281, 1991.

130. Aoyagi, N., Comparative studies of griseofulvin bioavailability among man and animals, Thesis, Kyoto University, 1986.

131. Lennernas, H. and Regardh, C.G., Regional gastrointestinal absorption of the beta-blocker pafenolol in the rat and intestinal transit rate determined by movement of 14C-polyethylene glycol (PEG) 4000, Pharm. Res., 10, 130, 1993.

132. Proano, M., Camilleri, M., Phillips, S.F., Thomforde, G.M., Brown, M.L., and Tucker, R.L., Unprepared human colon does not discriminate between solids and liquids, Am. J. Physiol., 260, G13, 1991.

133. Owen, R.L. and Jones, A.L., Epithelial cell specialization within human Peyer's patches: an ultrastructural study of intestinal lymphoid follicles, Gastroenterology, 66, 189, 1974.

134. Madara, J.L. Bye, W.A., and Trier, J.S., Structural features of and cholesterol distribution in M-cell membranes in guinea pig, rat, and mouse Peyer's patches, Gastroenterology, 87, 1091, 1984.

135. HogenEsch, H. and Felsburg, P.J., Ultrastructure and alkaline phosphatase activity of the dome epithelium of canine Peyer's patches, Vet. Immunol. Immunopathol., 24, 177, 1990.

136. Torres-Medina, A., Morphologic characteristics of the epithelial surface of aggregated lymphoid follicles (Peyer's patches) in the small intestine of newborn gnotobiotic calves and pigs, Am. J. Vet. Res., 42, 232, 1981.

137. Neutra, M.R., Phillips, T.L., Mayer, E.L., and Fishkind, D.J., Transport of membrane-bound macromolecules by M cells in follicle-associated epithelium of rabbit Peyer's patch, Cell Tissue Res., 247, 537, 1987.

138. Comes, J.S., Number, size and distribution of Peyer's patches in the human small intestine, Gut, 6, 225, 1965.

139. Faulk, W.P., McCormick, J.N., Goodman, J.R., Yoffey, J.M., and Fudenberg, H.H., Peyer's patches: morphologic studies, Cell Immunol., 1, 500, 1970.

140. Binns, R.M. and Pabst, R., Lymphoid cell migration and homing in the young pig: alternative immune mechanisms in action, in Migration and Homing of Lymphoid Cells, Husband, A., Ed., Chemical Rubber Company, Boca Raton, FL, 1988.

141. Pabst, R., Geist, M., Rothkotter, H.J., and Fritz, F.J., Postnatal development and lymphocyte production of jejunal and ileal Peyer's patches in normal and gnotobiotic pigs, Immunology, 64, 539, 1988.

142. HogenEsch, H., Honsman, J.M., and Feldburg, P.J., Canine Peyer's patches: macroscopic, light, microscopic, scanning electron microscopic and immunohistochemical investigations, in Recent Advances in Mucosal Immunology, Mestecky, J., McGhee, J.R., Bienenstock, J., and Ogra, P.L., Eds., Plenum, NY, 1987.

143. Proulx, P., Structure–function relationships in intestinal brush border membranes, Biochem. Biophys. Acta, 1071, 255, 1991.

144. Christiansen, K. and Carlsen, J., Microvillus membrane vesicles from pig small intestine. Purity and lipid composition, Biochem. Biophys. Acta, 647, 188, 1981.

145. Hauser, H., Howell, K., Dawson, R.M., and Bowyer, D.E., Rabbit small intestinal brush border membrane preparation and lipid composition, Biochem. Biophys. Acta, 602, 567, 1980.

146. Kawai, K., Fujita, M., and Nakao, M., Lipid components of two different regions of an intestinal epithelial cell membrane of mouse, Biochem. Biophys. Acta, 369, 222, 1974.

147. Bouhours, J.F., Glycosphingolipids and ceramide distribution in brush border and basolateral membranes of the rat mature intestinal cells [proceedings], *Arch. Int. Physiol. Biochem.*, 86, 847, 1978.

148. Forstner, G.G. and Wherrett, J.R., Plasma membrane and mucosal glycosphingolipids in the rat intestine, *Biochem. Biophys. Acta*, 306, 446, 1973.

149. Brasitus, T.A. and Dudeja, P.K., Small and large intestinal plasma membranes: structure and function, in *Lipid Domains and Their Relationship to Membrane Function*, Aloia, R.C., Curtain, C.C., and Gordon, L.M., Eds., Liss, NY, 1988.

150. Schwarz, S.M., Hostetler, B., Ling, S., Mone, M., and Watkins, J.B., Intestinal membrane lipid composition and fluidity during development in the rat, *Am. J. Physiol*, 248, G200, 1985.

151. Brasitus, T.A., Yeh, K.Y., Holt, P.R., and Schachter, D., Lipid fluidity and composition of intestinal microvillus membranes isolated from rats of different ages, *Biochem. Biophys. Acta*, 778, 341, 1984.

152. Hubner, C., Lindner, S.G., Stern, M., Claussen, M., and Kohlschutter, A., Membrane fluidity and lipid composition of rat small intestinal brush-border membranes during postnatal maturation, *Biochem. Biophys. Acta*, 939, 145, 1988.

153. Holmes, R. and Lobley, R.W., Intestinal brush border revisited, *Gut*, 30, 1667, 1989.

154. Bouwstra, J.A. and Honeywell-Nguyen, P.L., Skin structure and mode of action of vesicles, *Adv. Drug Deliv. Rev.*, 54 Suppl 1, S41, 2002.

155. Scheuplein, R.J., Skin permeation, in *The Physiology and Pathophysiology of the Skin*, Jarret, A., Ed., Academic Press, London, 1978.

156. Scheuplein, R.J. and Bronaugh, R.L., Percutaneous absorption, in *Biochemistry and Physiology of the Skin*, Goldsmith, L.A., Ed., Oxford University Press, NY, 1983.

157. Warner, R.R., Myers, M.C., and Taylor, D.A., Electron probe analysis of human skin: determination of the water concentration profile, *J. Invest. Dermatol.*, 90, 218, 1988.

158. Fartasch, M., Human barrier formation and reaction to irritation, *Curr. Probl. Dermatol.*, 23, 95, 1995.

159. Schatzlein, A. and Cevc, G., Nonuniform cellular packing of the stratum corneum and permeability barrier function of intact skin: a high-resolution confocal laser scanning microscopy study using highly deformable vesicles (transfersomes), *Br. J. Dermatol.*, 138, 583, 1998.

160. Schatzlein, A., and Cevc, G., Non-uniform cellular packing of the stratum corneum and permeability barrier function of intact skin: a high-resolution confocal laser scanning microscopy study using highly deformable vesicles (Transfersomes), *Br. J. Dermatol.*, 138(4), 583, 1998.

161. Chapman, S.J. and Walsh, A., Desmosomes, corneosomes and desquamation. An ultrastructural study of adult pig epidermis, *Arch. Dermatol. Res.*, 282, 304, 1990.

162. Swartzendruber, D.C., Manganaro, A., Madison, K.C., Kremer, M., Wertz, P.W., and Squier, C.A., Organization of the intercellular spaces of porcine epidermal and palatal stratum corneum: a quantitative study employing ruthenium tetroxide, *Cell Tissue Res.*, 279, 271, 1995.

163. Chang, F., Swartzendruber, D.C., Wertz, P.W., and Squier, C.A., Covalently bound lipids in keratinizing epithelia, *Biochem. Biophys. Acta*, 1150, 98, 1993.

164. Hadgraft, J., and Guy, R.H., Eds., *Transdermal Drug Delivery. Developmental Issues and Research Initiatives*, Marcel Dekker, NY, 1989.

165. Christophers, E., Wolff, H.H., and Laurence, E.B., The formation of epidermal cell columns, *J. Invest. Dermatol.*, 62, 555, 1974.
166. Odland, G.F., A submicroscopic granular component in human epidermis, *J. Invest. Dermatol.*, 34, 11, 1960.
167. Landmann, L., Epidermal permeability barrier: transformation of lamellar granule-disks into intercellular sheets by a membrane-fusion process, a freeze-fracture study, *J. Invest. Dermatol.*, 87, 202, 1986.
168. *Liposome Dermatics: Chemical Aspects of the Skin Lipid Approach,* Springer, Berlin, 1992.
169. White, R. and Walker, M., Thermotropic and lyotropic behaviour of epidermal lipid fractions, *Biochem. Soc. Trans.*, 18, 881, 1990.
170. Tsai, J.C., Guy, R.H., Thornfeldt, C.R., Gao, W.N., Feingold, K.R., and Elias, P.M., Metabolic approaches to enhance transdermal drug delivery. 1. Effect of lipid synthesis inhibitors, *J. Pharm. Sci.*, 85, 643, 1996.
171. Fartasch, M., Bassukas, I.D., and Diepgen, T.L., Disturbed extruding mechanism of lamellar bodies in dry noneczematous skin of atopics, *Br. J. Dermatol.*, 127, 221, 1992.
172. Potts, R.O. and Guy, R.H., Predicting skin permeability, *Pharm. Res.*, 9, 663, 1992.
173. Hou, S.Y., Mitra, A.K., White, S.H., Menon, G.K., Ghadially, R., and Elias, P.M., Membrane structures in normal and essential fatty acid-deficient stratum corneum: characterization by ruthenium tetroxide staining and x-ray diffraction, *J. Invest. Dermatol.*, 96, 215, 1991.
174. Van den Bergh, B.A., Swartzendruber, D.C., Bos-Van der Geest, A., Hoogstraate, J.J., Schrijvers, A.H., Bodde, H.E., Junginger, H.E., and Bouwstra, J.A., Development of an optimal protocol for the ultrastructural examination of skin by transmission electron microscopy, *J. Microsc.*, 187 (Pt 2), 125, 1997.
175. Downing, D.T., Lipid and protein structures in the permeability barrier of mammalian epidermis, *J. Lipid Res.*, 33, 301, 1992.
176. Swartzendruber, D.C., Studies of epidermal lipids using electron microscopy, *Semin. Dermatol.*, 11, 157, 1992.
177. Van Duzee, B.F., Thermal analysis of human stratum corneum, *J. Invest. Dermatol.*, 65, 404, 1975.
178. Gay, C.L., Guy, R.H., Golden, G.M., Mak, V.H., and Francoeur, M.L., Characterization of low-temperature (i.e., <65°C) lipid transitions in human stratum corneum, *J. Invest. Dermatol.*, 103, 233, 1994.
179. Ongpipattanakul, B., Burnette, R.R., Potts, R.O., and Francoeur, M.L., Evidence that oleic acid exists in a separate phase within stratum corneum lipids, *Pharm. Res.*, 8, 350, 1991.
180. White, S.H., Mirejovsky, D., and King, G.I., Structure of lamellar lipid domains and corneocyte envelopes of murine stratum corneum. An X-ray diffraction study, *Biochemistry*, 27, 3725, 1988.
181. Robson, K.J., Stewart, M.E., Michelsen, S., Lazo, N.D., and Downing, D.T., 6-Hydroxy-4-sphingenine in human epidermal ceramides, *J. Lipid Res.*, 35, 2060, 1994.
182. Stewart, M.E. and Downing, D.T., A new 6-hydroxy-4-sphingenine-containing ceramide in human skin, *J. Lipid Res.*, 40, 1434, 1999.
183. Whiting, D.A., Histology of normal hair, in *Atlas of Hair and Nails,* Hordinsky, M.K., Sawaya, M.E., and Scher, R.K., Eds., Churchill Livingstone, Philadelphia, 2000.

184. Greene, R.S., Downing, D.T., Pochi, P.E., and Strauss, J.S., Anatomical variation in the amount and composition of human skin surface lipid, *J. Invest. Dermatol.*, 54, 240, 1970.

185. Clarys, P. and Barel, A., Quantitative evaluation of skin surface lipids, *Clin. Dermatol.*, 13, 307, 1995.

186. Schaefer, H. and Redelmeier, T.E., *Skin Barrier: Principles of Percutaneous Absorption*, Basel, 1996.

187. Cev, G., Drug delivery across the skin, *Expert. Opin. Investig. Drugs*, 6, 1887, 1997.

188. Cross, S.E. and Roberts, M.S., Physical enhancement of transdermal drug application: is delivery technology keeping up with pharmaceutical development? *Curr. Drug Deliv.*, 1, 81, 2004.

189. Pitt, W.G., Husseini, G.A., and Staples, B.J., Ultrasonic drug delivery—a general review, *Expert Opin. Drug Deliv.*, 1(1), 37, 2004.

190. Alvarez-Roman, R., Merino, G., Kalia, Y.N., Naik, A., and Guy, R.H., Skin permeability enhancement by low frequency sonophoresis: lipid extraction and transport pathways, *J. Pharm. Sci.*, 92, 1138, 2003.

191. Merino, G., Kalia, Y.N., Delgado-Charro, M.B., Potts, R.O., and Guy, R.H., Frequency and thermal effects on the enhancement of transdermal transport by sonophoresis, *J. Control Release*, 88, 85, 2003.

192. Benson, H.A.E. and Mclnay, J.C., Topical nonsteroidal antiinflammatory products as ultrasound couplants: their potential in phonophoresis, *Physiotherapy*, 80, 74, 1994.

193. Byl, N.N., The use of ultrasound as an enhancer for transcutaneous drug delivery: phonophoresis, *Phys. Ther.*, 75, 539, 1995.

194. Mitragotri, S., Blankschtein, D., and Langer, R., Ultrasound-mediated transdermal protein delivery, *Science*, 269, 850, 1995.

195. Ueda, H., Ogihara, M., Sugibayashi, K., and Morimoto, Y., Difference in the enhancing effects of ultrasound on the skin permeation of polar and nonpolar drugs, *Chem. Pharm. Bull. (Tokyo)*, 44, 1973, 1996.

196. Mitragotri, S., Effect of therapeutic ultrasound on partition and diffusion coefficients in human stratum corneum, *J. Control Release*, 71, 23, 2001.

197. Boucaud, A., Montharu, J., Machet, L., Arbeille, B., Machet, M.C., Patat, F., and Vaillant, L., Clinical, histologic, and electron microscopy study of skin exposed to low-frequency ultrasound, *Anat. Rec.*, 264, 114, 2001.

198. Weaver, J.C., Electroporation: a general phenomenon for manipulating cells and tissues, *J. Cell Biochem.*, 51, 426, 1993.

199. Prausnitz, M.R., Edelman, E.R., Gimm, J.A., Langer, R., and Weaver, J.C., Transdermal delivery of heparin by skin electroporation, *Biotechnology* (NY), 13, 1205, 1995.

200. Prausnitz, M.R., A practical assessment of transdermal drug delivery by skin electroporation, *Adv. Drug Deliv. Rev.*, 35, 61, 1999.

201. Grassi, G., Dawson, P., Guarnieri, G., Kandolf, R., and Grassi, M., Therapeutic potential of hammerhead ribozymes in the treatment of hyper-proliferative diseases, *Curr. Pharm. Biotechnol.*, 5, 369, 2004.

202. Agostini, F., Dapas, B., Farra, R., Grassi, M., Racchi, G., Klingel, K., Kandolf, R., Heidenreich, O., Mercatahnti, A., Rainaldi, G., Altamura, F., Guarnieri, G., and Grassi, G., Potential applications of small interfering RNAs in the cardiovascular field, *Drug of the Future*, 31, 513, 2006.

203. Zewert, T.E., Pliquett, U.F., Langer, R., and Weaver, J.C., Transdermal transport of DNA antisense oligonucleotides by electroporation, *Biochem. Biophys. Res. Commun.*, 212, 286, 1995.

204. Regnier, V. and Preat, V., Localization of a FITC-labeled phosphorothioate oligodeoxynucleotide in the skin after topical delivery by iontophoresis and electroporation, *Pharm. Res.*, 15, 1596, 1998.

205. Vanbever, R., Prausnitz, M.R., and Preat, V., Macromolecules as novel transdermal transport enhancers for skin electroporation, *Pharm. Res.*, 14, 638, 1997.

206. Heller, R., Gilbert, R., and Jaroszeski, M.J., Clinical applications of electrochemotherapy, *Adv. Drug Deliv. Rev.*, 35, 119, 1999.

207. Riviere, J.E. and Heit, M.C., Electrically-assisted transdermal drug delivery, *Pharm. Res.*, 14, 687, 1997.

208. Banga, A.K., Bose, S., and Ghosh, T.K., Iontophoresis and electroporation: comparisons and contrasts, *Int. J. Pharm.*, 179, 1, 1999.

209. Chien, Y.W., Systemic delivery of peptide based pharmaceuticals by transdermal periodic iontotherapeutic system, in *Dermal and Transdermal Drug Delivery: New Insights and Perspectives,* Gurny, R. and Teubner, A. Eds., Wissenschaftliche Vierlags GmbEH, Struttgart, Germany, 1993.

210. Bidmon, H.J., Pitts, J.D., Solomon, H.F., Bondi, J.V., and Stumpf, W.E., Estradiol distribution and penetration in rat skin after topical application, studied by high resolution autoradiography, *Histochemistry*, 95, 43, 1990.

211. Turner, N.G. and Guy, R.H., Visualization and quantitation of iontophoretic pathways using confocal microscopy, *J. Investig. Dermatol. Symp. Proc.*, 3, 136, 1998.

212. Miller, L.L., Kolaskie, C.J., Smith, G.A., and Rivier, J., Transdermal iontophoresis of gonadotropin releasing hormone (LHRH) and two analogues, *J. Pharm. Sci.*, 79, 490, 1990.

213. Srinivasan, V., Su, M.H., Higuchi, W.I., and Behl, C.R., Iontophoresis of polypeptides: effect of ethanol pretreatment of human skin, *J. Pharm. Sci.*, 79, 588, 1990.

214. Bhatia, K.S. and Singh, J., Mechanism of transport enhancement of LHRH through porcine epidermis by terpenes and iontophoresis: permeability and lipid extraction studies, *Pharm. Res.*, 15, 1857, 1998.

215. Langkjaer, L., Brange, J., Grodsky, G.M., and Guy, R.H., Iontophoresis of monomeric insulin analogues *in vitro*: effects of insulin charge and skin pretreatment, *J. Control Release*, 51, 47, 1998.

216. Huang, Y.Y., Wu, S.M., and Wang, C.Y., Response surface method: a novel strategy to optimize iontophoretic transdermal delivery of thyrotropin-releasing hormone, *Pharm. Res.*, 13, 547, 1996.

217. Banga, A.K., Katakam, M., and Mitra, R., Transdermal iontophoretic delivery and degradation of vasopressin across human cadaver skin, *Int. J. Pharm.*, 116, 211, 1995.

218. Santi, P., Colombo, P., Bettini, R., Catellani, P.L., Minutello, A., and Volpato, N.M., Drug reservoir composition and transport of salmon calcitonin in transdermal iontophoresis, *Pharm. Res.*, 14, 63, 1997.

219. Guy, R.H., Delgado-Charro, M.B., and Kalia, Y.N., Iontophoretic transport across the skin, *Skin Pharmacol. Appl. Skin Physiol.*, 14 Suppl 1, 35, 2001.

220. White, P.J., Gray, A.C., Fogarty, R.D., Sinclair, R.D., Thumiger, S.P., Werther, G.A., and Wraight, C.J., C-5 propyne-modified oligonucleotides penetrate the epidermis in psoriatic and not normal human skin after topical application, *J. Invest. Dermatol.*, 118, 1003, 2002.

221. Singh, J., Gross, M., Sage, B., Davis, H.T., and Maibach, H.I., Regional variations in skin barrier function and cutaneous irritation due to iontophoresis in human subjects, *Food Chem. Toxicol.*, 39, 1079, 2001.
222. Meyer, B.R., Kreis, W., Eschbach, J., O'Mara, V., Rosen, S., and Sibalis, D., Transdermal versus subcutaneous leuprolide: a comparison of acute pharmacodynamic effect, *Clin. Pharmacol. Ther.*, 48, 340, 1990.
223. Tartas, M., Bouye, P., Koitka, A., Durand, S., Gallois, Y., Saumet, J.L., and Abraham, P., Early vasodilator response to anodal current application in human is not impaired by cyclooxygenase-2 blockade, *Am. J. Physiol. Heart Circ. Physiol.*, 288(4), H1668-73, 2005.
224. Doukas, A.G. and Flotte, T.J., Physical characteristics and biological effects of laser-induced stress waves, *Ultrasound Med. Biol.*, 22, 151, 1996.
225. Cross, S.E., Roberts, M.S., Physical enhancement of transdermal drug application: is delivery technology keeping up with pharmaceutical development? *Curr. Drug Deliv.*, 1(1), 81, 2004.
226. Lee, S., McAuliffe, D.J., Flotte, T.J., Kollias, N., and Doukas, A.G., Photomechanical transcutaneous delivery of macromolecules, *J. Invest. Dermatol.*, 111, 925, 1998.
227. Lee, S., Kollias, N., McAuliffe, D.J., Flotte, T.J., and Doukas, A.G., Topical drug delivery in humans with a single photomechanical wave, *Pharm. Res.*, 16, 1717, 1999.
228. Lee, S., McAuliffe, D.J., Mulholland, S.E., and Doukas, A.G., Photomechanical transdermal delivery of insulin *in vivo*, *Lasers Surg. Med.*, 28, 282, 2001.
229. Lee, S., McAuliffe, D.J., Kollias, N., Flotte, T.J., and Doukas, A.G., Photomechanical delivery of 100 nm microspheres through the stratum corneum: implications for transdermal drug delivery, *Lasers Surg. Med.*, 31, 207, 2002.
230. Gonzalez, S., Lee, S., Gonzalez, E., and Doukas, A.G., Rapid allergen delivery with photomechanical waves for inducing allergic skin reactions in the hairless guinea pig animal model, *Am. J. Contact Dermat.*, 12, 162, 2001.
231. Menon, G.K., Lee, S., McAuliffe, D.J., Kollias, N., and Doukas, A.G., *J. Invest. Dermatol.*, 114, 837, 2000.
232. Menon, G.K. and Elias, P.M., Morphologic basis for a pore-pathway in mammalian stratum corneum, *Skin Pharmacol.*, 10, 235, 1997.
233. Jacques, S.L., McAuliffe, D.J., Blank, I.H., and Parrish, J.A., Controlled removal of human stratum corneum by pulsed laser, *J. Invest. Dermatol.*, 88, 88, 1987.
234. Nelson, J.S., McCullough, J.L., Glenn, T.C., Wright, W.H., Liaw, L.H., and Jacques, S.L., Mid-infrared laser ablation of stratum corneum enhances *in vitro* percutaneous transport of drugs, *J. Invest. Dermatol.*, 97, 874, 1991.
235. Lee, W.R., Shen, S.C., Lai, H.H., Hu, C.H., and Fang, J.Y., Transdermal drug delivery enhanced and controlled by erbium: YAG laser: a comparative study of lipophilic and hydrophilic drugs, *J. Control Release*, 75, 155, 2001.
236. Bommannan, D.B., Tamada, J., Leung, L., and Potts, R.O., Effect of electroporation on transdermal iontophoretic delivery of luteinizing hormone releasing hormone (LHRH) *in vitro*, *Pharm. Res.*, 11, 1809, 1994.
237. Chang, S.L., Hofmann, G.A., Zhang, L., Deftos, L.J., and Banga, A.K., The effect of electroporation on iontophoretic transdermal delivery of calcium regulating hormones, *J. Control Release*, 66, 127, 2000.

238. Le, L., Kost, J., and Mitragotri, S., Combined effect of low-frequency ultrasound and iontophoresis: applications for transdermal heparin delivery, *Pharm. Res.*, 17, 1151, 2000.

239. Touitou, E., Dayan, N., Bergelson, L., Godin, B., and Eliaz, M., Ethosomes—novel vesicular carriers for enhanced delivery: characterization and skin penetration properties, *J. Control Release*, 65, 403, 2000.

240. Cevc, G., Schatzein, A., and Blume, G., Transdermal drug carriers: basic properties, optimization and transfer efficiencies in the case of epicutaneously applied peptides, *J.Control Release*, 36, 3, 1995.

241. Grassi, G., Grassi, M., Platz, J., Bauriedel, G., Kandolf, R., and Kuhn, A., Selection and characterization of active hammerhead ribozymes targeted against cyclin E and E2F1 full-length mRNA, *Antisense Nucleic Acid Drug Dev.*, 11, 271, 2001.

242. Grassi, G., Schneider, A., Engel, S., Racchi, G., Kandolf, R., and Kuhn, A., Hammerhead ribozymes targeted against cyclin E and E2F1 co-operate to down-regulate coronary smooth muscle cells proliferation, *J. Gene Med.*, 7, 1223, 2005.

243. Touitou, E., Shaco-Ezra, N., Dayan, N., Jushynski, M., Rafaeloff, R., and Azoury, R., Dyphylline liposomes for delivery to the skin, *J. Pharm. Sci.*, 81, 131, 1992.

244. Planas, M.E., Gonzalez, P., Rodriguez, L., Sanchez, S., and Cevc, G., Noninvasive percutaneous induction of topical analgesia by a new type of drug carrier, and prolongation of local pain insensitivity by anesthetic liposomes, *Anesth. Analg.*, 75, 615, 1992.

245. Touitou, E., Junginger, H.E., Weiner, N.D., Nagai, T., and Mezei, M., Liposomes as carriers for topical and transdermal delivery, *J. Pharm. Sci.*, 83, 1189, 1994.

246. Lauer, A.C., Lieb, L.M., Ramachandran, C., Flynn, G.L., and Weiner, N.D., Transfollicular drug delivery, *Pharm. Res.*, 12, 179, 1995.

247. Arunothayanun, P., Turton, J.A., Uchegbu, I.F., and Florence, A.T., Preparation and *in vitro/in vivo* evaluation of luteinizing hormone releasing hormone (LHRH)-loaded polyhedral and spherical/tubular niosomes, *J. Pharm. Sci.*, 88, 34, 1999.

248. Foong, V.C., Harsanyi, B.B., and Mezei, M., Biodispositions and histological evaluation of topically applied retinoic acid in liposomal cream and gel dosage forms, in *Phospholipids*, Hanning, I. and Peppeu, G., Eds., Plenum Press, NY, 1990.

249. Cevc, G., Transfersomes, liposomes and other lipid suspensions on the skin: permeation enhancement, vesicle penetration, and transdermal drug delivery, *Crit. Rev. Ther. Drug Carrier Syst.*, 13, 257, 1996.

250. Cevc, G. and Blume, G., New, highly efficient formulation of diclofenac for the topical, transdermal administration in ultradeformable drug carriers, Transfersomes, *Biochem. Biophys. Acta*, 1514, 191, 2001.

251. Cevc, G. and Blume, G., Lipid vesicles penetrate into intact skin owing to the transdermal osmotic gradients and hydration force, *Biochem. Biophys. Acta*, 1104, 226, 1992.

252. R&D Focus, in IMS Work Publication Limited, ed., 1989.

253. Dkeidek, I. and Touitou, E., Transdermal absorption of polypeptides, *AAPS Pharm. Sci.*, 1, s202, 1999.

254. Horwitz, E., Pisanty, S., Czerninski, R., Helser, M., Eliav, E., and Touitou, E., A clinical evaluation of a novel liposomal carrier for acyclovir in the topical treatment of recurrent herpes labialis, *Oral Surg. Oral Med. Oral Pathol. Oral Radiol. Endod.*, 87, 700, 1999.

255. Goodman, M. and Barry, B.W., Action of penetration enhancers on human stratum corneum as assessed by differential scanning calorimetry, in *Percutaneous Absorption: Mechanism—Methodology—Drug Delivery,* Bronaugh, R.L. and Maibach, H.I., Eds., Marcel Dekker, New York, 1989.

256. Touitou, E., Transdermal delivery of anxiolytics: *in vitro* skin permeation of midazolam maleate and diazepam, *Int. J. Pharm.*, 33, 37, 1986.

257. Hadgraft, J., Modulation of the barrier function of the skin, *Skin Pharmacol. Appl. Skin Physiol.*, 14 Suppl 1, 72, 2001.

258. Pilgram, G.S., Engelsma-van Pelt, A.M., Koerten, H.K., and Bouwstra, J.A., The effect of two azones on the lateral lipid organization of human stratum corneum and its permeability, *Pharm. Res.*, 17, 796, 2000.

259. Philips, C.A. and Michinak, B.B., Transdermal delivery of drugs with differing lipohilicities using azone analogs as dermal penetration enhancers, *J. Pharm. Sci.*, 84, 1427, 1995.

260. Investigational Drugs Database, Generic, 2002.

261. Marty, J.P., Transdermal 7-day HRT, *J. Menopause*, 2, 17, 2000.

262. Berner, B. and Liu, P., Alcohols in percutaneous penetration enhancers, in *Percutaneous Penetration Enhancers,* Smith, E.W. and Maibach, H.I., Eds., CRS Press, NY, 1995.

263. Touitou, E. and Fabin, B., Altered skin permeation of a highly lipophilic molecule: tetrahydrocannabinol, *Int. J. Pharm.*, 43, 17, 1988.

264. Ghanem, A.H., Mahmoud, H., Higuchi, W.I., Liu, P., and Good, W.R., The effects of ethanol on the transport of lipophilic and polar permeants across hairless mouse skin: methods/validation of a novel approach, *J. Control Release*, 59, 149, 1999.

265. Magee, P.S., Some new approaches to understanding and facilitating transdermal drug delivery, in *Percutaneous Penetration Enhancers,* Smith, E.W. and Maibach, H.I., Eds., CRS Press, NY, 1995.

266. Henry, S., McAllister, D.V., Allen, M.G., and Prausnitz, M.R., Microfabricated microneedles: a novel approach to transdermal drug delivery, *J. Pharm. Sci.*, 87, 922, 1998.

267. Zahn, J.D., Peshmukh, A.A., Pisano, A.P., and Liepmann, D., Conference Proceeding, 2001, pp. 503–506.

268. Kaushik, S., Hord, A.H., Denson, D.D., McAllister, D.V., Smitra, S., Allen, M.G., and Prausnitz, M.R., Lack of pain associated with microfabricated microneedles, *Anesth. Analg.*, 92, 502, 2001.

269. Figge, F.J.H. and Burnette, D.J., *Am. Practioner*, 3, 197, 1947.

270. Uchida, M., Jin, Y., Natsume, H., Kobayashi, D., Sugibayashi, K., and Morimoto, Y., Introduction of poly-L-lactic acid microspheres into the skin using supersonic flow: effects of helium gas pressure, particle size and microparticle dose on the amount introduced into hairless rat skin, *J. Pharm. Pharmacol.*, 54, 781, 2002.

271. Lin, M.T., Pulkkinen, L., Uitto, J., and Yoon, K., The gene gun: current applications in cutaneous gene therapy, *Int. J. Dermatol.*, 39, 161, 2000.

272. Schramm, J. and Mitragotri, S., Transdermal drug delivery by jet injectors: energetics of jet formation and penetration, *Pharm. Res.*, 19, 1673, 2002.
273. Burkoth, T.L., Bellhouse, B.J., Hewson, G., Longridge, D.J., Muddle, A.G., and Sarphie, D.F., Transdermal and transmucosal powdered drug delivery, *Crit. Rev. Ther. Drug Carrier Syst.*, 16, 331, 1999.
274. Dean, H.J., Fuller, D., and Osorio, J.E., Powder and particle-mediated approaches for delivery of DNA and protein vaccines into the epidermis, *Comp. Immunol. Microbiol. Infect. Dis.*, 26, 373, 2003.
275. Consoli, A., Capani, F., La Nava, G., Nicolucci, A., Prosperini, G.P., Santeusanio, G., and Sensi, S., Administration of semisynthetic human insulin by a spray injector, *Boll. Soc. Ital. Biol. Sper.*, 60, 1859, 1984.
276. Halle, J.P., Lambert, J., Lindmayer, I., Menassa, K., Coutu, F., Moghrabi, A., Legendre, L., Legault, C., and Lalumiere, G., Twice-daily mixed regular and NPH insulin injections with new jet injector versus conventional syringes: pharmacokinetics of insulin absorption, *Diabetes Care*, 9, 279, 1986.
277. Kerum, G., Profozic, V., Granic, M., and Skrabalo, Z., Blood glucose and free insulin levels after the administration of insulin by conventional syringe or jet injector in insulin treated type 2 diabetics, *Horm. Metab. Res.*, 19, 422, 1987.
278. Fuller, D.H., Simpson, L., Cole, K.S., Clements, J.E., Panicali, D.L., Montelaro, R.C., Murphey-Corb, M., and Haynes, J.R., Gene gun-based nucleic acid immunization alone or in combination with recombinant vaccinia vectors suppresses virus burden in rhesus macaques challenged with a heterologous SIV, *Immunol. Cell Biol.*, 75, 389, 1997.
279. Macklin, M.D., McCabe, D., McGregor, M.W., Neumann, V., Meyer, T., Callan, R., Hinshaw, V.S., and Swain, W.F., Immunization of pigs with a particle-mediated DNA vaccine to influenza A virus protects against challenge with homologous virus, *J. Virol.*, 72, 1491, 1998.
280. McCluskie, M.J., Brazolot Millan, C.L., Gramzinski, R.A., Robinson, H.L., Santoro, J.C., Fuller, J.T., Widera, G., Haynes, J.R., Purcell, R.H., and Davis, H.L., Route and method of delivery of DNA vaccine influence immune responses in mice and nonhuman primates, *Mol. Med.*, 5, 287, 1999.
281. Bustad, L.K., Pigs in the laboratory, *Sci. Am.*, 214, 94, 1966.
282. Bustad, L.K., and McClellan, R.O., Miniature swine: development, management, and utilization, *Lab. Anim. Care*, 18(2), 280, 1968.
283. Panepinto, L.M., Phillips, R.W., Wheeler, L.R., and Will, D.H., The Yucatan miniature pig as a laboratory animal, *Lab. Anim. Sci.*, 28, 308, 1978.
284. Lavker, R.M., Dong, G., Zheng, P.S., and Murphy, G.F., Hairless micropig skin. A novel model for studies of cutaneous biology, *Am. J. Pathol.*, 138, 687, 1991.
285. Panepinto, L.M. and Kroc, R.L., History, genetic origins and care of Yucatan miniature and micro pigs, *Lab. Anim.*, 24, 31, 1995.
286. Weinstein, G.D., Comparison of turnover time and keratinous protein fractions in swine and human epidermis, in *Swine in Biomedical Research*, Bustad LK, M.R., Ed., Seattle, 1966.
287. Monteiro-Riviere, N.A., Ultrastructure evaluation of the porcine integument, in *Swine in Biomedical Research*, Tumbleson, M.E., Ed., Plenum Press, New York, 1986.

288. Montagna, W. and Yun, J.S., The skin of the domestic pig, *J. Invest. Dermatol.*, 42, 11, 1964.

289. Montagna, W., The microscopic anatomy of the skin of swine and man, in *Swine in Biomedical Research*, Bustad LK, M.R., Ed., Seattle, 1966.

290. Meyer, W., Neurand, K., and Radke, B., Elastic fibre arrangement in the skin of the pig, *Arch. Dermatol. Res.*, 270, 391, 1981.

291. Meyer, W., Neurand, K., and Radke, B., Collagen fibre arrangement in the skin of the pig, *J. Anat.*, 134, 139, 1982.

292. Gray, G.M. and White, R.J., Glycosphingolipids and ceramides in human and pig epidermis, *J. Invest. Dermatol.*, 70, 336, 1978.

293. Meyer, W. and Neurand, K., The distribution of enzymes in the skin of the domestic pig, *Lab. Anim.*, 10, 237, 1976.

294. Forbes, P.D., Vascular supply of the skin and hair in swine, *Adv. Biol. Skin*, 9, 419, 1969.

295. Meyer, W. and Neurand, K.S.R., Die Haut der Haussaugetiere. 2. Ihre Bedeutung fur die dermatologische Forschumg: Hinweise zur speziellen Funktion einzelner Hautanteile, *Tierarztl Prax*, 6, 289, 1978.

296. Feldmann, R.J. and Maibach, H.I., Regional variation in percutaneous penetration of 14C cortisol in man, *J. Invest. Dermatol.*, 48, 181, 1967.

297. Bronaugh, R.L. and Maibach, H.I., Regional variations in percutaneous absorption, *in Percutaneous Absorption*, Dekker, New York, 1999.

298. Rougier, A., Lotte, C., and Maibach, H.I., *In vivo* relationship between percutaneous absorption and transepidermal water loss, in *Percutaneous Absorption*, Bronaugh, R.L. and Maibach, H.I., Eds., Dekker, New York, 1999.

299. Galey, W.R., Lonsdale, H.K., and Nacht, S., The *in vitro* permeability of skin and buccal mucosa to selected drugs and tritiated water, *J. Invest. Dermatol.*, 67, 713, 1976.

300. Gore, A.V., Liang, A.C., and Chien, Y.W., Comparative biomembrane permeation of tacrine using Yucatan minipigs and domestic pigs as the animal model, *J. Pharm. Sci.*, 87, 441, 1998.

301. Reifenrath, W.G., Chellquist, E.M., Shipwash, E.A., Jederberg, W.W., and Krueger, G.G., Percutaneous penetration in the hairless dog, weanling pig and grafted athymic nude mouse: evaluation of models for predicting skin penetration in man, *Br. J. Dermatol.*, 111 Suppl 27, 123, 1984.

302. Chow, C., Chow, A.Y., Downie, R.H., and Buttar, H.S., Percutaneous absorption of hexachlorophene in rats, guinea pigs and pigs, *Toxicology*, 9, 147, 1978.

303. Bartek, M.J., LaBudde, J.A., and Maibach, H.I., Skin permeability *in vivo*: comparison in rat, rabbit, pig and man, *J. Invest. Dermatol.*, 58, 114, 1972.

304. Reifenrath, W.G. and Hawkins, G.S., The weanling yorkshire pig as an animal model for measuring percutaneous penetration, in *Swine in Biomedical Research*, Tumbleson, M.E., Ed., Plenum Press, New York, 1986.

305. Marzulli, F.N. and Maibach H.I., Techniques for studying skin penetration, *Toxicol. Appl. Pharmacol.*, s76, 1969.

306. Bartek, M.J. and LaBudde, J.A., Percutaneous absorption, *in vitro*, in *Animals Models in Dermatology*, Maibach, H.I., Ed., Churchill Livingstone, Edinburgh, 1975.

307. Dick, I.P. and Scott, R.C., Pig ear skin as an *in-vitro* model for human skin permeability, *J. Pharm. Pharmacol.*, 44, 640, 1992.

308. de Lange, J., van Eck, P., Elliott, G.R., de Kort, W.L., and Wolthuis, O.L., The isolated blood-perfused pig ear: an inexpensive and animal-saving model for skin penetration studies, *J. Pharmacol. Toxicol. Methods*, 27, 71, 1992.

309. Van Rooij, J.G., Vinke, E., de Lange, J., Bruijnzeel, P.L., Bodelier-Bade, M.M., Noordhoek, J., and Jongeneelen, F.J., Dermal absorption of polycyclic aromatic hydrocarbons in the blood-perfused pig ear, *J. Appl. Toxicol.*, 15, 193, 1995.

310. Davies, H.W. and Trotter, M.D., Synthesis and turnover of membrane glycoconjugates in monolayer culture of pig and human epidermal cells, *Br. J. Dermatol.*, 104, 649, 1981.

311. Loveday, D.E., An *in vitro* method for studying percutaneous absorption, *J. Soc. Cosmet. Chem.*, 12, 224, 1961.

312. Tregear, R.T., Relative penetrability of hair follicles and epidermis, *J. Physiol.*, 156, 307, 1961.

313. Tregear, R.T., The permeability of mammalian skin to ions, *J. Invest. Dermatol.*, 46, 16, 1966.

314. Bronaugh, R.L., Stewart, R.F., and Congdon, E.R., Methods for *in vitro* percutaneous absorption studies. II. Animal models for human skin, *Toxicol. Appl. Pharmacol.*, 62, 481, 1982.

315. Marzulli, F.N. and Maibach, H.I., Relevance of animal models: the hexachlorophene story, in *Animal models in Dermatology*, Maibach, H.I., Ed., Churchill Livingstone, Edinburgh, 1975.

316. Gordon, G., Variety of the common or house mouse, *Zoologist*, 8, 2763, 1850.

317. Flanagan, S.P., 'Nude', a new hairless gene with pleiotropic effects in the mouse, *Genet. Res.*, 8, 295, 1966.

318. Pantelouris, E.M., Absence of thymus in a mouse mutant, *Nature*, 217, 370, 1968.

319. Gates, A.H., Arundell, F.D., and Karasek, M.A., Hereditary defect of the pilosebaceous unit in a new double mutant mouse, *J. Invest. Dermatol.*, 52, 115, 1969.

320. Stelzner, K.F., Four dominant autosomal mutations affecting skin and hair development in the mouse, *J. Hered.*, 74, 193, 1983.

321. Snell, G.D., Inheritance in the house mouse, the linkage relations of short-ear, hairless, and naked, *Genetics*, 16, 42, 1931.

322. Crew, F.A.E. and Mirskaia, L., The character "hairless" in the mouse, *J. Genet.*, 25, 17, 1931.

323. Chase, H.B. and Montagna, W., The development and consequences of hairlessness in the mouse, *Genetics*, 37, 537, 1952.

324. Howard, A., "Rhino", an allele of hairless in the house mouse, *J. Hered.*, 31, 467, 1940.

325. Haigh, J.M. and Smith, E.W., The selection and use of natural and synthetic membranes for *in vitro* diffusion experiments, *Eur. J. Pharm. Sci.*, 2, 311, 1994.

326. Ridout, G., Houk, J., Guy, R.H., Santus, G.C., Hadgraft, J., and Hall, L.L., An evaluation of structure–penetration relationships in percutaneous absorption, *Farmaco*, 47, 869, 1992.

327. Tsuruta, H., Skin absorption of organic solvent vapors in nude mice *in vivo*, *Ind. Health*, 27, 37, 1989.

328. Stoughton, R.B., Animal models for *in vitro* percutaneous absorption, in *Animal Models in Dermatology: Relevance to Human Dermatopharmacology and Dermatotoxicology*, Maibach H.I., Ed., Edinburgh, 1975.

329. Durrheim, H., Flynn, G.L., Higuchi, W.I., and Behl, C.R., Permeation of hairless mouse skin I: experimental methods and comparison with human epidermal permeation by alkanols, *J. Pharm. Sci.*, 69, 781, 1980.

330. Hinz, R.S., Lorence, C.R., Hodson, C.D., Hansch, C., Hall, L.L., and Guy, R.H., Percutaneous penetration of para-substituted phenols *in vitro*, *Fundam. Appl. Toxicol.*, 17, 575, 1991.

331. Kikkoji, T., Gumbleton, M., Higo, N., Guy, R.H., and Benet, L.Z., Percutaneous penetration kinetics of nitroglycerin and its dinitrate metabolites across hairless mouse skin *in vitro*, *Pharm. Res.*, 8, 1231, 1991.

332. Higo, N., Hinz, R.S., Lau, D.T., Benet, L.Z., and Guy, R.H., Cutaneous metabolism of nitroglycerin *in vitro*. II. Effects of skin condition and penetration enhancement, *Pharm. Res.*, 9, 303, 1992.

333. Halkier-Sorensen, L., Menon, G.K., and Elias, P.M., Thestrup-Pedersen, K., and Feingold, K.R., Cutaneous barrier function after cold exposure in hairless mice: a model to demonstrate how cold interferes with barrier homeostasis among workers in the fish-processing industry, *Br. J. Dermatol.*, 132, 391, 1995.

334. Behl, C.R., Flynn, G.L., Linn, E.E., and Smith, W.M., Percutaneous absorption of corticosteroids: age, site, and skin-sectioning influences on rates of permeation of hairless mouse skin by hydrocortisone, *J. Pharm. Sci.*, 73, 1287, 1984.

335. Hou, S.Y. and Flynn, G.L., Enhancement of hydrocortisone permeation of human and hairless mouse skin by 1-dodecylazacycloheptan-2-one, *J. Invest. Dermatol.*, 93, 774, 1989.

336. Kushla, G.P. and Zatz, J.L., Influence of pH on lidocaine penetration through human and hairless mouse skin *in vitro*, *Int. J. Pharm.*, 71, 167, 1991.

3 Rheology

3.1 INTRODUCTION

Rheology is a term formally introduced by Bingham in 1929 to designate the study of deformation and flow of matter [1]. It is evident from daily experience that materials can flow or deform under the action of body forces, i.e., their own weight or surface forces, applied over their boundaries (solid contours, free surface), and behave quite differently, both qualitatively and quantitatively. Indeed, the original definition of rheology seems to open a very large horizon spacing from highly rigid solids to low-density fluids. More precisely, the borders of this relatively young science are more restricted and marked by other more mature disciplines, belonging to the continuum mechanics family, such as the elasticity theory and the Newtonian fluid mechanics, which are more suitable to describe simple mechanical behaviors. In both cases only a physical property (elastic modulus or viscosity) is sufficient to describe the deformation or flow conditions produced by the stresses acting on the material. Just for their si mplicity such approaches hold only for a limited class of materials and under limiting conditions of flow and deformation. Many real substances exhibit more complex behaviors, which can be accommodated between the two extremes settled by the linear models proposed by Hooke and Newton. This is the reason why rheology emerged in the past twenties as an independent science to describe phenomena and to solve problems inaccessible by the classical approaches. In other words, the specific objective of rheology is the definition of constitutive equations or models suitable to account for the stress–strain relationships observed for real systems. Such equations, combined with the principles of continuum mechanics, can then be used positively in the resolution of flow and deformation problems which cannot be tackled with the classical approaches. In this sense rheology represents a natural expansion and a necessary integration of the continuum mechanics.

In addition, we must emphasize that since from its beginning rheology is strongly concerned with the structural phenomena governing the macroscopic behavior of materials under flow or deformation and determining their nonlinear properties. Real materials are characterized by a microstructure resulting from the way in which the elementary units (macromolecules, solid particles, droplets, micelles, etc.) are spatially distributed and interact among themselves. This is the case of concentrated polymeric and disperse systems, whose microstructure is frequently very complex and may undergo substantial modifications as a result of the imposed motion and deformation conditions. The peculiar and original aim of rheology may then be that of establishing a relationship between the microstructural processes and the macroscopic behavior. From this point of view, rheology overcomes the limits of continuum mechanics where microscopic or molecular phenomena and processes are ignored, and represents a new approach to the study of deformation and flow of matter.

Rheology deals with a wide variety of materials that have quite different molecular or microscopic structures and, hence, exhibiting quite different behaviors that can be described appropriately only by using more than one physical quantity and resorting to diverse and more complex constitutive equations than linear relationships. The classification of the rheological behaviors appears to be not a simple task, even because every material can behave differently under different conditions. Changes in applied stress or deformation conditions induce structural changes whose amplitude and kinetics depend on their intensity and time of evolution. Indeed, a given material can behave more similarly to an elastic solid or to a viscous liquid depending on the applied stress and on the time scale of the deformation process. As we will see later, the crucial factor is the ratio between two characteristic times which measure the velocity of change in structural conditions and applied deformation, respectively. Accordingly, we should replace the conventional classification between solid and liquid materials with a distinction between liquid-like and solid-like properties. All these considerations are relevant for the pharmaceutical sector, where the majority of the materials (ointments, creams, pastes, and gels) are viscoelastic, since they combine both viscous and elastic properties.

Consequently, the macroscopic behavior of a material depends not only on its internal structural conditions and external contours within or around which it flows or deforms, but also on intensity and temporal modes of action of driving forces. All these factors should be accounted for when the most appropriate rheological data is collected for process analysis and product performance in the pharmaceutical and other industrial sectors. By doing so, rheology can not only improve efficiency in processing but can also help formulators and end users in finding pharmaceutical products that are optimal for their individual needs. Pharmaceutical processes such as ingredient selection, preparations, material packaging, and shelf storage are often associated with complex flow conditions of materials. The application performance and, hence, the acceptance of the final product are also often dependent on its rheological behavior. A special case concerns the controlled release systems based on swellable gel matrices where the viscoelastic properties can play an essential and conditioning role in the drug release kinetics.

The present chapter is intended to orient the reader through the multi-disciplinary aspects of rheology, starting from basic concepts and material functions that are necessary to describe and classify the mechanical responses of materials, in particular when subjected to shear deformation and flow conditions. In the following sections different classes of materials, from simple liquids to more complex systems, will be examined to underline the connections between their structural features and rheological properties and, in conclusion, some significant examples of applications of rheology to the drug delivery systems will be illustrated.

3.2   VISCOSITY AND NEWTONIAN FLUIDS

Viscosity[1] is commonly perceived as "thickness," or resistance to pouring, and measures the internal resistance of a fluid to flow. Since ancient times viscosity has been the object of attention and speculations. In his work "De Rerum Natura" the Roman poet Lucretius observed that "olive oil is a thick liquid flowing more slowly than wine through a colander" adding an intuitive explanation of the stronger internal flow resistance of olive oil, 20 centuries before the official birth of rheology. More recently, in his "Principia" Newton touched on the flow of fluids and postulated that "the resistance which arises from the lack of slipperiness of the parts of the liquid, with other things equal, is proportional to the velocity with which the parts of the liquid are separated from one another" [2]. This first scientific definition of what we today call viscosity must be revised in order to arrive at the current version of the linear constitutive equation that governs the shear flow behavior of simple fluids. For this purpose, we must select an appropriate geometry to introduce the physical quantities that characterize both the kinematic and dynamic sides of shear flow. The simplest geometry for shear (laminar) flow conditions is represented in Figure 3.1.

The liquid is contained between two large parallel plates, each one of area A, separated by a distance d, and the upper plate moves with a constant velocity V in the direction x. In accordance with the original Newton's postulate, the tangential force F required to maintain the motion should be proportional to A and V, and inversely proportional to d:

$$\frac{F}{A} = \eta \frac{V}{d}, \tag{3.1}$$

where the constant of proportionality η is the fluid viscosity. F/A can be replaced by σ_{yx} which denotes the tangential stress (shear stress) acting in

---

[1]The word "viscosity" derives from the Latin word "viscum," the sticky substance named "bird-lime" made from mistletoe berries and smeared on branches to catch small birds.

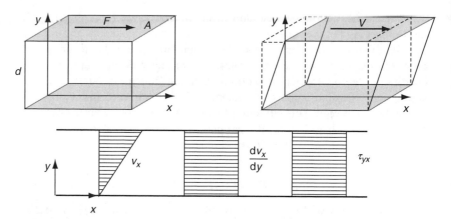

FIGURE 3.1 Shear flow in a plane geometry.

the x direction on a surface perpendicular to the y direction and has the same value in the whole space occupied by the liquid. V/d can be replaced by the velocity gradient dv_x/dy whose value is also constant; in this simple flow geometry it can be exchanged with the shear rate $\dot{\gamma}$, which corresponds to the time derivative of the deformation $\dot{\gamma}$. Then Equation 3.1 becomes

$$\sigma_{yx} = -\eta \frac{dv_x}{dy} = \eta \dot{\gamma}, \tag{3.2}$$

σ_{yx} represents also the momentum transfer from the liquid layer adjacent to moving boundary downward to the layers of progressively slower velocity. The negative sign underlines that the momentum transfer occurs in the opposite direction with respect to the velocity gradient.

Equation 3.2 is usually called Newton's law of viscosity and represents the simplest (linear) correlation, which can be established between the shear stress and the shear rate. On a macroscopic scale, Newton's law implies that the force F required to produce the motion is linearly proportional to the moving boundary velocity V and the liquid viscosity.

Similarly, for laminar flow in a pipe of radius R, Newton's law leads to a linear relation between the pressure drop ΔP over the whole pipe length L and the volumetric flow rate Q, i.e., to the Hagen–Poiseuille equation [3,4]:

$$\Delta P = \eta \frac{8L}{\pi R^4} Q. \tag{3.3}$$

Higher viscosity values imply higher inlet pressures to get a predetermined flow rate, or lower flow rates for a given inlet pressure. Equation 3.3 makes also evident about the important role of the flow geometry dimensions, in particular the strong dependence of pressure drop or flow rate on the tube radius, with all the factors constant.

The steady laminar flow conditions of Newtonian fluids in circular tubes as well as the Hagen–Poiseuille equation are easily derived from the only momentum balance, if the flow is incompressible and isothermal. The following linear radial distributions are derived for the tangential stress and, through Equation 3.2, for the velocity gradient (see Figure 3.2):

$$\sigma_{rz} = \frac{\Delta P}{L}\frac{r}{2},$$

(3.4)

$$\frac{dv_z}{dr} = -\frac{1}{\eta}\frac{\Delta P}{L}\frac{r}{2}.$$

(3.5)

If no slippage is assumed at the tube wall ($v_z = 0$ for $r = R$), a parabolic velocity distribution is obtained:

$$v_z = \frac{1}{\eta}\frac{\Delta PR^2}{4L}\left[1 - \left(\frac{r}{R}\right)^2\right].$$

(3.6)

Then Equation 3.6 leads to Equation 3.7 through a further step of integration:

$$Q = \int_0^{2\pi}\int_0^R v_z r\, d\vartheta\, dr = 2\pi\int_0^R v_z\, dr.$$

(3.7)

The previous relationships have been derived for fully developed laminar flow and hence can be correctly used only for long tubes (high L/R values) where the contribution of end effects is negligible. At the tube entrance and exit the flow is not unidirectional and parallel to the tube wall, so that also the radial component, v_r, must be accounted for. The axial velocity component, v_z, becomes practically independent of z and assumes the parabolic profile described by Equation 3.6 after a given distance from the tube inlet (entrance length) which depends on the flow intensity.

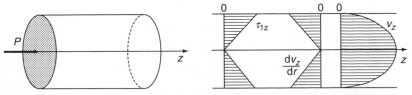

FIGURE 3.2 Tube flow: radial distributions of shearing stress, velocity gradient, and velocity.

Similar linear interrelations between the pressure drop and the flow rate can be derived for other internal flow geometries under steady, laminar, and isothermal conditions provided that the fluid is incompressible and Newton's law is valid. Also in the case of liquid agitation in a tank the applied torque T necessary to produce laminar flow conditions is linearly proportional to the impeller rotational velocity N and the proportionality factor is related to the liquid viscosity and the geometrical characteristics of the agitation system.

The Newton's law and the linear equations derived for the macroscopic quantities related to various laminar flow geometries (F vs. V, ΔP vs. Q, and T vs. N in the above examples) hold for simple liquids with low-molecular weight (below 5000). Accordingly, such fluids are called Newtonian.[2]

The current unit of viscosity in the SI system is Pa s (kg/m s or N/m^2 s) or, more conveniently, m Pa s for less viscous liquids and gases.[3] Indeed, viscosity values cover many orders of magnitude, more than other physical properties, from 10^{-5} Pa s (air and other gases in standard conditions) to 10^{12} Pa s (molten glasses) or even more (see Table 3.1).

Dividing η by the density ρ, we obtain the kinematic viscosity ν, which measures the momentum diffusivity and has the same units of mass and thermal diffusivities (m^2/s in the SI system). Hence, it can be conveniently used for comparing different transport processes, in particular when occurring simultaneously.[4]

TABLE 3.1
Viscosity Values of Newtonian Fluids (in Pa s)

Fluid	Viscosity	Fluid	Viscosity
Air (25°C, 1 atm)	2×10^{-5}	Liquid air (−192°C)	0.000173
Chloroform (20°C)	0.00058	Ethanol 20°C	0.0012
Water (20°C)	0.001	Propylene glycol (25°C)	0.040
Water (40°C)	0.0065	Propylene glycol (100°C)	0.0027
Aqueous glycerol sol (50%) (20°C)	0.006	Olive oil (20°C)	0.084
Glycerol (20°C)	1.41	Castor oil	0.98
White clover honey (13.7% water −25°C)	42	Glass (annealing point)	10^3
Corn syrup	10^3	Bitumen	10^9

[2]Also gases belong to the class of Newtonian fluids, but their macroscopic flow behavior can be derived only by combining the Newton's law and the equation of state suitable to describe their volumetric properties.

[3]Formerly, the widely used unit in the cgs system was the Poise (g/cm s). One Poise is equal to 0.1 Pa s.

[4]The dimensionless ratio between the kinematic viscosity ν and the thermal diffusivity α is called the Prandtl number Pr and measures the relative ease of momentum and energy transport in flow systems.

In general, the viscosity decreases with temperature for liquids, while it increases for low-pressure gases. Increasing the pressure over a liquid results in an exponential increase in viscosity, but the changes are appreciable only at high pressure, when compressibility of liquids cannot be ignored. In the temperature range from the freezing point to the normal boiling point, an Arrhenius-type equation generally holds:

$$\eta = A \exp\frac{B}{T},$$
(3.8)

where A and B are empirical constants. This simple relation is commonly referred to as the Andrade equation [5,6], even if it was firstly proposed by De Guzman [7]. A similar functional dependence was derived later by Eyring et al. [8] in their theoretical approach to rate processes, so that A and B can be calculated from molar volume and normal boiling point, even if with rough approximation. In general, for Newtonian liquids, the greater the viscosity, the stronger is the temperature dependence.

3.3 SHEAR-DEPENDENT VISCOSITY AND NON-NEWTONIAN FLUIDS

Many polymeric and disperse systems do not obey the Newton's law owing to their structural complexity and are referred to as non-Newtonian fluids. Their viscosity does not depend only on temperature and pressure but also, and often markedly, on flow intensity and duration. At constant shear rate the shear stress generally tends to a steady value $\sigma(\dot{\gamma})$, which depends on the imposed shear rate $\dot{\gamma}$. The ratio of the steady shear stress and shear rate is defined as steady shear viscosity and depends only on $\dot{\gamma}$ if the fluid behavior is reversible[5]: such a material function $\eta(\dot{\gamma})$ is conventionally used to classify the shear-dependent behavior of fluids. When a step change is imposed in the shear rate or shear stress conditions, steady state is often attained only after an appreciable transient period, whose duration depends on the change itself, i.e., on its amplitude and sign as well. This means that another classification must be taken into consideration, looking at the different time-dependent behaviors displayed by fluids when they are subjected to simple changes in shear conditions.

The classification of non-Newtonian fluids is traditionally based on the functional dependence of viscosity on shear rate, $\eta(\dot{\gamma})$. This choice has both cultural and practical origins, bound to the historical growth of the rheological

---

[5]Intense shear flow can sometimes lead to irreversible structural changes in disperse and polymeric systems on molecular or mesoscopic scale and then the same rheological behavior is irreversible depending on the previous mechanical history of the fluid sample.

TABLE 3.2
Shearing Rates Typical of Industrial Operations
and Product Applications

Operation	Shear Rate Range (s^{-1})
Sedimentation of fine particles in a liquid	10^{-6} to 10^{-4}
Leveling due to surface tension	10^{-2} to 10^{-1}
Draining under gravity	10^{-1} to 10^{1}
Polymer extrusion	10^{0} to 10^{2}
Chewing and swallowing of foods	10^{1} to 10^{2}
Dip coating	10^{1} to 10^{2}
Mixing and stirring of liquids	10^{1} to 10^{3}
Flowing in pipes	10^{0} to 10^{3}
Spraying and paint brushing	10^{3} to 10^{4}
Application of creams and lotions	10^{3} to 10^{5}
Pigment milling	10^{3} to 10^{5}
Paper coating	10^{5} to 10^{6}
Lubrication	10^{3} to 10^{7}

Source: Adapted from Lapasin, R. and Pricl, S., *Rheology of Industrial Polysaccharides: Theory and Applications*, Springer Verlag, 1995. Reprinted with permission from Springer Verlag.

knowledge, and to the important role played by viscosity in the analysis of many technological problems, respectively. Shear flow conditions are frequently encountered in many production processes and during product usage and applications; therefore, viscosity is (often reasonably) considered as the key rheological quantity. Table 3.2 contains a list of industrial operations and product applications where shearing conditions of quite different intensity can be encountered.

Indeed, since the beginning of rheology, researchers focused their interests on viscosity measurements, finding that many materials deviate from the Newtonian behavior, such as polymer solutions, emulsions, and dispersions. Only when more sophisticated instruments became available, their attention turned to determine also other rheological quantities, typical of shear flows, and, in more recent times, to characterize the behavior of several liquids, even of low viscosity, in elongational flow conditions. These aspects concerning other quantities or flows will be examined later.

In the majority of cases, the viscosity of disperse and polymeric systems decreases with increasing shear rate. Such behavior is called shear-thinning, even if the term pseudoplasticity is still used. Figure 3.3 illustrates the variation of viscosity with shear rate for hyaluronic acid solutions at different polymer concentrations.

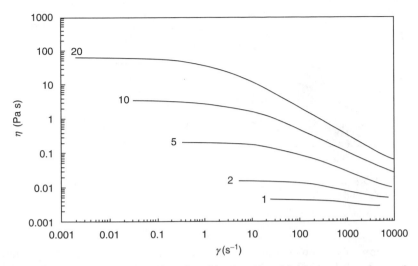

FIGURE 3.3 Shear-dependent viscosity of hyaluronic acid solutions at various polymer concentrations (1, 2, 5, 10, 20 g/L).

The curves indicate that the viscosity is constant in the limit of low-shear rates and tend to another but lower constant value at high-shear rates as well. These two Newtonian regions are usually indicated as lower and upper Newtonian plateaus, respectively. The corresponding viscosity values are called zero-shear rate viscosity η_0 and infinite-shear rate viscosity η_∞, and they cannot be easily measured, when the Newtonian plateaus are confined at very low- and high-shear rate values, respectively.

Less frequently, viscosity increases with increasing shear rate. Such a behavior is called shear-thickening, although the term dilatancy is also used, and is exhibited by concentrated suspensions of stabilized/deflocculated particles in a more or less extended range of $\dot{\gamma}$.

Other graphs are often used to represent the shear-dependent behavior, where shear stress is plotted vs. shear rate, using linear or logarithmic scales. In Figure 3.4, shear-thinning and shear-thickening behaviors are compared. As the shear rate increases, the stress also increases, but its dependence on $\dot{\gamma}$ is nonlinear, with shear stress increasing faster than the shear rate for shear-thickening fluids.

Finally, for several structured systems (concentrated suspensions of aggregated particles, physical gels, and others) η tends to diverge when $\dot{\gamma} \longrightarrow 0$. Correspondingly, in a linear plot of τ vs. $\dot{\gamma}$ the profile of the curve tends to a finite value τ_y, the *yield stress*, which is a peculiar feature of the plastic behavior (see Figure 3.4). This means that plastic materials can flow by

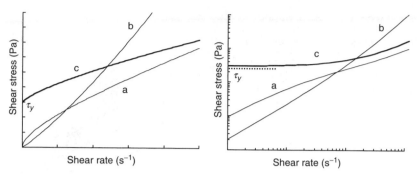

FIGURE 3.4 (a) Shear-thinning, (b) shear-thickening, and (c) plastic behaviors: τ vs. $\dot{\gamma}$ in linear and logarithmic graphs.

definition only if the critical stress τ_y is exceeded. Using logarithmic scales, differences between shear-thinning and plastic behavior appear more evidently in the low-shear region; moreover, in this diagram the τ_y value is individuated by the horizontal asymptotic branch of the curve. Obviously, τ_y can be estimated more reliably when more data are available at lower $\dot{\gamma}$.

Undoubtedly, the concept of yield stress is helpful for several practical purposes, since it serves to mark the onset of appreciable flow conditions and to distinguish zones of unsheared fluid within a given flow geometry. Nevertheless, the increasing use of more sophisticated controlled stress rheometers, suitable to explore low-shear regions, has made more evident that a Newtonian plateau characterized by high η_0 values can be individuated also in the case of concentrated dispersions and other structured fluids which are normally classified as plastic systems. For these systems the flow conditions are almost negligible at low stresses but change abruptly and become very intense when the applied stresses exceed a threshold value, which corresponds to the apparent yield stress. Thus, the diaphragm separating the two categories of plastic and shear-thinning behaviors practically vanishes. In addition, it is worthwhile to note that the apparent yield stress and the zero-shear rate viscosity are generally strictly related to each other within a class of similar systems and this implies an equivalent use of these quantities as guide parameters for formulation and flow control.

Several empirical equations are available to describe the different shear-dependent behaviors, part of them are founded on physical assumptions and considerations or convergent with theoretical treatments. They contain two or more adjustable parameters whose values can be determined only through comparison with the experimental data. Table 3.3 reports a list of equations which are most frequently used to describe shear-thinning (Equation 3.9 through Equation 3.11) or plastic behaviors (Equation 3.12 through Equation 3.14).

TABLE 3.3
Generalized Newtonian Fluid Models

Model	$\tau(\dot{\gamma})$	Parameters	Equation	Ref.
Ostwald–de Waele	$\tau = K\,\dot{\gamma}^n$	K, n	3.9	[10,11]
Cross	$\tau = \eta_\infty\dot{\gamma} + \dfrac{(\eta_0 - \eta_\infty)\dot{\gamma}}{1 + (\lambda\dot{\gamma})^{1-n}}$	$\eta_0, \eta_\infty, \lambda, n$	3.10	[12]
Carreau	$\tau = \eta_\infty\dot{\gamma} + \dfrac{(\eta_0 - \eta_\infty)\dot{\gamma}}{(1 + (\lambda\dot{\gamma})^2)^{(1-n)/2}}$	$\eta_0, \eta_\infty, \lambda, n$	3.11	[13]
Bingham	$\tau = \tau_y + \eta_\infty\,\dot{\gamma}$	τ_y, η_∞	3.12	[14,15]
Casson	$\tau^{1/2} = \tau_y^{1/2} + \eta_\infty^{1/2}\,\dot{\gamma}^{1/2}$	τ_y, η_∞	3.13	[16]
Herschel–Bulkley	$\tau = \tau_y + K\,\dot{\gamma}^n$	τ_y, K, n	3.14	[17,18]

3.4 SHEAR FLOWS

The knowledge of the shear dependence of viscosity is essential for the analysis and prediction of non-Newtonian flows since the local shear stress and, hence, also the viscosity can vary across the flow section in both internal and external flows. Equations and models describing the stress dependence of viscosity are more convenient to solve shear flow problems, since they be coupled more easily with the local stress conditions derived from mass and momentum balances.

Shear flows can be encountered within tubes and other closed geometries of constant flow section and in fluid films flowing over surfaces, and produced by internal and external driving forces (fluid weight and pressure forces) or moving boundaries. The plural action of driving forces and moving boundaries can lead to more complex flow patterns but the velocity distribution through the cross-section is still ordered on condition that flow is laminar. In the case of Newtonian fluids and simple geometries, the velocity distribution can be easily derived by combining the effects produced individually by the driving forces. Even if closed solutions of non-Newtonian flow problems can be obtained for simple geometries and relatively simple viscosity models [19], numerical methods are generally necessary to solve flow problems of greater complexity, due to the contours geometry and/or the rheological behavior of the fluid [20–22].

All shear flows can be ideally visualized as the relative sliding motion of rigid surfaces (*shearing surfaces*), whose form depends on the boundaries of the system. The distance between any two neighboring particles in a shearing surface is constant, and there is a constant separation between any two neighboring shearing surfaces. A special category is given by *viscometric flows*, which are usually selected for the rheological characterization of shear behavior in virtue of their simple geometry and kinematics (see Figure 3.5). Remarkable examples are steady tube flow, steady tangential annular flow,

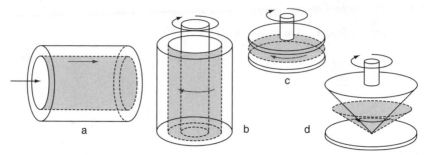

FIGURE 3.5 Geometries of viscometric flows: (a) tube, (b) coaxial cylinders, (c) parallel plate, (d) cone/plate.

steady cone/plate flow, and steady torsional flow. In all these cases, the lines of shear are also material lines: they are open trajectories in the case of the shear tube flow, and closed pathlines in the other three cases; as concerns the shearing surfaces, these are cylindrical, conical, and plane. Indeed, simple fundamental correlations can be established between the macroscopic quantities which are controlled and measured in the experimental tests, at least for Newtonian fluids, and this is why an absolute measure of viscosity can be performed using such instruments. The simplest case is given by the small angle cone/plate geometry, since the local shear stress and shear rate values are constant through the whole sample. Thus, the viscosity can be derived directly from the macroscopic quantities (torque T and angular velocity W) and the geometrical characteristics of the measuring device (angle ϑ_0 and radius R) for any liquid, no matter if Newtonian or not, using the following formula:

$$\eta = \frac{3T\vartheta_0}{2\pi R^3 W}. \tag{3.15}$$

Uniform conditions are approached in steady tangential annular flow only for small gaps between coaxial cylinders, whereas the shear rate distribution is always large in other geometries (parallel plates, capillaries). In the latter cases the correlation between the macroscopic quantities depends also on the rheological behavior of the fluid examined. Consequently, viscosity is derived directly from experimental tests only for Newtonian liquid, whereas appropriate data treatment is required for non-Newtonian samples. Detailed analyses of the various aspects related to rheometrical tests are reported elsewhere [9,19,23,24].

In the viscometric geometries of Figure 3.5, flow is internal or confined between two surfaces in relative motion. Shear conditions are encountered also in external flows for liquid films flowing over a surface owing to its own weight. Beyond a given distance from the entrance region (entrance length), the transversal distributions of both shear stress and velocity gradient in the flow section do not change in the flow direction, shear lines and material lines coincide and are rectilinear, while shearing surfaces are plane and cylindrical, respectively.

For Newtonian fluids the entrance length of tube flow can be calculated with satisfactory approximation from the following relation:

$$\frac{L_e}{D} = 0.035 \frac{D\langle v_z \rangle}{v}, \tag{3.16}$$

where D is the tube diameter, and $\langle v_z \rangle$ is the average velocity of the fluid. Equation 3.16 can be rewritten as

$$\frac{L_e}{D} = 0.035 \, \text{Re}, \tag{3.17}$$

where the Reynolds number Re corresponds to

$$\text{Re} = \frac{D\langle v_z \rangle \rho}{\mu}. \tag{3.18}$$

The magnitude of this dimensionless group gives a measure of the relative importance of the inertial and viscous forces in a fluid system. Experiments have shown that the laminar shear conditions are usually realized in tube flow of Newtonian liquids for Reynolds numbers up to about 2100.[6] Above that value, the inertial forces become predominant, the flow loses its order characteristics and becomes unstable: a transition to turbulent flow begins. There are eddies of fluid moving in all three coordinate directions, even though the bulk flow is in only one direction. The velocity profile (v_z vs. r) becomes flatter in the center region and steeper in the wall region than observed in laminar flow. In such conditions the Hagen–Poiseuille equation (Equation 3.3) no more holds and the pressure drop ΔP tends to be proportional to the square of the flow rate Q at very high Re and in rough pipes.

For other flow geometries the Reynolds number must be redefined by substituting D and $\langle v_z \rangle$ in Equation 3.18 with the characteristic linear dimension L and velocity V of the system, respectively. Also in these cases the Reynolds number provides a criterion to measure the flow intensity, to establish dynamic similarity between flow systems and to predict the laminar–turbulent transition, which can be sharp as for flow over a flat plate, or smooth

---

[6]Critical Reynolds numbers as high as 40,000 have been obtained in carefully controlled experiments, by eliminating any disturbance related to the entrance to the tube, wall roughness, or vibrations.

and extended over wide Re intervals as for flows over curved objects or within granular beds.

For non-Newtonian fluids the Reynolds number must be replaced by a generalized version proposed by Metzner and Reed or other dimensionless numbers in order to analyze the relative importance of the viscous and inertial forces, and then also to individuate the laminar–turbulent transition [25–28].

3.5 NOT ONLY VISCOSITY: NORMAL STRESS DIFFERENCES AND OTHER MATERIAL FUNCTIONS

The difference between Newtonian and non-Newtonian systems does not consist only in the shear dependence of viscosity. In shearing flow conditions the tensional state of a Newtonian liquid is totally given by the relevant tangential stresses, while other quantities must be accounted for in the case of non-Newtonian systems. Now, a better understanding of such differences can be achieved by examining how the local tensional state can be described in a deforming body.

3.5.1 TENSIONAL STATE

If we consider a generic point O and a small surface of area ΔS around it, all the interactions acting between the two sides of the material through the surface result into a force $\Delta \underline{F}$ which depends on the point location, and on the size and orientation of the area ΔS. The orientation of ΔS in space is defined by the *unit normal vector $\underline{n}$*, as indicated in Figure 3.6.

As ΔS tends to zero, the ratio $\Delta \underline{F}/\Delta S$ tends to the tension vector $\underline{t}$, which represents the force per unit area acting in O:

$$\lim_{\Delta S \to 0} \frac{\Delta \underline{F}}{\Delta S} = \underline{t}. \tag{3.19}$$

In the most general case, $\underline{t}$ varies with the unit vector $\underline{n}$ and the location of point O, and, hence, at any time t the tensional state in O is defined by all the

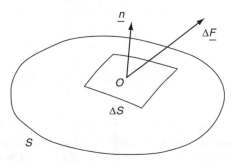

FIGURE 3.6 Stress vector definition.

pairs $(\underline{t}, \underline{n})$. Nevertheless, such a complex task is strongly simplified, since the tensional state can be univocally derived from only three independent pairs $(\underline{t}, \underline{n})$ relative to perpendicular planes.

Let $\underline{\delta}_1$, $\underline{\delta}_2$, $\underline{\delta}_3$ be the unit vectors in the direction of x_1, x_2, x_3 axes of a rectangular Cartesian reference frame, respectively. The stress vector $\underline{t}_1$, acting in O on a surface normal to the direction x_1, can be resolved into its components σ_{11}, σ_{12}, σ_{13} along the x_1, x_2, x_3 directions[7]:

$$\underline{t}_1 = \sigma_{11}\underline{\delta}_1 + \sigma_{12}\underline{\delta}_2 + \sigma_{13}\underline{\delta}_3. \qquad (3.20)$$

Similarly, the decomposition of the stress vectors $\underline{t}_2$ and $\underline{t}_3$, relative to the directions x_2 and x_3, leads to the components σ_{21}, σ_{22}, σ_{23} and σ_{31}, σ_{32}, σ_{33}, respectively. The stresses σ_{11}, σ_{22}, σ_{33} are termed *normal stresses*, the other components are called *shear or tangential stresses*. Figure 3.7 shows positive stress components on the faces of a block.[8]

The second-order *stress tensor* $\underline{\underline{\sigma}}$ associates the stress vectors $\underline{t}_1$, $\underline{t}_2$, and $\underline{t}_3$ with each coordinate direction according to

$$\underline{\underline{\sigma}} = \underline{t}_1\underline{\delta}_1 + \underline{t}_2\underline{\delta}_2 + \underline{t}_3\underline{\delta}_3. \qquad (3.21)$$

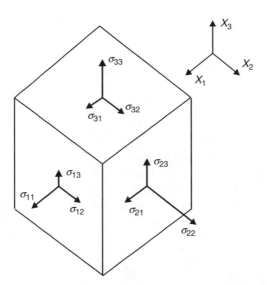

FIGURE 3.7 Stress components.

---

[7]The first suffix denotes the orientation of the normal to the surface and the second the direction of the resolution.
[8]According to the adopted sign convention, a positive σ_{ij} represents a tensile stress.

By substituting the expression for the generic stress vector t_i

$$t_i = \Sigma_j \sigma_{ij} \underline{\delta}_j \tag{3.22}$$

in Equation 3.21, we obtain

$$\underline{\underline{\sigma}} = \Sigma_i \Sigma_j \underline{\delta}_i \underline{\delta}_j \sigma_{ij}. \tag{3.23}$$

In the absence of body couples, the stress tensor is symmetric as a consequence of the balance of moments, so that

$$\sigma_{ij} = \sigma_{ji}. \tag{3.24}$$

This means that only six independent components of the stress tensor $\underline{\underline{\sigma}}$ are sufficient to determine the tensional state at every point, since the dependence of $\underline{t}$ on $\underline{n}$ can be defined through the stress tensor $\underline{\underline{\sigma}}$ according to

$$\underline{t} = \underline{\underline{\sigma}} \underline{n}. \tag{3.25}$$

At rest the tensional state of a fluid is isotropic and compressive, is determined by the *hydrostatic pressure P*:

$$\underline{t} = -P\underline{n} \tag{3.26}$$

and, hence, the stress tensor is simply given by

$$\underline{\underline{\sigma}} = -P\underline{\underline{\delta}}, \tag{3.27}$$

where $\underline{\underline{\delta}}$ is the *unit tensor*.

For a fluid at rest, pressure contributes to the normal stresses, but not to the shear stresses. In dynamic conditions the tensional state is more complex because of the additional contribution of surface forces due to deformation and flow, and in the most general case normal and shear stresses act on each element of the deforming body. The stress tensor is conveniently split into the isotropic pressure contribution and the *extra stress tensor* $\underline{\underline{\tau}}$, according to

$$\underline{\underline{\sigma}} = -P\underline{\underline{\delta}} + \underline{\underline{\tau}}, \tag{3.28}$$

$$\begin{aligned} \sigma_{ij} &= -P + \tau_{ij} \quad \text{for } i = j \\ \sigma_{ij} &= \tau_{ij} \quad \text{for } i \neq j, \end{aligned} \tag{3.29}$$

so that we can combine the analyses of the tensional state under both static and dynamic conditions. The six independent components of the symmetric stress tensor $\underline{\underline{T}}$ are related to the flow and deformation state of the material, and strictly associated to the rheological properties of the fluid through the constitutive equations.

3.5.2 DEFORMATION STATE

Let us examine the motion and deformation condition of a body. The velocity distribution can be defined through the function $\underline{v}(\underline{x}, t')$ or, alternatively, by the *velocity gradient tensor* $\underline{\underline{L}}$ (elsewhere designated as $\nabla\underline{v}$), the generic component of which may be written as

$$L_{ij} = \frac{\partial v_i}{\partial x_j}.$$ (3.30)

Even if the definition of the velocity field represents the final goal of the analysis and resolution of flow problems, more crucial to the understanding and definition of the rheological behavior of a fluid is the selection of meaningful quantities suitable for characterizing the local state of deformation and its change with time. For this purpose, the rate of deformation tensor is a useful quantity, as results from the following brief analysis of the deformation state.

Let us consider two neighboring material points P and Q separated by a small distance $d\underline{x}$ at time t, and by a distance $d\underline{r}$ at time t' (see Figure 3.8). $d\underline{r}$ and $d\underline{x}$ are related by a tensorial quantity, known as the *deformation gradient* $\underline{\underline{F}}$, whose generic component is given by

$$F_{ij} = \frac{\partial r_i}{\partial x_j}.$$ (3.31)

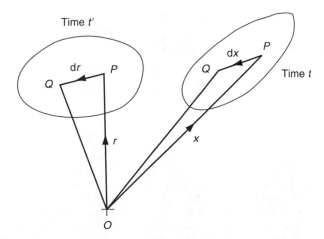

FIGURE 3.8 Motion of a body and configurations at times t' and t.

$\underline{\underline{F}}$ measures how the system changes its configuration in the surroundings around P at the given instant t'. In the most general case it can be expressed as the product of two contributions due to rigid rotation and pure deformation, as follows:

$$\underline{\underline{F}} = \underline{\underline{R}}\,\underline{\underline{U}},\qquad(3.32)$$

where $\underline{\underline{R}}$ and $\underline{\underline{U}}$ are the orthogonal *rotation tensor* and the *deformation tensor*, respectively. In order to measure the only local deformation, we must exclude the contribution due to rigid motion and this can be obtained with the symmetric tensor $\underline{\underline{C}}$, called *the Cauchy–Green tensor* and defined as[9]

$$\underline{\underline{C}} = \underline{\underline{F}}^{\mathrm{T}}\underline{\underline{F}}.\qquad(3.33)$$

By differentiating $\underline{\underline{C}}$ with respect to t', for $t' = t$, we obtain the *rate of deformation tensor* $\underline{\underline{\dot{\gamma}}}$, which corresponds to the sum of the velocity gradient and its transpose

$$\underline{\underline{\dot{\gamma}}} = \underline{\underline{L}} + \underline{\underline{L}}^{\mathrm{T}} = \nabla\underline{v} + \nabla\underline{v}^{\mathrm{T}}.\qquad(3.34)$$

The symmetric tensor $\underline{\underline{\dot{\gamma}}}$ describes the rate at which neighboring fluid particles move with respect to one another. A measure of the local rotation of the fluid is given by the antisymmetric *vorticity tensor* $\underline{\underline{\omega}}$, defined as[10]

$$\underline{\underline{\omega}} = \underline{\underline{L}} - \underline{\underline{L}}^{\mathrm{T}}.\qquad(3.35)$$

In shearing flow conditions the kinematics of the fluid can be expressed in scalar terms through the *shear rate* $\dot{\gamma}$, related to the second invariant $\mathrm{II}_{\dot{\gamma}}$ of the rate of deformation tensor $\underline{\underline{\dot{\gamma}}}$ as follows:

$$\dot{\gamma} = \sqrt{\frac{1}{2}\mathrm{II}_{\dot{\gamma}}} = \sqrt{\frac{1}{2}\sum_i\sum_j\dot{\gamma}_{ij}\dot{\gamma}_{ij}}.\qquad(3.36)$$

---

[9]In fact, by applying the rules of tensorial calculus, we can demonstrate that, by virtue of the orthogonality of $\underline{\underline{R}}$ and the symmetry of $\underline{\underline{U}}$, we have: $\underline{\underline{C}} = \underline{\underline{U}}^2$.

[10]There is considerable diversity in the literature regarding the definition of these quantities as well as their nomenclature. Frequently, factors of 1/2 are included on the right side of Equation 3.21 and Equation 3.22. In our opinion, the definition here adopted leads to a simpler analysis of flow conditions and formulation of the constitutive equations.

In the case of the simple plane flow geometry (see Figure 3.1) the rate of deformation tensor $\underline{\underline{\dot{\gamma}}}$ is given by

$$\underline{\underline{\dot{\gamma}}} = \dot{\gamma} \begin{vmatrix} 0 & 1 & 0 \\ 1 & 0 & 0 \\ 0 & 0 & 0 \end{vmatrix}.$$

3.5.3 NORMAL STRESS DIFFERENCES

In shear flow conditions the most general form of the stress tensor $\underline{\underline{\tau}}$ is

$$\underline{\underline{\tau}} = \begin{vmatrix} \tau_{xx} & \tau_{xy} & 0 \\ \tau_{yx} & \tau_{yy} & 0 \\ 0 & 0 & \tau_{zz} \end{vmatrix}.$$

For a Newtonian fluid the components of the stress tensor $\underline{\underline{\tau}}$ are linearly proportional to those of rate of deformation tensor $\underline{\underline{\dot{\gamma}}}$ and, hence, the tensional state is described by the only tangential stress τ_{yx} acting in the flow direction x (and its symmetric τ_{xy}). For non-Newtonian fluids the normal stresses (τ_{xx}, τ_{yy}, τ_{zz}) are not null and correspond to the dynamic pressure exerted by the fluid motion in the three directions. The only measurable quantities are the two normal stress differences N_1 and N_2, defined by

$$N_1 = \tau_{11} - \tau_{22}, \quad N_2 = \tau_{22} - \tau_{33}. \tag{3.37}$$

N_1 and N_2 are the first and the second normal stress differences, respectively. Consequently, two other material functions are to be considered in addition to the shear viscosity: they are the first normal stress coefficient Ψ_1 and the second normal stress coefficient Ψ_2, defined as

$$\Psi_1(\dot{\gamma}) = N_1(\dot{\gamma})/\dot{\gamma}^2, \quad \Psi_2(\dot{\gamma}) = N_2(\dot{\gamma})/\dot{\gamma}^2. \tag{3.38}$$

The first normal stress coefficient Ψ_1 is generally positive[11] and decreases with increasing shear rate. The second normal stress coefficient Ψ_2 is negative, and its absolute value is much smaller than Ψ_1, generally lower than 20% of Ψ_1. The effects of the second normal stress difference are frequently negligible, whereas the role of N_1 becomes more important at high-shear rates, because the stress ratio N_1/τ generally increases with increasing $\dot{\gamma}$.

---

[11]Negative N_1 values can be usually detected in an intermediate shear rate region in the case of concentrated crystalline polymeric solutions.

3.6   EXTENSIONAL FLOWS AND COMPLEX FLOWS

Another category of flows with simple kinematic characteristics is given by extensional or shearfree flows. Extensional flow conditions are frequently encountered in polymer processing operations, as fiber spinning and film blowing, where the flow field is bounded only by free surfaces and shearing components are absent.

In an orthogonal coordinate frame xyz the velocity field of all extensional flows is described by

$$\underline{v} = v_x(x)\underline{i} + v_y(y)\underline{j} + v_z(z)\underline{k}, \tag{3.39}$$

where $\underline{i}$, $\underline{j}$, and $\underline{k}$ are the unit vectors in the direction of x, y, and z axes, respectively. Each component of the velocity vector depends only on the corresponding coordinate and, consequently, the rate-of-deformation tensor has the following diagonal form:

$$\underline{\underline{\dot{\gamma}}} = \begin{vmatrix} \dot{\gamma}_{xx} & 0 & 0 \\ 0 & \dot{\gamma}_{yy} & 0 \\ 0 & 0 & \dot{\gamma}_{zz} \end{vmatrix}. \tag{3.40}$$

All shear components are zero ($\dot{\gamma}_{ij} = 0$, for $i \neq j$) and, hence, extensional flows are also named shearfree.

For incompressible fluids the velocity components are subjected to the following constraint deriving from the principle of matter conservation (continuity equation):

$$\frac{\partial v_x}{\partial x} + \frac{\partial v_y}{\partial y} + \frac{\partial v_z}{\partial z} = 0 \tag{3.41}$$

and, correspondingly, also the component of the rate-of-deformation tensor $\underline{\underline{\dot{\gamma}}}$ for extensional flow conditions must satisfy the following equation:

$$\dot{\gamma}_{xx} + \dot{\gamma}_{yy} + \dot{\gamma}_{zz} = 0. \tag{3.42}$$

Three main types of extensional flows can be individuated by selected combinations of $\dot{\gamma}_{ii}$ values which satisfy Equation 3.42.

The simple elongational flow (or uniaxial extensional flow) is described by the following velocity components:

$$v_x = \dot{\varepsilon}x, \quad v_y = -\frac{\dot{\varepsilon}}{2}y, \quad v_z = -\frac{\dot{\varepsilon}}{2}z, \tag{3.43}$$

where the elongation rate $\dot{\varepsilon}$ is sufficient to specify the whole velocity field. Such a flow condition is realized when the sample is extended along

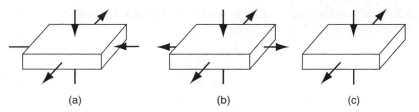

FIGURE 3.9 Extensional flows: (a) uniaxial, (b) biaxial, and (c) planar.

the direction of imposed deformation (x) and gets equally constricted along the two perpendicular directions (see Figure 3.9a) as in fiber spinning, where a cylindrical fluid element becomes longer and thinner as it is stretched.

The rate-of-deformation tensor then reduces to

$$\underline{\underline{\dot{\gamma}}} = \dot{\varepsilon} \begin{vmatrix} 2 & 0 & 0 \\ 0 & -1 & 0 \\ 0 & 0 & -1 \end{vmatrix}, \tag{3.44}$$

that is

$$\dot{\gamma}_{xx} = 2\dot{\varepsilon}, \quad \dot{\gamma}_{yy} = -\dot{\varepsilon}, \quad \dot{\gamma}_{zz} = -\dot{\varepsilon}. \tag{3.44'}$$

For any given value of $\dot{\varepsilon}$, the stress response of the material is expressed by the diagonal tensor

$$\underline{\underline{\tau}} = \begin{vmatrix} \tau_{xx} & 0 & 0 \\ 0 & \tau_{yy} & 0 \\ 0 & 0 & \tau_{zz} \end{vmatrix}, \tag{3.45}$$

where

$$\tau_{yy} - \tau_{zz} = 0. \tag{3.46}$$

Consequently, in the case of the uniaxial elongational flow the only measurable quantity is the normal stress difference $\tau_{xx} - \tau_{yy}$ (or $\tau_{xx} - \tau_{zz}$), and only one material function $\eta_E(\dot{\varepsilon})$, named elongational or extensional viscosity, is defined as

$$\eta_E(\dot{\varepsilon}) = \frac{\tau_{xx} - \tau_{yy}}{\dot{\varepsilon}} = \frac{\tau_{xx} - \tau_{zz}}{\dot{\varepsilon}}. \tag{3.47}$$

When the fluid is Newtonian, the direct proportionality among the components of the tensors $\underline{\underline{\tau}}$ and $\underline{\underline{\dot{\gamma}}}$ implies

$$\underline{\underline{\tau}} = \mu \underline{\underline{\dot{\gamma}}} = \mu \dot{\varepsilon} \begin{vmatrix} 2 & 0 & 0 \\ 0 & -1 & 0 \\ 0 & 0 & -1 \end{vmatrix}, \tag{3.48}$$

$$\tau_{xx} = 2\mu\dot{\varepsilon}, \quad \tau_{yy} = -\mu\dot{\varepsilon}, \quad \tau_{zz} = -\mu\dot{\varepsilon} \tag{3.48'}$$

From Equation 3.48 and Equation 3.48' we obtain Trouton's law, which holds whatever be the rate of strain

$$\eta_E = 3\mu. \tag{3.49}$$

For polymer solutions and melts the elongation viscosity η_E is dependent on the extensional rate $\dot{\varepsilon}$ and, for $\dot{\varepsilon} \rightarrow 0$, converges to a constant value, conforming to the Trouton's law:

$$\eta_E(\dot{\varepsilon})|_{\dot{\varepsilon} \rightarrow 0} = 3\eta(\dot{\gamma})|_{\dot{\gamma} \rightarrow 0}. \tag{3.50}$$

The biaxial extensional flows are characterized by the following kinematic characteristics:

$$v_x = \dot{\varepsilon}x, \quad v_y = \dot{\varepsilon}y, \quad v_z = -2\dot{\varepsilon}z. \tag{3.51}$$

The fluid element is equally stretched in both x- and y-directions, while it undergoes a contraction along the direction normal to the xy plane (see Figure 3.9b). Also in this case, for symmetry reasons

$$\tau_{xx} - \tau_{yy} = 0 \tag{3.52}$$

and the biaxial extensional viscosity η_{EB} is sufficient to characterize the rheological behavior:

$$\eta_{EB}(\dot{\varepsilon}) = \frac{\tau_{xx} - \tau_{zz}}{\dot{\varepsilon}} = \frac{\tau_{yy} - \tau_{zz}}{\dot{\varepsilon}}. \tag{3.53}$$

Finally, the velocity field in the planar elongational flows (see Figure 3.9c) is given by

$$v_x = \dot{\varepsilon}x, \quad v_y = 0, \quad v_z = -\dot{\varepsilon}z. \tag{3.54}$$

And the corresponding planar extensional viscosity η_{EP} is defined by

$$\eta_{EP}(\dot{\varepsilon}) = \frac{\tau_{xx} - \tau_{zz}}{\dot{\varepsilon}}. \tag{3.55}$$

This material function, like η_E, can be correlated with the shear viscosity, through the limiting condition

$$\eta_{EP}(\dot{\varepsilon})|_{\dot{\varepsilon}\to0} = 4\eta(\dot{\gamma})|_{\dot{\gamma}\to0}. \tag{3.56}$$

If the fluid is Newtonian, whatever be the rate of strain we can write

$$\eta_{EP} = 4\mu. \tag{3.57}$$

In all the extensional flows the relative motion of material particles is quite different than in shear flows. At constant $\dot{\varepsilon}$ the deformation of a fluid element increases exponentially with time and, hence, both the external configuration and the internal structure of the sample change much more significantly than in shear flows. This is the reason why extensional flows are classified as strong flows, whereas shear flows belong to the opposite class of weak flows. Moreover, the kinematic characteristics of extensional flows make the rheological characterization more difficult. One of the main problems in extensional rheometry is the difficulty in achieving steady flow, especially for low-viscosity liquids, while steady shear conditions can be achieved quite simply with different geometries by applying constant forces or at constant velocity of the moving boundary. Conversely, a constant elongation rate can be obtained by only moving the ends of a sample at an exponentially increasing velocity, and it is quite evident that in several practical cases the extensional flow can be maintained only over a limited period of time.

Shear and extensional flows have simple but opposite kinematics and represent the two extreme limits of a wide range of often complex flow patterns, which can occur also in simple geometrical arrangements. This is the case of entry flow through an abrupt circular contraction (see Figure 3.10). Shear flow is fully developed up to a given distance from the contraction, where shear and extensional components begin to coexist, and is encountered again after the entrance distance in the downstream tube. The flow pattern as well as the pressure drop in the contraction zone are strongly dependent on the rheological properties and the mean velocity of the fluid. A secondary-flow vortex or a stagnant zone may appear in the corner of the upstream tube in the presence of remarkable elastic or plastic properties, so increasing the entry pressure drop required to force the fluid through the contraction zone and to reach a fully developed velocity profile. Shear and extensional components coexist in all internal flows, when shape and/or dimensions of the contours change along the flow direction, in the close proximity of derivations and

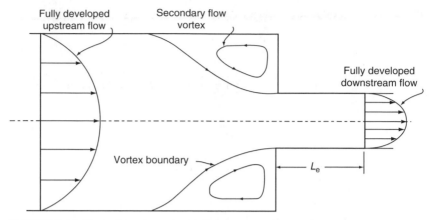

FIGURE 3.10 Entry flow through an abrupt contraction. (Adapted from Lapasin, R. and Pricl, S., *Rheology of Industrial Polysaccharides: Theory and Applications*, Springer Verlag, 1995. Reprinted with permission from Springer Verlag.)

bifurcations, or around obstacles present in the flow field. Numerical solutions of the entry-flow and other complex flow problems can be searched for by combining the equation of motion with an appropriate rheological model.

3.7   TIME-DEPENDENCE: THIXOTROPY AND VISCOELASTICITY

The dependence of viscosity and other material functions on the applied stress field and its intensity is governed by structural mechanisms which are concerned with the primary constitutive components (simple molecules, polymeric chains, dispersed particles) and, even more significantly when they are present, with the higher-order structures formed by molecules and particles in virtue of their attractive and/or repulsive mutual interactions.

The more sudden and strong are the imposed changes in the applied stress field, the more evident are the observed time-dependent rheological properties which are intrinsically related to the kinetics characteristics of the shear-induced structural processes.

When higher-order structures are absent, the response times of the material are controlled by the spatial (positional and orientational) rearrangements of primary components and also by their possible flow-induced deformation and extension, as for dispersed drops or dissolved polymeric chains. For concentrated dispersions and polymer solutions the topological interactions between the primary elements result into longer response times.

In the presence of aggregated particles or associated polymeric chains, the time-dependent behavior is mainly governed by the breakage and buildup of interparticle or interchain linkages. The amplitude and kinetics of these processes depend on the strength of attractive interactions and their cooperativity degree as well as on the amplitude of imposed stress variations. If reversible, the resulting time-dependent properties are classified as thixotropic and are often exhibited by concentrated suspensions of flocculated particles. A sudden increase in the imposed shear rate or shear stress causes a gradual decrease of the viscosity, while the opposite occurs for a sudden decrease of the controlled quantity, as illustrated in Figure 3.11.

Thixotropy is usually associated with shear-thinning, while the opposite type of behavior, involving a gradual increase of stress with time, is called antithixotropy, and is typical of highly concentrated particle suspensions, i.e., of shear-thickening systems.

Among the innumerous models and theories proposed to describe and elucidate the thixotropic behavior of disperse systems [29–31], a simple phenomenological approach can provide a satisfactory description of the thixotropic behavior, if elastic and other time-dependent effects are absent [32]. Such approach is based on the use of an arbitrary structural variable, often designated by λ, and implies the definition of two equations. The state equation correlates the variables τ, $\dot{\gamma}$, and λ, while the rate equation describes how the time variation of the structural variable, $d\lambda/dt$, due to the unbalance between structural buildup and breakdown processes, is controlled by $\dot{\gamma}$ and λ. However, in several other cases the time responses are more complex and a single structural parameter is not sufficient to account for the past rheological history, even in absence of appreciable elastic components [33].

More generally, time-dependent properties are viscoelastic and related to the time scales of relaxation processes induced by applied stresses or deformations in the system, on different length scales, from primary components to higher-order structures and the whole system as well. As the same term

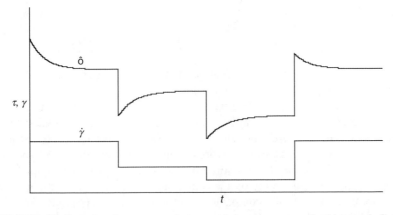

FIGURE 3.11 Stepwise shear rate variations and stress response of a thixotropic fluid.

suggests to our intuition, viscoelasticity represents a sort of hybrid from viscous and elastic behaviors which are the simplest mechanical archetypes and often assumed in the common language as synonyms of liquid and solid systems, respectively.

If a purely elastic solid is subjected to a given deformation, its tensional state instantaneously changes, internal stresses are proportional to the currently applied deformation at any instant. Hence, if deformation is kept constant, the tensional state does not change, too. If the flow of a purely viscous liquid is arrested, the deformation state remains unaltered while the internal stresses relax instantaneously and completely. Indeed, the real behaviors of materials differ from such simple idealized models, since at constant deformation stress relaxation occurs in more or less appreciable times. Very short relaxation times are peculiar to simple fluids of low-molecular weight (of the order of 10^{-12} s, as for water), while much longer times are typical of crosslinked polymers, or glassy systems (up to 10^{10} s, or more). In the daily practice, materials are classified as elastic solids or viscous liquids taking into account their mechanical response to low stresses over the ordinary time scale of minutes or seconds. However, if we apply a very wide range of stress over a wide spectrum of time, we are able to observe liquid-like properties in solids and solid-like properties in liquids. A drop of water falling on a surface may behave as an elastic solid within a very short time interval after the bump, while several years are needed to appreciate the flow properties of pitch or bituminous materials as demonstrated in the University of Queensland's pitch drop experiment [34].

Therefore, a given material can behave like a solid or a liquid depending on the time-scale of the deformation process. The distinction between liquid-like and solid-like behaviors can be based on the Deborah number which is defined as the ratio between the characteristic relaxation time of the material, λ, and the characteristic time, Λ, of the deformation process under observation

$$\text{De} = \frac{\lambda}{\Lambda}. \tag{3.58}$$

In a given process the rheological properties of a material can be easily classified as liquid-like if λ is much lower than Λ, or solid-like when λ is much higher than Λ, i.e., when the time it takes for the polymer to relax is very long compared to the rate at which deformation is accumulated. For a given material, the more pronounced are its solid-like properties, the faster is the deformation process. In other words, high Deborah numbers correspond to solid-like behavior and low Deborah numbers to liquid-like behavior. If De falls in the intermediate region the time scales of the material response and deformation process are comparable and hence the viscoelastic behavior cannot be ignored.

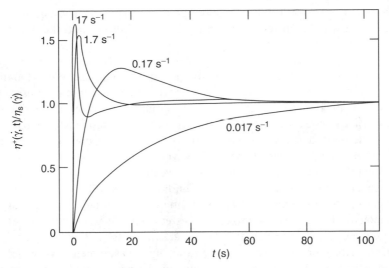

FIGURE 3.12 Transient responses (normalized viscosity vs. time) of a 1.5% poly-acrylamide solution in a water/glycerin mixture at 298 K at different shear rates. (Adapted from Lapasin, R. and Pricl, S., *Rheology of Industrial Polysaccharides: Theory and Applications*, Springer Verlag, 1995. Reprinted with permission from Springer Verlag.)

Figure 3.12 illustrates how the start-up response of a polymer solution changes also qualitatively, when the imposed shear flow condition is characterized by different $\dot{\gamma}$ values, and hence by different Λ values ($\Lambda = 1/\dot{\gamma}$). This example can be considered as paradigmatic of the transient behaviors of a viscoelastic system upon inception of flow, when the $\dot{\gamma}$ range is so large (here, three orders of magnitude) to encompass linear and nonlinear conditions. Viscosity data are reported in dimensionless terms: for each applied shear rate $\dot{\gamma}_0$ the ratio between the transient viscosity value $\eta^+(\dot{\gamma}, t)$ and the corresponding stationary value $\eta(\dot{\gamma})$ is plotted vs. time.

For very small shear rates, the shear stress (and the viscosity) approaches its steady-state value monotonically; for larger $\dot{\gamma}$, it goes through a maximum and then approaches the steady-state value, possibly after a stress undershoot condition, for still higher values of $\dot{\gamma}$. We can also observe that the maximum condition is not attained instantaneously; indeed, the time at which the maximum occurs decreases as $\dot{\gamma}$ is increased. These last considerations underline the neat differences between the viscoelastic and thixotropic behaviors. Indeed, the start-up responses of a thixotropic system are always characterized by an immediate stress overshoot followed by an exponential decay toward the steady state. In conclusion, a viscoelastic system can yield qualitatively different time-dependent responses, depending on the applied shear rate, i.e., for quite different De values. As $\dot{\gamma}$ increases, the characteristic time Λ $(= 1/\dot{\gamma})$ of the deformation process decreases; consequently, De decreases and the material behavior gradually shifts toward that typical of an elastic solid.

Conventionally, a distinction between thixotropy and viscoelasticity is made to underline the differences from a causal standpoint. While the time-dependent viscoelastic behavior arises because the response of stresses and strains in the fluid to changes in imposed strains and stresses, respectively, is not instantaneous, in a thixotropic fluid such response is instantaneous and the time-dependent properties arise purely because of changes in the structure of the fluid as a result of shear. However, we must emphasize that, even though it is convenient to distinguish between thixotropy and viscoelasticity as separate phenomena, many real fluids as weak gels and concentrated disperse systems exhibit both types of time-dependent behavior, even simultaneously. In general, the thixotropic effects prevail when the stresses and strain rates are relatively large, whereas at small deformations only the linear viscoelastic properties are manifested and their measurement is very useful as a physical probe of the microstructure of the system under examination. Conversely, the elastic components emerge quite evidently at higher shear flow conditions, more frequently for polymeric liquids. A number of effects and phenomena, such as die swelling in extrusion, reverse flow patterns in mixing tanks, or rod climbing of liquid in agitation, are manifestations of strong elastic components, firstly of high-primary normal stress differences.

3.8 LINEAR VISCOELASTICITY

3.8.1 Oscillatory Shear Test and Linear Viscoelastic Quantities

Small-amplitude oscillatory shear tests make possible to study the viscoelastic behavior of a system in a relatively extended range of kinematic conditions and, even more important, only perturbing its equilibrium structural state. Oscillatory motion can be realized by conventional rotational rheometers, rate- or stress-controlled, by imposing on one of the two boundary surfaces a displacement or a force, respectively, both sinusoidally varying with time.

Let us consider the unsteady motion of a fluid contained between two parallel plates, when the upper one undergoes a controlled sinusoidal motion of small amplitude in its own plane with frequency ω. The strain and shear rate distributions inside the sample are uniformly described by the following relationships:

$$\gamma = \gamma^0 \sin \omega t, \tag{3.59}$$

$$\dot{\gamma} = \gamma^0 \omega \cos \omega t = \dot{\gamma}^0 \cos \omega t, \tag{3.60}$$

where γ^0 and $\dot{\gamma}^0$ are the maximum strain and shear rate values during oscillating motion, respectively. The oscillatory shear stress response has the same frequency, but is shifted with respect to the strain

$$\tau = \tau^0 \sin(\omega t + \delta), \tag{3.61}$$

where τ^0 and δ are the maximum stress and the *phase shift*, respectively. Figure 3.13 gives a qualitative sketch of the response for viscoelastic fluids.
 The shear stress response can be split into two parts as follows:

$$\tau = \tau^0 \cos\delta \, \sin\omega t + \tau^0 \sin\delta \, \cos\omega t. \tag{3.62}$$

The first term is in phase with the strain and, hence, represents the elastic contribution, whereas the viscous one is given by the second term, which is in phase with the shear rate. In fact, in the two limiting cases $\delta = 0$ and $\delta = \pi/2$, we get the purely elastic and viscous behaviors, respectively,

$$\tau = \tau^0 \sin\omega t = G\gamma^0 \sin\omega t \quad (\text{for } \delta = 0), \tag{3.63}$$

$$\tau = \tau^0 \cos\omega t = \eta\dot{\gamma}^0 \cos\omega t \quad (\text{for } \delta = \pi/2). \tag{3.64}$$

Accordingly, Equation 3.62 can be written in two equivalent forms, by introducing two pairs of viscoelastic quantities, G', G'' and η', η'':

$$\tau = G'\gamma^0 \sin\omega t + G''\gamma^0 \cos\omega t, \tag{3.65}$$

$$\tau = \eta''\dot{\gamma}^0 \sin\omega t + \eta'\dot{\gamma}^0 \cos\omega t. \tag{3.66}$$

G' and G'' are called the storage modulus and the loss modulus, respectively, and, in general, vary with applied frequency ω. They represent the elastic and

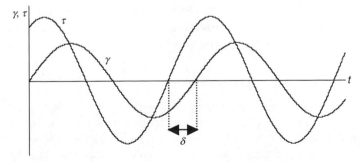

FIGURE 3.13 Small amplitude oscillatory shear tests: strain and stress vs. time.

viscous contributions to the properties of the fluid, and are related to the corresponding viscosity functions η' and η'' through

$$G' = \eta'' \omega, \qquad (3.67)$$

$$G'' = \eta' \omega. \qquad (3.68)$$

For a Newtonian fluid, the dynamic viscosity η' coincides with the shear viscosity and G' reduces to zero, whereas, for a purely elastic solid, we have $G'' = 0$ and the storage modulus is equal to the shear modulus G. Both the function pairs can be treated as real and imaginary parts of complex quantities, which measure the magnitude of the total response of the fluid. The complex viscosity η^* is defined by

$$\eta^* = \eta' - i\eta'' \qquad (3.69)$$

and, hence, we obtain the expression of the complex modulus G^*

$$G^* = i\omega \, \eta^* = G' + iG''. \qquad (3.70)$$

Its magnitude $|G^*|$ corresponds to the ratio between the maximum stress τ^0 and the maximum strain applied γ^0. At constant frequency, it does not depend on γ^0 and then the stress response is linearly proportional to the applied strain only if the oscillation amplitude is enough small, within the linear viscoelastic regime.

The linear viscoelastic behavior of a fluid is completely characterized when we know the frequency dependence of two functions, such as $|G^*|$ and the loss tangent, $\tan \delta = G''/G'$, or the dynamic moduli G' and G'', or any other combination of two quantities.

The frequency dependence of G' and G'' is normally plotted in a log–log diagram which is the mechanical spectrum of the system, almost an identity card of its linear viscoelastic behavior, where the weights of the elastic and viscous components as well as the relaxation modes are accurately recorded. This means that mechanical spectra obtained from small amplitude oscillatory tests represent a powerful tool to analyze the structural conditions of a system and offer a basis for molecular and structural interpretations.

3.8.2 LINEAR VISCOELASTIC MODELS

A simple way of understanding and describing the viscoelastic behavior is to make use of simple mechanical models, consisting of series and/or parallel combinations of linear elastic and viscous elements, i.e., springs and dashpots. The Maxwell model derives from the series combination of a Hookean spring of rigidity G, and a Newtonian dashpot of viscosity μ, as reported in Figure 3.14.

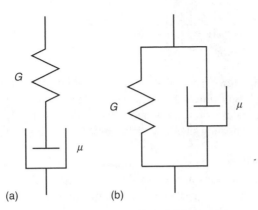

FIGURE 3.14 (a) The Maxwell model and (b) the Voigt model.

The response of such a mechanical model is strain-additive. In shear conditions the strain γ_{yx} and, hence, also the shear rate $\dot{\gamma}_{yx}$ produced by an applied stress τ_{yx} are calculated from the sum of both viscous and elastic contributions

$$\dot{\gamma}_{yx} = (\dot{\gamma}_{yx})_v + (\dot{\gamma}_{yx})_e = \frac{\tau_{yx}}{\mu} + \frac{1}{G}\frac{\partial \tau_{yx}}{\partial t}. \tag{3.71}$$

This equation defines the behavior of the simplest viscoelastic (Maxwellian) liquid. It can be rewritten as

$$\tau_{yx} + \lambda \frac{\partial \tau_{yx}}{\partial \tau} = \mu \dot{\gamma}_{yx}, \tag{3.71'}$$

where the relaxation time λ (equal to μ/G) weighs the importance of rate of change of stress.

The Newtonian fluid represents the limiting case of a Maxwellian fluid for $\lambda \longrightarrow 0$. Under steady flow conditions, Equation 3.71′ simplifies to the Newton's law. For rapid changes of the stress, the time-derivative term prevails and the behavior approximates that of a Hookean elastic solid.

Equation 3.71′ can be referred to every component of the stress tensor τ_{ij}, so that the more general version of the constitutive equation is

$$\tau_{ij} + \lambda_0 \frac{\partial \tau_{ij}}{\partial \tau} = \eta_0 \dot{\gamma}_{ij}, \tag{3.72}$$

where λ_0 and η_0 replace the previous λ and μ, respectively. By integrating Equation 3.72 over all the past times t' $(-\infty < t' < t)$, we obtain

$$\tau_{ij}(t) = \int_{-\infty}^{t} \left\{ \frac{\eta_0}{\lambda_0} \exp\left(-\frac{t-t'}{\lambda_0}\right) \right\} \dot{\gamma}_{ij}(t')dt', \tag{3.73}$$

where the function inside the braces is called the relaxation modulus. Accordingly, at any time t the stress state of the material can be interpreted as the result of the previous rheological history (defined by $\dot{\gamma}_{ij}(t')$), but with an exponentially vanishing memory of the past events, going backward in time ($t' \longrightarrow -\infty$).

The Maxwell model predicts a simple exponential stress response, when a constant shear rate $\dot{\gamma}_0$ is applied or the fluid is maintained at constant strain after its application. The growth and relaxation processes are controlled by λ_0:

$$\tau_{yx} = \eta_0 \dot{\gamma}_0 (1 - \exp(-t/\lambda_0)), \tag{3.74}$$

$$\tau_{yx} = \eta_0 \dot{\gamma}_0 \exp(-t/\lambda_0). \tag{3.75}$$

The stress relaxation following the sudden application of a strain γ_0 has the same form

$$\tau_{yx} = G\gamma_0 \exp(-t/\lambda_0). \tag{3.76}$$

The Maxwell model describes the behavior of a liquid-like material. In fact, when subjected to a constant nonzero stress τ^0, it undergoes a continuous deformation (creep flow)

$$\gamma_{yx} = \frac{\tau^0}{G} + \frac{\tau^0}{\eta_0}t = \gamma_0 + \dot{\gamma}_\infty t. \tag{3.77}$$

A solid-like viscoelastic behavior is described by the stress-additive Kelvin–Voigt model, resulting from the parallel connection of a Hookean spring and a Newtonian dashpot (see Figure 3.14). Now, the total stress comes from the sum of both viscous and elastic contributions

$$\tau_{ij} = (\tau_{ij})_v + (\tau_{ij})_e = \mu \frac{\partial \gamma_{ij}}{\partial t} + G\gamma_{ij}. \tag{3.78}$$

Here, the strain variation under constant stress τ_0 is given by

$$\gamma_{yx} = \frac{\tau_0}{G}\left(1 - \exp\left(-\frac{G}{\mu}t\right)\right) = \gamma_\infty\left(1 - \exp\left(-\frac{t}{\gamma_0}\right)\right). \tag{3.79}$$

The strain tends to a finite value γ_∞, as expected for a solid-like material. The rate of approach is governed by the retardation time $\lambda_0 \ (= \mu/G)$.

The liquid-like and solid-like character of the two models is well evidenced by the different relationships that characterize their response under oscillatory shear flow conditions.

In the case of the Kelvin–Voigt model, the dynamic viscosity and the storage modulus are constant and coincide with the model parameters μ and G. For the Maxwell model, we have

$$\eta' = \frac{\eta_0}{1 + (\lambda_0\omega)^2}, \quad \eta'' = \frac{\eta_0(\lambda_0\omega)}{1 + (\lambda_0\omega)^2}, \tag{3.80}$$

$$G' = G\frac{(\lambda_0\omega)^2}{1 + (\lambda_0\omega)^2}, \quad G'' = G\frac{\lambda_0\omega}{1 + (\lambda_0\omega)^2}. \tag{3.81}$$

We can see that the Maxwell model describes a progressive transition from viscous to elastic character of the fluid: increasing frequency ω results into η' decrease (from η_0 to 0), whereas the opposite trend is predicted for the storage modulus (from 0 to G). The transition coincides with the crossover condition ($G' = G''$) and occurs for $\lambda_0\omega = 1$, where the characteristic test time Λ ($= 1/\omega$) equals the relaxation time of the system (De = 1).

Even if the Maxwell and the Kelvin–Voigt models offer a simple and clear introduction to the linear viscoelasticity and its main features, they are not suitable to describe with enough approximation the more complex behaviors often exhibit by many disperse or polymeric systems. Starting from these simple archetypes, several other phenomenological models have been derived, following different criteria and strategies for their definition, i.e., by combining multiple linear viscous and elastic elements, changing the coordinate reference system, or resorting to more complex integrodifferential operators. A detailed review of the large family of linear viscoelastic constitutive equations is outside the scope of present book and we will illustrate only one of the most frequently used models.

Such a model is called generalized Maxwell model since it derives from the parallel combination of N Maxwell elements (see Figure 3.15). Accordingly, the total stress response is the sum of N contributions, each of which is given by the Maxwell equation (Equation 3.72).

The integral version of the model is

$$\tau_{ij}(t) = \int_{-\infty}^{t} \left\{ \sum_{k=1}^{N} \frac{\eta_k}{\lambda_k} \exp\left(-\frac{t - t'}{\lambda_k}\right) \right\} \dot{\gamma}_{ij}(t')\, dt'. \tag{3.82}$$

Here, the relaxation modulus is the sum of N exponential terms. The constitutive equation is defined by N relaxation times, λ_k and N viscosity constants, η_k (or, alternatively N modulus constants, $G_k = \eta_k\lambda_k$). The discrete set of G_k–λ_k (or η_k–λ_k) pairs is called the relaxation spectrum of a material and

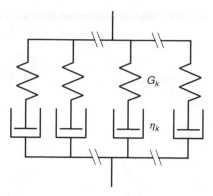

FIGURE 3.15 The generalized Maxwell model.

represents the most condensed form to summarize its time-dependent behavior in the linear viscoelastic regime.[12] If applied to steady shear flow conditions, the model predicts a Newtonian behavior as the simpler Maxwell equation does and the viscosity is given by

$$\eta_0 = \sum_{k=1}^{N} \eta_k.$$ (3.83)

The stress relaxation and growth processes predicted by the model are monotonic and symmetrical to each other, as for the Maxwell model, and result from the sum of the simple exponential contributions of each Maxwell element. The following expressions are derived for the stress relaxation and growth, respectively,

$$\tau_{yx} = \eta^- \dot{\gamma}_0 = \dot{\gamma}_0 \sum_k \eta_k \exp(-t/\lambda_k),$$ (3.84)

$$\tau_{yx} = \eta^+ \dot{\gamma}_0 = \dot{\gamma}_0 \sum_k \eta_k (1 - \exp(-t/\lambda_k)).$$ (3.85)

Similarly, the following expressions are derived for storage and loss moduli:

$$G' = \sum_k \frac{\eta_k \lambda_k \omega^2}{1 + (\lambda_k \omega)^2},$$ (3.86)

---

[12]The total number of parameters can be reduced by adopting empirical formulae or criteria for generating or scaling η_k and λ_k. Conversely, the model may be extended also to the case of an infinite number of Maxwell elements ($N \longrightarrow \infty$), so leading to a continuous relaxation spectrum $G(\lambda)$.

$$G'' = \sum_k \frac{\eta_k \omega}{1 + (\lambda_k \omega)^2}.$$ (3.87)

$G*$ and its components, G' and G'', are linear viscoelastic quantities derived from small amplitude dynamic shear tests. The analysis of the linear viscoelasticity can be performed also under oscillatory tensile conditions, so obtaining the complex Young's modulus $E*$ with its real (storage) E' and imaginary (loss) E'' components. In the more recent years several methods and instruments have been devised to measure the dynamic properties of viscoelastic materials, even for determining both complex moduli from one set of experiments on the same sample [35,36]. The relation between the tensile and shear moduli is

$$G* = \frac{E*}{2(1 + \nu)},$$ (3.88)

where the Poisson ratio ν is limited to values between 0 (completely compressible systems) and 0.5 (incompressible materials), between 0.4 and 0.5 for many materials of interest.[13] The macroscopic elastic properties and their interrelationships are strictly connected with the structural features of the material, which determine how the deformation field is spatially distributed, i.e., whether it is affine or not. In biopolymer networks and physical gels the degree of affinity of the deformation field increases down to the smallest length scales of the network in the case of high-crosslink density, molecular weight, and intrinsic rigidity [38].

3.9 RHEOLOGICAL PROPERTIES AND STRUCTURAL FEATURES

The Newtonian behavior is typical to simple fluids, such as pure compounds of low-molecular weight as well as their homogeneous mixtures whose components are compatible and do not give rise to supermolecular structures due to specific interactions. Newtonian fluids are also very dilute polymer solutions and disperse systems.

   Non-Newtonian properties appear if the molecular structure is intrinsically more complex as in the case of polymeric melts and blends, or when

---

[13]The Poisson ratio is defined as the negative transverse strain of a stretched or compressed body divided by its longitudinal strain. Ordinary materials contract laterally when stretched and expand laterally when compressed, and hence ν is positive (approximately 0.5 for rubbers and for soft biological tissues, for most metals between 0.25 and 0.35, from 0.1 to 0.4 for cellular solids such as typical polymer foams, nearly 0 for cork). However, negative Poisson's ratios are theoretically permissible and have been observed or realized in some auxetic materials, starting from novel foam structures [37].

higher-order structures are formed in a continuous phase by dissolved or dispersed components by virtue of their concentration and/or mutual inter-actions. A paradigmatic example is given by concentrated surfactant solutions which exhibit marked nonlinear properties inspite of the simple conformation of individual molecules. Polymeric concentrated solutions and gels, suspensions, emulsions, and foams display a wide variety of behaviors owing to the diverse distributions of solid particles, liquid droplets, and gas phase in a liquid phase.

The nonlinear behavior is related to the structural changes produced on different length scale by changes in the applied stress or deformation conditions. The structural conditions of polymeric and disperse systems depend on the intermolecular and surface forces which control the inter-actions within and between "particles" (molecules or macroscopic par-ticles). When polymeric chains and particles are individually distributed in the continuous phase, the viscosity is strictly related to their intrinsic features and spatial (positional and orientational) distribution in the flow field. If aggregates or associated structures are formed, the rheological behavior is essentially governed by flow-induced breakup and buildup of the higher-order structures.

In the following sections diverse classes of systems, from simple liquids to more complex fluids, will be examined with special attention to the interrelations between structural features and rheological properties.

3.9.1 SIMPLE FLUIDS AND HOMOGENEOUS MIXTURES

Viscosity values of Newtonian fluids may be quite different and cover several orders of magnitude. Even if there is no reliable theoretical basis for predicting liquid viscosities, some semiempirical estimation methods are available which are based on group contribution or other structural-sensitive parameters and then are valid for some homologous series. A comprehensive review is reported in Ref. [39].

Viscosity of homogeneous liquid mixtures is usually a nonlinear function of composition, so that it can be better described by resorting to $\log \eta$ and molar fraction as reference variables. The Grunberg–Nissan equation is widely used since it often provides a satisfactory correlation of experimental data inspite of its formal simplicity [40]. For a binary mixture, it gives

$$\log \eta = x_1 \log \eta_1 + x_2 \log \eta_2 + x_1 x_2 G_{12} \qquad (3.89)$$

and for multicomponent mixtures

$$\log \eta = \sum_i x_i \log \eta_i + \sum_i \sum_{j \neq i} x_i x_j G_{ij}, \qquad (3.90)$$

where x_i and η_i are the mole fraction and the viscosity of ith component, respectively, and G_{ij} is the interaction parameter between the components i and j. High G_{ij} values imply strong nonlinearity, while null values do not identify an ideal mixture condition, since a theoretical definition of ideal mixture viscosity does not exist differently from thermodynamic functions. G_{ij} values can be obtained from experimental data regression or calculated with the group contribution method proposed by Isdale et al. [41]. For aqueous systems and polar mixtures more accurate estimates are obtained with the Teja–Rice relation [42,43], which is based on the corresponding states principle. Details for both methods are reported in Ref. [39].

3.9.1.1 Salt Solutions

Concentrated and mixed salt solutions are often encountered in several contexts and then are worthy to be examined separately, in particular for their concentration effects. Up to now numerous experimental data of binary electrolyte solutions have been reported in the literature, but relatively few measurements are available for the mixed electrolyte solutions. Most theoretical investigations of viscosity have been focused on systems containing a single solute in dilute solutions [44]. In the limit of very low-ion concentrations ($c < 0.05$ mol/L) theoretical approaches correctly predict a square-root concentration dependence of the viscosity of ionic solutions [45]. For more concentrated solutions (up to 0.1–0.2 M) Jones and Dole proposed an empirical extension of such a simple model in order to account for intermolecular correlations [46]

$$\eta_r = \frac{\eta}{\eta_0} = 1 + Ac^{1/2} + Bc, \tag{3.91}$$

where η_r is the relative viscosity, c is the electrolyte molarity concentration (in mol/L), η and η_0 are the viscosities of the electrolyte solution and pure solvent (water), respectively. The term $Ac^{1/2}$ is identical to that obtained from the theoretical approaches, where A is a positive constant related to long-range Coulombic force interactions and mobilities of solute ions. B accounts for the interactions between the solvent and ions and the ion size, and its sign depends on the degree of solvent structuring introduced by the ions. Typically, large positive B values are found for strongly hydrated and structure-making (ordering) ions, while negative values are associated with structure-breaking (disordering) ions. For concentrated electrolyte solutions the concentration dependence of viscosity can be nonmonotonic, resulting from competition between various effects occurring in the ionic neighborhood, such as Coulombic interactions, size and shape of effects, alignment or orientation of polar molecules by the ionic field, and distortion of the solvent structure.

Extended versions of the Jones–Dole equation containing additional empirical terms were proposed to fit viscosity data at still higher concentrations. An example is given by the following expression [47]:

$$\eta_r = \frac{\eta}{\eta_0} = 1 + Ac^{1/2} + Bc + Dc^2 + Ec^{3.5}. \tag{3.92}$$

However, Equation 3.92 is not valid for systems containing more than one salt and cannot reproduce data for concentrated solutions (beyond ca. 1–3 M). Therefore, there is an evident need for theoretical models that would reproduce the viscosity of multicomponent solutions in the full ranges of concentration and temperature, or, alternatively, for simple predictive equations, which can make use of the available information on the binary electrolyte solutions and provide sufficiently accurate predictions for the mixed electrolyte solutions. Several efforts have been made in the last years in these directions, which are mainly based on coupling the Eyring's absolute rate theory with other theoretical models or empirical correlations [48–52].

3.9.2 SURFACTANT SOLUTIONS (FROM DILUTE TO MICELLAR AND CRYSTALLINE SYSTEMS)

Surfactant solutions exhibit quite different rheological behaviors owing to the polymorphic character of surfactants which can give rise to different microstructures, depending on concentration, temperature, and salinity of the system.

Surfactant molecules are composed of lyophilic groups chemically bonded to hydrophilic groups. A typical lyophilic group is a flexible carbon chain of short length (tail), while the shape of the hydrophilic group (head) is usually bulky. Different structural arrangements of surfactants are favored in water or oily phases owing to their amphiphilic nature: in an aqueous phase the tails tend to crowd together to minimize their exposure to water, while the opposite occurs in oil. Most of the surfactant molecules remain isolated until their concentration approaches a critical threshold (critical micellar concentration, CMC), above which almost all additional surfactant molecules self-assemble into aggregates (micelles of different shape, vesicles, bilayers). In a quiescent state such aggregates are dynamic structures resulting from the balance between continuous dissolution and reformation processes. The shape of surfactant aggregates can be predicted, even if approximately, for simple molecular shapes by referring to the molecular shape parameter and resorting to geometric packing arguments [53]. Conical molecules with bulky head groups form spherical micelles, cylindrical or slightly conical molecules with heads and tails of similar bulkiness form bilayers or vesicles, while wedge-shaped molecules with bulky tails form inverted micelles containing the heads in their interiors. Still depending on their shape, at high enough concentrations, aggregates can give origin to long-range structures, such as entangled solutions of giant wormy micelles, similar to concentrated ordinary polymer solutions, or crystalline phases of different positional and/or orientational order (nematic, hexagonal, smectic, cubic). It seems obvious to say that the rheological behavior is strongly dependent on the specific structural organization of the surfactant solution.

In contrast to the largely monodisperse spherical micelles, cylindrical micelles are polydisperse and their average length grows relatively rapidly with surfactant concentration. In their disordered state solutions of spherical micelles are low-viscosity liquids up to relatively high concentration (about 30% by volume). The increase of the relative viscosity with micellar volume fraction recalls that predicted for hard-sphere suspensions, but with higher viscosity values owing to the micellar shape fluctuations and electrostatic repulsions.

In the case of cylindrical micelles, they become longer and, hence, semiflexible with increasing surfactant concentration. As for ordinary polymer solutions, a dilute regime for such wormy micelles can be inferred directly from rheological experiments since at sufficiently low-surfactant concentration the viscosity is low and close to that of the solvent. It must be emphasized that the behavior of dilute solutions of wormy micelles is not quite simple, since they exhibit one of the most puzzling phenomena among the rich variety of shear-induced instabilities and transitions encountered in surfactant systems. Indeed, shear-thickening has been observed above a critical shear rate [54–62], because of the formation of a more viscous shear-induced phase, as confirmed by flow-birefringence [63–65] and SANS tests [55,56,59,65].

At high enough concentrations, worm-like micelles can entangle with each other, so forming a temporary network-like polymeric chains in concentrated solutions and melts. These giant micelles differ from ordinary polymers since they act like "living polymers" whose length distribution can vary reversibly in consequence of changes in concentration, temperature, salinity, and stress conditions as well. In fact, such wormy micelles are substantially broken, when subjected to strong flow conditions, but can reassemble themselves from their constituent molecules after cessation of flow. This imply that the theoretical approaches to the rheology of entangled polymers can be extended to entangled wormy micellar solutions, only accounting for the reversible processes of breakage and reconnection of micelles [66,67].

The mechanism of micelle breakage and reconnection depends on surfactant type and salinity. Figure 3.16 reports two limit structural configurations as illustrated by Candau et al. [68]: at low-electrolyte concentration, (a) the micelles are nonintersecting because of electrostatic repulsions and this structural feature is not altered by the breakage and reconnection processes, (b) while at high-salt concentration micelles become branched and intersect each other, forming a network whose junctions break and re-form. Other configurations are intermediate between these extremes. These brief observations

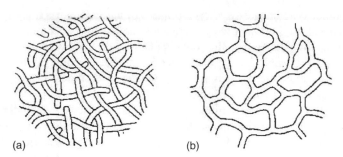

FIGURE 3.16 Structural configurations of wormy micelles: (a) entangled micelles, (b) interconnected micelles [68]. (Adapted from Walker, L.M. and Wagner, N.J., *J. Rheol.*, 38, 1525, 1994. With permission.)

underline the fundamental role of both salinity and concentration in determining the structural configurations and, hence, the rheological properties of surfactant systems, in particular of these concentrated solutions.

At high enough concentration, different types of liquid crystalline phases can appear in surfactant solutions, depending on the surfactant type and concentration, temperature, salinity, and the presence of other component such as cosurfactants [69,70]. Besides the distinction among nonionic, anionic, cationic, and zwitterionic surfactants, even more important factors are the presence of more tail or head groups and their relative dimensions, since the phase transitions are essentially governed by the space-filling constraints connected to such parameters. Different patterns of the composition–temperature phase diagrams can be encountered even when the analysis is restricted to the only category of aqueous solutions of one-tailed surfactant. Several structural transitions are produced by increasing surfactant concentration at constant temperature. At moderate temperatures, the most recurrent sequence implies the passage from disordered solution of spherical micelles to hexagonal packing of cylinders, to bicontinuous cubic phase, and to lamellar phase. The lamellar phases are very concentrated surfactant solutions that are generally persistent even at higher temperatures than the cubic phases.

Rheology of ordered surfactant systems is more complex and a consolidated basis of experimental observations and theoretical approaches is still lacking. Even if the flow behavior of surfactant nematics is similar to that of the corresponding polymeric systems, shear-thinning is associated with various flow-induced phenomena such as the formation of band textures, the onset of tumbling regime with increasing shear rate, the occurrence of multiple stress oscillations in start-up flow [71,72]. The flow curves of lamellar phases show a power-law shear-thinning region sometimes followed by a sharp viscosity drop toward Newtonian high-shear region [73,74]. Three different structural conditions are encountered with increasing shear rate: lamellar layers initially parallel to the shearing surfaces breakdown into multilamellar vesicles (onions) which are replaced by better or perfectly aligned layers at high shear [75]. The behavior of lamellar phases has received a lot of attention in the recent years [76–79] and many different shear effects have been reported, as referred in Ref. [80].

Bicontinuous cubic phases represent a class of highly ordered systems which exhibit elastic gel-like properties in the linear regime and apparently plastic behavior above a critical stress when the cubic phase ruptures along slip planes like other physical gels, as alginate or gelatin.

3.9.3 POLYMERIC SYSTEMS

The previous examination clearly demonstrates that marked non-Newtonian properties can appear also when simple units give origin to complex supramolecular structures as for surfactant molecules or micelles at high enough concentration. Conversely, complex structural units do not imply necessarily complex behaviors, as it can be easily evinced from the analysis of polymeric systems. Indeed, highly diluted solutions of polymer molecules, even of high-molecular weight, do not exhibit appreciable deviations from the Newtonian behavior. The rheological properties become important when topological constraints, associative interactions or chemical crosslinks between polymer chains lead to the formation of transient or permanent networks. Diverse structural subclasses can be distinguished within the wide family of polymeric systems, more than that for surfactant systems. Diversity factors are the intrinsic polymer characteristics, as the presence of one or more monomeric units, their number and sequence along the chain backbone, and the conformational properties which are strictly connected to the previous ones and to the nature of the solvent phase and the temperature conditions.

3.9.3.1 Flexible Chains and Ordinary Polymer Solutions

A convenient starting point for the analysis of polymeric systems is given by ordinary solutions where intermolecular chemical links or physical associations are absent. In an equilibrium state flexible macromolecules continuously change their conformation owing to Brownian forces and rotational motions of the large number of bonds in the backbone. The polymer conformation can be well represented by a random walk with a large number of steps or, equivalently, by a freely jointed chain with N_k rigid links, connected to each other at completely flexible joints, and whose length (Kuhn length) b_k corresponds to the effective random-walk stepsize. Then, in the case of Gaussian configuration distribution, the average end-to-end distance of the equivalent chain $\langle R^2 \rangle^{1/2}$ is given by $\sqrt{N_k} b_k$, whereas its fully extended length (contour length) L is equal to $N_k b_k$.[14] N_k is proportional to the polymer molecular weight M, while b_k increases with increasing chain stiffness, and corresponds to 5–10 times the monomer length l for typical flexible polymers. The b_k/l value is close to the characteristic ratio C_∞, more frequently used to measure the stiffness of polymer molecules. The polymer coil pervades an approximately spherical space whose radius is proportional to average end-to-end distance and then scales with $M^{1/2}$. This is strictly valid only for Gaussian random walk configuration and then it applies to polymer melts or concentrated solutions, but not to dilute solutions, except for theta conditions, depending on solvent–polymer pair and temperature.

Indeed, the polymer configuration in solution depends not only on the intrinsic flexibility of the polymer backbone but also on its environment. The solubility of a polymer in a solvent depends on relative magnitude of the always negative mixing entropy and the usually positive enthalpic term. For dilute solutions, the Flory–Huggins parameter χ accounts for polymer–solvent interactions and provides a measure of the solubility of the polymer [81,82]. In a good solvent ($\chi < 0.5$), the polymer segments preferably interact with the solvent and the polymer configuration is more expanded than the random walk one owing to excluded-volume interactions between any part of the chain and another: the effective volume fraction of the molecule increases; since the coil size is now proportional to $M^{3/5}$. In dilute solution with a poor solvent ($\chi > 0.5$) intramolecular interactions between polymer segments are favored and hence the molecule configuration is smaller than in a theta solvent (the coil radius scales with $M^{1/3}$), thus decreasing the effective volume fraction of the polymer. In the extreme case, the polymer molecules do not dissolve and then are dispersed in the solvent in form of aggregates. χ is temperature-sensitive, and the average coil size in dilute solutions is also sensitive to temperature. The theta condition ($\chi = 0.5$) occurs when the polymer coil contracts sufficiently to balance the expansion produced by excluded volume forces, thus recovering the random walk configuration.

Dilute solutions are those in which polymer molecules rarely overlap and, consequently, their rheology is governed solely by the dynamics and number of individual chains. When the concentration is increased, the polymer coils begin to overlap and become entangled with each other. The transition from dilute to concentrated systems is reached when the effective volume fraction pervaded by all the polymer molecules approaches 1. Accordingly the overlap concentration c^* decreases with increasing molecular weight M; it can be estimated from individual chain configuration and scales differently in theta

---

[14]The bonds of the real chain are not collinear even in the fully extended conformation and then the contour length is lower than the product Pl, where P is the number of monomers and l is the monomer length; for tetrahedral bonding angles $L \cong 0.82\,Pl$.

solvent (as $M^{-1/2}$) and in good solvent (as $M^{-4/5}$). In the latter case the dilute regime ends at lower $c*$.

Very dilute polymer solutions are Newtonian and their viscosity increases linearly with concentration for uncharged polymers or polyelectrolytes at constant ionic strength. For slightly higher concentrations, the solutions exhibit shear-thinning because polymer chains progressively orient and align with the flow direction as the shear rate is increased, thus reducing the viscous resistance. The viscosity remains constant only at low-shear rates (zero-shear viscosity η_0). The dependence of viscosity on concentration (at constant shear rate) can be better described by the Huggins equation where a quadratic term is added to account for second-order molecular interactions [83]

$$\eta_r = \frac{\eta}{\eta_s} = 1 + [\eta]c + k_H[\eta]^2 c^2, \tag{3.93}$$

where η_r is the relative viscosity (the solution viscosity η divided by the solvent viscosity η_s), c is the polymer concentration, and k_H is the Huggins constant. The rate of viscosity increase with increasing concentration is ruled by the intrinsic viscosity $[\eta]$ which can be extrapolated from data obtained for dilute systems as follows[15]:

$$[\eta] = \lim_{c \to 0} \frac{\eta_r - 1}{c} = \lim_{c \to 0} \frac{\eta_{sp}}{c}, \tag{3.95}$$

where η_{sp} is the specific viscosity. Only in excess of added salts, the specific viscosity of dilute aqueous solutions of polyelectrolytes linearly increases with increasing polymer concentration, as for nonionic polymers. As simple electrolyte concentration decreases, the viscosity increases and a more complex dependence upon polymer concentration is observed.

If derived from zero-shear-rate viscosity values, the intrinsic viscosity is related to the polymer configuration in equilibrium conditions. It depends only on the coil dimension and then increases with the polymer molecular weight. Several theoretical approaches[16] confirm the validity of the empirical relation suggested by Mark and Houwink to describe the dependence of $[\eta]$ on polymer molecular weight [85,86]

$$[\eta] = KM^a, \tag{3.96}$$

---

[15]Alternatively, the dependence of viscosity on concentration can be described with the Kraemer equation [84]:

$$\ln \eta_r = [\eta]c + k_k[\eta]^2 c^2. \tag{3.94}$$

[16]A detailed review of the theoretical approaches proposed for intrinsic viscosity is reported in Ref. [9].

where both parameters K and a depend on the polymer–solvent pair and temperature, and are related to the stiffness of the polymer. In particular, for a flexible random coil, a usually lies in the range 0.5–0.8 (from theta to good solvents), whereas it increases with increasing chain stiffness up to 1.8 for rigid rods.

In the case of polyelectrolytes, the Mark–Houwink parameter a increases with decreasing ionic strength I, since the chain is forced to a more extended and rigid conformation in order to minimize its electrostatic free energy. The influence of ionic strength on the polyelectrolyte chain stiffness can be derived from the analysis of viscosity data obtained at different constant I values. The intrinsic viscosity can be correlated with I as follows [9]:

$$[\eta]_0 = [\eta]_0^\infty + SI^{-1/2}, \tag{3.97}$$

where $[\eta]_0^\infty$ is the intrinsic viscosity at infinite ionic strength. The constant S could be used as a parameter of stiffness of the polyelectrolyte molecule resorting to theoretical considerations or empirical approaches [9].

Polymer solutions exhibit shear-thinning behavior and the concentration dependence of viscosity is stronger for lower shear rates. Therefore, the intrinsic viscosity is shear dependent, too, and the $[\eta]$–$\dot{\gamma}$ profile can be derived by extrapolating η_{sp}–c data sets obtained at different $\dot{\gamma}$. As evidenced by experimental observations and confirmed by molecular models, the degree of shear-thinning is much more pronounced in good solvents than in theta solvents, for semiflexible or rigid chains more than for flexible and more extendable macromolecules [9].

At higher concentrations intermolecular interactions become predominant and impact the rheology in a significant way. They may be viewed as localized temporary junctions (entanglements) which continually form and vanish among the chains in motion so acting to form a temporary network similar to crosslinked polymers [87], or as topological constraints which are put on the motion of each single molecule by the surrounding chains (see Figure 3.17). In the latter case the macromolecule is confined in a tube-like region, whose contours fluctuate with time, and the only way it can move is by reptating along the tube [88–91].

Now, the molecular weight and concentration of the polymer affect more significantly the rheology of the concentrated polymer solutions, which display marked shear-thinning and viscoelastic properties. Figure 3.18 clearly shows that concentrated solutions display high viscosities and very pronounced shear-thinning in the case of rigid chain conformation (scleroglucan), while deviations from the Newtonian behavior are detected only at high concentrations for flexible polymers (polyvinylpyrrolidone). In particular, the dependence of $\eta_{sp,0}$ on C is much stronger than for dilute solutions and this

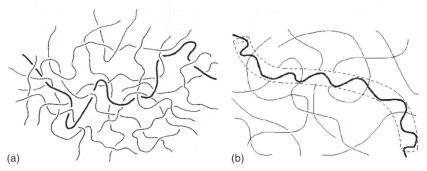

FIGURE 3.17 Physical representations of concentrated polymer systems: (a) entanglement network and (b) tube model.

difference can be clearly underlined by fitting data in dilute and concentrated regimes with the power-law type correlation

$$\eta_{sp,0} = aC^b. \tag{3.98}$$

Different values of the exponent b are obtained in the two regimes. For dilute solutions b normally varies between 1.1 and 1.6; the nonlinear dependence can be ascribed to the nonnegligible role played by the hydrodynamic interactions among the polymer coils. In the concentrated region the b values range from 1.9 to 5.6 in front of the theoretical value (3.75) derived by de Gennes from scaling arguments [92]. Even if the transition from dilute to concentrated solutions can be described by a continuous function $\eta_{sp,0}(C),$[17] encompassing both regimes, the discontinuous representation based on two power-law correlations is more convenient since it yields an objective criterion for determining, within the transition region, a single concentration value C^*, defined by the intersection of the two linear branches (see Figure 3.19).

Many efforts have been devoted to the definition of a more general representation, including the effects of both polymer molecular weight and

---

[17]A satisfactory fitting is generally provided by the Martin equation [93,94]:

$$\eta_{sp,0} = C[\eta]_0 \exp(K_M C[\eta]_0), \tag{3.99}$$

where $[\eta]$ is the zero-shear intrinsic vicosity, C is the polymer concentration, and K_M is a dimensionless parameter which depends on the polymer–solvent pair.

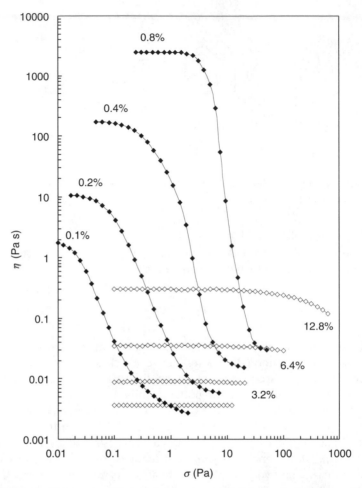

FIGURE 3.18 Flow curves of aqueous concentrated solutions of scleroglucan (◆) and polyvinylpyrrolidone (◇) at different polymer concentrations.

concentration. Viscosity data relative to flexible coil solutions can be reduced to a common master curve by using the following dimensionless quantities [95–97]:

$$\tilde{\eta} = \frac{\eta_{sp,0}}{C[\eta]_0}, \quad \tilde{C} = \frac{C}{\gamma}. \tag{3.100}$$

γ is a characteristic concentration, depending on the molecular weight M as follows:

$$\gamma = K_1 M^{-a_1}, \tag{3.101}$$

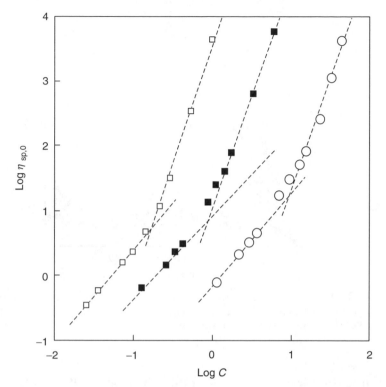

FIGURE 3.19 Zero-shear specific viscosity vs. concentration for ($\square$) guar gum, ($\blacksquare$) high-guluronate alginate in 0.2 M NaCl, and ($\bigcirc$) dextran. (Adapted from Lapasin, R. and Pricl, S., *Rheology of Industrial Polysaccharides: Theory and Applications*, Springer Verlag, 1995. Reprinted with permission from Springer Verlag.)

where the empirical parameter a_1 approximates the Mark–Houwink exponent in poor solvents. In such conditions $\widetilde{C}$ is equivalent to $C[\eta]_0$. In the field of polysaccharides, it is customary to use $\eta_{sp,0}$ and $C[\eta]_0$ as reference coordinates; accordingly, the transition from a dilute to a concentrated regime falls in a relatively narrow range of $C[\eta]_0$ values (2.5–4) and then an approximate criterion can be found for predicting C^* from $[\eta]_0$ data.

3.9.3.2 Rigid Chains and Crystalline Polymer Solutions

All the above considerations can be extended to rigid chains, because a neat distinction still exists between dilute and concentrated regimes. At very low concentration the viscosity dependence on polymer concentration and molecular weight is similar for both rigid and flexible polymers. In fact, the Brownian motion of each rod-like molecule is independent, and determined by the solvent viscosity up to a critical concentration, where the rotational and translational motions of each rod are severely restricted, due to entanglement

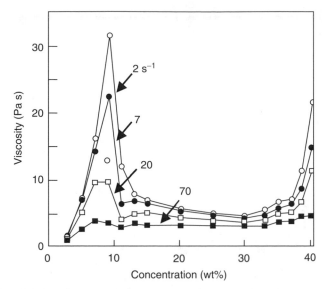

FIGURE 3.20 Viscosity vs. polymer concentration at various shear rates (poly-γ-benzylglutamate solutions in *m*-cresol). (Adapted from Walker, L.M. and Wagner, N.J., *J. Rheol.*, 38, 1525, 1994. With permission.)

formation. Nonetheless, the rotation of each given rod is still random at rest and the concentrated solution is isotropic. Above another critical concentration the random orientation is no longer allowed, eventually resulting in the appearance of an ordered phase of the liquid crystalline type.

Owing to the molecular characteristics and the liquid crystal formation, the viscosity behavior of rod-like polymer solutions in the concentrated regime is very different from that of flexible polymer solutions. A nonmonotonic increase of viscosity is observed with increasing polymer concentration. The viscosity curve goes through a maximum near the boundary concentration between the isotropic and biphasic regions, and then sharply decreases to increase again at higher C. The maximum is more pronounced and sharp at low-shear rates (see Figure 3.20). The profile of the curve is the consequence of two opposite effects, related to orientation effects and polymer concentration. Due to the orientation of their rods, the anisotropic domains have a lower viscosity than the isotropic regions. As polymer concentration increases, the viscosity of both phases tends to increase, but, contemporarily, also the concentration of anisotropic domains becomes higher.

According to the generalized behavior proposed by Onogi and Asada [98], three distinct regions can be distinguished in the flow curve of liquid crystalline polymer solutions (see Figure 3.21). In the central region the system displays either a constant viscosity or is, at least, less shear-thinning than in the subsequent

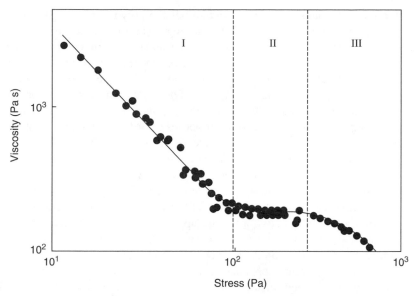

FIGURE 3.21 Viscosity vs. stress for 60% hydroxypropylcellulose in water [99]: flow curve subdivided in three regions according to Onogi and Asada [98]. (Adapted from Walker, L.M. and Wagner, N.J., *J. Rheol.*, 38, 1525, 1994. With permission.)

high-shear power-law region. Conversely, at low shear (yielding region) the behavior diverges from that of flexible polymers since it seems to be associated with an apparent yield stress. In this sense, an analogy could be established with the shear dependence of weak gels, highly filled polymer systems or concentrated dispersions, even if yield stress behavior could be ascribed to the piled poly-domain texture instead of the formation of a continuous three-dimensional network, deriving from macromolecule or particle aggregation.

For all isotropic polymer solutions the primary normal stress difference N_1 is always positive and increases with increasing shear rate, whereas three different regions can be distinguished in liquid crystalline polymers: at low-shear rates, N_1 depends linearly on $\dot{\gamma}$, in the intermediate range of shear rates, N_1 first decreases, becomes negative and, then, grows up again to positive values at higher shear rates. The negative N_1 zone is shifted to larger $\dot{\gamma}$ values as the polymer concentration is increased. Moreover, when subjected to sudden changes of shear flow conditions, liquid crystalline polymers generally exhibit peculiar time-dependent properties [9,70]. In most startup experiments and in other stepwise tests, damped oscillations are observed before the new steady state is reached, at a frequency inversely proportional to the imposed shear rate.

All these typical features of liquid crystal polymers are connected to the different specific structural processes which are involved in shearing conditions, i.e., rearrangement of molecular orientational distribution, relaxation of elastic distortion of domains, domain size evolution, and polydomain reorganization. They are characterized by quite different length and time scales and this can explain why the time-dependence of the viscous and elastic components are largely different. After the cessation of a previous shearing, the shear stress drops much more rapidly than the normal stress, and the complex viscosity decays exponentially with time scales of the order of hours, approaching "equilibrium" values, which depend markedly on the rate of applied preshear deformation.

Turning now to the viscoelastic properties of ordinary polymer solutions, they depend on the relative magnitude of their relaxation times and the reciprocal of the rate of deformation (i.e., the shear rate or frequency), there is more liquid-like behavior at low-shear rates (or frequency) and solid (or elastic)-like behavior at higher deformation rates. The relaxation time of a polymer chain reflects the time required to relax back to an equilibrium orientation and configuration after the application and subsequent removal of stress. The relaxation modes of a single chain are free in very dilute systems, only dependent on the molecular weight, while they are constrained by adjacent molecules in more concentrated solutions, so that the longest time of the relaxation spectrum increases with both polymer concentration and molecular weight. Very dilute solutions have enough time to relax to their equilibrium state within the usual deformation time scale, so that a fluid-like behavior with constant viscosity is observed, the viscous component of the shear modulus is remarkably larger than the elastic one. Conversely, at high-polymer concentration the deformation time scale can become much smaller than the relaxation time, and the systems show a solid-like behavior (with G' much larger than G'') over a large frequency range. For ordinary isotropic solutions the crossover condition shifts monotonically toward lower frequencies and larger modulus values with increasing polymer concentration (and/or molecular weight) (see Figure 3.22).

For liquid crystal polymers the influence of polymer concentration on the storage modulus resembles that exerted on the shear viscosity, since the G'–C curve displays a maximum value which is located at nearly the same concentration as the viscosity maximum. The decrease in the elastic modulus takes place when the degree of alignment in the polymer solution becomes very high.

All the above considerations developed for concentrated solutions can be extended to polymer melts, which can be simply considered as the limiting case in the class of entangled polymer systems. The effects of entanglements on the relaxation of polymer chains are similar to those observed for concentrated solutions, if the molecular weight exceeds a critical value M_c which plays the same role of overlap concentration for solutions. Above M_c the zero-shear rate viscosity and the longest relaxation time begin to rise as $M^{3.4}$, and a plateau appears in the frequency dependence of G', whose extension increases with increasing molecular weight. The value of the plateau modulus G_N^0 can be related to the density of entaglements so that the chain length between two consecutive entanglements can be estimated [70].

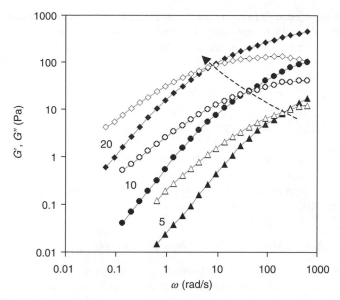

FIGURE 3.22 Mechanical spectra of hyaluronic acid solutions at different polymer concentrations: 5 g/L (triangles), 10 g/L (circles), 20 g/L (diamonds) (G': closed symbols, G'': open symbols).

3.9.3.3 Time Temperature Superposition Principle

Polymer solutions and melts belong to the class of thermorheologically simple fluids, since viscosity and other material functions are strong functions of temperature, but, more importantly, maintain their functional dependence on the shear rate or frequency. A change in temperature merely shifts the curves of $G'(\omega)$ and $G''(\omega)$ along the frequency axis without changing their shapes, as a consequence of the corresponding rigid shifting of the relaxation spectrum toward lower time values with increasing temperature. Experimental data determined at different temperatures can then be superposed so obtaining a generalized mechanical spectrum which covers a frequency window even more extended than that encompassed at any single temperature. This time–temperature superposition procedure makes then easier and more significant the analysis of temperature effects. The temperature dependence of the shift factor a_T required to obtain the superposition can be fit to an Arrhenius-type dependence similar to the Andrade equation or, more frequently, to the empirical WLF (Williams–Landel–Ferry) equation [100]

$$\log a_T = \frac{-c_1^0(T - T_0)}{c_2^0 + T - T_0},$$ (3.102)

in which c_1^0, c_2^0, and T_0 (a reference temperature) are constant. The flow curves obtained at different temperatures have similar shapes and then a master curve can be derived also for the viscosity function by plotting log η_r vs. log $\dot{\gamma}_r$, where the reduced variable are derived from experimental data $\eta(T)-\dot{\gamma}$ as follows:

$$\eta_r = \frac{\eta(T)}{a_T}, \quad \dot{\gamma}_r = a_T\dot{\gamma}. \tag{3.103}$$

The shift factor a_T corresponds also to the zero-shear viscosity ratio $\eta_0(T)/\eta_0(T_0)$ where T_0 is the reference temperature. This means that the temperature dependence of the zero-shear rate viscosity is provided by Equation 3.102.

3.9.3.4 Associative Polymer Solutions and Gels

Till now we have examined the rheological effects produced only by different chain conformations in dilute solutions and topological constraints due to entanglements or by molecular ordering and crystal domain formation in concentrated systems. Nevertheless, it must be underlined that in several other polymer systems physical or chemical intermolecular associations can occur and result into supramolecular structures that have different complexity and relaxation modes. Accordingly, a wide variety of rheological behaviors is displayed by associated polymeric systems. A first distinction must be made between chemical and physical associations, since covalent bonds can give origin to hard, irreversible, chemical gels of solid-like behavior, while physical associations produced by disparate forces such as Coulombic, van der Waals, dipole–dipole, hydrophobic interactions and hydrogen bonds are weak and reversible so that fluid properties can be still exhibited in appropriate stress conditions.

Chemical crosslinking of polymer molecules is accompanied by a progressive change in the rheological properties from those of a viscous liquid to those of an elastic solid. When the gel structure of permanent crosslinked network is completely formed, the elastic modulus can be related to the number of network strands per unit volume, ν, in accordance with the classical theory of rubber-like elasticity ($G_e = \nu RT$) on condition that the functionality of the network is sufficiently high. The analysis of partially cured or lightly crosslinked materials supplied the experimental basis for a better definition of the sol–gel transition (gel point) in substitution of the simpler and still widely used criterion ($G' = G''$). It has been observed that near the gel point the profiles of the storage and loss moduli are parallel straight lines over several frequency decades and their slope (in the log G'–G'' vs. log ω) depends on the stoichiometric ratio of crosslinker to precursor polymer molecules [101–103]. The power-law exponent derived from the

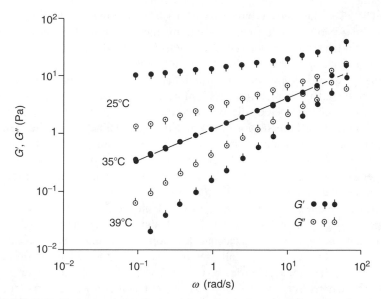

FIGURE 3.23 Temperature evolution of mechanical spectra for 1% w/w ι-carra-geenan in water [104]. (Adapted from Lapasin, R. and Pricl, S., *Rheology of Industrial Polysaccharides: Theory and Applications*, Springer Verlag, 1995. Reprinted with permission from Springer Verlag.)

frequency dependence of viscoelastic moduli (or, equivalently, from the time-dependence of the relaxation modulus) can be connected with the fractal scaling properties of gel clusters. Figure 3.23 shows that the temperature-driven sol–gel transition of an aqueous solution of ι-carrageenan can be individuated quite precisely from the mechanical spectra [104].

Turning now our attention to intermolecular physical associations, several factors may contribute and concur to determine the structural features of the polymeric system and the resultant rheological effects. The strength of phys-ical interactions is clearly important and significant effects can be expected from hydrogen bonds whose binding energies are one order of magnitude higher than van der Waals interactions, albeit much weaker than covalent bonds. However, the number and distribution of active sites along the chain backbone may be more important, since even weak physical bonds can cooperate locally thus giving rise to relatively strong intermolecular interactions.

In the case of telechelic polymers the associating hydrophobic groups are located only at the two ends of the hydrophilic chain and, then, in an aqueous environment they clump together to form very peculiar micellar structures, where the hydrophilic chains give origin to intramicellar loops as well as intermicellar bridges at high enough polymer concentration. Typical examples of telechelic polymers are hydrophobically modified ethoxylated urethanes

(HEUR). The rheological properties of associative polymers resemble those of ordinary solutions, even if an intermediate shear-thickening region is noticed in the flow curves of concentrated systems [105–112].

When multiple physical bonds of sufficient strength are more or less regularly spaced along the chain and the polymer concentration is high enough, the diffusive motion of each molecule is confined by entanglements and strongly conditioned by dissociation and reassociation of sticky junctions. This implies that the behavior of entangled sticky chains is characterized by longer relaxation times but it does not differ significantly from that of other entangled polymer systems. Shear-thickening in a polymer with multiple stickers can originate by a shear-induced change in the balance between intramolecular and intermolecular junctions.

More complex mechanisms of physical interaction are encountered in biopolymer systems, where the structural features of the chain play quite an important role [9]. Many polysaccharides adopt helical and ribbon-like shapes in the solid state and can partially retain such ordered forms in solution also by virtue of their relative intrinsic stiffness, when favorable interchain physical interactions act cooperatively to overcome the entropic drive to the disordered random coil state. This requires the alignment and interaction of two or more chain segments, usually characterized by extended regions of regular primary structure. A typical example is given by alginic acid and its salts. Alginic acid is a copolysaccharide containing three types of segments: one made up exclusively of α-D-mannuronic acid (the M-block), one of α-L-guluronic acid (the G-block), and the third in which the monomers approximate an alternate sequence (the MG-block). When divalent cations such as Ca^{2+} are introduced in sodium alginate solutions, Na^+ are replaced by Ca^{2+} ions which bind preferentially to G-blocks, creating egg-box ordered junction zones which can terminate with sharp kinks in the chain. If the guluronate block exceeds 20 residues, the process is highly cooperative, and can give origin to an expanded crosslinked network, where the extended junction zones are interconnected by more flexible chain segments. At sufficiently high polymer concentration, a physical gel is formed when chain–chain associations span the whole system. The network formation and its stability depend on the balance between junction zones, flexible segments, and kinks, in other terms between ordered and disordered conformations (i.e., between soluble and insoluble regions, or enthalpic and entropic contributions). Indeed, phase separation rather than gel formation is caused by an excess of intermolecular physical associations.

For alginates and pectins the junction zone formation is based on the ion-mediated egg-box mechanism, while in other cases, as for agar and carrageenan, it derives from double helix formation and/or helix–helix association, or from the persistence of rod-like triple helices in solution in the case of scleroglucan, or from disorder–local order transition and the lateral association of ordered chain sequences as for xanthan.

Looking at another important class of biopolymer systems, globular proteins can associate to form particulate networks made of ordered stringy structures or randomly branched aggregates. This is the case of ovalbumin and β-lactoglobulin whose textural and rheological properties are very sensitive to the environmental and preparation conditions. Small changes in pH or heating rate can produce remarkable changes in the rate and mode of aggregation, and then in the compactness of the particulate network, whose structural characteristics can be analyzed with fractal geometry criteria as for colloidal particle dispersions.

3.9.3.5   Weak and Strong Physical Gels

These few examples suffice to indicate that an astounding variety of physical interactions and association mechanisms can be envisaged, by virtue of the diversity in polymer conformations and, therefore, in the intermolecular spatial arrangements, which eventually lead to the formation of physical polymer gels. The rheological behavior of these systems lies in between those of chemical gels, where permanent covalent bonds exist between the polymer chains, and entanglement networks, formed by topological interchain interactions of brief lifetime. Indeed, the number and position of physical crosslinks can, and does, fluctuate with time and temperature as well as their strength may vary substantially from system to system. Furthermore, their spatial distribution within the network is usually more or less heterogeneous, since the interchain association mechanism often leads to an intricate hierarchy of structures, as illustrated before. Accordingly, the physical gels neatly differ from covalently crosslinked networks which are quite homogeneous over the entire distance scale (down to 10 nm).

A disputable but useful distinction between strong and weak gels can be made on the basis of the peculiar properties at small and, even more, at large deformations. For all the physical gels the extension of the linear viscoelastic regime is confined at lower strains than for ordinary solutions. However, the viscoelastic moduli of strong gels (as are usually denoted by agarose, gelatin, calcium alginate) become strain dependent at higher strains (≥ 0.2–0.25) than for weak gels. As for several concentrated particle dispersions, the critical strain derived from strain or stress sweep tests may be sensibly lower than 0.01. Moreover, the transition from linear to nonlinear regime is marked by a redistribution between the viscous and elastic components, so that the loss modulus initially increases with increasing strain. Such a behavior resembles that exhibited by flocculated suspensions, also as its concentration dependence is concerned, so that similar breakdown processes of the network can be invoked.

Another neat discrimination between strong and weak gels derives from their large deformation responses. Above a critical stress, strong gels undergo rupture with the formation of discontinuity surfaces across the sample, whereas homogeneous flow conditions without fracture are achieved for xanthan and other weak gels. In most cases the shear-dependent behavior of weak gels is apparently plastic: the low-shear rate Newtonian plateau is confined to very low-shear rates (10^{-4}–10^{-6} s^{-1}) and followed by a wide power-law region (with $n = 0.1$–0.2) which can be extended even over 6 shear rate decades. Neat changes in the time-dependent behavior occur entering the shear-thinning region owing to substantial breakdown of gel microdomains. Such thixotropic properties are characterized more suitably by means of stepwise procedures. The stress transients so obtained closely resemble those exhibited by flocculated dispersions, where structural buildup and breakdown processes regularly result from stepwise shear rate variations of opposite sign [9].

The elastic character of strong gels clearly emerges from the profile of dynamic moduli, since the storage modulus is almost independent of ω over a wide frequency range and its value is much greater than that of G'' (generally, much more than one order of magnitude). The mechanical spectra of weak gels show a slight frequency dependence with no apparent tendency to a Newtonian plateau as $\omega \longrightarrow 0$. In most cases the profiles of $G'(\omega)$ and $G''(\omega)$ are almost parallel and G' exceeds G'' over the ω range explored, in a lesser extent than for strong gels (see Figure 3.24). Accordingly, the relaxation modulus is satisfactorily described by a power-law time-dependence and this is another distinction between weak and strong gels. Indeed, such features

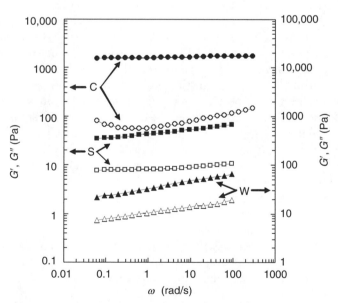

FIGURE 3.24 Mechanical spectra of aqueous physical gels: (C)—carbopol 980 2% (added with triethanolammine 0.15%), (S)—scleroglucan 1.2%, (W)—welan 0.8% (G': closed symbols, G'': open symbols).

closely resemble the behavior of chemically crosslinked polymer with strongly unbalanced stoichiometry at the sol–gel transition, since the exponent of the scaling law is far from 0.5 [103].

A strong dependence of the storage modulus of strong gels on polymer concentration is typically observed. More frequently, the storage modulus is a power-law function of the concentration with a power-law exponent greater than 2 [9]. This behavior is similar to the storage modulus-volume fraction (or concentration) relationship of attractive particle systems and highly entangled polymer solutions.

3.9.4 DISPERSE SYSTEMS

Physical weak gels represent a very peculiar class of polymer systems which introduces to the wide family of dispersions since their heterogeneous texture and the structural mechanisms governing the shear- and time-dependent properties in nonlinear regime resemble those of aggregated suspensions. Interaction forces between particle surfaces exert a fundamental role also on the structural characteristics of disperse systems and their mechanical responses, on condition that the particle size distribution covers, at least partially, the colloidal range. Indeed, dispersions of attractive, repulsive, and noninteracting colloidal particles display appreciable differences in their rheological behaviors.

Particle dimensions are important for the dispersion stability in both static and dynamic conditions. The tendency of large particles to settle under gravity increases linearly as the difference between phase densities $\Delta\rho$ increases and the viscosity of the continuous phase decreases in accordance with the Stokes equation:

$$v = \frac{2\Delta\rho r^2}{9\eta_s}.$$ (3.104)

In the case of fine particles, the oriented motion due to sedimentation is contrasted by the random Brownian motion produced by thermal forces. Accordingly, a criterion for static stability of spherical particles dispersed in a Newtonian liquid can be derived from the comparison between Brownian and gravitational forces

$$\frac{\dfrac{\Delta\rho g r^3}{kT}}{r} = \frac{\Delta\rho g r^4}{kT} < 1,$$ (3.105)

where r is the particle radius, and g is the gravity constant. Starting from Equation 3.105 an arbitrary border between colloidal and noncolloidal particles can be set around 1 μm. Under shear flow conditions coarser particles are subjected to other instability effects whose intensity is proportional to shear rate, as migration in the flow field, also contrasted by Brownian motion, or inertial effects, which are contrasted by liquid phase viscosity.

Smaller particle sizes mean more efficient action of Brownian motion against not only gravity but also hydrodynamic forces arising from the relative motion of particles to the surrounding fluid and the interactions between the particle surfaces.

Viscous stresses tend to produce an ordered configuration within the fluid in opposition to randomizing Brownian forces. The translational Peclet number, Pe, measures the relative efficacy of hydrodynamic and Brownian forces. For spherical particles it is given by

$$\text{Pe} \equiv \frac{6\pi\pi_s\dot{\gamma}r^2}{\dfrac{kT}{r}} = \frac{6\pi\pi_s\dot{\gamma}r^3}{kT}. \tag{3.106}$$

This dimensionless number can be also reconsidered as the ratio of two characteristic times t_D and t_f, which are associated with diffusional Brownian motion (derivable from the Stokes–Einstein hydrodynamic diffusion theory) and the shear motion (inversely proportional to shear rate

$$t_D = \frac{r^2}{D} = \frac{r^2}{\dfrac{kT}{6\pi\eta_s r}}, \tag{3.107}$$

$$t_f = \frac{1}{\dot{\gamma}}. \tag{3.108}$$

The spatial distribution of particles becomes more and more ordered as the shear rate increases, so offering minor resistance to motion. This is the only structural process governing the shear-thinning behavior of spherical noninteracting particles, as it will be illustrated hereafter.

Nonspherical particles are subjected also to periodic flow-induced rotation, depending on shear intensity and particle shape factor and contrasted by Brownian motion. Therefore, also a rotational Peclet number can be derived from the ratio of the two characteristic times related to rotational diffusion and shear motion. In the case of high-particle anisometry rotational Brownian diffusion is negligible and then hydrodynamic forces induce particle orientation even at low shear. Also this structural mechanism contributes, even significantly, to shear-thinning behavior of dispersions.

3.9.4.1   Colloidal Interactions and Dispersion State

For both effects Brownian and non-Brownian states of particles can be distinguished. Smaller particle size implies shorter mean interparticle distance and this is a fundamental reason why interparticle interactions play a primary role in determining the behavior of colloidal dispersions. These interactions, also named colloidal forces, result from the sum of the interactions between the individual molecules that make up the particles, and control if and how much the particles are either attracted or repulsed from each other and, consequently, affect the rheology of suspensions. The force F between two particles varies with the surface-to-surface separation distance D and is related to the interparticle potential energy function $W(D)$ by

$$F = -\frac{dW}{dD}. \tag{3.109}$$

Attractive and repulsive forces of different magnitude and range can contribute to $W(D)$ and control the microstructural condition of particle dispersions. Typically, the main contributions are given by excluded-volume, van der Waals and electrostatic forces, which are summed together in the DLVO theory [113,114]. Other specific contributions can derive from species present on the surface and/or in the continuous phase, as the steric and electrostatic repulsions due to adsorbed species (polymers and ions), the depletion and bridging flocculation, the hydrophobic and hydration effects.

Van der Waals attractive forces arise from dipole–dipole interactions in the molecules within each particle. Summing up the interactions between all the molecules in the particle, the van der Waals interaction energy between two identical spherical particles of radius r, is given approximately (when $D \gg r$) by [53]

$$W_{vdW} \approx -\frac{rA_H}{12D}, \tag{3.110}$$

where A_H is the Hamaker constant which is related to the dielectric constants and the refractive indices of the particles and the continuous phase. These attractive interactions (if not opposed by repulsive forces) lead to particle coagulation.

Many particle surfaces become charged in aqueous phases, the net surface charge is positive or negative at pH below or above the isoelectric point (IEP). A cloud of counterions shrouds each particle in order to maintain neutrality of the system. The surface charge together with the diffuse ion layer surrounding the charged particle constitutes the so-called double layer. When particles are pushed together, the counterion clouds begin to overlap and increase the counterion concentration in the space between the particles, so giving origin to an increase in osmotic pressure and, hence, to repulsive interactions. The thickness of the double layer (the Debye length) is inversely proportional to square root of the electrolyte concentration. At lower ionic strength (small counterion concentration) the particles are repulsive at large separation distances, so that the van der Waals attractive forces can be overwhelmed and the particle stabilization is ensured. Conversely, when sufficient salt is added, the range of the repulsive potential is decreased sufficiently so that the van der Waals attraction can dominate at larger separation distances and a secondary minimum appears. In such conditions the particles become weakly attractive and flocculate.

Colloidal dispersions can be stabilized also by either adsorbing or grafting polymer chains to the surface of particles, so creating a steric barrier to flocculation and originating repulsive forces which result from the two main contributions. The volume exclusion term arises from the compression and loss in configurational entropy of the adsorbed polymer layer on the approach of a second particle and is always repulsive. The mixing interaction term is repulsive only when the continuous phase represents a good solvent for the adsorbed polymer while poor solvents give rise to attractive mixing interactions. If compared electrostatic forces, steric interactions usually occur over short distances, which depend on the adsorbed layer thickness δ and, hence, on chain length of the adsorbed polymer, surface coverage, and thermodynamic solvent quality. For sufficiently large δ the van der Waals attraction can be overwhelmed and a well-dispersed suspension results, without the primary minimum in the interparticle potential profile. Both steric and electrostatic forces may contribute to the overall repulsion when polyelectrolyte chains are used as stabilizing additive and adsorbed on the particle surface (electrosteric stabilization).

Other interactions of polymeric-type lead to bridging and depletion flocculation. The former occurs in the case of high-molecular weight polymer adsorbed at low-surface coverage, when particles are held together by bridges of polymeric chains attached to two or more surfaces.

Depletion flocculation is typical of concentrated dispersions whose continuous phase contains nonadsorbing polymers [115]. If the separation of the particles is less than the gyration radius of the free polymer coil, polymer molecules are eliminated from the interparticle region in which they cannot exist without deformation. The depletion effect is induced by their osmotic pressure, due to the difference of polymer concentration between the inside and outside regions, so that solvent is displaced from the interparticle region to the bulk solution. As a result, an effective attractive force is produced between the colloidal particles driving them even closer together to phase separate or flocculate. Also in the case of polymer adsorption onto particles, the depletion effect can be produced by the excess polymer molecules after saturation adsorption.

Another short-range repulsive interaction may be observed when two surfaces are brought into proximate contact across a solvent medium, since the ordered layer of solvent molecules adsorbed on the particles restrict the further approach of two particles when the separation distance is comparable to the size of the layers. Conversely, long-range hydrophobic attractive interactions are often observed in aqueous dispersions when the particles are covered by an adsorbed monolayer of surfactant molecules with their hydrocarbon tails oriented outward to the aqueous phase.

3.9.4.2 Suspensions of Noninteracting Particles (Hard Sphere Model)

The simplest structural condition is that of noninteracting hard-sphere suspensions since repulsive excluded-volume interactions occur only when two particles come into contact. The interaction potential is null even at very short interparticle distance. Such a condition is encountered in real systems only if the always present van der Waals attractive interactions are negligible and this happens for very low values of the Hamaker constant, i.e., when the refraction indices of the disperse and continuous phases are close to each other.

At low concentrations the flow field around each sphere is independent of or only slightly dependent on the presence of neighboring spheres and, consequently, the behavior of sphere suspensions is Newtonian. For $\Phi < 0.05$–0.07 the viscosity can be predicted by the theoretical expression of Einstein [116,117]

$$\eta = \eta_s \eta_r = \eta_s(1 + 2.5\Phi), \tag{3.111}$$

where the viscosity results from the product of two contributions, the continuous phase viscosity η_s and the relative viscosity η_r which is linearly dependent on Φ.

A further term accounting for binary interactions must be added for $\Phi < 0.1$–0.15 according to Batchelor [118,119]

$$\eta = \eta_s(1 + 2.5\Phi + 6.2\Phi^2). \tag{3.112}$$

At higher concentrations the viscosity increases more rapidly than predicted by Einstein and Batchelor and an appreciable shear-thinning is observed for $\Phi > 0.25$–0.3. The spatial distribution of particles in flow conditions is governed by the Peclet number, which measures the balance between hydrodynamic and Brownian forces. As suggested by Krieger, a modified version of the Peclet number, corresponding to a reduced shear stress, can be obtained by replacing η_s with the suspension viscosity [120]

$$\mathrm{Pe_K} = \frac{\eta \dot{\gamma} r^3}{kT} = \frac{\sigma r^3}{kT} = \sigma_r. \tag{3.113}$$

By doing so, the relative viscosity can be expressed as a universal function of Φ and σ_r, that contains the dependence on shear stress, particle size, and temperature. An example is given in Figure 3.25 which concerns concentrated dispersions of polymethylmethacrylate spheres of different size in silicon fluids of different viscosity.

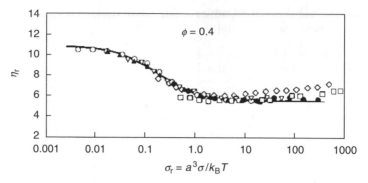

FIGURE 3.25 Relative viscosity vs. reduced shear stress for dispersions of poly-methylmethacrylate spheres of different size in silicon fluids of different viscosity [121]. (Adapted from Walker, L.M. and Wagner, N.J., *J. Rheol.*, 38, 1525, 1994. With permission.)

At constant volume fraction, experimental data of shear-thinning hard-sphere dispersions can be fitted with satisfactory approximation by the expression [122]

$$\eta_r = \eta_{r,\infty} + \frac{\eta_{r,0} - \eta_{r,\infty}}{1 + b\sigma_r}, \tag{3.114}$$

where $\eta_{r,0}$ and $\eta_{r,\infty}$ are the zero-shear-rate and infinite-shear-rate relative viscosities, respectively, and b is an adjustable parameter.

As the particle concentration increases, the behavior becomes more and more of shear-thinning with increasing difference between $\eta_{r,0}$ and $\eta_{r,\infty}$ ($< \eta_{r,0}$) as illustrated in Figure 3.26.

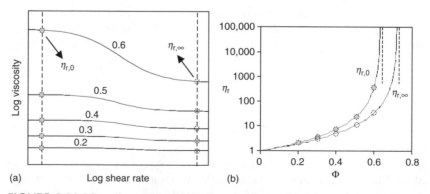

FIGURE 3.26 Monodisperse hard-sphere suspensions: (a) flow curves at different volume fractions (0.2–0.6); (b) relative viscosity vs. disperse phase volume fraction in the two limiting shear conditions.

The Φ dependences of both $\eta_{r,0}$ and $\eta_{r,\infty}$ can be suitably described by asymptotic relations as the Krieger–Dougherty equation [122]

$$\eta_r = \frac{\eta}{\eta_s} = \left(1 - \frac{\Phi}{\Phi_m}\right)^{-a}, \quad a = [\eta]\Phi_m, \qquad (3.115)$$

where both the maximum packing volume fraction Φ_m and the intrinsic viscosity $[\eta]$ are shear-dependent parameters. However, their product corresponding to the exponent a usually falls in a narrow interval so that almost equivalent fitting results are often obtained with the Quemada equation ($a = 2$) [123]. In the latter case the differences between the two limiting Newtonian viscosities are related only to the different values of $\Phi_{m,0}$ and $\Phi_{m,\infty}$ ($>\Phi_{m,0}$). For monomodal spherical particles they can be theoretically derived (for random packing $\Phi_{m,0} = 0.63$ and face-centered cubic packing $\Phi_{m,\infty} = 0.71$).

However, it must be underlined that at high-volume fractions the experimental determination of the two limiting Newtonian viscosities can be more difficult or disputable. At $\Phi > 0.4$–0.5 shear-thickening can occur gradually or abruptly at high-shear rates, beyond the shear-thinning region and hence the infinite-shear-rate viscosity becomes an ill-defined parameter. The onset of shear-thickening is generally caused by shear-induced microstructural changes in the suspension [124,125]. For Brownian hard-sphere suspensions it can be associated with an increase in the hydrodynamic contribution to the stress due to shear-induced cluster formation when hydrodynamic forces overcome Brownian repulsive interactions. The critical shear rate for shear-thickening onset is proportional to the separation between the particles and then it decreases rapidly with increasing the volume fraction, since the separation is inverse to the cubic of the volume fraction of particles. The critical shear rate strongly increases with decreasing particle radius and then can be less detectable for very fine particles. Moreover, at very high-volume fractions (above that of dense random packing, 0.63) and low-shear conditions, hard-sphere suspensions exhibit strong shear-thinning with apparent yield stress and very high-zero-shear-rate viscosity values which are difficult to determine.

The viscoelastic properties of all particulate suspensions are strongly dependent upon the volume fraction of particles as well as on the nature of the interactions between particles. At very high-volume concentrations of noninteracting particles, the storage modulus becomes independent of frequency, since the particles cannot rearrange themselves significantly within the time scale of the oscillatory perturbation. At lower volume both viscoelastic moduli are function of the frequency of oscillation.

The hard-sphere suspensions represent the simplest class of systems and an important starting point for the following analysis of the structural and rheological effects related to particle characteristics (size distribution and nonspherical shape) and to interparticle repulsive and attractive interactions.

3.9.4.3   Effects of Particle Size Distribution and Shape

When spherical particles of two or more differing sizes are mixed, the viscosity can be largely reduced keeping constant the total particle volume fraction Φ. The viscosity dependence on particle size distribution is more evident at sufficiently high concentration ($\Phi > 0.5$), not far from maximum packing condition, as illustrated in Figure 3.27 for bidisperse suspension of spherical particles of size ratio 5:1.

The viscosity minimum attained around a fraction of large particles equal to 0.6 is known as the Farris effect [126] and can be interpreted as the consequence of the packing of smaller particles into the voids between the larger ones. It must be emphasized that even a slight increase in Φ_m leads to strong viscosity reduction for concentrated dispersions. These considerations

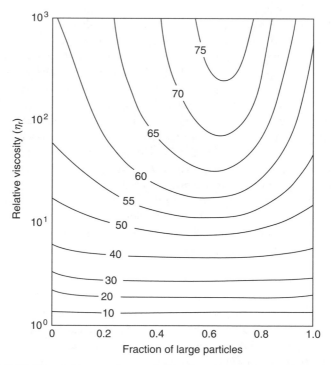

FIGURE 3.27 Bimodal suspensions: relative viscosity vs. large particle fraction at different disperse phase concentrations (volume percentage). (Adapted from Walker, L.M. and Wagner, N.J., *J. Rheol.*, 38, 1525, 1994. With permission.)

developed for bimodal or trimodal suspensions of hard spheres can be extended to well-dispersed bimodal colloidal suspensions, even if they are usually confirmed only at high-shear conditions where hydrodynamic forces prevail.

In the dilute regime the anisometric particles are sufficiently far apart from each other so that they are able to rotate freely without being impeded by neighboring particles. Such a condition is evidently restricted to very low-volume fractions (even below 0.01) for high-aspect ratios.[18] The periodic rotation of colloidal particles induced by shearing flow is contrasted by Brownian motion which tends to restore random orientation of particles and prevails at low shear. The transition from Brownian to non-Brownian behavior is then controlled by the rotational Peclet number, i.e., the ratio of shear rate to rotational diffusivity, and marks the beginning of shear-thinning regime. The Brownian behavior is generally shifted to low-shear rates, the rotational diffusivity is inversely proportional to the viscosity of the suspending medium and to the cubed longest dimension of the particles. The higher the aspect ratio p, the higher the viscosity and the shear-thinning character of the suspension.

At low concentration the Φ dependence of viscosity is described by a corrected version of the Einstein–Batchelor equation

$$\eta = \eta_s(1 + [\eta]\Phi + k_H\Phi^2), \qquad (3.116)$$

where both the intrinsic viscosity $[\eta]$ (>2.5) and the coefficient k_H (>6.2) are increasing functions of the aspect ratio. The intrinsic viscosity value is generally higher for rod-like particles than discs, for prolate than oblate spheroids of equivalent aspect ratio (p for prolate, $1/p$ for oblate).

As the aspect ratio increases, the maximum packing volume fraction Φ_m decreases so that the product $[\eta]\Phi_m$ does not vary appreciably. Hence, the Quemada equation can be used to describe the concentration dependence of viscosity with satisfactory approximation, in the place of the Krieger–Dougherty equation.

As outlined before, different types of interparticle interactions may be present in a suspension and their combined actions can give rise to diverse microstructural conditions and rheological responses. However, it must be underlined that, inspite of their diversity, any kind of interaction, either attractive or repulsive, will result in a viscosity increase with respect to the corresponding hard-sphere suspension with the same volume fraction and particle size distribution. Hereafter, we will examine the two classes of repulsive and attractive particle suspensions.

---

[18]The aspect ratio p of axisymmetric ellipsoidal particles (spheroids) is given by the ratio of the length of the particle along its axis of symmetry to the length perpendicular to this axis ($p > 1$ for prolate spheroids, $p < 1$ for oblate spheroids).

3.9.4.4 Suspensions of Stabilized Particles

The effects deriving from the presence of surface charges or steric-stabilization layers on the particle surface are similar and, hence, can be analyzed parallelly. Moreover the rheological behavior of repulsive particle systems is quite similar to that observed for hard spheres since it changes from Newtonian to shear-thinning and finally to apparently plastic with increasing volume fraction.

Sterically stabilized spherical particle can be generally treated as hard spheres on condition that the particle radius r is corrected for the adsorbed (or grafted) polymer layer of thickness Δ and, consequently, the effective volume fraction must be used with the Einstein–Batchelor equation

$$\Phi_{\text{eff}} = \Phi\left(1 + \frac{\Delta}{r}\right)^3. \tag{3.117}$$

Obviously, such a correction becomes important for colloidal dispersions with decreasing particle radius and with increasing volume fraction. If the adsorbed polymer is short and densely packed on the particle surface, the "effective hard sphere" approach can be extended up to high concentrations close to packing conditions, and the asymptotic model (Equation 3.115) can provide a good description of the concentration dependence of viscosity, by replacing the nominal volume fraction with effective one. Near the maximum packing, when the interparticle distance becomes less than twice the polymer layer thickness, repulsive forces become strong and result in a highly developed structure of particles. The behavior becomes apparently plastic since appreciable flow occurs only if sufficient stress is applied to overcome the interparticle forces.

In the case of long chains (high Δ value), the deformability of the polymer layer becomes significant at high Φ_{eff}, when the layers touch, begin to compress and interpenetrate each other. Particularly for colloidal particles, the "soft sphere" becomes the most appropriate reference model in the place of the "hard sphere" one. The decrease of the polymer layer thickness related to the compression produced by increasing volume fraction can be estimated by fitting the viscosity $\eta - \Phi_{\text{eff}}$ data with Equation 3.115 and using Δ as a fitting parameter.

The viscosity of a suspension of electrically charged particles is increased above that of hard spheres even at very dilute concentrations, because of the additional dissipation created by the distortion of the double layer produced by the flow of solvent around the particle (primary electroviscous effect) [127–129]. More important is the additional increase in viscosity caused by the overlapping double layers at higher concentrations (secondary electroviscous effect) [130–134]. Again, an effective volume fraction can be conveniently used accounting for the long-range repulsive interactions which tend to keep particles apart and then increase the effective particle diameter. Here the thickness of the "electrostatic layer," Δ, depends on electrolyte concentration, increasing with decreasing ionic strength. Electroviscous effects are more significant for fine colloidal particles, since the relative increase in the effective volume fraction due to the double layer can be sensibly large, and then a significant increment of viscosity can occur even at very low concentration and low-shear conditions. At low-ionic strength and higher concentration, the strong repulsive interactions due to the overlap of double layers result in the formation of a macrocrystalline lattice structure where each colloidal particle is caged by repulsive forces. The suspension thereby displays elastic properties at small deformation and plastic behavior.

Repulsive suspensions can exhibit yield stress when the volume fraction tends to the maximum packing condition and the range of the repulsive forces is greater than the interparticle separation distance. The repulsive forces hold every particle in a position midway between its neighbors and thermal motion may become insufficient to move the particles from this minimum energy position, if the magnitude and range of the forces is sufficient. The yield stress is the applied stress necessary to overcome the forces resisting deformation of the network in the repulsive systems.

Also for stabilized suspensions the shear thickening is associated with a microstructure transition from ordered layers to a disordered state [135–137]. Above a critical shear rate, the ordered structure breaks down because some particles are pushed out of the layered sheets or strings and interact more strongly with particles in adjacent layers, so leading to a dramatic increase in viscosity. Accordingly, the onset of shear-thickening is governed by the balance between hydrodynamic and repulsive forces. This means that it can be controlled by adjusting pH or salt concentration in the case of electrostatic particle stabilization; shear-thickening is shifted to higher critical shear rates by moving pH away from the isoelectric point (IEP).

Marked viscoelastic properties are observed at lower volume fractions than for hard sphere suspensions because the overlapping repulsive forces. At low Φ, G' increases with increasing frequency. The storage modulus becomes a weaker function of frequency with increasing volume fraction and a plateau region is observed at higher frequencies, where the particles are substantially frozen in their relative configuration by the multiple interactions with the adjacent particles.

3.9.4.5 Suspensions of Attractive Particles

When the ubiquitous van der Waals attraction is not contrasted by repulsive forces, strong attractive interactions occur between particles, which come into contact with each other in a deep primary minimum energy well and coagulate together, so making impossible any stable dispersion.

Weaker interparticle attractions occur when short-range repulsive forces are generated either by electrical double layers at very high-salt concentrations or by very thin adsorbed polymer layers. The short-range repulsion negates the van der Waals attraction only at small separation distances and a secondary minimum appears in the potential profile, so that flocculation takes place. Flocs are open and ramified structures which consist of more or less packed particles and can be usually described quite well in terms of the concepts of fractal geometry. Above a critical cut-off, the dimension of a floc containing n particles scales with n, the power-law exponent is inversely proportional to the fractal dimension D_f:

$$\frac{r_{gf}}{r} \approx n^{1/D_f}, \tag{3.118}$$

where r_g is the radius of gyration of the floc and r is the radius of the primary particles. Different floc density and formation kinetics can be ascribed to different mechanisms of aggregation [138]. Floc structures simulated in accordance with different cluster–cluster aggregation models (diffusion-limited cluster–cluster aggregation (DLCCA), ballistic cluster–cluster aggregation (BCCA), reaction-limited cluster–cluster aggregation (RLCCA)) are compared in Figure 3.28. In the case of diffusion-limited aggregation processes, whenever two particles or two clusters contact each other, they are combined irreversibly in the contacting configuration and the aggregation rate is limited solely by the time taken for clusters to collide via Brownian diffusion. Such a fast process leads to more open and porous flocs of lower fractal dimension ($D_f = 1.75$–1.8), whereas more interpenetrated and denser clusters ($D_f = 2.0$–2.2) are generated by the reaction-limited aggregation

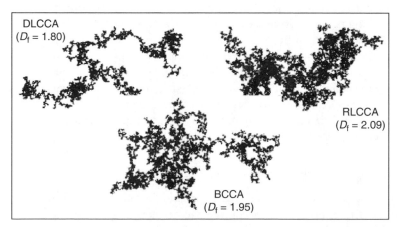

FIGURE 3.28 Floc structures derived from computer simulation using cluster–cluster aggregation models: DLCCA aggregation BCCA, RLCCA. (Adapted from Meakin, P., *J. Sol–Gel Sci. Technol.*, 15, 97, 1999. Reprinted with permission from Springer Verlag.)

mechanisms. In these cases the aggregation (reaction) rate is slower and limited by the probability of forming a bond upon collision of two clusters and then particle or clusters do not combine each time they come into contact, but continue on their random walks with many collisions before a pair of clusters finally become joined.

The particle flocs entrap part of the continuous phase and, then, their effective volume fraction, Φ_{eff}, is higher than that of the primary particles, Φ, so leading to higher viscosity than that of noninteracting ("hard sphere") or repulsive particle suspensions at identical concentration. In addition to the hydrodynamic effects, the flow properties of the flocculated suspensions are also strongly governed by interparticle interactions. Thus, in shear conditions flocs not only rotate, possibly deform and change their spatial distribution in the flow field but also break down to smaller aggregates and/or primary particles, when the applied stresses are high enough and hydrodynamic forces dominate the interparticle attractive interactions. The shear-thinning behavior of these systems can be mainly attributed to the progressive disruption of flocculated structures and the consequent decrease in the effective concentration of particles and flocs as the shear rate increases.

Shear-thinning is usually associated with thixotropic behavior, since the structural processes induced by shear are, at least partially, reversible, and flocculated structures are gradually rebuilt when the applied stress is reduced or removed. Brownian motion is the only driving force for particle aggregation in quiescent conditions and then the structural buildup and the thixotropic recovery of viscosity are faster for smaller particles.[19]

Highly porous flocs can fill up the available space even at low volume fraction Φ, thus forming a continuous three-dimensional particulate network, when the volume fraction exceeds the percolation threshold. The crossover from cluster–cluster aggregation to percolation takes place when the cluster size reaches a value given by

$$n \approx \Phi_f^D / (D_f - 3). \qquad (3.119)$$

This relation indicates that the percolation threshold can be attained for quite different structural conditions and particle volume fractions in dependence of the aggregation mechanism, i.e., of the fractal dimension D_f. The volume fraction at which a percolating sample-spanning network is formed is termed

---

[19]The self-similar fractal aggregate structures result from Brownian motion, a self-similar aggregation process which is qualitatively the same between large aggregates and between small aggregates. Shear-induced restructuring leads to more compact aggregates to which an increased fractal dimension is usually assigned. However, restructuring is in general not a self-similar process since it tends to occur at larger length scales before smaller ones. Consequently, the structure is no longer really fractal, but shows a change in mass scaling with length scale, the fractal dimension being apparently higher at larger length scales.

the gel point Φ_g [139] and may be as low as 0.10 to 0.15. The attractive particle networks exhibit marked plastic and elastic properties above the gel point and then at much lower volume fractions than either the repulsive or hard-sphere systems.

The interaction potential and spatial arrangement of particles within the isolated flocs or the three-dimensional network determines the strength of the aggregated structures and, hence, also the rheological behavior of the flocculated suspension at both low- and high-volume fractions.

According to the model proposed by Scales et al. [140], the yield stress of flocculated suspensions can be evaluated as summation of all pair interactions calculated from the interparticle DLVO forces. At a given volume fraction the maximum attraction and particle flocculation occur at the IEP where no charge exists on the surfaces of the particles and, hence, a maximum shear yield stress occur. As the pH is adjusted away from the IEP, the surface charge increases creating an electrical double layer repulsion that reduces the magnitude of the attractive forces and this results in lower yield stress. Stronger or weaker attraction between particles produces lower or higher gel points. The same model predicts a linear decrease of the yield stress with the square of the zeta potential.

Experimental data collected for several concentrated metal oxide suspensions confirm the model prediction and show that similar trends can be observed for viscosities throughout the entire shear rate range when pH is changed at a given volume fraction.

The storage modulus is greater than for systems of repulsive or noninteracting particles of the same size and for the same volume fraction. The more elastic character is due to the deep attractive energy minima that tend to maintain the particles in their rest configuration; G' increases with increasing depth of the potential well [141].

Volume fraction and particle size impact the rheology of flocculated suspensions in two ways, affecting the density of interparticle links and the microstructure of the suspension. At a given pH the yield stress increases very sharply as Φ increases and the particle radius r decreases

$$\tau_y = K \frac{\Phi^c}{r^2}, \tag{3.120}$$

where K is related with bond strength depending on material properties and system surface chemistry condition, while the power-law exponent c (approximately 4–5) can be correlated with the fractal dimension, which is associated with the interconnection and space-filling ability of the network microstructure [142,143]. Alternatively, the yield stress can be related with $(1 - \Phi)/\Phi$ or Φ/Φ_g [139,142].

While widening particle size distribution results into viscosity reduction of hard sphere dispersions at constant solid content, a polydisperse flocculated suspension exhibits a higher yield stress than the narrow size distributed suspension of the same volume average diameter. The effect of polydispersity on the yield stress can be better related with the surface area average diameter.

The storage modulus of attractive particle network can be related with the volume fraction via a power-law relationship. According to many researchers [139,141] the power-law exponent, which falls between 4 and 5 for many systems, can be related to the structure and interconnectivity of the particle network.

The rheological effects produced by bridging and depletion flocculation are similar to those deriving from neat attractive interaction potential [143]. Indeed, neutral polymers or polyelectrolytes can give a steric or electrosteric stabilizing effect when well adsorbed on the particle surface but can also produce bridging or depletion flocculation effects in other cases.

Bridging flocculation occurs when particles are held together by bridges of polymer molecules spanning between two or more particles. This is most likely to happen in the case of long chain polymers at low-surface coverage, usually at half surface coverage, particularly when polymers have opposite charges to the particle surface. Because of coexistence of the short-range steric repulsion, the role of polymeric bridging is often difficult to characterize with rheological techniques for concentrated suspensions.

When nonadsorbing polymers are present in the suspending medium, depletion flocculation could arise as the particles approach and the polymers are excluded from the gap due to an osmotic pressure gradient between the particles. Elastic, shear-thinning and plastic properties of the suspension increase substantially with increasing free polymer concentration. In the case of nonadsorbing polyelectrolytes, changes of the bulk concentration, the solution pH, and ionic strength can alter the chain dimensions and hence lead to different state of flocculation with different fractal dimension of the flocs.

3.9.4.6 Emulsions

The liquid nature of the disperse phase in emulsions introduces the important role that particle deformation can play in dispersion rheology and raises again the problem of system stability, which is more complex than for suspensions.

The intrinsic instability of emulsions is of thermodynamic nature. The free energy change associated with the formation of droplets and their dispersion in the outer phase is mainly due to the increment in the interfacial area, ΔA, and, in a lesser extent, to the entropy increase, ΔS_{config}

$$\Delta G_{formation} = \Gamma \Delta A - T \Delta S_{config}, \qquad (3.121)$$

where Γ is the interfacial tension. Stable conditions are attained only for very Γ values, when the entropic term prevails over the interfacial free energy contribution. This is the case of microemulsions, which can be formed by supplying very little mechanical energy.

Unless surfactants are present, the droplet dispersion will evolve toward two separate phases. Separation may occur through different mechanisms, such as droplet collision and fusion in the case of Brownian droplets, sedimentation or creaming, which is more significant for larger droplets, or Ostwald ripening, in the case of partial miscibility of the phases. Other processes of chemical or microbiological origin can contribute to system instability, as oxidation, hydrolysis, or microorganism growth.

Even if thermodynamically unstable, an emulsion can be rendered kinetically stable also for months, by using appropriate additives. Surfactants are normally used as emulsifiers which are adsorbed at the interface, so reducing the interfacial tension and favoring the formation of smaller droplets. Consequently, the disperse phase is more stable against droplet coalescence and gravity-driven sedimentation or creaming phenomena. Another solution for increasing stability consists in adding polymers (thickeners) which change the structural features and the rheological properties of the continuous phase, thus opposing the motion and mutual contacts of droplets. In the case of weak gel matrices, droplets are caged in the network meshes and very long stability can be reached.

Most emulsions contain droplets whose average size falls between 0.1 and 100 μm, and hence are partly affected by colloidal interactions. Accordingly, the droplet size distribution is an important parameter in determining the structural conditions of an emulsion and, then, its stability together with its rheological properties. It is directly connected with the preparation conditions, as the supplied mechanical energy and temperature, and the presence of additives; bimodal distribution is often observed, due to droplet flocculation or insufficient emulsifier during droplet formation.

The rheology of flocculated emulsions does not differ qualitatively from that of suspensions, since the shear- and time-dependent properties are mainly governed by the same floc breakup and aggregation processes. Conversely, peculiar shear-induced mechanisms must be accounted for when isolated droplets contribute individually to the behavior of the system with their deformation, breakup, and coalescence. The flow-induced deformation of an axisymmetric droplet is defined by the ratio $D = (L - B)/(L + B)$, where L and B are the lengths of its major and minor axes, respectively, or, alternatively, by the aspect ratio L/B. It increases with the viscosity ratio $M = \eta_d/\eta_c$, where η_d and η_c are the viscosities of the disperse and continuous phases, respectively, and, more markedly, with the capillary number Ca, defined as the ratio of viscous stresses ($\eta_s \dot{\gamma}$) to surface tensions related to the interfacial tension and droplet size (Γ/r). Droplet breakup occurs when the viscous stresses overcome the surface tension and D reaches a critical value, which increases with decreasing M. In nondilute systems both breakup and coalescence of droplets occur so that the size distribution of the disperse phase may not only depend on the shear conditions but also attain different steady states after different shear histories, and above a critical shear rate the steady-state average droplet size should be set by the balance between breakup and coalescence [70].

At very low concentration the behavior is Newtonian and the viscosity increases with the disperse phase volume fraction Φ in accordance with the Einstein equation (Equation 3.111), where the intrinsic viscosity $[\eta]$ can be calculated from the Taylor formula [144]

$$[\eta] = \frac{5M + 2}{2(M + 1)}. \tag{3.122}$$

If the disperse phase is very viscous ($M \longrightarrow \infty$), the droplets behave like hard spheres ($[\eta] \longrightarrow 2.5$). For $M \longrightarrow 0$, they are very deformable like bubbles and the relative viscosity is lower. In the dilute regime the Batchelor equation can be still used, by changing the coefficient of the quadratic term [145]. At higher concentrations the shear-thinning becomes more marked with increasing Φ. If the analysis of the concentration dependence is referred to the zero-shear rate relative viscosity ($\eta_{r,0} = \eta_0/\eta_s$), a satisfactory correlation can be provided by the empirical equation suggested by Pal which accounts for the maximum packing volume fraction Φ_m and the viscosity ratio M [146]

$$\eta_{r,0}^{1/K_1} = \exp\left(\frac{2.5\Phi}{1 - \Phi/\Phi_m}\right), \tag{3.123}$$

where

$$K_1 = \frac{0.4 + M}{1 + M}. \tag{3.124}$$

For $M \longrightarrow \infty$ it reduces to the Mooney equation [147] and the behavior of concentrated emulsions converge to that exhibited by concentrated hard-sphere suspensions. Higher Φ_m values are associated with lower M, since the viscosity tends to diverge at higher volume fractions. Now, the droplets are more deformable and can substantially modify their shape, when their concentration overcomes the random close packing condition (typically $\Phi_m \cong 0.63$ for mono-disperse hard particles) tending to a foam-like polyhedral cell configuration, where the continuous phase is confined to compressed thin films between deformed droplets. Not surprisingly, the shear-thinning behavior becomes apparently plastic with remarkable elastic components. Figure 3.29 shows how an increase in the disperse phase concentration leads to significant changes in both linear and nonlinear properties of concentrated inverse emulsions.

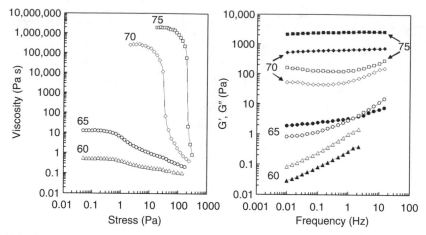

FIGURE 3.29 Shear viscosity and mechanical spectra of concentrated W/O emulsions (water/cyclomethicone, 0.5% NaCl, 2% cetyl dimethicone copolyol) at different disperse phase concentrations (%wt).

A plateau appears in the profile of the storage modulus G' vs. frequency and, correspondingly, the viscous component G'' shows a minimum. With increasing Φ the modulus plateau becomes more extended toward lower frequencies and its value increases, as it is strictly correlated to the osmotic pressure [148].

If the liquid phases are immiscible and surfactants are added to provide an adequate interfacial repulsion, coarsening of the droplet size distribution due to Ostwald ripening is negligible as well as droplet coalescence is suppressed. Consequently, the emulsion can remain stable for years also when osmotically compressed to form a biliquid foam.

All the considerations regarding the effects of particle size distribution and interparticle attractive interactions on suspension rheology can be extended to emulsions, even though the deformable and unstable character of droplets makes the analysis more complex. Traditional methods of emulsification typically lead to size distributions that have a large polydispersity, and the few tests performed on bimodal emulsions with different proportions of small and large droplets confirm the substantial reduction in viscosity observed also for suspensions, in particular at low shear and high concentration.

Attractive emulsions can show shear-thinning even at dilute Φ due to the breakup of flocs or aggregates as $\dot{\gamma}$ is increased. Indeed, the rheological properties must be referred to the effective volume fraction of the disperse phase Φ_{eff}. Even if droplet deformation plays a predominant role in the elasticity of compressed emulsions, attractions can significantly increase the plateau modulus for Φ_{eff} near and below the random packing condition in comparison with repulsive emulsions.

3.9.4.7 Microemulsions

Microemulsions are stable mixtures of oil, water, and surfactant, often in combination with a cosurfactant, whose disperse phase may be composed of isolated droplets at relatively low-inner phase concentration or form a porous sponge-like random network (bicontinuous state), depending the nature and proportions of components. Microemulsions composed of isolated droplets are low-viscosity Newtonian fluids, suitable for oral, parenteral, pulmonary, or even ocular delivery, while bicontinuous formulations may exhibit shear-thinning behavior. In a state diagram referred to oil, water, and surfactant system the domain of true microemulsions confines with other systems, such as emulsions, surfactant solutions, and crystalline gels, whose structural characteristics reflect into diverse rheological properties. Thus, viscosity measurements can help to define the more or less distinct borders in the state diagram as well as the transition from O/W to W/O systems within the microemulsions domain in conjunction with other complementary techniques [149].

The rheology of microemulsions can be tailor-made for topical administration by adding a polymeric thickener, which must be properly selected in order to impart marked shear-thinning without interacting appreciably with the surfactant system. In such a way, O/W microemulsions which are particularly suitable to guest drugs sparingly soluble in water, can be transformed in gels suitable for transdermal delivery [150].

3.9.4.8 Dispersions in Polymeric Matrices

These last considerations recall to our mind the contribution of polymeric phases to the rheology of dispersions. If solid particles are dispersed in a very viscous liquid, their diffusive motion is very slow and particle–particle interactions can be almost negligible, compared to viscous forces, so that the analysis of the solids contribution is relatively simple. The decrease of the relative viscosity with increasing shear rate is related to the ordering processes induced by viscous forces. The linear viscoelastic properties of filled polymers may differ from those of the matrix by an additional slow relaxation process which can be connected with the dynamics of the polymer chains adsorbed onto the particle surface. More often, the matrix and particles interact thermodynamically, through their interaction potential, and hydrodynamically, through flows mediated by the solvent and then the analysis is more complex due to the interplay of the structural and rheological contributions of the two phases.

Special attention must be paid to polymeric weak gel matrices, which can be profitably used to suit the rheology of dispersions to several industrial exigencies, normally ensuring easy manipulation and transportation as well as good stability. If disperse or soluble species are hosted in the matrix at low-concentration levels, as for active drugs in controlled release pharmaceutical systems, the rheological behavior of the final product is almost coincident with that of the matrix employed. If no interaction occurs between the components of the dispersions and the disperse phase is composed of stabilized very fine particles as for microemulsions, the latter can be reasonably hosted inside the network meshes.

Also when the disperse phase concentration is high as for many systems of practical interest, the behavior of the dispersion is essentially governed by the weak gel matrix even if the contributions of the structural mechanisms induced in both phases by changes in the shear flow field are interconnected and cannot be treated separately. At low shear the disperse phase is substantially caged within the tridimensional gel matrix and slightly contributes to dispersion viscosity. Only when a critical stress is exceeded, the matrix undergoes a significant breakup process and, hence, the disperse phase displays an increasing contribution to dispersion viscosity up to another critical stress above which the prevailing effects are due to structural rearrangements of the disperse phase in the flow field.

In other cases, the state of the disperse phase changes owing to the polymer addition. When the polymer is partially adsorbed on the particle surface, it contributes to improve the quality and stability of the dispersion, and then to reduce the viscosity. Contrary effects occur when the polymer chains interact with the particle surface forming bridges which increase the degree of connectivity of the network as well as the viscosity, the shear-thinning character and the elastic properties.

3.10 CONCLUSIVE CONSIDERATIONS ON RHEOLOGY APPLICATIONS TO DRUG DELIVERY SYSTEMS

As in several other scientific and industrial sectors rheology can play an essential role in the study and the development of drug delivery systems, not only for solving macroscopic problems connected with their preparation and application, but also, and even more significantly, to examine the structural features and processes that govern their behavior in flow conditions as well as under small deformation. Increasing interest has been manifested for rheology applications in the pharmaceutical field during the last 20 years as demonstrated by the corresponding growth of the scientific studies appearing in the relevant literature. An important contribution can be reasonably ascribed to the diffusion of more sophisticated instruments, suitable to perform experimental investigations of fundamental character, in particular dynamic tests at very small amplitude of stress or strain oscillation. This is a necessary requisite to detect the linear viscoelastic properties of polymeric hydrogels and other complex systems under near-equilibrium conditions, especially in the case of very fragile and weakly structured fluids.

Hydrogels are water-swellable networks composed of hydrophilic homopolymers or copolymers and are extremely useful in biomedical and pharmaceutical applications mainly due to their high-water content and viscoelastic properties, as well as their biocompatibility. Their swelling behavior can be ruled by different changes in their external environment (solvent concentration, pH, temperature, ionic strength), depending on the polymer type, and controlled by solvent diffusion or network relaxation in its kinetic aspects during solvent uptake [151], as illustrated in more detail in Chapter 6. Accordingly, the viscoelastic properties of hydrogels can play an important role for pharmaceutical applications, also as concerns the integrity of the drug delivery system during the lifetime of the application, and they can be modulated by changing the degree of crosslinking.

Most of the rheological studies have been devoted to sustain and address the development of novel delivery systems and to this purpose the analysis and understanding of the structural conditions and phenomena is fundamental. A typical example is given by bioadhesive formulations. Bioadhesivity is essential to localize a drug delivery device within the body, typically in the gastrointestinal tract, to enhance the drug absorption process in a site-specific manner. Bioadhesion research has primarily investigated the adhesion of thin film and microparticles based on hydrophilic polymers, to mucus gel. Hydrogels based on polyacrylic acid (PAA) are among the most accepted mucoadhesives. Various mechanical tests have been used in both *in vivo* and *in vitro* conditions to compare the adhesive strength of bioadhesive formulations and different mechanisms of adhesion have been discussed extensively in the literature as reviewed in Ref. [152]. Mucoadhesion is a complex process that involves bringing the polymer and mucus into intimate contact, followed by interpenetration and mechanical interlocking. The ability of mucoadhesive polymers to produce a large increase in the strain resistance when incorporated into a mucus gel, relative to when the mucus gel and the tested polymer are evaluated separately, is termed as rheological synergism and is often used as a measure of the strength of the mucoadhesive interaction. To this purpose, oscillatory tests are performed on polymer–mucin blends which simulate the interpenetration layer in the mucoadhesion process and empirical relations are used to calculate a synergism parameter from the values of the storage modulus at a fixed frequency (typically 1Hz) [153–156].

The relationships between synergism parameters and mucoadhesion are still matter of debate and object of controversies, primarily because the mucoadhesion testing itself is not well established for the gel materials. A positive rheological synergism means that the rheological response of a gel–mucin mixture is larger than the sum of the contributions from the gel and the mucin alone, but does not imply necessarily strong mucoadhesion. Conversely, elastic microgels tend to be very adhesive inspite of their low rank in rheological tests. Moreover, the rheology of mucin–polymer blends depends on polymer concentration, type of mucin, ionic strength, and nature of ions present. In the case of microgel particles, rheological measurements can give highly varying results, also depending on, the gap width of the measuring system [154]. A thorough investigation performed on the rheological properties of Pluronic–PAA copolymers showed that the fracture strength and the work of adhesion between the gels and the rat intestine correlated with the viscoelastic characteristics of gels such as the pseudoequilibrium modulus obtained in creep recovery tests and the loss angle measured at fixed oscillatory stress and frequency [157].

Crystalline liquid phases offer an enormous potential in the field of drug delivery. Cubic phases are excellent candidates for use as drug delivery matrices owing to their biodegradability, phase behavior, ability to deliver drugs of varying sizes and polarity, and to enhance the chemical and/or physical stability of incorporated drugs and proteins. However, the extremely high viscosity together with short release duration makes the cubic gel phases impossible to inject for parenteral administration so limiting their use to specific applications such as periodontal, mucosal, vaginal, and short acting oral and parenteral drug delivery [158]. Among the main types of liquid crystalline systems, mesophases with a lamellar structure that demonstrates the greatest similarity to the intercellular lipid membrane of the skin are primarily recommended for the development of a dermal dosage forms [159]. Rheological measurements can contribute to identify the lamellar phase, which is the most convenient for dermal and transdermal drug delivery [160].

Drug delivery to nasal or ocular mucosa for local treatment or systemic effect using gels is widely recognized. Among the hydrogel-forming polysaccharides, scleroglucan is considered of potential use for the design of controlled release devices owing to its mucoadhesive properties. At relatively low-polymer concentration aqueous scleroglucan solutions exhibit weak gel properties [161], which are almost insensitive to marked pH and temperature variations. Rheological studies are useful to detect the matrix structural changes produced by the inclusion of drugs which may modify also the drug-release profile [162].

Microemulsions can be profitably used in the pharmaceutical field as vesicular minicarriers for drug delivery systems, both for oral and topical formulations, because of high thermodynamical stability, easy preparation, and capacity of guesting drug molecules. O/W microemulsions are particularly suitable to guest drugs sparingly soluble in water, and can be destined for topical use on condition that their rheological properties are modified, since they are very fluid systems and cannot be applied directly on the skin. The solution is found by adding a hydrosoluble polymer suitable to impart proper rheological properties, without changing significantly the microemulsion state and affecting the release properties [150,163]. The use of microemulsions and closely related microemulsion-based systems as drug delivery vehicles is reviewed in Ref. [164].

An example of rheological analysis of the structural properties effecting the percutaneous absorption and stability in pharmaceutical organogels is reported in Ref. [165]. Organogels based on microemulsions, especially those stabilized by biosurfactants, have received particular attention, since they offer greater resistance to microbial contamination than aqueous-based formulations. They are generally more fluid than their hydrogel counterparts, but the incorporation of gelatin into W/O microemulsions can impart rheological properties comparable to those attainable using hydrogels. Consequently, these gelatin-containing microemulsion-based organogels (MBGs) can be employed as matrices suitable for transdermal drug delivery, with potential application in iontophoresis, because they are electrically conducting [166].

In situ-gelling systems that do not require organic solvents or copolymerization agents can be profitably used for various biomedical applications, including drug delivery, cell encapsulation, and tissue repair. These systems are injectable fluids that can be introduced into the body in a minimally invasive manner prior to gelling within the desired tissue, organ, or body cavity. Gelation can occur *in situ* by ionic crosslinking or after a change in pH or temperature. Particularly attractive are polymeric systems which undergo sol–gel transition *in situ* in response to temperature change, from ambient to body temperature. This is the case of cellulose derivatives, xyloglucan, chitosan/polyol salt combinations, *N*-isopropylacrylamide copolymers, poloxamers (Pluronics), poly(ethylene oxide)/(D,L-lactic acid-co-glycolic acid) (PEO/PLGA) copolymers, and thermosensitive liposome-based systems, as reviewed in Ref. [167]. Low-amplitude oscillatory shear (LAOS) tests performed with increasing temperature and/or in different isothermal conditions are fundamental to individuate the sol–gel transition and the viscoelastic properties before and after gelation.

Rheological studies have been carried out on poloxamer systems in view of their potential use as ocular vehicle for ophthalmic applications, vaginal formulations, or as stent coatings. In the former case, even if the sol–gel transition temperature can be modulated by changing the polymer concentration, adding a cosolvent or combining poloxamers analogs, it seems difficult to satisfy the rheological requisites related to phase transition temperature and dilution effects after administration to the eye [168,169]. No further improvement in the ocular retention was observed when adding a mucoadhesive polysaccharide, sodium hyaluronate, into the thermosetting gel due to the substantially decreased gel strength [169]. Also for vaginal applications an appropriate timely gelation and retention of *in situ*-gelling formulation is fundamental and even subtle differences in the composition might lead to remarkably different impacts of vaginal fluid dilution with dramatic differences in the viscoelastic properties [170].

An extensive rheological study combined with erosion tests showed that the viscoelastic properties of aqueous pluronic–alginate polymeric blends can be properly modulated in order to obtain promising candidates for active stent coating (endo-arterial paving technique), since erosion resistance of coating surface can be sensibly improved by alginate gelation after contact with a source of bivalent cations [171]. The addition of liposomes does not affect significantly the rheological properties in the final gel state, while the structural process leading to the gel formation starts at lower temperature and is accomplished over a broadened temperature interval.

A novel injectable *in situ*-gelling thermosensitive formulation, based on chitosan–β-glycerophosphate is proposed for tissue repair and sustained delivery, has been examined in Ref. [172]. Here the gelation rate and gel strength are slightly increased by the presence of the liposomes, whereas the release kinetics can be controlled by adjusting the liposome size and composition.

Oscillatory tests are particularly useful to detect the structural effects produced by shear flow on structured fluid. Potential substitutes for the vitreous body of the eye can be discriminated by comparing the mechanical spectra obtained before and after injection through a needle, a procedure that is essential for their use. Physically or lightly chemically crosslinked gels displayed negligible or only minor changes in their viscoelastic properties, thus resulting as the best potential candidates [173].

The development of topical products implies detailed rheological investigations in order to examine and solve the different problems connected to their preparation, application and final performance. For several formulations both continuous shear and viscoelastic measurements are needed to analyze and control manufacturing processes, to monitor changes during storage, as well as to establish relationships between rheological parameters and spreadability, sensory properties, adherence to skin, and drug release rate [174].

REFERENCES

1. Bingham, E.C., *The History of the Society of Rheology from 1924 to 1994*, 1944.
2. Newton, I.S., *Philosophiae Naturalis Principia Mathematica*, 1687.
3. Hagen, G.H.L., Ueber die bewegung des wassers in engen zylindrischen roehren, *Pogg. Ann.*, 46 (1), 423, 1839.
4. Poiseuille, J.L., *Comptes Rendus*, 12, 112, 1841; republished in an English translation by Herschel, W.H., in *Rheological Memoirs*, Easton, PA, 1940.
5. da C. Andrade, E.N., The viscosity of liquids, *Nature*, 125, 309, 582, 1930.
6. da C. Andrade, E.N., *Viscosity and Plasticity*, Chemical Publishing, New York, 1952.
7. De Guzman, J., Relation between fluidity and heat fusion, *Anales Soc. Espan. Fis. y Quim.*, 11, 353, 1913.
8. Eyring, H., Glasstone, S., and Laider, K.J., *Theory of Rate Processes*, New York, 1941.
9. Lapasin, R. and Pricl, S., *Rheology of Industrial Polysaccharides: Theory and Applications*, Springer Verlag, 1995.
10. Ostwald, W., Ueber die geschwindigkeitsfunktion der viscositaet disperse systeme, *Kolloid-Z.*, 36, 99, 1925.
11. de Waele, A., Viscometry and plastometry, *Oil Color Chem. Assoc. J.*, 6, 33, 1923.
12. Cross, M.M., Rheology of non-Newtonian fluids: a new flow equation for pseudoplastic systems, *J. Colloid Sci.*, 20, 417, 1965.
13. Carreau, P.J., *Rheological Equations from Molecular Network Theories*, University of Wisconsin, Madison, 1968.

14. Bingham, E.C., An investigation of the laws of plastic flow, *U.S. Bureau of Standards Bull.*, 13, 309, 1916.
15. Bingham, E.C., *Fluidity and Plasticity*, New York, McGraw-Hill, 1922.
16. Casson, N.A., A flow equation for pigment–oil suspensions of the printing ink type, in *Rheology of Disperse Systems,* Mills, C.C., Ed., Pergamon Press, New York, 1959, p. 84.
17. Herschel, W.H. and Bulkley, R., Konsistenzmessungen von gummi-benzol-lösungen, *Kolloid Z.*, 39, 291, 1926.
18. Herschel, W.H. and Bulkley, R., *Proc. ASTM*, 26, 621, 1926.
19. Bird, R.B., Armstrong, R.C., Hassager, O., *Dynamics of Polymeric Liquids,* Vol 1: *Fluid Mechanics*, 2nd edn., Wiley-Interscience, New York, 1987.
20. Crochet, M.J., Davies, A.R., and Walters, K., *Numerical Simulation of Non-Newtonian Flow*, Elsevier, Amsterdam, 1984.
21. Crochet, M.J., Numerical simulation of flow processes, *Chem. Eng. Sci.*, 42, 979, 2002.
22. Owens, R.G. and Phillips, T.N., *Computational Rheology*, Imperial College Press, London, 2002.
23. Walters, K., *Rheometry*, Wiley & Sons, New York, 1975.
24. Macosko, Ch.W., *Rheology: Principles, Measurements, and Applications*, Wiley, New York, 1994.
25. Metzner, A.B. and Reed J.C., Flow of non-Newtonian fluids—correlation of the laminar, transition and turbulent-flow regions, *AIChE J.*, 1, 434, 1955.
26. Govier, G.W. and Aziz, K., *The Flow of Complex Fluids in Pipes*, Kreiger Pub. Co., Malabar, 1982.
27. Heywood, N.I. and Cheng, D.C.-H., Comparison of methods for predicting head loss in turbulent pipe flow of non-Newtonian fluids, *Trans. Inst. Meas. Control*, 6, 33, 1984.
28. Chhabra, R.P. and Richardson, J.F., *Non-Newtonian Flow in the Process Industries*, Butterworth-Heinemann, Oxford, 1999.
29. Mewis, J., Thixotropy—a general review, *J. Non-Newtonian Fluid Mech.*, 6, 1, 1979.
30. Barnes, H.A., Thixotropy: a review, *J. Non-Newtonian Fluid Mech.*, 70, 1, 1997.
31. Mujumdar, A., Beris, A.N., and Metzner, A.B., Transient phenomena in thixotropic systems, *J. Non-Newtonian Fluid Mech.*, 102, 157, 2002.
32. Alessandrini, A., Lapasin, R., and Sturzi, F., The kinetics of the thixotropic behaviour in clay/kaolin aqueous suspensions, *Chem. Eng. Commun.*, 17, 13, 1982.
33. Dullaert, K. and Mewis, J., A model system for thixotropy studies, *Rheol. Acta*, 45, 23, 2005.
34. Edgeworth, R.R., Dalton, B.J., Parnell, T., The pitch drop experiment, *Eur. J. Phys.*, 5, 198, 1984.
35. Willis, R.L., Stone, T.S., and Berthelot, Y.H., An experimental–numerical technique for evaluating the bulk and shear dynamic moduli of viscoelastic materials, *J. Acoustic Soc. Am.*, 102, 3549, 1997.
36. Willis, R.L., Wu, L., and Berthelot, Y.H., Determination of the complex Young and shear dynamic moduli of viscoelastic materials, *J. Acoustic Soc. Am.*, 109, 611, 2001.
37. Lakes, R., Foam structures with a negative Poisson's ratio, *Science*, 235, 1038, 1987.

38. Head, D.A., Levine, A.J., and MacKintosh, F.C., Deformation of cross-linked semiflexible polymer networks, *Phys. Rev. Lett.*, 91, 108102–1, 2003.
39. Reid, R.C., Prausnitz, J.M., and Poling, B.E., *The Properties of Gases and Liquids*, McGraw-Hill, Singapore, 1988.
40. Grunberg, L. and Nissan, A.H., Mixture law for viscosity, *Nature*, 164, 799, 1949.
41. Isdale, J.D., Mac Gillivray, J.C., and Cartwright, G., Prediction of viscosity of organic liquid mixtures by a group contribution method, *Natl. Eng. Lab. Rept.*, East Kilbride, Glasgow, Scotland, 1985.
42. Teja, A.S. and Rice, P., A multifluid corresponding states principle for the thermodynamic properties of fluid mixtures, *Chem. Eng. Sci.*, 36, 1, 1981.
43. Teja, A.S. and Rice, P., Generalized corresponding states method for viscosities of liquid mixtures, *Ind. Eng. Chem. Fundam.*, 20, 77, 1981.
44. Horvath, A.L., *Handbook of Aqueous Electrolyte Solutions: Physical Properties, Estimation, and Correlation Methods*, Ellis Horwood, Chichester, UK, 1985.
45. Onsager, L. and Fuoss, R.M., Irreversible processes in electrolytes. Diffusion, conductance, and viscous flow in arbitrary mixtures of strong electrolytes, *J. Phys. Chem.*, 36, 2689, 1932.
46. Jones, G. and Dole, M., The viscosity of aqueous solutions of strong electrolytes with special reference to barium chloride, *J. Am. Chem. Soc.*, 51, 2950, 1929.
47. Zhang, H.L., Chen, G.-H., and Han, S.J., Viscosity and density of water + sodium chloride + potassium chloride solutions at 298.15 K, *J. Chem. Eng. Data*, 41, 516, 1996.
48. Hu, Y.-F., A new equation for predicting the density of multicomponent aqueous solutions conforming to the linear isopiestic relation, *Phys. Chem. Chem. Phys.*, 2, 2379, 2000.
49. Hu, Y.-F. and Lee, H., Prediction of viscosity of mixed electrolyte solutions based on the Eyring's absolute rate theory and the semi-ideal hydration model, *Electrochem. Acta*, 48, 1789, 2003.
50. Hu, Y.-F., Prediction of viscosity of mixed electrolyte solutions based on the Eyring's absolute rate theory and the equations of Patwardhan and Kumar, *Chem. Eng. Sci.*, 59, 2457, 2004.
51. Hu, Y.-F., Reply to Comments on 'Prediction of viscosity of mixed electrolyte solutions based on the Eyring's absolute rate theory and the equations of Patwardhan and Kumar', *Chem. Eng. Sci.*, 60, 3121, 2005.
52. Lencka, M.M., et al., Modeling viscosity of multicomponent electrolyte solutions, *Int. J. Thermophys.*, 19, 367, 1998.
53. Israelachvili, J., *Intermolecular and Surface Forces*, Academic Press, London, 1992.
54. Rehage, H. and Hoffmann, H., Shear induced phase transitions in highly dilute aqueous detergent solutions, *Rheol. Acta*, 21, 561, 1982.
55. Bewersdorff, H.W., et al., The conformation of drag reducing micelles from small-angle-neutron-scattering experiments, *Rheol. Acta*, 25, 642, 1986.
56. Hofmann, S., Rauscher, A., and Hoffmann, H., Shear induced micellar structures, *Ber. Bunsenges. Phys. Chem.* 95, 153, 1991.
57. Hu, Y., Wang, S.Q., and Jamieson, A.M., Rheological and flow birefringence studies of a shear-thickening complex fluid—a surfactant model system, *J. Rheol.*, 37, 531, 1993.

58. Hu, Y. and Matthys, E.F., Characterization of micellar structure dynamics for a drag-reducing surfactant solution under shear: normal stress studies and flow geometry effects, *Rheol. Acta*, 34, 450, 1995.
59. Schmitt, V., Schosseler, S., and Lequeux, F., Structure of salt-free wormlike micelles: signature by SANS at rest and under shear, *Europhys. Lett.*, 30, 31, 1995.
60. Hartmann, V. and Cressely, R., Simple salts effects on the characteristics of the shear thickening exhibited by an aqueous micellar solution of CTAB/NaSal, *Europhys. Lett.*, 40, 691, 1997.
61. Gamez-Corrales, R., et al., Shear-thickening dilute surfactant solutions: the equilibrium structure as studied by small-angle neutron scattering, *Langmuir*, 15, 6755, 1999.
62. Berret, J.-F., et al., Shear-thickening transition in surfactant solutions: new experimental features from rheology and flow birefringence, *Eur. Phys. J. E*, 2, 343, 2000.
63. Wunderlich, A.M., Hoffmann, H., and Rehage, H., Flow birefringence and rheological measurements on shear induced micellar structures, *Rheol. Acta*, 26, 532, 1987.
64. Oda, R., et al., Direct evidence of the shear-induced structure of wormlike micelles, *Langmuir*, 13, 6407, 1997.
65. Berret, J.-F., et al., Flow-structure relationship of shear-thickening surfactant solutions, *Europhys. Lett.*, 41, 677, 1998.
66. Cates, M.E., Repetition of living polymers: dynamics of entangled polymers in the presence of reversible chain-scission reactions, *Macromolecules*, 20, 2289, 1987.
67. Cates, M.E., Nonlinear viscoelasticity of wormlike micelles (and other reversibly breakable polymers), *J. Phys. Chem.*, 94, 371, 1990.
68. Candau, S.J., et al., Rheological behaviour of worm-like micelles: effect of salt content, *J. Phys. IV (France)*, 3, 197, 1993.
69. Larson, R.G., Monte Carlo simulations of the phase behavior of surfactant solutions, *J. Phys. II (France)*, 6, 1441, 1996.
70. Larson, R.G., *The Structure and Rheology of Complex Fluids*, University Press, New York, 1999.
71. Roux, D., et al., Shear induced orientations and textures of nematic wormlike micelles, *Macromolecules*, 28, 1681, 1995.
72. Berret, J.-F., et al., Tumbling behavior of nematic wormlike micelles under shear flow, *Europhys. Lett.*, 32, 137, 1995.
73. Roux, D., Nallet, F., and Diat, O., Rheology of lyotropic lamellar phases, *Europhys. Lett.*, 24, 53, 1993.
74. Bergenholtz, J. and Wagner, N.J., Formation of AOT/brine multilamellar vesicles, *Langmuir*, 12, 3122, 1996.
75. Diat, O., Roux, D., and Nallet, F., Effect of shear on a lyotropic lamellar phase, *J. Phys. II (France)*, 3, 1427, 1993.
76. Butler, P., Shear induces structures and transformations in complex fluids, *Curr. Opin. Colloid Interface Sci.*, 4, 214, 1999.
77. Richtering, W., Rheology and shear induced structures in surfactant solutions, *Curr. Opin. Colloid Interface Sci.*, 6, 446, 2001.

78. Mortensen, K., Structural studies of lamellar surfactant systems under shear, *Curr. Opin. Colloid Interface Sci.*, 6, 140, 2001.
79. Berni, M.G., Laurence, C.J., and Machin, D., A review of the rheology of the lamellar phase in surfactant systems, *Adv. Colloid Interface Sci.*, 98, 217, 2002.
80. Medronho, B., et al., Reversible size of shear-induced multi-lamellar vesicles, *Colloid Polym. Sci.*, 284, 317, 2005.
81. Flory, J.P., Thermodynamics of high polymer solutions, *J. Chem. Phys.*, 9, 660, 1941.
82. Huggins, M.L., Solutions of long chain compounds, *J. Chem. Phys.*, 9, 440, 1941.
83. Huggins, M.L., The Viscosity of dilute solutions of long-chain molecules. IV. Dependence on concentration, *J. Am. Chem. Soc.*, 64, 2716, 1942.
84. Kraemer, O., Molecular weights of celluloses, *Ind. Eng. Chem.*, 30, 1200, 1938.
85. Mark, H., *Der Feste Körper*, Leipzig, Hirzel, 1938.
86. Houwink, R., Zusammenhang zwischen viscosimetrisch und osmotisch bestimmten polymerisationsgraden bei hochpolymeren, *J. Prakt. Chem.*, 157, 15, 1940.
87. Graessley, W.W., The entanglement concept in polymer rheology, in *Advances in Polymer Science*, Heidelberg, Springer Berlin, 1974.
88. Edwards, S.F., The statistical mechanics of polymerized material, *Proc. Phys. Soc.*, 92, 9, 1967.
89. de Gennes, P.G., Repetition of a polymer chain in the presence of fixed obstacles, *J. Chem. Phys.*, 55, 572, 1971.
90. Doi, M. and Edwards, S.F., Dynamics of concentrated polymer systems, *J. Chem. Soc. Faraday II*, 74, 1789, 1978; Doi, M. and Edwards, S.F., Dynamics of concentrated polymer systems, *J. Chem. Soc. Faraday II*, 74, 1802, 1978; Doi, M. and Edwards, S.F., Dynamics of concentrated polymer systems, *J. Chem. Soc. Faraday II*, 74, 1818, 1978.
91. Doi, M. and Edwards, S.F., Dynamics of concentrated polymer systems, *J. Chem. Soc. Faraday II*, 75, 38, 1979.
92. de Gennes, P.G., Brownian motions of flexible polymer chains, *Nature*, 282, 367, 1979.
93. Martin, A.F., *American Chemical Society Meeting*, Memphis, 1942.
94. Spurlin, M., Martin, A.F. and Tennent, H.G., Characterization of cellulose derivatives by solution properties, *J. Polym. Sci.*, 1, 63, 1946.
95. Utracki, L.A. and Simha, R., Viscosity of polymer solutions: scaling relationships, *J. Rheol.*, 25, 329, 1981.
96. Simha, R. and Zakin, J.L., Solution viscosities of linear flexible high polymers, *J. Colloid Sci.*, 17, 270, 1962.
97. Utracki, L.A. and Simha, R., Corresponding states relations for moderately concentrated polymer solutions, *J. Polym. Sci.—Part A*, 1, 1089, 1963.
98. Onogi, S. and Asada, T., Rheology and rheo-optics of polymer liquid crystals, in *Rheology*, Astarita, G., Marrucci, G., Nicolais, G., Eds., Plenum Press, New York, 1980, pp. 127–147.
99. Walker, L.M. and Wagner, N.J., Rheology of region I flow in a lyotropic liquid–crystal polymer: the effects of defect texture, *J. Rheol.*, 38, 1525, 1994.
100. Williams, M.L., Landel, R.F., Ferry, J.D., The temperature dependence of relaxation mechanisms in amorphous polymers and other glass-forming liquids, *J. Am. Chem. Soc.*, 77, 3701, 1980.

101. Chambon, F. and Winter, H.H., Stopping of crosslinking reaction in a PDMS. Polymer at the gel point, *Polym Bull. (Berlin)*, 13, 499, 1985.
102. Winter, H.H. and Chambon, F., Linear viscoelasticity at the gel point of a crosslinking PDMS with imbalanced stoichiometry, *J. Rheol.*, 31, 683, 1987.
103. Scanlan, C. and Winter H.H., Composition dependence of the viscoelasticity of end-linked PDMS at the gel point, *Macromolecules*, 24, 47, 1991.
104. Cuvelier, G. and Launay, B., Frequency dependence of viscoelastic properties of some physical gels near the gel point, *Makromol. Chem. Macromol. Symp.*, 40, 23, 1990.
105. Witten, T.A., Associating polymers and shear thickening, *J. Phys. II (France)*, 49, 1055, 1988.
106. Jenkins, R.D., Silebi, C.A., and El-Aasser, M.S., In *Polymers as Rheology Modifiers*, Schulz, D.N. and Glass, J.E., Eds., *ACS Symp. Ser.*, 462, 1991, pp. 222–233.
107. Annable, T., et al., The rheology of solutions of associating polymers: comparison of experimental behavior with transient network theory, *J. Rheol.*, 37, 695, 1993.
108. Marrucci, G., Barghava, S., and Cooper, S.L., Models of shear-thickening behavior in physically cross-linked networks, *Macromolecules*, 26, 6483, 1993.
109. Hu, Y., Wang, S.Q., and Jamieson, A.M., Rheological and rheooptical studies of shear-thickening polyacrylamide solutions, *Macromolecules*, 28, 1847, 1995.
110. Tam, K.C., et al., Structural model of hydrophobically modified urethane-ethoxylate (HEUR) associative polymers in shear flows, *Macromolecules*, 31, 4149, 1998.
111. Berret, J.-F., et al., Nonlinear rheology of telechelic polymer networks, *J. Rheol.*, 45, 477, 2001.
112. Ma, S.X. and Cooper, S.L., Shear thickening in aqueous solutions of hydrocarbon end-capped poly(ethylene oxide), *Macromolecules*, 34, 3294, 2001.
113. Derjaguin, V. and Landau, L., Theory of stability of strongly charged sols and the adhesion of strongly charged particles in solutions of electrolytes, *Acta Physicochim. URSS*, 14, 633, 1941.
114. Vervey, J.W. and Overbeek, J.T.L., *Theory of Stability of Lipophobic Colloids*, Elsevier, Amsterdam, 1948.
115. Asakura, S. and Oosawa, F., Surface tension of high-polymer solutions, *J. Chem. Phys.*, 22, 1255, 1954.
116. Einstein, A., Eine neue bestimmung der molekuldimension, *Ann. Phys.*, 19, 289, 1906.
117. Einstein, A., Berichtigung zu meiner arbeit: eine neue bestimmung der molekuldimension, *Ann. Phys.*, 34, 591, 1911.
118. Batchelor, G.K., The stress system in a suspension of force-free particles, *J. Fluid. Mech.*, 41, 545, 1970.
119. Batchelor, G.K., The stress generated in a non-dilute suspension of elongated particles in pure straining motion, *J. Fluid. Mech.*, 46, 813, 1971.
120. Krieger, I.M., Rheology of monodisperse lattices, *Adv. Colloid Interface Sci.*, 3, 111, 1972.
121. Choi, G.N. and Krieger, I.M., Rheological studies on sterically stabilized dispersions of uniform colloidal spheres. Steady shear viscosity, *J. Colloid Interface Sci.*, 113, 101, 1986.

122. Krieger, I.M. and Dougherty, T.J., A mechanism for non-Newtonian flow in suspensions of rigid spheres, *Trans. Soc. Rheol.*, 3, 137, 1959.

123. Quemada, D., Rheology of concentrated disperse systems and minimum energy dissipation principle. I Viscosity–concentration relationship, *Rheol. Acta*, 16, 82, 1977.

124. Barnes, H.A., Shear-thickening ("dilatancy") in suspensions of nonaggregating solid particles dispersed in Newtonian liquids, *J. Rheol.*, 33, 329, 1989.

125. Zhou, Z., Scales, P.J., and Boger, D.V., Chemical and physical control of the rheology of concentrated metal oxide suspensions, *Chem. Eng. Sci.*, 56, 2901, 2001.

126. Farris, R.J., Prediction of the viscosity of multimodal suspensions from unimodal viscosity sata, *Trans. Soc. Rheol.*, 12, 281, 1968.

127. Russel, W.B., The rheology of suspensions of charged rigid spheres, *J. Fluid. Mech.*, 85, 209, 1978.

128. Sherwood, D., The primary electroviscous effect in a suspension of spheres, *J. Fluid. Mech.*, 101, 609, 1980.

129. Rubio-Hernández, F.J., Carrique, F., and Ruiz-Reina, E., The primary electroviscous effect in colloidal suspensions, *Adv. Colloid Interface Sci.*, 107, 51, 2004.

130. Krieger, I.M. and Eguiluz, M., The second electroviscous effect in polymer lattices, *Trans. Soc. Rheol.*, 20, 29, 1976.

131. Buscall, R., et al., Viscoelastic properties of concentrated lattices. Part I. Methods of examination, *J. Chem. Soc. Faraday Trans. I*, 78, 2873, 1982; Buscall, R., et al., Viscoelastic properties of concentrated lattices. Part 2. Theoretical analysis, *J. Chem. Soc. Faraday Trans. I*, 78, 2889, 1982.

132. Russel, W.B., Saville, D.A., and Schowalter, W.R., *Colloidal Dispersions*, Cambridge University Press, Cambridge, 1989.

133. Chen, L.B. and Zukoski, C.F., Flow of ordered latex suspensions: yielding and catastrophic shear thinning, *J. Chem. Soc. Faraday Trans.*, 86, 2629, 1990.

134. Ogawa, A., et al., Viscosity equation for concentrated suspensions of charged colloidal particles, *J. Rheol.*, 41, 769, 1997.

135. Hoffman, R.L., Discontinuous and dilatant viscosity behavior in concentrated suspensions. I. Observation of a flow instability, *Trans. Soc. Rheol.*, 16, 155, 1972.

136. Hoffman, R.L., Discontinuous and dilatant viscosity behavior in concentrated suspensions. II. Theory and experimental tests, *J. Colloid Interface Sci.*, 46, 491, 1974.

137. Boersma, W.H., Laven, J., and Stein, H.N., Shear thickening (dilatancy) in concentrated dispersions, *AIChE J.*, 36, 321, 1990.

138. Meakin, P., A historical introduction to computer models for fractal aggregates, *J. Sol–Gel Sci. Technol.*, 15, 97, 1999.

139. Channell, G.M. and Zukoski, C.F., Shear and compressive rheology of aggregated alumina suspensions, *AIChE J.*, 43, 1700, 1997.

140. Scales, P.J., et al., Shear yield stress of partially flocculated colloidal suspensions, *AIChE J.*, 44, 538, 1998.

141. Yanez, J,A., et al., Shear modulus and yield stress measurements of attractive alumina particle networks in aqueous slurries, *J. Am. Ceram. Soc.*, 79, 2917, 1996.

142. Zhou, Z., et al., The yield stress of concentrated flocculated suspensions of size distributed particles, *J. Rheol.*, 43, 651, 1999.

143. Zhou, Z., Scales, P.J., and Boger, D.V., Chemical and physical control of the rheology of concentrated, *Chem. Eng. Sci.*, 56, 2901, 2001.

144. Taylor, G.I., The viscosity of a fluid containing small drops of another fluid, *Royal Soc. London Ser. A*, 138, 41, 1932.

145. Choi, J.C. and Schowalter, W.R., Rheological properties of nondilute suspensions of deformable particles, *Phys. Fluids*, 18, 420, 1975.

146. Pal, R., Rheology of polymer-thickened emulsions, *J. Rheol.*, 36, 1245, 1992.

147. Mooney, M., The viscosity of a concentrated suspension of spherical particles, *J. Colloid. Sci.*, 6, 162, 1951.

148. Mason, T.G., Bibette, J., and Weitz, D.A., Elasticity of compressed emulsions, *Phys. Rev. Lett.*, 75, 2051, 1995.

149. Moulik, S.P. and Paul, B.K., Structure, dynamics and transport properties of microemulsions, *Adv. Colloid Interface Sci.*, 78, 99, 1998.

150. Coceani, N., Lapasin, R., and Grassi, M., Effects of polymer addition on the rheology of O/W microemulsions, *Rheol. Acta*, 40, 185, 2001.

151. Peppas, N.A., et al., Hydrogels in pharmaceutical formulations, *Eur. J. Pharm. Biopharm.*, 50, 27, 2000.

152. Peppas, N.A. and Sahlin, J.J., Hydrogels as mucoadhesive and bioadhesive materials: a review, *Biomaterials*, 17, 1553, 1996.

153. Madsen, F., Eberth, K., and Smart, J.D., A rheological examination of the mucoadhesive/mucus interaction: the effects of mucoadhesive type and concentration, *J. Contr. Rel.*, 50, 167, 1998.

154. Hagërström, H., Paulsson, M., and Edsman, K., Evaluation of mucoadhesion for two polyelectrolyte gels in simulated physiological conditions using a rheological method, *Eur. J. Pharm. Sci.*, 9, 301, 2000.

155. Riley, R.G., et al., An investigation of mucus/polymer rheological synergism using synthesised and characterised poly(acrylic acid)s, *Int. J. Pharm.*, 217, 87, 2001.

156. Hagërström, H. and Edsman, K., Limitations of the rheological mucoadhesion method: the effect of the choice of conditions and the rheological synergism parameter, *Eur. J. Pharm. Sci.*, 18, 349, 2003.

157. Bromberg, L., et al., Bioadhesive properties and rheology of polyether-modified poly(acrylic acid) hydrogels, *Int. J. Pharm.*, 282, 45, 2004.

158. Shah, J.C., Sadhale, Y., and Murthy Chilukuri, D., Cubic phase gels as drug delivery systems, *Adv. Drug Del. Rev.*, 47, 229, 2001.

159. Makai, M., et al., Structure and drug release of lamellar liquid crystals containing glycerol, *Int. J. Pharm.*, 256, 95, 2003.

160. Németh, Zs., et al., Rheological behavior of a lamellar liquid crystalline surfactant–water system, *Colloid. Surf. A*, 145, 107, 1998.

161. Grassi, M., Lapasin, R., and Pricl, S., A study of the rheological behavior of scleroglucan weak gel systems, *Carbohydr. Polym.*, 29, 169, 1996.

162. Francois, N.J., et al., Dynamic rheological measurements and drug release kinetics in swollen scleroglucan matrices, *J. Contr. Rel.*, 90, 355, 2003.

163. Valenta, C. and Schultz, K., Influence of carrageenan on the rheology and skin permeation of microemulsion formulations, *J. Contr. Rel.*, 95, 257, 2004.

164. Lawrence, M.J. and Rees, G.D., Microemulsion-based media as novel drug delivery systems, *Adv. Drug Del. Rev.*, 45, 89, 2000.

165. Pénzes, T., Csóka, I., and Eros, I., Rheological analysis of the structural properties effecting the percutaneous absorption and stability in pharmaceutical organogels, *Rheol. Acta*, 43, 457, 2004.
166. Kantaria, S., Rees, G.D., and Lawrence, M.J., Gelatin-stabilised microemulsion-based organogels: rheology and application in iontophoretic transdermal drug delivery, *J. Contr. Rel.*, 60, 355, 1999.
167. Ruel-Gariépy, E. and Leroux, J.-C., *In situ*-forming hydrogels—review of temperature-sensitive systems, *Eur. J. Pharm. Biopharm.*, 58, 409, 2004.
168. Edsman, K., Carlfors, J., and Petersson, R., Rheological evaluation of poloxamer as *in situ* gel for ophthalmic use, *Eur. J. Pharm. Sci.*, 6, 105, 1998.
169. Wei, G., et al., Thermosetting gels with modulated gelation temperature for ophthalmic use: the rheological and gamma scintigraphic studies, *J. Contr. Rel.*, 83, 65, 2002.
170. Chang, J.Y., et al., Rheological evaluation of thermosensitive and mucoadhesive vaginal gels in physiological conditions, *Int. J. Pharm.*, 241, 155, 2002.
171. Grassi, M., et al., Rheological properties of aqueous pluronic–alginate systems containing liposomes, *J. Colloid Interface Sci.*, 301, 282, 2006.
172. Ruel-Gariépy, E., et al., Thermosensitive chitosan-based hydrogel containing liposomes for the delivery of hydrophilic molecules, *J. Contr. Rel.*, 82, 373, 2002.
173. Dalton, P.D., et al., Oscillatory shear experiments as criteria for potential vitreous substitutes, *Poly. Gel. Netw.*, 3, 429, 1995.
174. Lippacher, A., Müller, R.H., and Mäder, K., Liquid and semisolid SLN dispersions for topical application: rheological characterization, *Eur. J. Pharm. Biopharm.*, 58, 561, 2004.

4 Mass Transport

4.1 INTRODUCTION

In order to exert a therapeutic effect, a drug should have high affinity and selectivity for its intended biological target (for example, a protein or a protein complex in or on a particular cells type) and also reach a sufficient concentration at that site [1]. In general, to match this goal, the drug has to be released from the delivery system, transported from the site of application to the site of action, biotransformed and finally eliminated (metabolism) from the body [2]. Thus, the whole process can be divided into the pharmaceutical phase (where drug liberation takes place) and the pharmacokinetic phase describing the course of the drug in the body and its metabolism. Undoubtedly, both the first and the second phases are strongly dependent on mass transport and that is why it plays a central role in determining delivery system reliability and effectiveness. Indeed, in the pharmaceutical phase, regardless the delivery system considered (in *passive preprogrammed systems*, release rate is predetermined and it is irresponsive to external biological stimuli; in *active preprogrammed systems*, release rate can be controlled by a source external to the body; in *active self-programmed systems*, release rate is driven by external physiological stimuli [3]) release kinetics depends on mass transport characteristics both in the case of diffusion and convection controlled processes. While in passive preprogrammed systems, release kinetics is usually controlled by drug or water diffusion in the polymeric network, in self-programmed systems, both diffusion and convection can play an important role. For example, Siegel [4], approaching insulin delivery by means of an implantable mechanochemical pump, makes use of both convection and diffusion. Indeed, this device is made up by a housing containing three chambers (Figure 4.1). While chamber I contains an insulin solution, chamber II contains aqueous fluid, and chamber III is made up by a glucose-sensitive swellable hydrogel separated by the external environment by means of a rigid membrane permeable only to small molecules and impermeable to large molecules such as plasma proteins. As soon as blood glucose concentration increases, glucose permeates through the rigid membrane causing further hydrogel expansion (diffusive mass transport). This, in turn, provokes the forward movement of the diaphragm and of the partition separating, respectively, chamber III from chamber II and chamber II from chamber I. Accordingly, chamber I volume reduces and insulin solution is sent in the

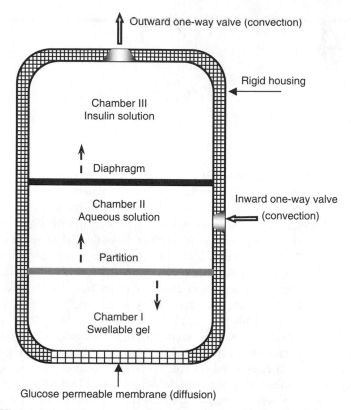

FIGURE 4.1 Schematic representation of an implantable insulin delivery system. When external environment concentration increases, glucose crosses the permeable membrane causing further gel swelling and forward motion of both diaphragm and partition (see *dashed* arrows). This, in turn, provokes insulin delivery through the outward one-way valve. If external glucose concentration decreases, gel shrinks, partition comes back (see *dashed* arrow) and water is recalled from the external environment in chamber II through the one-way inward valve. (Adapted from Siegel, R.A., *Pulsed and Self-Regulated Drug Delivery*, Kost, J., Ed., CRC Press, Boca Raton, FL, 1990.)

external environment (convective mass transport) through the outward one-way valve. When glucose concentration decreases, hydrogel shrinks, diaphragm recedes and chamber II absorbs liquid through the inward one-way valve so that movable partition does not move. This process works until the complete depletion of chamber I. Among the many other examples that can be mentioned, we can remember urea release from nonerodible pH-sensitive membranes [5], drug release from photochemically controlled microcapsules [6–11], verapamil (calcium channel blocker, indicated for the hyperstress treatment) release from a mixture of rapid and slow release beads coated by a polymeric film consisting in an hydroxypropylmethylcellulose/ethylcellulose blend [12], transdermal delivery systems [13] and colon targeted delivery systems [14].

Obviously, mass transport is very important also in the pharmacokinetic phase as compounds are absorbed in the body on condition that they cross cellular membranes and this takes place according to passive diffusion or owing to active mechanisms [15]. While passive diffusion occurs spontaneously, some active mechanisms require the support of external energy fronts as it happens for the ATP pump [16].

In this context, Chapter 4 aims to provide some fundamental thermodynamic concepts (that will be also used in other chapters) strictly linked to the mass transport, to discuss the mass conservation law, to show some important analytical solution of it, to focus the attention on diffusion coefficient, and to show the effect of particular boundary conditions (presence of stagnant layers and release from holed surface) on release kinetics.

4.2 THERMODYNAMIC CONCEPTS

Thermodynamically speaking, the part of the universe we are studying is called the *system* and *surrounding* is the remaining part of the universe. The system typically consists of a specified amount of chemical substance or substances, such as a given mass of a gas, liquid, or solid. On the basis of its relations with the surrounding, the system can be classified as *isolated* (neither matter nor energy can flow into or out of the system), *closed* (energy may be exchanged with the surrounding, but not matter), and *open* (both energy and matter can enter or leave the system). Additionally, the system can be defined *homogeneous* if it is characterized by uniform properties or *heterogeneous* if it is constituted by two or more phases (each one can be viewed as a homogeneous system). The thermodynamic energy U associated to the system descends from the energy of its atoms and molecules. This, in turn, depends on the electronic distribution within the material and on atomic or molecular motion, namely *translation* (the movement of individual molecules in space), *vibration* (the movement of atoms or groups of atoms with respect to each other within a molecule), and *rotation* (the revolution of molecules about an axis). While translation energy represents the kinetic part of U, vibrational, rotational, and electronic energy represent the internal part of U. Of course, U does not represent the whole system energy as kinetic energy (result of system motion as a whole) and potential energy (result of system position) constitute system mechanical energy and other forms of energy can be owned by the system as a whole (surface energy and electromagnetic energy among others. In this sense, now, we are dealing with *simple systems*). Accordingly, U can be seen as a measure of molecular and atomic distributions and motions, as well as the electronic distribution within atoms and molecules. It is also evident that U is an *extensive* system property as its value increases with system mass (proportional to the number of system atoms) even if the mean atoms, molecules, and electrons energy is the same. On the contrary, temperature T is an *intensive* system parameter as it is a measure of the mean atoms, molecules, and electrons energy and does not depend on system mass. The condition of no translational, vibrational, rotational, and electronic energy corresponds to the absolute zero equal to $-273.15°C$. Volume (V) and moles number (n_i) of each system component are other fundamental extensive system properties.

In order to describe a thermodynamic system, we need to define the walls separating it from the surrounding and defining the boundary conditions or these external bounds. For example, a rigid wall impedes V variations and this makes impossible work exchange between the system and the surrounding. In addition, while a semi-impermeable membrane can be restrictive with respect to the number of moles n_i of the ith component, an adiabatic wall is restrictive with respect to heat flux.

4.2.1 THERMODYNAMIC EQUILIBRIUM

Upon boundary conditions modification, system intensive and extensive properties vary and processes take place. When all system properties no longer change with time, the system is said to be in *thermodynamic equilibrium* and the central problem of thermodynamic consists in the determination of the new U, V, and n_i values. This problem is solved by means of the second thermodynamic principle enunciated in an axiomatic form according to which for each system there exists a function of the extensive parameters (U, V, n_i), called entropy (S) and defined for all system equilibrium conditions, characterized by the fact that (a) the values assumed by extensive parameters at equilibrium maximize S, (b) S value corresponding to a heterogeneous system is given by the sum of each phase entropy, (c) S is, mathematically speaking, a continuous, differentiable, and monotonic increasing function of U. This enunciation allows writing down the following general equations:

$$S = \sum_{j=1}^{m} S_j \tag{4.1}$$

$$S_j = S_j(U_j, V_j, n_i^j) \quad \forall j \text{ and } 1 < i < r \tag{4.2}$$

$$dS = 0, \quad d^2S < 0, \tag{4.3}$$

where S_j, U_j, and V_j represent, respectively, the entropy, the energy, and the volume of the jth system phase, m is phase number, r is system components number and n_i^j denotes the moles number relative to the ith component belonging to the jth phase. While Equation 4.1 expresses entropy additivity,

Equation 4.3 imposes the S maximum condition at equilibrium. On the basis of the above-mentioned S properties, the following equations can be obtained:

$$U = U(S, V, n_i) \quad \forall i \tag{4.4}$$

$$dU = \left(\frac{\partial U}{\partial S}\right)_{V,n_1,\dots,n_r} dS + \left(\frac{\partial U}{\partial V}\right)_{S,n_1,\dots,n_r} dV + \sum_{i=1}^{r} \left(\frac{\partial U}{\partial n_i}\right)_{S,V,n_j \neq n_i} dn_i \tag{4.5}$$

Equation 4.5 gives the opportunity of defining some important intensive parameters:

$$\left(\frac{\partial U}{\partial S}\right)_{V,n_1,\dots,n_r} \equiv T(S, V, n_1, \dots, n_r) \tag{4.6}$$

$$-\left(\frac{\partial U}{\partial V}\right)_{S,n_1,\dots,n_r} \equiv P(S, V, n_1, \dots, n_r) \tag{4.7}$$

$$\left(\frac{\partial U}{\partial n_i}\right)_{S,V,n_j \neq n_i} \equiv \mu_i(S, V, n_1, \dots, n_r), \tag{4.8}$$

where T and P are, respectively, system temperature and pressure while μ_i is the ith component chemical potential. Accordingly, Equation 4.5 can be rewritten as

$$dU = TdS - PdV + \sum_{i=1}^{r} \mu_i \, dn_i. \tag{4.9}$$

In addition, on the basis of Equation 4.6 through Equation 4.8, it follows:

$$\left(\frac{\partial S}{\partial U}\right)_{V,n_1,\dots,n_r} = \frac{1}{T} \tag{4.10}$$

$$\left(\frac{\partial S}{\partial V}\right)_{U,n_1,\dots,n_r} = -\left(\frac{\partial S}{\partial U}\right)_{V,n_1,\dots,n_r} \left(\frac{\partial U}{\partial V}\right)_{S,n_1,\dots,n_r} = -\frac{1}{T}(-P) = \frac{P}{T} \tag{4.11}$$

and finally

$$dS = \left(\frac{\partial S}{\partial U}\right)_{V,n_1,\dots,n_r} dU + \left(\frac{\partial S}{\partial V}\right)_{U,n_1,\dots,n_r} dV + \sum_{i=1}^{r} \left(\frac{\partial S}{\partial n_i}\right)_{S,V,n_j \neq n_i} dn_i \tag{4.12}$$

$$dS = \frac{dU}{T} + \frac{P}{T}dV - \sum_{i=1}^{r} \frac{\mu_i}{T} \, dn_i. \tag{4.13}$$

It is now useful expressing Equation 4.9 and Equation 4.13 explicitly in terms of entropy (S_j), energy (U_j), volume (V_j), temperature (T_j), pressure (P_j), and chemical potential (μ_i^j) referred to each system phase j:

$$dU = \sum_{j=1}^{m} dU_j = \sum_{j=1}^{m} \left(T_j dS_j - P_j dV_j + \sum_{i=1}^{r} \mu_i^j dn_i^j \right) \tag{4.14}$$

$$dS = \sum_{j=1}^{m} dS_j = \sum_{j=1}^{m} \left(\frac{dU_j}{T_j} + \frac{P_j}{T_j} dV_j - \sum_{i=1}^{r} \frac{\mu_i^j}{T_j} dn_i^j \right). \tag{4.15}$$

Remembering the definition of isolated system, the following relations hold:

$$dU = \sum_{j=1}^{m} dU_j = 0 \quad \Longrightarrow \quad dU_1 = -\sum_{j=2}^{m} dU_j \tag{4.16}$$

$$dV = \sum_{j=1}^{m} dV_j = 0 \quad \Longrightarrow \quad dV_1 = -\sum_{j=2}^{m} dV_j \tag{4.17}$$

$$dn_i = \sum_{j=1}^{m} dn_i^j = 0 \quad \Longrightarrow \quad dn_i^1 = -\sum_{j=2}^{m} dn_i^j \quad \forall i. \tag{4.18}$$

Consequently, system equilibrium conditions (Equation 4.3) become

$$dS = \sum_{j=1}^{m} \left(\left(\frac{1}{T_j} - \frac{1}{T_1} \right) dU_j + \left(\frac{P_j}{T_j} - \frac{P_1}{T_1} \right) dV_j \right.$$
$$\left. - \sum_{i=1}^{r} \left(\frac{\mu_i^j}{T_j} - \frac{\mu_i^1}{T_1} \right) dn_i^j \right) = 0. \tag{4.19}$$

As dS must be zero for any arbitrary set of dU_j, dV_j, and dn_i^j, we have the following equilibrium conditions:

$$T_1 = T_j = T \quad \forall j \quad \text{thermal equilibrium} \tag{4.20}$$

$$\frac{P_1}{T_1} = \frac{P_j}{T_j} \quad \Longrightarrow \quad P_1 = P_j = P \quad \forall j \quad \text{mechanical equilibrium} \tag{4.21}$$

$$\mu_i^1 = \mu_i^j \quad \Longrightarrow \quad \forall i, j \quad \text{chemical equilibrium.} \tag{4.22}$$

Equation 4.20 through Equation 4.22 express that an isolated, heterogeneous (simple) system made up by m phases and r components is in thermodynamic equilibrium when each phase is characterized by the same temperature and pressure, and the chemical potential of each component is the same in every phase system. On the basis of S and U properties, it can be demonstrated that the S maximum condition required for the system equilibrium translates into a U minimum condition and equilibrium conditions are always expressed by Equation 4.20 through Equation 4.22. Indeed, it is easy to verify that the substitution of these equations into Equation 4.14 yields $dU = 0$. Obviously, equilibrium conditions expressed by Equation 4.20 through Equation 4.22 descend from the implicit assumption that surface effects are negligible. When this is not the case, equilibrium conditions modify as discussed in Chapter 5 and Chapter 6.

4.2.2 GIBBS–DUHEM EQUATION

A function Y is said to be homogeneous of grade k with respect to its independent variables $X_1, X_2, \ldots, X_i, \ldots, X_r$, if the following relation holds whatever may be the value of the parameter λ:

$$Y(\lambda X_1, \lambda X_2, \ldots, \lambda X_i, \ldots, \lambda X_{r+2}) = \lambda^k Y(X_1, X_2, \ldots, X_i, \ldots, X_{r+2}). \quad (4.23)$$

An important property of homogeneous functions consists in the following relation:

$$\sum_{i=1}^{r+2} \left(\frac{\partial Y}{\partial (\lambda X_i)} \bigg|_{\lambda X_j} \frac{\partial (\lambda X_i)}{\partial \lambda} \bigg|_{X_i} \right) = \sum_{i=1}^{r+2} \left(\frac{\partial Y}{\partial (\lambda X_i)} \bigg|_{\lambda X_j} X_i \right) \underline{x} = k \lambda^{(k-1)} Y. \quad (4.24)$$

In particular, for $\lambda = 1$, Equation 4.24 represents the Euler equation. As U is a homogeneous function of grade 1 with respect to its independent variables (U is an extensive system property):

$$U(\lambda S, \lambda V, \lambda n_1, \ldots, \lambda n_i, \ldots, \lambda n_r) = \lambda U(S, V, n_1, \ldots, n_i, \ldots, n_r). \quad (4.25)$$

Euler equation for U becomes:

$$U = \left(\frac{\partial U}{\partial S} \right)_{V, n_1, \ldots, n_r} S + \left(\frac{\partial U}{\partial V} \right)_{S, n_1, \ldots, n_r} V + \sum_{i=1}^{r} \left(\frac{\partial U}{\partial n_i} \right)_{S, V, n_j \neq n_i} n_i \quad (4.26)$$

$$U = TS - PV + \sum_{i=1}^{r} \mu_i n_i. \quad (4.26a)$$

From Equation 4.26a, a relation between the intensive system parameters T, P, and μ_i can be found. Indeed, Equation 4.26a differentiation leads to

$$dU = TdS + SdT - PdV - VdP + \sum_{i=1}^{r} \mu_i \, dn_i + \sum_{i=1}^{r} n_i \, d\mu_i. \quad (4.27)$$

However, as this equation must be equal to

$$dU = TdS - PdV + \sum_{i=1}^{r} \mu_i \, dn_i, \quad (4.9)$$

it follows the well-known Gibbs–Duhem equation establishing a relation between the T, P, and μ_i

$$SdT - VdP + \sum_{i=1}^{r} n_i \, d\mu_i = 0. \qquad (4.28)$$

Accordingly, the degrees of freedom of a simple (homogeneous) system, composed by r components, is given by the sum of its intensive parameters (T, P, n_1, $n_2, \ldots, n_i, \ldots, n_r$ for a total of $r + 2$ variables) minus the Gibbs–Duhem equation for a net sum of $r + 1$. Consequently, a heterogeneous system made up by m phases is characterized by $m(r + 1)$ independent variables. Nevertheless, if equilibrium conditions hold between phases, $(m - 1)$ Equation 4.20 plus $(m - 1)$ Equation 4.21 plus $r(m - 1)$ Equation 4.22 can be written down so that system degrees of freedom F is given by

$$F = m(r + 1) - (m - 1)(r + 2) = r + 2 - m. \qquad (4.29)$$

4.2.3 Thermodynamic Potentials

As, sometimes, it cannot be so convenient to choose entropy S and volume V as independent variables, U partial Legendre transformations [17] enable to define other extensive thermodynamic potentials using different pairs of the four variables P, T, V, and S. Accordingly, the Helmholtz energy A is the U partial Legendre transformation substituting S with T:

$$A = A(T, V, n_1, n_2, \ldots, n_i, \ldots, n_r) \qquad (4.30)$$

$$A = U - TS \qquad (4.30')$$

$$dA = -SdT - PdV + \sum_{i=1}^{r} \mu_i \, dn_i. \qquad (4.30'')$$

Enthalpy H is the U partial Legendre transformation substituting V with P:

$$H = H(S, P, n_1, n_2, \ldots, n_i, \ldots, n_r) \qquad (4.31)$$

$$H = U + PV \qquad (4.31')$$

$$dH = TdS + VdP + \sum_{i=1}^{r} \mu_i \, dn_i. \qquad (4.31'')$$

Gibbs energy G is the U partial Legendre transformation substituting S with T and V with P:

$$G = G(T, P, n_1, n_2, \ldots, n_i, \ldots, n_r) \qquad (4.32)$$

$$G = U - TS + PV = H - TS = A + PV \qquad (4.32')$$

$$dG = -SdT + VdP + \sum_{i=1}^{r} \mu_i \, dn_i. \qquad (4.32'')$$

It can be demonstrated that the system equilibrium conditions requiring maximum S or minimum U also imply that A, H, and G get a minimum.

From the above relations regarding U, A, H, and G, the following identities hold:

$$T = \left(\frac{\partial U}{\partial S}\right)_{V, n_i} = \left(\frac{\partial H}{\partial S}\right)_{P, n_i} \qquad (4.33)$$

$$P = -\left(\frac{\partial U}{\partial V}\right)_{S, n_i} = -\left(\frac{\partial A}{\partial V}\right)_{T, n_i} \qquad (4.34)$$

$$V = \left(\frac{\partial G}{\partial P}\right)_{T, n_i} = \left(\frac{\partial H}{\partial P}\right)_{S, n_i} \qquad (4.35)$$

$$S = -\left(\frac{\partial G}{\partial T}\right)_{P, n_i} = -\left(\frac{\partial A}{\partial T}\right)_{V, n_i} \qquad (4.36)$$

$$\mu_i = \left(\frac{\partial U}{\partial n_i}\right)_{S, V, n_j} = \left(\frac{\partial H}{\partial n_i}\right)_{S, P, n_j} = \left(\frac{\partial G}{\partial n_i}\right)_{T, P, n_j} = \left(\frac{\partial A}{\partial n_i}\right)_{T, V, n_j}. \qquad (4.37)$$

Interestingly, Equation 4.37 clearly evidences that μ_i depends on composition, temperature, and pressure. As chemical potential has no immediate physical meaning, some auxiliary functions are needed. Consequently, Prausnitz et al. [18] introduced the concept of fugacity f_i and activity a_i of ith element according to the following relation:

$$\mu_i - \mu_i^0 = RT \ln\left(\frac{f_i}{f_i^0}\right) = RT \ln(a_i), \qquad (4.38)$$

where R is the universal gas constant while μ_i^0 and f_i^0 are, respectively, ith component chemical potential and fugacity in the reference state. a_i provides a measure of the difference between the component's chemical potential in the state of interest and that in its standard state.

4.3   KINETICS

Mass transport represents a reaction towards the perturbation of a system thermodynamic equilibrium state. Indeed, as soon as Equation 4.20 through

Equation 4.22 are no longer satisfied, in the attempt to restore a new equilibrium condition implying a maximum for S (Equation 4.3) or a minimum for U (Equation 4.14), system intensive and extensive parameters are modified and mass transport is a possible tool to perform this action. In order to have a precise physical idea of what happens after equilibrium perturbation, let us focus the attention on Figure 4.2 showing an isolated system (in thermodynamic equilibrium) made up by two phases (a, on the left; b, on the right) constituted by r components and separated by an ideal fixed membrane permeable to all components. Suppose now to increase the pressure in phase a so that $P_a > P_b$ with unchanged temperature ($T_a = T_b$) and components chemical potential ($\mu_i^a = \mu_i^b \ \forall i$). Among the many possible ways the system has to react to this perturbation, the simplest one consists in pressure reequilibration with no changes in temperature and chemical potential values [19]. This reequilibration implies mass motion from phase a to phase b through the permeable membrane until P_a again equals P_b. Mass flow due to a spatial pressure gradient is called *convection*. Analogously, system equilibrium can be broken by modifying (increasing concentration, for example) the chemical potential μ_i^a of only one component in phase a (component i) without altering all other parameters (temperature, pressure, and remaining components chemical potentials). Again, among the possible ways the system has to get a new equilibrium condition, the simplest one is to render μ_i^a equal to μ_i^b without altering all other parameters. Chemical potential reequilibration implies the movement of component i from phase a to phase b crossing the permeable membrane until $\mu_i^a = \mu_i^b$. In this case,

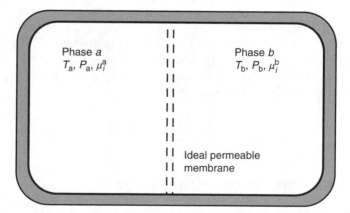

FIGURE 4.2 Isolated thermodynamic system made up by two phases a and b separated by an ideal membrane permeable to all r system components. T, P, and μ represent temperature, pressure, and chemical potential, respectively. (Adapted from Topp, E.M., *Transport Processes in Pharmaceutical Systems*, Amidon, G.L., Lee, I.L., and Topp, E.M., Eds., Marcell Dekker, New York, 2000, Chap. 1.)

however, mass transport is the result of the random, thermally induced, Brownian motion of i molecules as no pressure gradient exists to determine a convective flow. Mass flow induced by a chemical potential gradient is called *diffusion*.

4.3.1 MASS CONSERVATION LAW

According to the mass conservation law, mass can be neither created nor destroyed so that, in a fixed space region, mass can increase or decrease only by addition or subtraction from the surroundings, respectively. Although this law cannot be applied in the case of radioisotope decay or nuclear fission where mass converts into energy according to the well-known Einstein equation, in the typical pharmaceutical and engineering applications it holds and that is why it is widely used. In order to transform this law into equations, let us focus the attention on a generic flowing fluid composed by r components. In the above-mentioned fixed space region (that, for the sake of simplicity, can be thought as a parallelepiped (see Figure 4.3)), the conservation law translates into a mass balance that, for the generic ith fluid component can be expressed as

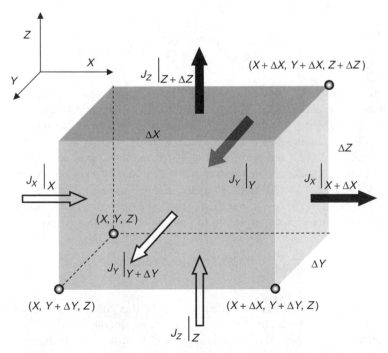

FIGURE 4.3 Fixed space region (parallelepiped) on which the mass balance is performed. J_X, J_Y, and J_Z represent, respectively, the X, Y, and Z mass flux components.

$$\begin{bmatrix} \text{rate of } i \text{ mass} \\ \text{change} \end{bmatrix} - \begin{bmatrix} \text{rate of } i \text{ gain} \\ \text{from surroundings} \end{bmatrix} \begin{bmatrix} \text{rate of } i \text{ loss} \\ \text{from surroundings} \end{bmatrix}$$
$$\pm \begin{bmatrix} \text{rate of } i \text{ gain or loss} \\ \text{by reaction} \end{bmatrix}, \qquad (4.39)$$

where the last term (called *generative term*) refers to a generic reaction that can, in principle, involves ith component and can cause i loss or gain depending on whether i is a reaction reactant or a product. Obviously, to solve the problem of mass transport, $(r-1)$ Equation 4.39 like equations plus an overall mass balance must be written as

$$\begin{bmatrix} \text{rate of total mass} \\ \text{change} \end{bmatrix} = \begin{bmatrix} \text{rate of total mass gain} \\ \text{from surroundings} \end{bmatrix} - \begin{bmatrix} \text{rate of total mass loss} \\ \text{from surroundings} \end{bmatrix},$$
$$(4.40)$$

where now, the generative term disappears as the chemical reaction can comport neither a net mass gain nor a net mass loss but only a fluid composition change. If ΔX, ΔY, and ΔZ represent our parallelepiped dimensions in the X, Y, and Z Cartesian axes directions, respectively, the first term of Equation 4.39 will be given by

$$\begin{bmatrix} \text{rate of } i \text{ mass} \\ \text{change} \end{bmatrix} = \frac{\partial C_i}{\partial t} \Delta X \, \Delta Y \, \Delta Z, \qquad (4.41)$$

where t is time, C_i is ith component concentration (mass/volume) in the volume $\Delta V = \Delta X * \Delta Y * \Delta Z$. The second term in Equation 4.39 can be computed remembering that the incoming mass flow is given by the sum of mass normally (i.e., in the direction of surface normal) entering the three surfaces ΔS_{YZ} (area $= \Delta Y * \Delta Z$ and positioned in $X = X$; normal X axis, see Figure 4.3), ΔS_{XZ} (area $= \Delta X * \Delta Z$ and positioned in $Y = Y$; normal Y axis, see Figure 4.3), and ΔS_{XY} (area $= \Delta X * \Delta Y$ and positioned in $Z = Z$; normal Z axis, see Figure 4.3). Accordingly, if J_X, J_Y, and J_Z represent the three Cartesian components of the mass flux vector J_i (referred to the ith component), we have

$$\begin{bmatrix} \text{rate of } i \text{ mass gain} \\ \text{from surroundings} \end{bmatrix} = \Delta Y \Delta Z J_{iX}|_X + \Delta X \Delta Z J_{iY}|_Y + \Delta X \Delta Y J_{iZ}|_Z, \quad (4.42)$$

where the subscripts X, Y, and Z denote that the flux components J_{iX}, J_{iY}, and J_{iZ} are evaluated, respectively, at X, Y, and Z. Analogous considerations can be drawn for the third term in Equation 4.39 with the only difference that, now, we are dealing with an exiting mass flow across the three surfaces ΔS_{YZ},

ΔS_{XZ}, and ΔS_{XY} positioned in $(X + \Delta X)$, $(Y + \Delta Y)$, and $(Z + \Delta Z)$, respectively (see Figure 4.3). Accordingly, it follows:

$$\begin{bmatrix} \text{rate of } i \text{ total mass loss} \\ \text{from surroundings} \end{bmatrix} = \Delta Y \Delta Z J_{iX}|_{X+\Delta X} + \Delta X \Delta Z J_{iY}|_{Y+\Delta Y} + \Delta X \Delta Y J_{iZ}|_{Z+\Delta Z}.$$

(4.43)

As the explicit expression of the fourth term in Equation 4.39 would require exact knowledge of the chemical reaction involving the ith component, for the moment, it will be simply indicated as $G_i*\Delta V$, where G_i indicates the mass of i generated or destroyed per unit time and volume in virtue of the chemical reaction. Finally, Equation 4.39 becomes

$$\frac{\partial C_i}{\partial t} \Delta V = \Delta Y \Delta Z (J_{iX}|_X - J_{iX}|_{X+\Delta X}) + \Delta X \Delta Z (J_{iY}|_Y - J_{iY}|_{Y+\Delta Y})$$

$$+ \Delta X \Delta Y (J_{iZ}|_Z - J_{iZ}|_{Z+\Delta Z}) + G_i \Delta V.$$

(4.44)

If both the left- and right-hand side terms are divided for ΔV and the limit ΔX, ΔY, and $\Delta Z \longrightarrow 0$ is considered, on the basis of partial derivates definition, it follows:

$$\frac{\partial C_i}{\partial t} = -\left(\frac{\partial J_{iX}}{\partial X} + \frac{\partial J_{iY}}{\partial Y} + \frac{\partial J_{iZ}}{\partial Z} \right) + G_i.$$

(4.45)

To make Equation 4.45 operative, the mass flux components J_{iX}, J_{iY}, and J_{iZ} must be specified. Remembering that in Section 4.3 it was observed that mass transport occurs according to *convection* (mass flow is caused by a pressure gradient) or *diffusion* (mass flow is caused by a chemical potential gradient), it is natural expressing J_{iX}, J_{iY}, and J_{iZ} in the following manner:

$$J_{iX} = C_i v_X - C_i B_i \frac{\partial \mu_i}{\partial X}, \quad J_{iY} = C_i v_Y - C_i B_i \frac{\partial \mu_i}{\partial Y},$$

$$J_{iZ} = C_i v_Z - C_i B_i \frac{\partial \mu_i}{\partial Z},$$

(4.46)

where v_X, v_Y, and v_Z are, respectively, the X, Y, and Z components of the fluid average ponderal velocity vector v, while B_i is a coefficient reflecting the mobility of the diffusing components [20]. While the terms $C_i * v_{X,Y,Z}$ clearly indicate the convection contribute, the terms containing the partial derivative of the chemical potential represent the contribution of diffusion (the sign minus indicates that diffusion takes place in the direction of the negative chemical potential gradient). Assuming ideal solution conditions, the contribution of diffusion can be replaced by the more usual Fick's law [21] according to which Equation 4.46 becomes

$$J_{iX} = C_i v_X - \rho D_{im} \frac{\partial \omega_i}{\partial X}, \quad J_{iY} = C_i v_Y - \rho D_{im} \frac{\partial \omega_i}{\partial Y},$$

$$J_{iZ} = C_i v_Z - \rho D_{im} \frac{\partial \omega_i}{\partial Z}, \tag{4.47}$$

where ρ is fluid density, D_{im} represents the diffusion coefficient of the ith component in the mixture (our fluid) while ω_i indicates the ith component mass fraction defined as the ratio between the i mass and the mixture mass contained in the volume ΔV when ΔX, ΔY, and $\Delta Z \longrightarrow 0$. Remembering the relation between C_i and ω_i:

$$\omega_i = \frac{C_i}{\rho}. \tag{4.48}$$

Equation 4.47 can be expressed explicitly in terms of C_i instead of ω_i:

$$J_{iX} = C_i v_X - D_{im} \left(\frac{\partial C_i}{\partial X} - \frac{C_i}{\rho} \frac{\partial \rho}{\partial X} \right) J_{iY} = C_i v_Y - D_{im} \left(\frac{\partial C_i}{\partial Y} - \frac{C_i}{\rho} \frac{\partial \rho}{\partial Y} \right)$$

$$J_{iZ} = C_i v_Z - D_{im} \left(\frac{\partial C_i}{\partial Z} - \frac{C_i}{\rho} \frac{\partial \rho}{\partial Z} \right). \tag{4.49}$$

Notably, diffusion depends also on density gradient and not only on concentration gradient. Recalling the definition of the nabla operator vector ∇ (dimensionally a length^{-1}) and its quadratic expression (dimensionally a length^{-2}) in Cartesian coordinates:

$$\nabla = \frac{\partial}{\partial X} \underline{i} + \frac{\partial}{\partial Y} \underline{j} + \frac{\partial}{\partial Z} \underline{k} \quad \nabla^2 = \frac{\partial^2}{\partial X^2} \underline{i} + \frac{\partial^2}{\partial Y^2} \underline{j} + \frac{\partial^2}{\partial Z^2} \underline{k}, \tag{4.50}$$

where $\underline{i}$, $\underline{j}$, and $\underline{k}$ are, respectively, the X, Y, and Z axes versors (unitary vectors defining direction and versus), Equation 4.45 can be written in a more compact form:

$$\frac{\partial C_i}{\partial t} = -(\nabla \cdot J_i) + G_i = -(\nabla \cdot C_i v) + \nabla \cdot (D_{im} \nabla \omega_i) + G_i, \tag{4.45'}$$

where $D_{im} \nabla \omega_i = D_{im} \frac{\partial \omega_i}{\partial X} \underline{i} + D_{im} \frac{\partial \omega_i}{\partial Y} \underline{j} + D_{im} \frac{\partial \omega_i}{\partial Z} \underline{k}$ while $\cdot$ indicates the scalar product between vectors. In the case of constant diffusivity D_{im}, Equation 4.45' becomes

$$\frac{\partial C_i}{\partial t} = -(\nabla \cdot J_i) + G_i = -(\nabla \cdot C_i v) + D_{im} \nabla^2 \omega_i + G_i. \tag{4.45''}$$

Apart from compactness, the advantage of expressing Equation 4.45 into Equation 4.45' form consists in the possibility of using it not only in Cartesian coordinates X, Y, and Z, but also in cylindrical and spherical coordinates once the proper ∇ and ∇^2 expression is considered [21]:

$$\nabla = \frac{1}{R}\frac{\partial}{\partial R}\left(R\frac{\partial}{\partial R}()\right)\underline{r} + \frac{1}{R}\frac{\partial}{\partial\theta}\underline{t} + \frac{\partial}{\partial Z}\underline{k} \quad \text{(cylinder)}, \tag{4.51}$$

$$\nabla^2 = \frac{1}{R}\frac{\partial}{\partial R}\left(R\frac{\partial}{\partial R}()\right)\underline{r} + \frac{1}{R^2}\frac{\partial^2}{\partial\theta^2}\underline{t} + \frac{\partial^2}{\partial Z^2}\underline{k} \quad \text{(cylinder)}, \tag{4.51'}$$

$$\nabla^2 = \frac{1}{R^2}\frac{\partial}{\partial R}\left(R\frac{\partial}{\partial R}()\right)\underline{r} + \frac{1}{R\sin(\theta)}\frac{\partial}{\partial\theta}(\sin(\theta))\underline{t} + \frac{1}{R\sin(\theta)}\frac{\partial}{\partial\phi}\underline{p} \quad \text{(sphere)}, \tag{4.52}$$

$$\nabla^2 = \frac{1}{R^2}\frac{\partial}{\partial R}\left(R^2\frac{\partial}{\partial R}()\right)\underline{r} + \frac{1}{R^2\sin(\theta)}\frac{\partial}{\partial\theta}\left(\sin(\theta)\frac{\partial}{\partial\theta}()\right)\underline{t} + \frac{1}{R^2\sin^2(\theta)}\frac{\partial^2}{\partial\phi^2}\underline{p} \quad \text{(sphere)}, \tag{4.52'}$$

where R is the radial coordinate (versor $\underline{r}$), θ is the tangential coordinate (versor $\underline{t}$), and φ is the angular coordinate (versor $\underline{p}$).

In order to write down the mass balance referred to the whole flowing fluid, it is sufficient to sum up Equation 4.45' for the r component remembering that

$$\sum_{i=1}^{r} C_i = \rho \tag{4.53}$$

$$\sum_{i=1}^{r} D_{i\text{m}}\nabla\omega_i - 0 \tag{4.54}$$

$$\sum_{i=1}^{r} G_i = 0. \tag{4.55}$$

Equation 4.53 states that fluid density is the sum of each component concentration, Equation 4.54 affirms that no net fluid motion can take place due to diffusion while Equation 4.55 makes clear that fluid mass cannot modify because of a whatever chemical reaction. Accordingly, the well-known *continuity equation* is found

$$\frac{\partial\rho}{\partial t} = -(\nabla \cdot \rho v). \tag{4.56}$$

Notably, this equation can be also derived starting from an overall mass balance similar to that expressed by Equation 4.45 assuming, $G_i = 0$, $J_{iX} = C_i\,v_X$, $J_{iY} = C_i\,v_Y$, and $J_{iZ} = C_i\,v_Z$.

To conclude this paragraph, it is important to remember that mass diffusion is not only due to the presence of the concentration gradient (see Equation 4.46 or Equation 4.47), but it can also happen because of temperature or pressure gradients. Indeed, as mass diffusion is because of the chemical potential gradient, which, in turn, is also a function of temperature and pressure (see Equation 4.8 and Equation 4.37), it is reasonable that mass diffusion takes place because of concentration, temperature, or pressure gradient. In addition, other gradients can be responsible for mass diffusion. Indeed, the presence of an electrical potential gradient provokes ions diffusion [19]. In this case, it is convenient to define the electrochemical potential $\tilde{\mu}_i$ as follows:

$$\tilde{\mu}_i = \mu_i + z_i F \psi, \tag{4.57}$$

where z_i is the net charge on species i, F is Faraday constant, and ψ is the electrostatic potential. Starting from this definition, after some manipulations, it is possible to derive the well-known Nerst–Plank equation:

$$J_i = -D_i \nabla C_i - z_i C_i \frac{F}{RT} \nabla \psi, \tag{4.58}$$

where J_i is the ith component flux, R and T are, respectively, universal gas constant and absolute temperature. It is clear that the first and second terms on the right represent, respectively, the Fickian and the electrostatic potential gradient contributions to diffusion.

4.3.2 MOMENTUM CONSERVATION LAW

Although r kinetics equations are now available, the problem of getting components profile concentration in the flowing fluid is not yet solved as the knowledge of the fluid velocity field is required. Indeed, Equation 4.45′ makes clear (see the term $(\nabla \cdot C_i \, v)$) that the time and space variation of the fluid velocity vector v must be known to get C_i space and time dependence $\forall \, i$. To solve this problem, recourse must be made to a particular force balance. It is well known that the velocity v of a body of mass m (for the sake of simplicity here thought of spherical shape) changes with time due to the action of all the forces acting on it according to Newton's law:

$$ma = m\frac{dv}{dt} = \sum F_i, \tag{4.59}$$

where a is body acceleration and F_i is the generic force acting on the body. In these terms, Equation 4.59 holds if m is a constant. When, on the contrary, m is time dependent, Equation 4.59 has to be generalized. Apart from relativistic effects taking place near speed of light, m can be time dependent also in much more common situations. At this purpose, let us focus the attention on a cubic vessel containing an amount M of sand and sliding (see Figure 4.4) on an inclined plane. If at time t_0 a hole is opened on the vessel back face, sand will progressively go out and M decreases with time causing a system mass m (sand plus vessel mass) decrease. In order to describe this situation, Equation 4.59 must be generalized asserting that the vector $\boldsymbol{p} = m\boldsymbol{v}$, called *linear momentum* or *simply momentum*, changes because of the action of all the forces acting on the vessel and because of sand leaving from the hole. Accordingly, subject of our attention is no longer body velocity ($\boldsymbol{v}$) but body momentum ($\boldsymbol{p} = m\boldsymbol{v}$). This scenario can be further complicated imagining that vessel motion takes place under a sand rain. Accordingly, if on one hand, m tends to decrease for sand escaping from the hole, on the other hand, m tends to increase due to sand coming in through the upper open vessel surface (see Figure 4.4). In conclusion, the time variation of system momentum $\left(\dfrac{\mathrm{d}\boldsymbol{p}}{\mathrm{d}t} = \dfrac{\mathrm{d}m\boldsymbol{v}}{\mathrm{d}t} \right)$ depends on the sum of forces acting on the body plus the incoming and escaping momentum due to in- and out-sand flow rates, respectively. This simple example can be used to understand what happens to a fluid element (that, again, can be thought as a parallelepiped of volume ΔV

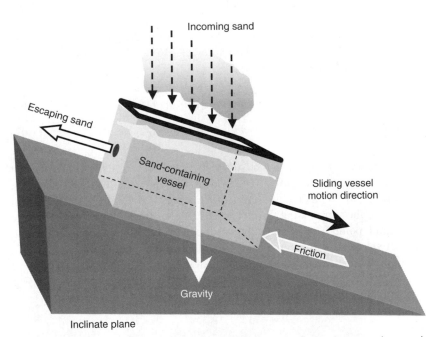

Incoming sand

Escaping sand

Sand-containing vessel

Sliding vessel motion direction

Friction

Gravity

Inclinate plane

FIGURE 4.4 Sliding vessel momentum depends on both the forces acting on it (gravity and friction) and on its mass variation due to escaping and incoming sand.

and dimensions ΔX, ΔY, and ΔZ) of a flowing fluid. Indeed, as in the vessel example, the variation of the fluid element momentum $(\Delta V \frac{\partial \rho v}{\partial t})$ is due to the action of all the forces acting on it, namely surface and body forces, plus the difference between the incoming and exiting momentum across parallelepiped surfaces. If the expression of surface and body forces is not so complicated, as shown later, much more care must be devoted to the expression of the inflow and outflow momentum. Indeed, as momentum is a vector and it is characterized by three components, it is convenient, for the moment, to limit our considerations to its X component per unit volume (the scalar quantity ρv_X can be thought as the momentum X component concentration per unit volume), being the following argumentations extensible to the other remaining two components, namely Y (ρv_Y) and Z (ρv_Z). The advantage of this strategy relies on the possibility of dealing with a balance regarding a scalar quantity (ρv_X) and not a vectorial one (ρv). Accordingly, as in to the case of the mass conservation law (paragraph 4.3.1), we are again dealing with the expression of a scalar quantity (ρv_X) flux through a parallelepiped. Consequently, the net balance through the two surfaces (ΔS_{YZ}) of normal X and placed in X and $X + \Delta X$, reads

$$\Delta Y \Delta Z([(\rho v_X)v_X]_X - [(\rho v_X)v_X]_{X+\Delta X}). \tag{4.60}$$

Similarly, the balances through the two surfaces (ΔS_{XZ}) of normal Y and placed in Y and $Y + \Delta Y$ and the two surfaces (ΔS_{XY}) of normal Z and placed in Z and $Z + \Delta Z$ read, respectively,

$$\Delta X \Delta Z([(\rho v_X)v_Y]_Y - [(\rho v_X)v_Y]_{Y+\Delta Y}), \tag{4.61}$$

$$\Delta X \Delta Y([(\rho v_X)v_Z]_Z - [(\rho v_X)v_Z]_{Z+\Delta Z}). \tag{4.62}$$

Thus, the sum of Equation 4.60 through Equation 4.62 constitutes the net balance of the momentum X component through parallelepiped surfaces. Now, let us focus the attention on surface forces acting on the fluid element and due to the surrounding fluid elements. As the stress (force/surface) state in a fluid point L is completely described if three stress vectors acting, respectively, on three mutual perpendicular planes passing through L [22] are known, nine scalars (three for each stress vector) are needed to know the stress state in L. In other words, we need the knowledge of the stress tensor $\underline{\tau}$, whose mathematical expression is

$$\underline{\tau} = \begin{bmatrix} \tau_{XX} & \tau_{YX} & \tau_{ZX} \\ \tau_{XY} & \tau_{YY} & \tau_{ZY} \\ \tau_{XZ} & \tau_{YZ} & \tau_{ZZ} \end{bmatrix}, \tag{4.63}$$

where τ_{ij} represents the generic tensor element corresponding to the component in the j direction of the stress vector acting on a plane of normal i. Thus, for example, τ_{XX}, τ_{YX}, and τ_{ZX} represent the X component of the stress vectors acting on a plane whose normal is the X, Y, and Z axis, respectively (see Figure 4.5). It is important to underline that for a resting fluid $\tau_{ij} = 0 \; \forall i, j$, this meaning that $\underline{\tau}$ is associated exclusively to fluid motion. In addition, it can be demonstrated [22] that $\underline{\tau}$ is symmetric, this reflecting in the following relation $\tau_{ij} = \tau_{ji}$ for $i \neq j$. In order to be homogeneous with the strategy adopted in writing Equation 4.60 through Equation 4.62, also in this case only the X component of the three stress vectors will be considered (namely, τ_{XX}, τ_{YX}, and τ_{ZX}). Accordingly, the sum of the X component of the surface forces reads (see Figure 4.5)

$$\Delta Y \Delta Z (\tau_{XX}|_X - \tau_{XX}|_{X+\Delta X}) + \Delta X \Delta Z (\tau_{YX}|_Y - \tau_{YX}|_{Y+\Delta Y})$$
$$+ \Delta X \Delta Y (\tau_{ZX}|_Z - \tau_{ZX}|_{Z+\Delta Z}). \tag{4.64}$$

Among surface forces, of course, the effect of hydrostatic pressure P has to be considered. The sum of the X component of surface forces due to P is given by

$$\Delta Y \Delta Z (P|_X - P|_{X+\Delta X}). \tag{4.65}$$

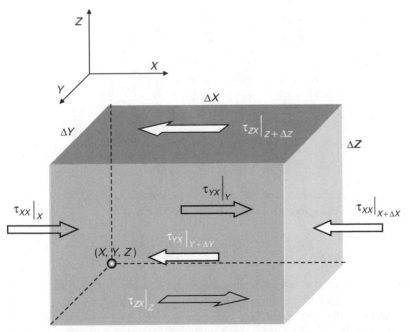

FIGURE 4.5 Representation of the stress tensor X components. Arrows indicate stresses caused by adjacent surrounding fluid elements.

Again, it is important to underline that this surface force does not depend on fluid motion and it is also present in a resting fluid. Finally, the most common body force acting on the fluid element under examination is gravity g, although other body forces due to the presence of electric and magnetic fields could be similarly considered. The X component of gravity force reads

$$\Delta X \Delta Y \Delta Z \rho g_X. \tag{4.66}$$

Equating the X component of the fluid element momentum variation $\left(\Delta V \frac{\partial \rho v_X}{\partial t}\right)$ to the sum of Equation 4.60 through Equation 4.62 and Equation 4.64 through Equation 4.66, dividing both the right- and left-hand side terms for ΔV and considering the limit ΔX, ΔY, and $\Delta Z \longrightarrow 0$, we get the X component of the generalized form of Equation 4.59:

$$\frac{\partial}{\partial t}(\rho v_X) = -\left(\frac{\partial}{\partial X}(\rho v_X v_X) + \frac{\partial}{\partial Y}(\rho v_X v_Y) + \frac{\partial}{\partial Z}(\rho v_X v_Z)\right)$$
$$-\left(\frac{\partial \tau_{XX}}{\partial X} + \frac{\partial \tau_{YX}}{\partial Y} + \frac{\partial \tau_{ZX}}{\partial Z}\right) - \frac{\partial P}{\partial X} + \rho g_X. \tag{4.67}$$

Analogously, for the Y and Z components, we have, respectively,

$$\frac{\partial}{\partial t}(\rho v_Y) = -\left(\frac{\partial}{\partial X}(\rho v_Y v_X) + \frac{\partial}{\partial Y}(\rho v_Y v_Y) + \frac{\partial}{\partial Z}(\rho v_Y v_Z)\right)$$
$$-\left(\frac{\partial \tau_{XY}}{\partial X} + \frac{\partial \tau_{YY}}{\partial Y} + \frac{\partial \tau_{ZY}}{\partial Z}\right) - \frac{\partial P}{\partial Y} + \rho g_Y, \tag{4.68}$$

$$\frac{\partial}{\partial t}(\rho v_Z) = -\left(\frac{\partial}{\partial X}(\rho v_Z v_X) + \frac{\partial}{\partial Y}(\rho v_Z v_Y) + \frac{\partial}{\partial Z}(\rho v_Z v_Z)\right)$$
$$-\left(\frac{\partial \tau_{XZ}}{\partial X} + \frac{\partial \tau_{YZ}}{\partial Y} + \frac{\partial \tau_{ZZ}}{\partial Z}\right) - \frac{\partial P}{\partial Z} + \rho g_Z. \tag{4.69}$$

As for the mass balance, also in this case it is possible to give a more compact expression of Equation 4.67 through Equation 4.69:

$$\frac{\partial}{\partial t}(\rho v) = -(\nabla \cdot \rho v v) - (\nabla \cdot \underline{\tau}) - \nabla P + \rho g, \tag{4.70}$$

where the left-hand side term is the fluid element momentum (per unit volume) time variation, while the first, second, third, and fourth term on the right are, respectively, the net momentum variation rate due to incoming and exiting mass, the sum of surface forces due to fluid motion and hydrostatic pressure, and the effect of gravitational field. Equation 4.70 is also referred to as *momentum conservation law* as it states that the rate of momentum variation is due to the sum of forces acting on the fluid element plus the net momentum balance due to the entering and exiting mass through fluid element surfaces.

The quantity ρvv is a tensor whose expression is

$$\rho vv = \rho \begin{bmatrix} v_X v_X & v_X v_Y & v_X v_Z \\ v_Y v_X & v_Y v_Y & v_Y v_Z \\ v_Z v_X & v_Z v_Y & v_Z v_Z \end{bmatrix}. \tag{4.71}$$

Remembering the definition of substantive derivative $D(-)/Dt$:

$$\frac{D}{Dt}(-) = \frac{\partial}{\partial t}(-) + (v \cdot \nabla(-)) \tag{4.72}$$

and after some manipulations, Equation 4.70 can be recast in the following form:

$$\rho \frac{D}{Dt}(v) = -(\nabla \cdot \underline{\tau}) - \nabla P + \rho g. \tag{4.70'}$$

Notably, this equation is formally equal to Equation 4.59 as it states that the product of mass (per unit volume) and acceleration is equal to the sum of all the forces acting on the fluid element. In other words, Equation 4.59 structure can be maintained also for the description of a fluid element motion provided that traditional acceleration is substituted by a "substantive" acceleration.

In order to render Equation 4.70 operative, it is necessary to define the relation between the stress tensor components and the velocity gradient. For a Newtonian fluid, in Cartesian coordinates we have

$$\tau_{XX} = -2\eta \frac{\partial v_X}{\partial X} + \left(\frac{2}{3}\eta - \kappa\right)(\nabla \cdot v), \tag{4.73}$$

$$\tau_{YY} = -2\eta \frac{\partial v_Y}{\partial Y} + \left(\frac{2}{3}\eta - \kappa\right)(\nabla \cdot v), \tag{4.74}$$

$$\tau_{ZZ} = -2\eta \frac{\partial v_Z}{\partial Z} + \left(\frac{2}{3}\eta - \kappa\right)(\nabla \cdot v), \tag{4.75}$$

$$\tau_{XY} = \tau_{YX} = -\eta \left(\frac{\partial v_X}{\partial Y} + \frac{\partial v_Y}{\partial X}\right), \tag{4.76}$$

$$\tau_{YZ} = \tau_{ZY} = -\eta \left(\frac{\partial v_Y}{\partial Z} + \frac{\partial v_Z}{\partial Y}\right), \tag{4.77}$$

$$\tau_{ZX} = \tau_{XZ} = -\eta \left(\frac{\partial v_Z}{\partial X} + \frac{\partial v_X}{\partial Z}\right), \tag{4.78}$$

where η is fluid shear viscosity, while κ is mass viscosity that for gas and liquid is usually close to zero and, thus, can be neglected [21]. Equation 4.73 through Equation 4.75 make clear that, for incompressible fluids ($(\nabla \cdot v) = 0$), τ_{xx}, τ_{yy}, and τ_{zz} are nonzero only when solid boundaries impose fluid acceleration or deceleration. This is the typical situation occurring when the fluid flows in a variable diameter tube.

Without going in further details, it is sufficient an inspection of Equation 4.73 through Equation 4.78 to clearly understand that the stress tensor, and thus velocity, depends on one important fluid property, namely shear viscosity. In turn, as shear viscosity is temperature dependent (this is true regardless we are dealing with Newtonian or more complex fluids; see Chapter 3), velocity, stress tensor and thus mass transport depend also on temperature. In addition, fluid density also depends on temperature. Accordingly, at least theoretically, the r mass balance equations (Equations 4.46) and the motion equation (Equation 4.70) must be coupled to an energy balance.

4.3.3 ENERGY CONSERVATION LAW

The fluid occupying the usual element of volume ΔV and dimensions ΔX, ΔY, and ΔZ possess an amount of energy that can be split into two contributes: the kinetic one, due to fluid motion, and the internal one due to molecules chaotic Brownian movements, defined U in paragraph 1.2. Accordingly, the energy balance on ΔV reads

$$
\begin{bmatrix} \text{rate of } internal \\ \text{and } kinetic \\ \text{energy increase} \end{bmatrix} = \begin{bmatrix} \text{rate of } internal \\ \text{and } kinetic \\ \text{energy gain} \\ \text{due to} \\ \text{convection} \end{bmatrix} - \begin{bmatrix} \text{rate of } internal \\ \text{and } kinetic \\ \text{energy loss} \\ \text{due to} \\ \text{convection} \end{bmatrix}
$$

$$
+ \begin{bmatrix} \text{net rate of} \\ heat \text{ received} \\ \text{due to} \\ \text{conduction} \end{bmatrix} - \begin{bmatrix} \text{power dissipated} \\ \text{against the} \\ \text{surrounding} \end{bmatrix} . \quad (4.79)
$$

This is nothing more than the first thermodynamic principle for an open system in a variable regime [21]. Of course, this balance is incomplete as, for example, some kinds of energy and mechanism of energy transport, such as nuclear, irradiative, and electromagnetic, have been neglected for the sake of simplicity. Nevertheless, its formulation can be easily modified to account for these other energy typologies. The rate of internal and kinetic energy increase (or decrease) can be expressed as

$$
\Delta X \Delta Y \Delta Z \frac{\partial}{\partial t} \left(\rho \hat{U} + \frac{1}{2} \rho v^2 \right), \quad (4.80)
$$

where $\hat{U}$ represents the fluid internal energy per unit mass while v is the local fluid velocity module. While the net internal and kinetic energy income due to convection is expressed by

$$\Delta Y \Delta Z \left[v_X \left(\rho \hat{U} + \frac{1}{2} \rho v^2 \right) \Big|_X - v_X \left(\rho \hat{U} + \frac{1}{2} \rho v^2 \right) \Big|_{X+\Delta X} \right] +$$

$$+ \Delta X \Delta Z \left[v_Y \left(\rho \hat{U} + \frac{1}{2} \rho v^2 \right) \Big|_Y - v_Y \left(\rho \hat{U} + \frac{1}{2} \rho v^2 \right) \Big|_{Y+\Delta Y} \right] +$$

$$+ \Delta X \Delta Y \left[v_z Z \left(\rho \hat{U} + \frac{1}{2} \rho v^2 \right) \Big|_Z - v_Z \left(\rho \hat{U} + \frac{1}{2} \rho v^2 \right) \Big|_{Z+\Delta Z} \right], \qquad (4.81)$$

the net energy income due to conduction reads

$$\Delta Y \Delta Z (q_X|_X - q_X|_{X+\Delta X}) + \Delta X \Delta Z (q_Y|_Y - q_Y|_{Y+\Delta Y}) + \Delta X \Delta Y (q_Z|_Z - q_Z|_{Z+\Delta Z}), \qquad (4.82)$$

where q_X, q_Y, and q_Z are, respectively, the X, Y, and Z components of the conduction energy flux $\boldsymbol{q}$ whose components can be expressed by the well-known Fourier law:

$$q_X = -k \frac{\partial T}{\partial X}, \quad q_Y = -k \frac{\partial T}{\partial Y}, \quad q_Z = -k \frac{\partial T}{\partial Z}, \qquad (4.83)$$

where k is fluid conductivity. In its motion, the fluid makes work per unit time (power) against body and surface forces. As, for the sake of simplicity, only gravity has been considered among the possible body forces, the fluid power dissipated for this reason is

$$-\rho \Delta X \Delta Y \Delta Z (v_X g_X + v_Y g_Y + v_Z g_Z), \qquad (4.84)$$

where the minus reminds that the work per unit time is made against the gravity field when $\boldsymbol{v}$ and $\boldsymbol{g}$ have opposite versus. The power dissipated due to surface forces regards pressure P and viscous forces (represented by the stress tensor $\underline{\tau}$) acting on fluid element faces. For what concerns P, we have

$$- \Delta Y \Delta Z ((P v_X)|_X - ((P v_X)|_{X+\Delta X}) - \Delta X \Delta Z ((P v_Y)|_Y - (P v_Y)|_{Y+\Delta Y})$$
$$- \Delta X \Delta Y ((P v_Z)|_Z - (P v_Z)|_{Z+\Delta Z}). \qquad (4.85)$$

Similarly, for viscous forces, the power dissipated reads

$$- \Delta Y \Delta Z \left((\tau_{XX} v_X + \tau_{XY} v_Y + \tau_{XZ} v_Z)|_X - (\tau_{XX} v_X + \tau_{XY} v_Y + \tau_{XZ} v_Z)|_{X+\Delta X} \right) +$$
$$- \Delta X \Delta Z \left((\tau_{YX} v_X + \tau_{YY} v_Y + \tau_{YZ} v_Z)|_Y - (\tau_{YX} v_X + \tau_{YY} v_Y + \tau_{YZ} v_Z)|_{Y+\Delta Y} \right) +$$
$$- \Delta X \Delta Y \left((\tau_{ZX} v_X + \tau_{ZY} v_Y + \tau_{ZZ} v_Z)|_Z - (\tau_{ZX} v_X + \tau_{ZY} v_Y + \tau_{ZZ} v_Z)|_{Z+\Delta Z} \right).$$

$$(4.86)$$

Substituting previous expressions into Equation 4.80 and dividing both members for ΔV in the limit ΔX, ΔY, and $\Delta Z \longrightarrow 0$, the equation energy is found

$$\frac{\partial}{\partial t}\left(\rho \hat{U} + \frac{1}{2}\rho v^2\right) = -\left[\frac{\partial}{\partial X}\left(v_X\left(\rho \hat{U} + \frac{1}{2}\rho v^2\right)\right) + \frac{\partial}{\partial Y}\left(v_Y\left(\rho \hat{U} + \frac{1}{2}\rho v^2\right)\right)\right.$$
$$+ \frac{\partial}{\partial Z}\left(v_Z\left(\rho \hat{U} + \frac{1}{2}\rho v^2\right)\right)\Bigg] +$$
$$- \left(\frac{\partial q_X}{\partial X} + \frac{\partial q_Y}{\partial Y} + \frac{\partial q_Z}{\partial Z}\right) + \rho(v_X g_X + v_Y g_Y + v_Z g_Z) +$$
$$- \left(\frac{\partial P v_X}{\partial X} + \frac{\partial P v_Y}{\partial Y} + \frac{\partial P v_Z}{\partial Z}\right) +$$
$$- \left[\frac{\partial}{\partial X}(\tau_{XX} v_X + \tau_{XY} v_Y + \tau_{XZ} v_Z) + \frac{\partial}{\partial Y}(\tau_{YX} v_X \right.$$
$$+ \tau_{YY} v_Y + \tau_{YZ} v_Z)$$
$$\left. + \frac{\partial}{\partial Z}(\tau_{ZX} v_X + \tau_{ZY} v_Y + \tau_{ZZ} v_Z)\right]. \qquad (4.87)$$

Obviously, also this equation can be rewritten in a more compact way

$$\frac{\partial}{\partial t}\left(\rho \hat{U} + \frac{1}{2}\rho v^2\right) = -\left(\nabla \cdot \left[\rho v\left(\hat{U} + \frac{1}{2}v^2\right)\right]\right) - (\nabla \cdot q) + \rho(v \cdot g)$$
$$- (\nabla \cdot Pv) - (\nabla \cdot [\underline{\tau} \cdot v]), \qquad (4.88)$$

where the term on the left represents the rate of internal and kinetic energy (per unit volume) increase (ore decrease), while the right terms are, respectively, the internal and kinetic energy (per unit volume) income rate due to convection and conduction minus the work (per unit time and volume) made by the fluid element against gravity, hydrostatic pressure, and viscous forces. Recalling the continuity and momentum equations, after some algebraic manipulations that are out of the aim of this chapter, Equation 4.88 can be rewritten in a more useful form, i.e., explicitly as function of temperature

$$\rho \hat{C}_p \frac{DT}{Dt} = -(\nabla \cdot q) - \left(\frac{\partial \ln(\rho)}{\partial \ln(T)}\right)_P \frac{DP}{Dt} - [\underline{\tau} : \nabla v], \qquad (4.88')$$

where $\hat{C}_p \left(= \left(\dfrac{\partial \hat{H}}{\partial T} \right)_P \right)$ is fluid specific heat per unit mass, $\hat{H}$ represents fluid enthalpy per unit mass, $-(\nabla \cdot \boldsymbol{q}) = k\nabla^2 T$ according to Fourier law and $[\underline{\tau}: \nabla \boldsymbol{v}]$ represents fluid heating due to viscous friction and its expression is

$$[\underline{\tau}: \nabla \boldsymbol{v}] = \left(\tau_{XX} \frac{\partial v_X}{\partial X} + \tau_{YY} \frac{\partial v_Y}{\partial Y} + \tau_{ZZ} \frac{\partial v_Z}{\partial Z} \right) +$$
$$\left(\tau_{XY} \left[\frac{\partial v_X}{\partial Y} + \frac{\partial v_Y}{\partial X} \right] + \tau_{XZ} \left[\frac{\partial v_X}{\partial Z} + \frac{\partial v_Z}{\partial X} \right] + \tau_{YZ} \left[\frac{\partial v_Y}{\partial Z} + \frac{\partial v_Z}{\partial Y} \right] \right). \quad (4.89)$$

In the case of a Newtonian fluid, the stress tensor elements are expressed by Equation 4.73 through Equation 4.78 and $[\underline{\tau}: \nabla \boldsymbol{v}]$ becomes

$$[\underline{\tau}: \nabla \boldsymbol{v}] = \eta \Phi_v, \quad (4.90)$$

$$\Phi_v = 2 \left[\left(\frac{\partial v_X}{\partial X} \right)^2 + \left(\frac{\partial v_Y}{\partial Y} \right)^2 + \left(\frac{\partial v_Z}{\partial Z} \right)^2 \right] + \left[\left(\frac{\partial v_X}{\partial Y} + \frac{\partial v_Y}{\partial X} \right)^2 + \left(\frac{\partial v_X}{\partial Z} + \frac{\partial v_Z}{\partial X} \right)^2 \right.$$
$$\left. + \left(\frac{\partial v_Y}{\partial Z} + \frac{\partial v_Z}{\partial Y} \right)^2 \right] - \frac{2}{3} \left[\frac{\partial v_X}{\partial X} + \frac{\partial v_Y}{\partial Y} + \frac{\partial v_Z}{\partial Z} \right]^2, \quad (4.91)$$

where Φ_v is the so-called *dissipation function*. Notably, Equation 4.88' states that the temperature of a fluid element moving with fluid flow is due to heat conduction $(-(\nabla \cdot \boldsymbol{q}))$, expansion effects $\left(-\left(\dfrac{\partial \ln (\rho)}{\partial \ln (T)} \right)_P \dfrac{DP}{Dt} \right)$, and viscous heating $(-[\underline{\tau}: \nabla \boldsymbol{v}])$.

4.3.4 FINAL CONSIDERATIONS

On the bases of what was discussed in Section 4.3.1 through Section 4.3.3, the concentration profile determination of the r elements constituting our flowing fluid requires to solve r equations expressing *mass conservation law* (Equation 4.45'), three equations expressing *momentum conservation law* (Equation 4.67 through Equation 4.69) and one equation expressing *energy conservation law* (Equation 4.88'). At this purpose, we need some more information regarding fluid properties dependence on pressure (P), temperature (T), and composition (C_i). In other words, we need the knowledge of the state equation for fluid density $(\rho = \rho(P, T, C_i))$, shear viscosity $(\eta = \eta(P, T, C_i))$, specific heat at constant pressure $(\hat{C}_p = \hat{C}_p(P, T, C_i))$, conductivity $(k = k(P, T, C_i))$, and diffusion coefficients $(D_i = D_i(P, T, C_i))$. Once the initial and boundary conditions are defined, the above-mentioned equations must be simultaneously solved to get solution. For the sake of simplicity, in this section, equations have been essentially derived and discussed referring to Cartesian coordinates system. However, due the vectorial form in which they have been presented, they can be easily adapted to cylindrical and spherical coordinates systems (that can be very useful in practical applications) once the expressions of the various operators are known in each particular coordinates system. The explicit expressions of these equations in cylindrical and spherical coordinates systems can be found in specific books [21].

4.4 FICK'S EQUATION

Despite the theoretical complexity connected to mass transport, fortunately, in the majority of pharmaceutical applications, the scenario can be considerably simplified as equations can be solved separately. In addition, in several cases, while energy balance is not required as isothermal conditions can be assumed, momentum conservation law becomes trivial as a uniform and constant velocity field realizes. Indeed, a typical drug delivery system is made up by three components: a matrix structure (that does not diffuse and, hence, its diffusion coefficient is zero), water (coming in from the external environment and moving inside matrix structure), and drug (that, usually, diffuses from the inner matrix to the external release environment). Accordingly, the problem of drug release can be matched considering only two equations of Equation 4.45′ type. Finally, when water absorption is not needed for drug release or it is very fast in comparison to drug diffusion, only one equation is required. Consequently, the equation under attention derives from Equation 4.45′ assuming a vanishing velocity field ($\boldsymbol{v} = 0$) and thus, according to mass conservation law (Equation 4.56), a constant and uniform system density so that $\nabla C_i = \nabla \omega_i$:

$$\frac{\partial C_i}{\partial t} = \nabla \cdot (D_{im} \nabla C_i) + G_i, \tag{4.92}$$

where Fick's law for the matter flux is assumed. For constant diffusivity, Equation 4.92 becomes

$$\frac{\partial C_i}{\partial t} = D_{im} \nabla^2 C_i + G_i. \tag{4.92′}$$

4.4.1 DIFFUSION COEFFICIENT

One of the most important parameters involved in Equation 4.92 (the others are connected to the generative term G_i) is surely the diffusion coefficient D_{im}. However, until now, nothing about this parameter has been specified if not a generic definition. It is now necessary to be more precise and to provide some more information on this topic. For this purpose, let us focus the attention on a fluid mixture composed by only two components named i and j, respectively [23], the following considerations are easily extensible to a mixture composed by r components. Once an appropriate frame of reference is chosen, this system may be described in terms of the *mutual diffusion coefficient* D_{ij} (diffusivity of i in j and vice versa). Unfortunately, however, unless i and j molecules are identical in mass and size, mobility of i is different with respect to that of j. This leads to a hydrostatic pressure gradient compensated by a bulk flow (convective contribution to species transport) of the i–j mixture. Consequently, the mutual diffusion coefficient is the combined result of the bulk flow and the molecules random motion. Accordingly, an *intrinsic diffusion coefficient* (D_i and D_j) can be defined to account only for molecules random motion. Finally, using radioactively labeled molecules, it is possible to observe the rate of diffusion of i in a mixture, composed by labeled and not labeled i molecules, where uniform chemical composition is attained. In doing so, the *self-diffusion coefficient* (D_i*) can be defined. It is possible to verify that, theoretically, both D_i and D_i* are concentration and temperature dependent. Indeed, the force f acting on an i molecule at point X is proportional to the i chemical potential gradient [20]

$$f \propto -\nabla \mu_i. \tag{4.93}$$

Consequently, the total force f_T acting on all molecules is proportional to

$$f_T \propto -C_i \nabla \mu_i, \tag{4.94}$$

where C_i represents i concentration. Assuming that the flux F_i is proportional to the total force, we have

$$F_i = -\frac{C_i}{\sigma_i \eta} \nabla \mu_i, \tag{4.95}$$

where $\sigma_i \eta$ is a resistance coefficient connected with diffusing molecules mobility [20,23]. On the basis of Equation 4.38 we have

$$F_i = -\frac{RT}{\sigma_i \eta} \frac{d(\ln(a_i))}{d(\ln(C_i))} \nabla C_i. \tag{4.96}$$

Accordingly, D_i expression is

$$D_i = \frac{RT}{\sigma_i \eta} \frac{d(\ln(a_i))}{d(\ln(C_i))}. \tag{4.97}$$

This equation clearly states that D_i is temperature and concentration dependent. Only in the case of an ideal solution, the concentration dependence disappears as a_i coincides with C_i. An analog dissertation developed on D_i^* leads to

$$D_i^* = \frac{RT}{\sigma_i \eta'} \tag{4.98}$$

as, obviously, now ideal solution conditions hold. If $\sigma_i \eta$ and $\sigma_i \eta'$ can be retained equal, we have

$$D_i = D_i^* \frac{d(\ln(a_i))}{d(\ln(C_i))}. \tag{4.99}$$

4.4.1.1   Diffusion in Solid or Solid-Like Medium

In the case of diffusion in a solid phase or in a solid-like phase such as gels (polymeric networks entrapping a solvent), it is convenient to assume the solid, stationary phase as a fixed reference and consider only diffusing solutes as mobile components [20]. In this light, the distinction between mutual and intrinsic diffusion coefficient is no longer needed and we can simply speak about the diffusion coefficient of a solute in a known solid or semisolid structure where Equation 4.97 is still valid. To further particularize the scenario, it is important to remember that, as mentioned before, a typical drug delivery system can be thought as a three components system (or, at least, it can be idealized as a three components system). The first one is the fixed structure that, usually, identifies with a three-dimensional polymeric network stabilized by the presence of junction zones (called crosslinks and that can be of physical or chemical nature) among different polymeric chains. Obviously, its diffusion coefficient is zero. The second component is a solvent unable to disassemble network structure because of the crosslinks presence. The last is the pharmacological active principle, namely the drug. Of course, for the sake of clarity, at this stage, the presence and the influence of possible excipients are neglected. As a consequence, we are interested in three diffusion coefficients, namely that of the solvent phase in the dry polymeric network D_s, that of drug in the pure solvent phase D_0 (this is the mutual drug diffusion coefficient in the pure solvent) and, finally, that of the drug in the polymeric network–solvent system D [24]. Indeed, although more complicated situations can take place as discussed in Chapter 7, usually, the interest is focused on dry polymeric network swelling by means of an external solvent and on the diffusion of drug from an already swollen polymeric network.

Regardless the diffusion coefficient we are dealing with ($D_s, D_0,$ and D), its evaluation is usually performed recurring to the *hydrodynamic, kinetic,* and *statistical mechanical* theories. While the first two approaches can be classified as *molecular theories*, as they are mathematical expressions of the system (liquid, polymeric solution, or crosslinked network) physical view at the molecular scale, the last one can be defined as *simulation theories* as they are based on atomistic simulations and they do not yield to a mathematical equation (or set of equations) for diffusion coefficient evaluation. While, theoretically, this last category is the most promising for what concerns diffusion coefficient prediction, due to huge computational resources needed, at present, it is applicable only to very simple molecules [25]. In addition, till now, no rigorous theoretical approach exists for the description of liquids and dense fluids; this is particularly true for polymeric solutions and swellable crosslinked polymeric networks [26]. This is the reason why it is usual to study liquids extending, in a semitheoretical way, the above-mentioned approaches [27].

According to the most famous hydrodynamic theory, the well-known Stokes–Einstein equation, D_0 can be evaluated according to

$$D_0 = \frac{KT}{6\pi\eta R_H},$$ (4.100)

where K is the Boltzman constant, T is absolute temperature, η is solvent (continuum medium) shear viscosity, and R_H is the solute molecules hydrodynamic radius. This equation holds for large, spherical shaped solute molecules immersed in a continuous solvent provided that the resulting solution is a diluted solution. For nonspherical polyatomic molecules (freely jointed Lennard–Jones (LJ) chain fluids), Reis et al. [28] propose to evaluate D_0 combining the kinetic theory of Chapman–Enskog model for dense fluid [29] with van der Waals mixing rules [30]. Accordingly, Chapman–Enskog equation for self-diffusion coefficient D_{00}^* is considered

$$D_0^* = \frac{D_{00}^*}{g(\rho)}, \quad D_{00}^* = \frac{3}{8\rho\sigma^2}\sqrt{\frac{KT}{\pi m}},$$ (4.101)

where $g(\rho)$ is the radial distribution function for hard spheres at the contact point, ρ is number density, D_{00}^* is the low-density diffusion coefficient, σ is the molecular diameter, and m is the molecular mass. Assuming that the volume of polyatomic molecules, containing N segments of mass m_{seg}, is given by $N\frac{4}{3}\pi\left(\frac{d}{2}\right)^3 = \frac{4}{3}\pi\left(\frac{\sigma}{2}\right)^3$, where d is segment diameter, yields to the conclusion that the corresponding molecular diameter $\sigma = N^{1/3}d$ [31]. Upon substitution in Equation 4.101, it follows:

$$D_0^* = \frac{D_{00}^*}{g(\rho)}, \quad D_{00}^* = \frac{3}{8\rho d^2 N^{\frac{2}{3}}}\sqrt{\frac{KT}{\pi m}} \quad m = Nm_{seg}.$$ (4.101′)

Finally, Reis et al. [26], working on 96 self-diffusion coefficient for chains length of 2, 4, 8, and 16, generalize Equation 4.101' for freely jointed LJ chain fluids:

$$D_0^* = \frac{D_{00}^*}{\left(\dfrac{g(\rho)}{R(\rho^*,T^*)F(N,\rho^*,T^*)} + \dfrac{0.04N^2}{(T^*)^{1.5}}\right)}, \quad D_{00}^* = \frac{3}{8\rho d^2 N^{\frac{2}{3}}}\sqrt{\frac{KT}{\pi m}}, \quad (4.102)$$

where $\rho^*(=\rho N\sigma^3)$ and $T^*(=KT/\varepsilon)$ are, respectively, the reduced density and temperature, σ and ε assume the meaning of LJ potential parameters, while R and F are, respectively, correction factors accounting for high density and chain connectivity and are defined by

$$R = \left(1 - \frac{\rho^*}{1.12(T^*)^{0.2}}\right)\left[1 + 0.97\sqrt{\rho^*} + 5.1(\rho^*)^2 + \right.$$
$$\left.\left(\frac{3.1\rho^* - 2.9\sqrt{\rho^*}}{(T^*)^{0.2}}\right)\right]e^{\left(-\frac{\rho^*}{2T^*}\right)},$$

$$(4.103)$$

$$\ln(F) = -0.018(N-1) - (1+1.05\sqrt{T^*})\left(\frac{N-1}{N}\right)\rho^* + \frac{2.09\left(\dfrac{N-1}{N}\right)\rho^*}{\sqrt{1 + \sqrt{\dfrac{T^*}{0.527}}}}.$$

$$(4.104)$$

Correctly, when $N=1$, $F=1$ (no connectivity at all) and when ρ^* is small (this happens for low-density systems and it is exalted at high temperature T^*) $R \longrightarrow 1$. At the same time, for $N=1$, the contribution of the term $(0.04 N^2/(T^*)^{1.5})$ becomes negligible. As a consequence, for $N=1$ and at low density, Equation 4.102 reduces to Equation 4.101, g is given by

$$g = \frac{1 - 0.5\theta}{(1-\theta)^3}, \quad \theta = (\pi/6)\rho N\sigma^3, \quad (4.105)$$

where θ is the packing fraction for disconnected spheres. Equation 4.103 through Equation 4.105 have been determined in the range $0.1 < \rho^* < 0.9$ and $1.5 < T^* < 4$. The extension of Equation 4.102 to binary mixtures for the estimation of the mutual diffusion coefficient D_0, implies assuming van der Waals mixing rules [30]

$$\rho^* = \rho_1 N_1 \sigma_1^3 + \rho_2 N_2 \sigma_2^3, \quad (4.106)$$

$$\varepsilon = \omega_1^2 \varepsilon_1 + \omega_2^2 \varepsilon_2 + 2\omega_1\omega_2\varepsilon_{12} \quad \varepsilon_{12} = \sqrt{\varepsilon_1\varepsilon_2} \quad (4.107)$$

$$N = \omega_1^2 N_1 + \omega_2^2 N_2 + 2\omega_1\omega_2 N_{12}, \quad N_{12} = \left(\frac{N_1^{1/3}\sigma_1 + N_2^{1/3}\sigma_2}{2\sigma_{12}}\right)^3,$$

$$\sigma_{12} = \frac{\sigma_1 + \sigma_2}{2} \tag{4.108}$$

$$m_{12} = \frac{m_1 m_2}{m_1 + m_2}, \tag{4.109}$$

where subscripts 1 and 2 refer, respectively, to solute and solvent and ω represents components mass fraction, replacing, in Equation 4.102, D_{00}^* with the mutual diffusion coefficient at law density D_{00}, given by

$$D_{00} = \frac{3}{8\rho\sigma_{12}^2 N_{12}^{2/3}}\sqrt{\frac{KT}{2\pi m_{12}}} \tag{4.110}$$

and assuming the following expression for g [32]:

$$g = \frac{1}{1 - \zeta_3} + \frac{3\sigma_1\sigma_2}{\sigma_1 + \sigma_2}\frac{\zeta_2}{(1 - \zeta_3)^2} + 2\left(\frac{\sigma_1\sigma_2}{\sigma_1 + \sigma_2}\right)^2\frac{\zeta_2^2}{(1 - \zeta_3)^3}, \tag{4.111}$$

where

$$\zeta_2 = \frac{\pi}{6}\rho\sum_{i=1}^{2}\omega_i N_i\sigma_i^2, \quad \zeta_3 = \frac{\pi}{6}\rho\sum_{i=1}^{2}\omega_i N_i\sigma_i^3. \tag{4.112}$$

According to the authors [28], this approach seems to be a promising tool for the correlation of mutual diffusion coefficients for both small–molecules systems and polymer–solvent systems.

The theoretical estimation of D must account for the fact that polymer chains slow down solute movements by acting as physical obstructions thereby increasing the path length of the solute, by increasing the hydrodynamic drag experienced by the solute and by reducing the average free volume per molecule available to the solute. Accordingly, in addition to *hydrodynamic* and *kinetic* theories, *obstruction* theory has been proposed [33]. Models based on obstruction theory assume that the presence of impenetrable polymer chains causes an increase in the path length for diffusive transport. Polymer chains, acting as a sieve, allow the passage of only sufficiently small solutes. Carman [34], schematizing the network as an interconnected bundle of tortuous cylindrical capillaries with constant cross section, demonstrates that D is given by

$$\frac{D}{D_0} = \left(\frac{1}{\tau}\right)^2, \tag{4.113}$$

where τ is tortousity, defined as the mean increase of the diffusion path due to the presence of obstructions. For solute molecule of the same size as polymer segments, Muhr and Blanshard [35], assuming a lattice model for the water–polymer hydrogel where polymer occupies a fraction φ of the whole site and, thus, solute transport occurs only within the free sites, suggest

$$\frac{D}{D_0} = \left(\frac{1-\varphi}{1+\varphi}\right)^2, \tag{4.114}$$

where φ represents also the polymer volume fraction. When solute molecules are much bigger than polymer segments, the Ogston approach has to be considered [36]. He assumes that solute diffusion occurs by a succession of directionally random unit steps whose execution takes place on condition that the solute does not meet a polymer chain. While solute is assumed to be a hard sphere, the crosslinked polymer is thought as a random network of straight long fibers of vanishing width. The unit step length coincides with the root-mean-square average diameter of spherical spaces residing between the network fibers. The resulting model is

$$\frac{D}{D_0} = e^{\left(-\frac{r_s+r_f}{r_f}\varphi^{1/2}\right)}, \tag{4.115}$$

where r_s and r_f are, respectively, solute and fiber radius. Phillips et al. [37] schematize the network as an ensemble of parallel fibers each one constituted by a series of nontangent consecutive spheres of diameter r_f. Applying the dispersional theory of Taylor to this network, they estimate the ratio D/D_0 in the case of solute diffusion perpendicular with respect to fibers direction. The result of this analysis is approximately given by

$$\frac{D}{D_0} = e^{(-\alpha\varphi^{1/2})}, \tag{4.116}$$

$$\alpha = 5.1768 - 4.0075\lambda + 5.4388\lambda^2 - 0.6081\lambda^3, \text{ where } \lambda = \frac{r_s}{r_f}. \tag{4.117}$$

Amsden [33] assumes that solute movement through the polymeric network is a stochastic process. Motion occurs through paths constituted by a succession of network openings large enough to allow solute molecule transit (openings must be larger than solute hydrodynamic radius). In addition, supposing that openings size distribution can be described by Ogston's equation relative to straight, randomly oriented polymer fibers [38], he gets

$$\frac{D}{D_0} = e^{\left(-\frac{\pi}{4}\left(\frac{r_s+r_f}{\bar{r}+r_f}\right)^2\right)},$$ (4.118)

where $\bar{r}$ is opening average radius that is related to the average end-to-end distance between polymer chains ξ according to

$$\bar{r} = 0.5\xi = 0.5k_s\varphi^{-0.5},$$ (4.119)

where k_s is a constant for a given polymer–solvent couple. In virtue of the straight polymer hypothesis, this model is applicable to network characterized by strong crosslinks typical of chemically crosslinked polymeric networks. Nevertheless, by using scaling laws for the description of the average distance between polymer chains, it is possible to render the model suitable also for weakly crosslinked network as it happens for physically crosslinked polymeric networks. In this case, indeed, mesh openings are neither constant in size nor location as, on the contrary, happens for strongly crosslinked polymeric network (see Chapter 7 for a more detailed discussion about chemically and physically crosslinked polymeric networks).

Hydrodynamic theory assumes that solute mobility, and thus its diffusion coefficient, depends on the frictional drag exerted by liquid phase molecules entrapped in the network [21]. In particular, polymer chains are seen as centers of hydrodynamic resistance as they reduce the mobility of the liquid phase and this, in turn, reflects in an increased drag effect exerted by liquid phase molecules on solute. The starting point of this theory is the Stokes–Einstein equation (Equation 4.100):

$$D_0 = \frac{KT}{f} = \frac{KT}{6\pi\eta R_H},$$ (4.100')

where f is the friction drag coefficient. Accordingly, this category of models is concerned with the calculation of f. For strongly crosslinked gels (rigid polymeric chains), Cukier [39] suggests

$$\frac{D}{D_0} = e^{\left(-\left(\frac{3\pi L_c N_A}{M_f \ln(L_c/2r_f)}\right)r_s\varphi^{1/2}\right)},$$ (4.120)

where L_c and M_f are, respectively, the polymer chains length and molecular weight, N_A is Avogadro number, and r_f is the polymer fiber radius. For weakly crosslinked gels (flexible polymeric chains), the same author proposes

$$\frac{D}{D_0} = e^{(-k_c r_s\varphi^{0.75})},$$ (4.121)

where k_c is a parameter depending on the polymer–solvent system.

In order to diffuse, a solute molecule must acquire the energy necessary to win the attraction forces exerted by surrounding solvent molecules. In this manner, it can jump into adjacent voids formed in the liquid space due to liquid molecules thermal motion. According to kinetic *Eyring theory* [21], the most important step is the first one, while, for the kinetic *free volume theory* [40], voids formation is the rate determining step. According to Eyring theory, the expression of the solute diffusion coefficient D_0 in a pure liquid reads

$$D_0 = \lambda^2 \underline{k}, \qquad (4.122)$$

where λ is the mean diffusive jump length and $\underline{k}$ is the jump frequency defined by

$$\underline{k} = \frac{KT}{\sqrt{2\pi m_r KT}} V_f^{-1/3} e^{\left(-\frac{\varepsilon}{KT}\right)}, \qquad (4.123)$$

where K is Boltzman constant, m_r is the solvent–solute couple reduced mass, V_f is the mean free volume available per solute molecule while ε is an energy per molecule representing the difference, between the energy molecule in the activated state and that at 0 K. The extension of Equation 4.123 and Equation 4.122 to the case of solute diffusion in a swollen polymeric network reads [35]

$$\frac{D}{D_0} = \left(\frac{\lambda'}{\lambda}\right)^2 \left(\frac{V_f}{V_f'}\right)^{\frac{1}{3}} e^{\left(\frac{\varepsilon - \varepsilon'}{KT}\right)}, \qquad (4.124)$$

where superscript refers to solvent–polymer properties. Unfortunately, the difficulty of parameters estimation makes this equation not so useful in practical applications.

According to the kinetic *free volume theory*, solute diffusion depends on jumping distance, solute thermal velocity and on the probability that there is an adjacent void (free volume) sufficiently large to host solute molecule. Physically speaking, the free volume of a solvent coincides with the difference between solvent volume (evaluated at fixed pressure and temperature) and the volume occupied by all its molecules perfectly packed to be tangent each other. The probability p_h that a sufficiently large void forms in the proximity of the diffusing solute molecule is given by

$$p_h = e^{\left(-\gamma \frac{V^*}{V_f}\right)}, \qquad (4.125)$$

where V_f is the mean free volume available per solute molecule, $V*$ is the critical local hole free volume required for a solute molecule to jump into and

γ is a numerical factor used to correct for overlap of free volume available to more than one molecule ($0.5 \leq \gamma \leq 1$). Accordingly, we have

$$D_0 \propto v_T \lambda e^{\left(-\gamma \frac{V^*}{V_f}\right)}, \qquad (4.126)$$

where v_T is solute thermal velocity and λ is jump length. Assuming negligible mixing effects, the free volume V_f of a mixture composed by solvent, polymer, and drug is given by

$$V_f = V_{fd}\omega_d + V_{fs}\omega_s + V_{fp}\omega_p, \qquad (4.127)$$

where V_{fd}, V_{fs}, and V_{fp} represent, respectively, drug, solvent, and polymer free volume, while ω_d, ω_s, and ω_p are, respectively, drug, solvent, and polymer mass fraction. Starting from Equation 4.126 and Equation 4.127, Muhr and Blanshard [35], for small polymer volume fraction φ, find the following relation:

$$\frac{D}{D_0} = e^{\left(\frac{1}{P - \frac{q}{\varphi}}\right)}, \qquad (4.128)$$

where P and q are φ independent parameters. Always for small φ values, Peppas and Reinhart [41], resorting to the free volume theory, arrive to

$$\frac{D}{D_0} = k_1 \left(\frac{\overline{M}_c - \overline{M}_c^*}{\overline{M}_n - \overline{M}_c^*}\right) e^{\left(-k_2 r_s^2 \left(\frac{\varphi}{1-\varphi}\right)\right)}, \qquad (4.129)$$

where k_1 and k_2 are two constants, $\overline{M}_c$ is the number average molecular weight between polymer crosslinks, $\overline{M}_n$ is the number average molecular weight of the uncrosslinked polymer, $\overline{M}_c^*$ is a critical molecular weight between crosslinks, φ is polymer volume fraction, and r_s is solute radius. Lustig and Peppas [42], introducing the idea of the scaling correlation length between crosslinks ζ, suppose that solute molecules can move inside the three-dimensional network only if $r_s < \zeta$. Accordingly, on the basis of the free volume theory and assuming $(1 - r_s/\zeta)$ as sieve factor, they suggest the following model:

$$\frac{D}{D_0} = \left(1 - \frac{r_s}{\zeta}\right) e^{\left(-Y\left(\frac{\varphi}{1-\varphi}\right)\right)}, \qquad (4.130)$$

where $Y = \gamma \pi \lambda r_s^2 / V_{fs}$. Consequently, Y represents the ratio between γV^* (critical local hole free volume required for a solute molecule to jump into times γ) and the average free volume per molecule of solvent. The same

TABLE 4.1
Diffusion Coefficient D_0 in Water and Radius r_s of Some Solutes

Solute	$D_0'' \times 10^6$ (cm²/sec)	T (°C)	r_s (Å)
Urea	18.1	37	1.9
Glucose	6.4	23	3.6
Theophylline	8.2	37	3.9
Sucrose	7.0	37	4.8
Caffeine	6.3	37	5.3
Phenylpropanolamine	5.5	37	6.0
Vitamin B_{12}	3.8	37	8.6
PEG 326	4.9	25	7.5
PEG 1118	2.8	25	13.1
PEG 2834	1.8	25	20.4
PEG 3978	1.5	25	24.5
Ribonuclease	0.13	20	16.3
Myoglobin	0.11	20	18.9
Lysozyme	0.11	20	19.1
Pepsin	0.09	20	23.8
Ovalbumin	0.07	20	29.3
Bovine Serum Albumin	0.06	20	36.3
Immunoglobulin G	0.04	20	56.3
Fibrinogen	0.02	20	107

Source: Adapted from Amsden, B., *Macromolecules*, 31, 8382, 1998.

authors suggest that, for correlation purposes, Y can be considered equal to 1. According to Amsden [33], free volume and hydrodynamic theories should be used to deal with weakly crosslinked networks, while for strongly crosslinked networks obstruction theory is more consistent with the experimental data.

Table 4.1, showing some examples of solute radius r_s and D_0 (in water) values, makes clear that, usually, the larger the solute molecule, the lower the corresponding D_0. However, it can be seen that chemical properties also play an important role as, for example, PEG 3978 is characterized by a D_0 that is one order of magnitude higher than those corresponding to smaller solutes (ribonuclease, myoglobin, lysozyme, and pepsin). In addition, Table 4.2 and Table 4.3 show the best fitting results obtained by considering three models deriving, respectively, from the hydrodynamic, free volume, and obstruction theory. While hydrodynamic (Equation 4.121) and free volume (Equation 4.130) models are tested on weakly (physically) crosslinked network, the obstruction one (Equation 4.118) is tested on strongly (chemically) cross-linked network. While in the first case (hydrodynamic and free volume models) polymer volume fraction φ ranges between 0 and 0.5, in the second

TABLE 4.2
Equation 4.121 and Equation 4.130 Best Fitting (Fitting Parameters k_c and k_2, Respectively) on Experimental Data Referred to Different Polymers and Solutes (Polymer Concentration φ Is the Independent Variable). Equation 4.121 Best Fitting Is Performed Assuming $r_s \ll \zeta$ and $Y = k_2 \times r_s^2$

Polymer	Solute	k_c (Å^{-1})	r_s (Å)	k_2 (Å^{-2})	r_s (Å)
		Hydrodynamic Theory (Equation 4.121)		Free Volume Theory (Equation 4.130)	
PAAM	Urea	1.12	1.9	0.774	1.9
	Sucrose	1.06	4.75	0.281	4.75
	Ribonuclease	0.55	16.3	0.060	16.6
	Bovin Serum Albumin	0.45	36.3	0.023	36.3
Dextran	Lysozyme	0.57	19.1	0.038	19.4
	Bovin Serum Albumin	0.58	36.3	0.021	36.3
	Immunoglobulin G	0.66	56.3	0.016	56.5
PVA	Vitamin B$_{12}$	0.62	8.7	0.061	8.7
	Lysozyme	0.40	19.1	0.044	19.4
PEO	Caffeine	0.88	5.25	0.179	5.25
PHEMA	Phenylpropanolamine	1.10	6.0	0.081	6.0

Source: Adapted from Amsden, B., *Macromolecules*, 31, 8382, 1998.
Note: PAAM, polyacrylamide; PVA, polyvinylalcohol; PEO, polyethyleneoxide; PHEMA, polyhydroxyethylmethacrylate.

TABLE 4.3
Equation 4.118 Best Fitting (Fitting Parameter k_s) on Experimental Data Referred to Different Polymers and Solutes (Polymer Concentration φ Is the Independent Variable). Fitting Is Performed Assuming $r_f = 8$ Å

Polymer	Solute	k_s (Å)	r_s (Å)
		Obstruction Theory (Equation 4.118)	
Alginate	Bovin Serum Albumin	5.73	36.3
Agarose	Myoglobin	11.63	18.9
	Bovin Serum Albumin	12.45	36.3

Source: Adapted from Amsden, B., *Macromolecules*, 31, 8382, 1998.

one, φ ranges in a smaller range ($0 < \varphi < 0.06$). As for Equation 4.121, Equation 4.130, and Equation 4.120 the correlation coefficient ranges, respectively, between 0.87 and 0.99, 0.75 and 0.99, and 0.77 and 0.98, a reasonably good data fitting is achieved for all models.

Duda et al. [45] apply the free volume theory to the polymer–solvent mixtures assuming temperature constant thermal expansion coefficients, no mixing effects (solvent and polymer specific volumes concentration independent), that solvent chemical potential μ_s is given by the Flory theory [44] (see also Chapter 7, Section 7.3.5.1):

$$\mu_s = \mu_s^0 + RT[\ln(1 - \varphi) + \varphi + \chi\varphi^2],\tag{4.131}$$

where μ_s^0 is the solvent chemical potential in the reference state and χ is the Flory interaction parameter, and that the following relations hold:

$$D_s = \frac{D_{ss}\rho_s}{RT}\left(\frac{\partial\mu_s}{\partial\rho_s}\right)_{T,P}\tag{4.132}$$

$$D_{ss} = D_{0s}e^{\left(-\gamma\frac{\omega_s V_s^* + \omega_p V_p^* \xi}{V_{FH}}\right)}, \quad D_{0s} = D_{0ss}e^{\left(-\frac{E}{RT}\right)},\tag{4.133}$$

where ρ_s, μ_s, ω_s, and V_s^* are, respectively, solvent density, chemical potential, mass fraction, and specific critical free volume, ω_p and V_P^* are, respectively, polymer mass fraction and specific critical free volume, D_{0ss} is a preexponential factor, γ is a numerical factor used to correct for overlap of free volume available to more than one molecule ($0.5 \leq \gamma \leq 1$), V_{FH} is the specific polymer–solvent mixture average free volume while ξ is the ratio between the solvent and polymer jump unit critical molar volume (for small solvents, the jump unit coincides with solvent molecule, while polymer jump unit coincides with the smallest polymer chain rigid segment that, sometimes, can be the monomeric unit). On the basis of these assumptions, they get

$$D_s = (1 - \varphi)^2(1 - 2\chi\varphi)D_{0s}e^{\left(-\frac{\omega_s V_s^* + \omega_p V_p^* \xi}{V_{FH}/\gamma}\right)},\tag{4.134}$$

where

$$\omega_s = \frac{\rho_s\varphi}{\rho_s(1 - \varphi) + \rho_p\varphi} \quad \omega_p = 1 - \omega_s\tag{4.135}$$

$$\frac{V_{FH}}{\gamma} = \frac{K_{11}}{\gamma}\omega_1(K_{21} + T - T_{g1}) + \frac{K_{12}}{\gamma}\omega_2(K_{22} + T - T_{g2}),\tag{4.136}$$

where Equation 4.136 parameters (K_{11}/γ, K_{12}/γ, ($K_{21}-T_{g1}$) and ($K_{22}-T_{g2}$)), for several polymer–solvent systems, can be found in literature [43,45,46]. Grassi et al. [47], studying drug release from swellable polyvynilpirrolidone, apply a simplified form of Equation 4.134 and they find, on data fitting basis, that water diffusion coefficient in the polymeric matrix ranges between 10^{-10} (dry state) and 10^{-7} cm^2/sec (swollen state). Although, nowadays, free volume theory is the most used approach, its predictive ability, over the entire range of polymer mass fraction (this means that we span from glassy to rubbery polymer), is limited to a qualitative data description rather than to a quantitative one [28]. The main reason for this relies on the fact that free volume theory does not account for the real mechanism of solvent diffusion, at least, in glassy systems. Indeed, atomistic simulation makes clear that diffusion takes place because of solvent molecule jump from a sort of polymer pocket to another one due to the formation of a neck between the two pockets. Neck opening occurs on a random fluctuation basis [25]. The "typical jump" model [48] is the result of a molecular theory accounting for this deduction. Of the two model variants, the linear elastic solid version seems the most promising. This theory, based on transition state approach, accounts for polymer and solvent interaction potentials and polymeric matrix elastic properties. In particular, it supposes that polymer behaves like linear elastic solid in the proximity of the deformation necessary for formation, with the same materials constants as the bulk (macroscopic) solid. The resulting expression for the diffusion coefficient is

$$D_s = \frac{1}{6}k_{jump}L^2, \quad k_{jump} = \frac{KT}{h}\frac{Q^+}{Q}e^{-\left(\frac{E_0}{KT}\right)}, \tag{4.137}$$

where k_{jump} and L are, respectively, jump rate and length, Q^+/Q is the ratio between the partition function of the transition and reactant state, E_0 is a critical energy, T is temperature while K and h are Boltzman and Planck constant, respectively. While this model yields good predictions for what concerns small nonpolar solvents, it fails for larger solvents.

4.4.2 FICKIAN KINETICS

If the generative term G_i in Equation 4.92 is zero and D_{im} is constant ($= D_{d0}$), the resulting equation reads

$$\frac{\partial C_i}{\partial t} = -D_{d0}\nabla\cdot(-\nabla C_i) = D_{d0}\nabla^2 C_i, \tag{4.138}$$

where for the sake of simplicity, C_i will be named simply C. This equation is also referred to as *Fick's second law* while Fick's first law states the

proportionality between mass flux and concentration gradient ($\boldsymbol{J} = - D_{d0}\nabla C_i$). Although Equation 4.138 is the result of drastic simplifications that rarely find a match in real experimental conditions, its solution provides useful didactic information about drug release kinetics, although referred to as an idealized situation. At this purpose, let us focus the attention on a slab, of cross-section area S and thickness L faced on one side to a release environment characterized by an infinite volume (this means that drug concentration is always 0 in the release environment. These are the so-called "*sink conditions*"). In addition, let us suppose that the slab is uniformly loaded by the drug at a concentration C_0 and that diffusion takes place only in the X direction. These conditions translate into the following mathematical equations:
initial conditions ($t = 0$)

$$C = C_0, \quad 0 < X < L \quad \text{uniform drug distribution in the slab,} \qquad (4.139)$$

boundary conditions ($t > 0$)

$$\left. \frac{\partial C}{\partial X} \right|_{X=0} = 0 \quad \text{no drug flux on the slab side in } X = 0, \qquad (4.140)$$

$$C = 0, \quad X = L \quad \text{infinite release environment volume.} \qquad (4.141)$$

Equation 4.138 solution, in the light of conditions expressed by Equation 4.139 through Equation 4.141, is given by [23]

$$M_t^+ = \frac{M_t}{M_\infty} = 1 - \sum_{n=0}^{n=\infty} \frac{8}{(2n+1)^2 \pi^2} e^{\left(-\frac{(2n+1)^2 \pi^2}{4} t^+\right)} \qquad (4.142)$$

$$M_\infty = SLC_0, \quad t^+ = \frac{t D_{d0}}{L^2}, \qquad (4.142')$$

where M_∞ is the drug amount released after an infinite time and t^+ is the dimensionless time. Equation 4.142 expresses that, in the so-called Fickian release, the ratio M_t/M_∞ increases according to $(t^+)^{0.5}$ for $0.2 < M_t/M_\infty < 0.6$ as shown in Figure 4.6. Of course, similar considerations can be done for the one-dimensional diffusion in a cylinder or in sphere. In particular, in the case of drug release from a cylinder of radius R_0 and length L (release takes place only in the radial direction), we have initial conditions ($t = 0$),

$$C = C_0, \quad 0 \le R < R_0 \quad \text{uniform drug distribution in the cylinder,} \quad (4.143)$$

boundary conditions ($t > 0$),

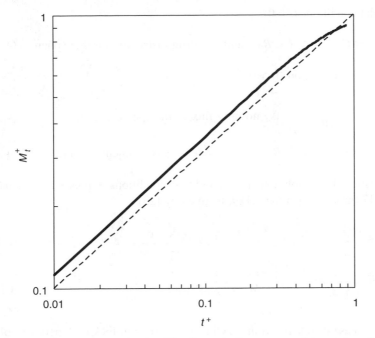

FIGURE 4.6 Fickian release kinetics. The dimensionless amount of drug released M_t^+ increases with the square root of the dimensionless time t^+ in the range $0.2 < M_t^+ < 0.6$ (release from slab. *Dashed line* represents a 0.5 sloped curve). In the case of release from cylinder and sphere, $M_t^+ \propto (t^+)^{0.45}$ and $M_t^+ \propto (t^+)^{0.43}$ would occur, respectively.

$$\left.\frac{\partial C}{\partial R}\right|_{R=0} = 0 \quad \text{no drug flux on the cylinder axis,} \tag{4.144}$$

$$C = 0 \quad R = R_0 \quad \text{infinite release environment volume.} \tag{4.145}$$

Equation 4.138 solution, in the light of conditions expressed by Equation 4.143 through Equation 4.145, is given by [23]

$$M_t^+ = \frac{M_t}{M_\infty} = 1 - \sum_{n=1}^{n=\infty} \frac{4}{R_0^2 \, \alpha_n^2} e^{(-\alpha_n^2 R_0^2 t^+)}, \tag{4.146}$$

$$M_\infty = \pi \, R_0^2 \, LC_0 \quad t^+ = \frac{t D_{d0}}{R_0^2}, \tag{4.146'}$$

where α_n are the positive roots of $J_0(R_0 \, \alpha_n) = 0$ and J_0 is the Bessel function. Analogously, in the case of drug release from a sphere of radius R_0 (release takes place in the radial direction), we have initial conditions ($t = 0$),

$$C = C_0, \quad 0 \leq R < R_0 \quad \text{uniform drug distribution in the sphere} \quad (4.147)$$

boundary conditions ($t > 0$),

$$\left. \frac{\partial C}{\partial R} \right|_{R=0} = 0 \quad \text{no drug flux in the sphere center,} \quad (4.148)$$

$$C = 0, \quad R = R_0 \quad \text{infinite release environment volume.} \quad (4.149)$$

Equation 4.138 solution, in the light of conditions expressed by Equation 4.147 through Equation 4.149, is given by [23]

$$M_t^+ = \frac{M_t}{M_\infty} = 1 - \sum_{n=1}^{n=\infty} \frac{6}{n^2 \pi^2} e^{(-n^2 \pi^2 t^+)}, \quad (4.150)$$

$$M_\infty = \frac{4}{3} \pi R_0^3 C_0 \ t^+ = \frac{t D_{d0}}{R_0^2}. \quad (4.150')$$

In the case of release from a cylinder or a sphere, Fickian kinetics resolves, respectively, in an M_t/M_∞ increases according to $(t^+)^{0.45}$ and $(t^+)^{0.43}$ for $0.2 < M_t/M_\infty < 0.6$ [49].

4.5   EFFECT OF INITIAL AND BOUNDARY CONDITIONS

It is important to underline that if an experimental M_t/M_∞ increase proportional to $(t^+)^n$ ($n = 0.5$, 0.45, and 0.43 for slab, cylinder, and sphere, respectively, when $0.2 < M_t/M_\infty < 0.6$) means that mass transport inside the release system obeys Fick's first law, different release kinetics are not automatically a proof of Fick's first law failure. Indeed, as it is still valid Fick's first law, particular initial and boundary conditions can yield to an M_t/M_∞ increase that does not follow $(t^+)^{0.5}$, $(t^+)^{0.45}$, and $(t^+)^{0.43}$ for slab, cylinder, and sphere, respectively. Consequently, it is necessary to distinguish between macroscopic and microscopic release kinetics. If, from the macroscopic viewpoint, release kinetics can be surely classified as non-Fickian when M_t/M_∞ is not proportional to $(t^+)^n$ ($n = 0.5$, 0.45, and 0.43 for slab, cylinder, and sphere, respectively, when $0.2 < M_t/M_\infty < 0.6$), microscopically, on the contrary, we are obliged to verify whether Fick's first law holds or not, regardless the macroscopic release kinetics. While in Chapter 7 and Chapter 9 the microscopic failure of Fick's first law will be examined, this section focuses the attention on some of the most important initial and boundary conditions able to yield macroscopic non-Fickian behavior, Fick's first law holding at the microscopic level. In particular, the effect of finite release environment volume, partition coefficient, initial drug distribution, stagnant layer, and release through a holed surface will be examined in the following.

4.5.1 FINITE RELEASE ENVIRONMENT VOLUME AND PARTITION COEFFICIENT

In several situations, both the hypothesis of an infinite release volume V_r and a unitary partition coefficient k_p at matrix/release environment interface do not hold. Accordingly, all other initial and boundary conditions are still valid, Equation 4.141, Equation 4.145, and Equation 4.149 (for slab, cylinder, and sphere, respectively) must be replaced by

$$V_r \frac{\partial C_r}{\partial t} = -SD_{d0}\nabla C \quad \text{matrix–fluid interface} \quad \text{finite } V_r \qquad (4.151)$$

$$C_r = C/k_p \quad \text{matrix–fluid interface} \quad \text{partitioning.} \qquad (4.152)$$

While Equation 4.151 states that the rate at which the solute leaves the matrix is always equal to that it enters the release volume, Equation 4.152 imposes that solute concentration in the release fluid C_r is not equal to the solute concentration at the matrix–release fluid interface. In the light of these new boundary conditions, Equation 4.138 solution becomes, for slab, cylinder and sphere, respectively [23],
slab

$$M_t^+ = \frac{M_t}{M_\infty} = 1 - \sum_{n=1}^{n=\infty} \frac{2\alpha(1+\alpha)}{1+\alpha+\alpha^2 q_n^2} e^{(-q_n^2 t^+)}, \qquad (4.153)$$

q_n nonzero positive roots of $\tan(q_n) = -\alpha q_n$, $\alpha = \dfrac{V_r}{SLK_p}$

$$M_\infty = \frac{SLC_0}{1+k_p SL/V_r}, \qquad t^+ = \frac{tD_{d0}}{L^2}, \qquad (4.153')$$

cylinder

$$M_t^+ = \frac{M_t}{M_\infty} = 1 - \sum_{n=1}^{n=\infty} \frac{4\alpha(1+\alpha)}{4+4\alpha+\alpha^2 q_n^2} e^{(-q_n^2 t^+)}, \qquad (4.154)$$

q_n nonzero positive roots of $\alpha q_n J_0(q_n) + 2J_1(q_n) = 0$, $\alpha = \dfrac{V_r}{\pi R_0^2 L k_p}$

$$M_\infty = \frac{\pi R_0^2 L C_0}{1+k_p \pi R_0^2 L/V_r}, \qquad t^+ = \frac{tD_{d0}}{R_0^2}. \qquad (4.154')$$

J_1 is the first order Bessel function,

sphere

$$M_t^+ = \frac{M_t}{M_\infty} = 1 - \sum_{n=1}^{n=\infty} \frac{6\alpha(1+\alpha)}{9 + 9\alpha + \alpha^2 q_n^2} e^{(-q_n^2 t^+)}, \qquad (4.155)$$

q_n nonzero roots of $\tan(q_n) = \dfrac{3q_n}{3 + \alpha q_n^2}, \qquad \alpha = \dfrac{3V_r}{4\pi R_0^3 k_p}$

$$M_\infty = \frac{(4/3)\pi R_0^3 C_0}{1 + k_p(4/3)\pi R_0^3/V_r}, \qquad t^+ = \frac{t D_{d0}}{R_0^2}. \qquad (4.155')$$

In order to better appreciate the effect of the finite release environment volume V_r, it is convenient to consider the dimensionless quantity M_{t0}^+ ($= M_t/M_0$) instead of the usual M_t^+ ($= M_t/M_\infty$). Indeed, this representation immediately gives an idea of the incomplete release of the initial drug load. Figure 4.7 shows, in the case of drug release from a slab, release kinetics profiles assuming different ratios R_v between matrix volume V_m ($= SL$) and release environment volume V_r. While for infinite V_r, the traditional Fickian

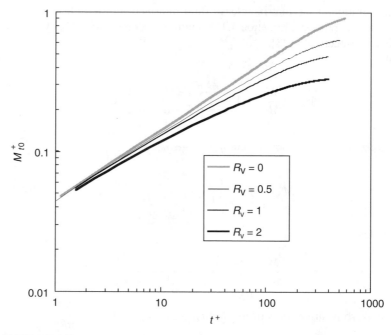

FIGURE 4.7 Effect of the release environment volume on drug release from a slab. For increasing values of the ratio R_v (= matrix volume/release environment volume) drug release kinetics detaches from the classical Fickian kinetics represented by the grey curve ($R_v = 0$). Indeed, sink conditions no longer hold. In addition, only a fraction of the initial drug load can be released.

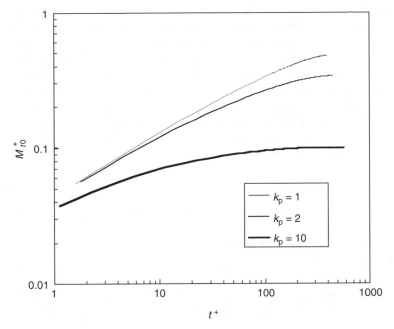

FIGURE 4.8 Effect of partitioning at the matrix/release environment interface assuming equality between matrix and release environment volumes ($R_v = 1$). Partition coefficient k_p increase exalts the deviation from the macroscopic Fickian release and it further reduces the initial drug load that can be released.

release is found ($M_{t0}^+ \propto t^{0.5}$), as soon as R_v increases, the deviation from Fickian release becomes evident, although Fick's law for diffusion holds at the microscopic level. In addition, the delivery system can no longer completely release its initial drug load and that is why (M_{t0}^+) does not tend to 1. As these effects become important for $R_v \geq 1$, their practical importance is related to particular administration routes such as ocular or implantable where a limited liquid release environment is easily met. If, additionally, drug partitioning phenomenon takes place at the matrix/release environment interface, macroscopic deviation from Fickian kinetics is exalted. Indeed, Figure 4.8 shows that, fixed $R_v = 1$, the increase of the partition coefficient k_p implies both a further deviation from Fickian release and a further decrease of the drug amount delivered (incomplete release). In particular, remembering Equation 4.153' and assuming $R_v = 1$ and $k_p = 10$, the drug amount released after an infinite time (M_∞) results in 9.1% of the total initial drug load SLC_0.

4.5.2 INITIAL DRUG DISTRIBUTION

The idea of a nonuniform drug distribution inside the matrix has been considered by many authors in the past and Lee [50,51] is the one who firstly rationalized this approach. The importance of this technique, indeed, relies on the possibility of properly controlling release kinetics without any chemical or physical modification of both the matrix and the drug. In this light, the possibility of providing analytical solutions for Fick's second law is very important for both the theoretical and practical point of view. In addition, these solutions assume also an interesting didactic valence for students. Accordingly, the boundary conditions set by Equation 4.140 and Equation 4.141, Equation 4.144 through Equation 4.145 and Equation 4.148 and Equation 4.149 are still valid for slab, cylinder, and sphere, respectively, the initial conditions must be replaced by

$$C = f(X), \quad 0 \leq X < X_0 \quad \text{slab}, \tag{4.156}$$

$$C = f(R), \quad 0 \leq R < R_0 \quad \text{cylinder, sphere.} \tag{4.157}$$

Fick's second law solution reads
slab

$$M_t^+ = \frac{M_t}{M_\infty} = 1 - \sum_{n=0}^{\infty} \frac{(-1)^{n+1}}{2n+1} I_{1(n)} e^{\left[\frac{-(2n+1)^2 \pi^2 t^+}{4}\right]} \Bigg/ \sum_{n=0}^{\infty} \frac{(-1)^{n+1}}{2n+1} I_{1(n)}, \tag{4.158}$$

where

$$I_{1(n)} = \int_0^1 f(\xi) \cos\left[(n+0.5)\pi\xi\right] d\xi, \qquad \xi = \frac{X}{L}, \qquad t^+ = \frac{D_{d0} t}{L^2}.$$

cylinder

$$M_t^+ = \frac{M_t}{M_\infty} = 1 - \sum_{n=1}^{\infty} \frac{I_{2(n)}}{\beta_n J_1(\beta_n)} e^{(-\beta_n^2 t^+)} \Bigg/ \sum_{n=1}^{\infty} \frac{I_{2(n)}}{\beta_n J_1(\beta_n)}, \tag{4.159}$$

where

$$I_{2(n)} = \int_0^1 \xi f(\xi) J_0(\beta_n \xi) d\xi \quad \beta_n \text{ roots of } J_0(\beta_n) = 0$$

$$\xi = \frac{R}{R_0}, \qquad t^+ = \frac{D_{d0} t}{R_0^2}.$$

sphere

$$M_t^+ = \frac{M_t}{M_\infty} = 1 - \left[\sum_{n=1}^{\infty} \frac{(-1)^{n+1}}{n} I_{3(n)} e^{(-n^2\pi^2 t^+)} \middle/ \sum_{n=1}^{\infty} \frac{(-1)^{n+1}}{n} I_{3(n)} \right],$$

(4.160)

where

$$I_{3(n)} = \int_0^1 \xi f(\xi) \sin(n\pi\xi) \, d\xi, \qquad \xi = \frac{R}{R_0}, \qquad t^+ = \frac{D_{d0} t}{R_0^2}.$$

For its practical relevance, and in the light of the qualitatively similar behavior found for the other geometries (slab, cylinder), it is sufficient to comment on the effect of initial drug distribution in the case of a sphere. In particular, we focus the attention on the stepwise distribution as it can properly approximate any other distribution provided that a sufficiently thin particle radius subdivision is considered. In this case, $I_{3(n)}$ analytical expression reads:

$$I_{3(n)} = \int_0^1 \xi f(\xi) \sin(n\pi\xi) \, d\xi = \sum_{i=1}^{N_s} C_i^+ \int_{\xi_{i-1}}^{\xi_i} \xi \sin(n\pi\xi) \, d\xi$$

$$= \sum_{i=1}^{N_s} \frac{C_i^+}{(n\pi)^2} [\sin(n\pi\xi_i) - n\pi\xi_i \cos(n\pi\xi_i) - \sin(n\pi\xi_{i-1})$$

$$+ n\pi\xi_{i-1} \cos(n\pi\xi_{i-1})],$$

(4.161)

where N_s represents the number of parts in which the radius R_0 has been subdivided in, C_i^+ ($= C_i/C_{max}$) is the dimensionless constant drug concentration occurring in $\xi_{i-1} < \xi < \xi_i$, $\xi_0 = 0$ (please note that $f(\xi)$ is now a stepwise function of concentration), and C_{max} is the maximum concentration value measured in the sphere. Obviously, $M_\infty = \sum_{i=1}^{N_s} \frac{4}{3}\pi C_i (R_i^3 - R_{i-1}^3)$.

Among the many existing stepwise distributions, the attention is focused on two particular kinds that find a match with real distributions. Indeed, they, approximately, represent the result of common techniques employed to partially deplete matrices [52] or they are the result of an imperfect drug loading [53]. Accordingly, the first kind is a one-step distribution characterized by a uniform concentration until $\xi = \xi_c$ (in particular, $\xi_c = 0.9, 0.75$, and 0.4 will be considered) and zero drug concentration in the remaining outer part (see insert in Figure 4.9). In the second kind, instead, radius R_0 is subdivided into N_s parts (five, in the specific case considered, see insert in Figure 4.11) each one characterized by different and constant drug concentration C_i^+. While the

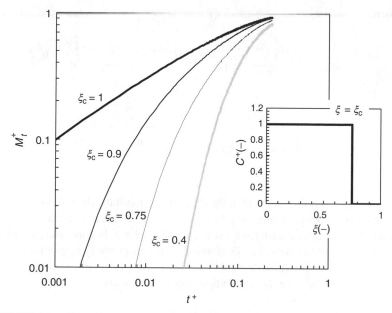

FIGURE 4.9 Effect of one-step initial drug distribution (see insert) on drug release kinetics from a sphere. As step position (ξ_c) detaches from one (uniform drug distribution), a neat macroscopic non-Fickian behavior appears.

increasing distribution represented by $C_i^+ = 0.0625, 0.125, 0.25, 0.5, 1$ can be the result of an imperfect drug loading, the decreasing distribution $C_i^+ = 1, 0.5, 0.25, 0.125, 0.0625$ can be the result of matrix depletion. Finally, for didactic reasons, a maximum shaped distribution ($C_i^+ = 0.25, 0.5, 1, 0.5, 0.25$) is considered. Figure 4.9 shows the trend of M_t^+ ($= M_t/M_\infty$) vs. t^+ in the case of uniform distribution (solid thick line) and three stepwise distributions of the first kind. It is evident how the release kinetics is highly influenced by the stepwise distribution and by its pattern. A soon as ξ_c decreases, a sensible reduction of drug release takes place and evident macroscopic non-Fickian behavior occurs. In addition, Figure 4.10 , showing the time variation of drug flux

$$\frac{d}{dt^+}\left(\frac{M_t}{M_\infty}\right) = \sum_{n=1}^{\infty}(-1)^{n+1}n\pi^2 I_{3(n)}e^{(-n^2\pi^2 t^+)} \bigg/ \sum_{n=1}^{\infty}\frac{(-1)^{n+1}}{n}I_{3(n)} \quad (4.162)$$

vs. t^+, evidences that smaller ξ_c corresponds to smaller flux variation. Indeed, in the case of uniform distribution, the flux spans from 0 to 25 while it spans from 0 to 16, 8, and 6 for $\xi_c = 0.9, 0.75$, and 0.4, respectively. Figure 4.11 reports the trend of M_t^+ ($= M_t/M_\infty$) vs. t^+ in the case of uniform distribution (solid thick line) and three different distributions of the second kind (increasing, decreasing, and maximum). Again, also in this case, drug distribution

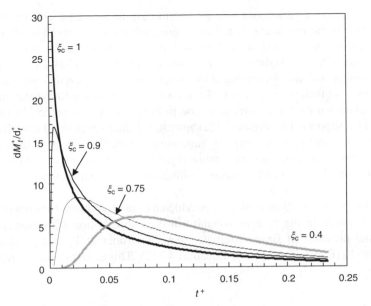

FIGURE 4.10 Effect of one-step initial drug distribution on time evolution of M_t^+ time (t^+) derivative. As step position (ξ_c) decreases, smoother M_t^+ time derivative occurs.

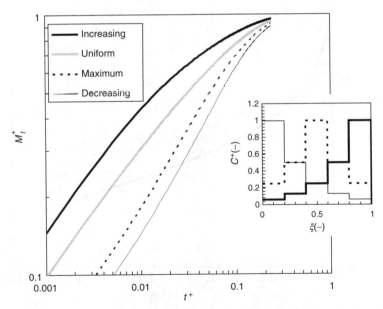

FIGURE 4.11 Effect of increasing, maximum, and decreasing initial drug distribution (see insert) on drug release kinetics from a sphere. Release kinetics is heavily influenced by the distribution features. Maximum and decreasing distribution yield to a macroscopic non-Fickian behavior.

pattern plays a paramount role in determining drug release characteristics. Indeed, in the increasing distribution, drug release is improved while in the decreasing distribution it is depressed and neat macroscopic non-Fickian behavior occurs. Maximum distribution, despite in the middle between the increasing and the decreasing distribution for what concerns concentration pattern, is much more similar to the decreasing distribution for what concerns cumulative drug release (see gray line in Figure 4.11). These considerations are also supported by Figure 4.12, showing the time variation of drug flux. If increasing distribution comports huge variation of flux (it spans over three decades), decreasing and maximum distributions imply much smaller flux variations. Uniform concentration collocates in between increasing and decreasing distributions.

Interestingly, Zhou et al. [54] provide analytical solutions for drug release from sphere in the case of nonuniform drug distribution (Equation 4.157) in the presence of finite external release environment V_r and partitioning ($k_p \neq 1$) (Equation 4.151 and Equation 4.152). This general solution reads

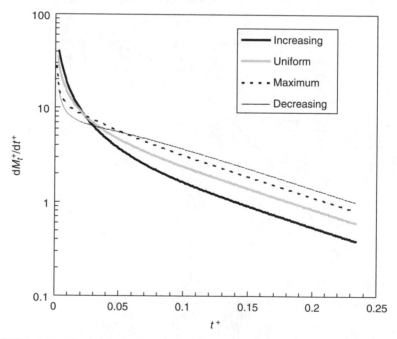

FIGURE 4.12 Effect of increasing, maximum, and decreasing initial drug distribution on time evolution of M_t^+ time (t^+) derivative. Smoother M_t^+ time derivative occurs for decreasing and maximum distribution.

$$M_t^+ = \frac{M_t}{M_\infty} = \sum_{n=1}^{\infty} A_n \operatorname{sen}(q_n)\left(1 - e^{-(q_n^2 t^+)}\right) \Big/ \sum_{n=1}^{\infty} A_n \operatorname{sen}(q_n), \qquad (4.163)$$

where

$$q_n \text{ are positive roots of } \tan(q_n) = \frac{3q_n}{3 + \alpha q_n^2}; \quad \alpha = \frac{3V_r}{4\pi R_0^3 k_p}, \qquad (4.164)$$

$$A_n = \frac{\displaystyle\int_0^{R_0} R(C(R) - C_\infty)[R_0 \operatorname{sen}(q_n R/R_0) - R \operatorname{sen}(q_n)]\, dR}{\displaystyle\int_0^{R_0} \operatorname{sen}(q_n R/R_0)[R_0 \operatorname{sen}(q_n R/R_0) - R \operatorname{sen}(q_n)]\, dR}, \qquad (4.165)$$

$$C_\infty = \frac{3}{R_0^3(\alpha + 1)} \int_0^{R_0} R^2 C(R)\, dR, \quad M_\infty = \frac{C_\infty V_r}{k_p} \quad M_0 = 4\pi \int_0^{R_0} R^2 C(R)\, dR \qquad (4.166)$$

and M_0 is the total drug amount contained in the matrix. In particular, for linear, quadratic, and sigmoidal distributions, Equation 4.163 becomes, respectively,
linear,

$$C(R) = C_{\max}\left(1 - \frac{R}{R_0}\right), \qquad (4.167)$$

$$M_t^+ = \frac{M_t}{M_\infty} = 1 - \sum_{n=1}^{n=\infty} \frac{24(1+\alpha)[(6 + 3q_n^2 + 2\alpha q_n^2)\operatorname{sen}(q_n) - 6q_n]}{(9 + 9\alpha + \alpha^2 q_n^2) q_n^4 \operatorname{sen}(q_n)} e^{(-q_n^2 t^+)}, \qquad (4.168)$$

$$C_\infty = \frac{C_{\max}}{4(\alpha + 1)}, \quad M_\infty = \frac{C_\infty V_r}{k_p}, \quad M_0 = \frac{\pi R_0^3}{3} C_{\max} \qquad (4.169)$$

quadratic,

$$C(R) = C_{\max}\left(1 - \left(\frac{R}{R_0}\right)^2\right), \qquad (4.170)$$

$$M_t^+ = \frac{M_t}{M_\infty} = 1 - \sum_{n=1}^{n=\infty} \frac{90(1+\alpha)^2\, e^{(-q_n^2 t^+)}}{(9 + 9\alpha + \alpha^2 q_n^2)\, q_n^2}, \qquad (4.171)$$

$$C_\infty = \frac{2C_{max}}{5(\alpha + 1)}, \qquad M_\infty = \frac{C_\infty V_r}{k_p}, \qquad M_0 = \frac{8\pi R_0^3}{15} C_{max} \qquad (4.172)$$

sigmoidal,

$$C(R) = C_{max} \left\{ \left[\left(\frac{R}{R_0}\right)^2 - 2 \right] \left(\frac{R}{R_0}\right)^2 + 1 \right\}, \qquad (4.173)$$

$$M_t^+ = \frac{M_t}{M_\infty} = 1 - \sum_{n=1}^{n=\infty} \frac{210\,(1+\alpha)\,[15(1+\alpha) - \alpha q_n^2]}{(9 + 9\alpha + \alpha^2 q_n^2)\, q_n^4} e^{(-q_n^2 t^+)}, \qquad (4.174)$$

$$C_\infty = \frac{8C_{max}}{35(\alpha + 1)}, \qquad M_\infty = \frac{C_\infty V_r}{k_p}, \qquad M_0 = \frac{32\pi R_0^3}{105} C_{max}. \qquad (4.175)$$

In the case of a stepwise distribution made up by N_s steps, the following A_n expression has to be considered in Equation 4.163:

$$A_n = C_{max} R_0 \frac{\displaystyle\sum_{i=1}^{N_s} (C_i^+ - C_\infty^+) \left\{ \frac{B_{ni}}{q_n^2} - \frac{\text{sen}(q_n)}{3}(\xi_i^3 - \xi_{i-1}^3) \right\}}{\displaystyle\sum_{i=1}^{N_s} \left\{ \frac{\xi_i - \xi_{i-1}}{2} - \frac{1}{4q_n}[\text{sen}(2z_{ni-1}) - \text{sen}(2z_{ni})] - B_{ni}\frac{\text{sen}(q_n)}{q_n^2} \right\}}, \qquad (4.176)$$

$$B_{ni} = \text{sen}\,(z_{ni}) - z_{ni}\cos\,(z_{ni}) - \text{sen}\,(z_{ni-1}) + z_{ni-1}\cos\,(z_{ni-1}), \qquad (4.176')$$

$$C_i^+ = \frac{C_i}{C_{max}}, \qquad z_{ni} = q_n\xi_i, \quad z_{ni-1} = q_n\xi_{i-1}, \quad \xi_i = \frac{R_i}{R_0}, \qquad (4.176'')$$

$$C_\infty^+ = \frac{C_\infty}{C_{max}} = \frac{1}{(\alpha + 1)}\sum_{i=1}^{N_s} C_i^+(\xi_i^3 - \xi_{i-1}^3), \qquad M_\infty = \frac{C_\infty V_r}{k_p}, \qquad (4.177)$$

$$M_0 = R_0^3 C_{max} \frac{4}{3}\pi \sum_{i=1}^{N_s} C_i^+(\xi_i^3 - \xi_{i-1}^3). \qquad (4.178)$$

As discussed above, in order to better appreciate the effect of the finite release environment volume V_r, it is convenient to consider the dimensionless quantity $M_{r0}^+ (= M_t/M_0)$ instead of the more usual $M_t^+ (= M_t/M_\infty)$. Indeed, this representation immediately gives an idea of the incomplete release of the initial drug load. It is also easy to verify that $M_{r0}^+ = (M_t/M_\infty) \times \alpha/(\alpha + 1)$. Figure 4.13 reports the combined effect of initial drug distribution (descending stepwise distribution; see insert in Figure 4.11) and finite release

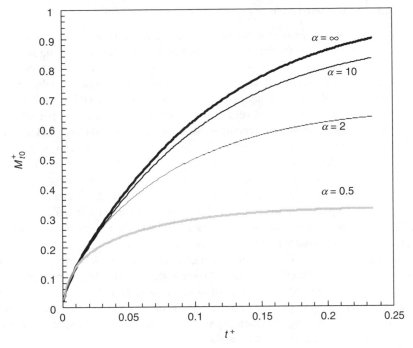

FIGURE 4.13 Effect of finite release environment volume assuming a decreasing stepwise initial drug concentration distribution (see insert in Figure 4.11). The effect of finite release environment volume is evaluated through the parameter α appearing in Equation 4.165.

environment. As α decreases ($\alpha = \infty$ corresponds to an infinite release environment volume) the fraction of the initial drug load that can be released decreases and, for very law α values (<1), the release kinetics typical of descending stepwise concentration is completely hidden.

To conclude this paragraph, it is important to underline that if the usefulness of analytical solutions to Equation 4.138 is out of discussion for evident reasons that include an easier data fitting, their use needs some care. Indeed, for analytical solutions involving the numerical determination of q_n, it is not true that the higher the summation terms (in Equation 4.163, for example), the more accurate the solution is. Practical test reveals that if q_n estimation is performed using the bisection method [55] with tolerance of 10^{-6}, a reasonable compromise between accuracy and time required for solution calculation fixes the optimal number of summation terms around eight. Of course, this problem is particularly important for small time as, for longer time, the rapid decay of the exponential makes the contribution of higher terms vanishing.

4.5.3 Stagnant Layer

Typically, the existence of a stagnant layer around release surfaces is due to insufficient stirring in the release environment or it is the result of releasing surfaces erosion because of chemical or physical reasons or it is due to a combination of both. Indeed, it is well known that, regardless of stirring conditions, at the solid–liquid interface a film of stagnant liquid forms [21]. Obviously, as film thickness depends on many factors such as liquid velocity field, surface roughness, and porosity, its extension is larger in the presence of weak velocity fields and rough surfaces. In addition, in the presence of erosion, film nature and extension can be heavily influenced by the matter leaving the releasing surfaces and moving into the bulk. Accordingly, if complete stagnant layer elimination is virtually impossible, its importance on drug release kinetics is determining only when it exerts a diffusive resistance comparable to those related, for example, to drug diffusion and dissolution inside the delivery system. This is the reason why the importance of stagnant layer becomes relevant in highly swollen physical gels that [56], typically, comport a moderate or weak resistance on drug molecules diffusion. In addition, another reason for studying the effect of stagnant layer is suggested by the common practice of deliberately coating the delivery system with a membrane that, often, represents the rate determining step of the whole release process [57]. In mathematical terms, the presence of the stagnant layer requires Fick's second law solution (Equation 4.138) on both the matrix and the stagnant layer

$$\frac{\partial C}{\partial t} = D_{d0} \nabla^2 C \quad \text{matrix,} \tag{4.179}$$

$$\frac{\partial C_{sl}}{\partial t} = D_{sl} \nabla^2 C_{sl} \quad \text{stagnant layer,} \tag{4.180}$$

where D_{sl} and C_{sl} represent, respectively, drug diffusion coefficient and concentration in the stagnant layer. In the majority of the situations, it can be assumed that stagnant layer formation is fast in comparison with drug release kinetics so that its initial drug concentration is zero. On the contrary, it is reasonable to assume that uniform drug concentration (C_0) realizes in the matrix: initial conditions ($t = 0$),

$$C = C_0 \quad \text{matrix,} \tag{4.181}$$

$$C_{sl} = 0 \quad \text{stagnant layer.} \tag{4.182}$$

Boundary conditions must account for finite release environment volume, for drug partitioning at the stagnant layer–release environment and matrix–stagnant layer interfaces and by zero flux condition in the matrix symmetry locus (sphere center, cylinder symmetry axis, slab symmetry plane, or slab surface opposite to the releasing one) boundary conditions ($t > 0$),

$$V_r \frac{\partial C_r}{\partial t} = -S \nabla C_{sl} \quad \text{finite } V_r\text{-stagnant layer/rel. env. interface,} \tag{4.183}$$

$$C_r = \frac{C_{sl}}{k_{psr}} \text{partitioning-stagnantlayer/releaseenvironmentinterface,} \tag{4.184}$$

$$C_{sl} = \frac{C}{k_{pms}} \quad \text{partitioning-matrix/stagnant layer interface,} \tag{4.185}$$

$$D_{d0} \nabla C = D_{sl} \nabla C_{sl} \quad \text{no drug accumulation,} \tag{4.186}$$

$$\nabla C = 0 \quad \text{no drug flux in matrix symmetry locus,} \tag{4.187}$$

where k_{psr} and k_{pms} represent, respectively, drug partition coefficient at the stagnant layer–release fluid interface and at the matrix–stagnant layer interface. Obviously, $k_p = k_{psr} \times k_{pms}$. Equation 4.186, simply states that no drug accumulation takes place at the matrix–stagnant layer interface. Indeed, it imposes that drug flux leaving the matrix is equal to that entering the stagnant layer. Unfortunately, if $D_{sl} \neq D_{d0}$ (and this is the most common case), the above set of equations (Equation 4.181 through Equation 4.187) cannot be analytically solved and a numerical solution is needed (for example, the control volume method [58]). Assuming, for a better understanding of stagnant layer effect, that $k_{psr} = k_{pms} = 1$ and that the release environment volume is infinite, Figure 4.14 shows the effect of the ratio R_{sm} between stagnant layer and matrix thickness (slab geometry) in the hypothesis of setting $D_{sl} = 0.1 \times D_{d0}$. If, as expected, when $R_{sm} = 0$ a macroscopic Fickian release occurs, for higher values, a clearly non-Fickian macroscopic kinetics develops and release curve slope considerably increases for both cases shown. Figure 4.15 reports the release curve behavior fixing $D_{sl} = D_{d0}$ and considering the same three different R_{sm} values. It is clear that the macroscopic non-Fickian character arises when $R_{sm} = 5\%$ and 20% but, now, it is less pronounced than what happened in Figure 4.14. Indeed, the macroscopic non-Fickian behavior is only due to the stepwise drug distribution in the matrix–stagnant layer system. The assumption $D_{sl} = D_{d0}$ makes, mathematically speaking, the matrix–stagnant layer system as a unique body characterized by a stepwise drug distribution as previously discussed (paragraph 4.5.2).

4.5.4 HOLED SURFACES

The necessity for a theoretical study on drug release from perforated membranes arises from the discovery that different release kinetics can be obtained from a simple tablet once it is coated by an impermeable perforated membrane [59,60]. Basically, this makes possible the improvement of the potentiality of traditional tablets by means of a relatively easy and low-cost

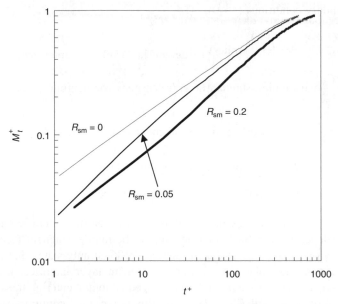

FIGURE 4.14 Effect of stagnant layer presence on drug release kinetics from a slab. R_{sm} indicates the ratio between stagnant layer and matrix thickness, while $D_{sl} = 0.1 \times D_{d0}$.

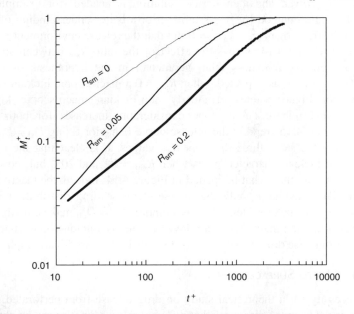

FIGURE 4.15 Effect of stagnant layer presence on drug release kinetics from a slab. R_{sm} indicates the ratio between stagnant layer and matrix thickness. The simulation is performed assuming that the diffusion coefficient in the stagnant layer is equal to that in the matrix ($D_{sl} = D_{d0}$).

manufacturing technique. In these holed systems, no drug diffusion occurs through the coating membrane and release kinetics is governed by holes characteristics. Indeed, the rate and duration of drug release can be easily adjusted by properly designing holes size and density. In addition, the study of drug release from holed surfaces can be very important in experimental techniques devoted to *in vitro* evaluation of diffusive properties of not self-sustaining matrices. Indeed, these systems need to be placed in a holed housing to study release characteristics [56].

Despite the great variety of holed surfaces that can be imagined, for the sake of clarity, it is convenient to refer to the scheme reported in Figure 4.16. Here, it is hypothesized that one of the two plane bases of a cylindrical matrix (radius r_m, height h_c) is put in contact with the release medium via a plane, impermeable, and circular membrane of radius r_m carrying all equal circular holes with radius r_h. In addition, we suppose that membrane thickness is very small with respect to cylinder height [24,61]. Accordingly, membrane geometrical characteristics are perfectly defined by its void fraction ϕ, equal to the ratio between holes area and membrane area, and by the ratio r_h/r_m. As a

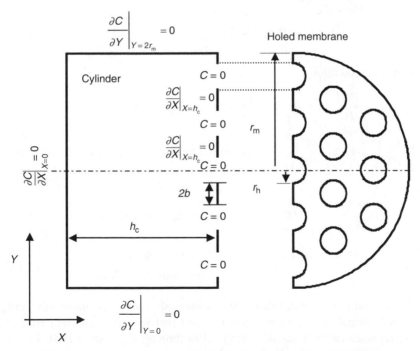

FIGURE 4.16 Schematic representation of the cylindrical matrix coated by an impermeable holed membrane on its bottom plane surface. This figure also shows the boundary conditions for Equation 4.189 solution in the hypothesis of reducing the three-dimensional problem to a simpler two-dimensional problem.

theoretical detailed analysis of drug release from this system would be very complicated by the presence of mixed boundary conditions (drug can leave the cylindrical matrix only through holes) requiring a three-dimensional solution scheme, it is more convenient to reduce this frame to a two-dimensional problem. As a consequence, the holed membrane is represented by a dashed straight line constituted by an alternation of sectors available for drug release (holes) and impermeable sectors (membranes) (see Figure 4.16). In this simplified picture, ϕ is defined as the ratio between holes length ($N_f 2r_h$) and membrane length ($2r_m$), where N_f is the number of holes. Drug release kinetics can be deduced by numerically solving (control volume method [58]) Fick's second law in two dimensions:

$$\frac{\partial C}{\partial t} = D_{d0}\left(\frac{\partial^2 C}{\partial X^2} + \frac{\partial^2 C}{\partial Y^2}\right),$$ (4.188)

where X and Y are Cartesian coordinates while D_{d0} and C are, respectively, drug diffusion coefficient and concentration in the cylindrical matrix. Equation 4.188 must undergo the following initial and boundary conditions (see also Figure 4.16):
initial conditions,

$$C(X, Y) = C_0 \quad 0 < Y < 2r_m,$$
$$0 < X < h \quad \text{uniform drug distribution,}$$ (4.189)

boundary conditions,

$$\left.\frac{\partial C}{\partial X}\right|_{X=0} = 0 \quad \text{impermeable coating on the back,}$$ (4.190)

$$\left.\frac{\partial C}{\partial Y}\right|_{Y=0} = 0; \quad \left.\frac{\partial C}{\partial Y}\right|_{Y=2r_m} = 0 \quad \text{impermeable coating on lateral surface,}$$ (4.191)

$$\left.\frac{\partial C}{\partial X}\right|_{X=h_c \text{ on the membrane}} = 0 \quad \text{impermeable membrane,}$$ (4.192)

$$C(X = h_c \text{ on the holes}) = 0 \quad \text{infinite release}$$
$$\text{environment volume.}$$ (4.193)

These conditions state that the cylindrical matrix is uniformly drug loaded (initial drug concentration C_0) and that the only releasing surfaces are represented by holes (Equation 4.190 through Equation 4.193). For the sake of simplicity, it is also assumed that the hypothesis of an infinite release environment volume holds (Equation 4.193), that the effect of the unavoidable stagnant layer in $X = h_c$ is negligible and that no matrix erosion takes place in correspondence of holes. In this context, two parameters among N_f, ϕ, and r_h/r_m, ($\phi = N_f r_h/r_m$) represent the independent designing parameters. Indeed, as depicted in Figure 4.16, it is assumed that, whatever be N_f, holes disposition on the membrane follows a symmetrical scheme with respect to cylinder axis (lying in $Y = r_m$).

Supposing that only one hole is present on the membrane, Figure 4.17 shows the effects induced by modifying ϕ. If for ϕ tending to 1, a typical macroscopic Fickian release takes place (thick solid line), for smaller values a macroscopic non-Fickian release appears (thin solid line). In particular, small ϕ variations below 1 reflect in a reduced amount of drug released without altering the dominant macroscopic Fickian character. On the contrary, larger ϕ reductions lead to a considerable increase of release curve slope approaching first order release kinetics. It is interesting to notice that the use of the simple, empirical $M_t^+ = kt^n$ equation for the fitting of simulation results obtained for increasing ϕ, leads to the conclusion that when $\phi = 1, 0.75, 0.5, 0.25$, and 0.1, $n = 0.5$, 0.532, 0.605, 0.689, and 0.764, respectively. Figure 4.18 and Figure 4.19 show the effect of N_f (or r_h/r_m) variation, once ϕ has been fixed. It is interesting to see that N_f increase corresponds to an increase in drug released, where the curve slope, is practically unaffected (this is particularly evident for small ϕ values as shown in Figure 4.19). This leads to the important

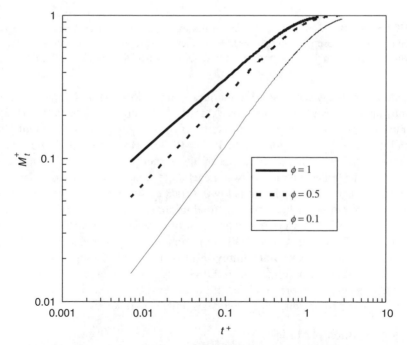

FIGURE 4.17 Effect of membrane void fraction ϕ assuming the existence of only one hole. The smaller ϕ, the more pronounced the macroscopic non-Fickian character of the release curve. The simulation is performed assuming a square system ($h_c = 2r_m$).

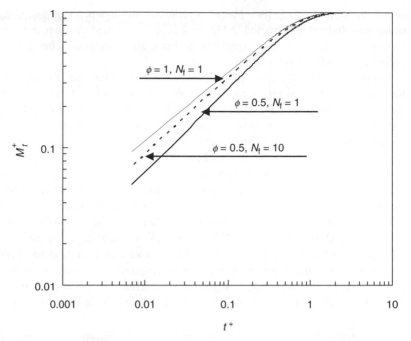

FIGURE 4.18 Effect of N_f (or r_h/r_m) variation, once ϕ has been fixed. Higher N_f favors drug release (see dotted line). $\phi = 1$ curve is reported as reference (Fickian kinetics). The simulation is performed assuming a square system ($h_c = 2r_m$).

conclusion that ϕ can be used to modulate curve slope, while N_f can be set to shift up and down the release curve. Figure 4.20 reports the surface concentration setting $\phi = 0.5$, $N_f = 3$ ($r_h/r_m = 0.167$), $h_c/2r_m = 1$, sink conditions and $t^+ = 0.0027$, corresponding to $M_t^+ = 0.0365$. Holes position is very well individuated by a huge reduction of the dimensionless concentration C^+ ($= C/C_0$). On the contrary, between each hole, concentration is considerably higher as, here, drug depletion is lower (practically, only the Y component of the concentration gradient exists). In addition, correctly, in proximity of the membrane ($X^+ = 1$) a general depression of concentration surface is clearly detectable. Finally, it is didactically important to verify that, once fixed ϕ and r_h/r_m, $h_c/2r_m$ ratio variation simply reflects in a rigid shifting of the release curve as witnessed by Figure 4.21 . Obviously, the thinner the system (lower $h_c/2r_m$ values), the more rapid depletion is. This is a direct consequence of the lower mean path the diffusing molecules must undertake.

4.5.4.1   Buchtel Model

Although the solution to the drug release problem from a perforated impermeable membrane can be performed according to the numerical procedure

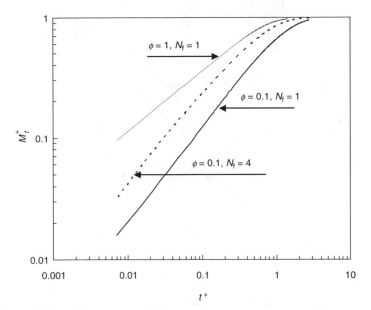

FIGURE 4.19 Effect of N_f (or r_h/r_m) variation, once ϕ has been fixed. Higher N_f favors drug release (see dotted line). $\phi = 1$ curve is reported as reference (Fickian kinetics). The simulation is performed assuming a square system ($h_c = 2r_m$).

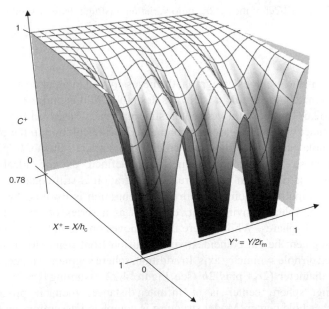

FIGURE 4.20 Dimensionless concentration surface $C^+(X^+, Y^+)$ assuming $\phi = 0.5$, $N_f = 3$ ($r_h/r_m = 0.167$), $h_c/2r_m = 1$, sink conditions and $t^+ = 0.0027$, corresponding to $M_t^+ = 0.0365$.

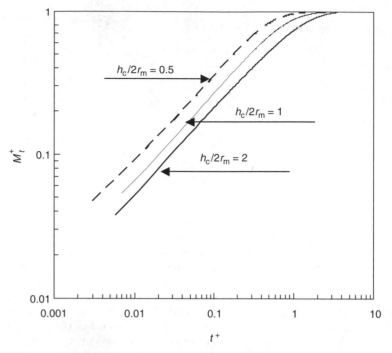

FIGURE 4.21 Effect of the $h_c/2r_m$ ratio variation on drug release assuming $\phi = 0.5$ and $N_f = 1$.

discussed in the previous paragraph, the necessity of a more agile and easy tool aimed to match this problem is always desirable. At this purpose, Pywell and Collet [62,63] proposed a very interesting approach based on the idea that, upon contact with the release environment, matrix depletion in the proximity of each hole occurs according to a precise geometrical scheme. In particular, they suppose that the diffusion front, defined as the ideal surface separating matrix region where drug concentration is still equal to its initial value from matrix region where drug concentration is lowered by diffusion towards the release environment, develops as a series of changing radius segments of spheres. While sphere center moves on the hole symmetry axis, sphere segment height coincides with diffusion front penetration distance h measured on hole symmetry axis. In addition, sphere segment wideness is equal to hole diameter ($2r_h$) plus $2h$ (see Figure 4.22 assuming $k = 1$). If, at the beginning, sphere center is at infinite distance, then, it progressively approaches hole center. Model structure is completed assuming, on the basis of some reasonable assumptions, that h is proportional to the square root of time. Despite the good Pywell and Collet intuition, they do not consider the

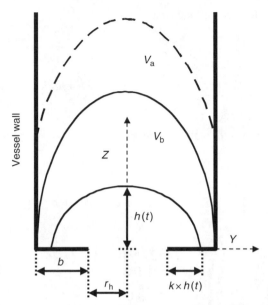

FIGURE 4.22 Time evolution of the diffusion front according to the Buchtel model. The front develops as a half rotational ellipsoid having height $h(t)$ and radius $(r_h + k \times h(t))$. When vessel wall or adjacent holes diffusion fronts meet, the diffusion front surface results from the intersection between the ellipsoid and a cylinder of radius $(r_h + b)$. In the Pywell and Collet approach, instead, the diffusion front surface is a segment of sphere having height h and cord $2(r_h + h)$. Accordingly, its shape is more squeezed than the Buchtel one and its center, lying on the Z axis, approaches hole surface as h increases. (From Pywell, E.J. and Collet, J.H., *Drug Dev. Ind. Pharm.*, 14, 2387, 1988.)

problem of diffusion front evolution after vessel wall or adjacent holes diffusion fronts matching. In addition, Equation 4.188 and Equation 4.189 numerical solution evidences that the diffusion front shape is better expressed by the following half rotational ellipsoid:

$$\frac{Z^2}{h^2} + \frac{Y^2}{(kh + r_h)^2} + \frac{X^2}{(kh + r_h)^2} = 1, \qquad Z \geq 0, \qquad (4.194)$$

whose Z–Y plane projection is depicted in Figure 4.22. The fact that k can be different from one increases approach generality. As soon as matrix depletion develops, h increases until the diffusion front reaches vessel walls or diffusion fronts competing to adjacent holes (for the sake of simplicity, we assume a circular, symmetric disposition of all equal holes). It is easy to verify that the half rotational ellipsoid volume reads

$kh \leq b$ (i.e., before vessel wall matching)

$$V_b = 4 \int_0^{kh+r_h} dY \int_0^{\sqrt{(kh+r_h)^2 - Y^2}} dX \int_0^{h\sqrt{1 - \frac{X^2 + Y^2}{(kh+r_h)^2}}} dZ = \frac{2}{3}\pi h(kh + r_h)^2, \qquad (4.195)$$

$kh > b$ (i.e., after vessel wall matching)

$$V_a = 4 \int_0^{b+r_h} dY \int_0^{\sqrt{(b+r_h)^2 - Y^2}} dX \int_0^{h\sqrt{1 - \frac{X^2 + Y^2}{(b+r_h)^2}}} dZ = \frac{2}{3}\pi h(kh + r_h)^2$$
$$\left(1 - \left(\sqrt{1 - \left(\frac{b+r_h}{kh+r_h}\right)^2}\right)^3\right).$$
$$\qquad (4.196)$$

The amount M_t of drug released from a membrane equipped by N_f holes is given by

$kh \leq b$

$$M_t = N_f\left(V_b C_0 - \int_0^h C(h')\,dV_b\,(h')\right), \qquad (4.197)$$

$kh > b$

$$M_t = N_f\left(V_a C_0 - \int_0^{b/k} C(h')\,dV_b\,(h') - \int_{b/k}^h C(h')\,dV_a\,(h')\right), \quad h \leq h_{fin}, \quad (4.198')$$

$$M_t = N_f\left(V_a C_0 - \int_0^{b/k} C(h')\,dV_b\,(h') - \int_{b/k}^{h_{fin}} C(h')\,dV_a\,(h')\right), \quad h > h_{fin}, \quad (4.198'')$$

where h_{fin} represents matrix thickness, h', spanning from 0 to h, is the integration variable, C_0 is the initial and uniform drug concentration in the matrix while $C(h')$ represents drug profile concentration that, on the basis of numerical simulations and literature indications [64], can be satis-factory approximated by the following parabolic expression:

$$C(h) = 2C_0 \frac{h'}{h} - C_0 \left(\frac{h'}{h}\right)^2. \tag{4.199}$$

This equation states that C is zero on hole surface (sink conditions in $h' = 0$) and that the parabolic profile gets its maximum in $h' = h$. For the relevance in practical applications, Equation 4.197 and Equation 4.198″ assume that h is always smaller than matrix thickness h_{fin} before vessel wall (or adjacent holes diffusion fronts) matching. If it were not the case, model formal variations are straightforward.

To make Equation 4.197, Equation 4.198′, and Equation 4.198″ operative, h time evolution must be defined. Starting points are the following mass balances:

$kh < b$

$$V_b(h) C_0 - \int_0^h C(h') dV_b(h') = \int_0^t \left(\pi r_h^2 D \frac{dC}{dh'}\bigg|_{h'=0}\right) dt, \tag{4.200}$$

$kh > b$

$$V_a(h) C_0 - \int_0^{b/k} C(h') dV_b(h') - \int_{b/k}^h C(h') dV_a(h') = \int_0^t \left(\pi r_s^2 D \frac{dC}{dh'}\bigg|_{h'=0}\right) dt, \tag{4.201}$$

where $\frac{dC}{dh'}\big|_{h'=0} = \frac{2C_0}{h}$. Both of them simply state that the drug amount that has left the matrix (left-hand side term) is equal to the drug flow integration over time (right-hand side term). Equation 4.201 integration yields to

$$\frac{2}{3} \pi C_0 \left(\frac{h^3 k^2}{10} + \frac{r_h h^2 k}{3} + \frac{r_h^2 h}{3}\right) = 2\pi r_h^2 DC_0 \int_0^t \frac{dt}{h}. \tag{4.202}$$

The derivation of both terms with respect to t leads to

$$\left(\frac{3}{10} h^2 k^2 + \frac{2}{3} r_h hk + \frac{1}{3} r_h^2\right) \frac{dh}{dt} = 3 \frac{D r_h^2}{h}. \tag{4.203}$$

Finally, the integration of this first order ordinary differential equation gives the relation between h and t:

$$\frac{3}{40} k^2 h^4 + \frac{2}{9} r_h k h^3 + \frac{1}{6} r_h^2 h^2 = 3 D r_h^2 t. \tag{4.204}$$

As, usually, h is small before vessel wall (or adjacent holes diffusion fronts) matching, only the quadratic term is important and the following simplification holds:

$$\frac{1}{6} r_h^2 h^2 \approx 3 D r_h^2 t \quad \Rightarrow h = \sqrt{18Dt}. \tag{4.205}$$

The same procedure, repeated for Equation 4.201, leads to a very complicated expression that can be roughly approximated by

$$h = \frac{r_h}{r_h + b} \sqrt{12Dt}. \tag{4.206}$$

It is easy to verify that Equation 4.206 is the result we would have if V_a were a cylinder of height h and radius $(r_h + b)$. As V_a tends to this cylinder volume for high h values, Equation 4.206 rigorously holds only after a long time. Reasonably, while before diffusion fronts matching h do not depend on hole radius, after contact, hole radius and holes half distance (b) become important. In both cases, anyway, in accordance with Lee [64], h square root dependence on time is found. Accordingly, for $h \leq b/k$, the model reads

$$M_t(h(t)) = N_f \frac{2}{3} \pi C_0 \left[\frac{k^2 h^3}{40} + \frac{r_h k h^2}{3} + \frac{r_h^2 h}{3} \right], \quad h = \sqrt{18Dt} \tag{4.207}$$

$$M_{t\infty} = N_f \pi h_{fin} C_0 (r_h + b)^2, \quad M_t^+ = \frac{M_t}{M_{t\infty}}. $$

For $h > b/k$, on the contrary, this model does not yield an analytical solution as the second integral in Equation 4.198′ and Equation 4.198″ needs a numerical solution. Nevertheless, its heaviness in terms of computational duties is much less than those required for Equation 4.188 numerical solution. In order to evaluate its characteristics, it is useful to compare Buchtel model with the two-dimensional-numeric solution of Equation 4.188. At this purpose, diffusion coefficient D, membrane void fraction ϕ, radius r_m, and holes number N_f are set. Then, the ratio b/r_h is calculated, for Buchtel, as

$$\phi = \frac{\pi r_h^2}{\pi (r_h + b)^2} \quad \Rightarrow \quad \frac{b}{r_h} = \frac{1 - \sqrt{\phi}}{\sqrt{\phi}} \tag{4.208}$$

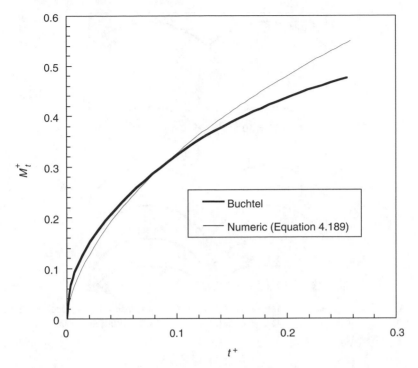

FIGURE 4.23 Comparison between Buchtel model (thick line) and numeric solution (thin line) Equation 4.189 assuming $D = 10^{-6}\,\text{cm}^2/\text{sec}$, $\phi = 0.5$, $r_m = 2263\,\mu\text{m}$ ($h_{\text{fin}} = 2263\,\mu\text{m}$) mm, 2 holes and $k = 0.5$. A satisfactory agreement holds up to $M_t^+ \approx 0.4$.

and for the numeric solution as

$$\phi = \frac{r_h}{(r_h + b)} \quad \Rightarrow \quad \frac{b}{r_h} = \frac{1 - \phi}{\phi}. \tag{4.209}$$

This asymmetric way of b calculation is made necessary by the intrinsic three and two dimensionality of Buchtel and numeric approach (Equation 4.188), respectively. Figure 4.23 shows the comparison between Buchtel (thick line) and two-dimensional-numerical solution (thin line) assuming $D = 10^{-6}$ cm^2/sec, $\phi = 0.5$, $r_m = 2263$ µm ($h_{\text{fin}} = 2263$ µm), two holes, and $k = 0.5$. While a satisfactory agreement occurs up to $M_t^+ \approx 0.4$, then, Buchtel underestimates the numeric solution. As this happens after that Buchtel diffusion front reaches cylinder height h_{fin} (Figure 4.24; compare the diffusion fronts dimensionless time and t^+ in Figure 4.23), we can argue that up to this condition Buchtel works properly and then it needs some improvements. Nevertheless, we can affirm that, in general, a qualitative agreement between Buchtel and two-dimensional-numerical solution takes place whatever the

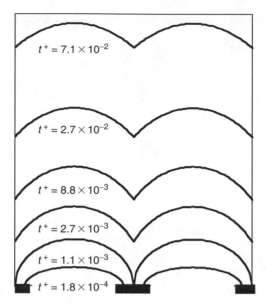

FIGURE 4.24 Time evolution of the diffusion front according to the Buchtel model in the case of two holes. The parameter used in this simulation is the same used in Figure 4.23.

values fixed for the geometrical parameters although this requires, sometimes, to modify the k value to make Buchtel diffusion front more similar to that calculated by numeric solution. It is also interesting to underline (see Figure 4.24) that while at the beginning diffusion front is represented by a rotational ellipsoid squeezed in the vertical direction (for $t^+ = 0$ it is a plane coinciding with hole surface), then, its squeezing occurs in the horizontal direction. Obviously, for long time, the diffusion front shape tends to be a plane and this will occur sooner for higher k values. Consequently, in data fitting, Buchtel is a two fitting parameters model (D and k).

Finally, we would like to conclude this paragraph with a curiosity. We named this model as Buchtel, in case of many holes, the diffusion front shape looks like the upper part of a typical Trieste cake named Buchtel (probably of Austrian origins).

4.6 CONCLUSIONS

The aim of this chapter was to present the principles on which mass transport and, thus, drug release, lie. Accordingly, the attention was initially focused on the definition of the intensive (pressure and temperature, for instance) and extensive (volume and thermodynamic potentials, for example) parameters characterizing a general thermodynamic system. On this basis, it was possible to define the conditions for the physical thermodynamic equilibrium that, basically, require same temperature, pressure (if surface phenomena are not important, see Chapter 5 and Chapter 6), and chemical potential in all system phases. The perturbation of this equilibrium, regardless of the reasons, implies a system reaction aimed to get a new equilibrium situation corresponding to a system minimum energy condition or system maximum entropy condition. Mass transport is simply one of the tools the system has to realize this target. In particular, while convective mass transport occurs due to the presence of a pressure gradient, diffusive mass transport takes place because of a chemical potential gradient that, in turn, can be caused by concentration and/or temperature gradients. Accordingly, the necessity of defining in mathematical terms the mechanisms ruling mass transport arises and mass, momentum, and energy conservation laws represent the result of this process. The simultaneous solution of these equations, combined with the proper initial and boundary conditions plus the constitutive equations describing system physical charac-teristics such as, density, viscosity, thermal properties, and diffusion coeffi-cients, allow the solution to mass transport problem. Fortunately, however, despite the complexity of the general frame hosting mass transport, in many situations, the problem can be considerably simplified as isothermal conditions and a pure diffusive mass transport can be assumed. Accordingly, only mass conservation law is necessary to match the problem of mass transport. If, additionally, diffusion coefficient can be retained time and concentration independent, mass transport problem reduces to the solution of Fick's second law. This theoretical approach can properly describe, at least with a reasonable approximation degree, what happens in many drug delivery systems. For this reason, Fick's second law analytical solutions for slab, cylindrical, and spher-ical matrices are proposed and discussed with the aim of establishing the macroscopic characteristics of Fickian release, namely a square root time dependence of the amount of drug released from slab and slightly different dependence in the case of cylinder and sphere. With this in mind, the effect of initial and boundary conditions on macroscopic release kinetics are evaluated. In particular, the effect of finite release environment volume, initial drug distribution, the presence of a stagnant layer surrounding the releasing surface and the presence of a holed impermeable membrane surrounding the releasing surface are considered. The most important finding is that the presence of the above-mentioned conditions can give origin to macroscopic non-Fickian re-lease kinetics although, at microscopic level, Fick's second law holds. Subse-quently, this analysis suggests possible tools to get different release kinetics aimed to improve system therapeutic efficacy and reliability. Finally, exact or approximate Fick's second law analytical solutions accounting for the above-mentioned initial and boundary conditions are provided in the light of improv-ing delivery systems designing.

REFERENCES

1. Camenish, G., Folkers, G., Waterbeemd van de, H., Review of theoretical passive drug absorption models: historical background, recent developments and limitations, *Pharm. Acta Helv.*, 71, 309, 1996.
2. Hildebrand, M., Pharmakokinetic und Arzneimittelentwicklung, *Deutske Apotheker*, Z.4, 23, 1994.
3. *Controlled Drug Delivery: Fundamentals and Applications*, Lee, V.H.L., Robinson, J.R., Eds., Marcel Dekker Inc., New York, Basel, 1987.
4. Siegel, R.A., pH-sensitive gels: swelling equilibria, kinetics and applications for drug delivery, *Pulsed and Self-regulated Drug Delivery*, Kost, J., Ed., CRC Press, Boca Raton, FL, 1990.
5. Ishihara, K., et al., pH induced reversible permeability control of the 4-carboxy acrylanilide-methyl methacrylate copolymer membrane, *J. Polym. Sci. Polym. Chem. Ed.*, 23, 2841, 1985.
6. Mathiowitz, E., et al., Photochemically rupture of capsules. I. A model system, *J. Appl. Polym. Sci.*, 26, 809, 1981.
7. Mathiowitz, E. and Cohen, M.D., Polyamide microcapsules for controlled release. I. Characterization of the membranes, *J. Membr. Sci.*, 40, 1, 1989.
8. Mathiowitz, E. and Cohen, M.D., Polyamide microcapsules for controlled release. II. Release characteristics of the microcapsules, *J. Membr. Sci.*, 40, 27, 1989.
9. Mathiowitz, E. and Cohen, M.D., Polyamide microcapsules for controlled release. III. Spontaneous release of azobenzene, *J. Membr. Sci.*, 40, 43, 1989.
10. Mathiowitz, E. and Cohen, M.D., Polyamide microcapsules for controlled release. 4. Effect of swelling, *J. Membr. Sci.*, 40, 67, 1989.
11. Mathiowitz, E. and Cohen, M.D., Polyamide microcapsules for controlled release. 5. Photochemical release, *J. Membr. Sci.*, 40, 67, 1989.
12. Lecomnte, F., et al., Polymers blends used for coating if multiparticulates: comparison of aqueous and organic coating techniques, *Pharm. Res.*, 21, 882, 2004.
13. Cardinal, J.R., et al., Use of asymmetric membranes in delivery devices, US Patent No. 5,612,059, 1997.
14. Macleod, G.S., et al., Selective drug delivery to the colon using pectin:chitosan: hydroxypropyl methylcellulose film coated tablets, *Int. J. Pharm.*, 187, 251, 1999.
15. Singer, S.J. and Nicolson, G.L., The fluid mosaic model of the structure of cell membranes, *Science*, 175, 720, 1972.
16. Rubino, A., Field, M., Shwachman, H., Intestinal transport of amino acid residues of dipeptides. I. Influx of the glycine residue of glycyl-L-proline across mucosal border, *J. Biol. Chem.*, 246, 3542, 1971.
17. Callen, H.B., *Thermodynamics*, John Wiley & Sons, Inc., New York, 1960, pp. 90–100.
18. Prausnitz, J.M., Lichtenhaler, R.N., Gomes de Azevedo, E., Classical thermodynamics of phase equilibria, in *Molecular Thermodynamics of Fluid-Phase Equilibria*, Prentice-Hall Inc., 1986, Chapter 2.
19. Topp, E.M., Principles of mass transfer, in *Transport Processes in Pharmaceutical Systems*, Amidon, G.L., Lee, I.L., Topp, E.M., Eds., Marcell Dekker, 2000, Chapter 1.
20. Flynn, G.L., Yalkowsky, S.H., Roseman, T.J., Mass transport phenomena and models: theoretical concepts, *J. Pharm. Sci.*, 63, 479, 1974.

21. Bird, R.B., Stewart, W.E., Lightfoot, E.N., *Transport Phenomena*, John Wiley & Sons, Inc., New York, London, 1960.
22. Dealy, J.M. and Wissbrun, K.F., Introduction to rheology, in *Melt Rheology and its Role in Plastics Processing*, Kluwer Academic Publishers, 1999, Chapter 1.
23. Crank, J., *The Mathematics of Diffusion*, 2nd ed., Clarendon Press, Oxford, 1975.
24. Grassi, M., Diffusione in matrici di idrogel polimerici per sistemi farmaceutici a rilsacio controllato, Thesis/Dissertation, University of Trieste (Italy), Department of Chemical Engineering, 1996.
25. Tonge, M.P. and Gilbert, R.G., Testing models for penetrant diffusion in glassy polymers, *Polymer*, 42, 501, 2001.
26. Reis, R.A., et al., Molecular dynamics simulation data of self-diffusion coefficient for Lennard–Jones chain fluids, *Fluid Phase Equilib.*, 221, 25, 2004.
27. Liu, H., Silvas, C.M., Macedost, E.A., Unified approach to the self-diffusion coefficients of dense fluids over wide ranges of temperature and pressure— hard-sphere, square-well, Lennard–Jones and real substances, *Chem. Eng. Sci.*, 53, 2403, 1998.
28. Reis, R.A., et al., Self- and mutual diffusion coefficient equation for pure fluids, liquid mixtures and polymeric solutions, *Chem. Eng. Sci.*, 60, 4581, 2005.
29. Chapman, S. and Cowling, T.G., *The Mathematical Theory of Non-Uniform Gases*, Cambridge University Press, Cambridge, 1970.
30. Grundke, E.W. and Henderson, D., Distribution functions of multicomponent fluid mixtures of hard spheres, *Mol. Phys.*, 24, 269, 1972.
31. Yu, Y.X. and Guang-Hua Gao, G.H., Self-diffusion coefficient equation for polyatomic fluid, *Fluid Phase Equilib.*, 166, 111, 1999.
32. Mansoori, G.A., et al., Equilibrium thermodynamic properties of the mixture of hard spheres, *J. Chem. Phys.*, 54, 1523, 1971.
33. Amsden, B., Solute diffusion within hydrogels. Mechanisms and models, *Macromolecules*, 31, 8382, 1998.
34. Carman, P.C., *Flow of Gases Through Porous Media*, Butterworths, London, 1956, pp. 45–50.
35. Muhr, A.H. and Blanshard, J.M.V., Diffusion in gels, *Polymer*, 23, 1012, 1982.
36. Ogston, A.G., Preston, B.N., Wells, J.D., On the transport of compact particles through solutions of chain-polymers, *Proc. R. Soc. Lond. Ser. A*, 333, 297–316, 1973.
37. Phillips, R.J., Deen, W.M., Brady, J.F., Hindered transport in fibrous membranes and gels: effect of solute size and fiber configuration, *J. Colloid Interface Sci.*, 139, 363, 1990.
38. Ogston, A.G., The spaces in a uniform random suspension of fibres, *Trans. Faraday Soc.*, 54, 1754, 1958.
39. Cukier, R.I., Diffusion of Brownian spheres in semidilute polymer solutions, *Macromolecules*, 17, 252, 1984.
40. Vrentas, J.S., et al., Prediction of diffusion coefficients for polymer–solvent systems, *AIChE J.*, 28, 279, 1982.
41. Peppas, N.A. and Reinhart, C.T., Solute diffusion in swollen membranes. Part I. A new theory, *J. Membr. Sci.*, 15, 275, 1983.
42. Lustig, S.R. and Peppas, N.A., Solute diffusion in swollen membranes. IX. Scaling laws for solute diffusion in gels, *J. Appl. Polym. Sci.*, 36, 735, 1988.
43. Duda, J.L. and Vrentas, J.S., Predictive methods for self-diffusion and mutual diffusion coefficient in polymer–solvent systems, *Eur. Polym. J.*, 34, 797, 1998.

44. Flory, P.J., *Principles of Polymer Chemistry*, Cornell University Press, Ithaca, NY, 1953.
45. Duda, J.L., et al., Prediction of diffusion coefficients for polymer–solvent systems, *AIChE J.*, 28, 279, 1982.
46. Hong, S.U., Prediction of polymer/solvent diffusion behaviour using free-volume theory, *Ind. Eng. Chem. Res.*, 34, 2556, 1995.
47. Grassi, M., Colombo, I., Lapasin, R., Drug release from an ensemble of swellable crosslinked polymer particles, *J. Contr. Rel.*, 68, 97, 2000.
48. Gray-Weale, A.A., et al., Transition-state theory model for the diffusion coefficients of small penetrants in glassy polymers, *Macromolecules*, 30, 7296, 1997.
49. Siepmann, J. and Peppas, N.A., Modeling of drug release from delivery systems based on hydroxypropyl methylcellulose (HPMC), *Adv. Drug Deliv. Rev.*, 48, 139, 2001.
50. Lee, P.I., Initial concentration distribution as a mechanism for regulating drug release from diffusion controlled and surface erosion controlled matrix systems, *J. Contr. Rel.*, 4, 1, 1986.
51. Lee, P.I., Novel approach to zero-order drug delivery via immobilized nonuniform drug distribution in glassy hydrogels, *J. Pharm. Sci.*, 73, 1344, 1984.
52. Lee, P.I., Diffusion-controlled matrix systems, in *Treatise on Controlled Drug Delivery*, Kydonieus, A., Ed., Marcel Dekker, Inc., 1992, Chapter 3.
53. Meriani, F., et al., *In vitro* nimesulide absorption from different formulations, *J. Pharm. Sci.*, 93, 540, 2004.
54. Zhou, Y., Chu, J.S., Wu, X.Y., Theoretical analysis of drug release into a finite medium from sphere ensembles with various size and concentration distributions, *Eur. J. Pharm. Sci.*, 22, 251, 2004.
55. Press, W.H., et al., *Numerical Recepies in FORTRAN*, 2nd ed., Cambridge University Press, Cambridge, USA, 1992.
56. Grassi, M., et al., Apparent non-fickian release from a scleroglucan gel matrix, *Chem. Eng. Commun.*, 155, 89, 1996.
57. Grassi, M., Membranes in drug delivery, in *Handbook of Membrane Separations: Chemical Pharmaceutical and Biotechnological Applications*, Sastre, A.M., Pabby, A.K., Rizvi, S.S.H., Eds., Marcel Dekker, 2006.
58. Patankar, S.V., *Numerical Heat Transfer and Fluid Flow*, McGraw-Hill/Hemi-Hemisphere Publishing Corporation, New York, 1990.
59. Kuu, W.Y. and Yalkowsky, S.H., Multiple-hole approach to zero-order release, *J. Pharm. Sci.*, 74, 926, 1985.
60. Kim, C.J., Release kinetics of coated, donut-shaped tablets for water soluble drugs, *Eur. J. Pharm. Sci.*, 7, 237, 1999.
61. Grassi, M., et al., Analysis and modeling of release experiments, in *Proceedings of the 22nd International Symposium on Controlled Release of Bioactive Materials*, Controlled Release Society, p. 364, 1995.
62. Pywell, E.J. and Collet, J.H., Theoretical rationale for drug release from non-idealised, planar, perforated laminates, *Drug Dev. Ind. Pharm.*, 14, 2387, 1988.
63. Pywell, E.J. and Collet, J.H., Drug release from polymeric, multi-perforated, laminated matrices, *Drug Dev. Ind. Pharm.*, 14, 2397, 1988.
64. Lee, P.I., Diffusional release of solute from a polymeric matrix. Approximate analytical solutions, *J. Membr. Sci.*, 7, 255, 1980.

5 Drug Dissolution and Partitioning

5.1 INTRODUCTION

Regardless the administration route, key factor for the success and reliability of a whatever formulation is drug bioavailability, defined as the rate and extent to which the active drug is absorbed from a pharmaceutical form and becomes available at the site of drug action [1]. Although metabolism and physiological factors highly affect drug absorption by living tissues, bioavailability strongly depends on drug permeability through cell membranes and drug solubilization in physiological fluids. Indeed, especially for what concerns oral formulations, if solubilization is the first absorption step, permeation is the second one (see also Chapter 2) as drug must dissolve in the physiological fluids and then it must cross cellular membranes. Although solubilization implies the drug dissolution process, permeation implies drug partitioning between a polar aqueous phase and an apolar phase (cellular membranes) unless active mechanisms rule drug permeation as discussed in Chapter 2. Indeed, for example, partitioning is the basis for the interpretation of passive absorption according to the two-step distribution model [2–4] interpreting permeability as the sum of two polar–apolar partitioning steps on both cell membrane sides (internal and external) as discussed later on. On the other hand, dissolution assumes relevant importance for poorly water-soluble drugs that, interestingly, represent more than 40% of the drugs recorded in the US Pharmacopoeia [5,6]. Indeed, most of them are optimized solely on the basis of pharmacological activity and not for what concerns bioavailability. In addition, this percentage is expected to increase with the advent of new biotechnology-based products (peptides and proteins) and the ever-increasing number of new compounds emerging from the discovery process that integrates combinatorial chemistry and high-throughput screening techniques [7]. Examples of commonly marketed drugs that are poorly soluble in water (less than 100 $\mu g/cm^3$) include analgesics, cardiovasculars, hormones, antivirals, immune suppressants, and antibiotics. Drugs with improved water solubility can be administered in a lower concentrated dose, with a reduction of local and systemic side-effects; this is crucial for drugs with important side-effect; profiles, such as antibiotics, antifungals, or antivirals. Moreover, improved dissolution means higher onset of action (i.e., faster drug therapeutic effect) that is particularly valuable for drugs intended to work immediately as required in pain, antianxiety, or antiemetic management [8].

The importance of both dissolution properties and permeability on drug delivery systems is very well emphasized by the biopharmaceutical classification system (BCS) [9] setting the regulatory framework for drugs classification. The BCS classification derives from the theoretical study led by Amidon and coworkers and devoted to simulate absorption from an ensemble of solid drug particles flowing in the intestinal lumen assuming constant permeability, plug flow conditions, and no particle–particle inter-action (aggregation). In particular, three dimensionless numbers ($Do =$ dose number, $Dn =$ dissolution number, and $An =$ absorption number) are individu-ated as ruling drug absorption. The theoretical analysis led on the basis of Do, Dn, and An, reveals that drug permeability and solubility are the key param-eters controlling drug absorption. Thus, drugs can be divided into high- or low-solubility–permeability classes and the expectations regarding *in vitro–in vivo* correlations more clearly stated. *Class I*, representing high-solubility–high-permeability drugs, contains drugs that are well absorbed and the rate limiting step to drug absorption is drug dissolution or gastric emptying if dissolution is very rapid. *Class II*, comprehending low-solubility–high-permeability drugs, is characterized by high *An* and low *Dn*. Accordingly, *in vivo* drug dissolution is the rate determining step in drug absorption (except at very high *Do*) and absorption is usually slower than for *class I*. Since the intestinal luminal contents and the intestinal membrane change along the intestine, and much more of the intestine is exposed to the drug, the dissolution profile along the intestine for a much greater time and absorption will occur over an extended period of time and pH. On the contrary, for *class III* drugs (high solubility–low permeability) permeability is the rate controlling step and both the rate and extent of drug absorption may be highly variable. Finally, *class IV* (low-solubility–low-permeability drugs) presents significant problems for effective oral delivery.

It is now clear that both drug dissolution and partitioning are key factors for understanding drug bioavailability and *in vitro* tests can provide useful information about it. In this light, Chapter 5 firstly deals with dissolution from a theoretical and experimental point of view evidencing its key param-eters such as solid surface properties, hydrodynamic of the solid–liquid interface, solid solubility, and drug stability in the dissolution medium. Then, with the same methodology, partitioning is matched devoting particular care to schemes approaching what happens *in vivo*. Of course, following the philosophy of this book, large use of mathematical modeling will be done and simultaneously dissolution and partitioning situations will be finally analyzed.

5.2 DISSOLUTION

The dissolution of a solid in a solvent is a rather complex process determined by a multiplicity of physicochemical properties of solute and solvent. Indeed, it can be considered as a consecutive process driven by energy changes [10] (Figure 5.1). The first step consists of the contact of the solvent with the solid surface (*wetting*), which leads to the production of a solid–liquid interface starting from solid–vapor one. The breakdown of molecular bonds of the solid (*fusion*) and passage of molecules to the solid–liquid interface (*solvation*) are the second and third steps, respectively. The final step implies the transfer of the solvated molecules from the interfacial region into the bulk solution (*diffusion*). Obviously, for performing each step, energy is required and the total energy required for solid dissolution is the sum of the energies relative to the four mentioned steps. Typically, the most important contribution to solid dissolution, in terms of energy required, is given by the *fusion* step. Although *solvation* and *diffusion* depend on solid–solvent chemical nature and on dissolution environment conditions (temperature and mechanical agitation, for example), *wetting* and *fusion* also depend on solid microstructure.

The classical interpretation of solid dissolution, the film theory [11], assuming a very large dissolution environment, suggests that, despite the

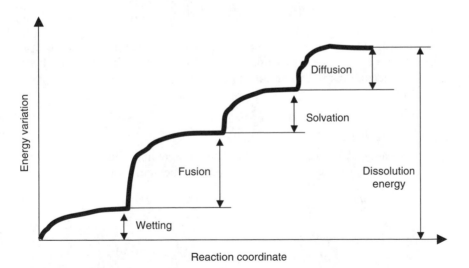

FIGURE 5.1 Dissolution can be considered as the sum of four consecutive energetic steps: solid wetting, crystal network destruction (fusion), solute molecule solvation, and diffusion in the bulk liquid phase (Adapted from Hsia, D.C., Kim, C.K., Kildsig, D.O., *J. Pharm. Sci.*, 66, 961, 1977.)

agitation, the solid surface is surrounded by a liquid stagnant film whose constant thickness h decreases with increasing solid surface–bulk liquid relative velocity (usually referred to as bulk liquid hydrodynamic conditions). Although the concentration of the solid molecules at the solid surface, C_s, is assumed to be equal to solid solubility in the dissolution medium, drug concentration C_b in the bulk solution is postulated homogeneous and smaller than C_s due to the mass transfer resistance exerted by the film. Indeed, it is assumed that inside the film, mass transfer occurs only due to diffusion. Accordingly, the drug mass conservation law assuming Fick's first law with constant diffusion coefficient for drug flux reads (see Equation 4.139):

$$\frac{\partial C}{\partial t} = D_{d0} \nabla^2 C, \tag{5.1}$$

where t is time, whereas D_{d0} and C are, respectively, drug diffusion coefficient and concentration in the stagnant layer. Postulating steady-state conditions ($\partial C / \partial t = 0$) (this assumption is more than reasonable in the light of usual h and D_{d0} values in drug dissolution) and assuming h to be very small in comparison with solid surface curvature radii (this means that solid surface can be considered as a plane), Equation 5.1 becomes

$$D_{d0} \frac{\partial^2 C}{\partial X^2} = D_{d0} \frac{\partial}{\partial X} \left(\frac{\partial C}{\partial X} \right) = 0. \tag{5.2}$$

In the light of the set boundary conditions (drug concentration equal to C_s in $X = 0$ and C_b in $X = h$), Equation 5.2 solution provides the following linear drug concentration in the film thickness:

$$C = (C_b - C_s)\frac{X}{h} + C_s, \quad \frac{\partial C}{\partial X} = \frac{(C_b - C_s)}{h}. \tag{5.3}$$

Accordingly, C_b (or m_b, the amount of drug present in the dissolution medium of volume V) increase in the bulk dissolution medium becomes

$$\frac{dm_b}{dt} = V\frac{dC_b}{dt} = S(-J) = SD_{d0}\frac{\partial C}{\partial X} = S\frac{D_{d0}}{h}(C_s - C_b), \tag{5.4}$$

$$\frac{dC_b}{dt} = \frac{S}{V}\frac{D_{d0}}{h}(C_s - C_b) = \frac{S}{V}k_d(C_s - C_b), \tag{5.4'}$$

where S is the solid surface area, J is drug flux, whereas k_d is the intrinsic drug dissolution constant expressed as a ratio of D_{d0} and h. Accordingly, k_d is strongly dependent on dissolution medium hydrodynamic conditions and this is the reason why some authors propose to define as "intrinsic dissolution constant" k_d value for very large (theoretically, infinite) difference between bulk liquid velocity and solid surface according to an extrapolation procedure [12]. Equation 5.4′ validity has been demonstrated on an experimental basis comparing it with a huge amount of data [13] and its solution, assuming that initial drug bulk concentration C_b is zero, is [14]:

$$C_b = C_s \left[1 - e^{\left(-k_d \frac{S}{V} t \right)} \right]. \tag{5.5}$$

Interestingly, Equation 5.4′ makes clear that dissolution is the result of kinetics (k_d) and thermodynamic equilibrium (C_s) contributions and both of them are very important in determining the dissolution rate. In addition, as this model holds after the development of a linear concentration profile in the limiting stagnant layer (steady-state assumption), the wetting step, influencing the very beginning of the dissolution process, is not explicitly accounted for. The effect of this kinetics aspect implicitly reflects in lower k_d values as discussed in Section 5.2.1. This brief introduction allows enucleating the fundamental aspects of dissolution and the topics that must be addressed to model it. Accordingly, in the light of Equation 5.4′ and of the energetic dissolution diagram shown in Figure 5.1, surface properties, hydrodynamic conditions, drug solubility, and drug stability in the dissolution medium will be discussed in detail in the following paragraphs.

5.2.1 SURFACE PROPERTIES

The separation zone between the two phases constituted by different matter aggregation state or between two immiscible phases in the same aggregation state represents the "interfacial phase." Accordingly, the possible existing interfacial phase regions that can occur are vapor–liquid, vapor–solid, liquid–liquid, liquid–solid, and solid–solid. Although, traditionally, matter properties are referred to as the bulk phase, sometimes, interfacial phase chemical and physical properties become relevant. Indeed, interfacial phase molecules experience different conditions with respect to those belonging to the homogeneous bulk phase. For example, in the liquid–vapor interface, whereas bulk phase molecules undergo attractive intermolecular forces of equal entity oriented in all the directions and with zero resultant, interfacial phase molecules undergo attractive forces whose resultant is directed toward the inner bulk phase. This depends on vapor molecules' low density (with respect to the liquid phase) that impedes to completely balance the attraction of the liquid molecules. Consequently, interfacial molecules are characterized by a separation distance bigger than that taking place in the bulk phase and thus, they show a higher surface energy (see Figure 5.2). In order to increase the liquid–vapor surface, some bulk phase molecules have to be driven in the interfacial phase region. This, of course, implies increasing molecules' energy and that is why proper energy has to be supplied by the external environment

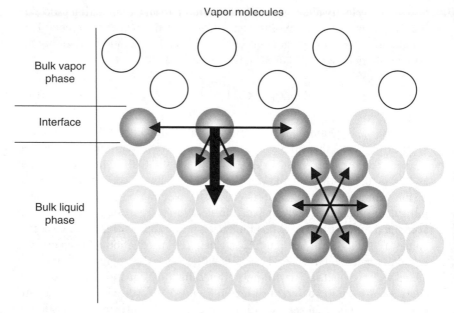

FIGURE 5.2 While in the bulk liquid phase the force resultant acting on the generic molecule is zero, this is no longer true for interfacial molecules. Indeed, due to the low density of vapor molecules, interfacial molecules force resultant is directed toward the liquid phase (see the *black big arrow*).

to make possible this transportation. The amount of energy supplied is the so-called *surface tension* [15], although, dimensionally, this is the energy per square unit or force per unit length or tension multiplied unit length. When this energy plays a relevant role, the thermodynamic system characterization performed in Chapter 4 must be implemented. Indeed, for an isolated system, the total internal energy U is given by

$$U = U^{V} + U^{\sigma}, \tag{5.6}$$

where U^{V} and U^{σ} represent, respectively, the volumetric and superficial contributions. Equation 5.6 can also be rewritten as

$$U = u^{V}V + u^{\sigma}\,AS, \tag{5.7}$$

where AS and V are, respectively, system surface area and volume, whereas u^{V} and u^{σ} are, respectively, the specific internal energy referring to the bulk and the interfacial phases. Of course, Equation 5.7 can be rewritten as

$$u = u^{V} + u^{\sigma}(AS/V). \tag{5.8}$$

Equation 5.8 evidences how the surface/volume ratio constitutes a very important parameter for the thermodynamic description of the system. Indeed, its value can signify the effect of the surface energy in comparison with the bulk energy.

5.2.1.1 Young's Equation

The contact angle of a liquid on a solid surface is defined as the angle comprised between the tangent to solid surface and the tangent to the liquid surface (see Figure 5.3) at the vapor–liquid–solid intersection. When a drop is put in contact with a solid substrate, in the absence of the gravity field, it assumes the shape minimizing the system free energy and finding an equilibrium condition between the solid–vapor (γ_{sv}), solid–liquid (γ_{sl}), and liquid–vapor (γ_{lv}) interfacial energies/square unit or forces/unit length or tension $*$ unit length. For a pure substance, at constant temperature and volume, the system Helmholtz energy A is function of

$$A = A(AS_{sv}, AS_{sl}, AS_{lv}), \tag{5.9}$$

where AS_{sv}, AS_{sl}, and AS_{lv} are, respectively, solid–vapor, solid—liquid, and liquid–vapor interfacial areas. Accordingly, the total differential A will be

$$dA = \left(\frac{\partial A}{\partial AS_{sv}}\right)_{T,V,AS_{sl},AS_{lv}} dAS_{sv} + \left(\frac{\partial A}{\partial AS_{sl}}\right)_{T,V,AS_{sv},AS_{lv}} dAS_{sl} + \left(\frac{\partial A}{\partial AS_{lv}}\right)_{T,V,AS_{sv},AS_{sl}} dAS_{lv}. \tag{5.10}$$

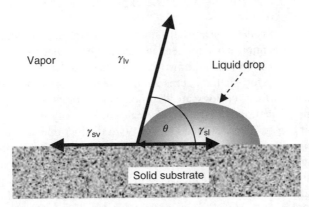

FIGURE 5.3 Solid–liquid contact angle θ and solid–vapor (γ_{sv}), solid–liquid (γ_{sl}) and liquid–vapor (γ_{lv}) surface tensions.

Remembering that

$$dAS_{sl} = -dAS_{sv} \left(\frac{\partial A}{\partial AS_i} \right)_{T,V,AS_{j \neq i}} = \gamma_i, \qquad (5.11)$$

Equation 5.10 can be rearranged as

$$dA = -\gamma_{sv} \, dAS_{sl} + \gamma_{sl} \, dAS_{sl} + \gamma_{lv} \, dAS_{lv}. \qquad (5.12)$$

Then, dividing for dAS_{sl}, and renumbering that at equlibrium $dA = 0$, we have

$$\left(\frac{\partial A}{\partial AS_{sl}} \right) = -\gamma_{sv} + \gamma_{sl} + \gamma_{lv} \left(\frac{\partial AS_{lv}}{\partial AS_{sl}} \right) = 0. \qquad (5.13)$$

But, at constant volume, the following relation holds [16]:

$$\left(\frac{\partial AS_{lv}}{\partial AS_{sl}} \right) = \cos \theta, \qquad (5.14)$$

where θ is the contact angle between the liquid and solid phases at the vapor–liquid–solid intersection. Consequently, Equation 5.13 can be rewritten as

$$\gamma_{lv} \cos(\theta) = \gamma_{sv} - \gamma_{sl}. \qquad (5.15)$$

This equation, known as Young's equation [17,18], represents the equilibrium of the three interfacial or surface tensions (see Figure 5.3) and the conditions for a complete wetting of a solid surface corresponding to $\theta = 0$. This means that the attraction forces between the liquid and the solid phases (adhesion forces) are equal or greater than those explicating inside the liquid phase (cohesion forces).

The above considerations strictly hold good when dealing with an ideal solid surface; this implies a homogeneous and perfectly plane surface without any irregularity. Unfortunately, however, real systems detach from this ideal scheme especially for what concerns solid surfaces. From a pure theoretical viewpoint, as surface tensions are thermodynamic properties of the solid–liquid couple, Equation 5.15 implies the existence of only one contact angle. However, it can be demonstrated [19] that two distinct contact angles exist: the advancing (θ_a) and receding (θ_r) contact angle. Although θ_a represents the contact angle appearing between a solid–liquid interface when the liquid phase is advancing on solid surface, θ_r is the contact angle appearing between a solid–liquid interface when the liquid phase is receding on solid surface. Due to solid surface irregularities, θ_a differs from θ_r, and this is the reason why, in general, experimental contact angle needs some corrections. In the case of surface roughness (Figure 5.4a), it is clear that the real contact angle θ_{re} is quite different from the apparent one θ_{ap}. Accordingly, defining the parameter r as follows:

$$r = \frac{A_{re}}{A_{ap}},\qquad(5.16)$$

where A_{re} is the real solid surface area and A_{ap} is the apparent solid surface area, the following relation holds [20]

$$\cos(\theta_{ap}) = r\cos(\theta_{re}).\qquad(5.17)$$

This equation allows the determination of θ_{re} once r and θ_{ap} are known.

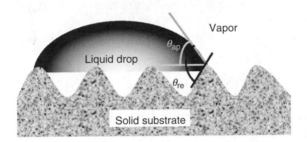

(a)

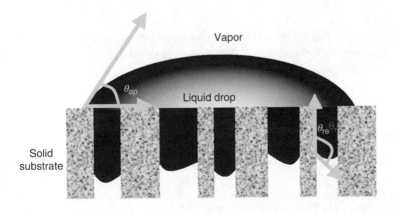

(b)

FIGURE 5.4 Apparent (θ_{ap}) and real (θ_{re}) solid–liquid contact angle in the case of rough (a) and porous (b) solid surfaces.

As clearly shown in Figure 5.4b, in the case of a porous surface also, θ_{ap} differs from θ_{re} and a correction is needed. As a consequence, assuming a plane and uniform solid surface, the Cassie–Baxter equation holds [21–23]

$$\cos(\theta_{ap}) = f_1 \cos(\theta_{re}) - f_2, \tag{5.18}$$

where f_1 and f_2 represent, respectively, the solid and void surface fraction. As f_1 and f_2 are proportional to solid porosity ε, Equation 5.18 can be rearranged as

$$\frac{\cos(\theta_{ap})}{1 - \varepsilon} = \cos(\theta_{re}) - \frac{\varepsilon}{1 - \varepsilon}. \tag{5.19}$$

Plotting $\cos(\theta_{ap})/(1 - \varepsilon)$ vs. $\varepsilon/(1 - \varepsilon)$ for a set of solid surfaces (substrates) with different porosities, it is possible to determine $\cos(\theta_{re})$ as the intercept of the straight line.

As real solid surfaces are both rough and porous, the general equation for real contact angle determination is

$$\cos(\theta_{ap}) = r\, f_1 \cos(\theta_{re}) - f_2. \tag{5.20}$$

When the solid surface is composed of a mixing of different phases each one showing its own contact angle and supposing that the extension of each surface phase is negligible in comparison with the macroscopic fluctuations of the liquid phase, the general form of the Cassie–Baxter equation holds

$$\cos(\theta_{ap}) = \sum_{i=1}^{n} f_i \cos(\theta_{rei}), \tag{5.21}$$

where i indicates the ith solid phase and f_i represents the solid surface fraction competing with the phase i. When only two phases constitute the solid surface, Equation 5.21 becomes

$$\cos(\theta_{ap}) = f_1 \cos(\theta_{re1}) + f_2 \cos(\theta_{re2}) \tag{5.22}$$

and, remembering that

$$f_1 + f_2 = 1, \tag{5.23}$$

we finally have

$$\cos(\theta_{ap}) = f_1(\cos(\theta_{re1}) - \cos(\theta_{re2})) + \cos(\theta_{re2}). \tag{5.24}$$

As Equation 5.24 represents a straight line in the variables $\cos(\theta_{ap})$ and f_1, performing different θ_{ap} measurements corresponding to different f_1 values, it is possible to indirectly determine θ_{re1} and θ_{re2} by resorting to the knowledge of Equation 5.24 slope $(\cos(\theta_{re1}) - \cos(\theta_{re2}))$ and intercept $\cos(\theta_{re2})$. Conversely, f_1 can be determined on the basis of θ_{ap}, θ_1, and θ_2.

5.2.1.2   Surface Tension Determination

Especially for organic solids, surface tension determination is usually based on the contact angle measurement. The starting point is the Young's equation (Equation 5.15) that establishes a relation between contact angle and the three surface tensions. Indeed, once $\cos(\theta)$ and γ_{lv} are experimentally measured [24] and γ_{sv} is theoretically determined, Young's equation allows the estimation of γ_{sl}. As a consequence, the key point of this method consists in γ_{sv} determination. For this purpose, different approaches have been developed in the past and the most important of them are discussed in the following. Basically, they can be divided into two main categories, namely the *surface tension components* and the *equation of state* [18].

The surface tension components strategy is based on the idea that surface tension is the sum of different contributes due to the different intermolecular forces. The first example of this category is that of Owens-Fowkes [24,25]. They start from the definition of work (per unit surface) of adhesion w_{AB} between two immiscible phases A and B:

$$w_{AB} = \gamma_A + \gamma_B - \gamma_{AB}, \qquad (5.25)$$

where γ_A and γ_B are, respectively, the surface energies per unit surface (or surface forces per unit length or surface tensions multiplied unit length) between the vapor and the A or B phase, whereas γ_{AB} represents the interfacial energy per unit surface (or surface forces per unit length or surface tensions multiplied unit length) between A and B. In other words, w_{AB}, is nothing more than the energy required to bring A (γ_A) and B (γ_B) molecules to the interface less the energy interaction across the interface (γ_{AB}). Of course, w_{AB} can be calculated on the basis of the potentials due to the presence of interaction forces such as dispersive, dipole–dipole, hydrogen bonding, and so on. This theory assumes that the most important interaction forces are the dispersive (i.e., London forces) and polar (dipole–dipole, hydrogen bonding) forces and the relative potentials ($\varepsilon_{AB}^d(r)$, $\varepsilon_{AB}^p(r)$) can be calculated according to

$$\varepsilon_{AB}^d(r) = \sqrt{\varepsilon_A^d \varepsilon_B^d} \qquad \varepsilon_{AB}^p(r) = \sqrt{\varepsilon_A^p \varepsilon_B^p}, \qquad (5.26)$$

where r indicates the distance, whereas ε_A^d, ε_B^d, ε_A^p, and ε_B^p are, respectively, the dispersion and polar forces potential referring to the phases A and B. Direct consequence of Equation 5.26 is

$$w_{AB} = \sqrt{w_A w_B}, \tag{5.27}$$

where, by definition, $w_A = 2\gamma_A$ and $w_B = 2\gamma_B$. Indeed, w_A and w_B are the energy of cohesion that corresponds to the energy required to detach in two part phases A and B, respectively. Accordingly, it follows:

$$w_{AB} = 2\left(\sqrt{\gamma_A^d \gamma_B^d} + \sqrt{\gamma_A^p \gamma_B^p} \right) = \gamma_A + \gamma_B - \gamma_{AB}, \tag{5.25'}$$

where γ_A^d, γ_B^d, γ_A^p, and γ_B^p are, respectively the dispersive and polar part of γ_A $(= \gamma_A^d + \gamma_A^p)$ and γ_B $(= \gamma_B^d + \gamma_B^p)$. If A is a solid phase and B a liquid phase, Equation 5.25' becomes

$$w_{sl} = 2\left(\sqrt{\gamma_{sv}^d \gamma_{sv}^d} + \sqrt{\gamma_{lv}^p \gamma_{lv}^p} \right) = \gamma_{sv} + \gamma_{lv} - \gamma_{sl}. \tag{5.28}$$

Remembering Young's equation (Equation 5.15), the model final form, in the solid–liquid case, reads

$$\gamma_{lv}(1 + \cos(\theta)) = 2\left(\sqrt{\gamma_{sv}^d \gamma_{sv}^d} + \sqrt{\gamma_{lv}^p \gamma_{lv}^p} \right). \tag{5.29}$$

Wu [26], following the same path, but assuming a harmonic mean for the definition of $\varepsilon_{AB}^d(r)$ and $\varepsilon_{AB}^p(r)$, gets

$$\gamma_{lv}(1 + \cos(\theta)) = \frac{4\gamma_{sv}^d \gamma_{lv}^d}{\gamma_{sv}^d + \gamma_{lv}^d} + \frac{4\gamma_{sv}^p \gamma_{lv}^p}{\gamma_{sv}^p + \gamma_{lv}^p}. \tag{5.30}$$

In both the Fowkes and the Wu model, γ_{lv} and θ, are experimentally known, γ_{lv}^d, γ_{lv}^p are known from literature, whereas γ_{sv}^d, γ_{sv}^p are the unknowns. Accordingly, their determination requires the simultaneous solution of two equations, Equation 5.29 or Equation 5.30 type referring to two different liquids on the same solid surface. As, now, the number of equations equates that of the unknowns, γ_{sv}^d, γ_{sv}^p can be determined jointly with γ_{sv} $(= \gamma_{sv}^d + \gamma_{sv}^p)$. The *acid–base* approach (also called *Lifshitz–van der Waals* (LW) approach) [27] can be considered an extension of Fowkes et al.'s approach as surface tension is divided into three components, namely the apolar (called LW) and the polar part (*ab*) that, in turn, is divided into the acid (+) and the base (−) part. Accordingly, for example, the solid–vapor surface tension γ_{sv} is given by

$$\gamma_{sv} = \gamma_{sv}^{LW} + \gamma_{sv}^{ab} = \gamma_{sv}^{LW} + 2\sqrt{\gamma_{sv}^{+}\gamma_{sv}^{-}}, \qquad (5.31)$$

where γ_{sv}^{LW}, γ_{sv}^{ab}, γ_{sv}^{+}, and γ_{sv}^{-} are, respectively, the apolar and polar parts that, in turn, are given by the acid and base parts. Accordingly, Equation 5.25' becomes

$$w_{sl} = 2\left(\sqrt{\gamma_{sv}^{LW}\gamma_{lv}^{LW}} + \sqrt{\gamma_{sv}^{+}\gamma_{lv}^{-}} + \sqrt{\gamma_{sv}^{-}\gamma_{lv}^{+}}\right) = \gamma_{sv} + \gamma_{lv} - \gamma_{sl}. \qquad (5.25'')$$

In the light of Young's equation (Equation 5.15), the final approach equation reads

$$\gamma_{lv}(1 + \cos\theta) = 2\left(\sqrt{\gamma_{sv}^{LW}\gamma_{lv}^{LW}} + \sqrt{\gamma_{sv}^{+}\gamma_{lv}^{-}} + \sqrt{\gamma_{sv}^{-}\gamma_{lv}^{+}}\right), \qquad (5.32)$$

whose unknowns are γ_{sv}^{LW}, γ_{sv}^{+}, and γ_{sv}^{-}, whereas γ_{lv}, and θ are experimentally determined and γ_{lv}^{LW}, γ_{lv}^{+}, and γ_{lv}^{-} are known from literature. As a consequence, determination of unknowns requires the simultaneous solution of three equations of Equation 5.32 type referring to three different liquids on the same solid. Table 5.1 reports the values of Wu approach (Equation 5.31) parameters (γ_{lv}^{d}, γ_{lv}^{p}) and the acid–base approach (Equation 5.32) parameters (γ_{lv}^{LW}, γ_{lv}^{+}, γ_{lv}^{-}) for some liquids commonly used in contact angle measurements.

According to Kwok and Neumann [18], the *equation of state* strategy is much more reliable in determining γ_{sv} than the *surface tension components*

TABLE 5.1
Liquid–Vapor Surface Tension γ_{lv} (20°C) and Relative Apolar ($\gamma_{lv}^{d} = \gamma_{lv}^{LW}$) and Polar ($\gamma_{lv}^{p} = \gamma_{lv}^{ab}$) Components ($\gamma_{lv}^{ab} = 2\sqrt{\gamma_{lv}^{+}\gamma_{lv}^{-}}$)

Liquid	γ_{lv} (mJ/m^2)	$\gamma_{lv}^{d} = \gamma_{lv}^{LW}$ (mJ/m^2)	$\gamma_{lv}^{p} = \gamma_{lv}^{ab}$ (mJ/m^2)	γ_{lv}^{+} (mJ/m^2)	γ_{lv}^{-} (mJ/m^2)
Water	72.8	21.8	51.0	25.5	25.5
Glycerol	64	34	30	3.92	57.4
Ethylene Glycol	48.0	29	19.0	1.92	47.0
Formamide	58	39	19	2.28	39.6
Dimethyl Sulfoxide	44	36	8	0.5	32
Chloroform	27.15	27.15	0	3.8	0
α-Bromonaphtalene	44.4	43.5	~0	—	—
Diiodomethane	50.8	50.8	~0	—	—

Source: From Leòn, V., Tusa, A., Araujo, Y.C., *Colloids Surf. A*, 155, 131, 1999. With permission from Elsevier.

strategy. The first equation of state is that based on Berthelot rule identifying with Equation 5.28 [18]. Combining Equation 5.25 and Equation 5.27, we get

$$w_{sl} = 2\sqrt{\gamma_{lv}\gamma_{sv}} = \gamma_{sv} + \gamma_{lv} - \gamma_{sl}. \tag{5.33}$$

Remembering Young's equation (Equation 5.15) and rearranging, it follows:

$$\gamma_{sv} = \gamma_{lv}\frac{(1 + \cos(\theta))^2}{4} \quad \text{or} \quad \cos(\theta) = 2\sqrt{\frac{\gamma_{sv}}{\gamma_{lv}}} - 1. \tag{5.34}$$

This equation holds good, approximately, for $|\gamma_{lv} - \gamma_{sv}| \le 35$ mJ/m². For higher differences, indeed, the geometric mean adopted (Equation 5.27) overestimates the potential $\varepsilon_{sl}(r)(=\varepsilon_{AB}(r))$ [18]. In order to improve Equation 5.34, Li [28] suggests the following correction for the expression of the work of adhesion w_{sl}:

$$w_{sl} = 2\sqrt{\gamma_{lv}\gamma_{sv}}e^{-\beta(\gamma_{lv}-\gamma_{sv})^2}, \tag{5.35}$$

where β is a model parameter that can be determined by means of the Newton's method [29] or, jointly with γ_{sv}, by means of data fitting. Notably, when γ_{lv} and γ_{sv} assume similar values, Equation 5.35 reduces to Equation 5.34. Inserting Equation 5.35 into Equation 5.25, remembering Young's equation (Equation 5.15) and rearranging, Li model finally reads

$$\cos(\theta) = \left(2\sqrt{\frac{\gamma_{sv}}{\gamma_{lv}}}\right)e^{-\beta(\gamma_{lv}-\gamma_{sv})^2} - 1. \tag{5.36}$$

As the application of this equation on many experimental data yields a mean β value equal to $12.47 * 10^{-5}$ (m²/mJ)², Equation 5.36 can be reasonably used to directly get γ_{sv} once γ_{lv} and $\cos(\theta)$ are known [18]. Following the same strategy yielding to Equation 5.36, but with a different correction factor for the work of adhesion, Kwok and Neumann propose [30]:

$$\cos(\theta) = 2\sqrt{\frac{\gamma_{sv}}{\gamma_{lv}}}(1 - \beta_1(\gamma_{lv} - \gamma_{sv})^2) - 1, \tag{5.37}$$

where γ_{sv} and β_1 can be determined by data fitting. Also in this case, as application of Equation 5.38 on many experimental data yields a mean β_1 value equal to 10.57×10^{-5} (m²/mJ)², it can be reasonably used to directly calculate γ_{sv} once γ_{lv} and $\cos(\theta)$ are known [18]. In alternative, Neumann proposes [31]:

$$\cos(\theta) = \frac{(0.015\gamma_{sv} - 2)\sqrt{\gamma_{sv}\gamma_{lv}} + \gamma_{lv}}{\gamma_{lv}(0.015\sqrt{\gamma_{lv}\gamma_{sv}} - 1)}. \tag{5.38}$$

Neumann and Kwok demonstrate the equivalence of Equation 5.36 and Equation 5.37.

It is now interesting to compare the abovementioned models on the γ_{sv} estimation relative to nimesulide, a typical nonsteroidal anti-inflammatory and poorly water-soluble (12 μg/mL at 25°C) drug belonging to class II according to the Amidon BCS [9]. For this purpose, three different liquids are considered: water (polar liquid; see Table 5.1), ethylene glycol (medium polar liquid; see Table 5.1), and di-iodomethane (apolar liquid; see Table 5.1). Contact angle measurements are made by means of a tensiometer (G10-Krüss GmbH, Hamburg, D) that automatically calculates contact angle on the basis of drop shape analysis (sessile drop configuration). Pure nimesulide tablets (1 cm diameter) are realized by the compression of approximately 0.5 g of nimesulide powder at 5 t. Table 5.2, reporting the results referring to two *surface tension components* approaches, i.e., Wu and acid–base models, makes clear that solid–vapor surface tension γ_{sv} according to the Wu model depends on the liquid couple considered. Indeed, it ranges from ~40 to ~51 mJ/m^2 and this would suggest considering the mean γ_{sv} obtained from the three possible couples derived from the use of three liquids. In addition, the results shown in Table 5.2 indicate that smaller γ_{sv} variations occur when a clear apolar liquid is used (at least when an apolar solid surface such as that of nimesulide is considered). Finally, although the polar component γ_{sv}^p increases, the apolar

TABLE 5.2
Nimesulide Solid–Vapor Surface Tension (γ_{sv}) Values According to Two Surface Tension Components Models

Liquid	θ (°)	Wu (Equation 5.31)			Acid-Base (Equation 5.33)
		γ_{sv} (mJ/m^2)	γ_{sv}^d (mJ/m^2)	γ_{sv}^p (mJ/m^2)	γ_{sv}^{LW} (mJ/m^2)
					50.3
Water	70 ± 5				
		40.3	21.6	18.6	$\sqrt{\gamma_{sv}^+}$ (mJ/m^2)$^{0.5}$
Ethylene glycol	35 ± 2				
					1.4
Water	70 ± 5				
		51.3	37.9	13.4	$\sqrt{\gamma_{sv}^-}$ (mJ/m^2)$^{0.5}$
Diiodomethane	17 ± 5				
					−0.17
Ethylene glycol	35 ± 2				
		48.6	42.6	6.0	γ_{sv} (mJ/m^2)
Diiodomethane	17 ± 5				
					49.9
Mean		**46.7**	**34.0**	**12.7**	

component γ_{sv}^d decreases with the increasing liquid couples mean polarity even if the predominance of γ_{sv}^d on γ_{sv}^p is always found regardless of the liquid couple considered. Finally, the γ_{sv} value estimated with the acid–base model does not differ too much from the mean coming from the Wu model. In addition, in this case, correctly, the predominance of γ_{sv}^d on γ_{sv}^p is clear even if it is much more accentuated than in the Wu case. On the other side, Table 5.3 reports the results referring to the *equation of states* approach. In particular, the Berthelot, Li, Kwok, and Neumann models are considered. The γ_{sv} determination is led by firstly considering the data coming from each liquid and then considering the data all together performing a data fitting. Although according to the Berthelot model, γ_{sv} increases with increasing liquid polarity, this does not happen with the other three models assuming, as suggested by the authors, $\beta = 12.47 \times 10^{-5}$ $(m^2/mJ)^2$ and $\beta_1 = 10.57 \times 10^{-5}$ $(m^2/mJ)^2$, for the Li and Kwok models, respectively. Of course, this is not surprising, as previously discussed, the Berthelot model does not hold when the difference $|\gamma_{lv} - \gamma_{sv}|$ is too high (>35 mJ/m^2) and this happens in the case of water (difference $= 46.8$ mJ/m^2). Basically, excluding the γ_{sv} value relative to water, the mean γ_{sv} value calculated according to the Berthelot approach does not substantially detach from the mean value coming from the other three models spanning from ~43 to 46 mJ/m^2. Interestingly, in this range, the γ_{sv} values coming from the Li, Kwok, and Neumann model best fitting on experimental data fall. In addition, the resulting β and β_1 are very similar to those suggested by Li and Kwok. In conclusion, although general conclusions

TABLE 5.3
Nimesulide Solid–Vapor Surface Tension (γ_{sv}) Values According to Different State Equations Models

Liquid	θ (°)	Berthelot (Equation 5.35) γ_{sv} (mJ/m²)	Li (Equation 5.37) γ_{sv} (mJ/m²)	Kwok (Equation 5.38) γ_{sv} (mJ/m²)	Neumann (Equation 5.39) γ_{sv} (mJ/m²)
Water	70 ± 5	33.2	42.0	41.3	48.1
Ethylene Glycol	35 ± 2	39.6	40.2	40.1	40.2
Diiodomethane	17 ± 5	48.6	48.6	48.6	48.6
Mean		40.4	43.6	43.4	45.7
Models best fitting results referring to the three liquids considered		β $(m^2/mJ)^2$ 18.4×10^{-5}	β_1 $(m^2/mJ)^2$ 17.0×10^{-5}		γ_{sv} (mJ/m²)
		γ_{sv} (mJ/m²) 44.6	γ_{sv} (mJ/m²) 44.6		46

Source: Contact angles θ are taken from De Simone, Studio dell'interazione polimero tensioattivo per l'incremento della bagnabilità di sistemi polimerici attivati impiegati per uso orale, Graduate Thesis, Dept. of Pharm. Sci, University of Trieste, Italy, 2003.

would require testing all models on many other drug solid surfaces, the indications coming from this brief analysis are that the most accurate γ_{sv} determination requires Li or Kwok equations fitting on experimental data referring to at least three liquids comprehending both polar and apolar liquids.

5.2.1.3 Wettability Effect on Drug Dissolution

A possible way for modeling the effect of wettability on drug dissolution consists in considering a mass transfer resistance $(1/k_m)$ at the solid–liquid interface and then solving the following equation:

$$\frac{\partial C}{\partial t} = D_{d0}\nabla^2 C \tag{5.1}$$

inside the stagnant layer with the following initial and boundary conditions: initial conditions

$$C = 0, \ \ 0 < X < h, \tag{5.39}$$

boundary conditions

$$D_{d0}\frac{\partial C}{\partial X}\bigg|_{X=0} = -k_m(C_s - C(X=0)), \tag{5.40}$$

$$\frac{\partial C}{\partial X}\bigg|_{X=h} = -\frac{V}{SD_{d0}}\frac{dC_b}{dt} = -\frac{V}{Shk_d}\frac{dC_b}{dt}, \ \ C_b = C(X=h). \tag{5.41}$$

Although Equation 5.39 simply states that, initially, the stagnant layer is drug empty, Equation 5.40 affirms that drug flux at the solid–liquid interface ($X=0$) depends on both k_m and the difference between drug solubility and actual local drug concentration. In other words, contrary to what is supposed by the film theory, Equation 5.40 states that drug concentration at the solid–liquid interface is not constantly equal to drug solubility but, starting from zero, it grows up to drug solubility and k_m rules this kinetics process. Accordingly, if high k_m values mean a fast interfacial drug concentration increase, this reflecting in good solid wettability, low k_m values are responsible for low interfacial increase, this translating in poor solid wettability. Consequently, for infinitely large k_m values, Equation 5.1 solution in the light of initial and boundary conditions expressed by Equation 5.39 through Equation 5.41 reduces to Equation 5.5. Finally, Equation 5.41 simply imposes that the drug amount leaving the stagnant layer is equal to that entering the bulk liquid. Assuming the following dimensionless variables:

$$t^+ = \frac{tD_{d0}}{h^2}, \ \ X^+ = \frac{X}{h}, \ \ C^+ = \frac{C}{C_s}. \tag{5.42}$$

Equation 5.1 and Equation 5.39 through Equation 5.41 become, respectively:

$$\frac{\partial C^+}{\partial t^+} = \frac{\partial}{\partial X^+}\left(\frac{\partial C^+}{\partial X^+}\right), \tag{5.1'}$$

$$C^+ = 0, \quad 0 < X^+ < 1, \tag{5.39'}$$

$$\left.\frac{\partial C^+}{\partial X^+}\right|_{X^+=0} = -\beta(1 - C^+(X^+ = 0)), \qquad \beta = \frac{k_m}{k_d}, \tag{5.40'}$$

$$\left.\frac{\partial C^+}{\partial X^+}\right|_{X^+=1} = -\alpha\frac{dC_b^+}{dt^+}, \qquad C_b^+ = C^+(X^+ = 1), \qquad \alpha = \frac{V}{Sh}, \tag{5.41'}$$

where β expresses the relative importance of wettability and dissolution, whereas α is the ratio between the dissolution medium volume and the volume occupied by the stagnant layer. Assuming, for example, $\alpha = 100$, Figure 5.5 shows Equation 5.1' numeric solution (control volume method [32]) corresponding to different β values ranging between 0.1 and infinity. While for an infinite β value (k_m is infinitely large), Equation 5.1' solution coincides with Equation 5.5 one, for lower β values, the entire dissolution

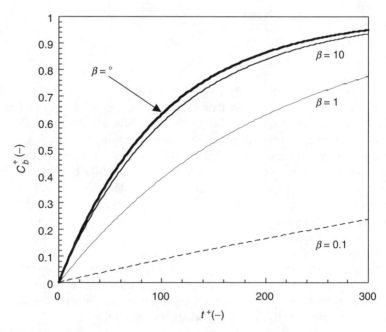

FIGURE 5.5 Effect of solid wettability on drug dissolution. The reduction of solid wettability (low β values) reflects into reduced dissolution rate. Accordingly, the dimensionless drug concentration C_b^+ increase vs. the dimensionless time t^+ is lowered.

process is considerably depressed. It is also easy to verify that Equation 5.5 can perfectly describe Equation 5.1′ solutions corresponding to different β values provided that smaller k_d values are assumed. In particular, referring to Figure 5.5, the curves characterized by $\beta = 10$, 1, and 0.1 correspond, respectively, to Equation 5.5 where k_d is equal to 0.91, 0.5, and 0.09 times the k_d value assumed for calculating the $\beta =$ infinity curve. Consequently, this demonstrates that Equation 5.5 implicitly accounts for solid wettability by means of lower k_d values. It is interesting to notice that this model yields an analytical solution under some reasonable simplifications. Indeed, Equation 5.1′ numerical solution, evidences how, despite the hypothesis of an interfacial drug concentration constantly equal to solubility does not hold, drug profile concentration in the stagnant layer approximately develops as a straight line shifting upward and maintaining a constant slope (this, of course, does not apply to the very beginning of the dissolution process). Accordingly, the linear concentration hypothesis leads to

$$-\alpha \frac{dC_b^+}{dt^+} = \frac{\partial C^+}{\partial X^+}\bigg|_{X^+=1} = \frac{\partial C^+}{\partial X^+}\bigg|_{X^+=0} = -\beta(1 - C^+(X^+=0)), \quad (5.43)$$

$$C^+(X^+=0) = \frac{C^+(X^+=1) + \beta}{1 + \beta} \quad C^+(X^+=1) = C_b^+ \quad (5.44)$$

From Equation 5.43 and Equation 5.44, it follows:

$$\frac{dC_b^+}{dt^+} = \frac{\beta/\alpha}{1 + \beta}(1 - C_b^+), \quad (5.45)$$

whose solution is

$$C_b^+ = 1 - e^{\left(-\frac{1}{\alpha} * \frac{1}{1+1/\beta} * t^+\right)} = 1 - e^{\left(-\frac{Sh}{V} * \frac{1}{1+k_d/k_m} * t^+\right)}. \quad (5.46)$$

Ultimately, this equation represents an approximated solution of Equation 5.1′, Equation 5.39′ through Equation 5.41′ and, correctly, if β approaches infinity (no wettability problems occur), it coincides with Equation 5.5. Due to the mathematical form of Equation 5.46, the simultaneous determination of k_d and k_m is not possible. However, remembering that, for sufficiently small times, Equation 5.46 reduces to the following linear expressions:

$$C_b^+ \approx \frac{S}{V} \frac{k_m k_d}{k_m + k_d} t \quad (5.47)$$

and, for very high k_m values (no wettability problems) it becomes

$$C_b^+ \approx \frac{S}{V} k_d t \quad k_m \to \infty, \tag{5.48}$$

the determination of both k_d and k_m becomes possible provided that a proper set of experimental data are available. For this purpose, we can focus the attention on the interesting paper of Chow and group [33]. The data shown in this paper deal with the effect of dopants on the dissolution rate of phenytoin (5,5-diphenylhydantoin) (DPH). In particular, doped DPH crystals are obtained by recrystallization from methanol with one of the following dopants: 3-acetoxymethyl-DPH (AMDPH), 3-propanoyloxymethyl-DPH (PMDPH), and 3-butanoyloxyymethyl-DPH (BMDPH). DPH doping provokes a change in crystal habit from acicular prisms to long thin plates, a decrease in melting enthalpy and entropy and an enhancement of intrinsic dissolution rate (IDR). Although the absorption of each dopant by DPH crystals is substantially comparable in magnitude under identical crystallization conditions, the amount of their adsorption on crystal surface is different as PMDPH is most extensively adsorbed and BMDPH the least. These aspects macroscopically translate into the surface properties shown in Table 5.4 reporting liquid–solid contact angles and solid–vapor surface tensions. While contact angles are measured by means of the sessile drop method (Rame–Hart type contact angle goniometer Model NRL-100), solid–vapor surface tension γ_{sv} is determined according to the sedimentation method [34]. This method is based on the concept that when no interactions occur between the solid and the liquid surface, the solid–liquid surface tension γ_{sl} is zero and, then, the liquid–vapor surface tension γ_{lv} equates the solid–vapor surface tension γ_{sv} (see Young's equation, Equation 5.15; $\cos \theta = 1$). Accordingly, in the case of a sedimentation experiment involving a particulate solid phase, the sedimentation volume would be minimal when the liquid phase γ_{lv} equates γ_{sv}. Practically, the authors, for each solidphase considered (pure DPH, APH–AMDPH, DPH–PMDPH, DPH–BMDPH, AMDPH, PMDPH, and BMDPH) evaluate the dependence of the sedimentation volume in liquid mixtures (1-propanol/water, 1,2-propanediol/water, 1-propanol/ethylene glycol, 1-octanol/1,3 propanediol) characterized by different compositions and, thus, by different liquid–vapor tension (25°C). The liquid mixture corresponding to the minimum value of the sedimentation volume indicates the solid–vapor surface tension. Table 5.4 shows, for all the solid surface considered, the contact angle with water and formamide jointly with the solid–vapor surface tension γ_{sv} calculated according to the sedimentation, Wu (Equation 5.30) and Li (Equation 5.36) methods. Wu approach is led considering the formamide–solid surface and water–solid surface contact angles reported in Table 5.4 and knowing that for formamide we have $\gamma_{lv}^d = 39$ mJ/m^2 and $\gamma_{lv}^p = 19$ mJ/m^2 [35], while for water we have $\gamma_{lv}^d = 22.1$ mJ/m^2, $\gamma_{lv}^p = 50.7$ mJ/m^2. Li approach is based on the

TABLE 5.4
Contact Angles and Solid–Vapor Surface Tension (γ_{sv}) (25°C) for Pure DPH, Dopants (AMDPH, PMDPH, BMDPH) and Doped DPH at Different Dopant Concentration in the Crystallization Solution (g/L)

Sample	θ_{Water} (°)	$\theta_{Formamide}$ (°)	γ_{sv} (mJ/m^2) Sedimentation	γ_{sv} (mJ/m^2) Wu (Equation 5.30)	γ_{sv} (mJ/m^2) Li (Equation 5.36)
DPH	65.8	47.5	42.4	43.21	42.45
AMDPH	68.5	52.1	44.1	40.75	39.60
PMDPH	66.5	50.1	42.9	42.09	40.59
BMDPH	45.0	33.2	50.9	55.70	51.20
DPH–AMDPH					
2 g/L	62.9	46.9	44.9	44.41	42.05
5 g/L	51.1	36.6	49.1	52.10	47.74
7 g/L	60.6	44.6	45.2	45.98	43.20
DPH–PMDPH					
2 g/L	62.4	46.4	43.4	44.75	42.30
5 g/L	60.9	44.6	44.4	45.86	43.30
7 g/L	62.6	43.8	43.7	45.51	44.32
DPH–BMDPH					
2 g/L	63.5	44.9	44.5	44.84	43.76
5 g/L	63.4	43.1	45.8	45.51	45.00
7 g/L	57.1	37.3	50.1	49.46	47.48

Source: From Chow, A.H.L., et al., *Int. J. Pharm.*, 126, 21, 1995. With permission from Elsevier.
Note: Three different approaches have been considered for the γ_{sv} calculation.

formamide–solid surface and water–solid surface contact angles reported in Table 5.4 and considering β as fitting parameter. It is clear that the presence of dopants modifies the solid–vapor surface tension and that this modification depends on the dopant kind and concentration. In general, the addition of dopant improves the γ_{sv} and this aspect is roughly proportional to dopant concentration. This behavior is particularly evident in the DPH–BMDPH case. Interestingly, the three methods considered for γ_{sv} evaluation provide similar results and, in general, the sedimentation values lie in between those of Wu and Li.

The effect of the abovementioned surface properties on drug dissolution can be now evaluated by means of the proposed model (Equation 5.47 and Equation 5.48) in the light of the intrinsic dissolution data provided by Chow and coworkers [33]. Practically, they provide, for pure DPH and for all other doped solids, the ratio R between the IDR corresponding to the generic doped solid and that of pure DPH (see Table 5.5). Accordingly, for DPH, we have $R = 1$. The determination of the k_d/k_m ratio for the different solids considered can be performed assuming that, in the DPH–AMDPH 5 g/L case, $R = 2.15$ is

TABLE 5.5
Dependence of R (Ratio Between the Intrinsic Dissolution Rate of the Generic Doped Solid and That of Pure DPH) On the Solid–Liquid Contact Angle (Water, 25°C) and the Corresponding Solid–Vapor γ_{sv} (Calculated According to the Li Equation) and Solid–Liquid Surface Tension γ_{sl} (Calculated According to Young's Equation)

Sample	R (°)	θ_{Water} (°)	γ_{sv} (mJ/m^2) Li (Equation 5.36)	γ_{sl} (mJ/m^2) Young's (Equation 5.15)
DPH	1	65.8	42.45	3.21
DPH–AMDPH				
2 g/L	1.42	62.9	42.05	2.42
5 g/L	2.15	51.1	47.74	1.17
7 g/L	1.95	60.6	43.20	1.90
DPH–PMDPH				
2 g/L	1.58	62.4	42.30	2.30
5 g/L	1.68	60.9	43.30	1.99
7 g/L	1.72	62.6	44.32	2.46
DPH–BMDPH				
2 g/L	1.76	63.5	43.76	2.67
5 g/L	1.77	63.4	45.00	2.65
7 g/L	1.91	57.1	47.48	1.35

so big that k_d/k_m is vanishing (k_m is very high in comparison to k_d and, thus, wettability does not slow down dissolution). Accordingly, on the basis of Equation 5.47 and Equation 5.48, R can be expressed by

$$R = 2.15 = R_{\text{DPH–AMDPH5}} = \frac{Sk_d}{V} \frac{V}{S} \frac{k_m + k_d}{k_m k_d} \Rightarrow \frac{k_d}{k_m^{\text{DPH–AMDPH5}}}$$

$$= R_{\text{DPH–AMDPH5}} - 1. \tag{5.49}$$

For all other solids considered, R can be expressed as

$$R = R_i = \frac{k_d k_m^i}{k_d + k_m^i} \frac{k_d + k_m^{\text{DPH–AMDPH5}}}{k_d k_m^{\text{DPH–AMDPH5}}}, \tag{5.50}$$

where R_i is the ratio referred to the generic solid while k_m^i and $k_m^{\text{DPH–AMDPH5}}$ refer, respectively, to the generic and to the DPH–AMDPH 5 g/L solid surface. Implicitly, Equation 5.50 assumes that k_d is the same for all the solid surfaces considered. After some simple algebraic manipulations, Equation 5.50 becomes

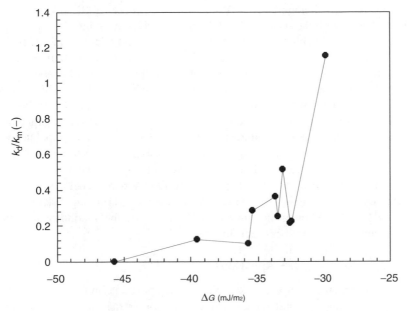

FIGURE 5.6 Trend of k_d/k_m vs. system energetic variation due to solid immersion $\Delta G = \gamma_{sl} - \gamma_{sv}$ at constant pressure and temperature.

$$\frac{k_d}{k_m^i} = \frac{R_{DPH-AMDPH5} - R_i}{R_i}. \tag{5.50'}$$

Figure 5.6 reports the trend of k_d/k_m vs. system energetic variation due to solid immersion $\Delta G = \gamma_{sl} - \gamma_{sv}$ at constant pressure and temperature. Interestingly, this figure shows that, in general, the more negative ΔG (wetting is, thermodynamically, more favorite) the lower k_d/k_m is and, thus, the higher k_m is as k_d can be retained constant for the different solids considered. In other words, when ΔG is highly negative, no wettability problems occur and, consequently, k_m assumes high values.

5.2.2 HYDRODYNAMICS

In order to make *in vitro* dissolution tests theoretically interpretable, many experimental setups have been considered [36]. This is the reason why a comprehensive and exhaustive detailed description of all of them is out of the aim of this chapter. Nevertheless, the most important need to be mentioned and some of them discussed in details in the light of practical implications. For example, Shah and coworkers [37] propose a magnet-driven rotating filter dissolution apparatus promoting stable flow of the dissolution medium near the dissolving surface. System hydrodynamics is studied by means of a convective diffusion model [38]. Vongirat and coworkers [39], modifying the Shah device with the introduction of a motor-driven rotating filter, realize a potentially predictable flow since they deal with a cylinder within a cylinder which represents, in the case of Newtonian fluids (see Chapter 3) an ordinary fluid mechanics problem. USP XXV IDR apparatus consists of a vessel containing the dissolution medium (900 ml of buffer pH 1.2 or pH 6.8 at $T = 37 \pm 0.5°C$) whose stirring is ensured by a rotating paddle. On the vessel bottom center a wax mould, carrying the dissolving disk, is fixed. Crane and coworkers [40] propose an interesting simplified approach to extrapolate system hydrodynamic and drug flux from the dissolving disk. Khoury and coworkers [36] study drug dissolution in the case of a stationary disk fixed in a rotating fluid (SDRF). Levich [13] performs a fundamental analysis of hydrodynamic and drug dissolution in the opposite case, i.e., a rotating disk in a fluid at rest. Finally, Peltonen and coworkers [41] approach the dissolution problem by means of a modified channel flow method.

5.2.2.1 Rotating Disk Apparatus

This experimental setup, also known as IDR apparatus, implies a drug disk fixed on a rotating shaft (ω is angular velocity) immersed in a very large dissolution medium (Figure 5.7). In order to greatly simplify the theoretical analysis, drug cylinder lateral surface is coated by a water impermeable membrane so that dissolution takes place only through the disk bottom plane surface. Shaft rotation provokes the dissolution medium movement driven by the perfect liquid adhesion to the rotating solid surface (see Figure 5.7). Accordingly, at the liquid–solid interface, dissolution fluid moves exclusively in a rotational (same angular velocity ω of the rotating disk) manner. As far as we detach from the interface, axial (v_y), tangential (v_ϕ), and radial (v_r) fluid velocity components are all nonzero and that is why the fluid approaches the solid surface according to spiral trajectory. Obviously, the radial velocity component is due to centrifugal forces. As the distance from the disk rotating surface increases, the tangential and radial velocity components become smaller and smaller so that for a distance greater than δ_0 they vanish and the only nonzero velocity component is the axial one. Practically, due to fluid continuity and perfect adhesion to the solid surface, the rotating disk recalls fluid from deeper dissolution environment and pushes it in the radial direction so that fluid stream lines describe a loop circulation (see Figure 5.7). Assuming isothermal, stationary, and Newtonian flow conditions and neglecting edge effects (disk radius much bigger than disk thickness), the description of the abovementioned physical frame requires the solution of the continuity (Equation 4.56) and momentum equation (Equation 4.70):

$$(\nabla \cdot \rho v) = 0, \tag{4.56}$$

$$(\nabla \cdot \rho vv) - (\nabla \cdot \underline{\tau}) - \nabla P + \rho \mathbf{g} = \mathbf{0}, \tag{4.70}$$

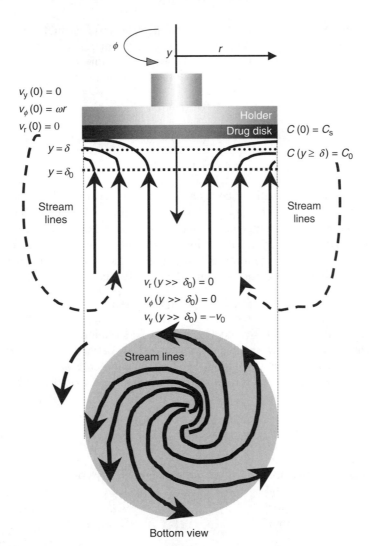

FIGURE 5.7 Schematic representation of fluid stream lines induced by drug disk rotation. The diffusion (δ) and the hydrodynamic (δ_0) boundary layers arise close to the solid surface.

provided that the following boundary conditions are attained:

$$v_r = 0; \quad v_\phi = \omega r; \quad v_y = 0, \quad y = 0 \text{ (disk surface)}, \qquad (5.51)$$

$$v_r = 0; \quad v_\phi = 0; \quad v_y = -v_0, \quad y \to \infty. \qquad (5.52)$$

While Equation 5.51 states that at the solid–liquid interface ($y = 0$) fluid moves only in the tangential direction (coherently with the solid surface),

Equation 5.52 imposes that very far from the solid surface, fluid moves only in the axial direction at constant velocity $-v_0$ (the minus reminds that fluid movement occurs upwards, i.e., in the negative y direction). Levich [13] demonstrates that the approximate analytical solution to this problem is

$$v_y \approx -0.89\sqrt{v\omega}, \quad y \to \infty, \tag{5.53}$$

$$v_y \approx -0.51 y^2 \sqrt{\omega^3/v}, \quad y \ll \sqrt{v/\omega}, \tag{5.54}$$

$$\delta_0 \approx 3.6\sqrt{v/\omega}, \tag{5.55}$$

where v is fluid kinematic viscosity and δ_0 may be regarded as the thickness of the hydrodynamic boundary layer on the disk surface. Within the boundary layer, the radial and tangential velocity components are not zero, while beyond that layer, only axial motion exists (see Figure 5.7). In order to definitively get a solution to mass transport from the rotating disk apparatus, stationary mass balance equation must be considered:

$$-(\nabla \cdot Cv) + \nabla \cdot (D\nabla C) = 0, \tag{4.46'}$$

where D is the drug diffusion coefficient in the dissolution medium, C is drug concentration, and v represents the convective field defined by Equation 5.51 through Equation 5.54. Assuming that C is function only of the distance y from disk surface and that it does not depend on either r or ϕ, Equation 4.46′ simplifies to

$$v_y(y)\frac{dC}{dy} = D\frac{d^2C}{dy^2}. \tag{5.56}$$

This equation must accomplish the following boundary conditions:

$$C(y = 0) = C_s, \tag{5.57}$$

$$C(y = \infty) = C_0, \tag{5.58}$$

where C_s and C_0 are, respectively, drug solubility and drug concentration in the dissolution medium bulk. While Equation 5.57 states that drug concentration at the solid–liquid interface (no wettability problems occur) equates drug solubility, Equation 5.58 assumes that drug concentration in the dissolution medium bulk is homogeneous and equal to C_0. To get Equation 5.56 solution, a double integration is needed:

$$\frac{dC}{dy} = a_1 \exp\left(\frac{1}{D}\int_0^y v_y(y^*)\,dy^*\right) \quad \text{1st integration,} \tag{5.59}$$

$$C = a_1 \int\limits_0^y \exp\left(\frac{1}{D} \int\limits_0^{y^\nabla} v_y(y^*) \, dy^*\right) dy^\nabla + a_2 \quad \text{2nd integration,} \qquad (5.60)$$

where a_1 and a_2 are constants to be determined on the basis of boundary conditions. While from Equation 5.57 it descends $a_2 = C_s$, Equation 5.58 requires

$$\frac{(C_0 - C_s)}{a_1} = \int\limits_0^y \exp\left(\frac{1}{D} \int\limits_0^{y^\nabla} v_y(y^*) \, dy^*\right) dy^\nabla. \qquad (5.61)$$

In the hypothesis of $v/D \gg 1$ (this is the typical situation met in drug dissolution experiments) and recalling dependence of v_y on y (Equation 5.53 and Equation 5.54), Levich demonstrates that the right-hand side term in Equation 5.61 is equal to $J_1 = 1.61166 \, D^{1/3} \, v^{1/6} \, \omega^{-1/2}$. Accordingly, we have $a_1 = (C_0 - C_s)/J_1$. As a consequence, the drug flux F leaving the disk surface is given by

$$F = -D \frac{\partial C}{\partial y}\bigg|_{y=0} = \frac{D(C_s - C_0)}{J_1} \exp\left(\frac{1}{D} \int\limits_0^{y=0} v_y(y^*) \, dy^*\right)$$

$$= (C_s - C_0) 0.62 D^{2/3} v^{-1/6} \omega^{1//2}. \qquad (5.62)$$

Finally, this equation allows the estimation of the diffusion boundary layer thickness δ:

$$\delta = \frac{D(C_s - C_0)}{F} = 1.61 \left(\frac{D}{v}\right)^{1/3} \sqrt{\frac{v}{\omega}} = 0.4772 \left(\frac{D}{v}\right)^{1/3} \delta_0. \qquad (5.63)$$

It is interesting to notice that, typically, δ is much smaller than δ_0. Indeed, in the very common case of dissolution in an aqueous medium at 37°C, $v = 7.02 \times 10^7 \, \text{m}^2/\text{sec}$ while D can be assumed equal to $10^{-9} \, \text{m}^2/\text{sec}$ (usually, drug diffusion coefficient in water at 37°C ranges between 10^{-9} and 10^{-8} m^2/sec). Accordingly, $\delta = 0.05 \, \delta_0$. Finally, one of the most important conclusions deriving from Levich's analysis consists in the independence of the diffusion boundary layer thickness δ on the distance from the rotation axis. In addition, this theory allows to express the intrinsic drug dissolution constant k_d (appearing in Equation 5.5) as a function of drug (D), dissolution medium (v) and hydrodynamics (ω) conditions:

$$k_d = \frac{D}{\delta} = 0.621 D^{2/3} v^{-1/6} \sqrt{\omega}. \qquad (5.64)$$

Before discussing an application example of this theory, it is interesting to focus attention on the hypotheses on which it relies. Apart from the already mentioned ones (isothermal, stationary, and Newtonian flow conditions, negligible edge effects and perfect liquid adhesion to the rotating solid surface), others, equally important hypotheses, need to be discussed. First of all, laminar flow conditions must be attained; this means that Reynolds number Re $((\omega R_D) * R_D/\nu)$, where R_D is disk radius) has to be lower than 10^4. Practically, assuming $R_D = 0.01$ m and $\nu = 7.02 \times 10^{-7}$ m^2/sec (water, 37°C), laminar flow conditions are attained up to $\omega = 68.06$ rad/sec (650 rpm). On the other side, the independence of δ on radial position holds only if local Re $((\omega r) * r/\nu)$, where r is local radial coordinate) exceeds 10^2.

Below this value, δ assumes smaller values than those predicted by Equation 5.63 so that a central zone exists where δ is smaller than on the remaining part of disk surface and, consequently, drug flux is higher. Assuming $\nu = 7.02 \times 10^{-7}$ m^2/sec (water, 37°C) and different ω and R_D values, Table 5.6 shows the ratio A_c/A_t between disk central surface where $Re < 10^2$ and, thus, Equation 5.63 no longer holds, and total disk surface. Assuming that Levich's hypotheses are accomplished if $A_c/A_t < 0.33$, it follows that when $R_D = 0.005$ m the minimum ω should be 100 rpm while this limit lowers to 75 rpm in the case of $R_D = 0.0075$ m while, for $R_D = 0.01$, no minimum rotating velocity problems occur. Finally, Levich's approach holds if the solid surface is perfectly smooth. Indeed, the presence of a depression on a surface (see Figure 5.8) causes turbulence if its depth d is comparable with the hydrodynamic boundary layer δ_0. In this case, even at small Re, separation of the streamlines above the depression occurs while inside it a reversed flow takes place. The turbulence zone spans not only in and above depression, but

---

TABLE 5.6
Dependence of A_c/A_t (Ratio Between Disk Central Surface (Radius R_C) Where $Re < 10^2$ and, thus, Equation 5.63 No Longer Holds, and Total Disk Surface) on Disk Radius R_D and Disk Angular Velocity ω

	$100 * R_D$(m) $\Rightarrow$		0.5	0.75	1.0
ω (rpm)	ω (rad/s)	$100 * R_C$ (m)	A_t/A_c	A_t/A_c	A_t/A_c
50	5.23	0.37	0.55	0.24	0.14
75	7.85	0.30	0.36	0.16	0.09
100	10.47	0.26	0.27	0.12	0.07
150	15.71	0.21	0.17	0.08	0.04
200	20.94	0.19	0.14	0.06	0.04
250	26.18	0.17	0.11	0.05	0.03

Note: Gray cells indicate unacceptable conditions for Levich's theory to hold.

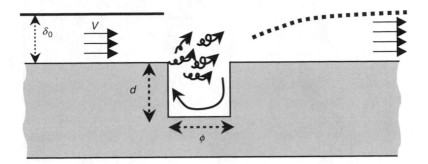

FIGURE 5.8 Effect of depressions on hydrodynamic boundary layer (δ_0). (From Grijseels, H., et al., *Int. J. Pharm.* 14, 299, 1983. With permission from Elsevier.)

also in the region immediately downstream from it [42]. Then, it is dampened gradually and the laminar boundary layer recurs. Obviously, δ_0 destruction implies also the diffusion layer δ destruction or thinning so that, locally, dissolution rate is improved. Experimental evidences [43] show that this improvement depends on liquid velocity and kinematic viscosity, depression shape, dimensions, and position. Accordingly, it is very hard to derive a general model able to comprehend all these aspects. Nevertheless, the huge work of Grijseels and coworkers allows the derivation of some useful guidelines [43]. First of all, they suggest that a solid surface can be defined smooth if the height of the highest surface irregularity is smaller than the following critical height h_c:

$$h_c = \alpha \, r \, Re^{-0.75}, \quad Re = (\omega r) * r / \nu, \tag{5.65}$$

where for a single, cylindrical, roughness element $\alpha = 7.865$ (for sharp elements, considerably smaller α values are found). Figure 5.9, showing Equation 5.65 trend vs. disk radial coordinate r for $\alpha = 1.1$ and different angular velocities ω, makes clear that the higher the ω, the smaller the h_c, regardless of r. In addition, h_c decreases as r increases. In conclusion, this simulation suggests that, in common experimental conditions, if roughness height is smaller than 20 μm, solid surface can be defined smooth.

Although turbulence appearance and, thus, dissolution rate increase, are proportional to both depression depth d and diameter ϕ, when $\phi \leq 1$ mm (this is the usual condition for drug disks obtained by compression) and $d/\phi \geq 1.5$, the effect of depth is negligible. In this hypothesis, the effect of surface depression on δ_0 (and, thus, on the dissolution rate) becomes significant if depression diameter ϕ is bigger than a critical value ϕ_c. On the basis of the work of Grijseels and coworkers [44], it is possible to derive a relation

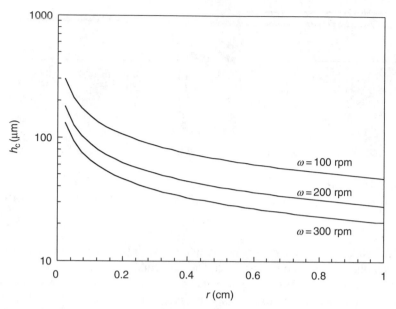

FIGURE 5.9 Dependence of the critical surface irregularity (h_c) vs. tablet radius (r) assuming three different tablet rotational speeds (ω) ($T = 37°C$, dissolution medium: water). If all surface irregularities are lower than h_c, surface can be defined smooth.

between the depression critical diameter ϕ_c, radial position r, liquid kinematic viscosity ν, and angular velocity ω:

$$\phi_c = \frac{2.744 \times 10^{-6}}{\nu^{1/4} \omega^{3/4} r^{1/2}}. \tag{5.66}$$

This relation establishes that ϕ_c decreases with increasing r, ν, and ω although with different exponents. This dependence is shown in Figure 5.10 assuming that dissolution takes place in water at 37°C ($\nu = 7.02 \times 10^{-7}$ m^2/sec) and that ω spans from 50 to 500 rpm. Interestingly, it comes out that if $\phi < 50$ μm, the effect of depressions is negligible whatever the radial position r and angular velocity ω considered are.

It is now useful to use Levich's approach to estimate theophylline mono-hydrated ($C_7H_8N_4O_2 \cdot H_2O$; Carlo Erba, Milano) diffusion coefficient D_{THEO} in water at 37°C [45]. A smooth cylindrical theophylline disk (5 mg, 1 cm diameter, compressed at 5 t) is attached, by means of liquefied paraffin, to a rotating stainless steel disk of the same diameter. The whole apparatus is immersed in a 150 cm^3 thermostatic water volume and disk motion is immediately started after immersion. Theophylline time concentration is continuously collected by means of a personal computer managing an UV detector (UV–Vis Spectrophotometer, Lambda 6, Perkin Elmer-USA) set at

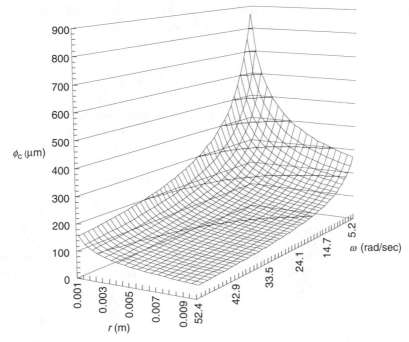

FIGURE 5.10 Dependence of the depression critical diameter ϕ_c on radial position r and angular velocity ω ($T = 37°C$; dissolution medium: water).

271 nm for 300 sec. Three different ω are considered (11.52, 15.71, and 18.32 rad/sec). Dissolution data are fitted according to Equation 5.5 knowing that theophylline solubility C_s in water at 37°C is 12495 ± 104 $\mu g/cm^3$. The only fitting parameter, k_d, results in $(1.87 \pm 0.001) \times 10^{-3}$, $(2.17 \pm 0.0012) \times 10^{-3}$, and $(2.37 \pm 0.02) \times 10^{-3}$ cm/sec, respectively, for $\omega = 11.52$, 15.71, and 18.32 rad/sec. These data are then fitted by means of Equation 5.64 to finally get $D_{THEO} = (8.2 \pm 0.6) \times 10^{-10}$ m^2/sec. The knowledge of D_{THEO} allows the determination of theophylline molecule hydrodynamic radius R_H on the basis of the Stokes–Einstein equation:

$$D_{THEO} = \frac{KT}{6\pi\eta R_H}, \qquad (4.100)$$

where K is the Boltzman constant, T is the absolute temperature (37°C), η is the water shear viscosity (6.97×10^{-4} Pa sec). The result we get, $R_H = 4$ Å, is perfectly in agreement with the estimation of theophylline van der Waals radius (3.7 Å) calculated according to molecular modeling techniques [46]. In addition, it is possible to further verify the result we got. Indeed, knowing that $R_H = 4$ Å and η (25°C) $= 8.94 \times 10^{-4}$ Pa sec, Stokes–Einstein equation allows to conclude that $D_{THEO(25°C)} - (6.2 \perp 0.4) \times 10^{-10}$ m²/sec which is in perfect agreement with what was found by De Smidt and group [47].

5.2.2.2 Fixed Disk Apparatus

In this case, a stationary disk is immersed in a fluid that, ideally, moves, very far from disk surface, in a solid-body rotation. When the rotating fluid approaches the stationary disk, it moves inward and it is discharged axially as reported in Figure 5.11. Accordingly, the hydrodynamic and diffusion layers δ_0 and δ are formed at the outermost edge of the disk. Assuming that a magnetic stirring bar provokes fluid motion, Khoury and group [36] suppose that mass flux F can be described by the following classical equation:

$$F = D\frac{(C_s - C_0)}{\delta} \tag{5.67}$$

although, theoretically, the use of this equation is questionable in the present physical frame. Their strategy is to derive a proper expression for δ in order to

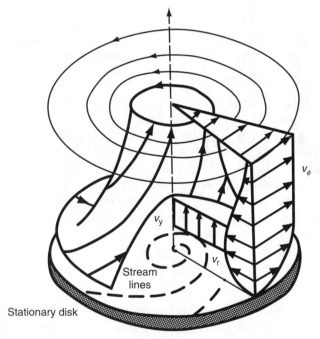

FIGURE 5.11 Velocity field components (v_y, v_r, and v_ϕ are, respectively the axial, radial, and tangential one) and flow stream lines occurring in the stationary disk apparatus (From Khoury, N., Mauger, J.W., Howard, S., *Pharm. Res.*, 5, 495, 1988. With kind permission of Springer Science and Business Media.)

build up a model able to describe experimental drug release rate Q. While very far from the solid surface an ideal frictionless rotational flow is supposed, within the hydrodynamic layer δ_0, a frictional effect occurs. Accordingly, the only two forces acting on the generic fluid element belonging to the hydrodynamic layer and rotating at distance r from disk center are centrifugal force f_c (radially directed) and force f_τ, due to wall shearing stress, forming an angle θ with the tangential component of velocity. Thus, at equilibrium we have

$$f_c = dm\,\omega^2 r = \tau\,ds\,sen(\theta) = f_\tau\,sen(\theta) \quad \text{radial component,} \qquad (5.68)$$

$$f_\tau \cos(\theta) = \tau\,ds\cos(\theta) \propto ds\,\eta\,\frac{\partial v}{\partial z} \approx ds\,\frac{\eta\omega r}{\delta_0} \quad \text{tangential component,} \quad (5.69)$$

where dm and ds are, respectively, element mass and surface, η is fluid viscosity and τ is the wall stress. Remembering that $ds = dV/\delta_0$, where dV is element volume, Equation 5.68 and Equation 5.69 become

$$\tau\,sen(\theta) = \rho\omega^2 r\delta_0 \quad \text{radial component,} \qquad (5.68')$$

$$\tau\cos(\theta) \propto \frac{\eta\omega r}{\delta_0} \quad \text{tangential component,} \qquad (5.69')$$

where $\rho = dm/dV$ (fluid density). Solving Equation 5.68' and Equation 5.69' for δ_0, we have

$$\delta_0 \propto \sqrt{(v/\omega)\,tg(\theta)}, \qquad (5.70)$$

where v is fluid kinematic viscosity. Recalling the relation between δ and δ_0 found by Levich in the rotating disk case ($\delta \propto (D/v)^{1/3}\delta_0$, Equation 5.63), we have

$$\delta \propto \left(\frac{D}{v}\right)^{1/3}\sqrt{\frac{v}{\omega}}. \qquad (5.71)$$

In order to render this approach more adherent to reality, the authors, modify Equation 5.71 so that δ is dependent on radial position r:

$$\delta \propto \left(\frac{D}{v}\right)^{1/3}\sqrt{\frac{v}{\omega}}g(r), \qquad (5.71')$$

where $g(r)$ is an r function to be defined on the basis of experimental data. Finally, the combination of Equation 5.67 and Equation 5.71 leads to

$$F \propto \frac{(C_s - C_0)\,D^{2/3}v^{-1/6}\omega^{1/2}}{g(r)} \qquad (5.72)$$

and Q becomes

$$Q = \int\limits_0^r 2\pi r F(r)\, dr = k_0(C_s - C_0)D^{2/3}\nu^{-1/6}\omega^{1/2}\int\limits_0^r \frac{r\, dr}{g(r)}, \qquad (5.73)$$

where k_0 is a constant to be determined on the basis of experimental data. Khoury and coworkers, working on hydrocortisone alcohol and hydrocortisone acetate tablets dissolution in water (25°C) and assuming $C_0 = 0$, find that $k_0 = 1.33$, while $\int_0^r \frac{r\, dr}{g(r)} = r^{3/2}$. Accordingly, $g(r) = (2/3)\sqrt{r}$.

5.2.2.3 USP Apparatus

The experimental setup, shown in Figure 5.12, consists in a drug disk fixed to a wax mould positioned at 3 mm from the vessel base. To ensure that drug dissolution takes place only from disk lateral surface, a wax coating is applied on the disk top surface. Dissolution medium consists of 900 mL of 0.1 N HCl solution stirred at 50 rpm (37°C). Drug concentration time course is measured by sampling the dissolution medium at desired time.

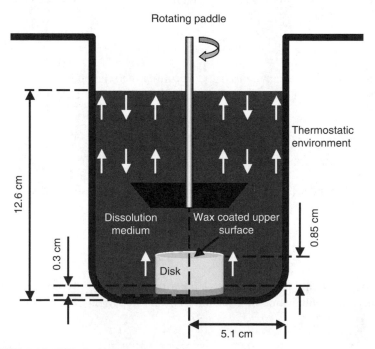

FIGURE 5.12 USP dissolution apparatus.

Crane and coworkers [40] approach the dissolution process according to the mass conservation law (with constant diffusivity and a zero generative term) in the diffusion layer surrounding the cylindrical solid surface:

$$\frac{\partial C}{\partial t} = -(\nabla \cdot C v) + D \nabla^2 C. \tag{4.46'}$$

The dissolution process is not explicitly accounted for in Equation 4.46′ but it appears in the boundary condition expressed by Equation 5.74. Indeed (see Figure 5.13), the amount of solid drug dissolved in the time interval dt (left-hand side part of Equation 5.74 and dotted rectangle in Figure 5.13) must be equal to the drug amount that left the diffusion layer δ in the same time interval (right-hand side of Equation 5.74):

$$(C_{d0} - C_s)(r(t) - r(t + dt)) = -D \frac{\partial C}{\partial r}\bigg|_{r=r(t)} dt, \tag{5.74}$$

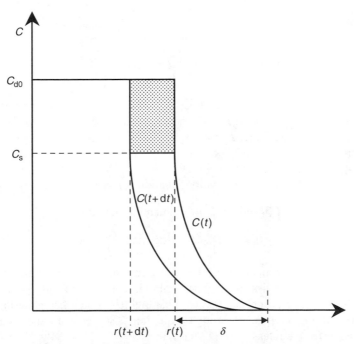

FIGURE 5.13 Schematic representation of solid dissolution according to Crane and coworkers [48]. r is the radial coordinate, C_s is drug solubility, C_{d0} is initial drug concentration while δ represents the diffusion boundary layer. (From Crane, M., et al., *Simulation Modelling Practice and Theory*, 12, 147, 2004. With permission from Elsevier.)

where C_{d0} is drug concentration in the dissolving disk (it would equal drug density if disk porosity were vanishing), C_s is drug solubility in the dissolving medium (it is implicitly supposed that drug concentration at the solid or liquid interface is equal to drug solubility in the liquid medium and, thus, no wettability problems occur), r is the radial coordinate and t is time. Drug concentration in the boundary layer always goes to zero to attain sink conditions. Consequently, it follows:

$$\frac{dr}{dt} = -\frac{D}{(C_{d0} - C_s)} \frac{\partial C}{\partial r}\bigg|_{r=r(t)}. \tag{5.75}$$

On the basis of Equation 5.75, an order of magnitude analysis demonstrates that Equation 4.46′ can be simplified as the term $\frac{\partial C}{\partial t}$ is negligible in comparison to $D\nabla^2 C$. Indeed, we have

$$\frac{\partial C}{\partial t} \approx \frac{C_s}{\delta} \frac{dr}{dt} \approx \frac{C_s}{\delta}\left(\frac{D}{C_{d0}} - C_s\right) \frac{C_s}{\delta} \approx \left(\frac{C_s}{\delta}\right)^2 \left(\frac{D}{C_{d0}} - C_s\right) \tag{5.76}$$

and, thus

$$\frac{\partial C}{\partial t}\bigg/ D\frac{\partial^2 C}{\partial r^2} \approx \left(\frac{C_s}{\delta}\right)^2 \left(\frac{D}{C_{d0} - C_s}\right)\bigg/ D\frac{C_s}{\delta^2} = \frac{C_s}{C_{d0} - C_s} = \frac{\lambda}{1 - \lambda}, \tag{5.77}$$

where $\lambda = C_s/C_{d0}$. As, in usual situations $\lambda \ll 1$, it is evident that $\partial C/\partial t$ is negligible in comparison to $D\nabla^2 C$ and the problem we are dealing with can be considered quasi-static. Accordingly, the equations ruling the entire process are

$$(\nabla \cdot v) = 0 \quad \text{continuity equation} \quad (\rho = \text{constant}), \tag{5.78}$$

$$(\nabla \cdot vv) = \nu\nabla^2 v \quad \text{momentum equation}, \tag{5.79}$$

$$(\nabla \cdot Cv) = D\nabla^2 C \quad \text{drug mass balance}, \tag{5.80}$$

where ν is fluid kinematic viscosity. Assuming that we are dealing with an infinitely long disk (disk radius a negligible in comparison with its height L) and that velocity v is constant (U_0) and parallel to cylinder lateral surface, the solution of Equation 5.78 through Equation 5.80 leads to the following expression for the release rate Q (mass/time):

$$Q = 4.26aD^{2/3}C_s\nu^{1/3}\left[\sqrt{\frac{U_0 L}{\nu}} + 0.42\frac{L}{a}\right]. \tag{5.81}$$

Remembering the equivalent Q expression:

$$Q = 2L\pi a C_s \frac{D}{\delta} = 2L\pi a C_s k_d, \qquad (5.82)$$

the δ and k_d expressions deriving from the Crane approach are

$$\delta = \frac{2\pi L}{4.26\left(\dfrac{\nu}{D}\right)^{1/3}\left[\sqrt{\dfrac{U_0 L}{\nu}} + 0.42\dfrac{L}{a}\right]}, \qquad (5.83)$$

$$k_d = \frac{4.26 D^{2/3}\nu^{1/3}\left[\sqrt{\dfrac{U_0 L}{\nu}} + 0.42\dfrac{L}{a}\right]}{2\pi L}. \qquad (5.84)$$

Interestingly, Equation 5.64 and Equation 5.84 predict the same k_d dependence on velocity and diffusion coefficient although a completely different dependence on fluid kinematic viscosity appears.

Crane and group, working on benzoic acid dissolution (C_s in 0.1 N HCl, 37°C, $= 4.55$ mg/mL; $D = 1.236 \times 10^{-9}$ m^2/sec; $U_0 = 0.0183$ m/sec, $\nu = 7.86 \times 10^{-7}$ m^2/sec; $L = 0.0085$ m; $a = 0.0065$ m), find that the predicted Q value according to Equation 5.81 ($= 2 \times 10^{-8}$ kg/sec) differs only by 18% from the experimental one ($= 2.43 \times 10^{-8}$ kg/sec). Of course, this is a good result especially in the light of model simplifying hypotheses (infinite long disk, ν constant and parallel to disk lateral surface). Undoubtedly, the elegance and practical importance of this approach are considerable. Nevertheless, it suffers for the necessity to independently determine U_0 according to a numerical simulation. Crane and group provide $U_0 = 0.0183$ m/sec when the paddle rotating speed is 50 rpm.

5.2.2.4 Channel Flow Method

Channel flow technique, more common in physical chemistry than in the pharmaceutical field, allows gaining very useful information on dissolution kinetics due to the well-defined hydrodynamics realized in the channel. Briefly, the body of the apparatus consists in a hydrophilic polymer that can withstand organic solvents such as acetone (see Figure 5.14). By means of a proper holder positioned on the top of the channel, the tablet faces to the flowing fluid and its alignment can be realized according to an adjusting screw. In order to guarantee a perfect sealing, a rubber O-ring is placed in between the top and bottom channel parts that are jointed together by means of six stainless steel screws. The tablet is positioned at the center of the channel to avoid edge effects. Aqueous solution flow rate, controlled by a peristaltic pump, can be

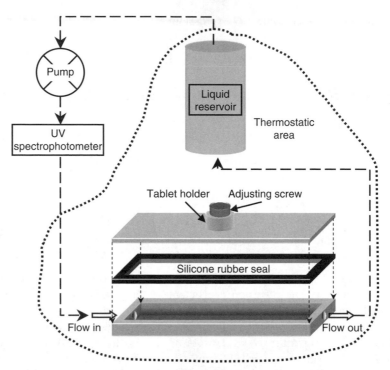

FIGURE 5.14 Schematic representation of the channel flow method apparatus. (From Peltonen, L., et al. *Eur. J. Pharm. Sci.*, 19, 395, 3003.)

varied between 100 and 900 mL/h. Drug concentration is measured by an on-line UV–Vis spectrophotometer, while temperature is controlled by submerging the solution reservoir and channel in a thermostatic bath.

In order to study dissolution kinetics in the channel flow method, Peltonen and group [41] assume that drug flux F across the diffusion boundary layer (δ) can be described by the following equation:

$$F = k_d(C_s^* - C), \qquad (5.85)$$

where k_d is the mass transfer coefficient, while C_s^* and C are, respectively, drug concentration at the solid–liquid interface and in the dissolution medium bulk. Implicitly, Equation 5.85 is based on the hypothesis that dissolution flux is constant over the tablet surface (this means that the diffusion layer thickness δ is constant all over the solid surface) and that the convective diffusion in the channel flow is taken into account through k_d that, of course, is not constant. At this purpose, the authors consider the relation suggested by Levich [13]:

$$k_d = \beta(Q_{FC})^{1/3}, \tag{5.86}$$

where β is a constant depending on channel geometry and on dissolving drug diffusion coefficient D, while Q_{FC} is solution flow rate. Assuming that drug dissolution can be represented by first-order reaction [48]:

$$F = k_f - k_b C_s^*, \tag{5.87}$$

where k_f and k_b are the forward (dissolution) and reverse (precipitation) rate constants, respectively, C_s^* can be eliminated to get F final expression:

$$F = \frac{k_d k_f (K - C)}{k_f + k_d K}, \tag{5.88}$$

where $K = k_f/k_b$ is the equilibrium constant of the dissolution reaction (solubility). Finally, a mass balance on the whole system allows the determination of drug bulk concentration C increase:

$$\frac{dC(t)}{dt} = \frac{FA}{V}, \tag{5.89}$$

where A is tablet area exposed to flow and V is the dissolution medium volume. Equation 5.89 solution leads to

$$C = K\left(1 - \exp\left(-\frac{k_d k_f}{k_f + k_d K}\frac{A}{V}t\right)\right). \tag{5.90}$$

If dissolution is very fast in comparison to drug diffusion through the diffusion layer ($k_f \longrightarrow \infty$), Equation 5.91 simplifies into

$$C = K\left(1 - \exp\left(-k_d\frac{A}{V}t\right)\right). \tag{5.91}$$

From the abovementioned equation, it is possible to derive an expression for the diffusion layer thickness δ:

$$k_d = \frac{D}{\delta} = \beta(Q_{FC})^{1/3} \rightarrow \delta = \frac{D}{\beta(Q_{FC})^{1/3}}. \tag{5.92}$$

Peltonen and group [41] apply this model to acetylsalicylic acid dissolution in buffer solutions (hydrochloric acid buffer at pH 1.2 and phosphate buffer at pH 6.8; $V = 100$ cm^3; $A = 0.636$ cm^2; $T = 37°C$) assuming K and β as fitting parameters. In particular, β is retained independent on Q_{FC} and pH so that a unique value applies to all the experiments performed. In addition, as fitting is insensitive to k_f, (dissolution is ruled by mass transfer), Equation 5.91 is considered in place of Equation 5.90. A good data fitting is achieved with $\beta = 0.8/\text{min}^{2/3}$; $K = 0.3$ mg/mL, pH $= 1.2$; $K = 0.57$ mg/mL, pH $= 6.8$. Remembering that, for acetylsalicylic acid $D = 7.7 \times 10^{-10}$ m^2/sec [49] and on the basis of Equation 5.92, it is possible to estimate the diffusion layer thickness δ dependence on Q_{FC} ($\delta = 5.8$ μm, $Q_{FC} = 180$ mL/h; $\delta = 4.6$ μm, $Q_{FC} = 350$ mL/h; $\delta = 4.0$ μm, $Q_{FC} = 540$ mL/h; $\delta = 3.7$ μm, $Q_{FC} = 700$ mL/h; both pH).

5.2.3 DRUG SOLUBILITY AND STABILITY

Up till now, it has been implicitly assumed that drug properties, such as solubility in the dissolution medium, do not change upon dissolution. But, it is well known that many organic compounds may exist in different crystalline structures such as polymorphs and solvated forms as it often happens in organic drugs [50]. Of course, temperature, pressure, and surrounding conditions (i.e., the existence of liquid phase in contact with the solid phase) establish which crystalline structure is the most stable one. Accordingly, upon dissolution (temperature and pressure remain constant) the contact with a new fluid phase (dissolution medium) can comport a solid phase transformation from one polymorphic phase (most stable configuration in absence of the dissolution medium) to another one (most stable configuration in presence of the dissolution medium). Consequently, the polymorphic transformation can take place in some dissolution media (solvents) and not in other. As, for thermodynamic reasons, a drug exhibiting crystal polymorphism should possess different activities depending on crystal structure [51], polymorphic transformations generally imply a variation of drug solubility and this aspect highly influences the entire dissolution process. A particularly interesting phase transformation is that involving the amorphous drug transformation into solid drug usually occurring in presence of aqueous dissolution media. Also in this case a considerable variation (decrease) of drug solubility can occur [52]. For example, Nogami and coworkers [50] study the dissolution of p-hydroxybenzoic acid and phenobarbital involving transformation to the respective hydrates formed during dissolution. Analogously, De Smidt and coworkers [47] consider the dissolution of anhydrous theophylline implying the formation of theophylline-hydrate with consequent solubility reduction while Di Martino and coworkers [53] study anhydrate and hydrate naproxen sodium dissolution. Finally, Higuchi and coworkers [54] focus the attention on the dissolution rate of two polymorphic forms of sulfathiazole and methylprednisolone.

Obviously, phase transformation is not the only physical phenomenon that can affect drug dissolution. Indeed, complex behaviors can arise from the dissolution of polyphase mixture made up by chemically different solid phases [55] or by two chemically equal polymorphs. Finally, the dissolution process can be also affected by intrinsic drug instability in the dissolution medium (due to particular pH conditions, for example) that can induce drug degradation. In this light, this section focuses the attention on the mathematical modeling of drug dissolution in case of phase transformation, dissolution of polyphase mixtures and drug dissolution in presence of drug degradation induced by the dissolution medium.

5.2.3.1 Dissolution and Phase Changes

As previously discussed, the contact with a fluid phase can induce solid polymorphic or anhydrous–hydrate transformation. Physically speaking, we can suppose that, initially, solid surface is entirely made up by, let us say, polymorph I (or anhydrous solid). Assuming that the drug release rate at interface is much faster than mass transport process across the diffusion layer, superficial phase transformation occurs until the whole dissolution surface is composed by polymorph II (or hydrated solid). This implies that solid solubility spans from the initial value C_{si}, competing to polymorph I (or anhydrous solid), to the final one C_{sf} competing to polymorph II (or hydrate solid). Supposing that the rate of crystal growth (polymorph II or hydrate solid) is proportional to the degree of supersaturation and is ruled by a first-order kinetics with respect to concentration, solubility variation can be described by the following equation [50]:

$$\frac{dC_s}{dt} = k_r(C_{sf} - C_s), \tag{5.93a}$$

where C_s is the time-dependent solid solubility and k_r is the recrystallization constant. Consequently, it follows:

$$C_s = C_{sf} + (C_{si} - C_{sf})e^{(-k_r t)}. \tag{5.93b}$$

If we further assume that phase transformation does not imply dissolution surface S variation and that drug diffusion coefficient in the diffusion layer δ is the same for the two polymorphic forms, the entire dissolution process can be described by inserting Equation 5.93b into Equation 5.4′:

$$\frac{dC_b}{dt} = \frac{S}{V}k_d(C_s(t) - C_b) = \frac{S}{V}k_d(C_{sf} + (C_{si} - C_{sf})e^{(-k_r t)} - C_b), \tag{5.94}$$

whose solution is

$$C_b = C_{sf}\left(1 - e^{\left(-\frac{k_d S}{V_r}t\right)}\right) + \frac{(C_{si} - C_{sf})}{\left(1 - \frac{k_r V_r}{k_d S}\right)}\left(e^{(-k_r t)} - e^{\left(-\frac{k_d S}{V_r}t\right)}\right), \quad (5.95)$$

where C_b is bulk concentration.

Due to the high parameters correlation, the direct use of Equation 5.95 for data fitting is not recommended and a more articulate model parameter determination is desirable. For this purpose, let us focus the attention on the work of De Smidt and coworkers [47] dealing with anhydrous theophylline dissolution performed according to the rotating disk method (Section 5.2.2.1). Indeed, theophylline can crystallize either in anhydrous or monohydrated form and the anhydrous form solubility exceeds that of the monohydrated one. In addition, upon contact with aqueous media, the metastable anhydrous form transforms into the more stable monohydrate form. Dissolution, taking place in buffer solution ($V_r = 100$ cm^3) pH 6.8 at 25°C, is performed compressing 750 mg anhydrous theophylline to get a disk characterized by a radius $R = 0.75$ cm. Rotational speed ω is equal to 54.4 rad/sec (520 rpm), while monohydrated theophylline solubility $C_{sf} = 6.125$ mg/cm^3. As sink conditions are attained (drug concentration is always very far from solubility, see Figure 5.15), Equation 5.94 can be simplified to

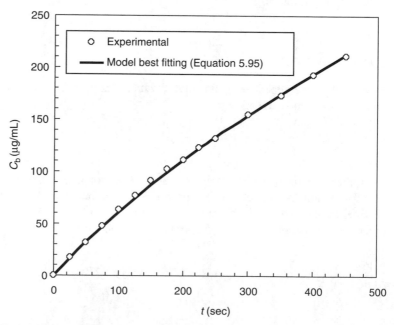

FIGURE 5.15 Comparison between experimental anhydrous theophylline dissolution (*symbols*) and model best fitting (solid line, Equation 5.95).

$$\frac{dC_b}{dt} = \frac{S}{V}k_d(C_s(t) - C_b) \approx \frac{S}{V}k_dC_s(t) = \frac{S}{V}k_d(C_{sf} + (C_{si} - C_{sf})e^{(-k_r t)}). \quad (5.94')$$

Accordingly, the simplified form of Equation 5.95 reads

$$C_b = \frac{k_d S}{V}\left(C_{sf}t + \frac{C_{si} - C_{sf}}{k_r}(1 - e^{(-k_r t)})\right), \quad (5.95')$$

k_d determination can be performed by equating the experimental slope of the $C_b(t)$ curve to the theoretical one (coming from Equation 5.95') for very long times ($t \longrightarrow \infty$):

$$\left.\frac{dC_b}{dt}\right|_{t\to\infty} = \frac{k_d S}{V}(C_{sf} + (C_{si} - C_{sf})e^{(-k_r t\to\infty)}) = \frac{k_d S}{V}C_{sf}$$

$$= 3.64 \times 10^{-4} \ (mg/cm^3/sec). \quad (5.96)$$

Consequently, we have $k_d = 3.36 \times 10^{-3}$ cm/sec that is in close agreement with 3.39×10^{-3} cm/sec, value descending from Equation 5.64 knowing that $v = 8.97 \times 10^{-3}$ cm^2/sec, $D = 6.2 \times 10^{-6}$ cm^2/sec (see Paragraph 5.2.2.1), and $\omega = 54.4$ rad/sec. The C_{si} determination follows an analogous path but with the difference that now we equate the initial ($t = 0$) experimental slope of the $C_b(t)$ curve with the theoretical one (coming from Equation 5.95'):

$$\left.\frac{dC_b}{dt}\right|_{t=0} = \frac{k_d S}{V}\left(C_{sf} + (C_{si} - C_{sf})e^{(-k_r t=0)}\right) = \frac{k_d S}{V}C_{si}$$

$$= 6.9 \times 10^{-4} \ (mg/cm^3/sec). \quad (5.97)$$

Thus, $C_{si} = 11.6$ mg/mL. Now, it is possible to fit experimental data with the whole model (Equation 5.95) in order to determine the last fitting parameter k_r. Figure 5.15 shows that data fitting (solid line) is very good with $k_r = 6 \times 10^{-3} s^{-1}$. In addition, a neat change in experimental release curve slope is visible after 100 sec to get a constant lower value after 250 sec and this behavior is due to the anhydrous–hydrate theophylline transformation with consequent solubility reduction.

5.2.3.2 Dissolution of Polyphase Mixtures

When a uniform, intimate, nondisintegrating mixture of two solid compounds A and B is put in contact with a dissolution medium (solvent), solids dissolution starts. The dissolution rate is proportional to the each solid solubility and diffusion coefficient in the boundary layer δ according to Equation 5.5. As dissolution proceeds, however, the solid–liquid interface depletes of the compound characterized by the higher dissolution rate (B) so that the interfacial solid phase is made up only by the other compound (A) (see Figure 5.16).

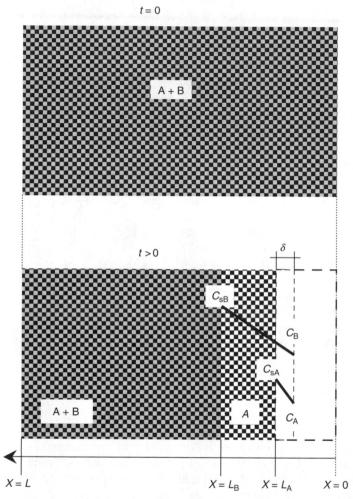

FIGURE 5.16 Upon dissolution ($t > 0$), the initially homogeneous matrix ($t = 0$) starts dissolving so that the solid–liquid front recedes. Due to its higher dissolution rate, solid B front recedes more rapidly than that of A and the dissolving fronts competing to the two solids do not coincide. While δ represents the diffusion boundary layer, C_{sA}, C_{sB}, C_A, and C_B are, respectively, A and B solubility and bulk concentration.

In other words, the receding speed of the solid B dissolution front is higher than that of A and, thus, the position of B front (L_B) is bigger than that of A (L_A). Obviously, the bigger the difference $L_B - L_A$, the lower the B dissolution rate, as B molecules have to cross an increasing distance given by the sum of $L_B - L_A$ and the diffusion layer δ. In order to model this physical frame, according to Higuchi suggestions [55], the following equation set holds:

$$\frac{dC_A}{dt} = \frac{S_A}{V} \frac{D_A(C_{sA} - C_A)}{\delta}, \tag{5.98}$$

$$\frac{dC_B}{dt} = \frac{S_B}{V} \frac{D_B(C_{sB} - C_B)}{\delta + (\tau/\varepsilon)(L_B - L_A)}, \tag{5.99}$$

$$L_A = \frac{VC_A(t)}{A_A S}, \tag{5.100}$$

$$L_B = \frac{VC_B(t)}{A_B S}, \tag{5.101}$$

where

$$S_A = \frac{S}{1 + R_{AB}}, \tag{5.102}$$

$$S_B = \frac{SR_{AB}}{1 + R_{AB}} \frac{1}{\varepsilon_s}, \tag{5.103}$$

$$A_A = \frac{R_{AB}}{1/\rho_B + R_{AB}/\rho_A}, \tag{5.104}$$

$$A_B = \frac{1}{1/\rho_B + R_{AB}/\rho_A}, \tag{5.105}$$

$$R_{AB} = \frac{m_A}{m_B}, \tag{5.106}$$

$$\varepsilon - \frac{\rho_A}{\rho_A + \rho_B R_{AB}}, \tag{5.107}$$

where S, S_A, and S_B are the whole solid dissolving surface and those competing to A and B, respectively, D_A and D_B are the A and B diffusion coefficient in the dissolving medium, respectively, V is the dissolution medium volume, C_A, C_{sA}, C_B, and C_{sB} are, respectively, A and B concentration and solubility in the dissolution medium, τ and ε are tortuosity and porosity of the B depleted zone (whose volume is $S * (L_A - L_B)$), ε_s is the surface porosity of the A solid–liquid interface, m_A, m_B, ρ_A, and ρ_B are component A and B mass and density, respectively, while A_A and A_B are component A and B mass per unit volume in the mixture. Equation 5.98 is nothing more than the particular expression of Equation 5.5, where the dissolving surface S is replaced by the surface competing to A (S_A) and the dissolution constant is $k_{dA} = D_A/\delta$. Indeed, assuming that no wettability problems occur, A molecules have to cross the diffusion layer δ to reach the dissolution medium bulk. Accordingly, Equation 5.98 analytical solution reads

$$C_A = C_{sA} \left[1 - e^{\left(-k_{dA} \frac{S_A}{V} t \right)} \right]. \qquad (5.108)$$

Implicitly, this equation assumes that A dissolution is not influenced by the presence of B and vice versa. Equation 5.99 is structurally identical to Equation 5.98 with the only, fundamental difference, that the dissolution constant k_{dB} is equal to $D_B / (\delta + (\tau/\varepsilon) (L_B - L_A))$. Indeed, B molecules must travel across the porous layer $(L_B - L_A)$, due to B molecules depletion, and then the diffusion layer δ to finally get the dissolution medium bulk. Obviously, the length of the path B molecules have to run, is bigger than the difference $(L_B - L_A)$ by virtue of the tortuosity τ ($\tau = 1$, no tortuosity) of the channels formed inside and on porosity ε (for no porous systems, $\varepsilon = 0$, obviously, $k_{dB} = 0$). Equation 5.100 and Equation 5.101 are the expressions, of A and B mass balances. Indeed, the amount of A and B molecules that left the solid phase, $L_A * S * A_A$, and $L_B * S * A_B$, respectively, equal to the amount present in the dissolution medium, namely $C_A(t) * V$ and $C_B(t) * V$, respectively. Due to the L_A and L_B dependence on C_A and C_B, respectively, Equation 5.99 does not yield an analytical solution and that is why a numerical solution is required [56]. This model, differs from the original Higuchi formulation [55] for what concerns the L_A and L_B definition (Equation 5.100 and Equation 5.101, respectively) and for the fact that Higuchi model holds only in sink conditions while this new formulation does not require sink conditions.

In order to verify the reliability of this model, we can focus the attention on Higuchi data [55] about the dissolution of benzoic acid and salicylic acid mixtures. Experiments are performed according to the rotating disk apparatus (disk diameter $2R = 0.95$ cm, rotational speed $\omega = 150$ rpm) where dissolution medium ($V = 400$ cm^3) is 0.1 N HCl at 30°C. Benzoic and salicylic acid solubility and density are, respectively, $C_{sB} = 4.2$ mg/cm^3, $\rho_B = 1.3$ g/cm^3, $C_s = 2.4$ mg/cm^3, and $\rho_S = 1.44$ g/cm^3. A severe strategy for model verification requires the determination of the dissolution constant, diffusion coefficient, and diffusion layer thickness relative to pure compounds. Then, on the basis of these parameters, it is possible to fit the model to experimental data relative to mixtures assuming τ and ε_s as the only fitting parameters. Model best fitting on pure component dissolution yields $k_{dB} = 3.5 \times 10^{-3}$ cm/sec and $k_{dS} = 3.4 \times 10^{-3}$ cm/sec. On the basis of Equation 5.64 it is also possible to determine benzoic acid and salicylic acid diffusion coefficients and diffusion layer thicknesses: $D_B = 1.62 \times 10^{-5}$ cm^2/sec, $\delta_B = 46$ μm, $D_A = 1.53 \times 10^{-5}$ cm^2/sec, and $\delta_A = 45$ μm. These values evidence that benzoic acid dissolution rate ($K_{dB} * C_{sB} = 14.8$ μg/sec cm^2) is higher than that of salicylic acid ($K_{dS} * C_{sS} = 7.8$ μg/sec cm^2). This means that in binary mixtures, benzoic acid identifies with compound B of Figure 5.16, while salicylic acid identifies with compound A of the same figure. Assuming the $D_A, D_B, \delta_A,$ and δ_B values coming from pure compounds dissolution, model is fitted on $R_{SA-BA} = 1, 3$

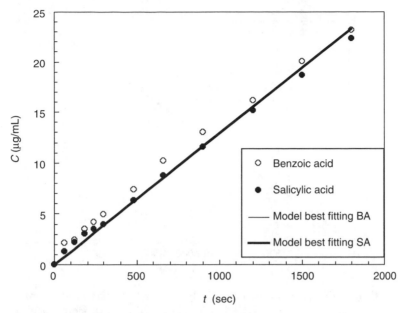

FIGURE 5.17 Comparison between model best fitting (*solid lines*) and experimental dissolution data referring to a 1:1 salicylic acid (*filled circles*)–benzoic acid (*open circles mixture*).

and 1/3 salicylic acid–benzoic acid mixtures where τ and ε_s are the only fitting parameters. Figure 5.17, Figure 5.18, and Figure 5.19 show model best fitting (solid line) on experimental data (symbols) referring, respectively, to the three different mixtures considered. It is clear that while in the $R_{\text{SA-BA}} = 1$ and 3 case model best fitting is good, in the $R_{\text{SA-BA}} = 1/3$ case, the fitting is not equally satisfactory although the trend of both salicylic acid and benzoic acid is correctly described. This is probably due to the fact that, in this case, the excess of benzoic acid, the faster dissolving solid in the mixture, could give origin to a complex dissolution surface that is not flat. Consequently, dissolution surfaces should be higher than those predicted by Equation 5.102 and Equation 5.103. Nevertheless, Figure 5.17, Figure 5.18, and Figure 5.19 substantially confirm model reliability in describing binary mixture dissolution of not interacting solids. Fitting parameters, referring to these three figures, are $\varepsilon_s = 0.53$ ($R_{\text{SA-BA}} = 1$), 0.67 ($R_{\text{SA-BA}} = 1/3$), and 0.52 ($R_{\text{SA-BA}} = 1/3$), while τ is always 1. Interestingly, while in first case, $R_{\text{SA-BA}} = 1$, ε and ε_s are similar ($\varepsilon = 0.52$), in the second case, $R_{\text{SA-BA}} = 3$, $\varepsilon < \varepsilon_s$ ($\varepsilon = 0.27$). In addition, as τ is always equal to 1, channels formed in the solid mixture should be mainly straight line shaped. Finally, Figure 5.20 shows the trend of the difference $L_{\text{BA}} - L_{\text{SA}}$ expressing the distance between benzoic acid

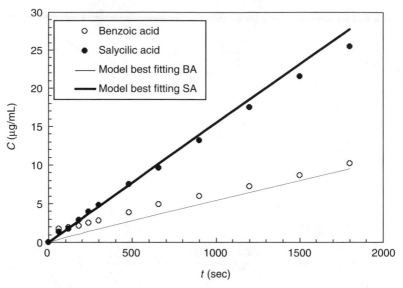

FIGURE 5.18 Comparison between model best fitting (*solid lines*) and experimental dissolution data referring to a 3:1 salicylic acid (*filled circles*)–benzoic acid (*open circles*) mixture.

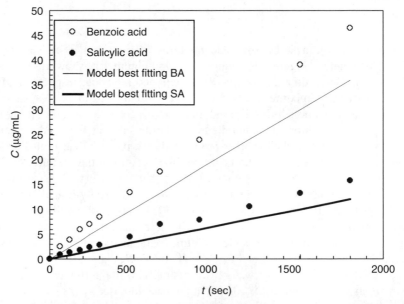

FIGURE 5.19 Comparison between model best fitting (*solid lines*) and experimental dissolution data referring to a 1:3 salicylic acid (*filled circles*)–benzoic acid (*open circles*) mixture.

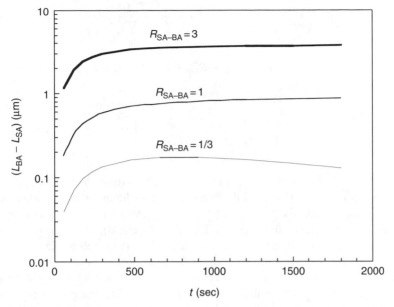

FIGURE 5.20 Temporal evolution of the distance between the benzoic acid front (L_{BA}) and the salicylic acid front (L_{SA}) for the three different mixtures considered ($R_{SA-BA} = 3, 1, 1/3$).

and salicylic acid dissolution fronts. While this difference increases with time (except for the last part of $R_{SA-BA} = 1/3$), it progressively reduces with decreasing the percentage of salicylic acid in the mixture.

5.2.3.3 Dissolution of Polyphase Mixtures: A Particular Case

A particular situation in the frame of not interacting polyphase mixtures dissolution is represented by the simultaneous dissolution of the crystalline and amorphous phase of the same drug. Although it could seem an artificial and, almost, a theoretical condition, it occurs more frequently than it would be expected. Indeed, it is not unusual that, due to the treatment processes, such as mechanical comminution and compression cycle, part of the crystalline drug undergoes the transformation to amorphous drug as it happens, for example, in the case of commercial micronized griseofulvin, a poorly water-soluble antibiotic drug. The mathematical modeling of this situation is made more complex by the fact that, in solution, it is not possible to distinguish between the drug coming from crystalline or amorphous phase dissolution. In addition, as discussed in Section 5.2.3.1, upon dissolution, amorphous drug undergoes recrystallization so that its solubility decreases with time. Accordingly, the model presented in Section 5.2.3.2 needs to be modified as follows:

$$V \frac{dC}{dt} = S_c k_{dc} (C_{sc} - C) + \frac{S_a(C_a(t) - C)}{\frac{1}{k_{da}} + \frac{\tau}{D\varepsilon}(L_a - L_c)}, \quad (5.109)$$

$$C_a = C_{sc} + (C_{sa} - C_{sc})e^{(-k_r t)}, \quad (5.109')$$

$$\frac{dM_c}{dt} = -SA_c \frac{dL_c}{dt} = -S_c k_{dc}(C_{sc} - C), \quad (5.110)$$

$$VC(t) = SL_c(t)A_c + SL_a(t)A_a \Rightarrow L_a(t) = \frac{VC(t)}{SA_a} - \frac{L_c A_c}{A_a}, \quad (5.111)$$

where C is drug concentration in the dissolution medium, S, S_a, and S_c are the whole dissolving surface and those competing to the amorphous and crystalline phases, respectively, k_{dc} and k_{da} are crystalline and amorphous phase dissolution constants (theoretically, they must be equal), respectively, C_{sc} and C_{sa} are, respectively, crystalline and amorphous drug solubilities, k_r is the recrystallization constant, D is drug diffusion coefficient in the dissolution medium, L_a and L_c are the position of the amorphous and crystalline diffusion fronts, respectively. M_c indicates the mass of crystalline phase. Finally, A_a and A_c represent the amount of amorphous and crystalline phase per unit volume and are defined by equations analogous to Equation 5.105 and Equation 5.104, respectively. Equation 5.109 states that the variation of drug concentration is due to the sum of two distinct contributes: that of the crystalline phase (Equation 5.109, first right-hand side term) and that of the amorphous phase (Equation 5.109, second right-hand side term). Obviously, as amorphous phase is characterized by a higher solubility (although decreasing from C_{sa} to C_{sc} according to the usual Equation 5.109'), this is the phase whose dissolution front recedes more rapidly and that is why it coincides with compound B of Figure 5.16. Equation 5.110 expresses the reduction of crystalline mass and, consequently, it allows the evaluation of the time increase of L_c. Finally, Equation 5.111 is an overall mass balance made up on the drug allowing the determination of the L_a time variation.

It is now of interest to verify approach reliability by applying this model on the intrinsic dissolution of commercial micronized griseofulvin that is composed of crystalline, nanocrystalline, and amorphous drug (for details see Paragraph 6.4). In order to reduce model degrees of freedom, the majority of its parameters have to be determined by means of independent experiments, while the remainder has to be considered as fitting parameters. Specifically, the ratio $\tau/D\varepsilon$ and the two dissolving surfaces (S_a, S_c) are considered fitting parameters, while C_{sc} is experimentally determined (11.9 ± 0.5 μg/cm^3), $\rho_a(1.333 \pm 0.001$ g/cm^3) and $\rho_c(1.495 \pm 0.001$ g/cm^3) are measured by helium pycnometry, amorphous griseofulvin solubility (C_{sa}), dissolution constants $k_{dc} = k_{da}$, and recrystallization constant k_r derive from an independent intrinsic dissolution experiment (sink condition) of completely amorphous

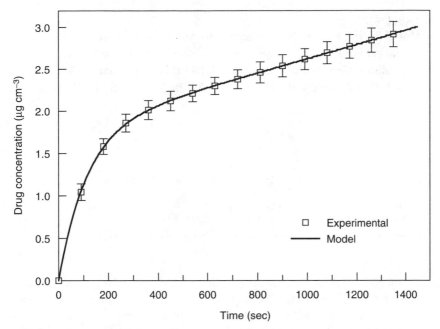

FIGURE 5.21 Constant surface area dissolution profile of the amorphous griseoful-vin. The solid line represents the model best fit Equation 5.95 on the experimental data (*symbols*). All measurements were done in triplicate. Standard deviations are indicated by error bars.

griseofulvin. Amorphous griseofulvin is prepared by quenching the melted raw material directly on the stainless steel disk used for the dissolution test. The quenching procedure leads to a completely amorphous material as confirmed by wide angle x-ray diffraction characterization. Figure 5.21 shows the comparison between model best fitting (Equation 5.95′, solid line) and dissolution data (symbols) referring to the completely amorphous griseofulvin ($V = 250$ cm^3, $S = 3.14$ cm^2). The good agreement shown, proved by the statistical F test ($F(3,16,0.005) = 85,464.9$), ensures fitting parameter values reliability (see Table 5.7). In the fitting strategy applied to the experimental dissolution profile of the native drug, the equilibrium solubility of the crystalline phase C_{sc} is considered as a fitting parameter. Indeed, since in this material the stable phase is made up of crystals in the nanometer size range, its equilibrium solubility value should be higher than that of the corresponding large crystals (see Section 6.4). In addition, its experimental determination is not so easy probably due to the contemporary occurrence of dissolution and recrystallization of both the amorphous and nanocrystalline phases. Dissolution experiments are performed using a smooth cylindrical griseofulvin disk (700 mg, 2 cm diameter, compressed at 2 t) attached, by means of liquefied

TABLE 5.7
Model Fitting Parameters Referring to Amorphous Griseofulvin (Equation 5.95′) and Multiphase (Amorphous, Nanocrystalline, Crystalline) Griseofulvin (Equation 5.109 through Equation 5.111) Dissolution Experiments (IDR) in pH 7.5 Buffer Solution at 37°C

Amorphous Griseofulvin Dissolution (Equation 5.95′)

C_{sa} ($=C_{si}$) ($\mu g/cm^3$)	k_d (cm/sec)	k_r (1/sec)
235 ± 2	$(3.36 \pm 0.03) \times 10^{-3}$	$(9.08 \pm 0.06) \times 10^{-4}$

Multiphase Griseofulvin Dissolution (Equation 5.109 through Equation 5.111)

C_{sc} ($\mu g/cm^3$)	S_a (cm^2)	S_c (cm^2)	$\tau/D\varepsilon$ (sec/cm^2)
60.2 ± 4.3	1.617 ± 0.082	1.551 ± 0.081	$(6.32 \pm 0.31) \times 10^{-5}$

microcrystalline wax, to a rotating stainless steel disk of the same diameter. The lateral surface of the disk is coated by a water impermeable membrane (PTFE). The whole apparatus is immersed in a 250 cm^3 of buffer medium (phosphate, pH 7.5) at 37°C and disk motion is immediately started after immersion (250 rpm). Drug concentration vs. the time is continuously collected by means of a pipette guide sampling probes system and a Watson Marlow peristaltic pump (mod. 505 V), at a flow rate of 30 cm^3/min, connected to a Perkin Elmer Lambda 20 spectrophotometer in a flow-through arrangement. The spectrophotometer is equipped with 10 mm flow-through cuvettes. The solution is continuously withdrawn through a polyethylene 10 μm filter at the inlet port of the probe. Spectrophotometer records drug concentration every 5 sec, at $\lambda = 295.3$ nm. Figure 5.22 shows model best fitting (solid line; Equation 5.109 through Equation 5.111) on dissolution data (symbols) while fitting parameters are shown in Table 5.7 (all other model parameters come from data fitting shown in Figure 5.21). Correctly, the C_{sc} value clearly shows that the solubility of the nanocrystals is higher than the equilibrium solubility of the recrystallized material but lower than that of the pure amorphous phase. These results seem to be in agreement with the thermodynamic prediction regarding the increase of the solubility of a crystalline solid with the reduction of the crystals size in the nanometer size range (see Chapter 6). From Table 5.7 it is also possible to determine the solid surface composition (in terms of amorphous and crystalline drug superficial abundance) on the basis of the fitting parameters S_a and S_c. Indeed, the amorphous surface fraction f_a is deduced by $f_a = S_a/(S_a + S_c)$ ($= 0.511 \pm 0.06$). As this amorphous fraction differs from that relative to

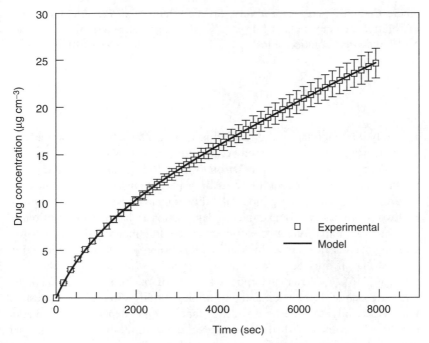

FIGURE 5.22 Constant surface area dissolution profile of the native, multiphase, griseofulvin. The solid line represents the model best fit Equation 5.109 through Equation 5.111 on the experimental data (*symbols*). All measurements were done in triplicate. Standard deviations are indicated by error bars.

the bulk one (0.1) measured by DSC and XRPD (see Section 6.4), we can conclude that surface properties differ from bulk properties.

5.2.3.4 Dissolution and Drug Degradation

Ancillary to the dissolution process, drug stability in the dissolution medium stands. Indeed, it is not so unusual that drug can undergo degradation in the dissolution medium due to, for example, pH conditions as it happens for some antibiotics. Accordingly, a proper interpretation of dissolution data requires to incorporate in the mathematical model (Equation 5.4′) a term related to drug degradation. Assuming a first-order degradation (one of the most common degradation kinetics), Equation 5.4′ becomes

$$\frac{dC_b}{dt} = \frac{S}{V} k_d (C_s - C_b) - \frac{k_{degr}}{V} C_b, \qquad (5.112)$$

where C_b and C_s are, respectively, drug concentration and solubility in the dissolution medium bulk, while k_{degr} is the degradation constant. Equation 5.112 analytical solution reads

$$C = \frac{C_s}{1 + \frac{k_{degr}}{Sk_d}} \left(1 - e^{-\left(\frac{Sk_d + k_{degr}}{V} t \right)} \right).$$

(5.113)

Basically, the importance of degradation is particularly heard in oral administration where drug bioavailability depends, among other factors, on drug partitioning between a polar (gastrointestinal fluids) and an apolar phase (cellular membrane). Accordingly, if drug partitioning is very fast in comparison to degradation, drug bioavailability is guaranteed. If, on the contrary, the two processes are comparable or degradation is faster than partitioning, drug bioavailability cannot be sufficient as the drug majority degrades before crossing cellular membrane. In addition, this frame is further complicated by the fact that, if drug degradation is pH dependent (in the physiological pH range (1.2–6), k_{degr} can undergo 3–4 order of magnitude variations), the relative speed of degradation and partitioning modifies in the gastrointestinal tract. Practically, this means that drugs that have pH-dependent degradation can be efficiently absorbed only in precise gastrointestinal tracts (the so-called "absorption windows") and not in others.

5.2.4 DISSOLUTION OF PARTICULATE SYSTEMS

Previous paragraphs deal with drug dissolution from wide surfaces (usually flat) typical of tablets as this is the best way of studying the dissolution process. Nevertheless, in common pharmaceutical practice, dissolution from particulate systems can be also of interest. Accordingly, in order to match this problem, it is convenient to consider the simplest situation, namely the dissolution of an ensemble of N, all equal, spherical particles in a very large dissolution medium so that sink conditions are attained. Accordingly, dissolution can be modeled by means of the following equations:

$$\frac{dC}{dt} = N \frac{4\pi R^2}{V} k_d C_s,$$

(5.114)

where C is drug concentration in the dissolution medium and R is the time-dependent particle's radius. This equation must be simultaneously solved with Equation 5.115, accounting for R reduction provoked by dissolution.

$$N \frac{dm}{dt} = N \frac{d}{dt} \left(\frac{4}{3} \pi R^3 \rho \right) = N 4\pi \rho R^2 \frac{dR}{dt} = -N 4\pi R^2 k_d C_s,$$

(5.115)

where m and ρ are, respectively, particle mass and density. The integration of Equation 5.115 leads to

$$R = R_0 - \frac{k_d C_s}{\rho} t, \qquad (5.116)$$

where R_0 is the initial particle radius corresponding to $m = m_0$. Model final form is found by inserting Equation 5.116 into Equation 5.114 and solving to finally get

$$C(t) = \frac{M_0}{V}(K^3 t^3 - 3K^2 t^2 + 3Kt) \qquad K = \frac{k_d C_s}{\rho R_0}, \qquad (5.117)$$

where M_0 $(= N * m_0)$ is the initial, undissolved, particles mass and V is the dissolution medium volume. Ultimately, this is the well-known Hixson–Crowell equation [57]. If, upon dissolution, a phase change happens at the solid–liquid interface (see Paragraph 5.2.3.1) and, consequently, drug solubility varies according to Equation 5.93b, the equation set comprising of Equation 5.114 and Equation 5.115 needs to be modified as follows:

$$\frac{dC}{dt} = N \frac{4\pi R^2}{V} k_d (C_{sf} - (C_{si} - C_{sf})e^{-k_r t}), \qquad (5.114')$$

$$\frac{dR}{dt} = -\frac{k_d}{\rho}(C_{sf} - (C_{si} - C_{sf})e^{-k_r t}), \qquad (5.115')$$

where C_{sf} and C_{si} indicate, respectively, final and initial drug solubility, while k_r is the recrystallization constant. Equation 5.114' and Equation 5.115' integration leads to

$$R(t) = R_0 - \frac{k_d C_{sf}}{\rho} t - \frac{k_d (C_{si} - C_{sf})}{\rho k_r}(1 - e^{-k_r t}), \qquad (5.118)$$

$$C(t) = \frac{M_0}{V}\left(1 - \left(\frac{R(t)}{R_0}\right)^3\right), \qquad (5.119)$$

it is easy to verify that in the limit $k_r \longrightarrow \infty$ (recrystallization is instantaneous), Equation 5.118 and Equation 5.119 transform into Equation 5.116 and Equation 5.117, respectively, where C_s is replaced by C_{sf}. Similarly, for $k_r \longrightarrow 0$ (recrystallization never happens), Equation 5.118 and Equation 5.119 transform into Equation 5.116 and Equation 5.117, respectively, where C_s is replaced by C_{si}.

One of the most elegant and powerful approaches about drug dissolution from an ensemble of polydisperse spherical particles in nonsink conditions is that of Pedersen and coworkers [58–61]. They assume that particles distribution can be described by the following log-normal distribution $f(\phi_0)$ [60]:

$$f(\phi_0) = \frac{N(\ln(\phi_0), \mu, \sigma)/\phi_0}{\displaystyle\int_{\phi_0^{\min}}^{\phi_0^{\max}} N(\ln(\phi_0), \mu, \sigma) \, d\phi_0/\phi_0} \qquad \phi_0^{\min} \le \phi_0 \le \phi_0^{\max}, \qquad (5.120)$$

$$N(\ln(\phi_0), \mu, \sigma) = \frac{1}{\sigma\pi\sqrt{2}} e^{-\left(\frac{\ln(\phi_0)-\mu}{\sigma\sqrt{2}}\right)_2}, \qquad \phi_0^{\min} = e^{(\mu - i\sigma)} \qquad \phi_0^{\max} = e^{(\mu + j\sigma)},$$

$$(5.120')$$

where $\phi_0 \, (= 2R_0)$ represents particles initial diameters spanning from $\phi_0^{\min}$ to $\phi_0^{\max}$, μ and σ are, respectively, initial particles distribution "mean" and "standard deviation" ($\sigma = 0$, monodisperse distribution) while "i" and "j" are the lower and upper truncation parameters (dimensionless) of the particles distribution accounting for the possible truncated nature of experimental distribution. In addition, defining the dissolution capacity coefficient as follows:

$$\alpha = \left(\frac{C_s V - M_0}{M_0}\right)^{1/3}, \qquad (5.121)$$

the authors finally get

$$C(t) = \frac{M_0}{V}\left(1 - \sum_{n=0}^{3} \frac{3!}{(3-n)!n!}(-G)^{(3-n)} \frac{F(T_2 - n\sigma) - F(T_1 - n\sigma)}{F(j - 3\sigma) - F(-i - 3\sigma)} e^{\frac{(n^2-9)\sigma^2}{2}}\right),$$

$$(5.122)$$

where

$$T_1 = \frac{\max(\ln(G), -i\sigma)}{\sigma} \qquad T_2 = \frac{\max(\ln(G), j\sigma)}{\sigma}, \qquad (5.123)$$

$$G = \frac{\alpha^3 K}{1+\alpha^3} t + \frac{K}{1+\alpha^3} \int_0^t \frac{M}{M_0} dt \qquad K = \frac{k_d C_s}{\rho R_0}, \qquad (5.124)$$

$$F(X) = \frac{1}{\pi\sqrt{2}} \int_{-\infty}^{X} e^{-\frac{u^2}{2}} du = \frac{1}{2}\left[\text{erf}\left(\frac{X}{\sqrt{2}}\right) + 1\right], \qquad (5.125)$$

where "erf" is the error function. The use of the "max" relationship (Equation 5.123) depends on the fact that, for t approaching infinite, C approaches M_0/V. In particular, the "max" condition on T_1 is related to the disappearance of the smallest particles, while the same condition for T_2 denotes the complete dissolution of the biggest particles. If the truncation parameters i and j are >3, particle distribution can be approximated by the ideal distribution $(i = j = \infty)$ and Equation 5.122 becomes

$$C(t) = \frac{M_0}{V} \left(1 - \sum_{n=0}^{3} \frac{3!}{(3-n)!n!} (-G)^{(3-n)} \left(1 - F\left(\frac{\ln(G)}{\sigma} - n\sigma \right) \right) e^{\frac{(n^2-9)\sigma^2}{2}} \right).$$

(5.122′)

If a monodisperse particles ensemble is considered $(\sigma = 0)$, Equation 5.122 and Equation 5.122′ become

$$C(t) = \frac{M_0}{V} \left(1 - \sum_{n=0}^{3} \frac{3!}{(3-n)!n!} (-G)^{(3-n)} \right) = \frac{M_0}{V} (1 - (1-G)^3)$$

$$= \frac{M_0}{V} (G^3 - 3G^2 + 3G).$$

(5.126)

If, in addition, sink conditions are attained $(\alpha \longrightarrow \infty)$, Equation 5.126 degenerates into the Hixson–Crowell equation (Equation 5.117).

Despite the apparent mathematical complexity, Equation 5.122, or Equation 5.122′, can be solved adopting standard numerical technique once it is transformed into an initial value problem. Indeed, letting $y = \int_{0}^{t} (M/M_0) \, dt$ and $y' = (M/M_0)$, remembering that

$$C(t) = \frac{M_0}{V} \left(1 - \frac{M}{M_0} \right)$$

and

$$h(t, y) = \sum_{n=0}^{3} \frac{3!}{(3-n)!n!} (-G)^{(3-n)} \frac{F(T_2 - n\sigma) - F(T_1 - n\sigma)}{F(j - 3\sigma) - F(-i - 3\sigma)} e^{\frac{(n^2-9)\sigma^2}{2}},$$

(5.127)

we have the following initial value problem:

$$C(t) = \frac{M_0}{V} (1 - y') \qquad y' = h(y,t) \qquad y(0) = 0 \qquad (5.128)$$

that can be solved, for example, with a fourth-order Runge–Kutta method [56].

5.3   PARTITIONING

A detailed study of drug partitioning between a polar aqueous phase and an apolar one is very important since some drug physicochemical properties and *in vivo* behavior can be determined on the basis of this phenomenon. In particular, the drug partition coefficient P, strictly connected to drug lipophilicity, has a paramount importance in predictive environmental studies [62] as it is used for the prediction of distribution among environmental compartments [63] in equations for the estimation of bioaccumulation in animals and plants [64] and in predicting the toxic effects of a substance [65]. Moreover, while lipophilicity encodes a wealth of structural information [66], especially in oral or parenteral administrations, drug bioavailability depends on P [67]. Additionally, the study of drug partitioning is also very important for what concerns drug release from disperse systems such as emulsions and microemulsions (see Chapter 8).

Drug transfer between the two liquid phases (phase 1, polar; phase 2, apolar) can be better understood recalling the physics of the liquid–liquid interface in the light of the Gibbs theory [24]. According to this theory, phase 1 and 2 properties are constant all over the respective volume except for a thin region containing the interface, whose position is not exactly defined. For example, in the case of water (phase 1) and *n*-octanol (phase 2), while water and oil concentration is always constant in the respective bulk region, it decreases approaching the interface to get a minimum value in the opposite phase as evidenced in Figure 5.23. This gives origin to two stagnant layers, sandwiching the interface, whose properties differ from those of phase 1 and 2 bulks. Consequently, drug transport from phase 1 toward phase 2 (and vice versa) can be explained by means of a three-step mechanism [4]: a diffusion controlled step toward the interface, a de– and resolvation step at the interface and, finally, a new diffusion controlled step (Figure 5.23). Indeed, the transfer from the first phase to the second one requires drug molecules to diffuse through the stagnant layer preceding the theoretical interface. Then, an energy step, allowing drug molecules passage from the first solvent to the second, occurs. Finally, a diffusion controlled step (diffusion through the second stagnant layer) away from interface develops [3,4]. Assuming that drug concentration assumes uniform values in the two bulk phases and in the two stagnant layers, the kinetics of this mechanism can be represented by means of the following consecutive first-order chemical reaction:

$$C_{\mathrm{w}} \underset{k_2'}{\overset{k_1'}{\rightleftharpoons}} C_{\mathrm{w}}^{\mathrm{ss}} \underset{k_2}{\overset{k_1}{\rightleftharpoons}} C_{\mathrm{o}}^{\mathrm{ss}} \underset{k_2''}{\overset{k_1''}{\rightleftharpoons}} C_{\mathrm{o}}, \qquad\qquad \text{Scheme (5.1)}$$

where C_{w} and C_{o} are, respectively, drug concentration in the water and oil bulk phase, $C_{\mathrm{w}}^{\mathrm{ss}}$ and $C_{\mathrm{o}}^{\mathrm{ss}}$ are, respectively, drug concentration in the water and

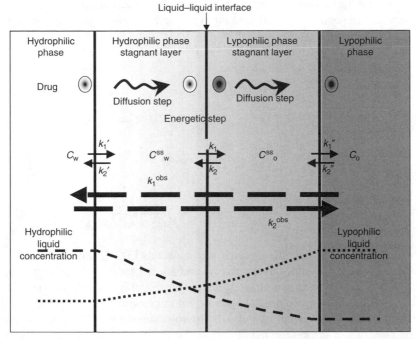

FIGURE 5.23 Drug partitioning between a hydrophilic and a lypophilic liquid phase can be viewed as a three-step mechanism: diffusion through the hydrophilic stagnant layer, drug desolvation and resolvation at the liquid–liquid interface and, finally, diffusion through the lypophilic stagnant layer.

oil stagnant layer while k_1', k_2', k_1, k_2, k_1'', and k_2'' are kinetics constants ruling each reaction. Scheme (5.1) mathematical translation reads

$$\frac{dC_w}{dt} = k_2' C_w^{ss} - k_1' C_w, \tag{5.129}$$

$$\frac{dC_w^{ss}}{dt} = k_1' C_w + k_2 C_o^{ss} - (k_1 + k_2') C_w^{ss}, \tag{5.130}$$

$$\frac{dC_o^{ss}}{dt} = k_1 C_w^{ss} + k_2'' C_o - (k_2 + k_1'') C_o^{ss}, \tag{5.131}$$

$$\frac{dC_o}{dt} = k_1'' C_o^{ss} - k_2'' C_o, \tag{5.132}$$

where t is time. For example, Equation 5.129 imposes that C_w variation depends on the difference between the drug flux coming from the water stagnant layer ($k_2' C_w^{ss}$) and that leaving the water bulk $-k_1' C_w$. Similarly,

Equation 5.130 through Equation 5.132 rule drug concentration variation in the water stagnant layer, oil stagnant layer, and oil bulk phase, respectively (Figure 5.23). Assuming $dC_w/dt = dC_o/dt = 0$ (steady-state treatment) [3], Equation 5.129 through Equation 5.132 reduce to

$$\frac{dC_w}{dt} = k_o^{obs} C_o - k_w^{obs} C_w, \tag{5.133}$$

where

$$C_o = \frac{M_0 - V_w C_w}{V_o}, \tag{5.134}$$

$$k_o^{obs} = \frac{k_2 \, k_2' \, k_2''}{k_2(k_2 + k_1'') + k_1'' k_1}; \qquad k_w^{obs} = \frac{k_1 \, k_1' \, k_1''}{k_2'(k_2 + k_1'') + k_1'' k_1}, \tag{5.135}$$

where k_o^{obs} and k_w^{obs} are the observed rate constants [3], V_o and V_w the apolar and polar phase volumes, respectively, and M_0 the total, and constant, drug amount present in the two-phase system. The reduction to only one kinetics equation has been made possible thanks to the introduction of the drug mass balance (made up on the two-phase system) expressed by Equation 5.134. Equation 5.133 can be formally expressed in a different way if apolar–polar (F_{ow}) and polar–apolar (F_{wo}) drug fluxes are, respectively, defined as follows:

$$F_{ow} = k_{ow} \, C_o, \qquad F_{wo} = k_{wo} \, C_w, \tag{5.136}$$

where k_{ow} and k_{wo} are the respective rate constants (dimensionally a velocity). Thus, we have

$$\frac{dC_w}{dt} = \frac{A F_{ow}}{V_w} - \frac{A F_{wo}}{V_w} = \frac{A k_{ow}}{V_w} C_o - \frac{A k_{wo}}{V_w} C_w, \tag{5.137}$$

where $k_o^{obs} = \frac{A k_{ow}}{V_w}$, $k_w^{obs} = \frac{A k_{wo}}{V_w}$ and A is the liquid–liquid interface area. Solution of Equation 5.137 and Equation 5.134 reads

$$C_w = \frac{k_{ow} M_0}{k_{wo} V_o + k_{ow} V_w} - \left(\frac{k_{ow} M_0}{k_{wo} V_o + k_{ow} V_w} - C_{wi} \right) e^{\left(-A \frac{k_{wo} V_o + k_{ow} V_w}{V_o V_w} t \right)}, \tag{5.138}$$

where C_{wi} is the initial drug concentration in the polar phase.

Although, usually, Equation 5.138 provides a satisfactory description of drug partitioning, it fails for sparingly soluble drugs in one or both phases. In this case, F_{ow} and F_{wo} expression must be modified to account for the finite drug solubility in both phases. For example, the following expression can be considered [68]:

$$F_{ow} = k_{ow} C_o \left(\frac{C_{sw} - C_w}{C_{sw}} \right) \qquad F_{wo} = k_{wo} C_w \left(\frac{C_{so} - C_o}{C_{so}} \right), \qquad (5.139)$$

where C_{so} and C_{sw} represent drug solubility in the apolar and polar phases, respectively. The advantage of this definition relies on the fact that, in contrast to Equation 5.136, it implies that F_{ow} or F_{wo} also vanishes when C_w or C_o approaches C_{sw} or C_{so}, respectively. This aspect is very important for sparingly soluble drugs in the water or oil phase. On the basis of this new flux definition, model expression becomes

$$\frac{dC_w}{dt} = C_w^2 a + C_w b + f, \qquad (5.137')$$

where

$$a = \frac{A}{V_o} \left(\frac{k_{ow}}{C_{sw}} - \frac{k_{wo}}{C_{so}} \right) \qquad f = Ak_{ow} \frac{M_0}{V_w V_o}, \qquad (5.140)$$

$$b = A \left(\frac{k_{wo}}{C_{so} V_w} \left(\frac{M_0}{V_o} - C_{so} \right) - \frac{k_{ow}}{C_{sw} V_o} \left(\frac{M_0}{V_w} + C_{sw} \right) \right). \qquad (5.141)$$

Depending on the a, b, and f values, Equation 5.137' solution reads

Case 1: $a > 0$ and $|b| < 2\sqrt{af}$

$$C_w = \left(\frac{\sqrt{f - b^2/4a}}{\sqrt{a}} \right) \frac{\left(\sqrt{f - b^2/4a} \right) \text{tg} \left(t\sqrt{a}\sqrt{f - b^2/4a} \right) + \sqrt{a}(C_{wi} + b/2a)}{\sqrt{f - b^2/4a} - \sqrt{a}(C_{wi} + b/2a) \text{tg}(t\sqrt{a}\sqrt{f - b^2/4a})}, \qquad (5.142)$$

Case 2: $a > 0$ and $|b| > 2\sqrt{af}$

$$C_w = \left(\sqrt{\frac{(b^2/4a) - f}{\sqrt{a}}} \right) \frac{1 + \frac{\sqrt{a}(C_{wi} + b/2a) - \sqrt{(b^2/4a) - f}}{\sqrt{a}(C_{wi} + b/2a) + \sqrt{(b^2/4a) - f}} e^{(2t\sqrt{(b^2/4a) - f}\sqrt{a})}}{1 - \frac{\sqrt{a}(C_{wi} + b/2a) - \sqrt{(b^2/4a) - f}}{\sqrt{a}(C_{wi} + b/2a) + \sqrt{(b^2/4a) - f}} e^{(2t\sqrt{(b^2/4a) - f}\sqrt{a})}} - \frac{b}{2a}, \qquad (5.143)$$

Case 3: $a = 0$

$$C_w = \left(\frac{f}{b} + C_{wi} \right) e^{(bt)} - \frac{f}{b}, \qquad (5.144)$$

Case 4: $a < 0$

$$C_w = \left(\frac{\sqrt{f - (b^2/4a)}}{\sqrt{-a}}\right) \frac{1 + \dfrac{\sqrt{-a}(C_{wi} + b/2a) - \sqrt{f - (b^2/4a)}}{\sqrt{-a}(C_{wi} + b/2a) + \sqrt{f - (b^2/4a)}} e^{(-2t\sqrt{f-(b^2/4a)}\sqrt{-a})}}{1 - \dfrac{\sqrt{-a}(C_{wi} + b/2a) - \sqrt{f - (b^2/4a)}}{\sqrt{-a}(C_{wi} + b/2a) + \sqrt{f - (b^2/4a)}} e^{(-2t\sqrt{f-(b^2/4a)}\sqrt{-a})}} - \frac{b}{2a}.$$

$$(5.145)$$

Although the model assumes different analytical expressions depending on the parameter values, its use for data fitting is made user friendly by embodying Equation 5.140 through Equation 5.145 into a proper routine that automatically selects the correct model expression (see the Appendix).

On the basis of this model (Equation 5.137′ and Equation 5.139), the drug partitioning dependence on the initial drug concentration C^0 in the two-phase system can be evaluated. Accordingly, let us define C^0 and the apparent partition coefficient P_a as follows:

$$P_a = \frac{C_o^{eq}}{C_w^{eq}} \quad C^0 = \frac{V_w C_{wi} + V_o C_{oi}}{V_w + V_o},$$

$$(5.146)$$

where C_o^{eq} and C_w^{eq} represent, respectively, the oil and water drug concentrations at equilibrium, while C_{oi} is the initial oil drug concentration. P_a dependence on C^0 can be deduced recalling the equilibrium condition ($F_{ow} = F_{wo}$) (see Equation 5.139) and the overall drug mass balance (Equation 5.134) allowing to express C_o as function of C_w. Consequently, it follows:

$$P_a = R_k \left(\frac{C_{so} - C^0(R_v + 1) + C_w^{eq} R_v}{(C_{sw} - C_w^{eq})\alpha}\right), \quad C_w^{eq} = -\frac{\beta}{2} \pm \sqrt{\left(\frac{\beta}{2}\right)^2 - \delta}, \quad (5.147)$$

where

$$\beta = -\left(\frac{C_{so} R_k}{(\alpha - R_k)R_v} + \frac{\alpha C_{sw}}{(\alpha - R_k)} + \frac{C^0(1 + R_v)}{R_v}\right); \quad \delta = \frac{\alpha C_{sw}(1 + R_v)}{(\alpha - R_k)R_v} C^0,$$

$$(5.148)$$

$$R_v = \frac{V_w}{V_o} \quad R_k = \frac{k_{wo}}{k_{ow}} \quad \alpha = \frac{C_{so}}{C_{sw}}.$$

$$(5.149)$$

The analysis of these equations reveals that if $R_k \neq \alpha$, P_a strongly depends (in an almost linear manner) on C^0, and, in particular, for $R_k > \alpha$ it decreases with C^0, while for $R_k < \alpha$ it increases with C^0. Only in the limiting case $R_k \longrightarrow \alpha$, P_a is concentration independent. In addition, whatever be R_k, $P_a \longrightarrow \alpha \, (= C_{so}/C_{sw})$

for increasing C^0 values. If, on the contrary, C^0 approaches 0 ($C_w^{eq} \approx 0$), P_a coincides with the partition coefficient P [15], its expression in terms of model parameters is:

$$P = \lim_{C^0 \to 0} (P_a) = R_k = \frac{k_{wo}}{k_{ow}}. \quad (5.150)$$

Interestingly, adopting the simplified approach (Equation 5.136), P_a would result as concentration independent and always equal to R_k.

Grassi and group [68] apply this approach (Equation 5.137' and Equation 5.139) to nimesulide and piroxicam (nonsteroidal anti-inflammatory drugs characterized by low-water solubility and pH-dependent solubility, see Table 5.8) partitioning in a water/n-octanol two-phase environment. Two different experimental conditions are studied. In the first case, polar phase (termed "water phase") consists of water buffered at pH 1.2 and saturated by n-octanol while the apolar phase (termed "oil phase") is represented by n-octanol saturated by water buffered at pH 1.2. Second case, differs from the first for the usage of water buffered at pH 7.5. In both cases water and oil phase presaturation is required to prevent oil migration into the water phase and vice versa. Indeed, this could probably affect drug solubility in the two

---

TABLE 5.8
Characteristics of the Four Different Kinds of Partition Experiments Performed

		Test 1	Test 2
Piroxicam	pH	1.2	7.5
	k_{wo} (cm/sec)	$(2.52 \pm 0.007) \times 10^{-3}$	$(3.63 \pm 0.02) \times 10^{-4}$
	k_{ow} (cm/sec)	$(2.85 \pm 0.02) \times 10^{-4}$	$(8.25 \pm 0.06) \times 10^{-4}$
	C_{sw} (µg/cm^3)	250 ± 3.47	2171 ± 4.8
	C_{so} (µg/cm^3)	1683 ± 57	1653 ± 38
	C_{wi} (µg/cm^3)	15.2 ± 0.2	21.8 ± 0.5
	$P(-)$	8.8	0.44
		Test 3	Test 4
Nimesulide	k_{wo} (cm/sec)	$(2.29 \pm 0.01) \times 10^{-3}$	$(1.1 \pm 0.003) \times 10^{-3}$
	k_{ow} (cm/sec)	$(2.85 \pm 0.2) \times 10^{-5}$	$(5.3 \pm 0.2) \times 10^{-5}$
	C_{sw} (µg/cm^3)	11.8 ± 0.5	104 ± 12
	C_{so} (µg/cm^3)	2789 ± 46	2702 ± 22
	C_{wi} (µg/cm^3)	8.5 ± 0.3	45 ± 2
	$P(-)$	80.4	20.7

Source: From Grassi, M., Coceani, N., Magarotto, L., *Int. J. Pharm.*, 239, 157, 2002. With permission from Elsevier.

phases during the partitioning experiment. Accordingly, piroxicam and nime-sulide solubilities (see Table 5.8) are measured in each of the presaturated phases at 37°C, this is the temperature of the partitioning tests. Water and oil phase volume V_w and V_o are equal to 150 and 50 cm^3, respectively, while the interfacial area $A = 34$ cm^2. Initially, the oil phase is drug free, while the aqueous phase is characterized by a drug concentration C_{wi} (see Table 5.8). Drug concentration decrease in the aqueous phase is monitored and recorded by means of an on-line UV spectrophotometer (UV spectrophotometer, Lambda 6/PECSS System, Perkin-Elmer Corporation, Norwalk, CT. Wavelength: piroxicam pH 1.2 = 354.2 nm; piroxicam pH 7.5 = 352.6 nm; nimesulide pH 1.2 = 300 nm; nimesulide pH 7.5 = 390 nm). Fluid recirculation is ensured by a peristaltic pump. Each test is performed in triplicate.

Model provides a very good data fitting in all the four experimental conditions considered (tests 1–4, see Table 5.8) as witnessed, for example, by Figure 5.24 referring to test 3 (nimesulide partition pH = 1.2; experimental data (open circles), model best fitting (solid line)). Interestingly, while pH reduction causes an order of magnitude decrease of nimesulide and piroxicam aqueous solubility (oil solubility is almost the same), k_{wo} and k_{ow} are not heavily affected by this variation for nimesulide while k_{wo} increases and k_{ow} reduces in the piroxicam case (see Table 5.8). This behavior can be ascribed to the fact that at pH = 1.2 piroxicam (weak acid, pKa = 5.1 [69]) is less

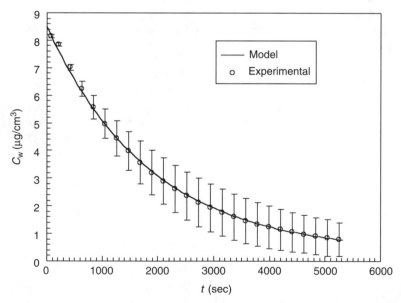

FIGURE 5.24 Comparison between model best fitting (*solid line*) and data (*symbols*) referring to nimesulide partitioning (pH = 1.2, $T = 37$°C). (From Grassi, M., Coceani, N., Magarotto, L., *Int. J. Pharm.*, 239, 157, 2002. With permission from Elsevier.)

dissociated than at pH $= 7.5$ and its dissociated form shows a higher affinity for the aqueous phase. On the basis of the determined k_{ow} and k_{wo} values, Table 5.8 shows the resulting nimesulide and piroxicam partition coefficient P. Correctly, these values, coincide with those determined resorting to equilibrium experiments. On the contrary, the use of Equation 5.138 leads to a nimesulide partition coefficient value (66.5) that differs from that experimentally determined and calculated according to the improved model. Finally, it is worth mentioning the border over which the use of the simplified approach (Equation 5.138) should be avoided. For this purpose and in order to generalize the following conclusions, let us consider the following dimensionless variables:

$$t^+ = k_{wo} \frac{tA}{V_o + V_w} \qquad C_w^+ = \frac{C_w}{C_{sw}}, \qquad (5.151)$$

where t^+ and C_w^+ are, respectively, dimensionless time and water drug concentration. Moreover, to be close to the experimental conditions previously examined (see Table 5.8, test 3 conditions), let us set $C_{so} = 2500$ $\mu g/cm^3$, $C_{sw} = 10$ $\mu g/cm^3$, $k_{ow} = 10^{-5}$ cm/sec; $k_{wo} = 10^{-3}$ cm/sec, $V_o = 50$ cm^3, $V_w = 150$ cm^3, $A = 34$ cm^2. Figure 5.25 shows the comparison between C_w^+ trend according to Equation 5.138 and the improved model in the case of $C_{oi}/C_{so} = 0.5$ (Equation 5.138, upper solid line; improved model, upper thin dashed line) and $C_{oi}/C_{so} = 0.4$ (Equation 5.138, lower solid line; improved model, lower thin dashed line). Besides the evident discrepancies arising between the two models, whatever the C_{oi}/C_{so} considered, it is interesting to notice that, in the $C_{oi}/C_{so} = 0.5$ case, Equation 5.138 erroneously predicts an exceeding of the drug solubility (C_w^+ exceeds 1). This meaningless prediction clearly reveals the unsuitability of Equation 5.138 in describing the oil–water partitioning of sparingly water-soluble drugs. Practically, Equation 5.138 can be used if initial drug concentration C^0 in the two-phases (Equation 5.146) does not exceed approximately 7% of system solubilization capacity S_C defined by

$$S_C = \frac{V_w C_{sw} + V_o C_{so}}{V_w + V_o}. \qquad (5.152)$$

5.3.1 THREE PHASES PARTITIONING

As mentioned at the beginning of this chapter, partitioning can have a paramount importance for what concerns drug effectiveness (alias, bioavailability), especially in oral administration, as the drug, moving from the aqueous gastroenteric fluid, has to cross the hydrophobic cell membranes to get blood circulation (the new aqueous environment). Accordingly,

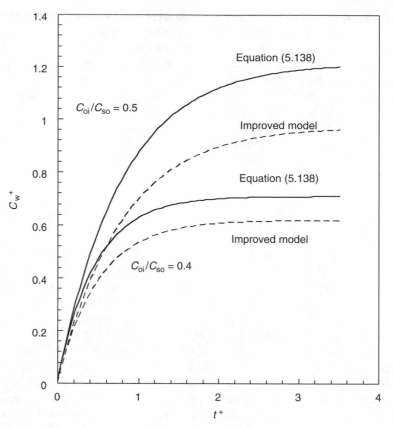

FIGURE 5.25 Comparison between Equation 5.138 (solid lines) and improved model (*dashed lines*) for two different values of the initial drug concentration in the oil phase (C_{oi}). (From Grassi, M., Coceani, N., Magarotto, L., *Int. J. Pharm.* 239, 157, 2002. With permission from Elsevier.)

partitioning plays a very important role and the experimental scheme previously discussed reveals to be very useful in providing some insight about the distribution phenomenon. However, this is not the unique interesting aspect of this technique. Indeed, by means of a simple implementation, it can represent a qualitative *in vitro* schematization of the *in vivo* conditions occurring in the gastroenteric environment, at least for what concerns drug distribution among hydrophilic and hydrophobic phases. It is sufficient to consider the scheme represented in Figure 5.26 where water 1 phase (representing the aqueous gastroenteric environment) is initially drug loaded, the oil (representing the epithelial cells) and water 2 (representing the blood circulation or cell contents) phases are initially drug free [70,71]. Thus, the attainment of the thermodynamic equilibrium conditions among the three phases implies that

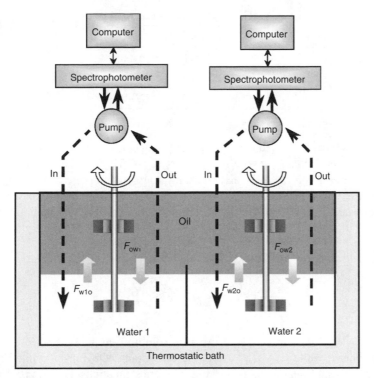

FIGURE 5.26 Schematic representation of the three phases partitioning apparatus. (Adapted from Sirotti, C., et al., *Proceedings of the CRS Winter Symposium and 11th International Symposium on Recent Advances in Drug Delivery Systems*, 3-3-2003, 51.)

the drug leaves water 1 to get water 2 passing through the oil phase. Mixing conditions inside each phase are ensured by impellers connected to rotating rods, whose rotational speed is set to prevent apolar–polar phases mixing at the interfaces. The temperature of the whole system is kept constant (37°C) by means of a surrounding thermostatic bath. Drug loading of water phase 1 is performed at time zero by means of a syringe injecting a small amount of drug solution. Of course, the solvent of the injected solution must be compatible with water and must be a good solvent for the drug so that a small solution volume is needed to get the desired concentration in water 1 without sensibly altering its volume. Drug concentration in the two aqueous phases can be measured by means of computer managed spectrophotometers each one belonging to a close loop comprehending a recirculation pump (Figure 5.26).

In the light of what was previously discussed (two-phase partitioning), the kinetics of drug partitioning can be described by the following equations [72]:

$$\frac{dC_{w1}}{dt} = \frac{A_1 F_{ow1}}{V_{w1}} - \frac{A_1 F_{w1o}}{V_{w1}}, \tag{5.153}$$

$$\frac{dC_{w2}}{dt} = \frac{A_2 F_{ow2}}{V_{w2}} - \frac{A_2 F_{w2o}}{V_{w2}}, \tag{5.154}$$

$$C_o = \frac{M_0 - C_{w1}V_{w1} - C_{w2}V_{w2}}{V_o}, \tag{5.155}$$

where t is time, C_o, C_{w1}, and C_{w2} represent drug concentration in the oil, water 1, and water 2 phases, respectively, V_o, V_{w1}, and V_{w2} are the volume of the three phases, M_0 is the total, and constant, drug amount present in the system while F_{ow1}, F_{ow2}, F_{w1o}, and F_{w2o}, indicate drug fluxes from the oil phase to the water 1 and 2 phases and vice versa, are defined according to Equation 5.139:

$$F_{owi} = k_{owi}C_o \left(\frac{C_{swi} - C_{wi}}{C_{swi}} \right), \tag{5.156}$$

$$F_{wio} = k_{wio}C_{wi} \left(\frac{C_{so} - C_o}{C_{so}} \right), \tag{5.157}$$

where $i = 1, 2$, C_{so} and C_{swi} are the drug solubility in the oil and in the two polar phases while k_{wio}, k_{owi}, are the rate constants. Obviously, while Equation 5.153 and Equation 5.154 represent kinetics equations, Equation 5.155 is the drug mass balance relative to the three phases. The determination of system equilibrium, i.e., the final drug concentration in the three phases, can be achieved considering that, at equilibrium, the time derivative appearing in Equation 5.153 and Equation 5.154 must vanish. Accordingly, at equilibrium, we have

$$C_{w1} = \frac{C_o}{P_1 + C_o \left(\frac{1}{C_{sw1}} - \frac{P_1}{C_{so}} \right)}, \tag{5.158}$$

$$C_{w2} = \frac{C_o}{P_2 + C_o \left(\frac{1}{C_{sw2}} - \frac{P_2}{C_{so}} \right)}, \tag{5.159}$$

where P_1 ($= k_{w1o}/k_{ow1}$) and P_2 ($= k_{w2o}/k_{ow2}$) are the partition coefficients between water 1/oil and water 2/oil phases, respectively (obviously, P_1 differs from P_2 only in the case of different aqueous phases). Inserting Equation 5.158 and Equation 5.159 into mass balance (Equation 5.155) leads to a cubic equation, whose solution allows the determination of the equilibrium C_o value, and, consequently, C_{w1} and C_{w2} equilibrium values. Model drug (nimesulide and piroxicam) and aqueous–oil phase are those previously used in the two-phases partitioning experiments. In particular, oil phase volume is $V_o = 100 \text{ cm}^3$, water 1 and water 2 volumes are $V_{w1} = 200 \text{ cm}^3$ and $V_{w2} = 100 \text{ cm}^3$, respectively, while the interfacial areas $A_1 = A_2 = 25 \text{ cm}^2$. At the beginning of the partitioning test, the oil phase and the water 2 phase are drug free, while water 1 phase is characterized by an initial drug concentration C_{w10} ($C_{w10}^{nime} = 54 \ \mu g/cm^3$, $C_{w10}^{pirox} = 52 \ \mu g/cm^3$). Drug concentration decrease in the water 1 phase and increase in the water 2 phase are monitored and recorded by means of an on-line UV spectrophotometer (Lambda 6/PECSS System, Perkin-Elmer Corporation, Norwalk, CT. Wavelength: piroxicam pH $1.2 = 354.2$ nm; nimesulide pH $7.5 = 390$ nm). Fluid recirculation is ensured by a peristaltic pump. Drug concentration in the oil phase is obtained from the total mass balance. Each test is performed in triplicate.

As an example, Figure 5.27 shows the very good comparison between model best fitting (solid lines) and the experimental data (symbols), in the case nimesulide partitioning at pH 7.5. Experimental data are represented by symbols, water 1 (triangles), oil (circles), and water 2 (squares). The experimental data standard errors, not reported for clarity reasons, reach a maximum value of 4%. The values of the rate constants, obtained as fitting parameters,

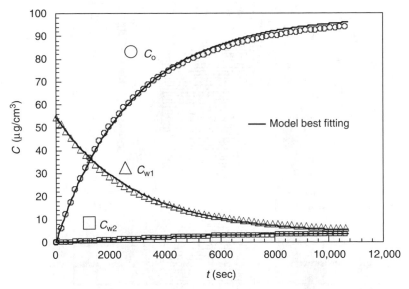

FIGURE 5.27 Comparison between model best fitting and experimental data, in the case of nimesulide partition (pH 7.5, 37°C). (Adapted from Sirotti, C., et al., *Proceedings of the CRS Winter Symposuim and 11th International Symposuim on Recent Advances in Drug Delivery Systems*, 3-3-2003, 51.)

are $k_{w1o} = k_{w2o} = 2.5 \times 10^3$ cm/sec, $k_{ow1} = k_{ow2} = 9.4 \times 10^{-5}$ cm/sec. Consequently, the partition coefficient, $P_{7.5}^{nime} = 27.1$, is very close to that previously determined according to the two-phases approach (see Table 5.8). Same results are found also for the other experimental conditions examined. Accordingly, this experimental–theoretical approach proves to be reliable and applicable to compare the behavior of different drugs, once their partition coefficient and solubility in the apolar and polar phases are known.

5.3.2 Partitioning and Permeation

It is now interesting to show just an example of how partitioning can affect drug permeability. For this purpose, among the great variety of models about permeability [73], let us consider the *two-step distribution model* of Van de Waterbeemd and group [3,4,71]. Basically, this model assumes that permeability is the result of two distribution steps. Indeed, the existence of an aqueous and an organic diffusion layer at each lipid–aqueous interface is postulated (see Figure 5.28). If the stagnant organic layers are assumed identical and supposing steady-state conditions, the permeation process depends on the observed forward (k_{13}) and backward (k_{24}) rate constants:

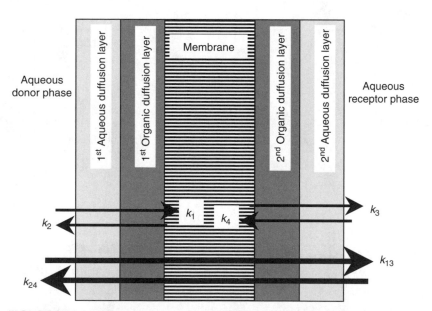

FIGURE 5.28 Membrane model (*two-step distribution model*) according to Waterbeemd and Jansen. This model assumes the existence of a double organic–aqueous stagnant layer at the membrane–aqueous phase interface. (Adapted from Waterbeemd van de, H., Jansen, C.A., Gerritsma, K.W., *Pharmaceutisch Weekblad*, 113, 1097, 1978.)

$$k_{13} = \frac{k_1 k_3}{k_2 + k_3} = \frac{k_{\text{org}} D_1}{(1 + \beta_1 D_1) + (1 + \beta_2 D_2)} \quad D_1 = \frac{C_{\text{org}}}{C_{\text{aq1}}} \quad \beta_1 = \frac{k_{\text{org}}}{k_{\text{aq1}}}, \quad (5.160)$$

$$k_{24} = \frac{k_4 k_2}{k_2 + k_3} = \frac{k_{\text{org}} D_2}{(1 + \beta_1 D_1) + (1 + \beta_2 D_2)} \quad D_2 = \frac{C_{\text{org}}}{C_{\text{aq2}}} \quad \beta_2 = \frac{k_{\text{org}}}{k_{\text{aq2}}}, \quad (5.161)$$

where k_1 and k_2 are the rate constants for the forward and reverse transport over the interface 1, k_3 and k_4 are the rate constants for the forward and reverse transport over the interface 2, D_1 and D_2 are the distribution coefficients at both sides of the membrane, k_{aq} and k_{org} are the diffusion rate constants in the stagnant aqueous layer and membrane, respectively ($k_{\text{org1}} = k_{\text{org2}} = k_{\text{org}}$), while $C_{\text{org}}, C_{\text{aq1}}$, and C_{aq2} are drug concentrations in the organic, aqueous 1, and aqueous 2 stagnant layer, respectively. Distribution coefficients (D) are used instead of partition coefficients (P) in order to account for potential ionization in the aqueous compartments at a particular pH. Assuming sink conditions in the aqueous receptor phase (only the mass transfer rate constant k_{13} is needed to describe membrane permeability) and remembering Fick's first law, the following relationship exists between k_{13} (1/sec) and the corresponding observed forward permeability coefficient $P_{e,13}$ (cm/sec) [74]:

$$k_{13} = P_{e,13} \frac{A}{V_A}, \quad (5.162)$$

where V_A is the solvent volume in the acceptor chamber and A is membrane surface. Combining Equation 5.160 and Equation 5.162, permeability P_e can be expressed as a function of partitioning properties:

$$P_{e,13} = P_e = \frac{k_{\text{org}} \frac{V_A}{A} D_1}{(1 + \beta_1 D_1) + (1 + \beta_2 D_2)}. \quad (5.163)$$

In order to render Equation 5.163 operative, we can remember that while, for a given system, A and V_A can be considered constant, k_{org} can be expressed by [75]

$$k_{\text{org}} = \frac{D_{\text{org}} A}{hV}, \quad (5.164)$$

where A and h are, respectively, the membrane surface area and thickness, V is acceptor chamber solvent volume and D_{org} is the solute diffusion coefficient in the solvent of the corresponding environment. For spherical particles, D_{org} can be approximated by the Stokes–Einstein equation:

$$D_{\text{org}} = \frac{KT}{6\pi\eta R_H}, \quad (4.100)$$

where K is the Boltzmann constant, T is the absolute temperature, η is the solvent (continuum medium) shear viscosity and R_H is the solute molecules hydrodynamic radius. Since membrane/water distribution coefficient (D_1 and D_2) are difficult to measure directly, they can be related to 1-octanol/water distribution coefficients (D_{oct}) via the well-known Collander [76,77] equation establishing a relation between lipophilicity of different solvents (A and B) with similar physical properties (in particular similar hydrogen-bonding capacity):

$$\log(D_A) = b\log(D_B) + \log(a). \tag{5.165}$$

Typical values for b are in the range 0.5–1.0, while $\log(a)$ varies between -3 and $+1$ [78].

In the light of Equation 5.164 and Equation 5.165, Equation 5.163 finally reads

$$P_e = \frac{k_{org}\dfrac{V_A}{A}a_1(D_{oct})^{b_1}}{(1+\beta_1 a_1(D_{oct})^{b_1}) + (1+\beta_2 a_2(D_{oct})^{b_2})} = \frac{a_1(D_{oct})^{b_1}}{(1+\beta_1'(D_{oct})^{b_1}) + (1+\beta_2'(D_{oct})^{b_2})}, \tag{5.166}$$

$$\beta_1' = \frac{k_{org}}{k_{aq1}}a_1 \qquad \beta_2' = \frac{k_{org}}{k_{aq2}}a_2. \tag{5.166'}$$

5.3.3 Log(P) Calculation

For its practical importance in permeability and solubility [79] determination, many authors devoted their research to the theoretical determination of P (or $\log(P)$). Although a detailed description of these models is out of the aim of this chapter, we believe that some of them can be briefly mentioned as a reference for the reader. Basically, for many approaches, $\log(P)$ results from the combined effects of "chemical groups" constituting molecular structure [80]. Sometimes these groups identify with hydrogen atoms substituents, or with molecular structure fragments (C, CH, CH_2, CH_3, OH, NH_2, and so on), or with molecule atoms. Other approaches, based on the well-known observation that a molecule is rarely a simple sum of its parts, try to evaluate $\log(P)$ on the basis of molecular properties.

Among the group contribution approaches, the Lipinski [81] implementation of Moriguchi [82] model can be considered. Basically, this method requires to count lipophilic atoms (all carbons and halogens with a multiplier rule for normalizing their contributions) and hydrophilic atoms (all nitrogen and oxygen atoms). Indeed, based on a collection of 1230 compounds, Moriguchi found that these two parameters alone account for 73% of the variance in the experimental $\log(P)$. To get model final result, some correction factors (four that increase the hydrophobicity and seven that increase the lipophilicity) need to be considered. The final equation accounts for 91% of the variance in the experimental $\log(P)$ of the 1230 compounds. The advantage of this approach consists in its easy programming and in the fact that it does not require a large database of parameter values. Another group contribution approach that is worth mentioning is CLOGP [80]. It simply states that $\log(P)$ of a solute can be estimated from the sum of the contributions of each fragment type, times the number of times it occurs plus the sum of factors (accounting for fragments interaction) times the number of times each factor occurs. Interestingly, fragmentation is not left to user discretion but it is well codified. This approach is even better than that of Lipinski [81]. Finally, the solvatochromic approach [83,84], essentially a molecular properties-based model, states that $\log(P)$ is the result of a weighted sum of solvatochromic parameters. In particular, they are solute molecular volume, solute polarity (or polarizability), solute hydrogen-bond acceptor strength, and solute hydrogen-bond donor strength.

5.4 DISSOLUTION AND PARTITIONING

To complete this panoramic discussion about dissolution and partitioning, it is necessary to discuss the simultaneous drug dissolution (in an aqueous environment) and partitioning in an organic phase. Indeed, not only this *in vitro* test can, at least qualitatively, yield some information about *in vivo* behavior, but it also allows checking the reliability of both the mathematical model and the experimental approach developed about dissolution and partitioning. Indeed, if theory and experiment are correct, it is possible to simulate the *in vitro* drug dissolution–partitioning behavior resorting to data coming from independent dissolution and partitioning experiments as shown later.

The experimental setup is nothing more than an extension of the IDR apparatus shown in Figure 5.7. A metallic rod, carrying a drug disk on its bottom part (Figure 5.29), is immersed in a two-phase environment constituted by the upper organic phase and the lower aqueous phase. To prevent solid drug–organic phase contact during metallic rod immersion, drug disk surface is coated with an impermeable membrane that is removed at the beginning of the experiment. In addition, drug disk lateral surface is coated by an impermeable membrane so that dissolution takes place only through the disk bottom surface. Aqueous and organic phase mixing are ensured, respectively, by rod rotation and by the fin fixed on the rod (Figure 5.29). Fin position and dimension must be studied to ensure a good mixing without causing liquid–liquid interfacial area increase due to the formation of undesirable waves. In this case, indeed, the interfacial surface would be very hard to determine and it would be dependent on rod rotational speed. Finally, in order to prevent aqueous solvent migration into the organic phase and vice versa during experiment, mutual presaturation of both phases is needed.

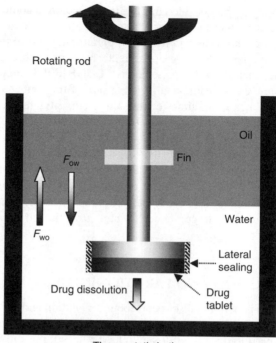

FIGURE 5.29 Schematic representation of the drug dissolution/partitioning apparatus.

Physically speaking, the mechanism ruling dissolution coupled with partitioning consists of drug leaving from solid surface, according to Levich's theory, followed by drug partitioning between the two phases as shown in Figure 5.29. This physical frame can be represented by the following set of differential equations:

$$\frac{dC_w}{dt} = \frac{k_d S}{V_w}(C_i(t) - C_w) - \frac{k_{degr}}{V_w}C_w - \frac{k_{wo}A_{wo}}{V_w}C_w\left(1 - \frac{C_o}{C_{so}}\right) + \frac{k_{ow}A_{wo}}{V_w}C_o\left(1 - \frac{C_w}{C_{sw}}\right),$$

$$(5.167)$$

$$\frac{dC_o}{dt} = \frac{k_{wo}A_{wo}}{V_o}C_w\left(1 - \frac{C_o}{C_{so}}\right) + \frac{k_{ow}A_{wo}}{V_o}C_o\left(1 - \frac{C_w}{C_{sw}}\right), \qquad (5.168)$$

where C_o and C_w are, respectively, drug concentration in the organic and aqueous phase, S is the dissolving solid surface area, k_d is the dissolution constant, V_w and V_o are, respectively, aqueous and organic phase volume, A_{wo} is the aqueous–organic interfacial area, k_{wo} and k_{ow} are kinetics constants, C_{so} and C_{sw} are, respectively, drug solubility in the organic and aqueous phases,

k_{degr} (degradation constant) accounts for possible drug degradation in the aqueous phase while $C_i(t)$ accounts for possible drug solubility variation at the solid drug–aqueous phase interface due to recrystallization (see Equation 5.109'):

$$C_i = C_w + (C_{wi} - C_w)e^{(-k_r t)},$$ (5.169)

where C_{wi} is drug solubility in the aqueous phase competing to less stable, more soluble, drug form (typically, $C_{wi} \gg C_w$). Basically, focusing the attention on its right-hand side, Equation 5.167 states that aqueous drug concentration varies with time due to drug dissolution (positive contribution), degradation (negative contribution), water–oil drug flux (negative contribution), and oil–water drug flux (positive contribution). Analogously, Equation 5.168 states that oil drug concentration varies with time due to water–oil drug flux (positive contribution) and oil–water drug flux (negative contribution). Due to the possible presence of drug degradation it is no longer possible to use a global mass balance to reduce to only one equation the differential system represented by Equation 5.167 and Equation 5.168. Accordingly, a numerical technique (for example, fourth-order Runge-Kutta method [56]) is needed to get model solution. If, on the contrary, the ratios C_o/C_{so} and C_w/C_{sw} tend to zero and no recrystallization takes place, the system of differential equation represented by Equation 5.167 and Equation 5.168 yields the following analytical solution:

$$C_w(t) = \frac{m}{z_1 z_2} + Ae^{z_1 t} + Be^{z_2 t},$$ (5.170)

$$C_o(t) = \frac{1}{\zeta}\left(e^{z_1 t}(Az_1 - \beta A) + e^{z_2 t}(Bz_2 - \beta B) - \alpha - \frac{\beta m}{z_1 z_2}\right),$$ (5.171)

where

$$\alpha = \frac{k_d S}{V_w}C_{sw}, \qquad \beta = -\left(\frac{k_{wo}A_{wo}}{V_w} + \frac{k_d S}{V_w} + \frac{k_{degr}}{V_w}\right), \qquad \zeta = \frac{k_{ow}A_{wo}}{V_w},$$ (5.172)

$$a = \frac{k_{ow}A_{wo}}{V_o} + \frac{k_{wo}A_{wo}}{V_w} + \frac{k_d S}{V_w} + \frac{k_{degr}}{V_w}, \qquad b = -\frac{k_{ow}A_{wo}}{V_w V_o}(k_d S + k_{degr}),$$ (5.172')

$$m = \frac{A_{wo}k_{ow}}{V_o}\frac{S k_d}{V_w}, \qquad z_1 = \frac{-a + \sqrt{a^2 - 4b}}{2}, \qquad z_2 = \frac{-a - \sqrt{a^2 - 4b}}{2},$$ (5.173)

$$B = \frac{m + \alpha z_2}{z_2(z_2 - z_1)}, \qquad A = \frac{m + \alpha z_1}{z_1(z_1 - z_2)}.$$ (5.174)

Interestingly, as $a > 0$ and $b < 0$, it follows $z_1 > 0$ and $z_2 < 0$. In addition, it is worth mentioning that A and B are defined to ensure $C_w(t = 0) = C_o(t = 0) = 0$, these reflecting usual experimental conditions.

To test this theoretical approach, we can refer to experimental data concerning piroxicam dissolution, partitioning and dissolution–partitioning at two different pH. In particular, as mentioned at the beginning of this paragraph, model parameters will be deduced by means of data fitting of independent dissolution and partitioning experiments and, then, model predictions will be compared with experimental data referring to drug dissolution in a two-phase environment. Regardless of the pH considered (1.2 and 7.5), the dissolution test is performed considering a drug tablet that has 1 cm diameter, rotational speed equal to 72 rpm while dissolution volume is 100 and 150 cm^3 for the pH 1.2 and 7.5 conditions, respectively. The use of unequal volumes is dictated by piroxicam aqueous solubility variation in relation to different pH conditions (37°C; pH 1.2, $C_{sw} = 250$ µg/cm^3; pH 7.5, $C_{sw} = 2171$ µg/cm^3; see Table 5.8). As in this case neither drug degradation nor drug recrystallization takes place, model expression identifies with Equation 5.5 and data fitting yields to $k_d = 1.6 \times 10^{-3}$ cm/sec and $k_d = 1.9 \times 10^{-3}$ cm/sec at pH = 1.2 and pH = 7.5, respectively. The rate constants connected to piroxicam partitioning are those determined in the "partitioning section" (Paragraph 5.3) and reported in Table 5.8 ($k_{wo} = 2.5 \times 10^{-3}$ cm/sec (pH = 1.2); 3.6×10^{-4} cm/sec (pH = 7.5); $k_{ow} = 2.8 \times 10^{-4}$ cm/sec (pH = 1.2); 8.3×10^{-4} cm/sec (pH = 7.5)). Based on the knowledge of k_d, k_{wo}, and k_{ow}, and knowing that aqueous and organic volumes are, at both pH, 150 and 50 cm^3, respectively, that aqueous–organic phase interfacial area is equal to 34 cm^2, that drug tablet diameter is 1 cm and that rotational speed is 72 rpm, it is possible to compare model prediction (due to piroxicam solubility values at both pH, model analytical form, Equation 5.170, can be used) and experimental data referring to simultaneous dissolution and partitioning. Figure 5.30 shows that the agreement between experimental data (symbols) and model prediction (solid line) is satisfactory although, in the pH = 1.2 case, model prediction underestimates experimental data. This could also be due to the fact that the rotational speed considered, in relation to drug tablet diameter, is a little bit low to ensure a full validity of Levich's approach (see Paragraph 5.2.2.1).

5.5   CONCLUSIONS

The aim of this chapter was essentially to match the problem of drug dissolution and partitioning from both an experimental and theoretical point of view. Accordingly, solid wetting, dissolution medium hydrodynamics, drug phase change upon dissolution, polyphase drug dissolution, drug degradation in the dissolution medium, and drug dissolution from a polydispersed

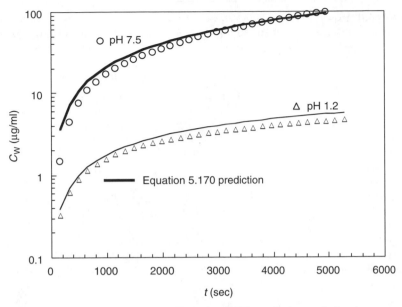

FIGURE 5.30 Comparison between Equation 5.170 prediction and piroxicam water concentration at two different pH.

particulate system have been discussed. In addition, the mechanisms ruling drug partitioning and their possible implications about drug permeation through cellular membranes have been considered. Finally, the simultaneous drug dissolution and partitioning have been matched. The theoretical frame presented serves to measure important drug physicochemical parameters and to predict, at least *in vitro*, drug behavior in the light of reducing and properly designing experimental trials. In addition, it is worth mentioning that, in a recent work by FDA researchers [69], the use of IDR test (or *disk intrinsic dissolution rate* test (DIDR)) has been successfully proposed for drug classification. According to the authors, this way of drug classification, which is based on a rate phenomenon instead of an equilibrium one, should be more closely correlated with *in vivo* drug dissolution dynamics than a classification based only on solubility. Consequently, they focus their attention on the robustness of DIDR procedure in relation to some experimental variables (compression force, dissolution volume, distance of compressed drug disk from the bottom of the dissolution vessel, and rotational speed). Assuming that solid surface area does not modify during dissolution and that sink conditions are always attained, experimental tests are performed varying compression force (range 600–5000 psi (4.13–34.47 MPa)), dissolution volume (range 225–900 cm^3), distance (range 0.5–1.5 in (1.27–3.81 cm)), and rotational speed (range 15–250 rpm (1.57–26.18 rad/sec)). Dissolution media

TABLE 5.9
Intrinsic Dissolution Rate "j" of Scarcely and Highly Soluble Drugs at $T = 37.4°C$ and Different pH

Drug	pK_a	j (mg/(min* cm²))		
		pH = 1.2	pH = 4.5	pH = 6.8
Low solubility				
Carbamazepine	—	0.025	0.024	0.029
Furosemide	3.9	0.0017	0.018	0.502
Griseofulvin	—	0.0026	0.0019	0.0022
Ketoprofen	3.5	0.016	0.062	0.567
Naproxen	4.2	0.0035	0.012	0.264
Piroxicam	5.1	0.022	0.0043	0.088
High solubility				
Atenolol	9.6	5.6	3.74	2.56
Cimetidine	6.8	7.3	2.9	1.07
Hydrochlorothiazide	7.9	0.119	0.124	0.113
Labetalol	7.4	1.03	2.88	0.76
Metoprolol	9.7	23.7	22.4	21.4
Nadolol	9.7	8.04	2.47	1.44
Nortriptyline·HCl	10.0	0.895	7.81	6.45
Propanolol	9.5	10.3	13.3	14.6
Ranitidine	—	46.1	47.9	43.1

Source: From Yu, L.X., et al., *Int. J. Pharm.*, 270, 221, 2004. With permission from Elsevier.
Note: Disk rotational speed = 100 rpm; dissolution medium volume = 900 cm³ for highly soluble drugs and 225 cm³ for scarcely soluble drugs; dissolution surface = 0.5 cm²; three repetitions for each condition; standard deviation is always less than 5% of the reported mean.

employed are 0.1 N HCl (pH = 1.2), 0.2 M acetate buffer (pH = 4.5), and 0.2 M phosphate buffer (pH = 6.8). Drug concentration is spectrophotometrically measured by sampling 4 cm³ from the dissolution environment volume and replacing it with an equal volume of pure solvent. IDR j is calculated according to the following equation:

$$j = \frac{V}{S}\frac{dC}{dt}, \qquad (5.175)$$

where V is the dissolution medium volume, S is the dissolving solid area, t is the time while C is the drug concentration in the dissolution medium. In this equation, dC/dt can be considered constant. Indeed, in sink conditions, Equation 5.5 can be approximated by

$$C = C_s\left[1 - e^{\left(-k_d\frac{S}{V}t\right)}\right] \approx C_s k_d \frac{S}{V} t \qquad (5.176)$$

and, thus

$$\frac{dC}{dt} = C_s k_d \frac{S}{V}. \qquad (5.177)$$

Table 5.9 shows the IDR of 15 model drugs (six scarcely soluble and nine highly soluble) at different pH and fixing to 100 rpm the rotational speed, $V = 900$ cm^3 for highly soluble and $V = 225$ cm^3 for scarcely soluble drugs, $S = 0.5$ cm^2 and $T = 37.4°$C. It is clear that highly soluble drugs show j values that are 3–4 orders of magnitude higher than that of scarcely soluble drugs. In addition, the pK_a value can considerably affect IDR. On the basis of this study, the authors conclude that DIDR is a robust technique for drug classification as the operative variables considered (compression force, dissolution volume, distance of compressed drug disk from the bottom) have no significant effect on j value, where the rotational speed is constant. In addition, they found that Equation 5.62 (with $C_0 = 0$) holds in the rotational speed range considered:

$$j = 0.62\, D^{2/3}\, \nu^{-1/6}\, \omega^{1//2}\, C_s. \qquad (5.62a)$$

According to Table 5.9, $j = 0.1$ mg/min/cm^2 should be the cut off value for drug classification into the high or low DIDR classes.

REFERENCES

1. *Medicinal Products for Human Use: Guidelines*, Pharmacos 4, Eudralex Collection, 2005, p. 234.
2. Waterbeemd van de, H. and Jansen, A., Transport in QSAR V: application of the interfacial drug transfer model, *Pharm. Weekbl. Sci. Ed.*, 3, 587, 1981.
3. Waterbeemd van de, H., et al., Transport in QSAR II: rate-equilibrium relationships and the interfacial transfer of drugs, *Eur. J. Med. Chem.*, 3, 279, 1980.
4. Waterbeemd van de, H., et al., Transport in QSAR IV: the interfacial transfer model. Relationships between partition coefficients and rate constants of drug partitioning, *Pharm. Weekbl. Sci. Ed.*, 3, 224, 1981.
5. Lipper, R.A., E pluribus product, *Mod. Drug Disc.*, 55, 1999.
6. Pace, S.N., et al., Novel injectable formulations of insoluble drugs, *Pharm. Tech.*, 116, 1999.
7. Wotton, P.K., Making the insoluble soluble, *Scrip. Mag.*, 12, 2001.
8. Bergese, P., Pharmaceutical nano-composite generated by microwaves induced diffusion (mind), Thesis/Dissertation, University of Brescia (Italy), 2002.
9. Amidon, G.L., et al., A theoretical basis for a biopharmaceutic drug classification: the correlation of *in vitro* drug product dissolution and *in vivo* bioavailability, *Pharm. Res.*, 12, 413, 1995.

10. Hsia, D.C., Kim, C.K., Kildsig, D.O., Determination of energy change associated with dissolution of a solid, *J. Pharm. Sci.*, 66, 961, 1977.
11. *Dissolution Technology*, The Industrial Pharmaceutical Technology Section of the Academy of Pharmaceutical Science, Washington D.C., 1974.
12. Nicklasson, M., Brodin, A., Sundelöf, L.O., On the determination of true intrinsic rate parameters from rotating disc experiments, *Int. J. Pharm.*, 15, 87, 1983.
13. Levich, V.G., *Physicochemical Hydrodynamics*, Prentice-Hall, 1962.
14. Banakar, U.V., *Pharmaceutical Dissolution Testing*, Marcel Dekker, New York, Basel, Hong Kong, 1991.
15. Martin, A., Swarbrick, J., Cammarata, A., *Physical Pharmacy*, Lea & Febiger, 1983, Chapter 16.
16. Good, R.J., A thermodynamic derivation of Wenzel's modification of Young's equation for contact angles together with a theory of hysteresis, *J. Am. Chem. Soc.*, 74, 5041, 1952.
17. Adamson, A.W. and Gast, A.P., *Physical Chemistry of Surfaces*, Wiley, 1997.
18. Young, T., *Miscellaneous Works*, Murray, London, 1855, p. 418.
19. Kwok, D.Y. and Neumann, A.W., Contact angle measurement and contact angle interpretation, *Adv. Colloid Interface Sci.*, 81, 167, 1999.
20. Adamson, A.W. and Gast, A.P., *Physical Chemistry of Surfaces*, Wiley, 1997, Chapter 10.
21. Cassie, A.B.D. and Baxter, S., Wettability of porous surfaces, *Trans. Faraday Soc.*, 40, 546, 1944.
22. Cassie, A.B.D., Contact angles, *Disc. Faraday Soc.*, 3, 11, 1948.
23. Canal, T., Colombo, I., Lovrecich, M., Lipidic composite materials: interface modifying agent effect on bulk and surface properties, *J. Contr. Rel.*, 27, 19, 1993.
24. Adamson, A.W. and Gast, A.P., *Physical Chemistry of Surface*, Wiley, New York, 1997.
25. Owens, D.K. and Wendt, R.C., Estimation of the surface free energy of polymers, *J. Appl. Polym. Sci.*, 19, 1741, 1969.
26. Wu, S.J., Calculation of interfacial tension in polymer systems, *Polym. Sci. Part C.*, 34, 19, 1971.
27. Good, R.J. and van Oss, C.J., The modern theory of contact angles and the hydrogen bond components of surface energies, in *Modern Approaches to Wettability: Theory and Applications*, Schrader, M. and Loeb, G., Eds., Plenum Press, 1992.
28. Li, D. and Neumann, A.W., A reformulation of the equation of state for interfacial tensions, *J. Colloid Interface Sci.*, 137, 304, 1990.
29. Swokowski, E.W., *Calculus with Analytic Geometry*, second ed., Prindle, Weber and Schmidt, Boston, 1979.
30. Kwok, D.Y. and Neumann, A.W., Contact angle interpretation in terms of solid surface tension, *Colloids Surf. A*, 161, 31, 2000.
31. Spelt, J.K. and Li, D., The equation of state approach to interfacial tensions, in *Applied Surface Thermodynamics*, Neumann, A.W. and Spelt, J.K., Eds., Marcel Dekker, 1996.
32. Patankar, S.V., *Numerical Heat Transfer and Fluid Flow*, McGraw-Hill/ Hemisphere Publishing Corporation, New York, 1990.
33. Chow, A.H.L., et al., Assessment of wettability and its relationship to the intrinsic dissolution rate of doped phenytoin crystals, *Int. J. Pharm.*, 126, 21, 1995.

34. Smith, R.P., et al., Approaches to determine the surface tension of small particles: equation of state considerations, *J. Colloid Interface Sci.*, 110, 521, 1986.

35. Leòn, V., Tusa, A., Araujo, Y.C., Determination of the solid surface tensions. The platinum case, *Colloids Surf. A*, 155, 131, 1999.

36. Khoury, N., Mauger, J.W., Howard, S., Dissolution rate studies from a stationary disk\rotating fluid system, *Pharm. Res.*, 5, 495, 1988.

37. Shah, A.C., Poet, C.B., Ochs, J.F., Design and evaluation of a rotating filter-stationary basket *in vitro* dissolution test apparatus I: fixed fluid volume system, *J. Pharm. Sci.*, 62, 671, 1973.

38. Nelson, K.J. and Shah, A.C., Convective diffusion model for a transport-controlled dissolution rate process, *J. Pharm. Sci.*, 64, 610, 1975.

39. Vongvirat, B., et al., Evaluation of a rotating filter apparatus: hydrodynamic characterization through modeling, *Int. J. Pharm.*, 9, 213, 1981.

40. Crane, M., et al., Simulation of the USP drug delivery problem using CFD: experimental, numerical and mathematical aspects, *Simulation Modelling Practice and Theory*, 12, 147, 2004.

41. Peltonen, L., et al., Dissolution testing of acetylsalicylic acid by a channel flow method-correlation to USP basket and intrinsic dissolution methods, *Eur. J. Pharm. Sci.*, 19, 395, 2003.

42. Grijseels, H. and de Blaey, C.J., Dissolution at porous interfaces, *Int. J. Pharm.*, 9, 337, 1981.

43. Grijseels, H., et al., Dissolution at porous interfaces II. A study of pore effects through rotating disc experiments, *Int. J. Pharm.*, 14, 299, 1983.

44. Grijseels, H., Crommelin, D.J.A., de Blaey, C.J., Dissolution at porous interfaces III. Pore effects in relation to the hydrodynamics at a rotating disc surface, *Int. J. Pharm.*, 14, 313, 1983.

45. Grassi, M., Colombo, I., Lapasin, R., Experimental determination of the theophylline diffusion coefficient in swollen sodium-alginate membranes, *J. Contr. Rel.*, 76, 93, 2001.

46. Coviello, T., et al., Structural and rheological characterization of Scleroglucan/boraxhydrogel for drug delivery, *Int. J. Biol. Macromol.*, 32, 83, 2003.

47. de Smidt, J.H., et al., Dissolution of theophylline monohydrate and anhydrous theophylline in buffer solutions, *J. Pharm. Sci.*, 75, 497, 1986.

48. Coles, B.A., et al., A hydrodynamic atomic force microscopy flow cell for the quantitative measurement of interfacial kinetics: the aqueous dissolution of salicylic acid and calcium carbonate, *Langmuir*, 14, 218, 1998.

49. Stojek, Z., Ciszkowska, M., Osteryoung, J.G., Self-enhancement of voltammetric waves of weak acids in the absence of supporting of electrolyte, *Anal. Chem.*, 66, 1507, 1994.

50. Nogami, H., Nagai, T., Yotsuyanagi, T., Dissolution phenomena of organic medicinals involving simultaneous phase changes, *Chem. Pharm. Bull.*, 17, 499, 1969.

51. Higuchi, W.I., Bernardo, P.D., Mehta, S.C., Polymorphism and drug availability II: dissolution rate behavior of the polymorphic forms of sulfathiazole and methylprednisolone, *J. Pharm. Sci.*, 56, 200, 1967.

52. Grassi, M., Colombo, I., Lapasin, R., Drug release from an ensemble of swellable crosslinked polymer particles, *J. Contr. Rel.*, 68, 97, 2000.

53. Di Martino, P., et al., Physical characterisation of nadroxen sodium hydrate forms, *Eur. J. Pharm. Sci.*, 14, 293, 2001.

54. Higuchi, W.I., Bernardo, P.D., Mehta, S.C., Polymorphism and drug availability. II. Dissolution rate behavior of the polymorphic forms of sulfathiazole and methylprednisolone, *J. Pharm. Sci.*, 56, 200, 1967.
55. Higuchi, W.I., Mir, N.A., Desai, S.J., Dissolution rates of polyphase mixtures, *J. Pharm. Sci.*, 54, 1405, 1965.
56. Press, W.H., et al., *Numerical Recepies in FORTRAN*, second ed., Cambridge University Press, Cambridge, USA, 1992.
57. Hixson, A. and Crowell, J., Dependence of reaction velocity upon surface and agitation: I theoretical consideration, *Ind. Eng. Chem.*, 23, 923, 1931.
58. Pedersen, V.P. and Myrick, J.W., Versatile kinetic approach to analysis of dissolution data, *J. Pharm. Sci.*, 67, 1450, 1978.
59. Pedersen, V.P. and Brown, K.F., General class of multiparticulate dissolution models, *J. Pharm. Sci.*, 66, 1435, 1977.
60. Pedersen, V.P. and Brown, K.F., Size distribution effects in multiparticulate dissolution, *J. Pharm. Sci.*, 64, 1981, 1975.
61. Pedersen, V.P. and Brown, K.F., Dissolution profile in relation to initial particle distribution, *J. Pharm. Sci.*, 64, 1192, 1975.
62. Finizio, F., Vighi, M., Sandroni, D., Determination of n-octanol/water partition coefficient (K_{ow}) of pesticide critical review and comparison of methods, *Chemosphere*, 34, 131, 1997.
63. Coehn, Y., et al., Dynamic partitioning of organic chemicals in regional environments: a multimedia screening-level modelling approach, *Environ. Sci. Technol.*, 24, 1549, 1982.
64. Briggs, G.G., Bromilov, R.H., Evans, A.A., Relationships between lipophilicity and root uptake and translocation of nonionized chemical by barley, *Pestic. Sci.*, 13, 495–504, 1982.
65. Calamari, D. and Vighi, M., Quantitative structure activity relationships in ecotoxicology: value and limitations, *Rev. Environ. Toxicol.*, 4, 1–112, 1990.
66. El Tayar, N., Testa, B., Carrupt, P.A., Polar intermolecular interactions encoded in partition coefficients: an indirect estimation of hydrogen-bond parameters of polyfunctional solutes, *J. Phys. Chem.*, 96, 1455, 1992.
67. Lung, S.H.S., Robinson, J.R., Lee, V.H., Parenteral products, in *Controlled Drug Delivery, Fundamental and Applications*, Robinson, J.R. and Lee, H.L., Eds., Marcel Dekker, 1987, Chapter 10.
68. Grassi, M., Coceani, N., Magarotto, L., Modelling partitioning of sparingly soluble drugs in a two phase liquid system, *Int. J. Pharm.*, 239, 157, 2002.
69. Yu, L.X., et al., Feasibility studies of utilizing disk intrinsic dissolution rate to classify drugs, *Int. J. Pharm.*, 270, 221, 2004.
70. Lippold, V.B.C. and Schneider, G.F., Zur optimierung der verfugbarkeit homologer quartarer ammoniumverbindunger, *Arzneim. -Forsch (Drug Res.)*, 25, 843, 1975.
71. Waterbeemd van de, H., Jansen, C.A., Gerritsma, K.W., Transport in QSAR, *Pharmaceutisch Weekblad*, 113, 1097, 1978.
72. Sirotti, C., et al., Mathematical modeling of sparingly soluble drugs partitioning in a three phase liquid system, in *Proceedings of the CRS Winter Symposium and 11th International Symposium on Recent Advances in Drug Delivery Systems*, 3-3-2003, 51.

73. Camenisch, G., Folkers, G., Waterbeemd van de, H., Shapes of membrane permeability–lipophilicity curves: extension of theoretical models with an aqueous pore pathway, *Eur. J. Pharm. Sci.*, 6, 321, 1998.
74. Flynn, G.L., Yalkowsky, S.H., Roseman, T.J., Mass transport phenomena and models: theoretical concepts, *J. Pharm. Sci.*, 63, 479, 1974.
75. Kubinyi, H., Quantitative structure–activity relationships: IV. Nonlinear dependence of biological activity on hydrophobic character: a new model, *Arzneim. Forsch. (Drug Res.)*, 26, 1991, 1976.
76. Collander, R., The distribution of organic compounds between iso-butanol and water, *Acta Chem. Scand.*, 4, 1085, 1950.
77. Collander, R., The partition of organic compounds between higher alcohols and water, *Acta Chem. Scand.*, 5, 774, 1951.
78. Leo, A., Hansch, C., Elkins, D., Partition coefficients and their uses, *Chem. Rev.*, 71, 525, 1971.
79. Yalkowski, S.H. and Valvani, S.C., Solubility and partitioning 1: solubility of nonelectrolytes in water, *J. Pharm. Sci.*, 69, 912, 1980.
80. Leo, A., Calculating LogPoct from structures, *Chem. Rev.*, 93, 1281, 1993.
81. Lipinski, C.A., et al., Experimental and computational approaches to estimate solubility and permeability in drug discovery and development settings, *Adv. Drug Deliv. Rev.*, 46, 3, 2001.
82. Moriguchi, I., et al., Simple method of calculating octanol/water partition coefficient, *Chem. Pharm. Bull.*, 40, 127, 1992.
83. Kamlet, M., et al., Linear solvation energy relationships. 23. A comprehensive collection of the solvatochromic parameters, PI, alpha, and beta, and some methods for simplifying the generalised solvatochromic equation, *J. Org. Chem.*, 48, 2877, 1983.
84. Kamlet, M., Abboud, J.-L., Taft, R., The solvatochromic comparison method. 6. The PI scale of solvent polarities, *J. Am. Chem. Soc.*, 99, 6027, 1977.

6 Dissolution of Crystallites: Effects of Size on Solubility

6.1 INTRODUCTION

The oral route is considered the best way of dosing drugs [1]. Most drugs orally administrated are absorbed by passive diffusion of the dissolved solid through the gastrointestinal (GI) cellular membranes. The oral bioavailability of a specific solid drug is thus determined by its solubility and permeability within the GI tract, which is an aqueous environment. For a given drug that dissolves into the GI tract, its molecular structure and the environment are the fixed parameters, and the solid phase dissolution becomes a key parameter for assessing the bioavailability and the onset of action of oral dosage forms [2,3]. Since an increasing number of newly developed chemical entities present poor water solubility, advances in enhancing oral bioavailability of poorly water-soluble drugs represent an actual challenge for pharmaceutical research, with the aims of improving therapeutic effectiveness and creating new market opportunities.

A classical formulation approach for such drugs is micronization, which means to increase the dissolution velocity by enlarging the specific surface area of the drug powder. Nowadays, many of the new drugs exhibit such a low solubility that micronization does not lead to a sufficiently high bioavailability. Consequently, a modern and very promising approach is to move from micronization to nanonization, which means, producing drug nanocrystals [4]. This microstructure manipulation can be achieved by forcing the crystalline state to rearrange into nanodimensional coherent domains, which are intertwined with a dramatic increase in the interface boundaries and a change in the drug melting temperature [5] and enthalpy [6]. The overall result is a relevant enhancement of the solubility and in turn of the bioavailability of the drug [5,7].

Over the last 10 years, nanoparticle engineering processes have been developed for pharmaceutical applications. In this approach, poorly water-soluble drugs are produced in the nanometer-size range where there is a large surface–volume ratio. Among all these processes aimed to obtain nanocrystals, one can differentiate between bottom-up and top-down technologies.

Bottom-up technologies start from the molecules that are dissolved and then precipitated by antisolvent, reactant, or evaporation methods. Top-down technologies are disintegration methods, which means wet milling or dry drug-carrier co-grinding processes. Some examples for precipitation techniques are hydrosols [8], high-gravity reactive precipitation [9], and polymer nanocomposites [10,11]. In the case of disintegration approaches, the various techniques can be divided into the following main groups: pearl or ball wet milling [12], high-pressure homogenization [13], and high-energy mechanical activation [14].

An outstanding feature of drug nanocrystals is the increase in apparent water saturation solubility and consequently an increase in the dissolution rate of the drug. This increase in the solubility seems to be related to an alteration of the solid microstructure induced by the manufacturing processes [15,16]. In fact, the above-mentioned technologies often lead to plastic deformation within the crystal lattice, creating disordered regions and surface outcrops of lattice defects (e.g., dislocations, stacking faults, and so on) [11]. For instance, fine mechanical comminution of crystals produces a reduction in the size of crystallites and an increase in the lattice defect density [17,18].

A number of studies have been undertaken in an attempt to measure the solubility of finely divided solids as a function of particle size. The well-known Ostwald–Freundlich thermodynamic equation indicates that the solubility increases with the decrease in the particle dimension and it also predicts that the nanoparticles are not stable in solution because of their high-solubility value. However, the published results do not seem to be in agreement with the theoretical predictions, probably as a result of some experimental difficulties involved in the experimental measurements. In this chapter, according to these arguments, we analyze, from a theoretical and experimental point of view, the importance of understanding the detailed mechanism of dissolution when attempts are made to experimentally measure the solubility of poorly water-soluble drugs. Moreover, the complex microstructure of the solid drugs is characterized and the influence of the phase composition of the surfaces of such solids is correlated with their experimentally measured apparent solubility.

6.2 THERMODYNAMIC CONCEPTS

To rigorously outline the properties of solids that determine their dissolution rate, we have to recall solubility theory. An excellent review of solubility theory can be found in Liu's book [19] about water-insoluble drugs. The subject is also discussed in many works about bioavailability of solid drugs.

Whenever the term solubility is employed, it is tacitly assumed that it is equilibrium solubility. In other words, solubility is defined as the concentration of the dissolved solid (the solute) in the solvent medium under conditions of complete thermodynamic equilibrium. These conditions are uniform temperature and pressure in the two phases, and constant composition in each phase. This definition is by no means easy to establish in practice since other variables, such as the particle size and phase composition of the solid, and the nature and composition of the solvent medium, make the experimental conditions difficult to control. It is also known that solids of higher energetic state (metastable amorphates) may have initially high apparent solubility and then reduced to a thermodynamically stable lower value over time. The time for such a reduction depends on several factors and could vary from seconds to months.

The role of thermodynamics in the context of this chapter is to provide (a) some of the thermodynamic principles that govern equilibrium between a solid phase and a liquid phase; (b) a basis for the construction of mathematical functions to represent experimental solubility data; and (c) enable useful thermodynamic quantities to be extracted from the same data.

6.2.1 IDEAL SOLUTIONS

Thermodynamics deals with a system, the object of experimental interest (see Section 4.2). In chemical terms, a system is made up of components. In physical terms, a system can contain a number of phases. A phase is a homogeneous part of a system, which implies that all intensive variables of the system (temperature, pressure, and composition) are uniform within it. The basic term used to describe any phase that contains more than one component is solution. A solution describes a liquid or solid phase containing more than one substance. For convenience, one of the substances, which is called the solvent, is treated differently from the other substances, which are called solutes. If the sum of the mole fractions of the solutes is small compared with unity, the solution is called a dilute solution.

Consider a system consisting of a completely miscible solution of solute (designated by subscript 2) and solvent in all proportions and assuming negligible solubility of the solvent in the solid phase; then, the equation of equilibrium is [20,21]

$$f_2^s = f_2^l \tag{6.1}$$

or

$$f_2^s = y_2 x_2 f_2^0, \tag{6.2}$$

where x_2 is the solubility (mole fraction), y_2 is the solid-phase activity coefficient, f_2^s, f_2^l, f_2^0 represent, respectively, the fugacity of the pure solid, of the solute in liquid solution, and in the standard state to which y_2 refers (see Equation 4.38).

From Equation 6.2 the solubility is

$$x_2 = \frac{f_2^s}{y_2 f_2^0}.$$ (6.3)

Thus, solubility depends not only on the activity coefficient ($a_2 = y_2 x_2$) but also on the ratio of the two fugacities. In other words, the solubility is not only a function of the intermolecular forces between solute and solvent but also depends on the fugacity of the standard state to which the activity coefficient refers and the fugacity of the pure solid.

The standard-state fugacity is arbitrary; the only thermodynamic requirement is that it must be at the same temperature as that of the solution. Although other standard states can be used, it is most convenient to define the standard-state fugacity as the fugacity of a pure, subcooled liquid at the same temperature as that of the solution and at some specific pressure. The ratio of the two fugacities can be readily calculated by the thermodynamic cycle shown in Figure 6.1. The molar Gibbs energy change for component 2 in going from a to d is related to the fugacities of solid and subcooled liquid by

$$\Delta g_{a \to d} = RT \ln \frac{f^l}{f^s},$$ (6.4)

where, for simplicity, the subscript 2 is omitted.

In addition, on the basis of Equation 4.32′, the Gibbs energy change is also related to the enthalpy and entropy changes by

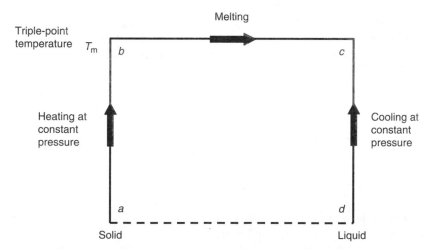

FIGURE 6.1 Thermodynamic cycle for calculating the fugacity of a pure supercooled liquid.

$$\Delta g_{a \to d} = \Delta h_{a \to d} - T \Delta s_{a \to d}. \tag{6.5}$$

The thermodynamic cycle shown in Figure 6.1 provides a method to evaluate the enthalpy and entropy changes given in Equation 6.5, as both enthalpy and entropy are state functions independent of the path, it is permissible to substitute the path $a \longrightarrow d$ by the alternate path $a \longrightarrow b \longrightarrow c \longrightarrow d$.

For the enthalpy change from a to d we have

$$\Delta h_{a \to d} = \Delta h_{a \to b} + \Delta h_{b \to c} + \Delta h_{c \to d}. \tag{6.6}$$

Equation 6.6 can be rewritten in terms of heat capacity c_p and enthalpy of melting Δh_m, neglecting the effect of pressure on the properties of solid and subcooled liquid:

$$\Delta h_{a \to d} = \Delta h_m(T_m) + \int_{T_m}^{T} \Delta c_p \, dT, \tag{6.7}$$

where $\Delta c_p = c_p(\text{liquid}) - c_p(\text{solid})$ and T_m is the melting temperature of the solid phase.

Similarly, for the entropy change from a to d

$$\Delta s_{a \to d} = \Delta s_{a \to b} + \Delta s_{b \to c} + \Delta s_{c \to d}, \tag{6.8}$$

which becomes

$$\Delta s_{a \to d} = \Delta s_m(T_m) + \int_{T_m}^{T} \frac{\Delta c_p}{T} \, dT. \tag{6.9}$$

At the melting point, the entropy of fusion Δs_m is

$$\Delta s_m = \frac{\Delta h_m}{T_m}. \tag{6.10}$$

Substituting Equation 6.5, Equation 6.7, Equation 6.9, and Equation 6.10 into Equation 6.4 and assuming that Δc_p is constant over the temperature range, $T \longrightarrow T_m$, we obtain

$$\ln \frac{f^l}{f^s} = \frac{\Delta h_m}{RT_m} \left(\frac{T_m}{T} - 1 \right) - \frac{\Delta c_p}{R} \left(\frac{T_m}{T} - 1 \right) + \frac{\Delta c_p}{R} \ln \left(\frac{T_m}{T} \right). \tag{6.11}$$

Equation 6.11 expresses the fugacity of the subcooled liquid at temperature T in terms of measurable thermodynamic properties.

An expression for the ideal solubility of a solid solute in a liquid solvent can be obtained by substituting Equation 6.11 into Equation 6.3 and assuming that $y_2 = 1$:

$$\ln x_2 = \frac{\Delta h_\mathrm{m}}{RT_\mathrm{m}}\left(\frac{T_\mathrm{m}}{T} - 1\right) - \frac{\Delta c_\mathrm{p}}{R}\left(\frac{T_\mathrm{m}}{T} - 1\right) + \frac{\Delta c_\mathrm{p}}{R}\ln\frac{T_\mathrm{m}}{T}. \tag{6.12}$$

Equation 6.12 immediately leads to some useful conclusions concerning the solubility of a solute in a liquid. Strictly, these conclusions apply only to ideal solutions but they are useful guides to describe the relationship between the physico-chemical properties of a crystalline solid and the solubility process.

For a given solid–solvent system, the solubility increases by (i) raising the experimental temperature, (ii) reducing the enthalpy of fusion Δh_m, and (iii) lowering the melting temperature of the solid solute T_m. These conditions can be achieved by modifying the drug microstructure. To demonstrate this, let us recall some basic principles of molecular structure and physical transformations of pure solid substances.

An amorphous phase may be defined with reference to a crystalline solid. The amorphous phase lacks the long-range order and the well-defined molecular packing of the (perfect) crystals, even if a short-range coherence exists, owing to intermolecular interactions among the nearest atoms [22]. As shown in Figure 6.2, the amorphous phase, because of disorder (which implies lower density and higher molecular mobility), has the highest enthalpy compared with the other solid phases, whereas the perfect crystalline phase has the lower one [22,23]. From the second law of thermodynamics (at constant pressure) and the definition of Gibbs free energy (see Section 4.2, Chapter 4), it follows that, at a given temperature T, $\Delta G = \Delta H - T\Delta S < 0$, where ΔG, ΔH, and ΔS are, respectively, the differences in Gibbs free energy, enthalpy, and entropy between the perfect crystalline and the amorphous phases. A lower Gibbs free energy of the perfect crystalline phase compared with the amorphous one indicates (i) that the amorphous phase is metastable and spontaneously tends to revert to a crystalline form, and (ii) that the amorphous phase has a higher activity.

Real solids enthalpy, represented in Figure 6.2 by the dashed line, lies in between the enthalpies of the amorphous and perfect crystalline phases. Actually, all the crystalline solids are imperfect in some sense, and may present polymorphism (different crystal packing). The higher enthalpy of a polymorphic phase arises from its less effective lattice packing or molecular conformation [24]. The crystal defects, instead, are associated with a local increase in disorder and constitute high-energy sites; thus, the total entropy of the crystalline phase increases with the density of lattice defects [25,26]. This scenario reflects the energy landscape model of solids [27], which regards the possible states of a solid as connected minima (corresponding to different molecular packing, conformations, and so on) of a multidimensional potential

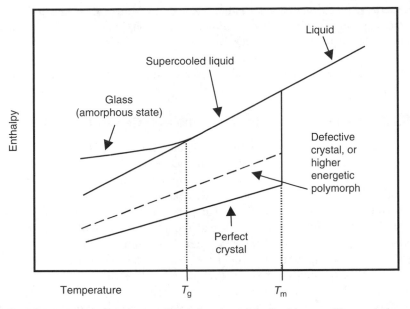

FIGURE 6.2 Enthalpy during cooling or heating of the liquid, crystalline, and glassy phases of a substance. T_m is the melting temperature, and T_g is the glass transition temperature. (Adapted from Elliott, S.R., *Physics of Amorphous Materials*, second ed., Longman Scientific & Technical, 1990.)

energy surface and suggests that the enthalpy of fusion of crystalline solids can be reduced by disorder.

On the other hand, recent experiments have shown that, at the nanoscale, the enthalpy of fusion, as well as the melting temperature, decreases with the size of the particles [28–30]. Moreover, a similar behavior was noticed for nanoparticles embedded in matrixes [5,31,32]. In these systems, the opposite thermal behavior has also been observed [33,34], showing that it depends on both particle size and particle–matrix interfaces. These studies indicate that the crystallite size, the enthalpy of fusion, and the melting temperature, which are independent of each other at the macroscale, are strongly dependent at the nanoscale.

In conclusion, according to Equation 6.12 and the earlier arguments, the solubility of a crystalline drug can be enhanced by (i) reducing the particle size, (ii) obtaining higher energetic polymorphs, and (iii) introducing defects. In other words, the key concept of solubilization of insoluble solids is to force the solid to assume a (metastable) microstructure characterized by short-range (nanoscale) coherence.

In the literature, it is reported that nanoparticles have considerably higher solubility than particles bigger than a micron [35,36]. Moreover, the increase in dissolution due to crystal lattice defects has been evidenced in a certain number of papers and reviewed by Burt and Mitchell [26] who experimentally verified a direct correlation between solubility and dislocation density. Finally, the transformation of a crystalline drug into one of its polymorphic modifications [37–39] or the amorphous form [40,41] is a well-established way to control the rate of drug solubility and dissolution.

6.2.2 MELTING TEMPERATURE DEPENDENCE ON CRYSTAL DIMENSION

The melting temperature of both free-standing and matrix-confined nanostructured materials deviates from the bulk value. Deviations are related to the curvature [42] and the nature [34,42–45] of the interfaces. The shifting of the melting point in nanosized materials was earlier predicted using the thermodynamic nucleation theory of the early 1900s (Gibbs–Thomson equation) [46,47]. Afterwards, this approach was reconsidered by more sophisticated treatments [48,49], which were then adapted to different melting mechanisms [50,51]. Accordingly, in this section we describe the melting of the crystallites with the thermodynamic approach developed by Brun et al. [49], which is based on Laplace and Gibbs–Duhem equations.

The variation of the triple-point temperature (drug melting temperature) depends on the solid–liquid–vapor interfaces curvature radii. The thermodynamic relationship between the triple-point temperature and the curvature radii of the three interfaces (solid–liquid, solid–vapor, and liquid–vapor) can be derived referring to the Gibbs–Duhem equation applied to the three volumetric phases and to the equilibrium conditions (chemical, thermal, and mechanical equilibrium) among the three volumetric phases and the three interfacial phases. The infinitesimal, reversible, variation of the internal energy E for a closed system made up of k components and three phases ("s" solid, "l" liquid, "v" vapor) is given by [52]

$$dE = dE^s + dE^l + dE^v + dE^{sv} + dE^{sl} + dE^{lv}, \tag{6.13}$$

$$dE^s = T^s \, dS^s + \sum_{i=1}^{k} \mu_i^s \, dn_i^s - P^s \, dV^s, \tag{6.14}$$

$$dE^l = T^l \, dS^l + \sum_{i=1}^{k} \mu_i^l \, dn_i^l - P^l \, dV^l, \tag{6.15}$$

$$dE^v = T^v \, dS^v + \sum_{i=1}^{k} \mu_i^v \, dn_i^v - P^v \, dV^v, \tag{6.16}$$

$$dE^{sv} = T^{sv} \, dS^{sv} + \sum_{i=1}^{k} \mu_i^{sv} \, dn_i^{sv} + \gamma^{sv} \, dA^{sv} + C_1^{sv} \, dc_1^{sv} + C_2^{sv} \, dc_2^{sv}, \tag{6.17}$$

$$dE^{sl} = T^{sl}\,dS^{sl} + \sum_{i=1}^{k} \mu_i^{sl}\,dn_i^{sl} + \gamma^{sl}\,dA^{sl} + C_1^{sl}\,dc_1^{sl} + C_2^{sl}\,dc_2^{sl}, \tag{6.18}$$

$$dE^{lv} = T^{lv}\,dS^{lv} + \sum_{i=1}^{k} \mu_i^{lv}\,dn_i^{lv} + \gamma^{lv}\,dA^{lv} + C_1^{lv}\,dc_1^{lv} + C_2^{lv}\,dc_2^{lv}, \tag{6.19}$$

where E^s, E^l, E^v, E^{sv}, E^{sl}, E^{lv} represent the solid, liquid, vapor, solid/vapor, solid/liquid, and liquid/vapor phase internal energy, respectively; S^s, S^l, S^v, S^{sv}, S^{sl}, S^{lv}, T^s, T^l, T^v, T^{sv}, T^{sl}, T^{lv} indicate entropy and temperature of the solid, liquid, vapor, solid/vapor, solid/liquid, and liquid/vapor phases, respectively; V^s, V^l, V^v, P^s, P^v, P^l are volume and pressure of the solid, liquid, and vapor phases, respectively; A^{sv}, A^{sl}, A^{lv}, γ^{sv}, γ^{sl}, γ^{lv} represent the solid/vapor, solid/liquid, and liquid/vapor interfacial area and surface tension, respectively; μ_i^s, μ_i^l, μ_i^v, μ_i^{sv}, μ_i^{sl}, μ_i^{lv}, n_i^s, n_i^l, n_i^v, n_i^{sv}, n_i^{sl}, n_i^{lv} represent, respectively, the chemical potential and the mole number of the ith component belonging to the solid, liquid, vapor, solid/vapor, solid/liquid, and liquid/vapor phases; c_1^{sv}, c_2^{sv}, c_1^{sl}, c_2^{sl}, c_1^{lv}, c_2^{lv} are the solid/vapor, solid/liquid, and liquid/vapor surface curvatures, respectively, while C_1^{sv}, C_2^{sv}, C_1^{sl}, C_2^{sl}, C_1^{lv}, C_2^{lv} are related constants.

Assuming a negligible contribution to the system internal energy E due to surface curvature variations [52] (this is strictly true for a sphere and for a plane), and considering the closed system hypothesis (system volume V, entropy S, and mole number n are constant),

$$dS = dS^s + dS^l + dS^v + dS^{sv} + dS^{sl} + dS^{lv} = 0, \tag{6.20}$$

$$dV = dV^s + dV^l + dV^v = 0, \tag{6.21}$$

$$dn = dn_i^s + dn_i^l + dn_i^v + dn_i^{sv} + dn_i^{sl} + dn_i^{lv} = 0, \tag{6.22}$$

the equilibrium condition ($dE = 0$) requires that

$$T^s = T^l = T^v = T^{sv} = T^{sl} = T^{lv} = T \quad \text{(thermal equilibrium)}, \tag{6.23}$$

$$\mu_i^s = \mu_i^l = \mu_i^v = \mu_i^{sv} = \mu_i^{sl} = \mu_i^{lv} = \mu_i \quad \text{(chemical equilibrium)}. \tag{6.24}$$

Accordingly, Equation 6.13 becomes

$$dE = -P^v\,dV^v - P^s\,dV^s - P^l\,dV^l + \gamma^{sv}\,dA^{sv} + \gamma^{sl}\,dA^{sl} + \gamma^{lv}\,dA^{lv} = 0. \tag{6.13'}$$

Referring to Equation 6.21 and observing that the following relation must hold between pressure drops across the solid/vapor, solid/liquid, and liquid/vapor interfaces:

$$(P^s - P^v) = (P^s - P^l) + (P^l - P^v), \tag{6.25}$$

the first part of Equation 6.13' becomes

$$-P^v\,dV^v - P^l\,dV^l + P^s\,(dV^v + dV^l) = dV^v(P^s - P^v) + dV^l(P^s - P^l)$$
$$= -dV^s(P^s - P^l) + dV^v(P^l - P^v).$$

The second part of Equation 6.13' can be modified bearing in mind the relation existing between surface tensions at the solid/liquid/vapor interfaces for a pure substance (contact angle is zero) (Young equation [52]):

$$\gamma^{lv} + \gamma^{sl} = \gamma^{sv}. \tag{6.26}$$

Consequently, we have

$$\gamma^{sv}\,dA^{sv} + \gamma^{sl}\,dA^{sl} + \gamma^{lv}\,dA^{lv} = \gamma^{sl}\,d(A^{sl} + A^{sv}) + \gamma^{lv}\,d(A^{lv} + A^{sv}).$$

Therefore, Equation 6.13' becomes

$$dE = -dV^s(P^s - P^l) + dV^v(P^l - P^v) + \gamma^{sl}\,d(A^{sl} + A^{sv})$$
$$+ \gamma^{lv}\,d(A^{lv} + A^{sv}) = 0 \tag{6.13'}$$

and we finally get the mechanical equilibrium conditions

$$(P^s - P^l) = \gamma^{sl}\,\frac{d(A^{sl} + A^{sv})}{dV^s} = \gamma^{sl}\,\frac{dA^s}{dV^s}, \tag{6.27}$$

and

$$(P^v - P^l) = \gamma^{lv}\,\frac{d(A^{lv} + A^{sv})}{dV^v} = \gamma^{lv}\,\frac{dA^v}{dV^v}. \tag{6.28}$$

These relations represent the Young–Laplace equation [52], which refers to the solid (A^s) and vapor (A^v) interfaces.

On the basis of the phase chemical and thermal equilibrium conditions (Equation 6.23 and Equation 6.24), Gibbs–Duhem equations for the three bulk phases (solid, liquid, and vapor), read

$$S^s\,dT + \sum_{i=1}^{k} n_i^s\,d\mu_i - V^s\,dP^s = 0, \tag{6.29}$$

$$S^l\,dT + \sum_{i=1}^{k} n_i^l\,d\mu_i - V^l\,dP^l = 0, \tag{6.30}$$

$$S^v \, dT + \sum_{i=1}^{k} n_i^v \, d\mu_i - V^v \, dP^v = 0. \tag{6.31}$$

Bearing in mind that we are dealing with only one component ($k=1$) and dividing Equation 6.29 through Equation 6.31 for n_1^s, n_1^l, and n_1^v, respectively, we have the Gibbs–Duhem equations expressed in molar quantities:

$$s^s \, dT + d\mu_1 - v^s \, dP^s = 0, \tag{6.29'}$$

$$s^l \, dT + d\mu_1 - v^l \, dP^l = 0, \tag{6.30'}$$

$$s^v \, dT + d\mu_1 - v^v \, dP^v = 0. \tag{6.31'}$$

Subtraction of Equation 6.30' from Equation 6.29' and Equation 6.31' from Equation 6.30' yields

$$(s^s - s^l) \, dT - v^s \, dP^s + v^l \, dP^l = 0, \tag{6.32}$$

$$(s^l - s^v) \, dT - v^l \, dP^l + v^v \, dP^v = 0. \tag{6.33}$$

Remembering that from the mechanical equilibrium conditions (Equation 6.27 and Equation 6.28) we have

$$dP^s = dP^l + d\left(\gamma^{sl} \frac{dA^s}{dV^s} \right), \tag{6.27'}$$

$$dP^v = dP^l + d\left(\gamma^{lv} \frac{dA^v}{dV^v} \right). \tag{6.28'}$$

Equation 6.32 and Equation 6.33 become, respectively,

$$dP^l = \left(\frac{s^s - s^l}{v^s - v^l} \right) dT - \left(\frac{v^s}{v^s - v^l} \right) d\left(\gamma^{sl} \frac{dA^s}{dV^s} \right) \tag{6.32'}$$

and

$$dP^l = \left(\frac{s^l - s^v}{v^l - v^v} \right) dT - \left(\frac{v^v}{v^l - v^v} \right) d\left(\gamma^{lv} \frac{dA^v}{dV^v} \right). \tag{6.33}$$

Equating the right-hand side of both equations and rearranging, we obtain

$$\left(\frac{s^s - s^l}{v^s - v^l} - \frac{s^l - s^v}{v^l - v^v} \right) dT = \left(\frac{v^v}{v^l - v^v} \right) d\left(\gamma^{lv} \frac{dA^v}{dV^v} \right) + \left(\frac{v^v}{v^s - v^l} \right) d\left(\gamma^{sl} \frac{dA^s}{dV^s} \right). \tag{6.34}$$

This is a general equation connecting triple-point temperature variation with interfacial properties. To make this equation more user friendly, we can suppose that v^{l} and v^{s} are negligible in comparison to v^{v} and $(s^{l} - s^{v})/v^{v} \ll (s^{s} - s^{l})/(v^{s} - v^{l})$. If we further define the difference $(s^{l} - s^{s})$ as the molar melting entropy ΔS_{f}, Equation 6.34 becomes

$$\Delta s_{m} dT = \left((v^{s} - v^{l}) d \left(\gamma^{lv} \frac{dA^{v}}{dV^{v}} \right) - v^{s} d \left(\gamma^{sl} \frac{dA^{s}}{dV^{s}} \right) \right). \tag{6.35}$$

Remembering that for spherical surfaces the following relations hold:

$$\frac{dA^{v}}{dV^{v}} = \frac{d(A^{lv} + A^{sv})}{dV^{v}} = -\left(\frac{2}{R^{lv}} + \frac{2}{R^{sv}} \right), \tag{6.36}$$

$$\frac{dA^{s}}{dV^{s}} = \frac{d(A^{sl} + A^{sv})}{dV^{s}} = \left(\frac{2}{R^{sl}} + \frac{2}{R^{sv}} \right), \tag{6.37}$$

where $R^{lv}, R^{sv},$ and R^{sl} are, respectively, the curvature radius of the liquid/vapor, solid/vapor, and solid/liquid interfaces, Equation 6.35 becomes

$$\Delta s_{m} dT = -\left((v^{s} - v^{l}) d \left(\gamma^{lv} \left(\frac{2}{R^{lv}} + \frac{2}{R^{sv}} \right) \right) + v^{s} d \left(\gamma^{sl} \left(\frac{2}{R^{sl}} + \frac{2}{R^{sv}} \right) \right) \right). \tag{6.38}$$

This equation establishes the dependence of the triple-point variation on surface properties in terms of surface tension and surface curvature radii. In the particular case, where only solid/liquid and liquid/vapor interfaces exist (this means that no contact exists between the solid and vapor phases as it happens in an isolated drug crystal melting), Equation 6.36 and Equation 6.37 become

$$\frac{dA^{v}}{dV^{v}} = \frac{d(A^{lv})}{dV^{v}} = -\left(\frac{2}{R^{lv}} \right), \tag{6.36'}$$

$$\frac{dA^{s}}{dV^{s}} = \frac{d(A^{sl})}{dV^{s}} = \left(\frac{2}{R^{sl}} \right), \tag{6.37'}$$

and Equation 6.38 reads

$$\Delta s_{m} dT = -\left((v^{s} - v^{l}) d \left(\gamma^{lv} \frac{2}{R^{lv}} \right) + v^{s} d \left(\gamma^{sl} \frac{2}{R^{sl}} \right) \right). \tag{6.39}$$

The estimation of R^{lv} and R^{sl} depends on the physical situation we are dealing with.

In order to simplify Equation 6.39, two situations can be considered. If the drug crystallites are dominant with respect to the amorphous boundaries, then they are tightly packed. In this case, we can assume that melting starts with each crystallite covered by a liquid skin, which means that $R^{lv} \approx R^{sl} = R^{nc}$ (R^{nc} is the radius of the drug nanocrystal). To the contrary, if the amorphous drug fraction is much larger than the crystalline one, then the crystallites melt dispersed in a liquid bath (we remember here that the drug amorphous phase becomes liquid as soon as its glass transition temperature is exceeded and this happens before nanocrystals melting), $R^{lv} \approx \infty$, and R^{sl} identifies with drug nanocrystal radius R^{nc}.

Accordingly, Equation 6.39 becomes

$$\Delta s_m \, dT = -2 \left((v^s - v^l) d \left(\frac{\gamma^{lv}}{R^{nc}} \right) + v^s d \left(\frac{\gamma^{sl}}{R^{nc}} \right) \right), \quad R^{lv} = R^{sl} = R^{nc} \quad (6.40)$$

$$\Delta s_m \, dT = -2 v^s d \left(\frac{\gamma^{sl}}{R^{nc}} \right). \quad R^{lv} = \infty; \quad R^{sl} = R^{nc} \quad (6.40')$$

The integration of Equation 6.40 and Equation 6.40', assuming $\Delta s_m = \Delta h_m / T$, ($\Delta h_m$ is the molar melting enthalpy), v^s, v^l, γ^{lv}, and γ^{ls} constant, leads, respectively, to

$$\int_{T_m}^{T} \frac{\Delta h_m}{T} \, dT = -2 \int_{0}^{1/R^{nc}} ((v^s - v^l)\gamma^{lv} + v^s \gamma^{sl}) d \left(\frac{1}{R} \right)$$

$$= -2 \frac{(v^s - v^l)\gamma^{lv} + v^s \gamma^{sl}}{R^{nc}}, \quad (6.41)$$

$$\int_{T_m}^{T} \frac{\Delta h_m}{T} \, dT = -2 \int_{0}^{1/R^{nc}} v^s \gamma^{sl} d \left(\frac{1}{R} \right) = -2 \frac{v^s \gamma^{sl}}{R^{nc}}. \quad (6.41')$$

Unfortunately, it is not possible to get an analytical integration of Equation 6.41 and Equation 6.41'; in fact, the left-hand side member Δh_m depends on both R^{nc} and T according to [51]:

$$\Delta h_m = \Delta h_m^{\infty} - \frac{3}{R^{nc}} (\gamma^{sv} v^s - \gamma^{lv} v^l) - \int_{T}^{T_m} \Delta c_p \, dT, \quad (6.42)$$

where T_m and Δh_m^{∞} are, respectively, the triple-point temperature and the melting enthalpy of the infinite drug crystal, and Δc_p is the difference between the liquid and the solid specific heat capacity. Assuming Δc_p temperature independent, Equation 6.42 becomes

$$\Delta h_{\rm m} = \Delta h_{\rm m}^{\infty} - \frac{3}{R^{\rm nc}}\,(\gamma^{\rm sv}\nu^{\rm s} - \gamma^{\rm lv}\nu^{\rm l}) - \Delta c_{\rm p}(T_{\rm m} - T). \tag{6.43}$$

Consequently, the functions $T(R^{\rm nc})$ and $\Delta h_{\rm m}(T)$ can be obtained only by numerically solving the system made up by Equation 6.41 (or Equation 6.41') and Equation 6.42.

It is interesting to note that if $\Delta s_{\rm m}$ is supposed to be temperature independent, the integration of Equation 6.39 yields the result found by Brun [49]:

$$\Delta T = \frac{-2}{\Delta s_{\rm m}}\left(\frac{(\nu^{\rm s} - \nu^{\rm l})\gamma^{\rm lv}}{R^{\rm lv}} + \frac{\nu^{\rm s}\gamma^{\rm sl}}{R^{\rm sl}}\right). \tag{6.44}$$

To date, few works on the melting of drug nanocrystals, free-standing or matrix-confined, have been published [5,10,31]. One of the attempts to investigate the melting behavior of nanostructured drugs embedded into a cross-linked polymeric matrix was made by Bergese et al. [53], who demonstrated the depression of the melting point of the nanocrystalline drugs with respect to the bulk drug.

In the light of their results, we decided to adopt the same thermodynamic model but taking into account the dependence of the melting entropy on both the temperature and the interfaces curvature radii. In the experiments, Nimesulide (a nonsteroidal, anti-inflammatory drug) was considered as the model drug for its low water solubility. Cross-linked polyvinylpyrrolidone (PVPCLM) (a water-insoluble powdered material obtained by popcorn polymerization) was used as the carrier. The physical mixture of raw materials (in a weight ratio equal to 1:3) was cold co-grinded for 2 h and 4 h by a high-energy mechanical activation (HEMA) patented process [17] in a planetary mill (Pulverisette 7, Fritsch GmbH, D). The HEMA process enables producing polymeric nanocomposites made of PVPCLM microparticles in which nanocrystals and molecular clusters of drug are confined [54,55]. Pure Nimesulide and co-grounded systems were characterized by means of a power-compensated differential scanning calorimeter (DSC) Pyris-1 (Perkin Elmer Corp., Noewalk, CT). The samples to be analyzed by DSC were put into aluminum pans (6–7 mg of nanocomposite and about 2 mg of raw drug) and then scanned under a N_2 stream of 20 cm^3/min at a heating rate of 10°C/min. In order to exclude polymorphic transformation induced by the co-grinding process, the pure drug and composite systems were characterized by x-ray powder diffraction (XRPD) technique. The measurements were performed with a Philips X'Pert PRO diffractometer in a θ/θ Bragg–Brentano geometry. Cu Kα radiation ($\lambda = 1.541$ Å), generated by a sealed x-ray tube (45 kV × 40 mA), and a real-time multiple strip detector (X'Celerator by Philips) were used for all the XRPD measurements. The powder samples were prepared, after gentle grinding, in a back loading sample holder and the XRPD patterns collected in continuous mode with a scanning speed of 0.0102°/sec. In order

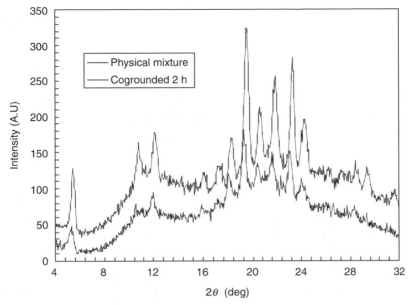

FIGURE 6.3 Comparison between the x-ray diffraction patterns relative to the Crosspovidone/ Nimesulide (3/1, w/w) physical mixture (thin line) and the 2 h co-grounded system (thick line).

to minimize preferred orientation effects, a sample spinner was used in all measurements. The evaluation of the weight fraction of the crystalline drug contained in the nanocomposites was done by quantitative XRPD. We adopted a single-line method that was specifically developed for largely amorphous composites [56].

Figure 6.3 shows the comparison between the x-ray diffraction patterns relative to the physical mixture and the same mixture after 2 h of co-grinding. It is possible to notice that no substantial shifting of the typical Nime-sulide diffraction peaks [57] occurs. As analog diffraction patterns could be obtained for the 4 h co-grinding time, we can affirm that no polymorphs are generated because of grinding. The measured weight fractions of the crystal-line drug embedded in co-grounded systems at 2 and 4 h were 0.32 and 0.11, respectively.

Figure 6.4 reports the thermogram relative to pure Nimesulide. It is clearly evident with the melting event taking place at 148.7°C and character-ized by a melting enthalpy Δh_m^∞ equal to 109 J/g. In Figure 6.5, the thermo-grams relative to the co-grounded systems (2 h, thick line; 4 h thin line) are shown. While in the first case melting temperature T is equal to 118.0°C and system melting enthalpy Δh_m is equal to 5.1 J/g, in the second case (4 h) melting temperature T is equal to 112.0°C and system melting enthalpy Δh_m

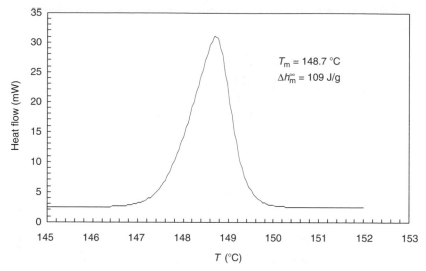

FIGURE 6.4 Pure Nimesulide thermogram performed by differential scanning calorimeter. Melting temperature is equal to 148.7°C and melting enthalpy is equal to 109 J/g.

is equal to 2.5 J/g. This result was expected because the Nimesulide embedded into PVPCLM is nanostructured, and it is in agreement with previous observations [11,57].

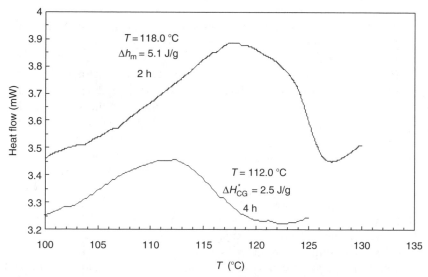

FIGURE 6.5 2 h co-grounded system (thick line) and 4 h co-grounded system (thin line) thermograms performed by differential scanning calorimeter.

Evaluating the crystallite size $(2R^{nc})$ of the drug from the melting data, Equation 6.41' and Equation 6.42 are numerically solved. The choice of Equation 6.41', instead of Equation 6.41, for the calculations is *a posteriori* motivated by the fact that as we found small values of the crystalline fractions for both systems (see the results previously shown), we can assume that nanocrystal melting happens in an infinite amount of liquid drug (liquid–vapor radius R^{lv} is infinite and solid–liquid radius R^{sl} identifies with drug nanocrystal radius R^{nc}).

The thermodynamic quantities required by the models were experimentally determined. The melting point depression of the embedded Nimesulide, ΔT, the bulk Nimesulide-specific enthalpy of fusion, Δh_m^∞, and Δc_p were obtained by DSC analysis. The density of solid and liquid Nimesulide, ρ^s and ρ^l, were measured by helium picnometry and calculated using the group contribution method [58]. The values obtained were 1.49 g/cm^3 and 1.34 g/cm^3, respectively. The solid–vapor, the liquid–vapor, and the solid–liquid interfacial energies, γ^{sv}, γ^{lv}, and γ^{sl}, were evaluated by contact angle experiments (see Section 5.2.1.2, Chapter 5) and the parachor method [58]. The calculated values were $\gamma^{sv} = 57.57 \times 10^{-3}$ J/m^2, $\gamma^{lv} = 44.26 \times 10^{-3}$ J/m^2 and $\gamma^{sl} = 1.31 \times 10^{-3}$ J/m^2. Figure 6.6 shows the reduction of crystal melting

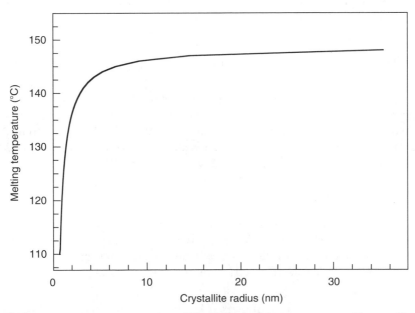

FIGURE 6.6 Theoretical reduction of Nimesulide melting temperature with crystallite radius.

temperature T_m with crystal radius R^{nc} descending from the numerical solution of Equation 6.41' and Equation 6.42. Although for crystal radii greater than approximately 6 nm, no sensible T_m reductions are detectable, for smaller radii, melting temperature reduction is very pronounced, falling to 112°C for $R^{nc} = 0.7$ nm. On the basis of these results, the thermodynamic model estimated the mean crystallite diameter of the confined Nimesulide nanocrystals equal to 1.64 nm and 1.40 nm in the 2 and 4 h co-grounded systems, respectively. To be more precise, the estimated value is the most probable value (mode) of the volumetric distribution of the crystallite diameters. Actually, when we consider the melting temperature of nanocrystals, we should refer to a distribution of the melting temperature corresponding to the temperature interval of the experimental DSC endotherm.

6.2.3 SOLUBILITY DEPENDENCE ON CRYSTAL DIMENSION

A number of studies have attempted to measure the solubility of finely divided solids (the successful expression of nanoparticles was still not coined) as a function of particle size. For example, studies by Banker and Rhodes [59] have shown that the micronized form of Griseofulvin reaches an apparent solubility of approximately 30 $\mu g/cm^3$, whereas the crystalline form of the drug reaches a value of less than 20 $\mu g/cm^3$. The resulting expression for the excess solubility of small spherical solid particles of density ρ, often referred to as the Ostwald–Freundlich relation [60,61]:

$$\rho \nu \frac{RT}{M} \ln\left(\frac{S_r}{S_\infty}\right) = \frac{2\gamma^{sl}}{r}. \tag{6.45}$$

In Equation 6.45, S_r is the solubility of particles of radius size r, S_∞ is the normal solubility value of an infinite plane crystal surface, γ^{sl} is the solid–liquid interfacial tension, and ν represents the number of moles of ions formed from one mole of electrolyte. Based on vapor condensation analogies, Equation 6.45 can be derived from a liquid–vapor system [62].

6.2.3.1 Liquid–Vapor System

Briefly, the work required in the creation of a surface area $d\sigma$ in a liquid of surface tension γ^{lv} is $\gamma^{lv} d\sigma$. For a spherical drop of radius r, the work to increase the radius by dr is thus $8\pi\gamma^{lv} r\, dr$, so that the force opposing deformation is $4\pi r^2\, dr$. This corresponds to an excess mechanical pressure

$$P = \frac{8\pi\gamma_{lv} r}{4\pi r^2} = \frac{2\gamma^{lv}}{r} \tag{6.46}$$

exerted on the fluid within the drop. Equation 6.46 is the Laplace equation.

From thermodynamics, it is shown that an imposed pressure P on an incompressible fluid raises its vapor pressure from the normal value p^0 to p, given by

$$RT \ln\left(\frac{p}{p^0}\right) = Pv = \frac{PM}{\rho}, \tag{6.47}$$

where v is the molar volume of the liquid and M and ρ are its molecular weight and density. Eliminating the mechanical pressure P from Equation 6.46 and Equation 6.47 gives

$$\rho = \frac{RT}{M} \ln\left(\frac{p}{p^0}\right) = \frac{2\gamma^{lv}}{r}. \tag{6.48}$$

Here r is positive and there is thus an increased vapor pressure. In the case of water, p/p^0 is about 1.001 if r is 10^{-4} cm, 1.011 if r is 10^{-5} cm, and 1.114 if r is 10^{-6} cm.

Equation 6.48 is frequently called the Kelvin equation: this relation is also referred to in the literature by the names of Gibbs–Thomson and Gibbs–Kelvin relation. The model provides a ready explanation for the ability of vapor to supersaturate; thus, the vapor pressure p over a concave meniscus must be less than the saturation pressure p^0 at the same temperature. This implies that a vapor will be able to condense to a liquid in a pore of a solid structure, even when its relative pressure is less than unity [63].

The Kelvin equation should also apply to crystals immersed in a liquid. Therefore,

$$\rho = \frac{RT}{M} \ln\left(\frac{p}{p^0}\right) = \frac{2\gamma^{sl}}{r}, \tag{6.49}$$

where γ^{sl} is the solid–liquid surface tension. As an actual crystal will be polyhedral in shape and may well expose faces of different surface tension, the question is what value of γ^{sl} and r should be used. As the Wulff theorem states that γ_i^{sl}/r_i ($i =$ face index) is invariant for all faces of an equilibrium crystal, we can choose the face we prefer.

Equation 6.49 can be generalized as

$$RT \ln\left(\frac{a}{a^0}\right) = \frac{2\gamma^{sl}M}{\rho r} \approx RT \ln\left(\frac{S_r}{S_\infty}\right), \tag{6.50}$$

where a and a^0 represent, respectively, the activity of the crystal of radius r and that of the crystal characterized by an infinite radius; S_r and S_∞ represent, respectively, the solubility of the crystal of radius r and that of the crystal characterized by an infinite radius. In principle, then, small crystals should have higher solubility in a given solvent than should large ones. Practically, significant solubility improvements can be achieved only when the crystal radius is, approximately, below 20 nm.

In the case of sparingly soluble salt that dissociates into v^+ positive ions M and v^- negative ions A, the solubility of S is given by

$$S = \frac{(M)}{v^+} = \frac{(A)}{v^-} \tag{6.51}$$

and the activity coefficient of the solute is given by

$$a = (M)^{v^+}(A)^{v^-} = S^{(v^+ + v^-)}(v^+)^{v^+}(v^-)^{v^-}, \tag{6.52}$$

if activity coefficients are neglected. On substituting into Equation 6.50, we have

$$\rho \frac{RT}{M}(v^+ + v^-) \ln\left(\frac{S_r}{S_\infty}\right) = \frac{2\gamma^{sl}}{r}. \tag{6.53}$$

A complicating factor is that of the electrical double layer presumably present [64], which gives the equation:

$$\rho \frac{RT}{M} \ln\left(\frac{S_r}{S_\infty}\right) = \frac{2\gamma^{sl}}{r} - \frac{q^2}{8\pi\varepsilon r^4}, \tag{6.54}$$

supposing the particle to possess a fixed double layer of charge q. Equation 6.45 was also derived thermodynamically, from a solid–liquid system, by Enüstün and Turkevich [65].

6.2.3.2 Solid–Liquid System

Considering a single crystal of surface area A at constant temperature and pressure in equilibrium with the solution phase, a simple extension of Equation 4.32″ [52], when the system possesses an appreciable surface area and it is composed by one substance, is given by

$$dG = \mu_0 \, dn + \gamma^{sl} \, dA \tag{6.55}$$

or

$$\mu = \mu_0 + \gamma^{sl} \frac{dA}{dn}, \tag{6.56}$$

where γ^{sl} is the mean Gibbs surface free energy and μ_0 is the bulk chemical potential of the crystal.

Since the molar surface area is NA and the molar volume $v = NV = M/\rho$, where N is the number of particles per mole and ρ is the bulk density, Equation 6.56 can be rewritten as

$$\mu = \mu_0 + \frac{M}{\rho}\gamma^{sl}\frac{dA}{dV}. \tag{6.57}$$

If r is any characteristic dimension of the solid (crystal radius) and supposing that the system remains isomorphic, that is, the shape is preserved when its size is changed, the surface area A and volume V of the single crystal are, respectively, $A = k_1 r^2$ and $V = k_2 r^3$; thus,

$$\frac{dA}{dV} = \frac{2\beta}{3r} \tag{6.58}$$

then, Equation 6.57 can be rewritten as

$$\mu = \mu_0 + \frac{2}{3}\gamma^{sl}\frac{M\beta}{\rho r}. \tag{6.59}$$

In Equation 6.59, the surface effect is expressed in terms of $M\beta/\rho r$, where β is a constant geometric factor for a given shape of the crystal (k_1/k_2).

In the case of an electrolyte, its chemical potential in the solution is given by

$$\mu' = \mu'_0 + RT \ln a, \tag{6.60}$$

where μ'_0 is the standard chemical potential and a the mean activity of the electrolyte in the solution. When the crystal is sufficiently large, the saturation activity (a) assumes its normal value a_∞ given by

$$\mu_0 = \mu'_0 + RT \ln a_\infty. \tag{6.61}$$

Equating μ and μ' as required by the condition for equilibrium between the two phases (crystal and solution), the generalized equation for the solubility as a function of crystal size is then obtained:

$$\rho\frac{RT}{M}\ln\left(\frac{a}{a_0}\right) = \frac{2}{3}\beta\frac{\gamma^{sl}}{r}. \tag{6.62}$$

For spherical crystals $\beta = 3$, Equation 6.62 becomes equivalent to Equation 6.45 and Equation 6.53.

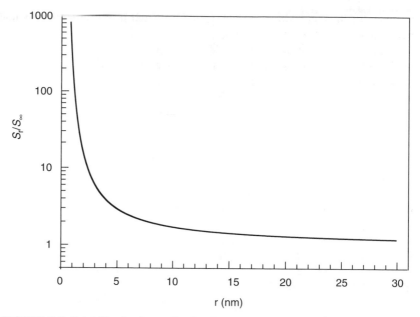

FIGURE 6.7 Solubility S_r of crystals of radius r, compared with the solubility of large crystals S_∞.

The Ostwald–Freundlich relation shows that the solubility of the solid at a given temperature decreases monotonically with an increase in the crystal size. The practical significance of this effect can be examined by plotting (as shown in Figure 6.7) the solubility ratio (s_r/s_∞) against the crystal radius (r) for the Nimesulide (molecular weight of 308.51 and a solid density of 1.49 g/cm^3). The solid–liquid interfacial free energy was evaluated by contact angle and liquid surface tension methods [66]. The calculated values were $\gamma^{sv} = 57.57$ mJ/m^2, $\gamma^{lv} = 44.26$ mJ/m^2, and $\gamma^{sl} = 32.06$ mJ/m^2 at 298 K. The predicted solubility ratio increases steeply as the radius is reduced from 10 to 1 nm; thus, the size range encompasses that of the critical nucleus (single unit cell) to that of about 10 unit cell. Actually, the reference crystal structure of the known form of Nimesulide, obtained from the Cambridge Structure Database (CSD), points out the following crystallographic parameters: (i) monoclinic system, (ii) cell volume of 2.769 nm^3, and (iii) cell parameters of $a = 3.366$ nm, $b = 0.513$ nm, $c = 1.601$ nm, $\alpha = \gamma = 90°$, and $\beta = 93.37°$. Consequently, considering the radius of a sphere with the same volume as the unit cell, the mean radius of one unit cell is 0.87 nm.

6.3 EXPERIMENTAL METHODS AND LIMITATIONS

The use of the Ostwald–Freundlich equation for describing the effect of crystal size on solubility has been the subject of serious criticism [67,68] but it is still popular in many current papers and books [69,70]. The expected experimental difficulties involved in applying this model are:

1. Finely divided crystals possessing lattice defects (amorphous regions) cannot be used, as their surface characteristics can be changed by manufacturing processes [71].
2. Small amounts of impurities can substantially change the solubility.
3. In a polydisperse powder system, a phenomenon known as Ostwald ripening occurs because a smaller particle, with a higher solubility, will dissolve in the unsaturated solution that is saturated with respect to a larger particle of lower solubility. Thus, larger particles will grow at the expense of smaller particles, and the concentration of the saturated solution will decrease asymptotically [72].

The time for attaining an equilibrium state is also a very important issue; several days or weeks are the typical choice in a careful experimental procedure. Therefore, variations in the experimental period may become a reason for discrepancy in the solubility data.

The effect of the amount of excess solids and (using the knowledge of the amorphous content of the material) the role that the drug microstructure can play in the apparent solubility of a partially crystalline drug is described in the following section. Experiments were performed according to the method classified as direct analytical method [73] in which the solubility of the micronized Griseofulvin (a poorly water-soluble antibiotic) was measured, under isothermal–isobaric conditions. The suspensions of known dispersion concentration (from 0.5 to 30 mg/cm^3) of the drug were prepared using an appropriate amount of buffer medium (pH 7.5). The suspensions were stirred on a magnetic stirring board for the desired duration in a temperature-controlled bath at 37°C. Then, they were filtered through a 0.2 μm Sartorius (regenerated cellulose) membrane before determining the concentration. All the glassware and filters were prewarmed in a thermostatic chamber for at least 1 h to avoid precipitation during the filtration. Finally, the concentrations of Griseofulvin were assayed spectrophotometrically (Perkin Elmer Lambda 20) at a wavelength of 295.3 nm; each value is the mean of at least three experiments.

To verify the existence of disordered phases in the solid-state structure of the native drug, a well-ordered structural powder was prepared by a double crystallization process (recrystallized Griseofulvin), using acetonitrile as solvent and an annealing treatment at 100°C in a saturated chamber with inert gas (N_2). Before the solubility measurement, the microstructural characteristics of Griseofulvin were investigated using DSC and XRPD. Thermal analyses were carried out with a power-compensated DSC Pyris-1 (Perkin Elmer). The samples to be analyzed by DSC were placed in aluminum pans (1–3 mg of drug) and then scanned under a N_2 stream of 20 cm^3/min at a heating rate of 10°C/min. Thermal analyses of native and recrystallized

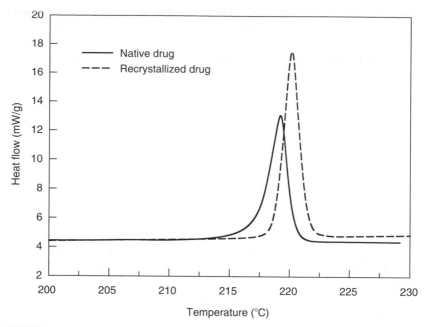

FIGURE 6.8 Differential scanning calorimeter profiles of the native and recrystallized Griseofulvin.

Griseofulvin are reported in Figure 6.8. The DSC profile of the native drug (continuous line) evidences a single endothermic event that can be ascribed to the melting of the crystalline phase. It follows that, for the native material, both the melting temperature, which is $218.9 \pm 0.3°C$, and the melting enthalpy, which is 115.5 J/g, are significantly lower than the values of the recrystallized one, which instead are $220.9 \pm 0.1°C$ and 124.4 J/g, respectively (dashed profile). The degree of disorder φ_A of the raw Griseofulvin was then calculated from the crystalline weight fraction content, obtained by dividing the normalized melting enthalpy of the native material Δh_m^n by that of the well-ordered recrystallized one Δh_m^R:

$$\varphi_A = 1 - \frac{\Delta h_m^n}{\Delta h_m^R}. \tag{6.63}$$

Therefore, the resulting degree of disorder of the tested material was 0.13.

This result was expected because the commercial Griseofulvin used for this study was subjected to preliminary size reduction treatment (mechanical comminution process) as stated by the supplier. As a result of mechanical treatment of the crystals (such as milling), the individual particles of the solid material increase their lattice disorder producing a partially amorphous solid. The extent of such transformation was confirmed by quantitative x-ray analysis (QXRD). The XRPD experiments were performed on a Philips X'Pert PRO diffractometer in θ/θ Bragg–Brentano geometry. Cu Kα radiation ($\lambda = 1.541$ Å), generated by a sealed x-ray tube (45 kV $\times$ 40 mA), and a real-time multiple strip detector (X'Celerator, Philips) were used for all the XRPD experiments. The powder samples were prepared, after gentle grinding, in a back loading sample holder and the XRPD patterns were collected in continuous mode with a scanning speed of 0.0102°/sec. To minimize preferred orientation effects, a sample spinner was used in all measurements.

To quantitatively track the crystalline fraction of the native material, we adopted a development of the QXRD analysis based on the diffraction–absorption method described by Zevin and Kimmel [74] and successively applied to the partially amorphous pharmaceutical composites [56].

In the hypothesis of a powder sample with randomly oriented grains measured in a Bragg–Brentano geometry, the basic equation that relates the integrated intensity of a phase diffraction peak to the phase abundance in the specimen is

$$I_{ij} = \frac{K_e K_{ij} c_j}{\rho_j \mu^*}, \tag{6.64}$$

where c_j is the weight fraction of crystalline phase j, K_e a constant for a particular experimental system, K_{ij} a constant for each diffraction peak i from the crystal structure of phase j, ρ_j the density of phase j, and μ^* is the mass absorption coefficient of the specimen [75]. By considering the sum of all the diffraction peaks i, belonging to phase j, we obtain equation

$$\sum_i I_{ij} = \frac{K_e \sum_i K_{ij} c_j}{\rho_j \mu^*}. \tag{6.65}$$

Applying Equation 6.65 to a pure crystalline phase j ($c_j = 1$), we have

$$\sum_i (I_{ij})_0 = \frac{K_e \sum_i K_{ij}}{\rho_j \mu_j^*}. \tag{6.66}$$

The diffraction–absorption method is based on the calculation of the crystalline phase abundance c_j obtained combining Equation 6.65 and Equation 6.66:

$$c_j = \frac{\sum_i I_{ij}}{\sum_i (I_{ij})_0} \frac{\mu^*}{\mu_j^*}. \tag{6.67}$$

In the case of a single crystalline phase and a single species, that mean $\mu^* = \mu_j^*$, Equation 6.67, becomes

$$c = \frac{\sum_i I_i}{\sum_i (I_i)_0}.$$ (6.68)

In Equation 6.68, $\sum_i I_i$ represent the full profile integrated intensities of the native material and $\sum_i (I_i)_0$ is the full profile integrated intensities of the recrystallized material (pure crystalline phase). Therefore, if the drug is present in amorphous and crystalline phases within the native material, the ratio between the two integrated intensities will give the crystalline weight fraction content of the raw Griseofulvin.

The XRPD patterns of both the commercial (continuous line) and the recrystallized Griseofulvin (dashed line) are shown in Figure 6.9. In the figure, the match of the Bragg peaks evidences that the crystallographic phase of the two materials coincide; this allowed us to point out that no phase transformation occurred during the drug recrystallization process.

The refined values of $\sum_i (I_i)$ and $\sum_i (I_i)_0$ were derived from the full profile fitting [56] of the two patterns, and the residual crystalline drug weight fraction (c) of the native drug was estimated to be 0.91, leading to a degree of disorder (φ_A) of 0.09. As expected, the two methods lead to the same result confirming the biphasic nature of the tested raw material.

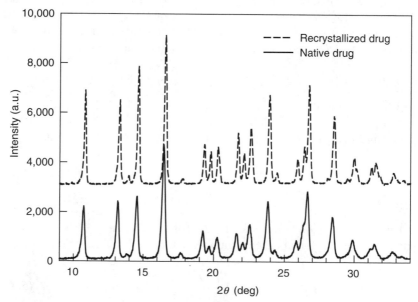

FIGURE 6.9 X-ray diffraction patterns of the native and recrystallized Griseofulvin.

Traditionally, the term microstructure has been used to describe the crystallite size and disorder or defects in the polycrystalline materials [76]. The methods commonly employed for the assessment of microstructure have been based on the x-ray diffraction line breadth measurements. The parameters most widely used as an indicator of the microstructure are full-width at half maximum (FWHM) and the integral breadth of x-ray diffraction peaks [77,78]. Theoretically, the line of the powder diffraction pattern should be sharp for a polycrystalline material containing sufficiently large and strain-free crystallites. However, in actual practice, one does not find diffraction lines with zero breadth because of the combined effects of instrumental and microstructural factors that tend to broaden the pure diffraction lines.

Microstructural factors include the crystallite size and imperfections in the crystalline lattice. Numerous studies have reported various techniques for extracting the crystallite size and microstrain properties of specimens producing broadened x-ray diffraction profiles [79]. In this context, the microstructural features of the native Griseofulvin were determined simultaneously with the refinement of the structural parameters with the Rietveld method [80] and using the Warren–Averbach method [81] of separating the crystallite size and microstrain. The Warren–Averbach approach employs the Fourier coefficient of the intrinsic diffraction profiles of two or more orders of well-resolved reflections. Accordingly, the purpose of this further characterization was to extract the crystallite size and microstrain parameters from the x-ray diffraction peaks of the commercial Griseofulvin to attempt a correlation between the microstructural characteristics of the solid phase and the related apparent solubility value.

The computer program MAUD [82], based on the Rietveld method, was used for the whole-pattern profile fitting structure refinement and to extract the microstructural parameters with the Warren–Averbach analysis. The standard material used for modeling the instrumental resolution was the recrystallized Griseofulvin. The Rietveld refinement was first carried out with the data collected on this reference material to obtain the appropriate instrumental parameters. The observed and calculated diffraction patterns obtained from the Rietveld refinement of the native Griseofulvin are reported in Figure 6.10, with the difference plot shown at the bottom.

The final results of the Warren–Averbach analysis can be summarized with two distribution functions: (i) the surface-weighted size distribution of the crystallite, as shown in Figure 6.11, and (ii) the root mean square strain (RMSS) distribution as reported in Figure 6.12. The mean area-weighted size value of 89.8 nm and the mean microstrain of 2.5×10^{-3} clearly show that in the single particles of the native material crystalline domains coexist, in the nanometer-size range, with the lattice disorder induced by the mechanical comminution process.

The crystallite size data represent the size of the crystal domains (coherent diffraction) which, however, are not necessarily the same as the particle sizes.

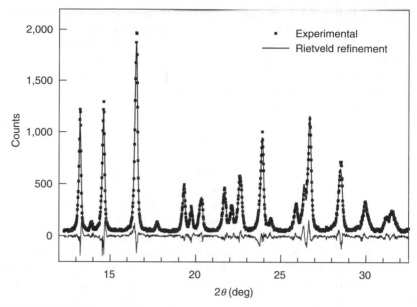

FIGURE 6.10 Observed and calculated x-ray diffraction patterns of the native Griseo-fulvin.

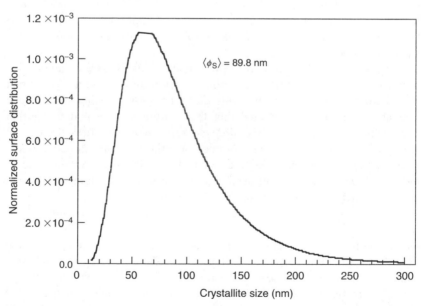

FIGURE 6.11 Surface-weighted size distribution of the crystallites in the native Griseofulvin.

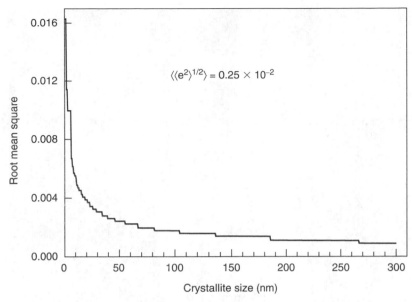

FIGURE 6.12 Root mean square strain distribution of the native Griseofulvin.

In the present case, it appears that the crystallite sizes obtained by x-ray diffraction analysis are different from the actual Griseofulvin particle sizes. The results of the scanning electron microscopy (SEM) analysis shown in Figure 6.13 demonstrate this difference: the measured crystallite sizes are smaller than the observed particle sizes. Actually, as it appears from the SEM image of native material, the morphology of the powder particles is variegate and shows a particle size distribution ranging from 0.5 to 5 μm.

It is now possible to conclude that the native polycrystalline Griseofulvin is made up of particles containing a significant weight fraction of nanocrystals (about 0.9), which have a mean crystallite size less than 100 nm, separated by amorphous interfacial regions (about 0.1 weight fraction).

The impact of the amount of excess solids on the dynamic solubility of the native Griseofulvin in buffer solution is shown in Figure 6.14. The increase in the solid excess leads to an increase in the drug concentrations in solution. The concentrations reached constant values after 24 h, in all the cases, and no degradation products were found in the HPLC analysis.

If the powder particles are partially crystalline and partially amorphous, as previously pointed out, a composite solid–liquid interface made up of crystalline and amorphous phases will be generated in the dispersion. The classical interpretation of a noninteracting polyphase mixtures dissolution (see Section 5.2.3.3, Chapter 5) assumes that the processes of the simultaneous

FIGURE 6.13 Scanning electron microscopy of the native Griseofulvin.

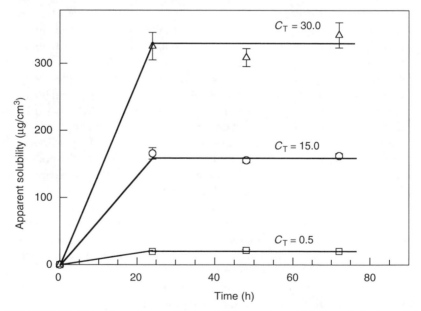

FIGURE 6.14 Time dependence of Griseofulvin apparent solubility. C_T is the solid amount added (mg/cm^3). The standard deviations are shown as error bars, although most of the bars are hidden in the symbols.

dissolution of the crystalline (stable form) and amorphous (metastable form) phases and of the crystal growth (crystallization to a more stable form) are both diffusion controlled. The surface areas available for the dissolution and growth will then be rate-determining factors for the extent of the drug concentration in solution. Therefore, the observed plateau supersaturation will be the region at which the dissolution rate of the two surface phases and the recrystallization rate from the supersaturated solution (crystals growth) are balanced. Moreover, its value is determined by the relative surface areas of the two phases and their kinetic constants.

The mean value of the drug concentration measured at 24 h, for each amount of solid added, was chosen as the apparent solubility value. In Figure 6.15, the phase-solubility diagrams of both the commercial and recrystallized Griseofulvin are shown.

The capacity of any system to form solutions has limits imposed by the phase rule of Gibbs (Equation 4.29):

$$F + P = C + 2 \tag{6.69}$$

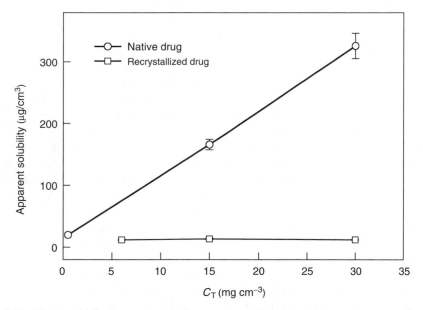

FIGURE 6.15 Phase-solubility diagrams of the native (open circles) and recrystallized (open squares) Griseofulvin. C_T is the solid amount added (mg/cm^3). The standard deviations are shown as error bars, although most of the bars are hidden in the symbols.

where F is the number of degree of freedoms in a system consisting of C components with P phases. For a system of two components and two phases (e.g., solid and liquid) under the pressure of their own vapor and at constant temperature, there is no degree of freedom ($F = 0$) and so there is a definite reproducible equilibrium solubility. Accordingly, neither the amount of excess solid nor the size of the particles present will change the position of the equilibrium. As can be clearly seen in the phase-solubility diagram (Figure 6.15), the well-ordered stress-free drug shows a reproducible solubility value of 11.9 $\mu g/cm^3$, independent of the solid amount added. On the contrary, the excess amount of solid affects the apparent solubility of the native drug; that is, the solubility of a material containing metastable phases (stressed drug) can be dramatically increased with the increase in the amount of excess solid.

It has been reported [70] that if the particle surface is disordered to some degree, the apparent solubility will be a function of the amount of drug dispersed in the medium. Under such a circumstance, the solid dispersion concentration will determine the total amount of disordered surface area present in the medium. The observed linear relationship for the phase-solubility diagram of the partially crystalline material (see Figure 6.15) seems to emphasize the critical role of the available disordered surface area for dissolution. Therefore, the existence of an unexpected metastable solubility of the native Griseofulvin should be induced by either (i) the presence of microstrains (mechanically induced lattice disorder) or (ii) the crystalline domains in the nanometer-size range or (iii) their combined effects. This phenomenon is caused by higher-energy particles resulting from a high density of crystal defects and a crystallites dimension in the nanometer-size range.

While it is straightforward to determine the equilibrium solubility of a phase that is stable with respect to conversion, the measurement of solubility of metastable phases that are susceptible to conversion does not seem to be a trivial matter. In this context, it is much more convenient to approach solubility determination through a kinetic experiment rather than equilibrium. In an attempt to measure the solubility of the amorphous and nanocrystalline phases, we developed a method based on the measurement of the simultaneous dissolution of crystalline and amorphous phases of the same drug (see Section 5.2.3.3, Chapter 5). The dissolution model takes into account the composite nature of the solid surface and the superficial phase transformation occurring for the amorphous fraction. This implies that the variation of drug concentration in the medium is due to the sum of two distinct contributions: that of the nanocrystalline phase and that of the amorphous phase.

The results obtained from the dynamic experiments show that the solubility of the nanocrystals (60.2 ± 4.3 $\mu g/cm^3$) is higher than the equilibrium value of the recrystallized material (11.9 ± 0.5 $\mu g/cm^3$) but lower than that of the pure amorphous solid (235.3 ± 2.0 $\mu g/cm^3$). These data confirm the thermodynamic prediction regarding the increase in the solubility of the crystalline drugs with the reduction of their dimensions in the nanometer-size range. Furthermore, from the same experiments it was possible to derive the material surface composition in terms of amorphous and nanocrystalline surface abundance. The amorphous surface fraction was 0.511 ± 0.026; the surface composition, showing a surface excess of the disordered phase, does not reflect the bulk composition (about 0.1) measured by DSC and XRPD techniques.

The dynamic dissolution approach, described in Chapter 5, appears to be most useful in the determination of solubility values for metastable phases, such as nanocrystallites and amorphates, in dissolution media in which they spontaneously convert into a more stable phase.

6.4 CONCLUSIONS

The aim of this chapter was to present some fundamental thermodynamic concepts strictly linked to the solubility of a solid in a solvent. Accordingly, assuming some crude approximations related to the equilibrium process, the ideal solubility of a solid solute in a liquid solvent was initially discussed. Moreover, the well-known relationship describing crystal solubility as a function of particle size was derived taking into account both the liquid–vapor and the solid–liquid interfaces. In addition, the effects of the amount of excess solids and the influence of solid microstructure on the apparent solubility of a model drug have been investigated. Finally, a much more convenient strategy to approach dynamic solubility measurement for a multiphase solid material was shown.

The theoretical thermodynamic frame presented serves to point out important solid physico-chemical properties affecting the solubility process. More precisely, either the lowering of the melting temperature or the reducing of the enthalpy of fusion of the solid leads to an increase in the solubility. The melting point depression of nanosized materials has been investigated and the proposed thermodynamic model was experimentally demonstrated and discussed.

Owing to the lack of long-range order, characteristic of the crystalline phase, an amorphous solid is in a metastable equilibrium state with its surroundings and shows metastable equilibrium solubility higher than that of the stable solid (well-ordered crystal). On the other hand, the crystalline solid, with dimensions in the nanometer-size range, should also be considered as the metastable phase in the light of the solubility process. In fact, the Ostwald–Freundlich relation predicts that nanocrystals are not stable in solution because of the high solubility value. It can be deduced that the slurries of nanocrystals are not stable thermodynamically but they are in a dynamically stable state. For solids with low recrystallization rates, the metastable states can be maintained for long periods. When experiments are carefully executed, a variation in solubility with particle size is rarely observed for freshly precipitated solids. On the contrary, the apparent solubility of stressed matcrials, made up of amorphous and nanocrystalline phases, increases with an increase in the solid amount added. The Ostwald–Freundlich equation is thermodynamically based, whereas dissolution is a dynamic process although determined by thermodynamic factors. The proposed experimental approach, based on the dissolution model of a biphasic mixture, helps in measuring solubility values for the two metastable phases: amorphous and nanocrystalline.

REFERENCES

1. Devane, J., Oral drug delivery technology: Addressing the solubility/permeability paradigm, *Pharm. Tech.*, 22, 68, 1998.
2. U.S. Department of Health and Human Services, Food and Drug Administration, Center for Drug Evaluation and Research, *Waiver of In Vivo Bioavailability and Bioequivalence Studies for Immediate-Release Solid Oral Dosage Forms Based on Biopharmaceutics Classification System*, 2000 (Internet site: http://www.fda.gov/cder/guidance/index.htm).
3. Amidon, G.L., Lennernäs, H., Shah, V.P., and Crison, J.R., A theoretical basis for a biopharmaceutic drug classification: The correlation of *in vitro* drug product dissolution and *in vivo* bioavailability, *Pharm. Res.*, 12(3), 413, 1995.
4. Petravale, V.B., Date, A.A., and Kulkarni, R.M., Nanosuspensions: A promising drug delivery strategy, *J. Pharm. Pharmacol.*, 56, 827, 2004.
5. Carli, F. and Colombo, I., Physical state of drug loaded into silica gel carriers, *Acta Pharm. Jugosl.*, 38, 361, 1988.
6. Liang, L.H., Zhao, M., and Jiang, Q., Melting enthalpy depression of nanocrystals based on the surface effect, *J. Mater. Sci. Lett.*, 21, 1843, 2000.
7. Grassi, M., Colombo, I., and Lapasin, R., Drug release from an ensemble of swellable crosslinked polymer particles, *J. Contr. Rel.*, 68, 97, 2000.
8. Gassmann, P., List, M., Schweitzer, A., and Sucker, H., Hydrosols—alternatives for the parenteral application of poorly water soluble drugs membranes, *J. Pharm. Biopharm.*, 40, 64, 1994.
9. Chen, J.F., Wang, Y.H., Guo, F., Wang, X.M., and Zheng, C., Synthesis of nanoparticles with novel technology: High-gravity reactive precipitation, *Ind. Eng. Chem. Res.*, 39, 948, 2000.
10. Carli, F., Colombo, I., Magarotto, L., and Torricelli, C., Influence of polymer characteristics on drug loading into Crospovidone, *Int. J. Pharm.*, 33, 115, 1986.
11. Bergese, P., Alessandri, I., Colombo, I., Coceani, N., and Depero, L.E., Microstructure and morphology of Nimesulide/Crospovidone nanocomposites by Raman and electron microscopies, *Composites Part A*, 36, 443, 2005.
12. Liversidge, K.C., Cundy, J., Bishop, D., and Czekai, D., Surface modified drug nanoparticles, Patent [5,145,684], 1991, US.
13. Müller, R.H., Becker, R., Kruss, B., and Peters, K., Pharmaceutical nanosuspensions for medicament administration as system of increased saturation solubility and rate of solution, Patent [5,858,410], 1998, US.
14. Lovrecich, M., Supported drugs with increased dissolution rate, and a process for their preparation, Patent [5,354,560], 1991, US.

15. Müller, R.H., Jacobs, C., and Kayser, O., Nanosuspensions as particulate drug formulations in therapy. Rationale for development and what we can expect for the future, *Adv. Drug Del. Rev.*, 7, 3, 2001.
16. Choi, W.S., Kim, H.I., Kwak, S.S., Chung, H.Y., Yamamoto, K., Oguchi, T., Tozuka, Y., Yonemochi, E., and Terada, K., Amorphous ultrafine particle preparation for improvement of bioavailability of insoluble drugs: Grinding characteristics of fine grinding mills, *Int. J. Miner. Process.*, 74S, S165, 2004.
17. Manca, D., Coceani, N., Magaratto, L., Colombo, I., and Grassi, M., High-energy mechanochemical activation of active principles, in *Proceedings of the Symposium GRICU*, Ischia (Napoli), SI 229, 2004.
18. De Giudici, G., Biddau, R., D'Incau, M., Leoni, M., and Scardi, P., Dissolution of nanocrystalline fluorite powder: An investigation by XRD and solution chemistry, *Geach. Cosmoch. Acta*, 69 (16), 4073, 2005.
19. Neau, S.H., Solubility theory, in *Water-Insoluble Drug Formulation*, Liu, R., Ed., Interpharm Press, Denver, Colorado, USA, 2000, Chapter 2.
20. Atkins, P.W., Solutions, in *Physical Chemistry*, sixth ed., Oxford University Press, Oxford, England, 1998, Chapter 4.
21. Prausnitz, J.M., Lichtenhaler, R.N., and Gomes de Azevedo, E., Solubility of solids in liquid, in *Molecular Thermodynamics of Fluid-Phase Equilibria*, Prentice-Hall Inc., Englewood Cliffs, New York, USA, 1986, Chapter 9.
22. Elliott S.R., *Physics of Amorphous Materials*, second ed., Longman Scientific & Technical, Burnt Mill, Harlow, Essex, England, 1990.
23. Craig, D.Q.M., Royall, P.G., Kett, L.V., and Hopton, L.M., The relevance of the amorphous state to pharmaceutical dosage forms: Glassy drugs and freeze dried systems, *Int. J. Pharm.*, 179, 179, 1999.
24. Vippagunta, S.R., Brittain, H.G., and Grant, D.J.W., Crystalline solids, *Adv. Drug Del. Rev.*, 48, 3, 2001.
25. Swalin, R.A., *Thermodynamics of Solids*, second ed., Wiley-Interscience Publication, John Wiley & Sons, New York, USA, 1972.
26. Burt, H.M. and Mitchell, A.G., Crystal defects and dissolution, *Int. J. Pharm.*, 9, 137, 1981.
27. Stillinger, F.H. and Weber, T.A., Packing structures and transition in liquid and solids, *Science*, 225, 983, 1984.
28. Sheng, H.W., Lu, K., and Ma, E., Melting and freezing behaviour of embedded nanoparticles in ball-milled 10 WT% M ($M = $ In, Sn, Bi, Cd, Pb) mixture, *Acta Mater.*, 46, 5195, 1998.
29. Bottani, C.E., Li Bassi, A., Stella, A., Cheyssac, P., and Kofman, R., Investigation on confined acoustic phonon of tin nanoparticles during melting, *Europhys. Lett.*, 56 (3), 386, 2001.
30. Jiang, Q., Zhang, Z., and Wang, Y.W., Thermal stability of low dimensional crystals, *Mater. Sci. Eng. A-Struct.*, 286 (1), 139, 2000.
31. Carli, F., Colombo, I., and Magarotto, L., Thermal analysis of drug loaded polymeric carriers, *Proc. the 13th Int. Symp. Cont. Rel. Bio. Mat.*, Norfolk, Virginia, USA, August, 3–6, 192, 1986.
32. Jackson, C.L. and Mckenna, G.B., The melting behaviour of organic materials confined in a porous solids, *J. Chem. Phys.*, 93 (12), 9002, 1990.
33. Jiang, Q., Zhang, Z., and Li, J.C., Melting thermodynamics of nanocrystals embedded in a matrix, *Acta Mater.*, 48 (20), 4791, 2000.

34. Lu, K. and Jin, Z.H., Melting and superheating of low-dimensional materials, *Curr. Opin. Solid. Mater. Sci.*, 5, 39, 2001.
35. Carstensen, J.T. and Musa, M.N., Dissolution rate patterns of log-normal distributed powders, *J. Pharm. Sci.*, 61 (2), 223, 1972.
36. Mermet, J.M., Kellner, R., Otto, M., and Widmer, H.M., *Analytical Chemistry*, Wiley-VHC, Weinheim, Germany, 1998.
37. Brittain, H.G., Ed., *Polymorphism in Pharmaceutical Solids*, Volume 95, Marcel Dekker Inc., New York, USA, 1999.
38. Heleblian, J.K. and McCrone, W.C., Pharmaceutical applications of polymorphism, *J. Pharm. Sci.*, 58, 911, 1969.
39. Neelima, V.P. and Suryanarayanan, R., Polymorphism in anhydrous theophylline—implications on the dissolution rate of theophylline tablets, *J. Pharm. Sci.*, 86, 1256, 1997.
40. Hancock, B.C. and Zografi, G., Characteristic and significance of the amorphous state in pharmaceutical systems, *J. Pharm. Sci.*, 86, 1, 1997.
41. Yu, L., Amorphous pharmaceutical solids: Preparation, characterization and stabilization, *Adv. Del. Rev.*, 48, 27, 2001.
42. Zhao, M., Zhou, X.H., and Jiang, Q., Comparison of different models for melting point change of metallic nanocrystals, *J. Mater. Res.*, 16, 3304, 2001.
43. Allen, G.L., Gile, W.W., and Jesser, W.A., The melting temperature of microcrystals embedded in a matrix, *Acta Metall.*, 28, 1695, 1980.
44. Shi, F.G., Size dependent thermal vibrations and melting in nanocrystals, *J. Mater. Res.*, 9, 1307, 1994.
45. Sanz, N., Boudet, A., and Ibanez, A., Melting behavior of organic nanocrystals grown in sol–gel matrices, *J. Nanop. Res.*, 4, 99, 2002.
46. Jackson, C.L. and McKenna, G.B., The melting behavior of organic materials confined in porous solids, *J. Chem. Phys.*, 93, 9002, 1990.
47. Adamson A.W. and Gast A.P., *Physical Chemistry of Surfaces,* sixth ed., Wiley-Interscience Publication, John Wiley & Sons, New York, USA, 1997, Chapter IX.
48. Couchman, P.R. and Jesser, W.A., Thermodynamic theory of size dependence of melting temperature in metals, *Nature*, 269, 481, 1977.
49. Brun, M., Lallemand, A., Quinson, J.F., and Eyraud, C., Changement d'etat liquide–solide dans les milieux poreux. II Etude theorique de la solidification d'un condensat capillaire, *J. Chimie Phys.*, 70, 979, 1973.
50. Peters, K.F., Cohen, J.B., and Chung, Y.W., Melting of Pb nanocrystals, *Phys. Rev. B*, 57, 13430, 1998.
51. Zhang, M., Efremov, M.Y., Schiettekatte, F., Olson, E.A., Kwan, A.T., Lai, S.L., Wisleder, T., Greene, J.E., and Allen, L.H., Size-dependent melting point depression of nanostructures: Nanocalorimetric measurements, *Phys. Rev. B*, 62, 10548, 2000.
52. Adamson A.W. and Gast A.P., *Physical Chemistry of Surfaces*, sixth ed., Wiley-Interscience Publication, John Wiley & Sons, New York, USA, 1997, Chapters II, III and X.
53. Bergese, P., Colombo, I., Gervasoni, D., and Depero, L.E., Melting of nanostructured drugs embedded into a polymeric matrix, *J. Phys. Chem. B*, 108, 15488, 2004.
54. Dobetti, L., Cadelli, G., Furlani, D., Zotti, M., Cerchia, D., and Grassi, M., Reliable experimental setup for the drug release from a polymeric powder loaded by means of co-grinding, *Proc. 28th Int. Symp. on Contr. Rel. of Bioact. Mat.*, San Diego (CA), USA, June 23–27, #6059, 2001.

55. Meriani, F., Coceani, N., Sirotti, C., Voinovich, D., and Grassi, M., *In vitro* Nimesulide absorption from different formulations, *J. Pharm. Sci.*, 93, 540, 2004.
56. Bergese, P., Colombo, I., Gervasoni, D., and Depero, L.E., Assessment of X-ray diffraction–absorption method for quantitative analysis of largely amorphous pharmaceutical composites, *J. Appl. Cryst.*, 36, 74, 2003.
57. Bergese, P., Bontempi, E., Colombo, I., Gervasoni, D., and Depero, L.E., Microstructural investigation of Nimesulide–Crospovidone composites by X-ray diffraction and thermal analysis, *Composites Sci. Technol.*, 63, 1197, 2003.
58. Van Krevelen, D.W., *Properties of Polymers,* third ed., Elsevier Science, Amsterdam, Nederland, 1997.
59. Banker, G.S. and Rhodes, C.T., *Modern Pharmaceutics*, Marcel Dekker, New York, USA, 1979, p. 172.
60. Ostwald, W., On the supposed isomerism of red and yellow mercury oxide and the surface tension of solid substances, *Z. Phys. Chem.*, 34, 495, 1900.
61. Freundlich, H., *Colloid and Capillary Chemistry*, E.P. Dutton and Company Inc., New York, USA, 1923, p. 153.
62. Adamson A.W. and Gast A.P., *Physical Chemistry of Surfaces,* sixth ed., Wiley-Interscience Publication, John Wiley & Sons, New York, USA, 1997, Chapters II and III.
63. Gregg, S.J. and Sing, K.S.W., *Adsorption, Surface Area and Porosity,* second ed., Academic Press, New York, USA, 1982, Chapter III.
64. Buckely, H.E., *Crystal Growth*, Wiley, New York, USA, 1951.
65. Enüstün, B.V. and Turkevich, J., Solubility of fine particles of strontium sulfate, *J. Am. Chem. Soc.*, 82, 4502, 1960.
66. Wu, S., Calculation of interfacial tension in polymer system, *J. Polymer Sci. C*, 34, 19, 1971.
67. Buckton, G. and Beezer, A.E., The relationship between particle size and solubility, *Int. J. Pharm.*, 82, R7, 1992.
68. Bikerman, J.J., *Physical Surfaces*, Academic Press, New York, USA, 1970, p. 216.
69. Jinno, J., Kamada, N., Miyake, M., Yamada, K., Mukai, T., Odomo, M., Toguchi, H., Liversidge, G.G., Higaki, K., and Kimura, T., Effect of particle size reduction on dissolution and oral absorption of a poorly water-soluble drug, cilostazol, in beagle dogs, *J. Contr. Rel.*, 111, 56, 2006.
70. Carstensen, J.T., *Advanced Pharmaceutical Solids*, Swarbrick, J., Ed., Marcel Dekker, Inc., New York, USA, 2001, Chapter 3.
71. Mosharraf, M. and Nyström, C., Apparent solubility of drug in partially crystalline systems, *Drug Dev. Ind. Pharm.*, 29 (6), 603, 2003.
72. Madras, G. and McCoy, B.J., Growth and ripening kinetics of crystalline polymorphs, *Crys. Gro. Des.*, 3 (6), 981, 2003.
73. Cohen-Adad, R. and Cohen-Adad, M.T., Solubility of solids in liquid, in *The Experimental Determination of Solubility*, Hefter, G.T. and Tomkins, R.P.T., Eds., John Wiley & Sons Ltd, Chichester, England, 2003, Chapter 4.1.
74. Zevin, L.S. and Kimmel, G., Methodology of quantitative phase analysis, in *Quantitative X-ray Diffractometry*, Mureinik, I., Ed., Springer Verlag, New York, USA, 1995, Chapter 4.
75. Alexander, L.E. and Klug, H.P., Basic aspects of X-ray absorption in quantitative diffraction analysis of powder mixture, *Anal. Chem.*, 20, 886, 1948.

76. Brandon, D. and Kaplan, W.D., The concept of microstructure, in *Microstructural Characterization of Materials*, John Wiley & Sons, Chichester, England, 1999, Chapter 1.
77. Klug, H.P. and Alexander, L.E., Crystallite size and lattice strains from line broadening, in *X-ray Diffraction Procedures*, second ed., Wiley-Interscience Publication, John Wiley & Sons, New York, USA, 1974, Chapter 9.
78. Langford, J.I., Line profile and sample microstructure, in *Industrial Applications of X-Ray Diffraction*, Chung, F.H. and Smith, D.K., Eds., Marcel Dekker Inc., New York, USA, 2000, Chapter 33.
79. Snyder, R.L., Fiala, J., and Bunge, H.J., Eds., Part I, Fundamentals of defect analysis in diffraction, in *Defect and Microstructure Analysis by Diffraction, IUCr Monographs on Crystallography*, Volume 10, Oxford University Press Inc., New York, USA, 1999.
80. Young, R.A., Ed., *The Rietveld Method, IUCr Monographs on Crystallography*, Volume 5, Oxford University Press Inc., New York, USA, 1993.
81. Warren, B.E. and Averbach, B.L., The effect of cold-work distortion on X-ray patterns, *J. Appl. Phys.*, 21, 595, 1950.
82. Lutterotti, L., Maud, Materials analysis using diffraction, general diffraction–reflectivity analysis program mainly based on the Rietveld method. http://www.ing.unitn.it/~maud/index.html, 2006.

7 Drug Release from Matrix Systems

7.1 INTRODUCTION

According to the release behavior, controlled release systems (CRS) can be classified into passively preprogrammed, actively preprogrammed, and actively self-programmed systems [1]. Although in the first category (passively preprogrammed system) the release rate is predetermined and it is irresponsive to external biological stimuli, in the second category (the actively preprogrammed system), the release rate can be controlled by a source external to the body as in the case of insulin and urea deliveries [1,2]. It is clear that the last category represents the new generation of delivery systems (regardless of the administration route) as self-programmed delivery systems will be able to autonomously regulate release kinetics in response to external stimuli such as the concentration of a fixed analyte [2,3]. While 15 years ago the majority of CRS fell into the first category, at present the importance of the last two categories has considerably increased [2]. Indeed, only the last two categories can fully accomplish the modern concept of therapeutic treatment the aim of which is to increase drug effectiveness and patient compliance, two variables strictly related to the administration frequency, and side effects connected to dosage. In this framework, both the "designing step" and the use of new materials are becoming more and more significant for the realization of CRS. Consequently, CRS requires knowledge of engineering, chemistry, pharmacy, and medicine merging in an attempt to realize more and more effective and reliable delivery systems [4]. Accordingly, mathematical models, a typical engineering tool, turn out to be very useful in predicting CRS behavior or in measuring some important related parameters, such as the drug diffusion coefficient. In addition, as the mathematical model is nothing more than a "mathematical metaphor of some aspects of reality" [5] (that, in this case, identifies with drug release), this approach requires a clear knowledge of the ensemble of phenomena ruling release kinetics.

Among the great variety of CRS, matrix systems, defined as the three-dimensional network containing the drug and other substances such as solvents and excipients, can benefit enormously from the cooperation of engineers, chemists, pharmacists, and medical doctors. Indeed, for instance, polymer science, surface, bioadhesion and thrombogenic properties, drug–matrix interactions, drug physicochemical and therapeutic characteristics, and biocompatibility merge together to determine the final reliability and effectiveness of a matrix-based delivery system. Accordingly, this chapter first deals with the description of matrice structures and the mechanisms ruling drug release from this kind of delivery systems. Then, based on this physical frame, the development of mathematical models aimed to describe release kinetics from different matrix systems is presented including a usually briefly mentioned topic such as drug release from the ensemble of polydisperse spherical matrices.

7.2   MATRICES

One of the most common approaches to get controlled release is to embed a drug in a hydrophobic or hydrophilic matrix to speed up or to reduce drug release kinetics depending on the final therapeutic target. Although wax, polyethylene, polypropylene, and ethylcellulose usually constitute hydrophobic matrices, hydrophilic matrices are generally made up by hydroxypropylcellulose, hydroxypropylmethylcellulose, methylcellulose, sodium carboxymethylcellulose and, in general, polysaccharides [6,7]. In addition, poly(vinyl alcohol), acrylates, methacrylates, polylactic acid, and polyglycolic acid cannot be forgotten [7]. Matrix-based delivery systems find a great variety of applications in the pharmaceutical field especially in oral administration. Although a detailed description of all these applications is out of the aim of this chapter, the analysis of the most recent ones can be interesting. For example, in regenerative medicine, matrices (hydrogels) are used as three-dimensional support for growth of cells as in the case of stem cells isolated from the human bone marrow [8] and chondrocytes [9–10]. Owing to a particularly soft loading process, matrices (hydrogels) are also suitable to release proteins and oligonucleotides, high-molecular weight and fragile drugs, which can easily undergo denaturation or degradation during the formulative process. For example, triblock copolymer poly(ethylene glycol) (PEG)–poly(lactic-co-glycolic acid) (PLGA)–PEG is used to accelerate the wound healing process by releasing *in situ* a protein that accelerates the wound healing process [11,12]. Matrix systems are also used to deliver antimicrobial [13], antifungal [14], antiherpes simplex type 2 (HSV-2) [15], and paclitaxel, a drug preventing *in situ* regrowth of the neoplastic tissue after surgical removal [16].

Matrices can be prepared by mixing the drug, in the form of a thin powder, with the prepolymer. Then, the whole mixture is poured into the polymerization reactor. Alternatively, matrix can be structured in advance and then put in contact with a highly concentrated drug solution able to swell the matrix (solvent swelling technique). Solvent removal is achieved, for example, by means of physical treatments [17]. Another approach relies on the mechanical energy supplied to the drug–carrier (usually a polymer constituting a matrix network) couple by cogrinding. In this manner, it is possible to load a drug into a matrix network avoiding the use of solvents, whose elimination from the final formulation can represent a very expensive and delicate step [18]. In addition, supercritical fluids can represent a profitable tool to achieve drug loading inside matrices [19–21]. Indeed, if on one hand supercritical fluids show a density approaching that of liquids (this usually implies a good solubility with respect to drug or solvents used in the solvent swelling technique), then on the other hand they are characterized by low viscosity, typical of gases. Consequently, they easily and efficiently swell the matrix (bringing the drug inside the matrix network or extracting solvents) and they can be then removed by a simple pressure decrease. Indeed, as a pressure decrease provokes the transition from the supercritical condition to the gas one, the matrix can be easily devoided by the supercritical solvent without dragging out the drug. Finally, tablets represent the simplest and the most traditional way of preparing a matrix-based delivery system as their formulation requires to compress, in a proper ratio, a carrier (usually a polymer), the drug, and various excipients.

7.2.1 MATRIX TOPOLOGY

Because of the great variety of typologies and materials that can be used, it is not easy to provide a comprehensive physical picture of all kinds of existing matrices. Nevertheless, in an attempt to give a rigorous description without losing generality and for the sake of simplicity, we can think of matrices as constituted by three different phases: (a) continuous, (b) shunt, and (c) dispersed [22]. These phase types have to be further classified into primary, secondary, tertiary, etc., on the basis of the spatial relationships of each phase to the other phases present. A primary continuous phase pervades the whole matrix and, depending on its composition relative to the compositions of associated phases, it may provide an uninterrupted diffusion path for a solute or it represents an inaccessible zone acting as a mere supporting structure. A primary shunt phase interests the whole matrix and identifies with a macrochannel structure (pores). If the shunt phase is not present, the matrix is defined as nonporous whereas it will be classified as microporous (pores diameter: 10–100 nm), mesoporous (pores diameter: 100–1000 nm), macroporous (pores diameter: 1–100 μm), or superporous (pores diameter: 10–1000 μm) in the opposite case. Although in the last two cases the pore diameter is much larger than the molecular size of hypothetical diffusing molecules, in the first two cases this is no longer true [23]. Finally, dispersed phases are embedded in continuous or shunt phases.

For their importance and wide use in the pharmaceutical field, it is worth particularizing the above-mentioned theoretical frame to polymeric matrices. In this case, usually, the shunt phase is not present (even if it is not always true as discussed later on) whereas a primary continuous phase (usually a liquid phase) is trapped in a swollen solid secondary continuous phase constituted by high-molecular weight molecules dispersed and collocated to form a continuous three-dimensional polymeric network pervading the whole system [24]. In very low swelling matrices, however, the polymeric phase can assume the role of primary phase due to its massive prevalence on the other (liquid) continuous phase. Regardless of which one is the primary phase, the ensemble of these two continuous phases can be thought as a coherent system, with the mechanical characteristics in between those of solids and liquids. The presence of crosslinks (polymer–polymer junctions) between polymeric chains hinders polymer dissolution in the liquid phase, which can only swell the network. This structure is roughly similar to that of a sponge filled by a liquid phase. Nevertheless, this is a particular sponge as, in the case of strong crosslinks (typically chemical covalent bonds), the network does not modify with time. When, on the contrary, weak crosslinks prevail (typically physical interactions such as Coulombic, van der Waals, dipole–dipole, hydrophobic, and hydrogen bonding interactions), polymeric chains are not so rigidly connected to each other and the similarity with the sponge is no longer so pertinent. Indeed, though crosslink density (number of crosslinks per unit volume) is constant with time (in static conditions), the Brownian motion of chains and segment of chains makes the distribution of crosslinks time dependent. As a consequence, though average dimensions of network meshes do not modify, each mesh can modify, thus resembling a statistical network. Obviously, this kind of network can easily undergo erosion owing to polymer–polymer junction weakness. This physical frame is made more complex by the fact that the whole structure can be constituted by an ensemble of small matrix domains embedded in a continuum, usually represented by a polymer solution as it occurs for Carbopol [25,26]. In addition, it is well known that many charged polysaccharides, i.e., alginate, pectate, gellan-gum, carrageenans, can form inhomogeneous matrix structures (hydrogels) in the presence of uni- or multivalent cations [27,28]. Indeed, at microscopic level, they show clusters with different and high-crosslink density dispersed in a less crosslinked medium [29–31]. Finally, the simultaneous presence of two networks (interpenetrating structures), formed by two different polymers, can further complicate the scenario. Typically, these systems are produced by an initial swelling of a monomer reacting to form a second intermeshing network structure [32–35]. Obviously, the choice of the polymer depends on the final CRS administration route (examples include oral, ophthalmic, rectal, vaginal, and subcutaneous) and on different factors such as matrix swelling degree, biodiversity, biocompatibility, interactions with drug, excipients, and mechanical properties.

Polysaccharides, in particular glucans and xanthan, represent typical examples of physically crosslinked matrices [24]. Indeed, interchained physical interactions allow the formation of junction zones where the regular coupling of chains belonging to different polymeric chains takes place (Figure 7.1). The long chain segments departing from these junction zones can form, with other chains, additional junction zones so that a polymeric

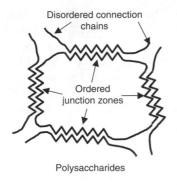

Disordered connection chains

Ordered junction zones

Polysaccharides

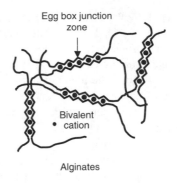

Egg box junction zone

Bivalent cation

Alginates

Spaghetti-like configuration

FIGURE 7.1 Examples of crosslink structure in physically crosslinked polymeric networks.

three-dimensional network can be built up. Obviously, in general, these are weak networks (even if network strength strongly depends on polymer concentration) that can easily undergo erosion leading to structure destruction. Alginates, on the contrary, represent a typical example of strong physical networks (matrices) [27]. In this case, junction zones resemble egg boxes for their characteristic shape [36] (Figure 7.1). Indeed, bivalent cations take part in interchain bonds between guluronic residues that, in their planar conformation, give rise to a bond cavity for cations. In particular, bivalent cations interact with oxygens of COO^- groups and some OH groups of guluronic residues belonging to different chains [37]. Sequences of these cavities form the above-mentioned egg boxes (junction zones). Affinity series of bivalent cations is [38]

$$Pb > Cu > Cd > Ba > Sr > Ca > Co, \quad Ni, Zn > Mn.$$

For many biological applications, Cu^{++} and Ca^{++} represent the cations of choice for the formation of interchain junction zones. Strontium and barium are used in hydrogels designed to contain cells [39]. Crosslinked polyvinyl-pyrrolidone (PVP, C_6H_9NO) represents a particular example of strong physically crosslinked matrix. Indeed, the popcorn polymerization used to get the polymeric matrix [40] gives rise to a very complex topological structure made up by entangled polymeric chains giving rise to a *spaghetti*-like structure (Figure 7.1). The fact that this matrix never dissolves in proper solvents (like water, for example) leads to the conclusion that the above-mentioned spaghetti-like structure is stabilized by a very small amount of chemical crosslinks [41].

Among the many examples of chemically crosslinked matrices, temperature responsive gels can be mentioned. These are N-alkylacrylamides with N-isopropylacrylamide (NIPA), N-N-Diethylacrylamide (DEAAm), and their copolymers, where a variety of crosslinkers ranging from N,N'-methylenebis (acrylamide) and ethylene glycol dimethacrylate to N-methylolacrylamide can be mentioned [42]. Peppas and group [43], for example, use poly(N-isopropylacrylamide *-co-methacrylic acid) gels crosslinked by ethylene glycol dimethacrylate for the delivery of antithrombotic agents such as heparin. In addition, different polysaccharides such as methylcellulose, hydroxypropyl-methylcellulose, hydroxypropylcellulose, and carboxymethylcellulose can be crosslinked with divinyl sulfone [44], adipic, sebacic, or succinic acid [45]. Finally, PEG and poly(hydroxyethyl methacrylate) (PHEMA) are other common polymers that can undergo chemical crosslinking [28].

Although the above-discussed examples mainly deal with nonporous structures, typical examples of porous matrices are represented by super-porous hydrogels (SPH), superabsorbent polymers (SAP) [46], some kinds of hydrophobic matrices, and tablets. SAP and SPH are crosslinked hydro-philic polymers with the ability to absorb considerable amounts of water or aqueous fluids (10–1000 times of their original weight or volume) in relatively short periods of time due to their wide porous structure generated by the decomposition of a bicarbonate compound during polymerization (Figure 7.2). To prevent the formation of heterogeneous structures associated with the violent exothermic reaction involved in polymerization, hydrophilic monomer (typically acrylamide, salts of acrylic acid, and sulfopropyl acrylate) is diluted in water by gently mixing at room temperature. Then, the neutralization step, required for ionic monomers, is followed by crosslinker addition. Glacial acetic acid or acrylic acid is subsequently added to obtain foam during polymerization. In addition, to get homogeneous SPH, a surfac-tant (such as triblock copolymers (polyethylene oxide–polypropylene oxide–polyethylene oxide)) is needed to stabilize the foam. Although in the case of SPA, polymerization is obtained through both redox and thermal systems, SPH are produced only through a redox mechanism. Accordingly, oxidant and reductant (and thermal initiator) are added to the monomer solution under gentle mixing. Finally, an acid-dependent foaming agent (typically sodium bicarbonate) is added. SPA and SPH products are then dehydrated (heating for SPA, alcohol and heating for SPH) to get a solid, brittle porous material, which is white in color because of a heterogeneous combination of polymer and pores. The final product can be ground into a particle shape (like super-absorbent particles), sliced into absorbent sheets, or machined into any shape and size. Obviously, pore characteristics and porosity can be varied by

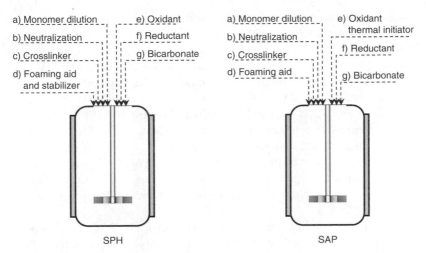

FIGURE 7.2 Schematic representation of steps involved in the production of SPH and SPA. (From Omidian, H., Rocca, J.G., Park, K., *J. Contr. Rel.*, 102, 3, 2005. With permission from Elsevier.)

properly modulating the production parameters so that various grades of products can be prepared spanning from nonporous up to superporous structures. Usually, swollen SPH show very poor mechanical properties and that is why their use in many practical applications turns out to be very difficult. Accordingly, new types of SPH have been proposed to overcome this serious limitation and two new generations of SPH have been realized. Although the first generation consists of fragile fast and high swelling ratio systems, the second generation is the fast and medium swelling systems characterized by improved mechanical properties. Finally, their good mechanical properties make the third generation suitable also for pharmaceutical and biomedical applications. To get second SPH generation, a crosslinked hydrophilic polymeric matrix (CHPM), which is able to absorb monomer solution, crosslinker, initiator, and remaining components, is added to monomer solution. Upon polymerization, CHPM serves as the local point of physical crosslinking (or entanglement) of the formed polymer chains. During the polymerization process, each CHPM particle acts as an isolated individual reactor in which crosslinking polymerization occurs. The final result is a structure made up by an ensemble of CHPM units connected to each other by polymeric chains. Third SPH generation, on the contrary, relies on the addition of a hybrid agent (HA) that can be crosslinked after monomer polymerization and crosslinking. Typically, HA is a water-soluble or water-dispersible polymer that can form a crosslinked (chemical or physical) structure pervading the primary SPH polymeric structure yielding to an interpenetrating network. Examples of hybrid agents are polysaccharides (sodium alginate, pectin, and chitosan) or synthetic water-soluble hydrophilic polymers such as poly(vinyl alcohol). Examples of third SPH generation are acrylamide-based structures in the presence of sodium alginate crosslinked by calcium ions and poly (acrylamide-co-acrylic acid) in the presence of polyethyleneimine [23].

Some hydrophobic matrices are essentially realized by melt extrusion of a mixture containing a low-melting binder such as stearic acid, the drug and hydrophilic excipients such as PEG 6000 and lactose [47,48]. Upon contact with an aqueous release environment, the progressive dissolution of hydrophilic compounds and drug leads to the formation of a porous matrix whose skeleton is represented by the insoluble low-melting binder. Obviously, matrix topological structure (porosity, tortuosity, etc.) changes with time unless the dissolution of hydrophilic excipients can be retained instantaneous and the drug content is low.

Tablets, obtained by simple compression of a proper mixture typically constituted by a polymer, a drug, and eventual excipients, represent one of the most traditional ways of getting a CRS. Nevertheless, the resulting structure is, from the release point of view, very complex. Indeed these systems are characterized by a changing porosity spanning from the initial one, referred to as the dry tablet, to a vanishing value corresponding to the swollen condition caused by the presence of an external aqueous release environment.

Finally, an interesting category of porous systems is represented by inorganic matrices that are emerging as a new category of host or guest devices [49]. Owing to their biological stability and their release properties depending on pore size, topology, and drug-pore surface interactions [50], there is a significant and increasing interest in these potential carriers. Examples of this type of carriers include synthetic zeolites [50], silica xerogels [51], and porous ceramics [52]. According to the IUPAC definition [53], inorganic porous materials are classified according to the pore diameter. Consequently, microporous (<2 nm), mesoporous ($2–50$ nm), and macroporous (>50 nm) categories are formed (although this classification differs from that given earlier about polymeric matrices [23], in the following, IUPAC classification will be used for porous inorganic matrices, whereas the other one will be followed for polymeric matrices). Zeolites probably represent the most famous example of inorganic porous materials used in drug delivery systems. Unfortunately, however, their application is limited by the small pores diameter. Indeed, for pore diameter lesser than 2 nm, release rates are slow and the incorporation of some interesting molecules is very difficult if not impossible due to the dimension of molecules. As a consequence, pore diameter increase has been one of the main goals in zeolite chemistry [53]. This is the reason why mesoporous materials were prepared by means of intercalation of layered materials such as double hydroxides, metal (titanium, zirconium) phosphates, and clays. Unfortunately, however, they show a broad mesopore size distribution spanning from the microscale to the mesoscale. On the contrary, a particular class of porous silica material, generally

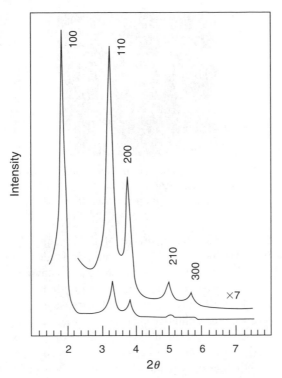

FIGURE 7.3 X-ray diffraction pattern of high-quality calcined MCM-41. (From Ciesla, U., and Schuth, F., *Microporous Mesoporous Mater.*, 27, 131, 1999. With permission from Elsevier.)

named as MCM-41, represents the first example of mesoporous material showing ordered pore arrangement coupled with a very narrow pore size distribution (average diameter 3 nm) and surface area up to 1000 m^2/g [54]. An evolution of MCM-41 is represented by highly ordered hexagonal silica structure named SBA-15, synthesized by using commercially available block–copolymer surfactants in strong acid media and characterized by mesopores of, approximately, 6 nm diameter [55].

MCM-41 is characterized by a honeycomb structure resulting from a hexagonal packing of unidimensional cylindrical pores. The x-ray diffraction pattern of MCM-41 typically shows 3–5 reflections between $2\theta = 2°$ and 5° (Figure 7.3), although samples with more reflections have also been reported [56,57]. The reflections are due to the ordered hexagonal array of parallel silica tubes and can be indexed assuming a hexagonal unit cell as (100), (110), (200), and (300). Since the materials are not crystalline at the atomic level, reflections at higher angles are not observed. In addition, the strong decrease of the structure factor at higher angles would render these reflections weak in any case. Despite the clear existence of focus problems, Figure 7.4 reports a

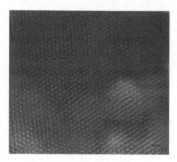

FIGURE 7.4 MCM-41 transmission electron micrograph. Hexagonally arranged 4.0 nm sized pores can be detected. (From Ciesla, U., and Schuth, F., *Microporous Mesoporous Mater.*, 27, 131, 1999. With permission from Elsevier.)

transmission electron microscopy image of the hexagonal arrangement of a uniform, 4 nm sized pores MCM-41 sample. It is also important to note that most MCM-41 not only shows ordered arrangements as reported in Figure 7.4 but also disordered zones constituted by lamellar and fingerprint structures [58]. MCM-41 surface area and pore size distribution can be determined by different techniques. Although the method of choice for surface area consists in the absorption of probe molecules such as N_2, O_2, and Ar, pore size distribution can be determined according to geometrical considerations [59], thermodynamic (basically Kelvin equation [60]), and freezing point depression [61]. Pore wall surface properties can be studied by adsorption of molecules on the surface and by using Fourier transform infrared (FTIR) analysis. Indeed, the absorption of polar or apolar molecules reveals the hydrophilic or hydrophobic character of a surface. The adsorption of cyclohexane, benzene, and water demonstrated the relatively hydrophobic character of siliceous MCM-41 [53]. The original MCM-41 synthesis was carried out in water under alkaline conditions and, similarly to zeolite syntheses, organic molecules (surfactants) function as templates forming an ordered organic–inorganic composite material. By means of calcinations, the surfactant is removed, leaving the porous silicate network. Nevertheless, in contrast to zeolites, the templates are not single organic molecules but liquid–crystalline self-assembled surfactant molecules. The formation of the inorganic–organic composites is based on electrostatic interactions between the positively charged surfactants and the negatively charged silicate species. The formation mechanism of MCM-41 can be explained by means of the "liquid–crystal templating" theory of Beck and group [62] who suppose that two main pathways exist for the porous structure formation (Figure 7.5). Although in pathway 1, the liquid–crystal phase is intact before the silicate species are added, in pathway 2 the addition of the silicate results in the ordering of the subsequent silicate-encased surfactant micelles. The existence of one mechanism rather than the other depends on surfactant properties,

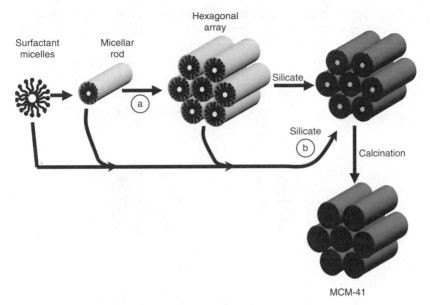

Hexagonal
array

Surfactant   Micellar
micelles   rod

(a)

Silicate

Silicate

(b)

Calcination

MCM-41

FIGURE 7.5 Two possible pathways for the formation of MCM-41: (a) liquid–crystal initiated; (b) silicate-initiated. (From Ciesla, U., and Schuth, F., *Microporous Mesoporous Mater.*, 27, 131, 1999. With permission from Elsevier.)

namely its water concentration and the presence of ions [63]. Indeed, surfactant concentration can span from the critical micelle concentration up to concentrations where liquid crystals are formed.

In the light of pharmaceutical applications, it is interesting to note that mesoporous silica-based materials can be produced in the form of films, fibers, spheres, and monoliths. As examples it is worth mentioning the application of MCM-41 for the controlled release of ibuprofen (anti-inflammatory drug, low water solubility and small molecular size) [49] and the applications of SBA-15 for the controlled release of amoxicillin (penicillin-like antibiotic) [64] and gentamycin (antibiotic for bacterial infections caused by staphylococci) [65].

7.2.2 RELEASE MECHANISMS

Despite their great variety (as discussed in the previous section), in general, it is possible to extrapolate the release mechanism of different matrices to that of polymeric (nonporous) matrices once proper simplifications are made. Indeed, drug release from polymeric matrices is, probably, the most complex one as it can imply simultaneous drug and polymer deep structural modifications. Accordingly, matrix swelling and erosion, drug dissolution (recrystallization) and diffusion, drug–matrix structure interactions, drug distribution and concentration inside the matrix, matrix geometry (cylindrical, spherical, etc.)

and matrix polydispersion in the case of delivery systems made of an ensemble of mini matrices [66,67] are the most important aspects of drug release from polymeric matrices (Figure 7.6). Except for some rare cases, such as ophthalmic matrices, due to dosing and system physical and chemical stabilities, matrix systems are stored in dry, shrunken state, i.e., without containing any liquid phase. In these conditions, the drug, present in the dry matrix network in the form of microcrystals, nanocrystals, or amorphous

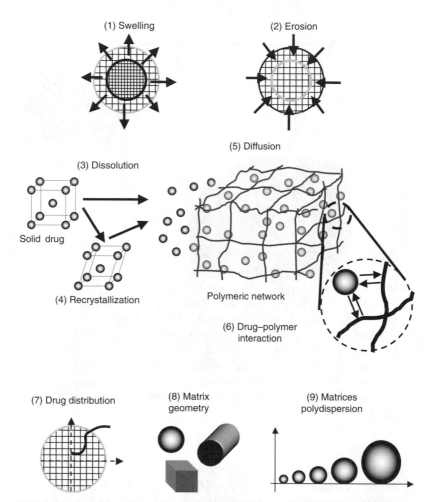

FIGURE 7.6 The most important phenomena affecting drug release from polymeric matrices are: (1) matrix swelling; (2) erosion; (3) drug dissolution; (4) recrystallization; (5) drug diffusion; (6) drug/matrix structure interactions; (7) drug distribution and concentration inside the matrix; (8) matrix geometry (cylindrical, spherical, etc.); and (9) matrices polydispersion in the case of delivery systems made up by an ensemble of mini matrices.

state, cannot diffuse through the network meshes or matrix pores [68]. Upon contact with the release fluids (water or physiological media), matrix structure can undergo swelling with consequent drug dissolution. In the case of polymeric matrices made of polymers showing a glass transition temperature higher than room temperature, the swelling process implies also the transition from the glassy dry state to the rubbery swollen one. Indeed, when the local solvent concentration exceeds a threshold value, polymeric chains unfold so that the glassy–rubbery polymer transition occurs and a gel-like layer, surrounding the matrix dry core, begins to appear [69–71]. This transition implies a molecular rearrangement of polymeric chains that tend to reach a new equilibrium condition as the old one was altered by the presence of the incoming solvent [72]. The time required for this rearrangement depends on the relaxation time t_r of the given polymer/solvent system which, in turn, is a function of both solvent concentration and temperature. If t_r is much lower than the characteristic time of diffusion t_d of the solvent (defined as the ratio of the square of a characteristic length and of the solvent diffusion coefficient at equilibrium), then solvent adsorption may be described by means of Fick's law with a concentration-dependent diffusion coefficient. On the contrary, if t_r is much greater than t_d, then a Fickian solvent adsorption with constant diffusivity takes place. However, in both cases, the diffusion of the drug molecule in the swelling network may be described by Fick's law with a nonconstant diffusion coefficient, and the macroscopic drug release is defined as Fickian. When $t_r \approx t_d$, solvent adsorption does not follow Fick's law of diffusion [73–75]. In such a case, the macroscopic drug release is called anomalous or non-Fickian [76]. Thus, solvent absorption and drug release depend also on the polymer/solvent couple viscoelastic properties [77].

Polymer erosion can take place because of chemical and physical reasons. Under particular physiological conditions, the hydrolysis of water-labile bonds incorporated in the polymer can eventually cause chain breaking. Moreover, erosion can also be due to enzyme attack and chemical reactions at particular polymeric chain sites [6]. On the contrary, in physically crosslinked matrices, erosion is usually due to chain disentanglement, induced by the matrix swelling fluid and the hydrodynamic conditions imposed in the release environment. Obviously, polymer characteristics play a very important role in erosion kinetics as extensively documented by the review of Miller-Chou and Koenig [78]. Erosion can be of two different kinds [6,79]: (a) surface or heterogeneous erosion; and (b) bulk or homogeneous erosion. Although in the first case only the outer parts of the matrix are affected by erosion, in the second case, the phenomenon also affects the polymeric bulk phase. Homogeneous erosion is usually due to a rapid swelling fluid uptake by the matrix system and consequent polymeric network degradation due to chemical reasons as discussed earlier. On the contrary, surface erosion can be caused by both "chemical" (in this case the swelling fluid uptake is slower) and "physical" reasons, namely hydrodynamic conditions imposed in the release environment. On the basis of the relative chain cleavage and solvent diffusion rate, a system can be defined as "surface erosion" or "bulk-erosion". In particular, if solvent diffusion is much slower than polymer degradation, we are dealing with a surface erosion system. Surface erosion, also referred to as polymer dissolution in the case of physically crosslinked matrices, can be further subdivided into two categories on the basis of the amorphous or semicrystalline nature of the polymer. Indeed, though in the case of amorphous polymers only chain disentanglement is necessary for erosion, in the case of semicrystalline polymers, crystal unfolding precedes chains disentanglement, thus resulting in slower erosion kinetics [80].

The glassy–rubbery transition enormously increases polymer chain mobility, so that network meshes enlarge and the drug can dissolve and diffuse through the developing gel layer. Although, in general, the dissolution of microcrystals (also referred to as macrocrystals in the following) does not show particularly interesting aspects, dissolution of nanocrystals and amorphous drug exhibits a peculiar behavior as discussed in Chapter 5. Indeed, as solubility depends on crystal size [60,81], both of them, due to their small dimension (amorphous state can be considered as a crystal of vanishing dimensions), are characterized by a higher solubility in aqueous medium with respect to the infinitely large crystal, here identifiable with microcrystal (for organic drugs, significant solubility improvements due to reduced dimensions should take place for crystals of radius smaller than 10–20 nm). Unfortunately, however, nanocrystals and amorphous state do not represent a thermodynamically stable condition and in the presence of the incoming solvent, they tend to merge to form the original macrocrystal. Accordingly, upon dissolution, nanocrystals and amorphous state tend to come back to the macrocrystal condition and yet the dissolved drug undergoes a recrystallization with unavoidable solubility reduction [82]. Dissolved drug recrystallization can take place inside the matrix and in the release environment, and this phenomenon can be characterized by different recrystallization constants [83]. Despite this situation, the average (in time) solubility and thus drug bioavailability are neatly increased, and this explains the use of the term "activated drug" for matrix systems containing nanocrystals or amorphous state [18]. Interestingly, provided that a proper polymer choice is made, in the dry, shrunken state, nanocrystals and amorphous state can be fully stabilized by the polymeric network that hinders the recombination of the drug molecules into macrocrystals. This action is mainly due to both the polymer–drug interactions and the physical presence of the polymeric chains. Indeed, drug macrocrystals can form on the condition that the dry network meshes are sufficiently wide [83]. It is worth mentioning that solubility variation of a drug upon dissolution can also take place because of a metastable crystalline form transformation into a more stable crystalline one. Typically, the metastable and the more stable forms are characterized by different solubilities as it happens for the anhydrous/hydrated transformation of theophylline [84], p-hydroxybenzoic acid, phenobarbital [85], and for the polymorphic transformation of sulfathiazole and methylprednisolone [86].

Of course, drug diffusion through the swelling network depends on matrix topology and on the relative size of network meshes and drug molecules, and all these reflect on the value of the drug diffusion coefficient as discussed in Chapter 4. Nevertheless, also physical and chemical interactions of polymer–drug can play an important role as nicely documented by Singh and coworkers [87,88]. Drug adsorption–desorption phenomena can take place on polymer chains due to, for example, electrostatic effects as it happens for charged polypeptide and antibiotics in collagen matrices [89]. Additionally, drug–polymer interactions can be expected to take place in several other cases [90] and this aspect becomes particularly important in gel molecular imprinting technology [91,92]. Indeed, this technology makes possible the recognition of a template molecule by a polymeric network on the basis of the chemical or physical interactions between the template molecule itself and particular arrangements of chemical groups fixed on the polymeric network. In other words, the polymeric network is equipped with specific sites able to interact with the template molecule according to an adsorption–desorption mechanism. Remembering that, for biological applications, noncovalent interactions with template molecules are preferred to get an easy binding/nonbinding process, the above-mentioned adsorption–desorption mechanism becomes very important. In addition, as imprinted polymeric networks will be important components of self-programmed delivery systems and biosensors, it highlights the importance of the adsorption–desorption mechanism in the designing of future delivery systems [3,93]. For example, hydroxyethylmethacrylate(HEMA)-based polymers were imprinted to recognize hydrocortisone [94], methacrylic acid (MAA) was used as a functional monomer for the controlled release of propanolol (β-blocker), tetracycline, and sulfasalazine (prodrug) [95] and other imprinted polymers have been profitably employed as biosensors in the detection of nandrolone [96], okadaic acid [97], caffeine [98], paracetamol [99], and tetracycline hydrochloride [100].

Drug diffusion can also be heavily affected by a particular aspect of matrix topology. Indeed, complex network topology, resulting from a high internal disorder degree, can confer a fractal character to matrix network [101]. If wide network meshes are defined as accessible sites for the diffusing drug and small network meshes (beside that, obviously, polymeric chains) as forbidden sites, the entire network can be seen as a percolative network. If forbidden sites approach a threshold value [102–104], the percolative network becomes a fractal structure. It can be demonstrated that diffusion in percolative (fractal) networks differs a lot from diffusion in nonfractal networks and release kinetics [101–105].

Drug distribution in the matrix (i.e., drug concentration profile) can heavily affect drug release kinetics. Lee [106–108] demonstrated that very different release kinetics can be achieved by properly selecting uniform, sigmoidal, steps, or parabolic drug distribution.

As said before, the above-discussed release mechanisms regarding nonporous polymeric matrices can be particularized for the other matrices considered in Section 7.2.1. Indeed, for example, in the case of hydrophobic matrices realized by melt extrusion of a mixture containing a low-melting binder, the drug and hydrophilic excipients, the key factors ruling drug release are only drug dissolution and diffusion in the liquid-filled matrix pores once a rapid dissolution of hydrophilic excipients is assumed. In the case of porous inorganic materials, conversely, release kinetics is essentially ruled by drug/pore surface interactions and drug diffusion through liquid-filled pores if we assume that mesoporous matrix can be efficiently made wet by the release environment fluid. In superporous polymeric matrices, the dominant mechanism ruling release kinetics is represented by drug dissolution in liquid-filled macropores whereas drug diffusion through pores should exert a minor role due to the width of the pores.

Although release kinetics from (hydrophilic) tablets depends on the same ingredients as discussed for nonporous polymeric matrices, in this case the scenario is more complicated. Indeed, the presence of excipients (such as inert fillers), the high drug–polymer ratio, and tablet compression process considerably affect tablet porosity that obviously differs a lot from the intrinsic porous character of the dry or swollen polymeric structure. In addition, tablet wettability is strongly affected by all the above-mentioned aspects. Accordingly, a detailed microscopic description of the phenomena involved in tablet swelling would lead to a very complex and uncertain scenario. As a consequence, also in the light of a possible mathematical modeling, a simpler and clearer view of the entire release mechanism needs a macroscopic point of view rather than a microscopic one. When hydrophilic tablets interact with aqueous media (water, buffers, physiological fluids, and so on), both polymer hydration and dissolution of soluble components take place [109]. The dissolution fluid penetrates into the tablet according to the local porosity and polymer physical properties. As soon as the local solvent concentration exceeds a threshold value, the glassy–rubbery transition occurs, thus, the polymeric chains start unfolding and a gel layer surrounding the tablet begins to appear [69,70]. Meanwhile, the amount of drug at the tablet surface dissolves giving rise to a burst effect in the release profile of the system, more or less pronounced depending on the drug solubility and the polymer hydration rate [110]. The glassy–rubbery transition enormously increases polymer chain mobility, so that drug molecules can diffuse throughout. Macroscopically, polymer hydration and drug dissolution/diffusion give rise to the formation of three moving fronts (Figure 7.7): the swelling front, the erosion front, and the diffusion front. The erosion front separates the release environment from the matrix (it moves outward when swelling kinetics is predominant in the erosion process, whereas it moves inward in the opposite case) and its position depends on the combination of release environment hydrodynamic conditions and matrix crosslinking strength. In other words, matrix erosion is a function of the tensions applied by the release

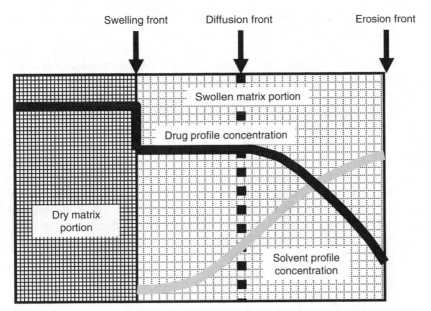

FIGURE 7.7 Polymer hydration and drug dissolution/diffusion give origin to the formation of three moving fronts: the swelling front, the erosion front, and the diffusion front.

environment and network connectivity. The swelling front, separating the dry glassy core from the swollen matrix portion, moves inward and its speed depends on polymer/solvent characteristics (namely, on the viscoelastic properties of the polymer/solvent couple) and on matrix porosity. Additionally, in the case of sparingly soluble drugs, a third front, the diffusion front, can appear between the outer portion of the swollen matrix, where the drug is completely dissolved, and the inner part, where the drug is not yet dissolved despite the rubbery state of the polymer [111,112]. The release kinetics working after the initial burst is heavily ruled by the gel layer thickness, which in turn depends on the relative position of the erosion front and the swelling front.

7.3 MATHEMATICAL MODELING

Once the physical aspects of a given phenomenon have been properly understood, mathematical modeling becomes possible. Obviously, there is no biunivocal correspondence between the phenomenon and the relative mathematical model. Indeed, the final model mathematical form heavily depends on the different hypotheses and assumptions on which it relies. Accordingly, the number of models associated with a given phenomenon depends only on the fantasy of researchers. At the same time, a model able to describe all the aspects of a given phenomenon does not exist but its existence, in the majority of the situations, is not needed. Indeed, greater the model generality, more difficult are its expression and practical applicability. In addition, it is important to remember that despite many aspects that constitute the phenomenon, their relative importance is not always the same. In other words, although in some cases polymer swelling plays the most important role, in other cases, erosion or drug dissolution can be the key aspect in determining release kinetics behavior. Usually, closer to reality the model is, the more complex it is. This is the reason why the so-called empirical models are represented by simple equations as their aim is just to describe the macroscopic behavior of a phenomenon regardless of its microscopic aspects. Nevertheless, they can be very useful, at least when the study of a phenomenon is made for the first time or when a qualitative comparison among different data series is needed as it often happens in laboratory practice. A typical example of an empirical model describing release kinetics from matrices is the Power law equation [113]:

$$\frac{M_t}{M_\infty} = Kt^n, \tag{7.1}$$

where M_t is the amount of drug released until time t, M_∞ is the amount of drug released after an infinite time, K is a constant, and n is the exponent characterizing the release process. If Fickian diffusion takes place, n is equal to 0.5, 0.45, and 0.43 for a thin film, a cylinder, and a sphere, respectively [114]. When n exceeds these thresholds, non-Fickian release takes place. Indeed, Equation 7.1 is an approximation of the solution of Fick's law when the ratio M_t/M_∞ ranges between 0.2 and 0.6. Equation 7.1 has to be used with great care for the classification of the release nature on the basis of n values. Indeed, it is well known that, despite Fickian diffusion in the matrix, boundary conditions such as the presence of a stagnant layer or a net faced on the release surface give rise to a release kinetics characterized by n greater than 0.5 as also discussed in Chapter 4 [77,115,116]. Peppas and Sahlin [117] couple the Fickian diffusional contribution with the non-Fickian one in a linear manner:

$$\frac{M_t}{M_\infty} = k_1 t^n + k_2 t^{2n}, \tag{7.2}$$

where k_1 and k_2 are constants related, respectively, to the Fickian and non-Fickian diffusional contribute. Grassi and colleagues [118] studying theophylline release from different kinds of hydroxypropylmethylcellulose tablets (HPMC; Methocel E5 and K100 M Premium), propose a semiempirical model based on the fact that release kinetics is, in this case, substantially determined by drug dissolution and drug diffusion through the gel layer surrounding the dry tablet core. The physical frame on which the model relies is similar to that proposed by Higuchi et al. [119] for the dissolution of a solid

$t = 0$

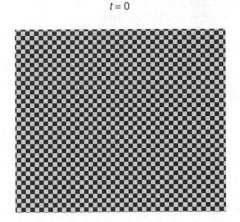

Phase 1 (solid drug) + Phase 2 (dry polymer)

$t > 0$

Gel layer thickness

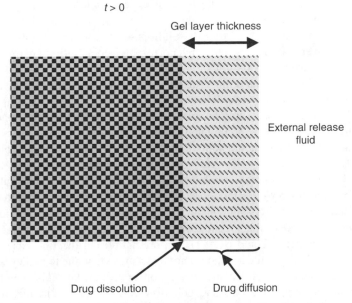

External release
fluid

Drug dissolution   Drug diffusion

FIGURE 7.8 At the beginning ($t = 0$), the system can be viewed as a uniform dispersion of two noninteracting solid phases. Upon contact with the external release fluid, drug (phase 1) starts dissolving, while polymer (phase 2) swells giving origin to a gel layer reducing drug delivery in the external release fluid.

biphasic system (see Figure 7.8). Accordingly, the tablet is seen as a uniform dispersion of two noninteracting solid phases whose dissolution kinetics depends on the respective dissolution constant and surface composition (the dissolution area available for each phase depends on the respective mass fraction). Though phase 1 coincides with the solid drug, phase 2 is the polymer. Accordingly, upon contact with the external release environment fluid, phase 1 (drug) starts dissolving whereas phase 2 (polymer) starts swelling. Consequently, gel layer formation slows down drug delivery and release kinetics depends on the sum of two resistances in series, namely drug dissolution and drug diffusion through the time-dependent gel layer thickness. Gel layer thickness is the result of two counteracting phenomena: polymer swelling, causing an increase in gel thickness, and polymer erosion, determining a decrease in gel thickness. As a consequence, the time evolution of the gel layer thickness, proportional to gel resistance R to drug delivery, shows a peak connecting the initial increase with the final decrease. Obviously, the peak may be more or less pronounced and broad and may appear in the early stage or after a very long time so that sometimes (highly crosslinked or entangled gels) it never appears and a final plateau is encountered. Mathematically speaking, any thing that describes the above-discussed gel thickness time evolution could be used to account for the gel layer resistance R. For the sake of simplicity and in the light of experimental data, two particular limiting situations are examined. In the first case, gel layer formation is assumed to be slower than drug dissolution so that, at the beginning (time zero), gel layer resistance is zero (this corresponds to an infinite permeability value of gel layer with its thickness vanishing) and the only drug delivery resistance coincides with solid drug dissolution. In the second case, on the contrary, gel layer formation is assumed to be very fast (ideally, instantaneous) in comparison with drug dissolution so that gel layer thickness reaches a maximum value at the beginning and, then, it decreases due to erosion. These two limiting situations should apply to the case of gels made of high-molecular weight polymers (first case) and low molecular weight polymers (second case). Indeed, it seems reasonable that the formation of a gel constituted by a lower molecular weight polymer is more rapid than the formation of a chemically equal gel constituted by a higher molecular weight polymer. In addition, although in the second case (low polymer molecular weight) the erosion phenomena should exert an important role, in the first case (high-molecular weight polymer) it should not be so important due to higher content of entanglements present in the gel. In order to translate this scenario in mathematical terms, the classical equation describing the dissolution of a solid drug in sink conditions is considered [114]

$$\frac{dC_r}{dt} = -\frac{k_d A}{V_r} C_s, \qquad (7.3)$$

where t is time, C_r and C_s are drug concentration and solubility in the release environment, respectively, A is the release surface area, V_r is the release environment volume, and k_d is the dissolution constant. Now, the effect of the gel layer resistance R and the tablet composition at the swelling front have to be properly embodied in Equation 7.3. Therefore, bearing in mind that the global resistance to diffusion of a multilayered membrane is given by the sum of the resistance of each layer [120] and $1/k_d$ can be thought of as the drug dissolution resistance, Equation 7.3 can be rewritten as

$$\frac{dC_r}{dt} = -\frac{x_d A}{V_r} \frac{C_s}{\left(\frac{f}{k_d} + R\right)}, \tag{7.4}$$

where x_d is the drug mass fraction at the swelling front and f is a parameter accounting for the fact that, due to gel presence, the drug intrinsic dissolution constant k_d will be lower than that relative to pure solid drug dissolution [22]. x_d accounts for the fact that drug release also depends on the effective drug dissolution area at the swelling front. Assuming that R is mainly affected by gel layer thickness the first (slow gel layer formation; high-molecular weight polymer) and second cases (fast gel layer formation; low molecular weight polymer) can be represented, respectively, by the following equations [121–123]:

$$R = B(1 - \exp^{-bt}), \tag{7.5}$$

$$R = Be^{-bt}, \tag{7.6}$$

where B and b are two model parameters to be determined by data fitting (obviously, gel permeability P is equal to $1/R$). Although Equation 7.5 implies an increase in gel resistance starting from zero up to a final plateau value ($R(t=\infty)=B$), Equation 7.6 implies an instantaneous build up of gel resistance ($R(t=0)=B$) followed by an exponential reduction due to erosion.

To complete the model, the time variation of the release area A has to be estimated. For this purpose A is identified with the geometrical area competing the diffusion front (in doing so it was assumed that the crucial point for drug delivery is the area at the diffusion front rather than tablet geometrical area, coinciding with the area of the erosion front) [124]. Moreover, for the sake of simplicity, it assumed that the ratio K between the height h and the radius R_a of the cylinder delimited by the diffusion front does not change with time and remains equal to that of the nonswollen tablet. Accordingly, A modifies with time in the following way:

$$A = 2\pi(1 + K)R_a^2(t). \tag{7.7}$$

In other words, the problem of the time dependence A is shifted on the R_a, which can be determined by means of the following mass balance:

$$M_0 = V_r C_r(t) + \pi K R_a^3(t)C_0, \tag{7.8}$$

where M_0 and C_0 are, respectively, the initial drug amount and concentration in the tablet. In writing Equation 7.8, it is implicitly assumed that the drug amount contained in the tablet portion delimited by the erosion and diffusion front is negligible. The rearrangement of Equation 7.8 leads to

$$R_a = \sqrt[3]{\frac{M_0 - V_r C_r(t)}{\pi K C_0}}. \tag{7.9}$$

Inserting Equation 7.9 into Equation 7.7 yields

$$A = (\alpha - \beta C_r(t))^{2/3}, \tag{7.10}$$

$$\alpha = \frac{(2\pi(1+K))^{3/2} M_0}{\pi K C_0}, \quad \beta = \frac{(2\pi(1+K))^{3/2} V_r}{\pi K C_0}. \tag{7.10'}$$

Now, Equation 7.10 can be embodied into Equation 7.4 to give the final expression of the model differential form, whose solution is

$$C_r^+ = \frac{C_r}{M_0/V_r} = 1 - \left(1 - \frac{2(1+K)x_d}{3M_0^{1/3}} \left(\frac{\pi}{K^2 C_0^2}\right)^{1/3} F(t)\right)^3, \tag{7.11}$$

where, for the first case (slow gel layer formation; high-molecular weight polymer), $F(t)$ is given by

$$F(t) = C_s \left[\frac{t}{B + 1/fk_d} + \frac{\ln(1 + B(1 - e^{-bt})fk_d)}{(B + 1/fk_d)^b}\right], \tag{7.12}$$

whereas it becomes

$$F(t) = fk_d C_s \left[t + \ln\left(\frac{Be^{-bt} + 1/(k_d f)}{B + 1/fk_d}\right)\Big/b\right] \tag{7.13}$$

for the second case (fast gel layer formation; low molecular weight polymer). It is interesting to note that, regardless of $F(t)$ expression, Equation 7.11 shows a flat behavior (its time derivatives vanish, see Equation 7.4, Equation 7.7, and Equation 7.9) approaching $C_r^+ = 1$ condition, as required by the physics of the problem under study. Finally, some considerations are needed about drug solubility C_s. Indeed, it may happen that upon dissolution the drug undergoes a phase transition reflecting into a modification of its solubility [85]. To generalize the model for this purpose, it is sufficient to insert the correct $C_s(t)$ function in Equation 7.4.

Figure 7.9 shows the comparison between model best fitting (solid lines) and the experimental data (symbols) referring to K100 and E5 matrices. As it

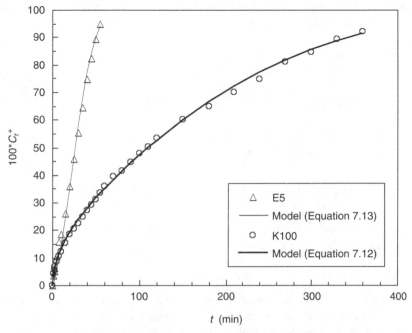

FIGURE 7.9 Model best fitting (solid lines) on theophylline release data from E5 (triangles) and K100 (circles) hydroxypropylmethylcellulose tablets. (Adapted from Moneghini, M., et al., *Drug Dev. Indus. Pharm.*, 2005.)

can be seen, a good data description is achieved only when Equation 7.12 and Equation 7.13 are considered in the model for the description of K100 and E5 data, respectively. Accordingly, in the E5 case it can be supposed to have rapid formation, all around the matrix dry core, of a very weak gel layer more similar to a viscous aqueous polymer solution rather than to a structured (physical) hydrogel. Due to the liquid-like nature of this layer and the superimposed hydrodynamic conditions of the release environment, it becomes increasingly thinner with time due to polymer disentanglement and subsequent migration in the release environment. Although the low E5 molecular weight justifies this interpretation, the high K100 molecular weight supports the idea of stronger gel layer all around the tablet. In addition, this layer should take a longer time to develop and it should be almost insensitive to erosion induced by release environment hydrodynamic conditions. Fitting procedure, carried out by knowing that $M_0 = 75$ mg, $C_0 = 565$ mg/cm^3, $x_d = 0.5$, $K = 0.1538$, $R_a(0) = 0.65$ cm, $C_s = 12.5$ mg/cm^3, and $k_d = 0.188$ cm/min (25°C) [125], yields the following fitting parameters values: (K100) $B = 50 \pm 2$, $b = 0.052 \pm 0.008$, $f = 0.52 \pm 0.1$; (E5) $B = 16 \pm 3$, $b = 0.083 \pm 0.01$, $f = 1$. In the case of partially coated tablets [126], where drug diffusion can take place only through the top planar tablet surface, the model can be adapted to account for this new situation with the advantage that nonsink conditions can also be considered. Accordingly, Equation 7.3 becomes

$$\frac{dC_r}{dt} = -\frac{x_d A}{V_r} \frac{(C_s - C_r)}{\left(\dfrac{f}{k_d} + R\right)}.$$ (7.14)

Remembering that, now, A is constant and considering Equation 7.5 for the R time evolution, model solution is

$$C_r = C_s \left(1 - (\exp^{bt}(1 + Bfk_d(1 - \exp^{-bt})))^{-\frac{Ax_d}{bV_r(B+1/(fk_d))}}\right).$$ (7.15)

Also in this particular case, the model (Equation 7.15) proves to yield good experimental data fitting [126].

7.3.1 DIFFUSION-CONTROLLED SYSTEMS

Among all the factors presented in Section 7.2.2, now, drug release kinetics is supposed to be ruled only by drug diffusion inside the matrix. This means that matrix erosion does not take place (or, at least, it develops very slowly in comparison with the drug release phenomenon), drug dissolution is virtually instantaneous or only dissolved drug is present in the matrix, matrix swelling does not take place or it is very fast or very slow in comparison to the release mechanism, drug–polymer interactions are negligible and drug concentration is uniform throughout the whole matrix. Although some of these systems have been presented and discussed in Chapter 4 in relation to solution of Fick's law, here some particular cases are considered.

Fu and colleagues [127], studying drug release from diffusion-controlled cylindrical matrices, furnish an analytical solution of Fick's law for considering mass transfer in three dimensions:

$$\frac{M_t}{M_\infty} = 1 - \frac{8}{h^2 r^2} \sum_{m=1}^{\infty} \alpha_m^{-2} \exp(-D\alpha_m^2 t)^* \sum_{n=0}^{\infty} \beta_n^{-2} \exp(-D\beta_n^2 t),$$ (7.16)

$$J_0(r\alpha_m) = 0, \quad \beta_n = \frac{(2n+1)\pi}{2h},$$ (7.16')

where M_t and M_∞ are, respectively, the amount of drug released until time t and after an infinite time, h denotes the half-length and r denotes the radius of the cylinder. D is the constant diffusion coefficient, α and β are defined in Equation 7.16', J_0 is a zero-order Bessel function and m and n are integers. Equation 7.16 holds for uniform initial drug distribution in the matrix and sink condition in the release environment.

The problem of drug release from diffusion-controlled cylindrical matrices can also be matched recurring to an approximate numerical solution. This is a way of reducing mass transport in two dimensions (radial and axial) to a one-dimensional problem [128]. Contrary to Equation 7.16, this approximate solution holds for nonsink conditions also. Supposing a constant drug diffusion coefficient D and a negligible gel density variation due to the diffusion process, Fick's second law, in a two-dimensional cylindrical coordinates system, can be written as

$$\frac{\partial C}{\partial t} = \frac{D}{R}\frac{\partial}{\partial R}\left(R\frac{\partial C}{\partial R}\right) + D\frac{\partial^2 C}{\partial Z^2}, \tag{7.17}$$

where t is time, C is the drug concentration (mass/volume) in the cylinder, R and Z are the radial and axial axes, respectively. This equation must satisfy the following initial and boundary conditions.

Initial conditions:

$$C(Z,R) = C_0, \quad -Z_c \le Z \le Z_c, \quad 0 \le R \le R_c, \tag{7.18}$$

$$C_r = 0. \tag{7.19}$$

Boundary conditions:

$$C(Z, R_c, t) = C(\pm Z_c, R, t) = k_p \, C_r(t), \tag{7.20}$$

$$\left.\frac{\partial C}{\partial R}\right|_{R=0} = 0, \quad -Z_c < Z < Z_c, \tag{7.21}$$

$$V_r C_r(t) = \pi R_c^2 2 Z_c C_0 - 2\int_0^{Z_c}\int_0^{R_c} C(Z,R,t)\,2\pi R\,\mathrm{d}R\,\mathrm{d}Z), \tag{7.22}$$

where $2Z_c$ and R_c are, respectively, the cylinder height and radius, C_0 is the initial drug concentration in the cylinder, C_r and V_r are the drug concentration and the volume of the release medium, and k_p is the drug partition coefficient between the cylindrical gel and the environmental release fluid. Equation 7.18 and Equation 7.19 state that the release environment is initially drug free, whereas the cylinder is uniformly loaded by a drug concentration C_0. Equation 7.20 expresses the partitioning condition at the cylinder/release fluid interface, Equation 7.21 ensures that no mass transport in radial direction takes place on the cylinder axis whereas Equation 7.22 is a drug-mass balance for the cylinder/release fluid system, which allows to state the relation between C_r and $C(Z, R, t)$. The equation set comprising of Equation 7.17 through Equation 7.21 can be numerically solved by means of the control volume method [129] even if fitting this equation to experimental data can be very time-consuming (on a Pentium IV 1.7 GHz personal computer, 225 (15×15 grid) control volume elements represents the upper limit for data fitting). The strategy adopted to perform two-dimensional to one-dimensional transformation is based on the idea of finding the correction factor to be applied to the one-dimensional solution to make it coincide with the two-dimensional one, provided the same boundary and initial conditions are considered. In order to optimize the numerical solution procedure (control volume method [129]) of the one-dimensional problem, the cylinder is sub-divided into cylindrical shells (computational elements), characterized by the same ratio K between the height ($2Z$) and the corresponding radius (R) (Figure 7.10). Consequently, the volume of the generic cylindrical shell is given by

$$dV = d(2Z\pi R^2) = d(K\pi R^3) = 3\pi K R^2 dR, \quad K = \frac{2Z_c}{R_c} = \frac{2Z}{R}. \quad (7.23)$$

Supposing a uniform drug concentration inside the cylindrical shell, Fick's second law reads as

$$-dV\frac{\partial C}{\partial t} = 2\pi(R + dR)2(Z + dZ)N_{R+dR} - 4\pi RZ N_R$$
$$+ 2\pi(R + dR)^2 N_{Z+dZ} - 2\pi R^2 N_Z, \quad (7.24)$$

where N_R and N_{R+dR} are, respectively, the radial matter fluxes evaluated in R and $R + dR$, whereas N_Z and N_{Z+dZ} are, respectively, the axial matter fluxes evaluated in Z and $Z + dZ$ (note that the factor 2 in the axial balance is related to the fact that drug diffusion takes place through both the two cylindrical shell bases). Recalling that

$$N_Z = -\frac{\partial C}{\partial Z}\Big|_Z = -\frac{2}{K}\frac{\partial C}{\partial R}\Big|_R = \frac{2}{K}N_R, \quad N_{Z+dZ} = -\frac{\partial C}{\partial Z}\Big|_{Z+dZ}$$
$$= -\frac{2}{K}\frac{\partial C}{\partial R}\Big|_{R+dR} = \frac{2}{K}N_{R+dR}, \quad (7.25)$$

the final expression of Fick's second law on this particular kind of computational elements is

$$\frac{\partial C}{\partial t} = \frac{2}{3}\left(\frac{2 + K^2}{K^2}\right)\frac{1}{R^2}\frac{\partial}{\partial R}\left(R^2 D \frac{\partial C}{\partial R}\right). \quad (7.26)$$

Of course, the above-described mathematical treatment, makes the solution of Equation 7.26 different from the solution of Equation 7.17 as, for instance, when $K = 2$, Equation 7.26 reduces to the equation describing the one-dimensional

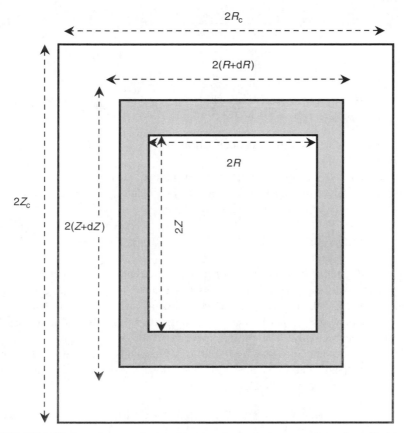

FIGURE 7.10 In order to optimize the numerical solution procedure, the cylindrical matrix is subdivided into cylindrical shells (computational elements), characterized by the same ratio K between the height $(2Z)$ and the corresponding radius (R). $2Z_c$ and R_c represent, respectively, matrix height and radius. (From Coviello, T., et al., *Biomaterials*, 24, 2789, 2003. With permission from Elsevier.)

radial diffusion in a sphere [76]. The problem is now to find out a correction factor making the two solutions coincide, provided that Equation 7.26 satisfies the initial and boundary conditions expressed by Equation 7.18 through Equation 7.19 and Equation 7.20 through Equation 7.22, respectively. It is possible to numerically verify that the solutions of Equation 7.17 and Equation 7.26 are approximately equal if a proper correcting factor $f(K)$ is introduced in Equation 7.26. The f dependence on K can be determined considering the dimensionless expression of Equation 7.17 and Equation 7.26:

$$\frac{\partial C^+}{\partial t^+} = \frac{1}{R^+}\frac{\partial}{\partial R^+}\left(R^+\frac{\partial C^+}{\partial R^+}\right) + \frac{\partial^2 C^+}{\partial (Z^+)^2}, \tag{7.17'}$$

$$\frac{\partial C^+}{\partial t^+} = f(K)\frac{2}{3}\left(\frac{2+K^2}{K^2}\right)\frac{1}{(R^+)^2}\frac{\partial}{\partial R^+}\left((R^+)^2\frac{\partial C^+}{\partial R^+}\right), \qquad (7.26')$$

where

$$t^+ = tD/R_c^2, \quad C^+ = C\frac{V_r + V_g k_p}{V_r C_{r0} + V_g C_0}, \quad R^+ = R/R_c, \quad Z^+ = Z/R_c, \quad (7.27)$$

where C_{r0} is the initial drug concentration in the release medium and V_g the cylinder volume.

Solutions of Equation 7.17′ through Equation 7.26′ for a discrete set of different K values ($0.5 < K < 16$) practically coincide provided that the correcting factor is given by the following equation:

$$f(K) = 0.797(1 - e^{-(0.677(K+0.993))}) \quad (0.5 < K < 16). \qquad (7.28)$$

For $K \geq 6$, f becomes constant and equal to 0.8, whereas, for lower K values, it progressively decreases. This analysis is limited to the $0.5 < K < 16$ range because for $K < 0.5$ and $K > 16$, the two-dimensional diffusion problem can be approximated to a one-dimensional totally axial or totally radial diffusion problem, respectively. The most important result of this approach is that Equation 7.26 and Equation 7.17 solutions practically coincide if, all other parameters and conditions are the same, in Equation 7.26 the diffusion coefficient considered D_{1D} is given by

$$D_{1D} = f(K)D_{2D} \quad (0.5 < K < 16), \qquad (7.28')$$

where D_{2D} is the true diffusion coefficient of the drug in the cylinder. Accordingly, once data fitting is performed by means of Equation 7.26 and D_{1D} has been determined, the calculation of D_{2D} is straightforward. Interestingly, a different choice of the computational shells shape would reflect only on the analytical expression of the correcting factor f as the constant factor multiplying the term

$$\frac{1}{(R^+)^2}\frac{\partial}{\partial R^+}\left((R^+)^2\frac{\partial C^+}{\partial R^+}\right)$$

in Equation 7.26′ would be different.

Figure 7.11 shows the rather good agreement between the Equation 7.26 best fitting and experimental data ($37°C$; distilled water pH $= 5.4$) referring to theophylline release from a cylindrical scleroglucan hydrogel cross-linked by borax (moles of borax $=$ moles of repeating unit of scleroglucan; polymer concentration in the final gel $= 0.7\%$ (w/v), theophylline concentration in the gel $C_0 = 4486$ μg/cm^3) [128]. The gel (height $= 1.0$ cm,

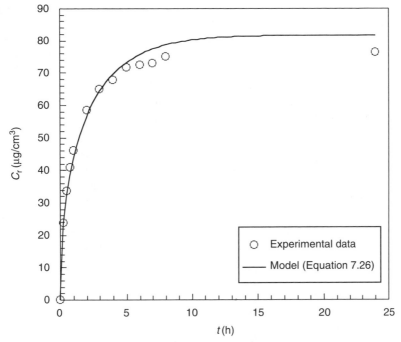

FIGURE 7.11 Comparison between model best fitting (Equation 6.26—solid line) and experimental data (theophylline release from cylindrical scleroglucan hydrogel crosslinked by borax. (From Coviello, T., et al., *Biomaterials*, 24, 2789, 2003. With permission from Elsevier.)

diameter $= 2.2$ cm), suspended in a thin web, is put in the release environment ($V_r = 200$ cm^3) and drug concentration is measured by means of a spectrophotometer (271 nm). Assuming $k_p = 1$, the resulting diffusion coefficient is $D = (7.4 \pm 2.1) \times 10^{-6}$ cm^2/s which is, correctly, lower than that in pure water ($D_w = (8.2 \pm 0.6) \times 10^{-6}$ cm^2/s [125].

Although the simplified approach now discussed can be applied only in the case of constant drug diffusivity inside the matrix, the equation set comprised of Equation 7.17 through Equation 7.22 can also be used to study drug release from inhomogeneous cylindrical matrices provided that the proper drug diffusion coefficient dependence on spatial coordinates is known. Indeed, it is well known [30] that many hydrogel matrices (especially those constituted of polymers), at microscopic level, are inhomogeneous. They show clusters with variable but high crosslink densities, dispersed in a medium at low crosslink densities. Consequently, the drug diffusion coefficient will be strongly position dependent as (see Chapter 4) higher the crosslink density, lower the drug diffusion coefficient. For this purpose, Coviello and colleagues [130] show an interesting case of myoglobin release from inhomogeneous cylindrical hydrogels made up of carboxylated scleroglucan (Sclerox) crosslinked by Ca^{++}. Figure 7.12 shows myoglobin (van der Waals radius $= 21$ Å) release in distilled water (200 cm^3) at 37°C from hydrogels (height $= 2$ cm; radius $= 0.6$) characterized by different polymer concentration c_p, (2%, 3%, and 4% w/v) and a drug concentration equal to 12821 $\mu g/cm^3$. It is clear that release kinetics is strongly dependent on polymer concentration and that whatever be the c_p, a clear two-phase behavior is detectable as a rapid concentration increase is followed by slow release kinetics. In principle, this behavior could also be attributed to a drug–polymer interaction or to the inhomogeneous gel nature induced by the preparation technique. Indeed, polymer crosslinking takes place because of the presence of divalent calcium ions crossing the dialysis membrane, which separates polymer solution from the divalent ions solution. Accordingly, hydrogel macromolecules closer to the two dialysis membranes should be more crosslinked than those in the inner part. Nevertheless, the authors show that intrinsic hydrogel inhomogeneity is the real cause of this behavior. In particular, they assume the existence of two different values of the diffusion coefficient: lower in the highly

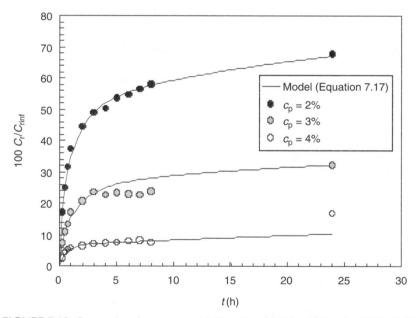

FIGURE 7.12 Comparison between model (Equation 7.17) best fitting (solid line) and myoglobin release data from inhomogeneous cylindrical hydrogels made up of carboxylated scleroglucan (Sclerox) crosslinked by Ca^{++}. Three different polymer concentrations (c_p) are considered (2% w/v, filled circles, 3% w/v gray circles, 4% open w/v circles). (From Coviello, T., et al., *Biomaterials*, 24, 2789, 2003. With permission from Elsevier.)

TABLE 7.1
Model Best Fitting Parameters Relative to Myoglobin Release from Sclerox Hydrogels

c_p (w/v %)	$F(2,11)$	D_h (cm^2/s)	D_l (cm^2/s)	V_{ci}/V_g (%)
2	3056	1×10^{-5}	1×10^{-7}	46
3	886	4.5×10^{-6}	8.0×10^{-9}	73
4	248	8.0×10^{-6}	3.0×10^{-9}	87

Source: From Coviello, T., et al., *J. Contr. Rel.*, 102, 643, 2005. With permission from Elsevier.
Note: *F* statistic expresses fitting goodness.

crosslinked polymeric clusters (D_l) and higher in the embedding, low cross-linked, surrounding medium (D_h). Due to the practical difficulty of knowing the exact conformation and distribution of the clusters in the gel, for the sake of simplicity, they assume that such clusters are all together joined to form an inner homogenous cylinder coated by a layer with a higher drug diffusion coefficient. Data fitting reveals that though myoglobin diffusion coefficient (D_h) in the low crosslinked gel zone (external coating) is not heavily affected by c_p (see Table 7.1), the diffusion coefficient D_l competing to the highly crosslinked gel zone (the inner volume, given by the sum of all the clusters whose volume, is V_{ci}) undergoes an abrupt reduction passing from $c_p = 2\%$ to $c_p = 3\%$ and 4%. At the same time, the volume of the highly crosslinked clusters increases passing from 46% to 87% of the whole gel volume V_g. As a consequence, the reduction of the myoglobin release kinetics is essentially due to the increase of cluster domains volume and crosslink density. Interestingly, the release of a much smaller solute as theophylline (van der Waals radius = 3.7 Å) is not significantly influenced by the polymer concentration increase.

7.3.2 DIFFUSION AND MATRIX/DRUG INTERACTIONS CONTROLLED SYSTEMS

The first example of a mathematical model dealing with drug/matrix interactions is represented by a modification [131] of the Higuchi equation [132]. Then, Crank [76] proposes to treat the adsoption–desorption process as a first-order reversible reaction. Subsequently, Muhr and Blanshard [133] assume that drug flux J can be given by the sum of two terms:

$$J = -D_b \nabla C_b - D_f \nabla C_f, \tag{7.29}$$

where D_b, D_f and C_b, C_f are, respectively, the diffusion coefficient and the concentration of the bound and free drug, whereas ∇ is the nabla operator.

Once the relationship between C_b, C_f and the values of D_b, D_f is known, it is possible to predict the measured drug diffusion coefficient D defined by

$$J = -D\nabla(C_b + C_f). \tag{7.30}$$

If D_b can be set at zero (more than reasonable assumption) and C_b can be related to C_f by the following relation (holding on condition of a very rapid adsorption–desorption kinetics in comparison to pure diffusion):

$$C_b = \alpha(C_b + C_f), \tag{7.31}$$

where α is position and concentration independent, the entire process can be modeled by the following equation:

$$J = -(1 - \alpha)D_f \nabla(C_b + C_f). \tag{7.32}$$

All these approaches are based on the hypothesis that the equilibrium between free and bound drug is reached instantaneously. If this approximation holds for low drug concentration in the matrix, it becomes less plausible when most of the binding sites are occupied [87].

Singh and coworkers [87] deviate from this approximation assuming that the binding interaction between the drug and the polymeric chains can be described by a Langmuir isotherm. The amount M_b of drug bound per unit mass of polymer is ruled by the following equation:

$$C_p \frac{\partial M_b}{\partial t} = r_a - r_d, \tag{7.33}$$

where C_p is the polymer concentration in the matrix and t is time. r_a and r_d are the adsorption and desorption rates, respectively, and are defined as follows:

$$r_a = k_a C_f (M_b^{\max} - M_b) C_p, \tag{7.34}$$

$$r_d = k_d M_b C_p, \tag{7.35}$$

where k_a and k_d are the adsorption and desorption constants, respectively, M_b represents the mass of drug bound per unit mass of polymer, $M_b^{\max}$ is the polymer binding capacity (maximum value of M_b), C_f is the free drug concentration inside the matrix. Obviously, $C_b = C_p \times M_b$. To account also for the case of poorly soluble drug in the liquid eventually pervading the matrix, Grassi and colleagues [134], generalize Equation 7.35 to get

$$r_d = k_d^* M_b C_p (C_s - C_f), \tag{7.35'}$$

where C_s is free drug solubility in the matrix pervading fluid. It is worth mentioning that Equation 7.34 states that adsorption rate vanishes when no free drug is present ($C_f = 0$) or when the binding sites are all occupied ($M_b = M_b^{max}$). Analogously, Equation 7.35' states that desorption rate vanishes when no bound drug is present ($M_b = 0$) or when saturation is reached in the matrix pervading fluid ($C_f = C_s$). Thus, the mass-balance equation describing the time and spatial free drug concentration C_f in the matrix is given by

$$\frac{\partial C_f}{\partial t} = D_f \nabla^2 C_f - C_p \frac{\partial M_b}{\partial t}. \tag{7.36}$$

In writing Equation 7.36 it has been implicitly assumed that D_f is constant and this implies that drug molecules dimension is very small in comparison with matrix meshes. If it were not the case, a position/concentration-dependent D_f should be considered. To complete the model expression, boundary and initial conditions have to be set up (one-dimensional problem).
Initial conditions:

$$C_b + C_f = C_0, \tag{7.37}$$

$$C_r = 0. \tag{7.38}$$

Boundary conditions:

$$\frac{\partial C_f}{\partial X}\Big|_{X=0} = 0, \tag{7.39}$$

$$M_0 = C_0 V_m = S\left(\int_0^L (C_f + C_b)dX\right) + V_r C_r, \tag{7.40}$$

where X is the abscissa, C_0 is the initial drug concentration in the matrix, V_m, S, and L are, respectively, matrix volume, cross section, and length, whereas V_r is the release environment volume. If Equation 7.37 establishes a uniform initial drug concentration, Equation 7.38 sets the initial drug concentration to zero in the release environment. Although Equation 7.39 implies that the free drug flux in $X = 0$ is equal to zero (obviously, it is assumed that bound drug cannot diffuse), Equation 7.40 is an overall mass balance needed to establish a connection between C_r and C_f ($X = L$). Assuming that, at the beginning, an equilibrium condition holds between C_b and C_f, it is possible to evaluate $C_b(t = 0) = C_{b0}$ and $C_f(t = 0) = C_{f0}$. Accordingly, equating Equation 7.34 to Equation 7.35' yields

$$C_{f0} = \frac{-b \pm \sqrt{b^2 - 4ac}}{2a}, \tag{7.41}$$

$$a = (k_a - k_d^*)(V_r + V_m), \tag{7.42}$$

$$b = (V_r + V_m)k_d^*C_s + V_m k_a C_{bmax} - M_0(k_a - k_d^*), \qquad (7.42')$$

$$c = M_0 k_d^* C_s, \qquad (7.42'')$$

$$C_{b0} = C_{bmax} \frac{k_a C_{f0}}{k_d^* C_s + C_{f0}(k_a - k_d^*)}, \qquad (7.43)$$

where, obviously, only the positive root of Equation 7.41 has to be considered.

On the basis of this model, the implicit expression for the concentration-dependent matrix/release environment partition coefficient k_p is

$$k_p = \frac{C}{C_r} = \frac{C_b + C_f}{C_r}. \qquad (7.44)$$

It is also interesting to note that k_p becomes concentration independent when the adsorption–desorption kinetics is much faster than the diffusion process (equilibrium is always attained between the bound and free drug molecules within the gel), free drug concentration C_f is always far from drug solubility C_s in fluid pervading the matrix and C_b is low in comparison with C_{bmax}. Indeed, from Equation 7.34 and Equation 7.35', it follows that

$$r_a = k_a C_f(C_{bmax} - C_b) \approx k_a C_f C_{bmax}, \quad C_b \ll C_{bmax}, \qquad (7.45)$$

$$r_d = k_d^* C_b(C_s - C_f) \approx k_d^* C_b C_s, \quad C_f \ll C_s. \qquad (7.46)$$

Assuming that the adsorption–desorption phenomenon is always at equilibrium, r_a is equal to r_d and thus

$$\frac{C_f}{C_b} \approx \frac{k_d^* C_s}{k_a C_{bmax}}. \qquad (7.47)$$

Remembering the expression of the partition coefficient k_p (see Equation 7.44) and that, at equilibrium, C_f has to be equal to C_r, it descends that

$$k_p = 1 + \frac{C_b}{C_r} = 1 + \frac{C_b}{C_f} \approx 1 + \frac{C_{bmax} k_a}{k_d^* C_s} = \text{constant}. \qquad (7.48)$$

Finally, it is interesting underlying that if the adsorption–desorption process is very fast in comparison with diffusion, C_f and C_b are in equilibrium and are related by the following reaction:

$$C_b = \frac{k_a C_{bmax} C_f}{k_a C_s + C_f(k_a - k_d^*)}. \qquad (7.49)$$

The insertion of Equation 7.49 into Equation 7.36 and rearranging, yields

$$\frac{\partial C_f}{\partial t} = D_e \nabla^2 C_f,$$ (7.50′)

$$D_e = D \frac{\left[C_s k_d + C_f (k_a - k_d^*) \right]^2}{k_d^* k_a C_{bmax} C_s + \left[C_s k_d^* + C_f (k_a - k_d^*) \right]^2}.$$ (7.50′)

This means that the entire process is now described by a Fickian law where an effective concentration-dependent diffusion coefficient D_e has to be considered.

Defining the following dimensionless parameters:

$$R = \frac{C_{d0}}{C_s}$$ (7.51)

$$C^+ = \frac{C}{C_\infty}, \quad C_\infty = \frac{M_0}{V_r + V_m (1 + C_{bmax}/C_s)} \quad t^+ = \frac{tD}{L^2}$$ (7.52)

and assuming

$$K^* = \frac{k_a}{k_d^*} = 1, \quad \varphi^2 = \frac{k_d^* L^2}{D} = 0.5, \quad \beta = \frac{C_{bmax}}{C_{d0}} = 10,$$ (7.53)

it is possible to evaluate some model characteristics. In the case of drug permeation through a drug–polymer interacting membrane, Figure 7.13 clearly reveals that, as drug solubility decreases (R increases), the desorption rate r_{des} decreases and, thus, drug diffusion is lowered. It is interesting to note that the original Singh model works only in the limit $R \approx 0$.

Singh and coworkers [88] apply the above-mentioned model (in the limit $R = 0$) to the case of poly-L-lysine (PLL) (positively charged polypeptide) release from succinylated (SC) or phosphonylated (PC) purified acid collagen matrices. The reasons for choosing PLL consist in its simple structure only positively charged residues, its hydrophilic nature (consequently, binding interactions are substantially mediated by electrostatic forces), its easy labeling, and its small size (15–30 kDa), this preventing hindered diffusion in the collagen matrix (final collagen concentration in the matrix ranges between 22 and 26 mg/cm^3). M_b^{max} and $K (= k_a/k_d)$ are calculated by applying to experimental equilibrium data the Langmuir adsorption isotherm, descending from the equilibrium condition $r_a = r_d$:

$$C_b = \frac{K C_{bmax} C_f}{1 + K C_f}.$$ (7.54)

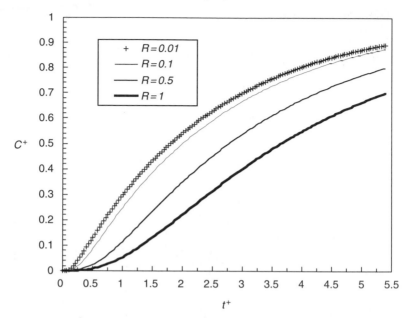

FIGURE 7.13 Model prediction (Equation 7.36) about drug permeation through a drug–polymer interacting membrane for different values of the parameter R. (From Grassi, M., Lapasin, R., Carli, F., *Proceedings of the 27th International Symposium on Controlled Release of Bioactive Materials*, Controlled Release Society, 7-9-2000. With permission from Controlled Release Society.)

Electron spin resonance (ESR) is used to study the labeled PLL (probe) and collagen binding at equilibrium (adsorption isotherm measurements) in PBS (phosphate buffered saline: 20 mM sodium phosphate, 130 mM sodium chloride, pH adjusted to 7.4). Experimental data evidence that though no interactions take place between the probe and the native collagen ($M_b^{max} = 0$ μg/mg; $K = k_a/k_d = 0$ cm^3/mg), they are clearly evident in the case of SC ($M_b^{max} = 60$ μg/mg; $K = 0.5$ cm^3/mg, $s_g = 51\%$; $M_b^{max} = 180$ μg/mg; $K = 0.6$ cm^3/mg, $s_g = 83\%$. $s_g =$ lysine groups on collagen modified during derivatization) and PC ($M_b^{max} = 240$ μg/mg; $K = 1.1$ cm^3/mg, $s_g = 22\%$). Indeed, though native collagen has no net charge under physiological conditions, SC and PC are characterized by a net negative charge. In particular, if in the case of SC, binding capacity increases with degree of succinylation, a stronger interaction is seen with PC due to its higher valence (-2 in HPO$_3^{2-}$ PC groups; -1 in SC groups). In addition, experimental data reveal that a reduction of NaCl concentration in PBS (pH = 7.4) reflects in an increase of the binding strength. Indeed, the lower the NaCl concentration, the smaller the shield effect exerted by salt ions. In particular, in the case of SC ($s_g = 83\%$), M_b^{max} and K assume, respectively, 0, 180, 380 μg/mg and

0, 0.6, 1.0 cm^3/mg when NaCl is equal to 500, 130, 0 mM. Finally, kinetic experiments reveal that, in the case of SC ($s_g = 83\%$), $k_a = 0.04$ and 0.04 cm^3/(mg s), $k_d = 0.06$ and 0.08/s in PBS and phosphate buffer (NaCl not present), respectively. Interestingly, on the basis of equilibrium and kinetic parameters, the model is able to provide a good data prediction assuming that the adsorption–desorption process is very fast in comparison with the diffusion processs. In other words, in this particular case, only equilibrium parameters (M_b^{max}, K) are really important for data prediction. The same group [89] also studied the release of gentamycin, a hydrophilic antibiotic having 3–4 primary amine groups present per molecule that ionize in solution, from SC and PC matrices. Again, the hypothesis of an instantaneous adsorption-desorption holds.

Kuijpers and coworkers [135], matching the problem of prosthetic valve endocarditis caused by the adherence of bacteria to the valve or to tissue at the site of implantation, studied *in situ* release of antibacterial proteins. In particular, lysozyme (a small, 14.4 kDa, cationic antibacterial protein) was chosen as model drug whereas crosslinked gelatin constituted the matrix (gelatin A extracted from porcine skin and processed by acidic pretreatment; gelatin B extracted from bovine skin and processed by alkaline pretreatment). The authors found that higher the amount of free carboxylic acids in the matrix (gelatin B), lower the lysozyme release kinetics due to the existing electrostatic interactions. In this case also, the hypothesis of an instantaneous adsorption–desorption process holds.

Brasseur and colleagues [136] studied the interaction between hemato-porphyrin (HP) and polyalkylcyanoacrylate (polyisobutylcyanoacrylate (PIBCA) and polyisohexylcyanoacrylate (PIHCA)) nanoparticles in the frame of photodynamic therapy. Indeed, porphyrins are macromolecules with inherent photosensitizing properties that are highly taken up and retained by tumor tissues. Local illumination of malignant tumors with red light in a few days following intravenous administration of porphyrins results in the production of singlet oxygen with damage to biomolecules and subsequent tumor cells death. The authors observed that both in the case of PIBCA and PIHCA, the absorption efficiency of HP on nanoparticles decreases drastically as the pH of the incubation medium increased from 5.9 to 7.4. This evidence is explained observing that the degree of ionization of drug heavily affects the adsorption–desorption process and the best condition for adsorption occurs when drug is in its nonionized form. The use of Langmuir isotherm (Equation 7.54) reveals that $M_b^{max} = 30.8$ μg/mg and $K = 0.134$ cm^3/mg in the case of PIBCA at pH 6.2.

Kwon and coworkers [137] found that an important electrostatic interaction takes place between positively charged polypeptides or proteins and albumin–heparin or albumin microspheres prepared as ion exchange gels. In particular, human lysozyme (positively charged protein) was chosen as model drug whereas heparin (negatively charged polyelectrolyte) was incorporated into albumin microspheres (heparin concentration $= 10\%$ w/w) to increase the electrostatic interaction with the positively charged lysozyme. Adsorption isotherms were carried out at 25°C in PBS (28 mM potassium phosphate and 39 mM disodium phosphate) pH 7.0. Although the authors interpreted adsorption isotherms by means of the Freundlich equation:

$$C_b = K_f C_f^{1/n}, \tag{7.55}$$

where K_f is the adsorption constant (ratio between the adsorption and desorption constants) and n is a model parameter, data interpretation according to the Langmuir approach (Equation 7.54) can also be performed. Statistically speaking, data fitting according to the Freundlich or Langmuir equation is equal. In the presence of heparin (albumin–heparin microspheres) fitting parameters are $K_f = 350$ mg/cm^3, $1/n = 0.72$ and $M_b^{max} = 370$ μg/mg and $K = 5.2$ cm^3/mg for the Freundlich and Langmuir approaches, respectively. On the contrary, when heparin is lacking in the microspheres, $K_f = 30$ mg/cm^3, $1/n = 0.24$ and $M_b^{max} = 60$ μg/mg and $K = 7.8$ cm^3/mg for the Freundlich and Langmuir approaches, respectively. These results clearly show that the presence of heparin improves the electrostatic interactions with the positively charged lysozyme a lot. Interestingly, neither the Freundlich nor the Langmuir approach is able to satisfactorily describe adsorption isotherms of chicken egg lysozyme on albumin–heparin microspheres.

On a completely different drug side, Frenning and Strømme [138] observed binding interactions between NaCl and agglomerated micronized cellulose (microcrystalline cellulose Avicel PH 101, FMC, USA). Indeed, they experimentally noted that NaCl release from microcrystalline tablets is characterized by a delay time that cannot be attributed to matrix swelling or tablet disintegration time due to its large value. For this purpose, they built a mathematical model conceptually equivalent to that described by Singh and colleagues (Equation 7.36) apart from the presence of the dissolutive term, the drug diffusion coefficient dependence on concentration and a somewhat different expression of r_a and r_b:

$$\frac{\partial C_f}{\partial t} = \nabla(D_f \nabla(C_f)) - \frac{\partial C_b}{\partial t} + K_t(C_s - C_f), \tag{7.56}$$

where K_t is drug dissolution constant, C_s is drug solubility in the liquid pervading the matrix and $K_t (C_s - C_f)$ is the dissolution term that is discussed in detail in the next section (Section 7.3.4). This model relies on the following hypotheses: (1) the tablet contains a large number of drug crystals, of approximately the same size and shape, dispersed into an insoluble matrix; (2) in contact with water the tablet breaks up into a number N of approximately spherical tablet fragments; (3) liquid absorption and tablet disintegration are much faster than drug dissolution. Accordingly, the initial state is character-ized by virtually complete liquid absorption, matrix swelling, and disintegra-tion but negligible drug dissolution. (4) The surrounding liquid is well mixed, so that its concentration is independent of the space coordinates. The authors assume the following D_f concentration dependence:

$$D_f = D_{f0}\left[(1 - \varepsilon)e^{(-bC_f)} + \varepsilon\right], \tag{7.57}$$

where D_{f0}, ε, and b are model parameters. In addition, they assume that the adsorption–desorption process is very fast (C_f and C_b are always in equili-brium) and can be represented by the following Langmuir adsorption isotherm:

$$C_b = \frac{C_{bmax} (KC_f)^\delta}{1 + (KC_f)^\delta}, \tag{7.58}$$

where δ is the heterogeneity parameter, a measure of the width of the distribution of adsorption energies. It assumes values between zero (wide distribution) and unity (narrow distribution). It is not worthwhile mentioning that Equation 7.58 implies the following expression for r_a and r_d:

$$r_a = k_aC_f(C_b^{max} - C_b)^{(1/\delta)} \tag{7.59}$$

$$r_d = k_d(C_b)^{(1/\delta)}. \tag{7.60}$$

On the basis of Equation 7.58, the explicit expression of the adsorption–desoprtion term in Equation 7.56 ($\partial C_b/\partial t$) is

$$\frac{\partial C_b}{\partial t} = KC_{bmax}\delta \frac{(KC_f)^{\delta-1}}{(1 + (KC_f)^\delta)^2} \frac{\partial C_f}{\partial t}. \tag{7.61}$$

Accordingly, the final model expression is

$$\left(1 + KC_{bmax}\delta\frac{(KC_f)^{\delta-1}}{(1+(KC_f)^\delta)^2}\right) \frac{\partial C_f}{\partial t} = \nabla(D_f(C_f)\nabla(C_f)) + K_t(C_s - C_f). \tag{7.62}$$

This equation is numerically solved in a one-dimensional (radial) spherical coordinates system as the tablet is assumed to immediately disrupt into N_p, all identical, swollen spherical particles (total swollen volume $= 0.8$ cm^3; polymer concentration in the swollen particles $C_p = 285$ mg/cm^3). Additionally, it is supposed that, at the beginning, the drug is present in the particles in a solid, undissolved, form and its distribution concentration is uniform (122 mg/cm^3). Two boundary conditions are considered:

$$\left.\frac{\partial C_f}{\partial r}\right|_{r=0} = 0, \tag{7.63}$$

$$C_f(t, R_p) = k_p C_r, \tag{7.64}$$

where r is the radial coordinate, R_p is particle radius, k_p the partition coefficient and C_r is drug concentration in the release environment. Although Equation 7.63 expresses an obvious symmetry condition in the particle center, Equation 7.64 imposes the partitioning condition at the particle–release environment interface. Remembering that $C_r = m_r / V_r$ (V_r ($= 9.35$ cm^3) is the release environment volume whereas m_r is the drug amount dissolved in it), Equation 7.64 becomes

$$\frac{dC_f(t, R_p)}{dt} = \frac{k_p}{V_r} \frac{dm_r(t)}{dt} = -\frac{k_p N_p 4\pi R_p^2}{V_r} \frac{\partial C_f(t, R_p)}{\partial r}. \tag{7.65}$$

Model best fitting provides a very good data description and model parameters, for what concerns the adsorption–desorption process, are $\delta \approx 1$, $M_b^{max} = 39$ μg/mg, $K = 16.4$ cm^3/mg, and $k_p = 0.25$. M_b^{max} and K values lead to the conclusion that the amount of NaCl bound on the polymeric chain is not very large but the adsorption process is much more pronounced than the desorption one. Accordingly, NaCl–microcrystalline cellulose interaction cannot be neglected in the release kinetics.

It is useful to end this section remembering that the above-discussed adsorption–desorption mechanism can be also applied to study molecularly imprinted polymers (see Section 7.2.2), basis of the self-regulated delivery system, the future of drug delivery field. Indeed, imprinted polymer–drug binding affinity can be quantified by means of the receptor–template dissociation constant K_d, that can be defined starting from the typical macromolecular–template complex $[MT]$ reaction [95]:

$$[M] + [T] \underset{k_r}{\overset{k_f}{\rightleftharpoons}} [MT], \tag{7.66}$$

where $[M]$ is the macromolecular binding sites concentration, $[T]$ is the template concentration whereas k_f and k_r represent the forward and reverse rate constants, respectively. The equilibrium binding constant or association constant, K_a, and the equilibrium dissociation constant, K_d, are expressed by

$$K_a = \frac{k_f}{k_r} = \frac{1}{K_d} = \frac{[MT]}{[M][T]}. \tag{7.67}$$

Equation 7.67 simply descends from the equilibrium condition requiring that the forward reaction rate R_f:

$$R_f = k_f [M][T] \tag{7.68}$$

equates the reverse one R_r:

$$R_f = k_r[MT]. \qquad (7.69)$$

Finally, Kim and colleagues [139], working on L- and D-Fmocs of tryptophan enantiomers on Fmoc-L-Trp imprinted polymer, found that experimental isotherm data can be conveniently described by bi- and tri-Langmuir adsorption isotherms (namely, the sum of two and three, Equation 7.54, terms each one characterized by different K and C_{bmax}).

7.3.3 DIFFUSION AND CHEMICAL REACTION CONTROLLED SYSTEMS

A particular kind of matrix is represented by pendant chain systems where the drug is bound to polymer backbone through covalent bonds [6]. Drug is gradually detached by means of bonds' hydrolytic or enzymatic cleavage and the resulting release kinetics depends on both cleavage rate and drug diffusion through the polymeric network.

The synthesis of pendant group systems can be performed according to two different strategies [6]. The first one implies active agent conversion to a polymerizable derivative that subsequently polymerizes to afford the macromolecular combination. In the second one, on the contrary, bioactive agent is chemically bound to preform synthetic or natural polymer backbone with the addition of proper spacers preventing from reaction failure due to steric hindrance problems. Ester, amide, orthoester, urethane, anhydride, and carbonate are typical hydrolyzable or biodegradable bonds employed in pendent chain systems [140]. Copolymers of N-(2-hydroxypropyl) methacrylamide (HPMA) are probably the most developed pendant group systems as anticancer agents that have already entered early clinical trials. Indeed, it is important to underline that, despite their versatility, these systems are rarely marketed as they are considered as new chemical entities and the path needed to get FDA approval is arduous and expensive.

Mathematical modeling of pendant chain systems release kinetics is not a trivial task due to the complexity of the phenomenon. Indeed, drug release from these systems requires external medium diffusion in the matrix, drug–polymer bond cleavage, drug diffusion through the polymeric network and spreading into external medium [141]. Accordingly, in the drug mass balance equation, an additional term $(\partial C_b/\partial t)$, accounting for the chemical reaction (enzymatic or hydrolytic), must be embodied:

$$\frac{\partial C_f}{\partial t} = \nabla(D_f \nabla(C_f)) - \frac{\partial C_b}{\partial t}, \qquad (7.70)$$

where C_f and C_b represent, as usual, free and bound drug concentration, respectively, t is time and D_f is free drug diffusion coefficient. The problem consists of the explicit expression of $\partial C_b/\partial t$ whose analytical form depends on the particular chemical reaction scheme. In addition, for both the cases of enzymatic or hydrolytic reaction, the availability of enzymes and water (external medium) must be accounted for by proper mass balance equations. As a consequence, it is not possible to derive a general model accounting for both diffusion and chemical reaction, but each particular case has to be matched independently. For example, in the case of hydrolytic cleavage, assuming a water-soluble drug, a homogeneous degradation, that diffusion process is very fast in comparison with chemical reaction, which can be represented by the following scheme:

$$P - Dr + \text{cleavage agent} \xrightarrow{k} P + Dr, \qquad (7.71)$$

where P is the polymer backbone and Dr is the drug, the release rate R_D depends only on chemical cleavage and its expression is

$$R_D = kC_0 e^{-kt}, \qquad (7.72)$$

where k is the hydrolysis constant rate and C_0 is the initial drug concentration. More complex situations (heterogeneous degradation) are considered by Tani and coworkers who study the case of matrices made up of both hydrophobic and hydrophilic parts [142]. These authors, assuming that drug diffusion is not the rate-determining step, distinguish three different situations based on the relative importance of hydrolysis rate R_h and water penetration rate R_w. If $R_h \gg R_w$, the hydrolysis occurs at the interface and zero-order release occurs until drug depletion:

$$R_D = R_w C_0. \qquad (7.73)$$

When $R_h \approx R_w$, assuming R_h as constant and independent of drug concentration, R_D increases linearly until surface drug concentration becomes 0:

$$R_D = R_h R_w\, t. \qquad (7.74)$$

Then, it becomes constant as in the case $R_h \gg R_w$. In particular, if R_h is first order, then

$$R_D = R_w\, C_0(1 - e^{-kt}). \qquad (7.75)$$

Finally, if $R_h \ll R_w$, Equation 7.72 is again found. To have a simpler match between model and experimental data, it is convenient to transform Equation 7.72 through Equation 7.75 in terms of M_t/M_∞ where M_t is the amount of drug released until time t and M_∞ is the drug released after an infinite long time. In this context, remembering that, dimensionally, R_D is (mass/(volume×time)) and that M_∞ can be approximated by $V_0 \times C_0$, V_0 is the initial matrix volume, it is given as follows:

$$M_t = \int_0^t V_0 R_D \, dt. \qquad (7.76)$$

Accordingly, M_t/M_∞ becomes

$$\frac{M_t}{M_\infty} = R_w t, \quad R_h \gg R_w, \qquad (7.73')$$

$$\frac{M_t}{M_\infty} = R_w \left(t - \frac{1 - e^{-kt}}{k} \right), \quad R_h \approx R_w, \quad R_h \text{ first order}, \qquad (7.75')$$

$$\frac{M_t}{M_\infty} = (1 - e^{-kt}), \quad R_h \ll R_w. \qquad (7.72')$$

Interestingly, Equation 7.75' degenerates into Equation 7.73' for very large k values. Additionally, R_w can be calculated as the ratio of the squared water front speed v_w^2 and the water diffusion D_w coefficient in the matrix. An interesting example of a more sophisticated model dealing with diffusion and reaction has been recently proposed by Abdekhodaie and Wu [143].

Shah and colleagues [144] provide an interesting validation of the above-discussed model studying the release of a series of substituted benzoic acids covalently linked with 2-hydroxyethyl methacrylate (HEMA). In particular, they focus the attention on p-nitrobenzoic acid, benzoic acid, p-methoxy benzoic acid, p-amino benzoic acid, and 2,4-dinitrobenzoic acid. The preparation of the various 2-methacryolyl ethyl esters (monomeric drug derivative) is followed by the bulk polymerization carried out in a test tube filled with HEMA, monomeric drug derivative (ratio 10:1 w/w) and 0.8% t-butyl hydroperoxide as initiator. The reaction is carried out at 60°C for 4 h and then at 80°C for 12 h. The cylindrical matrix (thickness of 1 mm) is isolated by breaking the tube. Release tests are performed in alkaline medium. In the case of p-nitrobenzoic acid, release data are well described assuming that $R_h \gg R_w$, this means that M_t/M_∞ is represented by Equation 7.73'. This result is absolutely reasonable remembering that the presence of a nitro group considerably increases the rate of hydrolysis so that the $R_h \gg R_w$ hypothesis sounds good. Interestingly, the fitting value of R_w (1.9×10^{-5}/s) is very close to that calculated (2.8×10^{-5}/s) resorting to the measured values of v_w (2.2×10^{-6} cm/s) and D_w (1.7×10^{-7} cm^2/s). When benzoic acid is considered, the $R_h \approx R_w$ hypothesis holds and M_t/M_∞ is given by Equation 7.75'. Knowing that $R_w = v_w^2/D_w = 1.8 \times 10^{-5}$/s, data fitting yields $k = 7.6 \times 10^{-5}$/s. In this case, the k/R_w ratio is approximately equal to 4. Due to electron donating nature of their substituents, the hydrolysis rate of p-amino benzoic acid and p-methoxy benzoic acid has to be much smaller than the rate of benzoic acid. Consequently, they fall in the $R_h \ll R_w$ case and Equation 7.72′ applies. Data fitting yields $k = 1.4 \times 10^{-7}/\text{s}$ and $k = 6.2 \times 10^{-7}/\text{s}$ for p-amino benzoic acid and p-methoxy benzoic acid, respectively. The comparison of these results with R_w ($= v_w^2 / D_w$) data ($1.9 \times 10^{-5}/\text{s}$ for both of them) reveals that the k/R_w ratio is approximately equal to 7.4×10^{-3} and 32×10^{-3} for p-amino benzoic acid and p-methoxy benzoic acid, respectively. This means that, now, R_w is, respectively, 135 and 30 times k. Interestingly, 2,4-dinitrobenzoic acid does not follow the Tani model as diffusion is the release rate-determining step. The same situation also occurs in the tributyltin carboxylate release from epoxy resin based pendant chain matrices [145]. In both the latter cases, Equation 7.70 must be used to understand experimental data.

7.3.4 DIFFUSION-AND DISSOLUTION-CONTROLLED SYSTEMS

In these matrices both drug diffusion and dissolution concur in determining release kinetics characteristics, whereas all other aspects are apparently less important. Apart from inorganic matrices, usually, organic matrices are composed of an insoluble skeleton (such as stearic acid) hosting the drug and some hydrophilic excipients (such as lactose). Upon contact with the aqueous release environment, hydrophilic excipients dissolve, conferring a changing porous topology to the matrix. In turn, the drug dissolves in the liquid-filled channels made free by hydrophilic excipients whose dissolution is, generally, faster than that of the drug. Then, the drug diffuses throughout the matrix channels (without crossing their wall) to get the release environment. The phenomenon complexity inferred by the changing porosity makes any attempt of a detailed matrix description in terms of Euclidean geometry very hard. Accordingly, an effective way to proceed is to assume an instantaneous hydrophilic compounds dissolution in the outer matrix layer, a negligible matrix porosity variation and to define a mean effective drug diffusion coefficient D_e accounting for matrix topology (porosity, ε, and tortuosity, τ) characterizing drug molecules motion in the cylindrical matrix according to the following equation:

$$D_e = \frac{D_w \varepsilon}{\tau}, \tag{7.77}$$

where D_w is the drug diffusion coefficient in the liquid filling the pores [146]. It is sometimes desirable [147] to incorporate into this expression a partition coefficient, K_p, for possible solute adsorption on pore walls and a restriction coefficient, K_r, accounting for hindered diffusion, and defined by

$$K_r = (1 - \lambda)^2, \quad \lambda = \frac{r_d}{r_p}, \tag{7.78}$$

where r_p and r_d are, respectively, pore and drug radii. Accordingly, Equation 7.77 becomes

$$D_e = D_w \frac{\varepsilon}{\tau} K_p K_r. \tag{7.79}$$

To account for the noninstantaneous hydrophilic excipients dissolution in the matrix inner part, two different D_e values can be distinguished: D_{ee} (higher) related to the matrix surface and the just-underneath-the surface thin layer (thickness h), D_{ei} (lower) related to the inner part of the matrix. Indeed, the lower D_{ei} value accounts for the later dissolution of the inner part excipients. Additionally, of course, it is supposed that matrix density does not vary upon drug release. Despite the presence of a stagnant layer surrounding the cylinder and its theoretical importance (as discussed in Chapter 4), it is assumed that its resistance is negligible in comparison with the other resistances (namely, drug diffusion and dissolution). On these basis, and on the assumption that no mass diffusion takes place in the angular direction due to matrix cylindrical symmetry, the whole drug release process can be schematically represented by the following two-dimensional (radial and axial) equation:

$$\frac{\partial C}{\partial t} = \frac{1}{R} \frac{\partial}{\partial R} \left[R D_e \frac{\partial C}{\partial R} \right] + \frac{\partial}{\partial Z} \left[D_e \frac{\partial C}{\partial Z} \right] - \frac{\partial C_d}{\partial t}, \tag{7.80}$$

$$\frac{\partial C_d}{\partial t} = -K_t(C_s - C), \tag{7.81}$$

where C and C_d are, respectively, the concentrations of the dissolved and undissolved drug fractions, K_t is the dissolution constant and C_s is the drug solubility in the release fluid. Equation 7.80 illustrates the differential drug mass balance, whereas Equation 7.81 illustrates the drug dissolution process occurring upon drug contact with the release fluid [148–153]. Equation 7.80 can be numerically solved (control volume method [129]) with the following initial conditions:

$$C_r = 0, \tag{7.82}$$

$$C(R, Z) = 0, \quad 0 < R < R_c, \quad -Z_c < Z < Z_c, \tag{7.83}$$

$$C_d(R, Z) = C_{d0}, \quad 0 < R < R_c, \quad -Z_c < Z < Z_c \tag{7.84}$$

and boundary conditions:

$$\left. \frac{\partial C}{\partial R} \right|_{R=0} = 0, \quad -Z_c < Z < Z_c, \tag{7.85}$$

$$C_r = \frac{C}{k_p} \text{ cylinder surface,} \tag{7.86}$$

$$M_0 = V_r C_r + N_c \int_{-Z_c}^{Z_c} \int_0^{R_c} [C(R,Z) + C_d(R,Z)] 2\pi R \, dR \, dZ, \tag{7.87}$$

where C_{d0} is the initial undissolved drug concentration, C_r is the drug concentration in the release environment, k_p is the drug partition coefficient, and N_c is the number of identical cylinders considered in the release experiments. It is important to note that, regardless the number of identical cylinders considered (1, 2, 3, ...), Equation 7.87 can be rewritten assuming that $N_c = 1$ and dividing V_r for the real number of cylinders used. Equation 7.82 through Equation 7.87 express the same conditions already discussed for Equation 7.18 through Equation 7.22 apart from Equation 7.84 that states the uniformity of the undissolved drug concentration in the cylinder at the beginning and Equation 7.87 where, now, the amount of undissolved drug must be considered in the overall mass balance. Figure 7.14 schematically represents the cylindrical delivery system status (a) at the beginning of the release experiments and (b) for $t > 0$. Although in the beginning all the drug is in the undissolved condition (filled circles) at the initial concentration C_{d0}, later on, in the outer matrix region only dissolved drug exists (blank circles) and only in the inner part we find incompletely dissolved drug at the original concentration (blank circles) or at lower concentration (gray circles: undissolved drug $C_d < C_{d0}$).

This theoretical approach is successfully used to study monohydrate theophylline release from ram extruded cylindrical matrices made up of stearic acid (low-melting binder), monohydrated lactose, and PEG 6000 (hydrophilic excipients) [48]. To verify the effect of components on drug release rate, different compositions are examined (Table 7.2). In particular, the percentage range for each component is 10%–75% for monohydrate theophylline, 25%–90% for stearic acid, 0%–15% for monohydrate lactose, and 0%–15% for PEG 6000. The dissolution test is performed according to the USP 24 method I. A dissolution rotating basket apparatus SOTAX (AT 7 Smart, Basel, CH) is employed with a stirring rate of 100 rpm and maintained at $37 \pm 0.1°C$. Samples of extrudates are dissolved in 900 mL of dissolution medium (freshly demineralized water; sink conditions always attained). The aqueous solution is filtered and continuously pumped to a flow cell in spectrophotometer Perkin–Elmer (Mod. Lambda 35, Monza, Italy) and absorbance values are recorded at the maximum wavelength of the drug (271 nm). PEG 6000, stearic acid, and lactose did not interfere with the UV analysis. Although D_{ee}, D_{ei}, K_t, and h are real model fitting parameters, in the attempt to greatly reduce the computational time required for data fitting (this model requires a numerical solution) h and K_t are set according to the

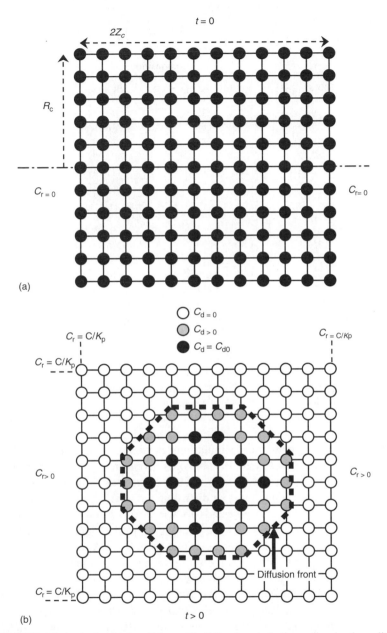

FIGURE 7.14 At the beginning ((a), $t = 0$) all the drug is in the undissolved condition (filled circles) at the initial uniform concentration C_{d0}. Upon contact with the release environment ((b), $t > 0$), outer matrix region depletes (only dissolved drug exists (blank circles)) and only in the inner part not completely dissolved drug at the original concentration (blank circles) or at lower concentration (gray circles: undissolved drug $C_d < C_{d0}$ can be seen. (From Grassi, M., et al., *J. Contr. Rel.*, 92, 275, 2003. With permission from Elsevier.)

TABLE 7.2
Extruded Matrices Composition

N°	Composition			
	$100^* \omega_{THEO}$	$100^* \omega_{SACID}$	$100^* \omega_{LACT}$	$100^* \omega_{PEG}$
1	10.05	69.19	10.38	10.38
2	29.09	54.55	8.18	8.18
3	48.13	39.40	5.98	5.98
4	67.17	25.25	3.79	3.79
5	52.46	25.05	11.24	11.24
6	38.37	45.19	8.22	8.22
7	24.28	65.32	5.20	1.04
8	10.18	85.45	2.18	2.18
9	37.81	54.01	0.08	8.10
10	35.94	51.34	5.03	7.70
11	34.06	48.66	9.97	7.30
12	32.19	45.99	14.92	6.90
13	37.81	54.01	8.10	0.08
14	35.94	51.34	7.7	5.03
15	34.06	48.66	7.30	9.97
16	32.19	45.99	6.90	14.92
17	35.00	50.00	7.50	7.50

Source: From Grassi, M., et al., *J. Contr. Rel.*, 92, 275, 2003. With permission from Elsevier.
Note: ω_{THEO} is the theophylline mass fraction; ω_{SACID} is the stearic acid mass fraction; ω_{LACT} is the monohydrated lactose mass fraction; and ω_{PEG} is the polyethylenglycol mass fraction.

following strategy. Numerical simulations carried out assuming test 17 conditions (this is the pivot system; Table 7.2) reveal that the predicted release curve remarkably changes when at least 9% of the total drug contained in the cylinder belongs to the higher drug diffusion coefficient (D_{ee}) zone. It is easy to demonstrate that this corresponds to $h = 58$ μm. Similarly, $h = 115$ μm corresponds to 18% of the total drug belonging to the higher drug diffusion coefficient (D_{ee}) zone. Moreover, it is found that when K_t exceeds 0.1 cm/s, the rate determining step is not dissolution but diffusion, and further K_t increases do not reflect in substantial drug release curve modifications. On the contrary, when $K_t < 0.1$ cm/s data fitting yields higher D_{ei} and D_{ee} values but data description is not completely satisfactory (this is evaluated on the basis of the F test, see Chapter 1). Accordingly, data fitting is performed by fixing discrete h values (0, 58, 115 μm, and so on), setting $K_t = 0.1$ cm/s, and letting D_{ei} and D_{ee} free to be modified. Figure 7.15, reporting the time variation of C_r^+ (ratio between C_r and its value after a very long time $C_{r\infty}$), shows a good agreement between model best fitting (solid line) and experimental data (symbols) relative to some of the systems considered. The fitting

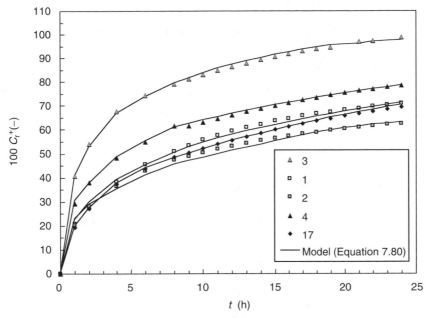

FIGURE 7.15 Comparison between model best fitting (Equation 6.80; solid line) and experimental data referring to theophylline release from system 1 (open squares), 2 (gray squares), 3 (gray triangles) 4 (filled triangles), and 17 (filled diamonds). In this case the attention is focussed on the effect of theophylline composition variation on release kinetics. (From Grassi, M., et al., *J. Contr. Rel.*, 92, 275, 2003. With permission from Elsevier.)

is performed with the knowledge that release volume $V_r = 900$ cm^3, cylinder height $2Z_c = 0.5$ cm, cylinder radius $R_c = 0.15$ cm, $M_0 = 26.4$ mg (this corresponds to $N = 2$ cylinders) and fixing $K_t = 0.1$ cm/s. In the case of the pivot system (17), this fitting yields $D_{ee} = 2 \times 10^{-6}$ cm^2/s, $D_{ei} = 4.2 \times 10^{-7}$ cm^2/s and $h = 58$ μm (see Table 7.3 for other systems results). It is interesting to note that, despite the small h value, data fitting would not be satisfactory had the existence of the external layer of thickness (characterized by a much higher diffusion coefficient) been neglected. In particular, the model would not be able to properly fit both the initial (0–4 h) and the final parts of the release curve since it could not describe simultaneously the initial burst effect and the following part characterized by a slower release kinetics. This particular kind of release kinetics can be explained because in the outer part of the matrix, drug release is mainly ruled by drug dissolution, and drug diffusion plays a less important effect, due to the thin diffusion layer. In the inner part, in contrast, drug diffusion is the limiting step. As a consequence, drug release is considerably slowed down. The presence of theophylline on the cylinder surface is confirmed by an XPS analysis. On the basis of the calculated effective diffusion coefficient (D_{ei}, D_{ee}) and the theophylline diffusion

TABLE 7.3
Theophylline Amount (M_0), Fitting Parameters (K_t, = 0.1 cm/s D_{ei}, D_{ee}, h), Tortuosity (τ_i, τ_e), Initial (ε_1) and Final (ε_2) Porosity of the 17 Systems Studied

N°	M_0 (mg)	D_{ei} (cm²/s)	D_{ee} (cm²/s)	H (µm)	ε_1 (−)	ε_2 (−)	τ_i (−)	τ_e (−)
1	24.6	3.40×10^{-8}	1.40×10^{-7}	115	0.16	0.24	49.1	11.9
2	23.0	1.25×10^{-7}	1.10×10^{-6}	115	0.13	0.38	16.8	1.9
3	24.8	1.90×10^{-6}	1.00×10^{-5}	115	0.10	0.53	1.4	0.3
4	26.5	1.80×10^{-7}	7.00×10^{-6}	115	0.07	0.68	17.1	0.4
5	22.1	6.00×10^{-6}	1.00×10^{-5}	115	0.20	0.69	0.6	0.4
6	16.8	9.00×10^{-7}	2.50×10^{-6}	58	0.14	0.47	2.8	1.0
7	19.0	8.50×10^{-8}	6.50×10^{-7}	58	0.05	0.25	14.5	1.9
8	23.8	8.00×10^{-9}	5.00×10^{-8}	115	0.03	0.11	65.2	11.7
9	27.7	4.00×10^{-7}	2.00×10^{-6}	58	0.07	0.39	4.7	0.9
10	27.0	5.30×10^{-7}	1.25×10^{-6}	115	0.11	0.41	4.0	1.7
11	26.6	1.75×10^{-7}	1.50×10^{-6}	115	0.14	0.43	13.5	1.6
12	25.1	2.80×10^{-7}	1.60×10^{-6}	115	0.18	0.46	9.3	1.6
13	27.0	2.00×10^{-7}	2.20×10^{-6}	58	0.06	0.38	9.0	0.8
14	27.3	3.10×10^{-7}	1.60×10^{-6}	58	0.10	0.41	6.8	1.3
15	25.7	4.00×10^{-7}	1.50×10^{-6}	115	0.15	0.44	6.0	1.6
16	23.5	3.60×10^{-7}	2.00×10^{-6}	115	0.19	0.46	7.4	0.4
17	26.4	4.20×10^{-7}	2.00×10^{-6}	58	0.12	0.42	5.3	1.1

Source: From Grassi, M., et al., *J. Contr. Rel.*, 92, 275, 2003.

coefficient D_w in the dissolution medium ($T = 37°C$; 8.2×10^{-6} cm²/s), matrices tortuosity τ can be computed according to Equation 7.79 assuming that $K_p = K_r = 1$ (see Table 7.3). Notably, of course, as matrix porosity ε varies with time because of the hydrophilic excipients (lactose, PEG) and theophylline dissolution, two distinct ε values are considered. Accordingly, ε is estimated at the beginning (ε_1) of the release process (theophylline is not dissolved and excipients are considered totally dissolved) and at the end (ε_2), when the matrix consists of stearic acid alone. Hence, external (τ_e) and internal (τ_i) tortuosities are evaluated averaging the values corresponding to ε_1 and ε_2 (Table 7.3). It is interesting to note that though tortuosity is not so important (from the release point of view) for what concerns the external layer (τ_e), it is much more important in the inner matrix space.

The use of this model also helps in gaining some information about the effect of matrix composition on release kinetics. Indeed, the inspection of D_{ee} and D_{ei} values coming from data fitting (see Table 7.3) evidences that though release kinetics is essentially slowed down by the increase of stearic acid in the matrix, the increase of theophylline and PEG determines an improvement in the release kinetics. Lactose, on the contrary, does not seem to affect the release process remarkably.

Finally, it is worth mentioning that mathematical modeling can also provide information that would not be easy to observe from an experimental viewpoint. Indeed, it is possible to theoretically calculate the diffusion front position and velocity as time passes. Assuming that diffusion front position (whose resolution depends on the grid mesh used for the numerical solution) is chosen such that grid points inside the matrix region delimited by its contour are characterized by $C_d > 0$, whereas outer points are characterized by $C_d = 0$, Figure 7.16 shows its time evolution. It is clear that it undergoes recession as time passes (1 min, 1, 32, 64, and 96 h; solid line). In addition, the dotted line (right ordinate axis), indicating the radial diffusion front velocity V_{df} (the same considerations can be made considering the axial directions) as a function of radial position (which, in turn, is a function of time), shows that V_{df} sharply decreases in the first 8 h and then becomes approximately constant after 32 h (the experiment lasted 24 h).

The above-described approach accounts for drug dissolution directly in the transport equation (Equation 7.80) and diffusion front position is a posteriori deduced on the basis of the undissolved drug concentration profile as discussed in Figure 7.14 and Figure 7.16. In addition, this approach leads to the conclusion that a real surface identifying with the diffusion front does not exist. The diffusion front can then be identified with a narrow region where undissolved and dissolved drug coexist and where a sharp variation of undissolved drug concentration takes place. Traditionally, however, the way of approaching diffusion/dissolution-controlled systems is to ideally assume the existence of a sharp diffusion front moving inside the matrix and to consider the proper macroscopic mass balance equation jointly with Fick's equation for diffusion in the undissolved drug free zone where pseudosteady state (PSS) is supposed. The advantage of this approximated approach clearly reflects the possibility of getting analytical solutions to the problem of release kinetics. On the contrary, the disadvantage consists of the inexplicit use of the solid drug dissolution constant K_t. Nevertheless, at least theoretically, it is possible to establish a link between the two approaches. At this purpose, from Equation 7.81 it descends that

$$\frac{\partial m_d}{\partial t} = dV \frac{\partial C_d}{\partial t} = -k_d dS (C_s - C), \qquad (7.81')$$

where m_d is the amount of solid drug contained in the volume dV, k_d is the drug intrinsic dissolution constant, and dS is the dissolution area. In Chapter 5 it has been already discussed that, in the case of intrinsic dissolution rate experiments (IDR), k_d is defined as the ratio between the drug diffusion

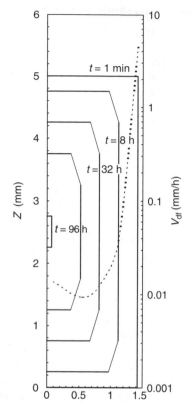

FIGURE 7.16 Model prediction of the temporary evolution of the diffusion front position (solid line) inside the cylindrical matrix (Z, R axial, and radial coordinates, respectively) assuming test 17 parameters. Dotted line indicates the diffusion front receding speed (V_{df}) in the radial direction. (From Grassi, M., et al., *J. Contr. Rel.*, 92, 275, 2003. With permission from Elsevier.)

coefficient in the environmental release liquid and the thickness of the unstirred liquid layer surrounding the solid surface. It is clear that this definition becomes a little bit unclear in the case of solid drug dissolution inside a swollen matrix. Although a wide experimental campaign should be needed, it can be argued that the value of k_d appearing in Equation 7.81' would be of the same order of magnitude of the value calculated according to the IDR method when disk rotating speed approaches zero. Indeed, k_d value rapidly decreases with hydrodynamic wetting liquid rotational speed. Alternatively, k_d can be simply thought of as a mass transfer coefficient and it can be estimated according to usual dimensionless relations applying to mass transfer problems [154]. Whatever the choice, however, it is reasonable to suppose that k_d inside the matrix is position (or time) dependent. This is also supported by the fact that, as shown later, k_d is directly proportional to diffusion front velocity that is not constant as proved by numerical simulation.

Dividing both members of Equation 7.81′ for dV, Equation 7.81″ is achieved:

$$\frac{\partial C_d}{\partial t} = -k_d A_V(C_s - C) = -K_t(C_s - C), \quad K_t = k_d A_V, \tag{7.81″}$$

where A_V is the solid drug dissolution area per matrix unit volume. Supposing that solid drug is present in the matrix in the form of monosized spherical particles of radius r_p, the number N_p of particles contained in the matrix is

$$N_p = \frac{C_{d0} V_M}{(4/3)\,\pi r_p^3 \rho_D}, \tag{7.88}$$

where ρ_D is drug density and V_M is matrix volume. Remembering that $A_v = 4\pi\, r_p^2\, N_p/V_M$, it follows a possible A_V estimation:

$$A_V = \frac{3C_{d0}}{r_p \rho_D}. \tag{7.89}$$

Equation 7.89 clearly shows that, theoretically, A_V is time dependent as r_p decreases with particles dissolution. Accordingly, the assumption of a constant A_V value holds only if r_p variation is moderate and this takes place at the beginning of the release experiment or if release environment is rapidly saturated by the released drug. As a thumb rule, if solid drug particles are very small ($r_p \leq 1$ μm; this corresponds to very high A_V values), A_V can be considered time independent. If this is not the case, A_V time dependence should be accounted for according to, for example, the following considerations. A small modification of Equation 7.81′ leads to

$$\frac{\partial m_d}{\partial t} = N_p \frac{\partial}{\partial t}\left(\frac{4}{3}\pi r_p^3 \rho_D\right) = -N_p 4\pi r_p^2 k_d(C_s - C). \tag{7.90}$$

Proper simplifications yield to the final form:

$$\frac{\partial}{\partial t}(r_p) = -\frac{k_d}{\rho_D}(C_s - C). \tag{7.91}$$

Obviously, Equation 7.91 must be coupled with Equation 7.80 and Equation 7.81.

Now it is possible to establish a link between the two approaches mentioned earlier. Indeed, the amount dm of drug dissolved due to a diffusion front movement equal to dx (this takes place in the time interval dt) is given by

$$dm = C_{d0}S\nu\,dt, \tag{7.92}$$

where ν is the diffusion front speed. At the same time, dm must be equal to (assuming sink conditions for dissolution)

$$dm = Sx_D k_d C_s\,dt, \tag{7.93}$$

where x_D is the surface drug mass fraction on S (practically, x_D can coincide with the initial drug bulk mass fraction). Equating Equation 7.92 and Equation 7.93, it follows the relation between k_d and v:

$$k_d = \frac{C_{d0}}{C_s x_D}\nu = \frac{K_t}{A_V}. \tag{7.94}$$

It is now interesting to discuss some mathematical models based on the diffusion front approach. The first example in this field is that of Higuchi [155]. His model, initially conceived for planar systems, was then extended to different geometries and porous systems [114]. The model is based on the hypotheses that (1) initial drug concentration in the matrix is much higher than drug solubility, (2) drug diffusion takes place only in one dimension (edge effects must be negligible), (3) solid drug particles are much smaller than system thickness, (4) matrix swelling and dissolution are negligible, (5) drug diffusivity is constant, and (6) perfect sink conditions are always attained in the release environment. Accordingly, model expression is given by

$$M_t = A\sqrt{D(2C_0 - C_s)C_s t}, \quad C_0 > C_S, \tag{7.95}$$

where M_t is the amount of drug released until time t, A is the release area, D is the drug diffusion coefficient, C_0 is the initial drug concentration in the matrix whereas C_s is drug solubility. Interestingly, this model shows an M_t square root dependence on time as exact Fick's solution predicts when the amount released is less than 60% [76]. A very elegant and recent improvement of this approach is that of Zhou and Wu who studied the release of salicylic acid from anisotropic matrices made of stearyl alcohol (54.9% w/w), lactose (21.9% w/w), and PVP (10.8% w/w) [156]. These matrices do not undergo swelling or erosion and, due to the compression procedure used to produce them, they show different diffusion properties in the radial and axial directions. The authors assume that matrix cross section can be subdivided into three regions indicated, respectively, as Ω_1, Ω_2, and Ω_3 (see Figure 7.17) where diffusion is ruled by the following equation:

$$\frac{\partial C}{\partial t} = \frac{D_{er}}{R} \frac{\partial}{\partial R} \left[R \frac{\partial C}{\partial R} \right] + D_{ea} \frac{\partial}{\partial Z} \left[\frac{\partial C}{\partial Z} \right], \tag{7.96}$$

where D_{er} and D_{ea} are, respectively, the effective drug diffusion coefficient in the radial and axial directions. Additionally, it is supposed that drug diffusion essentially takes place in radial direction in Ω_1, in the axial direction in Ω_2 and in both the radial and the axial directions in Ω_3. Hence, solute

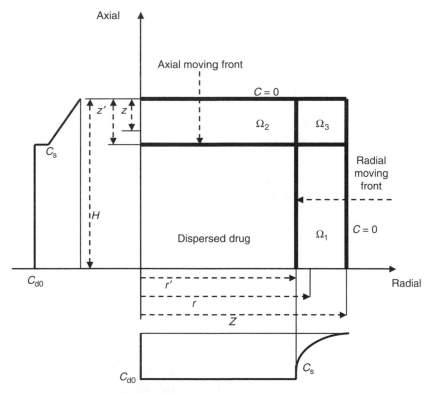

FIGURE 7.17 Schematic diagram for the radial and axial moving fronts jointly with drug concentration profile in the tablet. (From Zhou, Y., et al., *Biomaterials.*, 26, 945, 2005. With permission from Elsevier.)

concentration distribution can be described by a one-dimensional planar model, one-dimensional cylindrical model and by a two-dimensional hollow cylindrical model in Ω_1, Ω_2, and Ω_3, respectively. Assuming PSS condition to hold ($\partial C/\partial t = 0$ in Equation 7.96) and sink conditions at the tablet/release environment interface, drug concentration distribution in Ω_1, Ω_2, and Ω_3 is given by, respectively,

$$C(r) = C_s \frac{\ln(r/R)}{\ln(r'/R)} \quad \text{in } \Omega_1, \tag{7.97}$$

$$C(r) = C_s \frac{z}{z'} \quad \text{in } \Omega_2, \tag{7.98}$$

$$C(r) = C_s \frac{z}{z'} \frac{\ln(r/R)}{\ln(r'/R)} \quad \text{in } \Omega_3, \tag{7.99}$$

where R is the tablet radius, r and z are, respectively, the radial and axial coordinates whereas r' and z' denote the position of the diffusion front in radial and axial directions, respectively.

On the basis of Fick's law, the rate dm_r/dt of drug leaving the tablet in the radial direction across a cylindrical surface equal to $1*2\pi r'$ is given by

$$\frac{dm_r}{dt} = -2\pi r' D_{er} \frac{dC}{dr} = \frac{-2\pi D_{er} C_s}{\ln(r'/R)}. \tag{7.100}$$

Analogously, the rate dm_a/dt of drug leaving the tablet in the axial direction across a unit area, is given by

$$\frac{dm_a}{dt} = -D_{ea} \frac{dC}{dz} = \frac{D_{ea} C_s}{z'}, \tag{7.101}$$

where m_r and m_a can be expressed by means of the following mass balances:

$$m_r = \pi(R^2 - r'^2)C_{d0} - \int_{r'}^{R} 2\pi r C_s \frac{\ln(r/R)}{\ln(r'/R)} dr$$
$$= \pi \left[(R^2 - r'^2)C_{d0} + C_s \left(r'^2 - \frac{r'^2 - R^2}{2\ln(r'/R)} \right) \right], \tag{7.102}$$

$$m_a = z'(C_{d0} - C_s/2). \tag{7.103}$$

Differentiation of Equation 7.102 and Equation 7.103 with respect to time, yields

$$\frac{dm_r}{dt} = \frac{dm_r}{dr'}\frac{dr'}{dt} = \pi\left[2r'(C_s - C_{d0}) - \frac{C_s}{2}\left(\frac{2r'}{\ln(r'/R)} - \frac{r'^2 - R^2}{r'\ln^2(r'/R)}\right)\right]\frac{dr'}{dt},$$

(7.104)

$$\frac{dm_a}{dt} = \frac{dm_a}{dz'}\frac{dz'}{dt} = \left(C_{d0} - \frac{C_s}{2}\right)\frac{dz'}{dt}.$$

(7.105)

Equating Equation 7.100/Equation 7.104, Equation 7.101/Equation 7.105 and integrating, respectively, between r' and R and between 0 and z', the implicit relations linking time t and diffusion front position in the radial (r') and axial directions (z') are found:

$$t(r') = \frac{1}{2D_{er}}\int_{r'}^{R}\left[2\left(1 - \frac{C_{d0}}{C_s}\right)r\ln(r/R) - r\left(1 - \frac{1}{2\ln(r/R)} + \frac{R^2}{2r^2\ln(r/R)}\right)\right]dr,$$

(7.106)

$$t(z') = \frac{2C_{d0} - C_s}{2D_{ea}C_s}\int_{0'}^{z'}z\,dz = \frac{2C_{d0}/C_s - 1}{4D_{ea}}z'^2.$$

(7.107)

To account for the two singularities arising in Equation 7.106 ($r' = R$; $r' = 0$), the authors simply suggest to assume $r' = R - 0.00001$ and $r' = 0.0001$, respectively.

As at time t diffusion front is in r' and z' in the radial and axial directions, respectively, the right-hand side term of Equation 7.106 must be equal to the one of Equation 7.107. Accordingly, it is given as follows:

$$z' = \left\{\frac{4D_{ea}\left(2\frac{C_{d0}}{C_s} - 1\right)}{\left(2\frac{C_{d0}}{C_s} - 1\right)D_{er}}\int_{r'}^{R}\left[r\ln(r/R) - r\left(1 - \frac{1}{2\ln(r/R)} + \frac{R^2}{2r^2\ln(r/R)}\right)\right]dr\right\}^{1/2}.$$

(7.108)

In addition, the total amount M_r of drug released from Ω_1 region is given by

$$M_r = \pi(H - z')\left[C_{d0}(R^2 - r'^2) - \frac{2C_s}{\ln(r'/R)}\int_{r'}^{R}r\ln(r/R)\,dr\right]$$

(7.109)

$$= \pi(H - z')\left[\left(C_{d0} + \frac{C_s}{2\ln(r'/R)}\right)(R^2 - r'^2) - C_s r'^2\right],$$

where H is tablet half-thickness. The total amount M_a of drug released from Ω_2 region is given by

$$M_a = \left(C_{d0} - \frac{C_s}{2} \right) z' \pi r'^2. \tag{7.110}$$

The total amount M_c of drug released from Ω_3 region is given by

$$M_c = \pi \left[C_{d0}(R^2 - r'^2)z' - \frac{2C_s}{z' \ln (r'/R)} \int_0^{z'} z \, dz \int_{r'}^R r \ln (r/R) \, dr \right]$$

$$= \pi z' \left[\left(C_{d0} + \frac{C_s}{4 \ln (r'/R)} \right) (R^2 - r'^2) - \frac{C_s r'^2}{2} \right]. \tag{7.111}$$

As in the above derivation, only half of the cylinder is considered, the total amount M_t of drug released from the tablet is

$$M_t = 2(M_a + M_r + M_c). \tag{7.112}$$

Equation 7.106, Equation 7.108 through Equation 7.112 represent general solutions for the proposed problem assuming sink conditions, $C_{d0}/C_s \geq 3$, $r' > 0$ and $z' < H$ (see Figure 7.17).

A typical characteristic of these models is that despite a relatively simple expression for M_t, problems arise in the diffusion front position evaluation. Indeed, unfortunately, Equation 7.106 solution needs a numerical solution. Nevertheless, once this problem is solved, Equation 7.108 and Equation 7.112 yield model solution. Obviously, the difficulty arising in solving Equation 7.106 is less important than that involved in the numerical solution of Equation 7.80. Interestingly, if the dissolved drug amount in the tablet can be neglected ($C_{d0} \geq 10 \, C_s$), model solution becomes

$$t(r') = \frac{C_{d0}}{C_s D_{er}} \left[\frac{r'^2}{2} \ln (r'/R) + \frac{1}{4}(R^2 - r'^2) \right], \tag{7.106'}$$

$$t(z') = \frac{C_{d0}}{2 D_{ea} C_s} z'^2, \tag{7.107'}$$

$$z' = \left\{ \frac{2 D_{ea}}{D_{er}} \left[r'^2 \ln (r'/R) + \frac{1}{4}(R^2 - r'^2) \right] \right\}^{1/2}, \tag{7.108'}$$

$$M_r = \pi C_{d0}(R^2 - r'^2)(H - z'), \tag{7.109'}$$

$$M_a = C_{d0} z' \pi r'^2, \tag{7.110'}$$

$$M_c = \pi C_{d0}(R^2 - r'^2)z', \tag{7.111'}$$

$$M_t = 2\pi C_{d0}\left[H(R^2 - r'^2) + z'r'^2\right]. \tag{7.112'}$$

Authors prove the reliability of this model by measuring D_{ea} and D_{er} from independent experiments and then comparing model prediction and experimental data. In particular, they perform release experiments from partially coated tablets (top/bottom plane surfaces or lateral cylindrical surface; PVC adhesive tape (20 MILPVC)) to realize a totally one-dimensional release in the axial or radial direction. Solute release from the tablets is determined in a USP II apparatus containing 750 mL of pH 7.2 PBS at 37°C with 100 rpm stirring rate. Salicylic acid concentration in the dissolution medium is assayed by means of a spectrophotometer (wavelength of 294 nm). Fitting of exact solution of Fick's law on these experimental data yields $D_{ea} = 1 \times 10^{-6} \text{ cm}^2/\text{s}$ and $D_{er} = 8 \times 10^{-7} \text{ cm}^2/\text{s}$ underlying a moderate matrix anisotropy. On the basis of these values it is possible to compare model prediction and experimental data referred to salicylic acid release from uncoated tablet. Figure 7.18 clearly shows that the agreement is decisively satisfactory.

7.3.5 DIFFUSION-, DISSOLUTION-, AND SWELLING-CONTROLLED SYSTEMS

Systems falling in this category are typically represented by polymeric matrices characterized by covalent crosslinks or strong physical crosslinks as it happens for alginates and crosslinked PVP. This means that erosion and drug–polymer interactions play a negligible role at least before drug release has been completed. Indeed, it cannot be excluded (and sometimes it is desirable) that matrix undergoes erosion after a prolonged exposure to the release environment conditions. In addition, *in vivo*, the action of enzymes and other factors can lead to matrix erosion (surface or bulk). Finally, matrix/drug interactions are supposed to be negligible.

To provide a better understanding, this section discusses separately the key factors of this category systems, namely swelling equilibrium, swelling kinetics, and drug diffusion/dissolution.

7.3.5.1 Matrix Swelling Equilibrium

Swelling properties of polymeric matrices can affect both the drug-release kinetics and the drug-loading properties, hence, also the modes of preparation and use of the release system [157]. The swelling behavior is strongly dependent on the number of intermolecular junctions per unit of volume, namely the crosslink density ρ_x, and on the state of the interactions between the solvent molecules and the polymeric chain segments. In many situations, the latter feature of the gel network can be simply characterized by the Flory interaction parameter χ. For neutral polymers, the amount of absorbable

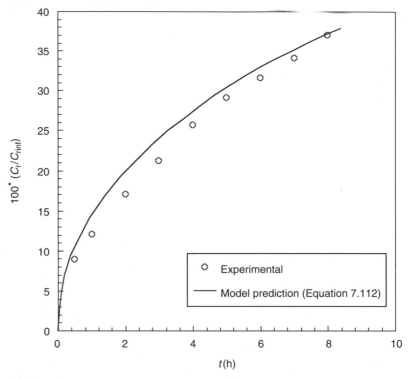

FIGURE 7.18 Comparison between model prediction (solid line, Equation 7.112) and experimental release data (open circles) from uncoated tablet. (From Zhou, Y., et al., *Biomaterials.*, 26, 945, 2005. With permission from Elsevier.)

solvent depends on both the solvent chemical affinity for the polymer and the elastic properties of the swollen polymeric network which, in turn, mainly depend on the number of intermolecular bonds, namely the crosslink density. In the case of charged polymers, the swelling equilibrium of the polymeric matrix is more complicated [158,159] as it heavily depends also on the ionic strength. Indeed, crosslinked polyelectrolyte chains tend to assume more extended conformations in pure water, whereas such an extension is usually hindered in an aqueous salt solution where electrostatic interactions occur between the polymer charges and the mobile ions present in solution [160,161]. The swelling equilibrium conditions of a polymeric gel can be obtained from the Gibb's free energy by imposing the condition of zero osmotic pressure [158,162,163]:

$$\pi = -\frac{\Delta\mu}{V_1} = \frac{\mu_1^{\mathrm{g}} - \mu_1^{\mathrm{l}}}{V_1} = 0, \tag{7.113}$$

where π is the osmotic pressure, μ_1^g and μ_1^l are the chemical potentials of the solvent in the matrix (g) and the liquid (l) phases, respectively, and V_1 is the molar volume of the solvent. Although other approaches can be followed [164–166], the osmotic pressure can be written according to the Flory–Rehner theory [158,167]:

$$\pi = \pi_{\text{mix}} + \pi_{\text{elas}} + \pi_{\text{ion}} + \pi_{\text{el}}, \tag{7.112}$$

where π_{mix} is the mixing free energy term, π_{elas} is the elastic contribution, connected with the deformation of polymeric network, π_{ion} is the ionic contribution, due to the difference in ion concentration between the gel and the liquid phases, and π_{el} is the electrostatic contribution deriving from the repulsive effects between equal charges present in the network. Even though more sophisticated approaches might be followed to express the mixing term [163,168], the Flory–Huggins expression can be used:

$$\pi_{\text{mix}} = -\frac{RT}{V_1}\left[\ln\left(1 - \phi_2\right) + \phi_2 + \chi\phi_2^2\right], \tag{7.113}$$

where ϕ_2 is the polymer volume fraction, R is the universal gas constant, T is the temperature, and χ is the Flory interaction parameter, related to the difference between the free energies of polymer–polymer, and polymer–solvent interactions. In the most general case, the elastic term could be calculated by assuming that the real structural conditions of the polymeric network are somewhat intermediate between two opposite ideal limits, corresponding to the affine and the phantom networks, respectively [161,163,169]. According to the phantom theory, the intermolecular constraints can freely fluctuate within the polymeric network and such an assumption leads to the following expression:

$$\pi_{\text{elas}}^{\text{phantom}} = -RT\rho_x\left(\frac{\phi_2}{\phi_2^0}\right)^{\frac{1}{3}}, \tag{7.114}$$

where ρ_x is the crosslink density (mol/cm^3), and ϕ_2^0 is the polymer volume fraction at the reference state. If dry polymer is chosen as reference state, $\phi_2^0 = 1$. Conversely, the fluctuations are hindered in the affine polymeric network, and such a structural condition yields

$$\pi_{\text{elas}}^{\text{affine}} = -RT\rho_x\left[\left(\frac{\phi_2}{\phi_2^0}\right)^{\frac{1}{3}} - 0.5\left(\frac{\phi_2}{\phi_2^0}\right)\right]. \tag{7.115}$$

For a real system lying between the phantom and affine network condition, we may write

$$\pi_{elas} = \pi_{elas}^{phantom}(1 - F) + \pi_{elas}^{affine} F, \tag{7.116}$$

where F ranges from 0 to 1, depending on the degree of interpenetration of the polymeric network and on its dilatation ratio $(\phi_2^0/\phi_2)^{1/3}$ [163,169,170]. Obviously, the affine network condition ($F = 1$) is more suitable to represent the structural conditions and the intermolecular constraints of a chemically crosslinked polymer network. Equation 7.114 and Equation 7.115 are derived on condition that the Gaussian statistic distribution could be assumed valid for the distance between two consecutive crosslinks. Such an assumption seems to be reasonable only for limited chain extensions. Indeed, for very extended swelling conditions, the polymeric chains closely approach their fully stretched conformation and behave as nonlinear elastic connectors so that a more complex distribution should be considered, namely the Langevin one [171,172]. The resulting expression of the elastic term is, then

$$\pi_{elas} = -\frac{RT}{3} \rho_x (\phi_2 \phi_{2c})^{\frac{2}{3}} (\phi_{2\,min})^{-\frac{1}{3}} \Im, \tag{7.117}$$

where ϕ_{2c} is the polymer volume fraction in the reference state, ϕ_{2min} is the minimum polymer volume fraction attainable by the swollen gel (all the polymeric chains are completely stretched) and $\Im$ is defined by

$$\Im = \beta + \frac{\beta^2 \sinh(\beta)\left(\left(\dfrac{\phi_{2\,min}}{\phi_2}\right)^{\frac{1}{3}} + \beta^{-1} - \coth(\beta)\right)}{\sinh^2(\beta) - \beta^2}, \tag{7.118}$$

where β is defined by

$$\beta = L^{-1}\left(\frac{\phi_{2min}}{\phi_2}\right), \tag{7.119}$$

where L^{-1} is the inverse of the Langevin function whose approximate form, according to Warner [173], is

$$\beta \approx \frac{3\left(\dfrac{\phi_{2min}}{\phi_2}\right)^{\frac{1}{3}}}{1 - \left(\dfrac{\phi_{2min}}{\phi_2}\right)^{\frac{2}{3}}}. \tag{7.120}$$

The ionic contribution to the osmotic pressure is derived from the Donnan equilibrium theory [158,159,161,163]. In the presence of a solution, the concentration of mobile ions within the gel phase is larger than in the external phase because of the charges located on the polymer chains. This difference gives rise to the following ionic contribution:

$$\pi_{ion} = RT \left[jn \frac{\phi_2}{V_u} - \nu(c_s^* - c_s) \right],$$ (7.121)

where n is the number of ionizable groups present on the monomer, j is the ionization degree (fraction of monomers carrying the ionizable groups), V_u is the monomer molar volume c_s^* and c_s are the salt concentrations in the external solution and in the gel phase, respectively, $\nu(= \nu_+ + \nu_-)$ is the sum of the positive and negative valences of the dissociating salt. To render Equation 7.121 operative, it is necessary expressing c_s as a function of known or measurable quantities. For this purpose, it is sufficient equating electrolyte activities in the gel and solution phase assuming, for the sake of simplicity, that activities can be substituted by concentrations [158]. Accordingly, in the case of monovalent ions present both in solution and on the

polymeric chains ($\nu = 2$), we have $c_s = 0.5 \left(-\frac{nj\phi_2}{V_u} + \sqrt{\left(\frac{nj\phi_2}{V_u}\right)^2 + 4(c_s^*)^2} \right).$

Consequently, Equation 7.121 becomes

$$\pi_{ion} = -RT \left[2c_s^* - \sqrt{\left(\frac{nj\phi_2}{V_u}\right)^2 + 4(c_s^*)^2} \right].$$ (7.122)

Finally, the electrostatic term originating from the repulsive effects between equal charges present in the polymeric network could be calculated by the Ilavsky approach [171], but its contribution is usually negligible in comparison with the ionic term [163]. In conclusion, for not highly swellable matrices, the equilibrium swelling conditions is

$$\frac{\Delta\mu}{RT} = \ln(1 - \phi_2) + \phi_2 + \chi\phi_2^2 + \rho_x V_1(\phi_2^{\frac{1}{3}} - 0.5\phi_2) + V_1$$

$$\times \left[2c_s^* - \sqrt{\left(\frac{nj\phi_2}{V_u}\right)^2 + 4(c_s^*)^2} \right] = 0.$$ (7.123)

For highly swellable matrices, the equilibrium condition reads

$$\frac{\Delta\mu}{RT} = \ln(1 - \phi_2) + \phi_2 + \chi\phi_2^2 + \frac{1}{3}\rho_x V_1(\phi_2\phi_{2c})^{\frac{2}{3}}(\phi_{2\,min})^{\frac{-1}{3}}\Im$$

$$+ V_1\left[2c_s^* - \sqrt{\left(\frac{nj\phi_2}{V_u}\right)^2 + 4(c_s^*)^2}\right] = 0. \tag{7.124}$$

If salt concentration c_s^* in the swelling medium is zero, Equation 7.124 becomes

$$\frac{\Delta\mu}{RT} = \ln(1 - \phi_2) + \phi_2 + \chi\phi_2^2 + \frac{1}{3}\rho_x V_1(\phi_2\phi_{2c})^{\frac{2}{3}}(\phi_{2\,min})^{\frac{-1}{3}}\Im - V_1\frac{nj\phi_2}{V_u} = 0. \tag{7.124'}$$

Coviello and colleagues [174], studying drug permeation through crosslinked (1,6-hexanedibromide crosslinker) sclerox (oxidized scleroglucan) membranes, gives an example of Equation 7.124 application. In particular, she fixes polymer concentration to 1.6% (w/v), two different crosslinking ratios r ($r = $ (equivalents of reagent)/(equivalents of polymer)) (0.5, 1), two temperatures (37°C, 7°C) and two NaCl concentrations in the swelling agent (water) (0, 0.154 M) (see Table 7.4). A useful strategy applicable to calculate χ and ρ_x is to collect the swelling equilibrium data in presence (ϕ_{2s}) and in absence (ϕ_{2w}) of an external salt (same temperature), and then solving, with respect to χ and ρ_x, Equation 7.124 and Equation 7.124':

TABLE 7.4
Swelling Equilibrium Characteristics of Crosslinked (1,6-Hexanedibromide Crosslinker) Sclerox (Oxidized Scleroglucan) Hydrogels

Parameters	Tests 1a–1b	Tests 2a–2b	Tests 3a–3b	Tests 4a–4b
R	1.0	0.5	1.0	0.5
T (°C)	37	37	7	7
c_s (mol/dm^3) (Test a)	0.0	0.0	0.0	0.0
c_s (mol/dm^3) (Test b)	0.154	0.154	0.154	0.154
V_1 (cm^3/mol)	18.12	18.12	18.0	18.0
N	2	2	2	2
J	1	1	1	1
V_u (cm^3/mol)	288.8	288.8	288.8	288.8
$\phi_{2w} \times 10^2$	1.10 ± 0.35	0.27 ± 0.01	1.60 ± 0.26	0.42 ± 0.02
$\phi_{2s} \times 10^2$	2.86 ± 0.39	0.99 ± 0.01	1.95 ± 0.03	1.07 ± 0.03
$\phi_{2min} \times 10^2$	0.895	0.255	1.57	0.425
$\rho_x \times 10^4$	2.10 ± 0.20	0.90 ± 0.02	2.30 ± 0.70	1.10 ± 0.04

Source: From Coviello, T., et al., *Biomaterials*, 22, 1899, 2001. With permission from Elsevier.

$$\chi = \frac{\frac{E(A+C)}{B} - F - D}{\left(\phi_{2w}^2 - \frac{E}{B}\phi_{2s}^2\right)}, \quad \rho_x = -\frac{A + C + \chi\phi_{2s}^2}{B}, \quad (7.125)$$

where

$$A = \ln(1 - \phi_{2s}) + \phi_{2s}, \quad B = \frac{1}{3}V_1(\phi_{2s}\phi_{2\,min})^{\frac{2}{3}}(\phi_{2c})^{\frac{2}{3}},$$

$$C = V_1\left(2c_s^* - \sqrt{\left(\frac{nj\phi_{2s}}{V_u}\right)^2 + 4c_s^{*2}}\right), \quad (7.126)$$

$$D = \ln(1 - \phi_{2w}) + \phi_{2w}; \quad E = \frac{1}{3}V_1(\phi_{2w}\phi_{min})^{\frac{2}{3}}(\phi_{2c})^{\frac{2}{3}}, \quad F = -V_1\frac{nj\phi_{2w}}{V_u}.$$

$$(7.127)$$

Although ϕ_{2min} experimental determination is not an easy task, it can be demonstrated that an exact knowledge of its value may not be strictly necessary and only an approximate evaluation is needed. For this purpose, let us focus the attention on the swelling equilibrium characterizing test 1a and 1b (Table 7.4). The solution to system of equations, Equation 7.124 through Equation 7.124′, is obtained by considering different ϕ_{2min} values and assuming $V_1 = 18.1\ cm^3/mol$, $n = 2$ and $j = 1$ (each monomer carries two univalent ions), calculating the monomer molar volume ($V_u = 288.8\ cm^3/mol$) by means of the solubility parameters [175], and knowing that $\phi_{2w} = (1.1 \pm 0.35) \times 10^{-2}$ and $\phi_{2s} = (2.86 \pm 0.4) \times 10^{-2}$ (see also Table 7.4). As Figure 7.19 clearly reveals, the unknown χ and ρ_x simultaneously assume a physically consistent value only when $0.82 \times 10^{-2} < \phi_{2min} < 0.97 \times 10^{-2}$. Moreover, though χ strongly depends on ϕ_{2min}, ρ_x is virtually independent and could be evaluated, for instance, by choosing the mean ϕ_{2min} value characterizing the above-mentioned interval. Accordingly, on condition to renounce to get the χ value, ρ_x can be calculated by choosing $\phi_{2min} = 0.895 \times 10^{-2}$. The solution to the system of equations, Equation 7.124 through Equation 7.124′, repeated for the other experimental conditions, leads to similar conclusions, namely the virtually independence of ρ_x on ϕ_{2min} in the interval where both ρ_x and χ assume physically consistent values. Consequently, the mean ϕ_{2min} values can be selected (see Table 7.4). It is worth mentioning that these ϕ_{2min} values accomplish a reasonable topological condition requiring that higher r values correspond to higher ϕ_{2min} values at both temperatures (37°C and 7°C) (see Table 7.4). Moreover, given an r value, higher is the temperature, lower is ϕ_{2min}. On this basis, the ρ_x calculation is performed for each experimental condition examined and the results are reported in Table 7.4. Notably, at each temperature, the calculated ρ_x value coming from the test characterized by a higher r value is, approximately, two times the value coming from the test

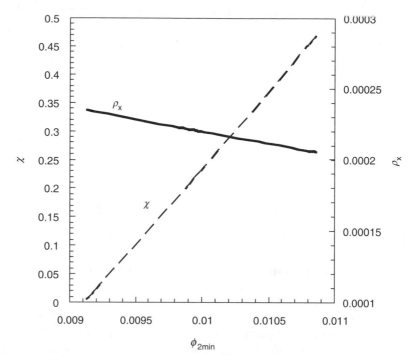

FIGURE 7.19 Although the Flory interaction parameter χ is strongly dependent on ϕ_{2min}, the network crosslink density ρ_x is virtually independent on ϕ_{2min}. In addition, ρ_x and χ simultaneously assume a physically consistent value only when $0.82 \times 10^{-2} < \phi_{2min} < 0.97 \times 10^{-2}$. (From Coviello, T., et al., *Biomaterials*, 22, 1899, 2001. With permission from Elsevier.)

characterized by a lower r value. Moreover, tests 1a–1b and tests 3a–3b give similar ρ_x values as it happens for tests 2a–2b and tests 4a–4b.

Mantovani and group [157], dealing with different kinds of crosslinked sodium starch glycolate matrices (sodium salt of a poly-α-glucopyranose, prepared by crosslinking (sodium trimetaphosphate) and by carboxymethylation of potato starch), make use of Equation 7.123 to get ρ_x and χ. In particular, the solution strategy implies to get experimental data relative to matrix–liquid water (ϕ_{2L}) and matrix–vapor water (ϕ_{2v}) equilibrium performed at the same temperature. In the case of matrix–vapor equilibrium, Equation 7.123 becomes [176–180]

$$\frac{\Delta\mu}{RT} = \ln(1 - \phi_{2v}) + \phi_{2v} + \chi\phi_{2v}^2 + \rho_x V_1(\phi_{2v}^{\frac{1}{3}} - 0.5\phi_{2v})$$

$$- V_1\left[2c_s + \frac{nj\phi_{2v}}{V_u}\right] - \ln\left(\frac{P}{P_0}\right) = 0, \tag{7.123'}$$

where P/P_0 is the relative vapor pressure. The simultaneous solution of Equation 7.123 and Equation 7.123 yields problem solution:

$$\chi = \frac{\frac{E(A+C)}{B} - F - D}{\left(\phi_{2v}^2 - \frac{E}{B}\phi_{2L}^2\right)}; \quad \rho_x = -\frac{A + C + \chi\phi_{2L}^2}{B}, \tag{7.128}$$

where

$$A = \ln(1 - \phi_{2L}) + \phi_{2L}; \quad B = V_1(\phi_{2L}^{1/3} - 0.5\phi_{2L}),$$

$$C = V_1\left(2c_s^* - \sqrt{\left(\frac{nj\phi_{2L}}{V_u}\right)^2 + 4c_s^{*2}}\right), \tag{7.129}$$

$$D = \ln(1 - \phi_{2v}) + \phi_{2v} + \ln\left(\frac{P}{P_0}\right), \quad E = V_1(\phi_{2v}^{1/3} - 0.5\phi_{2v})^{\frac{2}{3}}(\phi_{2c})^{\frac{2}{3}},$$

$$F = -V_1\left(2c_s + \frac{nj\phi_{2v}}{V_u}\right). \tag{7.130}$$

Table 7.5 shows the χ and ρ_x values for the four sodium starch glycolate kinds considered knowing that $P/P_0 = 0.97$, $V_1 = 18.1$ cm^3/mol, $V_u = 376.4$ cm^3/mol, $nj = 0.75$, $c_s^* = 10^{-7}$ mol/cm^3 and different values for c_s calculated on the basis of the residual NaCl mass (residual presence of the salt in the matrix is due to the crosslinking and carboxymethylation reactions). Interestingly, the ρ_x knowledge allows the determination of M_c, the molecular weight of the chain segment comprised between two consecutive crosslinks. Indeed, the following relation [159] holds:

$$M_c = \left(\frac{\rho_x}{\rho_p} + \frac{2}{M_n}\right)^{-1}, \tag{7.131}$$

TABLE 7.5
Characteristics of the Four Sodium Starch Glycolates Considered (CLV, LOW PH, V 17, and EXPLOTAB)

Parameters	CLV	LOW PH	V 17	EXPLOTAB
ρ_p (g/cm^3)	1.465	1.495	1.456	1.452
ϕ_{2L}	0.03858	0.0455	0.00545	0.0305
ϕ_{2v}	0.31	0.32	0.34	0.34
c_s (mol/cm^3)	4.4×10^{-4}	1.2×10^{-4}	6.0×10^{-4}	5.1×10^{-4}
χ	0.58	0.45	0.67	0.64
ρ_x (mol/cm^3)	2.2×10^{-4}	2.9×10^{-4}	6.0×10^{-5}	1.8×10^{-4}

Source: Adapted from Mantovani, F., et al., *Fluid Phase Equilibria*, 167, 63, 2000.

where ρ_p is the polymer density and M_n is the polymer molecular weight before crosslinking. For all the four sodium starch glycolate kinds considered (CLV, LOW PH, V17 and EXPLOTAB), the M_n value reported in the technical bulletin provided by the manufacturer [181] is 10^6. Accordingly, M_c is equal to 6455, 5156, 23 180, 7910 for the CLV, LOW PH, V17, and EXPLOTAB samples, respectively. The corresponding number of monomeric units between two consecutive crosslinks is 10.5, 8.2, 37.7, 12.9, for CLV, LOW PH, V17, and EXPLOTAB, respectively. According to a molecular simulation (DTMM software), it is possible to conclude that the end-to-end distance of the sodium starch glycolate monomer is equal to 0.892 nm. Assuming that (i) each branch between two consecutive crosslinks can be modeled as a freely jointed chain composed of monomer units and (ii) all the monomers comprised in the branch are perfectly aligned so that the chain configuration is fully extended, mesh size values should be 10.8, 9.5, 47.5, and 13.9 nm, for the CLV, LOW PH, V17, and EXPLOTAB, respectively. Obviously, this simplified calculation leads to overestimated mesh sizes since the actual configuration of the polymeric branches, resulting from the balance between the intrinsic constraints and the tension state produced by the network swelling, is likely to be less extended.

The above-mentioned examples show how to estimate χ and ρ_x exclusively resorting to equilibrium data. However, a more traditional way is to use Equation 7.123 or Equation 7.124 to calculate χ once ρ_x has been determined by means of mechanical experiments. Indeed, resorting to rubber elasticity and linear viscoelastic theory, it can be demonstrated that the Young (E) and shear (G) modulus of a swollen polymeric matrix can be expressed by [158,182]

$$E = 3(RT\rho_x)\,\phi_2^{-2/3}, \quad G = (RT\rho_x)\,\phi_2^{-\frac{2}{3}}, \tag{7.132}$$

where R is the universal gas constant and T is the temperature. Although for swollen matrices behaving like a solid, E (or $G = E/3$) can be estimated by matrix compression or elongation (or shear deformation) experiments, for viscoelastic matrices (where matrix response to a solicitation is not instantaneous but it develops in time), E (or G) identifies with the pure matrix elastic character E_0 ($G_0 = E_0/3$). In the case of relaxation tests, it follows [183]

$$\sigma(t) = \frac{\varepsilon_0}{t_1}\left[E_0 t_1 + \sum_{i=1}^{n} \eta_i e^{-\frac{E_i}{\eta_i}t}\left(e^{\frac{E_i}{\eta_i}t_1} - 1\right)\right], \tag{7.133}$$

where σ is the normal tension, ε_0 is the deformation, η_i and E_i represent generalized Maxwell model parameters (see Chapter 3) and t_1 is the time required to get the desired deformation ε_0. Obviously, when $t_1 \longrightarrow 0$ (instantaneous compression), Equation 7.133 becomes

$$\sigma(t) = \varepsilon_0 \left[E_0 + \sum_{i=1}^{n} E_i \mathrm{e}^{-\frac{E_i}{\eta_i}t} \right], \qquad (7.133')$$

where η_i/E_i can be seen as relaxation time. Equation 7.133 and Equation 7.133' clearly evidence that the normal stress approaches $\varepsilon_0 E_0$ after a complete relaxation. Pasut [39] applied Equation 7.133 on experimental data referring to the relaxation behavior of different types of sodium alginate cylinders characterized by polymer concentration ranging from 1% to 4% w/w. Correspondingly, he found that ρ_x ranges between 3×10^{-8} and $2 \times 10^{-6} \, \mathrm{mol/cm^3}$.

7.3.5.2 Matrix Swelling Kinetics

Modeling the drug release from a swellable matrix implies formulating the relevant mass balance and solving the corresponding constitutive equation for the flux of both the swelling fluid (entering the matrix) and the drug (leaving it), respectively. Whereas a suitable choice can be the classical Fick's law as the constitutive equation of the drug, in the case of the swelling fluid, recourse must be made to an equation which can account for the complex phenomena governing the flux of swelling fluid entering the matrix, and particularly the viscoelastic properties of the swellable matrix.

Several approaches have been proposed to interpret and model the swelling fluid diffusion in a glassy polymer matrix. For example, Joshi and Astarita assumed a time-dependent composition at the polymer/swelling fluid interface [184], whereas other authors imposed a swelling propagation law at the swelling front [185]. In an interesting series of papers, Cohen firstly generalized Crank's idea [76] of a time-dependent swelling fluid diffusion coefficient [186], and then supposed that the swelling fluid flux could be properly described by coupling the concentration and stress gradients, where the stress depends on concentration and time via a Maxwell-like viscoelastic relationship [187]. Later on, Cohen and coworkers developed and improved the above-mentioned approach by modifying the stress dependence on time and concentration [188–190]. The existence of a stress-related convective contribution to the swelling fluid flux was postulated by Frisch [191] and taken into consideration also by other authors [192,193]. According to Adib and Neogi [194] and Camera-Roda and Sarti [195], the swelling fluid diffusion flux may depend on the history of the swelling fluid concentration gradient. Singh and Fan [196] developed a generalized mathematical model for the simultaneous transport of a drug and a solvent in a planar glassy polymer matrix. The drug diffuses out of the matrix which is undergoing macromolecular chain relaxation and volume expansion due to solvent absorption from the environment into the matrix. The swelling behavior of the polymer is characterized by a stress-induced drift velocity term v. The change of volume due to the relaxation phenomenon is assumed instantaneous. The model incorporates convective transport of the two species induced by volume expansion and stress gradient. However, it was developed only for planar systems, and not for cylindrical matrices. The most general theory about the swelling fluid diffusion in a viscoelastic polymer matrix was developed by Lustig [197] who assumed that the swelling fluid flux should depend on several driving forces such as temperature gradient, species' inertial and body forces, chemical potentials, and stress gradient. One of the main advantages of this theory is that all the material properties taken into account may be measured so that no phenomenological coefficient must be introduced [198]. The predictions provided by the Lustig model are in semiquantitative agreement with the experimental results [199].

Despite its lower generality, we believe that the Camera-Roda and Sarti equation [200] can be used to model non-Fickian diffusion. This model avoids the determination of a great number of experimental information normally requested by other, more general models. Obviously, it implies the use of some phenomenological parameters which are not, however, empirical in nature, since all the parameters of the Camera-Roda and Sarti model do posses a well-defined physical meaning. Basically, these authors suppose that swelling fluid (SF) flux J may be expressed as the sum of two terms: J_f, characterized by a zero relaxation time and representing the Fickian contribution to the global flux, and J_r, characterized by a nonzero relaxation time and representing the non-Fickian contribution to the global flux, respectively. Accordingly, the global flux can be expressed as

$$J = J_f + J_r, \tag{7.134}$$

where

$$J_f = -D_f \nabla C, \tag{7.135}$$

$$J_r = -D_r \nabla C - \tau \frac{\partial J_r}{\partial t}, \tag{7.136}$$

where C is the swelling fluid concentration, τ is the relaxation time of the given polymer/SF system (in doing so, it is implicitly assumed that system viscoelastic behavior is represented by the simple Maxwell model constituted by the series of a hookean spring and viscous dashpot; see Chapter 3), D_f is the diffusion coefficient relative to the Fickian flux, D_r is the diffusion coefficient relative to the non-Fickian flux and t is the time. Under these assumptions, the balance equation governing the SF adsorption phenomenon is given by

$$\frac{\partial C}{\partial t} = -\nabla J. \tag{7.137}$$

On the basis of the free volume theory, Camera-Roda and Sarti (CRS) [200] assume, for D_f, D_r, and τ, the following functional dependencies on C:

$$D_f(C) = D_0, \tag{7.138}$$

$$D_f(C) = D_{eq} \exp[g(C - C_{eq})] - D_0, \tag{7.139}$$

$$\tau(C) = \tau_{eq} \exp[k(C_{eq} - C)], \tag{7.140}$$

where D_0 is the diffusion coefficient of the swelling fluid in the dry matrix $(C = 0)$, D_{eq} is the diffusion coefficient of the swelling fluid in the swollen matrix $(C = C_{eq})$, C_{eq} is the swelling fluid concentration in the polymeric matrix at equilibrium, g and k are two adjustable parameters, and τ_{eq} is the matrix relaxation time at equilibrium. Assume dealing with a one-dimensional problem (SF diffusion takes place only in the X direction), Equation 7.137 has to be solved under the following initial conditions:

$$J_r = 0, \quad 0 \le X \le L_0, \tag{7.141}$$

$$C = 0, \quad 0 \le X \le L_0, \tag{7.142}$$

$$C = 0.8C_{eq}, \quad X \le L_0, \tag{7.143}$$

and boundary conditions:

$$J = J_f = J_r = 0, \quad X = 0, \tag{7.144}$$

$$\tau\frac{dC}{dt}\bigg|_{X = L(t)} = C_{eq} - C, \tag{7.145}$$

where L_0 is the dry matrix thickness, $L(t)$ is the matrix thickness at time t (matrix thickness increases due to swelling) and τ is the matrix relaxation time defined by Equation 7.140. Although in most cases SF equilibrium concentration holds at the matrix/external environment interface, to infer a wider generality to the model, it is assumed that a relaxation process takes place even at that boundary. Equation 7.143 and Equation 7.144 rule this relaxation; suppose that a reasonable C starting value may be $0.8\,C_{eq}$ and that its increase rate is proportional to the difference $C_{eq} - C$ via $1/\tau$ [200].

Obviously, the SF income implies matrix volume increase. In the case of a cylinder absorbing the SF only from one base, volume increase and shape variation follow a two-step mechanism. Cylinder swelling takes place only in the X direction (anisotropic swelling) until the swelling fluid concentration in the matrix's innermost part is lower than a threshold value C^*. As soon as this threshold C^* is overcome, swelling can occur in all directions (isotropic swelling). Subdividing the polymeric matrix into V_c slices (volume elements), assuming volume additivity and a linear variation of C between two consecutive volume elements, the thickness ΔXi of the ith volume element is given by the following expressions:

$$\Delta X_i = \Delta X_0 \frac{2\rho_s}{2\rho_s - (C_i + C_{i-1})} \quad \text{(anisotropic swelling)} \quad (7.146)$$

and

$$\Delta X_i = \Delta X_0 \sqrt[3]{\frac{2\rho_s}{2\rho_s - (C_i + C_{i-1})}} \quad \text{(isotropic swelling)}, \quad (7.147)$$

where ρ_s is the SF density, C_i is the SF concentration inside the ith matrix volume element, C_{i-1} is the SF concentration inside the $(i-1)$th matrix volume element and ΔX_0 is the value assumed by ΔX_i when $C_i = C_{i-1} = 0$ [201,202]. Indeed, for what concerns anisotropic swelling, the thickness ΔX_i of the ith matrix volume element may be obtained firstly by making a SF mass balance on the volume element itself:

$$\Delta M_i = \int_0^{\Delta X_i} S\left(C_{i-1} + x\frac{C_i - C_{i-1}}{\Delta X_i}\right) dx = 0.5\, S\Delta X_i(C_i - C_{i-1}), \quad (7.148)$$

where ΔM_i is the SF amount contained in the ith matrix volume element and S is the constant cross-sectional surface of the sample. Supposing volume additivity, which means neglecting the volume variation due to mixing, the following volume balance holds:

$$S\Delta X_i = S\Delta X_0 + \frac{\Delta M_i}{\rho_s}, \quad (7.149)$$

where $S\Delta X_0$ is the volume of the element in absence of SF. Substituting Equation 7.148 in Equation 7.149 and solving the resultant equation for ΔX_i, Equation 7.146 is found. Isotropic swelling implies a simultaneous ΔX_i and S increase. According to Li and Tanaka [203] the swelling phenomenon is composed of two consecutive steps. The first consists of an anisotropic swelling in the X direction, the second is an instantaneous sample shape readjustment taking place at constant volume. For particular sample shape like cylinders, Li and Tanaka demonstrated that the above-mentioned steps are responsible for a sample volume increase obeying the following equation:

$$\frac{R_i}{R_0} = \frac{\Delta X_i}{\Delta X_0}, \quad (7.150)$$

where R_i and R_0 are the radius of the sample cross-sectional surface during swelling and in the dry state, respectively. From Equation 7.150 it follows that, for cylinders, we have

$$\frac{S_i}{S_0} = \left(\frac{R_i}{R_0}\right)^2 = \left(\frac{\Delta X_i}{\Delta X_0}\right)^2, \tag{7.151}$$

where S_i and S_0 are the surface of the element cross section during swelling and in the dry state, respectively. For the isotropic step, the SF mass balance and the volume balance are, respectively:

$$\Delta M_i = \int_0^{\Delta X_i} S_i \left(C_{i-1} + x\frac{C_i - C_{i-1}}{\Delta X_i}\right) dx = 0.5 \; S_i \Delta X_i (C_i - C_{i-1}), \tag{7.152}$$

$$S_i \Delta X_i = S_0 \Delta X_0 + \frac{\Delta M_i}{\rho_s}. \tag{7.153}$$

Consequently, it follows as

$$\Delta X_i = \Delta X_0 \frac{2\rho_s}{2\rho_s - (C_i + C_{i-1})} \frac{S_0}{S_i}. \tag{7.154}$$

Inserting Equation 7.151 into Equation 7.154 and solving the resultant expression for ΔX_i, Equation 7.147 is found.

To better understand the effects of all parameters on the swelling fluid uptake, Equation 7.138 through Equation 7.140 can be conveniently recast in dimensionless terms as follows:

$$D_f^+ = 1, \tag{7.138'}$$

$$D_r^+ = R_d \exp\left[g(C - C_{eq})\right] - 1, \tag{7.139'}$$

$$\tau^+(C) = \frac{De \exp\left[k(C_{eq} - C)\right]}{R_d}, \tag{7.140'}$$

where De is the Deborah number, defined as

$$De = \frac{\tau_{eq} D_{eq}}{L_{eq}^2}, \quad t^+ = \frac{t D_0}{L_{eq}^2}, \tag{7.155}$$

where L_{eq} is the matrix characteristic length (thickness in a one-dimensional swelling case) at the swelling equilibrium and R_d is given by

$$R_d = \frac{D_{eq}}{D_0}. \tag{7.156}$$

Higher the R_d, more pronounced the non-Fickian character of the sorption process is. Obviously, due to the nonlinearities coming from Equation 7.139 and Equation 7.140 this model has to be numerically solved [129,204].

An inspection of Figure 7.20 clearly reveals that when De is equal to 0 or infinite, a Fickian swelling fluid uptake takes place. Indeed, the sorption curves show a linear dependence on $(t^+)^{0.5}$ for $0 < 100\ M_t^+ < 60$. On the contrary, setting $De = 1$, a typical non-Fickian uptake is observed. Particularly, in this last case, the trend of M_t^+ vs. $(t^+)^{0.5}$ shows an overshoot, with respect to the equilibrium value, followed by damped oscillations. Although the overshoot is experimentally detectable [205–207], the subsequent oscillations are unrealistic and represent a mathematical drawback of the model. Interestingly, R_d increase reflects in an exaltation of the non-Fickian character of

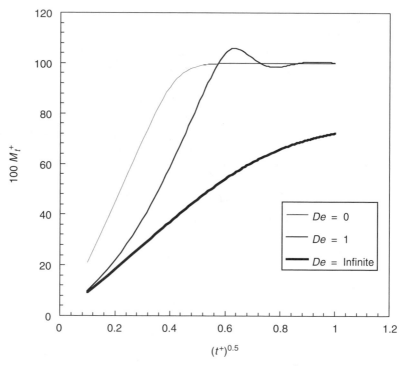

FIGURE 7.20 Dimensionless amount of solvent absorbed (M_t^+) vs. the square root of the dimensionless time t^+ for different Deborah numbers De assuming $R_d = 10$, $C_{eq} = 0.2$ g/cm^3, $g = 14$ cm^3/g and $k = 12$ cm^3/g. (From Grassi, M., Lapasin, R., Pricl, S., *Chem. Eng. Commun.*, 169, 79, 1998. Reproduced by permission of Taylor & Francis, Inc., *http://www.taylorandfrancis.com*.)

the sorption curve (for example, oscillation's amplitude increase). Fortunately, the oscillation drawback can be easily overcome by generalizing the model with the addition of more than one component for the relaxation flux J_r [208]. This means that the Maxwell model is too simple to provide a satisfactory description of the linear viscoelastic properties of a real polymeric matrix and more complicated models must be used, e.g., the generalized Maxwell model constituted by a parallel combination of spring-dashpot series, each of which is characterized by a relaxation time τ_i (see Chapter 3):

$$\tau_i = \eta_i/E_i, \tag{7.157}$$

where η_i and E_i represent, respectively, dashpot fluid viscosity and spring constant. Thus, to improve the Camera-Roda and Sarti model (CRS), the relaxation flux contribution J_r^+ may be given by

$$J_r^+ = \frac{\sum_{i=1}^{i=N} J_{ri}^+ \sigma_i}{\sum_{i=1}^{i=N} \sigma_i}, \tag{7.158}$$

$$J_{ri}^+ = -D_r^+ \nabla C^+ - \tau_i^+ \partial(J_{ri}^+)/\partial(t^+), \tag{7.159}$$

where τ_i^+ is the dimensionless form of τ_i and σ_i is the weighting factor of the ith Maxwell element. τ_i^+ analytical expression is

$$\tau_i^+(C^+) = (De_i/R_d) \exp(k\, C_{eq}(1 - C^+)), \tag{7.160}$$

$$De_i = \tau_{eqi} D_{eq}/L_{eq}^2, \tag{7.161}$$

where τ_{eqi}, D_{eq}, and L_{eq} are the ith relaxation time, the solvent diffusion coefficient, and the thickness of the swollen matrix, respectively. Figure 7.21 reports a comparison between the sorption curve predicted by the CRS model and those derived from its improved version with $N = 2$ (with two different combinations of τ_{eqi} and, hence, of De_i). The simulation is performed (one-dimensional swelling) by setting the mean Deborah number De_m equal to 5.005 for all the curves, $R_d = 100$ (high non-Fickian character) and $\sigma_1 = \sigma_2 = 100$. De_m is defined by

$$De_m = \sum_{i=1}^{i=N} De_i \sigma_i \Big/ \sum_{i=1}^{i=N} \sigma_i. \tag{7.162}$$

The two sets of Deborah number are $De_1 = 0.01$, $De_2 = 10$, and $De_1 = 1$, $De_2 = 9.01$, respectively. The sorption curves (M_t^+ vs. $t^{+0.5}$) calculated from the improved version of the CRS model do not exhibit damped oscillations, and their shapes depend on the set of De_i.

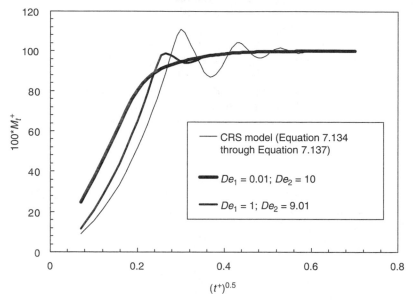

FIGURE 7.21 Comparison between CRS model prediction and those coming from its improved form considering two different Deborah numbers (De_1, De_2). Mean Deborah number is, in all cases, equal to 5.005. Simulations are performed assuming $R_d = 100$, $C_{eq} = 0.2$ g/cm^3, $g = 14$ cm^3/g, and $k = 12$ cm^3/g.

Figure 7.22 shows the simulation results obtained in the case of eight Maxwell elements ($De_1 = 0.01$; $De_{i+1} = 5$ De_i) and for three different weight distributions, ranging from the box-type (d_3) ($\sigma_i = 100$ for all i) to the edge-type (d_1) ($\sigma_i = 100$, 2.9, 0.023, 10^{-3}, 3.7×10^{-7}, 1.34×10^{-9}, 4.8×10^{-12}, 1.72×10^{-14}) passing through d_2 ($\sigma_i = 100$, 43, 29, 23, 19, 16, 14, 12). Although a soft overshoot is detectable (d_3 and d_2), no damped oscillation can be observed and a gradual shifting toward slower sorption kinetics can be noted passing from d_1, a typical liquid-like matrix swelling ($De_m = 0.011$), d_2 ($De_m = 48$) to d_3, a typical solid-like matrix swelling ($De_m = 122$). The advantage of this approach is in the possibility of transferring polymer/solvent rheological characteristics into the mass balance equation. Indeed, as discussed in Chapter 3, frequency sweep test provides the weight distribution (d), the number and the values of the relaxation times τ_{eqi} (or, De_i) relative to the polymer/solvent system under examination (relaxation spectra). Obviously, one strong approximation of this approach consists in assuming that all the relaxation times decrease with solvent concentration according to an exponential law (see Equation 7.160). A possible improvement of the model implies the determination of Equation 7.160 parameter (De_i and k) from experimental relaxation spectra corresponding to different solvent concentrations.

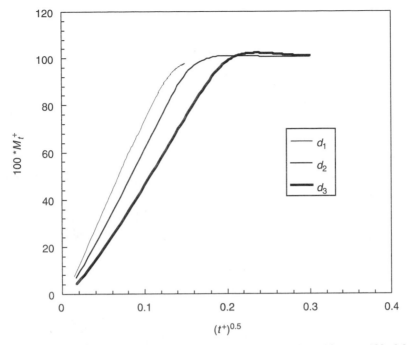

FIGURE 7.22 Effect of different Deborah number distributions (d_1: $\sigma_i = 100$, 2.9, 0.023, 10^{-3}, 3.7×10^{-7}, 1.34×10^{-9}, 4.8×10^{-12}, 1.72×10^{-14}; d_2: $\sigma_i = 100$, 43, 29, 23, 19, 16, 14, 12; box-type d_3: $\sigma_i = 100$ for all i) on the dimensionless amount of absorbed solvent M_t^+. Simulations are performed assuming $R_d = 100$, $C_{eq} = 0.2$ g/cm^3, $g = 14$ cm^3/g, and $k = 12$ cm^3/g.

Due to the presence of many fitting parameters, a proper way to test model reliability is to fit, for example, the swelling data and then compare predicted and experimental deswelling data. Although Colombo and colleagues [209] proved model reliability in describing swelling/deswelling cycle of thermo-reversible N-isopropylacrylamide copolymer matrices, Grassi and group [210] got interesting results by applying this model to thermoreversible dimethylaminoethylmethacrylate-co-acrylamide crosslinked (methylenebisa-crylamide) systems undergoing a considerable swelling equilibrium increase during passage from 40°C to 20°C. Figure 7.23 reports system swel-ling/deswelling kinetics in terms of the swelling ratio S_R(g/g) vs. time t (minutes). It can be seen that model fitting (increasing part of the solid curve) of swelling data (symbols) is quite good and that the subsequent model deswel-ling prediction (decreasing part of the solid curve) is reasonable too. Assuming $g = 14$ cm^3/g and $k = 12$ cm^3/g, knowing that $C_{eq}(20°C) = 0.98$ g/cm^3, that in the initial system water concentration in the membrane is 0.95 g/cm^3 (matrix swelling equilibrium at 40°C) model fitting yields to $\tau_{eq} = 19$ s,

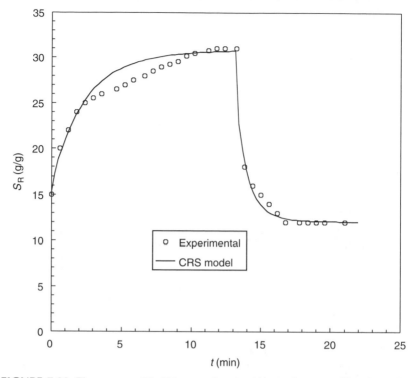

FIGURE 7.23 Thermoreversible *N*-isopropylacrylamide copolymer swelling deswelling kinetics (S_R is the swelling ratio whereas t is time). Although the CRS model has been fitted on swelling data (increasing part of the solid curve), the subsequent curve portion represents model prediction of deswelling kinetics. Both model data (open circles) fitting and prediction are satisfactory. (From Colombo, I., et al., *Fluid Phase Equilib.*, 116, 148, 1996. From Grassi, M., et al., *J. Membrane Sci*, 152, 241, 1999. With permission from Elsevier.)

$Deq = 5.1 \times 10^{-6}$ cm^2/s, and $D_0 = 6.45 \times 10^{-7}$ cm^2/s (this corresponds to $R_d = 12$; $De = 0.01$). It is now of great importance to stress the fact that, although the determined fitting parameters assume reasonable values, their exact values are not significant, as they may strongly depend on the adjustable parameters g and k, fixed according to Camera-Roda and Sarti. The most important finding concerns the existence of a reasonable set of model parameters by means of which the model can properly describe the sorption–desorption phenomena. Finally, it is interesting to point out that as De is very small, there is no need to use the implemented version (namely that with more than one relaxation time) of the Camera-Roda and Sarti model. Indeed, when R_d is about 10, this implementation is necessary only for De values ranging between 1 and 10.

7.3.5.3   Drug Dissolution and Diffusion

Although in previous sections (Section 7.3.1 through Section 7.3.5) only drug–polymeric network interactions were considered, now, also drug/incoming solvent interactions must be accounted for. Indeed the incoming swelling solvent provokes meshes enlargement, exerts a drag action on drug molecules and allows drug dissolution when its local concentration overcomes a threshold value. Accordingly, a bulk flow contribution to the global drug flux should be considered jointly with a position-dependent drug diffusion coefficient. In addition, it must be considered that drug dissolution is not instantaneous in the whole matrix but it firstly takes place in the outer parts and only later it interests the inner part of the matrix.

If the swelling kinetics is slow, the drag effect (and, thus, the bulk flow contribution) can be neglected and only drug diffusion coefficient D_d variation plays an important role. As widely discussed in Chapter 4, even if D_d depends on both the local swelling fluid concentration C and on drug concentration C_d, it can be assumed [149] that only the former plays an important role. As discussed in Chapter 4, many expressions are available for the D_d dependence on C and one of them is that descending from the free volume theory [211,212]:

$$\varphi = 1 - C/\rho_s, \tag{7.163}$$

$$\frac{D_d}{D_0} = Ze^{\left(-a\frac{\varphi}{1-\varphi}\right)} \tag{7.164}$$

where D_0 is the drug diffusion coefficient in the pure solvent, Z and a are two adjustable parameters related to network mesh size and drug molecule radius, respectively, while ρ_s is solvent density. Drug dissolution can be accounted for by Equation 7.81:

$$\frac{\partial C_{dd}}{\partial t} = -K_t (C_s - C_d), \tag{7.81}$$

where C_d and C_{dd} are, respectively, the dissolved and solid local drug concentration, C_s is the drug solubility in the swelling fluid and K_t is the drug dissolution constant. As upon dissolution (see Chapter 5), a phase transition variation can take place, Equation 7.81 needs to be implemented. This transition, although not affecting the drug diffusion coefficient, alters drug solubility and a suitable model for this modification may be that proposed by Nogami [82]:

$$\frac{dC_s}{dt} = K_r(C_s^f - C_s), \tag{7.165}$$

where K_r is the crystallization constant and C_s^f is the drug solubility at the end of the transition. Equation 7.165 solution leads to

$$C_s = (C_s^i - C_s^f)\, e^{(-K_r t^*)} + C_s^f, \tag{7.166}$$

where C_s^i is drug solubility at the beginning of the transition and t^* is defined by

$$t^* = t - t_t, \tag{7.167}$$

where t_t is the time needed to get a local solvent concentration greater than a given threshold value (for example, 0.04 C_{eq}). Obviously, t_t is zero at the matrix/release environment surface and it progressively increases moving inside the matrix. Accordingly, in a one-dimensional situation, the mass balance referred to the drug reads as

$$\frac{\partial C_d}{\partial t} = \frac{\partial}{\partial X}\left(D_d \frac{\partial C_d}{\partial X} - \frac{D_d C_d}{\rho_m}\frac{\partial \rho_m}{\partial X}\right) + K_t\left(\left((C_s^i - C_s^f)e^{(-K_r t^*)} + C_s^f\right) - C_d\right), \tag{7.168}$$

where ρ_m is matrix density and $D_d C_d/\rho_m\, \partial \rho_m/\partial X$ accounts for the effect of matrix density variation (due to swelling) on concentration gradient as discussed in Chapter 4. Supposing that the amount of drug contained in the system is sufficiently low to be neglected whereas computing matrix density, ρ_m, dependence on local solvent concentration is given by [72]

$$\rho_m(X) = \cfrac{4\rho_p + \cfrac{\rho_s(3C_X + C_{X-dX})}{2\rho_s - (C_X + C_{X-dX})} + \cfrac{\rho_s(3C_X + C_{X+dX})}{2\rho_s - (C_X + C_{X+dX})}}{4 + \cfrac{3C_X + C_{X-dX}}{2\rho_s - (C_X + C_{X-dX})} + \cfrac{3C_X + C_{X+dX}}{2\rho_s - (C_X + C_{X+dX})}}, \tag{7.169}$$

where ρ_p is polymer density, whereas C_{X-dX}, C_X, and C_{X+dX} represent swelling fluid concentration at the abscissa $X - dX$, X, and $X + dX$, respectively. Equation 7.169 assumes that the drug amount contained in the system is sufficiently low to be neglected whereas computing matrix density and ρ_m is given by

$$\rho_m(X) = \cfrac{\Delta p(X) + \Delta m(X)}{\cfrac{\Delta p(X)}{\rho_p} + \cfrac{\Delta m(X)}{\rho_s}}, \tag{7.170}$$

where $\Delta p(X)$ and $\Delta m(X)$ represent, respectively, the amount of polymer and solvent contained in a matrix volume element positioned at the abscissa X, having section S and thickness dX. Moreover, solvent concentration profile between two consecutive matrix volume elements is assumed linear.

Before proceeding with this discussion, it is interesting to note that so far it was implicitly assumed that the drug was present in the matrix exclusively in one of its possible three forms, namely macrocrystals, nanocrystals, or amorphous. However, as it is well known that these three phases can coexist inside a polymeric matrix [213,214], Equation 7.81 needs to be applied to each form. Accordingly, Equation 7.168 becomes

$$\frac{\partial C_d}{\partial t} = \frac{\partial}{\partial X}\left(D_d \frac{\partial C_d}{\partial X} - \frac{D_d C_d}{\rho_m}\frac{\partial \rho_m}{\partial X}\right) - \frac{\partial C_{dd}^a}{\partial t} - \frac{\partial C_{dd}^{nc}}{\partial t} - \frac{\partial C_{dd}^{mc}}{\partial t}, \quad (7.168')$$

where

$$\frac{\partial C_{dd}^a}{\partial t} = -K_t^a(C_s(t^*) - C_d) \quad C_s(t^*) = (C_a - C_{mc})\,e^{(-K_r t^*)} + C_{mc}, \quad (7.81^{III})$$

$$\frac{\partial C_{dd}^{nc}}{\partial t} = -K_t^{nc}(C_s(t^*) - C_d) \quad C_s(t^*) = (C_{nc} - C_{mc})\,e^{(-K_r t^*)} + C_{mc},$$

$$(7.81^{IV})$$

$$\frac{\partial C_{dd}^{mc}}{\partial t} = -K_t^{mc}(C_{mc} - C_d), \quad (7.81^V)$$

where C_{dd}^a, C_{dd}^{nc}, and C_{dd}^{mc} are, respectively, the concentration of undissolved amorphous, nanocrystalline, and macrocrystalline drug, K_t^a, K_t^{nc}, K_t^{mc} and C_a, C_{nc}, C_{mc} are the respective dissolution constants and the solubility. Although the dissolution constants can be different (indeed they depend on surface area, see Equation 7.94), limiting solubility (C_a, C_{mc}) and the recrystallization constant must be equal. Obviously, as widely discussed in Chapter 6, C_{nc} lies in between C_a and C_{mc}, its values depend on nanocrystal's radius.

An example of application of Equation 7.134 through Equation 7.140 and Equation 7.168 in the case of completely dissolved drug (the dissolution term in Equation 7.168 is always zero) is reported by Grassi and coworkers [210] who studied, in a side-by-side apparatus, hydrocortisone permeation through thermoreversible crosslinked dimethylaminoethylmethacrylate-co-acrylamide membranes. The peculiarity of these systems consist in an abrupt increase of C_{eq} during transition from 40°C to 20°C. Accordingly, by properly playing on the temperature, it is possible to modulate hydrocortisone efflux from the membrane as the higher C_{eq}, the lower the diffusive resistance offered by membrane (network meshes are larger). Figure 7.24 shows a good agreement between model best fitting (solid line) and experimental permeation data (symbols) in relation to a series of temperature step variation between 20°C and 40°C. Using the "solvent" parameters found in Section 7.3.5.2 ($g = 14\ \text{cm}^3/\text{g}$,

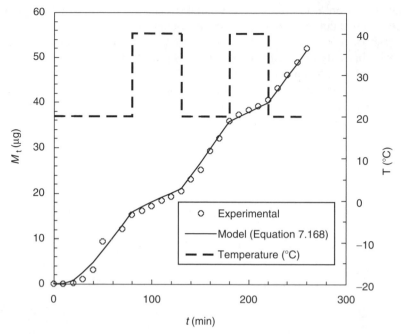

FIGURE 7.24 Model (Equation 7.168) best fitting on hydrocortisone permeation through thermoreversible *N*-isopropylacrylamide copolymer membrane undergoing temperature step variations. (From Grassi et al., *J. Membrane Sci.*, 152, 241, 1999. With permission from Elsevier.)

$k = 12$ cm^3/g, $C_{eq}(20°C) = 0.98$ g/cm^3, $C_{eq}(40°C) = 0.95$ g/cm^3, $\tau_{eq} = 19$ s, $Deq = 5.1 \times 10^{-6}$ cm^2/s, and $D_0 = 6.45 \times 10^{-7}$ cm^2/s) and assuming $\rho_s = 1$ g/cm^3, model best fitting yields $a = 59$ and $D_0 Z = 3.6 \times 10^{-6}$ cm^2/s. In the first period (0–80 min), no swelling/deswelling phenomena take place as the environmental temperature is kept at a constant value (20°C). The cumulative permeated hydrocortisone amount M_t vs. time t has a nonlinear trend as the inner membrane concentration profile has not completely been developed. Indeed, at 20°C, the membrane thickness is not negligible (0.1 cm) and a linear drug concentration profile takes several minutes to develop inside the membrane. The subsequent periods exhibit a linear M_t trend vs. time indicating complete development of the membrane inner concentration profile. This is due to the fact that the "swelling/deswelling time" is small compared to the "drug permeation time" which depends mainly on the hydrocortisone diffusion coefficient in the membrane.

7.3.6 EROSION-CONTROLLED SYSTEMS

A very interesting example dealing with mathematical models describing erosion induced by chemical reasons is the one reported by Thombre, Joshi and Himmelstein [215–217]. These authors focused their attention on polymer degradation from poly(orthoester)-based delivery systems containing an acid-producing species to accelerate matrix hydrolysis. Assuming a slab geometry and perfect sink conditions, they suppose that initially, water (A) moves in the matrix hydrolyzing the acid generator compound (B, for example, an acid anhydride). Then, the generated acid (C) catalyzes polymer degradation according to two steps: (i) formation of an unstable intermediate ester (D^*) and (ii) reaction with water to final polymer degradation products (E):

$$A + B \rightarrow C, \quad C + D \rightarrow D^*, \quad D^* + A \rightarrow E. \qquad (7.171)$$

The diffusion of the compounds due to the degradation process is ruled by the following equation:

$$\frac{\partial C_i}{\partial t} = \frac{\partial}{\partial X}\left(D_i(X, t)\frac{\partial C_i}{\partial X}\right) + v_i, \quad i = A, B, C, E. \qquad (7.172)$$

where C_i and D_i are the concentration and diffusion coefficients of the diffusing species, respectively, X is the space coordinate, v_i is the net sum of synthesis and degradation rate of species i while t is the time. To account for the increasing permeability of matrix due to polymer degradation, the following law for the species diffusion coefficient is considered:

$$D_i = D_i^0 e^{\left(\frac{\mu\left(c_D^0 - c_D\right)}{c_D^0}\right)}, \qquad (7.173)$$

where D_i^0 is the diffusion coefficient of species i in the polymer when the polymer is not hydrolyzed whereas μ is a constant. The model of Thombre, Joshi and Himmelstein yields a rather good agreement with experimental data reported in the literature [79].

　　Monte Carlo-based approaches proved useful to properly describe drug release and polymer dissolution both in case of surface and bulk erosion [79]. Zygourakis [218–220] developed the first example of drug release modeling from surface erosion polymeric matrices. Basically, the matrix is schematized by a two-dimensional grid whose nodes can be occupied by polymer, drug, filler, or can be void (pores). To simulate a drug or a polymer dissolution, each drug or polymer node is assigned a "life expectancy" representing the time needed for instantaneous dissolution after coming in contact with the incoming release fluid (solvent). In this manner, obviously, outer drug or polymer nodes dissolve earlier than inner nodes. To account for the fact that matrix compounds are characterized by different dissolution velocities, "life expectancy" is set proportional to each compound dissolution constant. Accordingly, slow dissolution compounds are characterized by higher "life expectancy" with respect to high-dissolution ones. Although this model is able to describe well, polymer dissolution, it does not account for species diffusion (solvent, drug, polymer degradation products, and so on). Probably, the most general model using Monte Carlo methods is the one proposed by Göpferich and Langer [221,222] that can be applied also to bulk erosion systems [223]. This model accounts for species diffusion, crystallization of polymer degradation products and microclimate pH effects besides, of course, matrix erosion. Analogous to Zygourakis model, matrix cross sections are represented by two-dimensional grids whose nodes can represent amorphous or crystalline polymer. In particular, two types of copolymers [1,3-*bis*(*p*-carboxyphenoxy)propane (CPP); sebacic acid (SA)] were considered [*p*(CPP-SA) 20:80 and *p*(CPP-SA 50:50)] [220]. "Life expectancy" for noneroded nodes are sampled randomly from first-order Erlang distributions with a direct Monte Carlo approach [224], whereas, crystalline nodes are assigned longer lifetimes with respect to those of amorphous nodes. Interestingly, this model is able to calculate matrix porosity $\varepsilon(t, x)$ evolution on time t and space x during erosion knowing, at each time step, the number of uneroded and eroded nodes. On this basis, monomeric species (SA and CPP) diffusion can be modeled, in one dimension, by means of the following equation:

$$\frac{\partial}{\partial t}[C(x,t)\varepsilon(X,t)] = \frac{\partial}{\partial x}\left[D(x,t)\varepsilon(X,t)\frac{\partial C(x,t)}{\partial x}\right] + k_{\mathrm{d}}m(x,t)[C_{\mathrm{s}} - C(x,t)],$$

$$(7.174)$$

where C is the diffusion species concentration in the void nodes (pores), D is the effective diffusion coefficient [146], k_{d} and m_{d} are the dissolution rate constant and mass of suspended crystallized species, whereas C_{s} is species solubility. It is worth mentioning that though the product $C^{*}\varepsilon$ indicates the amount of compound per unit volume of matrix, C represents the concentration referred to only matrix void space (matrix pores). In addition, though the first right-hand side term of Equation 7.174 represents the usual diffusion process driven by concentration gradient, the second term accounts for monomer dissolution. Indeed, the authors assume that, due to limited solubility, SA and CPP monomers instantaneously crystallize upon polymer degradation. Accordingly, the subsequent slower crystal dissolution has to be accounted for. In particular, for CPP monomer solubility, the authors suggest a functional dependence on pK_{a} and SA local concentration. Model fitting yields good agreement with experimental data except for CPP release from p(CPP-SA) 50:50 probably because the model assumes that most parameters are time independent. On the contrary, model prediction dealing with matrix mass reduction is very good for both copolymers [79].

Peppas and Mallapragada modeled drug release from semicrystalline polymers assuming that the key point is crystal dissolution [225,226]. These authors assume a first-order dependence on solvent concentration for what concerns crystal dissolution kinetics. Accordingly, the changes of the crystalline portion volume fraction v_{2c} during time are ruled by the following equation:

$$\frac{\partial v_{2c}}{\partial t} = -k_1 v_1 H(v_{2c}), \tag{7.175}$$

$$\left(v_{2c} = \frac{crystalline\ polymer\ volume}{polymer\ volume} \right) \tag{7.176}$$

where k_1 is the unfolding constant calculated according to energy considerations [215] and v_1 is water volume fraction. The Heavyside function is used to prevent v_{2c} to become negative as crystal dissolution occurs for an extended period. Amorphous polymer volume fraction v_{2a} is ruled by a traditional diffusion term and an additional term accounting for crystalline–amorphous polymer transformation during the unfolding process:

$$\frac{\partial v_{2a}}{\partial t} = \frac{\partial}{\partial x} \left[D_1 \frac{\partial v_{2a}}{\partial x} \right] + k_1 v_1 H(v_{2c}), \tag{7.177}$$

where D_1 is water diffusion coefficient depending exponentially on v_1. Drug volume fraction v_d is ruled by the following equation:

$$\frac{\partial v_d}{\partial t} = \frac{\partial}{\partial x} \left[D_d \frac{\partial v_d}{\partial x} \right], \tag{7.178}$$

where D_d is drug diffusion coefficient through semicrystalline polymer. It depends on both v_{2c} and polymer tortuosity τ according to

$$D_d = D_a \left(\frac{1 - v_{2c}}{\tau} \right), \tag{7.179}$$

where D_a is the drug diffusion coefficient through purely amorphous polymer. This model well reflects experimental evidences referring to metronidazole release from PVA matrices. Narasimham and Peppas [227], modifying the original model of Harland and colleagues [228], studied water and drug diffusion in a polymeric matrix made up of, amorphous, uncrosslinked, linear polymer undergoing chain disentanglement. Assuming slab geometry, a three-component system is considered, indicating water as component 1, polymer as component 2, and drug as component 3. The model considers two moving boundaries, R and S (R is the glassy–rubbery interface and S is the rubbery–solvent interface) and defines (S–R) as the gel layer thickness.

Quasiequilibrium conditions at the rubbery–solvent interface enable the use of the Flory–Rehner theory [167] to calculate the water (and drug) volume fractions at this interface. The disentanglement rate of the polymer is taken as the ratio between the polymer radius of gyration and its repetition time. An important contribution of this work is the presence of a diffusion boundary layer adjacent to the rubber–solvent interface, through which the disentangled chains (and the drug) have to diffuse. The quasisteady-state model solution is given by

$$-\frac{(S-R)}{B} - \frac{A}{B^2} \ln\left\{1 - \frac{B}{A}(S-R)\right\} = t, \qquad (7.180)$$

where

$$A = D(v_{1,eq} - v_1^*)\left(\frac{v_{1,eq}}{v_{1,eq} + v_{d,eq}} + \frac{1}{v_1^* + v_d^*}\right)$$

$$+ D_d(v_d^* - v_{d,eq})\left(\frac{v_{d,eq}}{v_{1,eq} + v_{d,eq}} + \frac{1}{v_1^* + v_d^*}\right), \qquad (7.181)$$

$$B = \frac{k_d}{v_{1,eq} + v_{d,eq}}, \qquad (7.182)$$

where D and D_d are, respectively, the diffusion coefficients of the solvent and the drug, v_1^* and v_d^* are, respectively, the characteristic concentrations of solvent and drug, whereas $v_{1,eq}$ and $v_{d,eq}$ are the equilibrium concentrations of solvent and drug, respectively. k_d is the disentanglement rate of the polymer chains and is calculated using reptation theory [229, 230]. The fraction of drug released reads as

$$\frac{M_d}{M_{d,\infty}} = \frac{v_{d,eq} + v_d^*}{2l}(\sqrt{2At} + Bt), \qquad (7.183)$$

where l is slab half-thickness whereas M_d and $M_{d\infty}$ represent, respectively, the amount of drug released at time t and after an infinite time. This model was successfully tested on different experimental situations [226].

Ju and coworkers [69,231,232] develop a comprehensive model describing the swelling/dissolution behaviors and drug release from cylindrical HPMC matrices. The most important feature of this model lies in defining the polymer disentanglement concentration $\rho_{p,dis}$, below which chain detachment from swollen network occurs. For HPMC matrices, the authors find that the following relation between $\rho_{p,dis}$ and the polymer molecular weight M holds as

$$\rho_{p,dis} = 0.05^*(M/96000)^{-0.8}. \qquad (7.184)$$

In turn, the model predicts the following scaling laws:

$$\frac{M_{\mathrm{p}}(t)}{M_{\mathrm{p}\infty}(t)} \propto M^{1.05}, \quad \frac{M_{\mathrm{d}}(t)}{M_{\mathrm{d}\infty}(t)} \propto M^{0.24}, \tag{7.185}$$

where $M_{\mathrm{p}}(t)$ and $M_{\mathrm{p}\infty}(t)$ are, respectively, the amount of eroded polymer at time t and after an infinitely long time, whereas $M_{\mathrm{d}}(t)$ and $M_{\mathrm{d}\infty}(t)$ are, respectively, the amount of released drug at time t and after an infinitely long time.

Assuming that a stagnant layer between the erosion front and the release environment arises (this is absolutely reasonable dealing with erosion systems [115]), that compounds diffusion takes place only in the radial direction r and that a matrix anisotropic expansion takes place, the model mathematical expression is

$$\frac{\partial \rho_i}{\partial t} = -\frac{\partial \rho_i}{\partial r}\frac{dr}{dt} + \frac{1}{r}\frac{\partial}{\partial r}\left(rD_i\rho_{\mathrm{t}}\frac{\partial w_i}{\partial r}\right) - \rho_i\frac{dV/V}{dt}, \tag{7.186}$$

where ρ_i and D_i are, respectively, concentration (mass/volume) and diffusivity of species i, w_i is species i mass fraction (mass i/mass total), t is time, V is matrix volume, and ρ_{t} is the local matrix density. The first right-hand side term is a convective contribute arising from the moving boundary. Indeed, Murray and Landis assumed that, in a swelling matrix, an elementary mass portion situated at radius r, moves with a velocity equal to dr/dt [233]. The second term is a usual Fickian contribute accounting also for matrix density variation due to swelling. It is clear that the authors implicitly suppose that drug dissolution is much faster than diffusion so that the model applies only for water-soluble drugs. In the case of the poorly water-soluble compounds, an additional generative term accounting for drug dissolution should be added. Finally, the third term is a generative contribute accounting for concentration variation due to local matrix volume changes induced by the swelling process.

According to the NMR measurements of Gao and Fagerness [234], the following concentration dependencies of the diffusivities are assumed:

$$\frac{D_{\mathrm{s,w}}}{D_{\mathrm{w,0}}} = k'_{\mathrm{w}}e^{(k_{\mathrm{w}}w_{\mathrm{w}})}, \tag{7.187}$$

$$\frac{D_{\mathrm{d}}}{D_{\mathrm{d,0}}} = e^{(-k_{\mathrm{d,p}}w_{\mathrm{p}}-k_{\mathrm{d,1}}w_{\mathrm{1}}-k_{\mathrm{d,d}}w_{\mathrm{d}})}, \tag{7.188}$$

where $D_{\mathrm{s,w}}$ and D_{d} are the self-diffusion coefficient of water and the mutual diffusion coefficient of the drug. The subscripts w, d, and p refer to water, drug, and polymer, respectively. D_{i0} represents the diffusivity of the species i in a dilute solution, whereas k'_{w}, k_{w}, and $k_{\mathrm{d,i}}$ are constants. For water and polymer, the mutual diffusion coefficient, D_{w}, is related to the self-diffusion coefficient of water, $D_{\mathrm{s,w}}$ and polymer, as follows [235]:

$$D_{\text{w}} = D_{\text{s,w}}\phi_{\text{p}} + D_{\text{s,p}}\phi_{\text{w}} \approx D_{\text{s,w}}\phi_{\text{p}}, \tag{7.189}$$

where ϕ_i is the volume fraction of the species i. Figure 7.25 and Figure 7.26 show the comparison between model predictions (lines) and experimental data (triangles) referring to HPMC (Figure 7.25) and adinazolam mesylate (ADM; Figure 7.26) release from two different HPMC tablets (HPMC K100LV, MW $\approx$ 30 kDa; HPMC K4M, MW $\approx$ 90 kDa) composed by 35% HPMC, 62% lactose, and 2.5% ADM [231]. Despite the fact that only radial diffusion is considered, the model provides rather good predictions.

Probably the most advanced model for what concerns drug release from amorphous polymers undergoing swelling and erosion is that developed by Siepmann and coworkers [236–238]. This model, dealing with HPMC matrices, can be considered an extension of the Ju's work as it holds also for poorly soluble drugs and it considers species diffusion in both radial and axial directions. Nevertheless, one of the most important differences with the Ju's work consists in the fact that, now, polymer dissolution is accounted for resorting to the reptation theory [239]. When water concentration exceeds a critical threshold, $C_{1,\text{crit}}$, surface polymer chains start to disentangle and

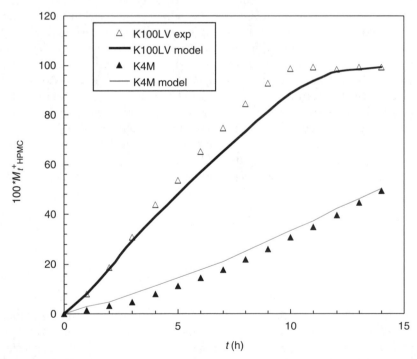

FIGURE 7.25 Comparison between model (Equation 7.186) predictions (lines) and experimental data (triangles) referring to HPMC release from two different HPMC tablets (HPMC K 100LV, MW $\approx$ 30 kDa; HPMC K4M, MW $\approx$ 90 kDa). (Adapted from Ju, R.T., et al., *J. Pharm, Sci.*, 84, 1464, 1995.)

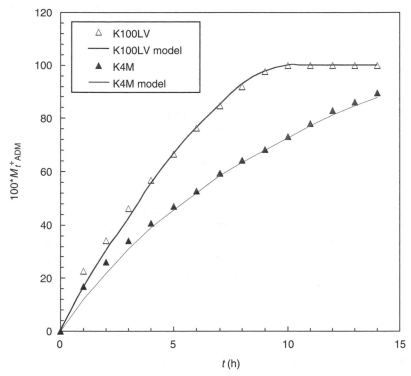

FIGURE 7.26 Comparison between model (Equation 7.186) predictions (lines) and experimental data (triangles) referring to adinazolam mesylate (ADM) release from two different HPMC tablets (HPMC K 100LV, MW $\approx$ 30 kDa; HPMC K4M, MW $\approx$ 90 kDa). (Adapted from Ju, R.T., et al., *J. Pharm, Sci.*, 84, 1464, 1995.)

diffuse through the unstirred layer into the bulk fluid. Polymer mass loss per unit area is ruled by the following equation:

$$M_{pt} = M_{p0} - k_{diss} A_t t, \qquad (7.190)$$

where A_t is the surface area of the device at time t, k_{diss} the dissolution rate constant whereas M_{pt} and M_{p0} are the dry polymer matrix at time t and $t = 0$, respectively. Water and drug diffusion are described using Fick's second law of diffusion for cylindrical geometry, taking into account axial and radial mass transport and concentration-dependent diffusivities:

$$\frac{\partial c_k}{\partial t} = \frac{1}{r} \left\{ \frac{\partial}{\partial r} \left(r D_k \frac{\partial c_k}{\partial r} \right) + \frac{\partial}{\partial z} \left(r D_k \frac{\partial c_k}{\partial z} \right) \right\}, \qquad (7.191)$$

where c_k and D_k represent concentration and diffusion coefficients of the diffusing species ($k = 1$, water; $k = 2$, drug), respectively, whereas r and z are, respectively, the radial and axial coordinates. The authors, in the light of the Fujita theory [240], choose the following equation for the diffusion coefficient dependence on the local water concentration:

$$D_k = D_{kcrit} e^{\left(-\beta_k \left(1-\frac{c_1}{c_{1,crit}}\right)\right)},$$ (7.192)

where β_1 and β_2 are dimensionless constants, whereas $D_{1,crit}$ and $D_{2,crit}$ are, respectively, water and drug diffusion coefficients at the matrix/release medium interface, where polymer chain disentanglement occurs. Supposing ideal mixing conditions (this means that no volume variations occur on drug, polymer, and water mixing), the total matrix volume at any instant is given by the sum of each component volume. Equation 7.191 is numerically solved imposing that at $t = 0$ the matrix is completely dry and that drug concentration is uniform. Moreover, $c_{1,crit}$ is calculated from the polymer disentanglement concentration [69,231,239] and drug concentration is assumed to be zero in the release environment (perfect sink conditions). This model has been successfully tested on theophylline release from different HPMC matrices [237].

7.3.7 POLYDISPERSE SYSTEMS

Many pharmaceutical systems essentially consist of a polymeric carrier hosting the active agent (drug) inside its three-dimensional network. Especially in the case of oral administration, they are often prepared as particulate systems since these forms present remarkable advantages over the single unit devices. The easier dispersion inside the stomach reflects into an appreciable reduction of the local drug concentration which is usually responsible for gastric irritation [241]. Moreover, they are very versatile and this is another reason for their wide use. In spite of the considerable importance of such a technical solution for the preparation of drug release systems, not many mathematical modeling attempts have been made. The main reason, probably, relies on the polydispersity character of particulate systems that obliges to subdivide the continuous particle size distribution into N classes, each one comprehending particles showing the same diameter. This, in turn, requires the simultaneous solution of the proper mass transport equations for each class as, in principle, drug release from each particle class is influenced by the presence of the other classes. Obviously, this reflects in a much more complex solution technique to be adopted.

Dappert and Thies [242] were the first, followed by Gross [243] and Hoffman and coworkers [244], who dealt with drug release from microcapsules, to develop an interesting approach essentially based on the idea that the cumulative amount of drug release pertaining to the ensemble of polydisperse particles can be deduced from the cumulative release of each particle class. Interestingly, they also clearly demonstrate that release kinetics from the particles ensemble differs from that of the single class [243,245]. This model assumes that drug concentration in the release environment is always vanishing so that release kinetics from the generic particle class is not influenced by other classes. The main drawback of this acute approach is that it is impossible to establish a direct link between release kinetics and fundamental parameters such as drug diffusion coefficient inside the particles. On the other hand, the main advantage of this approach relies on its applicability also to nonspherical particles.

Harland and group [153] studied drug release from porous, nonswellable, and nonerodable polymeric particles in both sink and nonsink conditions. Although they provided very useful analytical solutions for particular conditions and they considered the effect of mean particle radius variations, did not renounce the monodisperse hypothesis. Chang and Himmelstein [246], dealing with nonswelling and nonerodable polymeric particles, focus their attention on the effect of the drug dissolution process and initial drug distribution on release kinetics. Also in this case useful analytical solutions are provided but polydispersity is not considered. Abdekhodaie and Cheng [146] studied drug release from particles, but as before, they assumed monodispersity. Wu and group [247], although assuming particle monodispersity, deviate from previous work as, for the first time to our knowledge, they do not account for concentration homogeneity in the release environment. Accordingly, neglecting eventual particle swelling/erosion and supposing that drug dissolution is very fast in comparison with drug diffusion, they demonstrated the influence of particle positioning inside the release environment.

Ritger and Peppas [112,248], studying drug release from nonswellable and nonerodable particles, provide an interesting analytical solution (the only one we could find connecting macroscopic release with fundamental parameters) to the problem of drug release from an ensemble of polydisperse particles:

$$\frac{M_t}{M_\infty} = 1 - \frac{6}{\pi^2} \sum_{i=1}^{N} \left(\omega(r_i)^* \sum_{k=1}^{\infty} \frac{1}{k^2} \exp\left(-\frac{D k^2 \pi^2 t}{r_i^2} \right) \right), \qquad (7.193)$$

where r_i is the radius characterizing to the ith particle class, D is the drug diffusion coefficient inside particles, N is the number of particle classes, whereas $\omega(r_i)$ represents the weight fraction of ith particle class. This model holds if Fick's law with constant D is valid for each class, initial drug concentration inside particles is uniform, drug dissolution is instantaneous and if release environment concentration is always vanishing (sink condition attained). Indeed, Equation 7.193 implies that release kinetics relative to the ith class is not influenced by the presence of other classes and, consequently, the whole drug amount released at time t (M_t) is the sum of the amounts released from all the classes. The same authors, in the same hypotheses apart from assuming a case II release kinetics from each class (case II situation implies an M_t linear dependence on time when the release occurs from a plane sheet and this is typical of systems whose release kinetics is regulated by matrix swelling), propose the following equation:

$$\frac{M_t}{M_\infty} = 1 - \sum_{i=1}^{N}\left(\omega(r_i)\left[1 - \frac{k_0}{C_0 r_i}t\right]^3\right),$$
(7.194)

where k_0 is the model parameter defined as case II relaxation constant and C_0 is the initial drug concentration in the particles. When $N = 1$, case II release kinetics from a sphere of radius r_1 is obtained. In Equation 7.194, it remains unclear whether r_i is to be referred to swollen, nonswollen particles, or to an average value between them.

To our knowledge, the most recent examples of mathematical models accounting for particles polydispersity are those of Grassi and coworkers [83] and Zhou and co-workers [249]. Although both of them assume concentration uniformity in the release environment and nonerodable particles, Zhou further assumes nonswellable particles and instantaneous drug dissolution in the particles (diffusion-controlled systems). Nevertheless, these two models are conceptually equivalent and their novel character mainly consists in accounting for the mutual influence exerted by particles for what concerns release kinetics. Accordingly, their mathematical expression comes out from applying to each particle class, the equations discussed in this mathematical modeling section solvent uptake:

Mass balance equation:

$$\frac{\partial C_i}{\partial t} = \frac{1}{R_i^2}\frac{\partial}{\partial R_i}[R_i^2 J_i], \quad i = 1, 2, \ldots, N,$$
(7.195)

$$J_i = J_{fi} + J_{ri}, \quad i = 1, 2, \ldots, N,$$
(7.196)

$$J_{fi} = -D_0\frac{\partial C_i}{\partial R_i}, \quad i = 1, 2, \ldots, N,$$
(7.197)

$$J_{ri} = -D_r\frac{\partial C_i}{\partial R_i} - \tau\frac{\partial J_{ri}}{\partial t}, \quad i = 1, 2, \ldots, N,$$
(7.198)

$$D_r = D_{eq}e^{(g[C_i - C_{eq}])} - D_0, \quad i = 1, 2, \ldots, N,$$
(7.199)

$$\tau = \tau_{eq}e^{(f[C_{eq} - C_i])}, \quad i = 1, 2, \ldots, N,$$
(7.200)

$$R_{ij} = \sqrt[3]{R_{ij-1}^3 + (R_{ij}^3(0) - R_{ij-1}^3(0))\frac{\rho_s}{\rho_s - C_{ij}}},$$
$$i = 1, \ldots, N; \quad j = 1, \ldots, V_c$$
(7.201)

Initial conditions:

$$C_i = 0. \quad 0 \le R_i \le R_{\mathrm{pi}}. \quad i = 1, 2, \ldots, N, \tag{7.202}$$

$$C_i = 0.8 C_{\mathrm{eq}}. \quad R_i \le R_{\mathrm{pi}}. \quad i = 1, 2, \ldots, N, \tag{7.203}$$

$$J_{\mathrm{fi}} = J_{\mathrm{ri}} = 0. \quad 0 \le R_i \le R_{\mathrm{pi}}. \quad i = 1, 2, \ldots, N. \tag{7.204}$$

Boundary conditions:

$$J_{\mathrm{fi}} = J_{\mathrm{ri}} = 0, \quad R_i = 0, \quad i = 1, 2, \ldots, N, \tag{7.205}$$

$$\tau \frac{dC_i}{dt} = C_{\mathrm{eq}} - C_i, \quad R_i = R_{\mathrm{pi}}, \quad i = 1, 2, \ldots, N. \tag{7.206}$$

Drug release:
  Mass balance equation

$$\frac{\partial C_{\mathrm{di}}}{\partial t} = \frac{1}{R_i^2} \frac{\partial}{\partial R_i} \left[D(C_i(R_i)) \frac{\partial C_{\mathrm{di}}}{\partial R_i} R_i^2 \right] - \sum_{m=1}^{3} \frac{\partial C_{\mathrm{ddi}}^m}{\partial t}, \quad i = 1, 2, \ldots, N, \tag{7.207}$$

$$\frac{D}{D_0} = A e^{\left(-\frac{B\varphi_i}{1 - \varphi_i} \right)}, \quad \varphi_i = 1 - \frac{C_i(R_i)}{\rho_{\mathrm{s}}}, \tag{7.208}$$

$$\frac{dM_{\mathrm{c}}}{dt} = V_{\mathrm{r}} K_{\mathrm{rb}} (C_{\mathrm{r}} - C_{\mathrm{mc}}), \quad i = 1, 2, \ldots, N, \tag{7.209}$$

Initial conditions

$$C_{\mathrm{r}} = 0, \tag{7.210}$$

$$C_i(R_i) = 0, \quad 0 < R_i < R_{\mathrm{pi}}. \quad i = 1, 2, \ldots, N, \tag{7.211}$$

$$C_{\mathrm{ddi}}^m(R_i) = C_{\max} \left(1 - \frac{1 - e^{-\left[0.5 \left(\frac{1 - R_i^+}{1 - R_{\mathrm{f}}^+} \right)^2 \right]}}{1 - e^{-\left[0.5 \left(\frac{1}{1 - R_{\mathrm{f}}^+} \right)^2 \right]}} \right), \quad R_i^+ = \frac{R_i}{R_{\mathrm{pi}}}, \tag{7.212}$$

$$R_{\mathrm{f}}^+ = \frac{R_{\mathrm{fi}}}{R_{\mathrm{pi}}}, \quad i = 1, 2, \ldots, N,$$

$$M_{\mathrm{c}} = 0. \tag{7.213}$$

Boundary conditions:

$$\frac{\partial C_{di}}{\partial R_i} = 0, \quad R_i = 0, \quad i = 1, 2, \ldots, N, \tag{7.214}$$

$$C_r = \frac{C_i}{k_p}, \quad R_i = R_{pi}, \quad i = 1, 2, \ldots, N, \tag{7.215}$$

$$M_0 = V_r C_r + M_c + \sum_{i=1}^{N} N_{pi} \int_0^{R_{pi}} \left[C_{di}(R_i) + \sum_{m=1}^{3} C_{ddi}^m(R_i) \right] 4\pi R_i^2 dR_i, \tag{7.216}$$

where R_{pi} indicates the ith particles class radius, N_{pi} is the number of particles belonging the ith class, R_i is the radial coordinate relative to ith class, C_i and C_{di} represent, respectively, solvent and dissolved drug concentration in R_i, C_{ddi}^1, C_{ddi}^2, and C_{ddi}^3 are, respectively, the not dissolved concentrations of amorphous, nanocrystalline, and macrocrystalline drug in R_i, M_c is the amount of drug re-crystallized in the release environment, K_{rb} is the drug recrystallization constant in the release environment (it can differ from K_r, the drug recrystallization constant inside the swelling matrix), k_p is the partition coefficient, V_r and C_r are, respectively, release environment volume and drug concentration in it, M_0 is the whole drug amount contained in the drug loaded particles. For the sake of simplicity, in writing the solvent uptake equations only one relaxation time is considered, although their generalization to more than one relaxation time is immediate. Similarly, in the drug mass balance (Equation 7.207), the effect of the matrix density gradient has been neglected as its importance is not so relevant for typical polymeric matrix–based delivery systems. In addition, the dissolution of undissolved drug (C_{ddi}^1, C_{ddi}^2, and C_{ddi}^3) is ruled by Equation 7.81^{III}, Equation 7.81^{IV}, and Equation 7.81^{V}. Equation 7.201 expresses particle radius increase due to swelling. In particular, R_{ij} is the radius of the jth volume element in which the generic particle of the ith class can be subdivided into (V_c is the total volume element number per class); R_{ij-1} represents the volume element radius adjacent to R_{ij}, $R_{ij}(0)$ and $R_{ij-1}(0)$ represent the R_{ij} and R_{ij-1} values when the local solvent concentration C_{ij} is zero. This equation assumes that solvent concentration in each volume element is constant and that volume additivity holds as supposed in Equation 7.147. Although Equation 7.209 indicates the increase of drug amount recrystallized in the release environment, Equation 7.212 assumes that drug distribution inside the matrix is not uniform but has a sigmoidal shape showing the inflection point in $R_i = R_f$ and the maximum, C_{max}, in R_{pi} as suggested by Lee [108]. In principle, C_{ddi}^1, C_{ddi}^2 and C_{ddi}^3 can be characterized by different sigmoidal profiles concentrations.

Finally, Equation 7.216, as an overall mass balance ensuring that the sum of drug mass present in the release environment and in the particles is always constant and equal to M_0, serves for the C_r determination as it closes the balance between equations and unknowns. Indeed, from a mathematical viewpoint, numerical model solution (control volume method [129]) requires to solve N times a linearized form of the nonlinear system represented by the discretization of Equation 7.195 (its dimension coincides with the number V_c of control volumes in which the generic particle of the jth class is subdivided to calculate the internal solvent profile concentration). Subsequently, to get drug concentration profile inside the particles, we have to solve a linear system of algebraic equations (coming from Equation 7.207 discretization) whose dimension increases with V_c. Indeed, each particle class implies V_c control volumes and $(V_c + 2)$ unknown concentrations, and hence the total number of unknowns is $N^*(V_c + 2) + 1$, where 1 indicates the drug concentration in the release environment. The unknown number equals the number of the equations at disposal: (N^*V_c) Equation 7.207, N Equation 7.214, N Equation 7.215 and the total mass balance represented by Equation 7.216. The numerical method adopted easily fits the necessity of solving the differential Equation 7.206 in the above-mentioned linear system by considering a finite difference approximation. Nevertheless, this fact suggests excluding too large integration time steps Δt to avoid solution accuracy problems. Accordingly, we set $V_c = 10$, $N_c = 10$, and $\Delta t = 0.5$ s. Of course, one of the main assumptions of this model is that solvent concentration profile is not affected by drug presence so that solvent equations (Equation 7.195 through Equation 7.206) can be solved independently from drug equations (Equation 7.207 through Equation 7.216).

Finally, it is worth mentioning that any initial particle size distribution (this usually coincides with the dry particle size distribution) can be chosen. For example, in virtue of its mathematical power, the Weibull one [250] can be a good choice:

$$\frac{V}{V_0} = 1 - e^{\left(-\left(2\frac{R_p - R_{min}}{\eta}\right)^\delta\right)}, \tag{7.217}$$

where R_p and R_{min} are the generic particle radius and the minimum particle radius, respectively, η and δ are two parameters regulating the Weibull size distribution, whereas V_0 and V are the total volume occupied by the ensemble of polymeric particles and the volume occupied by those particles having a radius lower than or equal to R_p, respectively. Equation 7.217 discretization allows the calculation of R_{pi} and N_{pi}:

$$R_i^u = R_{min} + i\frac{R_{max} - R_{min}}{N}, \quad i = 0, 1, \dots, N, \tag{7.218}$$

$$R_{pi} = \frac{R_i^u + R_{i-1}^u}{2}, \quad i = 1, 2, \dots, N, \tag{7.219}$$

$$N_{pi} = \frac{V_0}{\frac{4}{3}\pi R_{pi}^3} \left(e^{\left(-\left(2\frac{R_{i-1}^u - R_{min}}{\eta}\right)^\delta\right)} - e^{\left(-\left(2\frac{R_i^u - R_{min}}{\eta}\right)^\delta\right)} \right), \quad i = 1, 2, \ldots, N, \quad (7.220)$$

where R_{max} is the maximum particle radius of the distribution.

Grassi and coworkers [83] check the correctness and the reliability of this model studying MAP (medroxyprogesterone acetate; antitumoral drug; Carlo Erba Milano, Italy) and TEM (Temazepam, benzodiazepin derived; sedative and hypnotic; Carlo Erba Milano, Italy) release from the same polymeric powder (crosslinked PVP, Kollidon CLM-BASF, Germany). MAP and TEM are chosen as model drugs because the former is present in the PVP particles in a completely macrocrystalline form whereas the latter is present in a completely amorphous form as demonstrated by x-rays and differential scanning calorimeter analysis. Accordingly, TEM should undergo an evident recrystallization process upon dissolution whereas MAP should not. In both cases, drug-loading process consists of swelling PVP by CH_2Cl_2 drug solution under continuous mixing, drying to constant weight in a vacuum oven (Vuototest, Mazzali, Italy) at 30°C for 3 h, deaggregating with a 45 mesh sieve and homogeneously mixing (Turbula, Bachofen, Switzerland). The dry particle size distribution is measured by mercury porosimetry (Mod. 2000 Carlo Erba Strumentazione, Italy) and applying the method of Mayer and Stone [251]. Release experiments are carried out in a well-stirred thermostatic (37°C) environment containing phosphate buffer solution (pH = 7.5). A peristaltic pump (40 cm^3/min) guarantees the solution motion to the UV analyzer and the return to the release environment (wavelength $\lambda = 247$ nm for MAP; wavelength $\lambda = 313$ nm for TEM). To avoid particle entrainment two filters were inserted in the circulation circuit.

In order to reduce model degrees of freedom, the majority of its parameters have to be determined by means of independent experiments or by resorting to the pertinent theory, whereas the remainders have to be considered fitting parameters. Nevertheless, due to the high number and the practical impossibility of *a priori* knowing all of them, some model parameters must be reasonably fixed. The reasonability of this choice is *a posteriori* checked by verifying that the parameters coming from fitting procedure assume physically consistent values. Nevertheless, it can be affirmed that, whatever be the choice, drug profile release cannot be successfully explained by a model if some of the model parameters, such as those related to the particle size distribution or the drug concentration profile inside the particles, do not assume their true value. Accordingly, it is found that dry PVP particle size distribution can be described by the following Weibull distribution ($\eta = (5.96 \pm 0.025) \times 10^{-4}$cm, $\delta = 3.88 \pm 0.08$; $R_{min} = 3.13 \times 10^{-5}$cm, $= R_{max} = 7.0 \times 10^{-4}$ cm). $C_{eq} = 0.31 \pm 0.01$ g/cm^3 and drug solubilities of both drugs in the swelling fluid ($C_{scTEM} = 224 \pm 8$ μg/cm^3; $C_{scMAP} = 3.43 \pm 0.2$ μg/cm^3)

TABLE 7.6
Values of the Preset and Fitting Parameters Relative to MAP and TEM Release

Parameters		MAP	TEM
$C_{p,eq}$ (g/cm^3)	Input parameters	0.31 ± 0.1	0.31 ± 0.1
f (cm^3)	Parameters	12	12
A (−)		1	1
D_{ds} (cm^2/s)		6.95×10^{-6}	7.13×10^{-6}
C_0 (μg/cm^3)		202,500	102,800
C_{sc} (μg/cm^3)		3.43 ± 0.2	224 ± 8
R_{max} (cm)		7.0×10^{-4}	7.0×10^{-4}
R_{min} (cm)		3.13×10^{-5}	3.13×10^{-5}
η (cm)		$(5.96 \pm 0.025) \times 10^{-4}$	$(5.96 \pm 0.025) \times 10^{-4}$
δ (−)		3.88 ± 0.08	3.88 ± 0.08
K_p (−)		1	1
V_0 (cm^3)		0.494	0.535
V_r (cm^3)		100	50
C_{sa} (μg/cm^3)	Fitting parameters	—	8649
τ_{eq} (s)		0.1	0.35
g (cm^3/g)		5	51.75
D_0 (cm^2/s)		10^{-8}	10^{-10}
D_{eq} (cm^2/s)		10^{-7}	10^{-7}
B (−)		0.71	2
K (L/s)		5.7	0.1025
K_r (L/s)		—	0.007
K_{rb} (L/s)		—	0.0121

Source: From Grassi, M., Colombo, I., Lapasin, R., *J. Contr. Rel.*, 68, 97, 2000. With permission from Elsevier.

are experimentally measured. The drug diffusion coefficients in the pure solvent ($D_{dsTEM} = 7.13 \times 10^{-6}$ cm^2/s; $D_{dsMAP} = 6.95 \times 10^{-6}$ cm^2/s) are calculated according to the well-known Stokes–Einstein equation [154] whereas Equation 7.207 parameter A is set to 1. The partition coefficient k_p for both drugs is set to 1, whereas Equation 7.200 parameter f is set in accordance with Camera-Roda and Sarti [200] ($f = 12$ cm^3/g). The remaining are fitting parameters. Table 7.6 shows the values of the preset and fitting parameters relative to MAP and TEM release. Figure 7.27 (MAP release; 600 mg corresponding to $V_0 \approx 0.494$ cm^3 characterized by an initial MAP concentration $C_{0MAP} = 202.5$ mg/cm^3) shows a satisfactory model data fitting with the following values of the adjustable parameters $D_{eq} = 10^{-7}$ cm^2/s, $D_0 = 10^{-8}$ cm^2/s, $\tau_{eq} = 0.1$ s, $g = 5$ cm^3/g, $K = 5.7$ 1/s, and $B = 0.71$ and considering a uniform drug distribution inside the particles at the beginning of the release experiments.

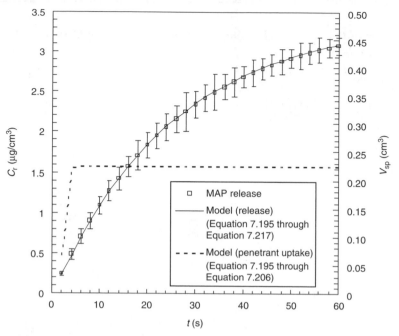

FIGURE 7.27 Comparison between model best fitting (solid line) and medroxyproges-terone acetate (MAP) release data (open circles) from polydispersed crosslinked Poly-rinylpyrrolidone particles. While vertical bars indicate standard error, dotted line represents model prediction about the volume of absorbed solvent V_{sp}. (From Grassi, M., Colombo, I., Lapasin, R., *J. Contr. Rel.*, 68, 97, 2000. With permission from Elsevier.)

It can be seen that the release kinetics is mainly governed by the dissolution constant K rather than by the characteristics of both the solvent inward diffu-sion (D_{eq}, D_0, τ_{eq}, g) and drug outward diffusion (B). Indeed, the simulation reveals that swelling extremely fast, is completed within 5 s only (see the dotted line in Figure 7.27). In such a case K_r, K_{rb} are not determined since no recrystallization occurs upon MAP dissolution. Quite different is the situation illustrated in Figure 7.28 where the fitting curve is compared with the experi-mental data for the TEM release (658 mg corresponding to $V_0 \approx 0.535$ cm^3 characterized by an initial TEM concentration $C_{0TEM} = 102.8$ mg/cm^3). The effects produced by the recrystallization manifests through the appearance of a maximum in the release curve (oversaturation) due to the progressive reduction of the TEM solubility both in the swelling particle and in the release environment. Even if there is a slight divergence between the calculated curve and the experimental data at longer times ($t > 200$ s), the model provides a reasonable fitting also in this complicated condition, with the following parameters: $D_{eq} = 10^{-7}$ cm^2/s, $D_0 = 10^{-10}$ cm^2/s, $\tau_{eq} = 0.35$ s, $g = 51.75$ cm^3/g, $K = 0.1025$ 1/s, $B = 2.0$, $C_{saTEM} = 8649$ µg/cm^3, $K_r = 0.007$ 1/s,

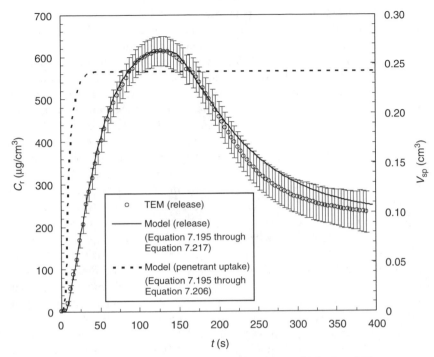

FIGURE 7.28 Comparison between model best fitting (solid line) and Temazepam (TEM) release data (open circles) from polydispersed crosslinked polyvinylpyrrolidone particles. While vertical bars indicate standard error, dotted line represents model prediction about the volume of absorbed solvent V_{sp}. (From Grassi, M., Colombo, I., Lapasin, R., *J. Contr. Rel.*, 68, 97, 2000. With permission from Elsevier.)

$K_{rb} = 0.0121$ 1/s and a uniform drug distribution inside the particles at the beginning of the release. Figure 7.28 also shows that PVP swelling equilibrium is reached in about 50s, which is 10 times slower with respect to the MAP case.

Grassi and coworkers [47] also tested this model in a simpler situation to check its reliability in predicting drug release behavior from an ensemble of nonswellable spherical particles once the release behavior of particles classes is known. In particular, they focus the attention on paracetamol (PA) (ACEF, Fiorenzuola D'Arda, Piacenza-Italy) release from stearic acid (SA) (Galeno, Milano-Italy) and monohydrate lactose (LA) (Pharmatose 200 mesh, Meggle, Wasserburg-Germany) particles obtained by melt extrusion. Upon contact with the release environment ($V_r = 900$ cm^3; buffer solution 0.2 M NaCl/0.2 M HCl, pH 1.2, 37°C; USP 24 rotating basket apparatus, Mod. DT-1, Erweka, Italy, stirring rate of 100 rpm; UV drug concentration determination at 245 nm), lactose rapidly dissolves and paracetamol release occurs due to dissolution and diffusion through the buffer-filled matrix channels

formed by lactose dissolution (supposed instantaneous respect to the paracetamol one). Due to the hydrophobic nature of the insoluble stearic acid
skeleton, matrix swelling does not occur and model equations reduce to those
devoted to drug (Equation 7.207 through Equation 7.216). In addition, as
paracetamol does not undergo re-crystallization, Equation 7.209 and Equation
7.213 disappear whereas the dissolution term in Equation 7.207 reduces to
only one term (macrocrystals dissolution). Due to the complex matrix topology, drug diffusion coefficient is substituted by the effective diffusion
coefficient, as discussed in Section 7.3.4. Finally, due to particles preparation
technique, a uniform initial drug concentration C_{d0} inside particles is assumed. After granulating, pellets (composition: SA:LA:PA = 20:20:60
(w/w)%) undergo a separation process by means of a vibrating apparatus
(Octagon 200, Endecotts, London-UK) and a set of sieves to get five classes:
ω_{2000}, ω_{1250}, ω_{800}, ω_{630}, and $\omega_{residual}$. Accordingly, ω_{2000} is composed of
particles whose radius ranges between 3000 and 2000 μm, ω_{1250} is composed
of particles whose radius ranges between 2000 and 1250 μm, ω_{800} is composed of particles whose radius ranges between 1250 and 800 μm, ω_{630} is
composed of particles whose radius ranges between 630 and 800 μm, and,
finally, $\omega_{residual}$ is composed of particles whose radius is smaller than 630 μm.
For the sake of simplicity, each fraction is supposed to be characterized by a
mean diameter ϕ_m, regardless of the fact that it is surely polydispersed.
Accordingly, ϕ_m, the effective diffusion coefficient D_e and the dissolution
constant K_t represent model fitting parameters. Despite different fitting strategies can be adopted to get adjustable parameters, we believe that the best
choice is to fit the model on the experimental data referring to ω_{2000} class with
D_e, K_t, and ϕ_m (that must be >2000 μm) as fitting parameters. Then, based on
the known D_e and K_t, all other classes of data are fitted assuming ϕ_m as
unique fitting parameter (D_e and K_t cannot vary among the five classes). The
reason for this approach relies on the fact that, as clearly shown in Figure 7.29,
the hypothesis of perfect spherical particles is fully attained for ω_{2000} whereas it
is a poor approximation for the smallest classes (ω_{630} and $\omega_{residual}$). Figure 7.30
shows the satisfactory agreement between model best fitting (solid line) and the
experimental data (symbols) referring to ω_{2000}, ω_{1250}, and ω_{800} fractions. The
ω_{2000} fraction fitting is performed knowing that the initial value of the undissolved drug concentration C_{d0} is equal to 741.53 mg/cm^3, paracetamol solubility C_s is equal to 18 mg/cm^3, and the amount of particles considered W_0 is
equal to 15.7 mg ($M_0 = 9.4$ mg). Fitting parameter values obtained
($D_e = 1.7 \times 10^{-4}$ cm^2/min, $K_t = 12$ cm/min, and $\phi_m = 2500$ μm) underline
the fact that the dissolution process is fast and, consequently, the release
process is mainly ruled by diffusion. Figure 7.30 also reports the comparison
between model best fitting (solid lines) and the experimental data (symbols)
referring to ω_{1250} and ω_{800} fractions. This fitting is performed assuming the
same conditions set for the ω_{2000} fraction ($D_e = 1.7 \times 10^{-4}$ cm^2/min and
$K_t = 12$ cm/min), except for W_0 that is equal to 18.1 mg ($M_0 = 10.86$ mg)

FIGURE 7.29 SEM pictures of pellets size fractions: 2000 μm (a); 1250 μm (b); 800 μm (c); 630 μm (d); <630 μm (residual) (e). (From Grassi, M., et al., *J. Contr. Rel.* 88, 381, 2003. With permission from Elsevier.)

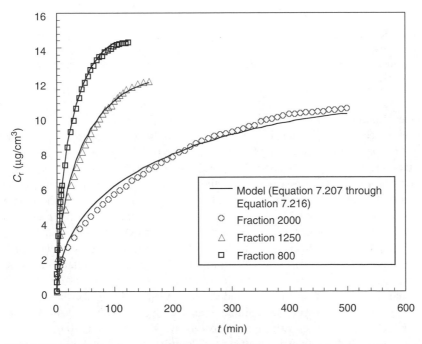

FIGURE 7.30 Comparison between model best fitting (solid line) and experimental paracetamol release from 2000 μm fraction (circles), 1250 μm fraction (triangles), and 800 μm fraction (squares). (Adapted from Grassi, M., et al., *J. Contr. Rel.*, 88, 381, 2003. With permission form Elsevier.)

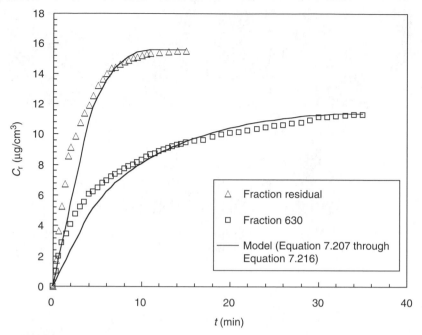

FIGURE 7.31 Comparison between model best fitting (solid line) and experimental paracetamol release from 630 μm fraction (triangles) and residual (<630 μm) (squares). (Adapted from Grassi, M., et al., *J. Contr. Rel.*, 88, 381, 2003. With permission from Elsevier.)

and 21.5 mg ($M_0 = 12.9$ mg) for the ω_{1250} and ω_{800} fractions, respectively. As it can be seen, the agreement is even better than in the ω_{2000} fraction case and the fitting parameter is $\phi_m = 1300$ μm and $\phi_m = 1025$ μm for ω_{1250} and ω_{800} fractions, respectively. These values (ϕ_m) are absolutely reasonable and this is not surprising as, also in the case, the hypothesis of spherical particles is reasonably attained (see Figure 7.29). On the contrary, Figure 7.31 shows that model best fitting (solid lines), especially at the beginning, does not provide a satisfactory description of the release data (symbols) referring to ω_{630} and $\omega_{residual}$ fractions. This is certainly due to lack of particles spherical in shape. Model best fitting is performed assuming the same conditions relative to ω_{2000} fraction ($D_e = 1.7 \times 10^{-4}$ cm^2/min and $K_t = 12$ cm/min), except for W_0 that is equal to 17 mg ($M_0 = 10.2$ mg) and 23.3 mg ($M_0 = 14$ mg) for the ω_{630} and $\omega_{residual}$ fractions, respectively. Fitting parameter ϕ_m is equal to 540 and 315 μm for the ω_{630} and $\omega_{residual}$ fractions, respectively. It is interesting to note that in the ω_{630} case, the model, in an attempt to describe experimental data, interprets the particle as spheres having a diameter lower than the physically acceptable value of 630 μm; this is a clear evidence for the failure of the spherical particle hypothesis. The most important proof of model reliability is given by Figure 7.32

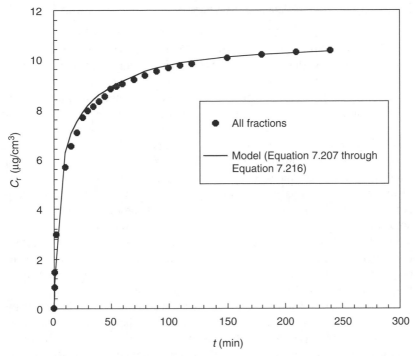

FIGURE 7.32 Comparison between model prediction (solid line) and experimental paracetamol release relative to a system composed of all pellets fractions (particle size composition: $\omega_{2000} = 0.153$, $\omega_{1250} = 0.159$, $\omega_{800} = 0.210$, $\omega_{630} = 0.160$, and $\omega_{residual} = 0.319$ (circles). (Adapted from Grassi, M., et al., *J. Contr. Rel.*, 88, 381, 2003.)

reporting paracetamol release data (open circles) from the ensemble of the five classes (w/w particle size distribution: $\omega_{2000} = 0.153$, $\omega_{1250} = 0.159$, $\omega_{800} = 0.210$, $\omega_{630} = 0.160$, and $\omega_{residual} = 0.319$) and model prediction (solid line) assuming $V_r = 5000 \text{ cm}^3$, $C_{d0} = 741.53 \text{ mg/cm}^3$, $W_0 = 88.8 \text{ mg }(M_0 = 53.3$ mg), $D_e = 1.7 \times 10^{-4} \text{ cm}^2/\text{min}$, $K_t = 12 \text{ cm/min}$, $\phi_{m2000} = 2500 \text{ μm}$, $\phi_{m1250} = 1300 \text{ μm}$, $\phi_{m800} = 1025 \text{ μm}$, $\phi_{m630} = 540 \text{ μm}$, and $\phi_{mResidual} = 315$ μm. Undoubtedly, the description is fully satisfactory and the model can be now considered a good tool to predict drug release from an ensemble of different classes characterized by different particle size distribution. Finally, as in the case of theophylline release from cylindrical matrix discussed in Section 7.3.4, it is possible to estimate particles tortuosity τ. Indeed, based on Equation 7.79, D_e, matrix porosity ε (beginning condition: paracetamol not dissolved and lactose totally dissolved, $\varepsilon_1 = 0.17$; final condition: paracetamol and lactose totally dissolved, $\varepsilon_2 = 0.74$) and knowing that paracetamol diffusion coefficient in water, D_w, is equal to $7.1 \times 10^{-4} \text{ cm}^2/\text{min}$ [47], τ value averaged on ε_1 and ε_2 value can be calculated and its value is 1.9.

7.3.8 Diffusion in Fractal Matrices

Fractal objects are interesting and intriguing structures theorized for the first time by Mandelbrot [252] who is considered "the father of fractals." The concept of fractals (the word fractal means fraction form and comes from the latin verb *frangere* (*frango, frangis, fregi, fractum, frangere*) meaning to fracture, break) relies on Mandelbrot's observation concerning the independence of fractal objects forms from scale at which they are observed. Indeed, he defines the fractal as a shape built with *autosimilar* (from the latin words *auto = self + similes = same*), but smaller shapes and forms. In turn, these smaller shapes are composed of even smaller *autosimilar* shapes which are also made up in the same way. Roughly speaking, this means that, regardless of the magnification used, a microscopic analysis of a fractal object would always reveal similar images. Although only theoretical fractals are truly autosimilar, naturally occurring forms show autosimilarity only in a certain range. Nevertheless, by means of the fractal geometry it is possible to theoretically recreate naturally occurring forms (coastlines, clouds, mountains, trees, and so on) with an incredible grade of similitude, and that is why fractal geometry is also called the geometry of nature.

Coming back on a more theoretical plane, we would like to emphasize that fractal geometry is based on the notion of dimension that, for a points set A, can be defined as the exponent d relating the number of boxes (segments, squares, cubes, or iper-cubes in one-, two-, three-, or n-dimensional space, respectively), $N(\eta)$ necessary to cover A, with boxes norm η (it is a characteristic length). In the R^n space, the Euclidean norm coincides with the distance of the point $X(x_1, x_2, \ldots, x_n)$ from coordinates system origin for η tending to zero:

$$N(\eta) = \frac{L(\eta)}{\eta^n} = \lambda \eta^{-d}, \quad \eta \to 0, \tag{7.221}$$

where n is the dimension of the space containing A and $L(\eta)$ is the measure of A length, surface, volume, or iper-volume. Daily experience tells us that for nonfractal (Euclidean) objects like simple segments ($n = 1$), squares ($n = 2$), or cubes ($n = 3$), d is equal to 1, 2, and 3, respectively, and this is the reason why Euclidean objects measure L ($= \lambda$) does not depend on boxes norm η. For fractal objects, on the contrary, this is no longer true and, thus, d can be viewed as the A fractal dimension. For example, it has been demonstrated [252] that the length $L(\eta)$ of a coastline, a typical natural occurring fractal contained in a $n = 1$ space, tends to infinite for η tending to zero and this trend is given by Equation 7.221 with $d < 1$. As a consequence, A is a fractal points set, when $d < 1$, 2, and 3, for $n = 1$, 2, and 3 respectively.

In physical and applied literature, it is customary to define the fractal dimension of a structure by means of the so-called mass–radius relation. If $m(r)$ represents the structure mass contained in the sphere of radius r and centered in the mass center, assuming a constant specific density (homogeneous fractals), Equation 7.221 is equivalent to

$$m(r) \propto r^d. \tag{7.222}$$

Accordingly, fractal structure density $\rho(r)$ results in a decreasing function of position:

$$\rho(r) = \frac{m(r)}{V(r)} \propto \frac{m(r)}{r^n} \propto r^{d-n}, \tag{7.223}$$

where $V(r)$ is structure volume ($n = 3$), surface ($n = 2$), and length ($n = 1$).

As transport phenomena are influenced by topological properties like connectivity, presence of closed loops or dead ends, it is reasonable that fractal media alter the usual rules of diffusion. For this purpose, let us focus the attention on the random walk of a particle moving in a fractal medium. Assuming X_0 as particle initial position, it can be demonstrated that particle mean square displacement $\langle r^2(t) \rangle$ depends on time t according to the following law:

$$\langle r^2(r) \rangle \propto t^\beta \quad \text{for } t \to \infty, \quad \beta = \frac{d_s}{d}, \tag{7.224}$$

where d_s is called fracton dimension [253]. Although for nonfractal structures $\beta = 1$, fractal media are characterized by $\beta < 1$. It is interesting to note that when $\beta = 1$, Equation 7.224 represents nothing more than Fick's law for diffusion (see Section 4.4.2) as it states that, for t approaching infinity, the mean displacement $\langle r(r) \rangle$ is proportional to the square root of time. This, in turn, implies that the mean diffusing molecules mass M_t is proportional to the square root of time, and this is exactly the solution of Fick's second law (see Equation 4.139). The fact that $\beta < 1$ in fractal media witnesses how diffusional propagation is hindered by geometrical heterogeneity. In addition, though β establishes the relation between diffusion characteristics and medium topology, it may happen that fractal media showing the same fractal dimension are not characterized by the same diffusive properties, that is, they are associated to different fracton dimensions d_s.

All these considerations make clear that the description of diffusion in fractal media can be undertaken, provided that medium topology is properly accounted for. Of course, different possibilities exist. One of them consists in adopting the classical differential mass balance (see Section 4.3.1), assuming that both diffusion coefficient D and fractal object differential volume dV depend on d. In particular, remembering that, by definition $\langle r^2(r) \rangle \propto D(r)t$, it follows $D(r) \propto r^2/t$ and, in the light of Equation 7.224, we finally have

$$D(r) \propto \frac{r^2}{t} \propto \frac{r^2}{r^{2/\beta}} \quad D(r) \propto r^{-2\left(\frac{d}{d_s}-1\right)} = r^{-\theta}, \quad D(r) = \alpha_0 r^{-\theta}. \tag{7.225}$$

Equation 7.225 shows that D decreases with r and reflects Equation 7.223 content concerning the reduction of fractal media density with r. In addition, remembering that, for Euclidean (i.e., not fractal) one- ($n = 1$), two- ($n = 2$), or three- ($n = 3$) dimensional objects the length, surface, or volume differentials are proportional to ($nr^{n-1} \, dr$), for fractal statistical radially symmetric objects we can assume [254]

$$dV(r) = ldr^{d-1} \, dr, \qquad (7.226)$$

where l is a prefactor related to Mandelbrodt's definition of lacunarity [252]. Equation 7.226 is nothing more than the Euclidean approximation for the volume scaling of a fractal structure, and thus, $V(r)$ has the dimension [length]^d. Relying on Equation 7.225 and Equation 7.226, O'Shaughnessy and Procaccia [255] get a macroscopic model for diffusion in fractal structures:

$$\frac{\partial C(t,r)}{\partial t} = \frac{\alpha_0}{r^\omega} \frac{\partial}{\partial r} \left(r^\nu \frac{\partial C(t,r)}{\partial r} \right), \quad \omega = d - 1, \quad \nu = \omega - \theta. \qquad (7.227)$$

Assuming an infinite medium and an initial concentration distribution showing a δ-Dirac form localized in $r = 0$, Equation 7.227 solution reads

$$C(t,r) = C_0 t^{-\kappa} e^{-\frac{r^\xi}{Bt}}, \quad \xi = 2\frac{d}{d_s}, \quad \kappa = 0.5d_s, \quad B = \alpha_0 \xi^2, \qquad (7.228)$$

where C_0 is a normalization constant depending on initial conditions [105].

Another possibility for the solution of diffusion in fractal media is to consider the traditional Euclidean mass balance (Section 4.3.1) and imposing that diffusion can take place only in some space zones (those belonging to the fractal structure) whereas others are forbidden. At this purpose, the use of percolation lattices is very significant. A percolation lattice is a spatially uncorrelated random distribution of empty and occupied sites characterized by a percolation probability p representing the frequency of empty sites (site where diffusion can occur). Percolation theory [103] affirms that there exists a critical percolation probability p_c above which transport is allowed. In other words, for $p \geq p_c$, there exists a connected cluster of empty sites (the so-called infinite cluster) so that a diffusing molecule can cross the entire lattice (it can be verified that, for two-dimensional square lattices, $p_c \approx 0.593$ [103]). Interestingly, just above p_c, a highly heterogeneous complex structure showing fractal properties for what concerns diffusion takes place. Accordingly, once the infinite cluster has been built up by means of a proper algorithm [256], in the case of two-dimensional lattices, mass balance equation

$$\frac{\partial C}{\partial t} = D \left(\frac{\partial^2 C}{\partial X^2} + \frac{\partial^2 C}{\partial X^2} \right) \qquad (7.229)$$

can be numerically solved only on empty sites. This implies Equation 7.229 discretization on the lattice (now thought as a grid of mesh size ΔL and composed, respectively, by h and k sites (squares) in the X and Y directions) according to, for example, the following explicit algorithm [257]:

$$C_{ij}^{t+\Delta t} = C_{ij}^{t} + \gamma', \sum_{(h,k)\in(i,j)} J_{hk\to i,j}^{t}, \qquad (7.230)$$

where t is the time, Δt is the time step, $\gamma' = D\,\Delta t/\Delta L$, (i,j) are generic lattice site coordinates, $I(i,j)$ is the set of the first four nearest neighbors of (i,j), $C_{ij}^{t+\Delta t}$ and C_{ij}^{t} are, respectively, diffusing molecules concentration in (i,j) at time $(t+\Delta t)$ and t, whereas $J_{hk \longrightarrow i,j}^{t}$ is the matter flux at time t:

$$J_{hk\to i,j}^{t} = \frac{C_{hk}^{t} - C_{ij}^{t}}{\Delta L}. \qquad (7.231)$$

Indicating with $\mathcal{M}$ the infinite cluster, its presence on the lattice can be accounted for by the following characteristic function $\chi(i,j)$:

$$\chi(i,j) = \begin{cases} 1, & (i,j) \in \mathcal{M} \\ 0, & (i,j) \notin \mathcal{M} \end{cases}. \qquad (7.232)$$

Empty sites (where diffusion can take place) are characterized by $\chi = 1$, while forbidden sites correspond to $\chi = 0$. Remembering that Equation 7.230 is defined only on empty sites and that $J_{hk \longrightarrow i,j}^{t}$ is zero in forbidden sites, the combination of Equation 7.230 and Equation 7.232 leads to the model final form:

$$C_{ij}^{t+\Delta t} = C_{ij}^{t} + \gamma \sum_{(h,k)\in I(i,j)} \chi(h,k)[C_{hk}^{t} - C_{ij}^{t}], \quad \gamma = \frac{D\Delta t}{\Delta L^{2}}. \qquad (7.233)$$

Notably, Equation 7.233 satisfies the conservation principle, that is,

$$\sum_{(i,j)} \chi(i,j) \sum_{(h,k)\in I(i,j)} \chi(h,k)[C_{hk}^{t} - C_{ij}^{t}] = 0, \qquad (7.234)$$

which is a fundamental requisite for the applicability of finite difference scheme deriving from mass conservation. The stability criteria applied to Equation 7.233 require that $\gamma < 0.5$ [76]. Interestingly, this approach can be easily modified to perform more sophisticated numerical techniques (e.g., implicit algorithms) and to account for more complex situations such as the presence of two phases inside $\mathcal{M}$. Indeed, in this case, the $\chi = 1$ set would be composed of two subsets χ_1 and χ_2 characterized, respectively, by two different values, D_1 and D_2, of the diffusion coefficient.

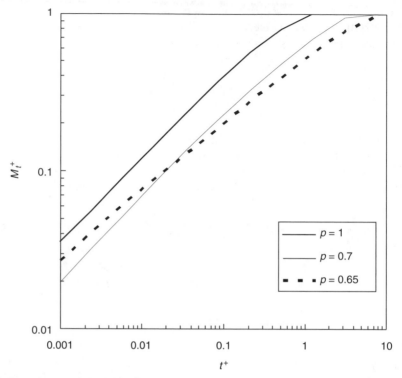

FIGURE 7.33 For high percolation probability ($p > 0.7$), mass transport through a two-dimensional percolation lattice shows a typical fickian character as the dimensionless mass released M_t^+ scales as $t^{0.5}$. When, on the contrary, p approaches the critical percolation value $p_c = 0.593$, medium complexity (fractal nature) hinders mass transport and M_t^+ scales as $t^{0.38}$. (Adapted from Adrover, A., Giona, M., and Grassi, M., *J. Membr. Sci.*, 113, 21, 1996.)

Assuming sink conditions in the release environment and a uniform initial distribution of diffusing molecules concentration, Equation 7.233 can be numerically solved to yield the results shown in Figure 7.33. It is evident that for $0.7 < p < 1$, a typical Fickian release occurs as the dimensionless amount of drug released M_t^+ is proportional to the square root of dimensionless time t^+. On the contrary, when p approaches p_c, M_t^+ depends on time according to a lower (<0.5) exponent. In particular, for $p = 0.65$, we have $M_t^+ \propto t^{0.38}$. These results evidence how anomalous diffusion in fractal, disordered media is associated with the reduction of the exponent ruling the M_t^+ dependence on time.

7.4 CONCLUSIONS

The aim of this chapter was to expose and discuss the phenomena ruling drug release (mass transport) from matrix systems, one of the most popular tools to get controlled release. Accordingly, a panoramic illustration of CRS is followed by the definition and discussion of matrix systems. In particular, the attention is focused on matrix topology and release mechanisms in an attempt to give a purely physical view of our matter of discussion. At this stage, no equation is used to avoid overshadowing concepts with the mathematical formalism. Subsequently, on the contrary, mathematics is largely used to translate the physical frame into equations, the core step of mathematical modeling. Nevertheless, paragraphs are structured to first show mathematical modeling and then to see how models match the experiments. In this manner, the reader who is uninterested in mathematical details, once knowing the problem and the hypotheses on which the model relies, can simply move to experimental data/model comparison. In this light, diffusion-controlled systems, diffusion and matrix/drug chemical–physical interaction controlled systems, diffusion- and dissolution-controlled systems, swelling-controlled systems, erosion-controlled systems, and fractal systems are considered. Finally, particular attention is devoted to drug release from delivery systems consisting of an ensemble of polydisperse spherical particles. Indeed, despite the importance of this kind of delivery system in real applications, to our knowledge, not many modeling efforts have been carried out till now. Obviously, although mathematical models can be always improved according to a natural evolution process (Darwin's concept of the survival of the fittest), it is important to remember that the search for the general, unifying model may not be the more convenient and useful way. We believe that, at least for industrial purposes, the best model is the simplest one able to reasonably describe the experimental data set.

REFERENCES

1. *Controlled Drug Delivery: Fundamentals and Applications*, Lee, V.H.L. and Robinson, J.R., Eds., Marcel Dekker, New York, Basel, 1987.
2. Kost, J. and Langer, R., Responsive polymeric delivery systems, *Adv. Drug Deliv. Rev.*, 46, 125, 2001.
3. Langer, R. and Peppas, N., Advances in biomaterials, drug delivery, and bionanotechnology, *AIChe. J.*, 49, 2990, 2003.
4. Dziubla, T.D., et al., Implantable drug delivery devices, in *Biomimetic Materials and Design*, Lowman, A.M. and Dillon, A.K., Eds., Marcell Dekker, 2002.
5. Israel, G., Balthazar, van der, Pol e il primo modello matematico del battito cardiaco, in *Modelli Matematici nelle Scienze Biologiche*, Freguglia, P., Ed., Edizioni Quattro Venti, 1998.
6. Kydonieus, A., Ed., *Treatise on Controlled Drug Delivery*, Marcell Dekker, New York, 1992.
7. Wise, L.D., Ed., *Handbook of Pharmaceutical Controlled Release Technology*, Marcell Dekker, New York, Basel, 2000.
8. Hishikawa, K., et al., Gene expression profile of human mesenchymal stem cells during osteogenesis in three-dimensional thermoreversible gelation polymer, *Biochem. Biophys. Res. Commun.*, 317, 1103, 2004.

9. Au, A., et al., Evaluation of thermoreversible polymers containing fibroblast growth factor 9 (FGF-9) for chondrocyte culture, *J. Biomed. Mater. Res. A*, 69, 367, 2004.

10. Au, A., et al., Thermally reversible polymer gel for chondrocyte culture, *J. Biomed. Mater. Res. A*, 67, 1310, 2003.

11. Lanning, D.A., et al., Myofibroblast induction with transforming growth factor-beta 1 and beta 3 in cutaneous fetal excisional wounds, *J. Pediatr. Surg.*, 35, 183, 2000.

12. Sidhu, G.S., et al., Curcumin enhances wound healing in streptozotocin induced diabetic rats and genetically diabetic mice, *Wound Repair Regen.*, 7, 362, 1999.

13. Wang, J., Bacterial vaginosis, *Prim. Care Updat. Ob. Gyns.*, 7, 181, 2000.

14. Chang, J.Y., et al., Prolonged antifungal effects of clotrimazole-containing mucoadhesive thermosensitive gels on vaginitis, *J. Contr. Rel.*, 82, 39, 2002.

15. Roy, S., et al., Thermoreversible gel formulations containing sodium lauryl sulfate or n-lauroylsarcosine as potential topical microbicides against sexually transmitted diseases, *Antimicrob. Agents Chemother.*, 45, 1671, 2001.

16. Ruel-Gariepy, E., et al., A thermosensitive chitosan-based hydrogel for the local delivery of paclitaxel, *Eur. J. Pharm. Biopharm.*, 57, 53, 2004.

17. Hsieh, D., Ed., *Controlled Release Systems: Fabrication Technology*, CRC Press, Boca Raton, Florida, 1987.

18. Meriani, F., et al., *In vitro* nimesulide absorption from different formulations, *J. Pharm. Sci.*, 93, 540, 2004.

19. Debenedetti, P.G., Tom, J.W., Yeo, S.D., Application of supercritical fluids for the production of sustained delivery devices, *J. Contr. Rel.*, 24, 27, 1993.

20. Falk, R., et al., Controlled release of ionic compounds from poly(-lactide) micro-spheres produced by precipitation with a compressed antisolvent, *J. Contr. Rel.*, 44, 77, 1997.

21. Ghaderi, R., Artursson, P., Carlfors, J., A new method for preparing biodegrad-able microparticles and entrapment of hydrocortisone in DL-PLG microparticles using supercritical fluids, *Eur. J. Pharm. Sci.*, 10, 1, 2000.

22. Flynn, G.L., Yalkowsky, S.H., Roseman, T.J., Mass transport phenomena and models: theoretical concepts, *J. Pharm. Sci.*, 63, 479, 1974.

23. Kim, D. and Park, K., Swelling and mechanical properties of superporous hydrogels of poly(acrylamide-co-acrylic acid)/polyethylenimine interpenetrat-ing polymer networks, *Polymer*, 45, 189, 2004.

24. Lapasin, R. and Pricl, S., *Rheology of Industrial Polysaccharides, Theory and Applications*, Chapman & Hall, London, 1995.

25. Barry, B.W. and Meyer, M.C., The rheological properties of carbopol gels. Continuous shear and creep properties of carbopol gels, *Int. J. Pharm.*, 2, 1, 1979.

26. Ketz, R.J., Prud'homme, R.K., Graessly, W.W., Rheology of concentrated microgel solutions, *Rheol. Acta*, 27, 531, 1988.

27. Skjak-Braek, G., Alginates: biosyntheses and some structure–function relation-ships relevant to biomedical and biotechnological applications, *Biochem. Soc. Trans.*, 20, 27, 1992.

28. Hoffman, A.S., Hydrogels for biomedical applications, *Adv. Drug Deliv. Rev.*, 54, 3, 2002.

29. Clark, A.H. and Ross-Murphy, S.B., Structural and mechanical properties of biopolymer gels, *Adv. Polym. Sci.*, 83, 57, 1987.

30. Crescenzi, V., Dentini, M., Coviello, T., Solution and gelling properties of polysaccharide polyelectrolytes, *Biophys. Chem.*, 41, 61, 1991.
31. Burchard, W., Structure formation by polysaccharides in concentrated solution, *Biomacromolecules*, 2, 342, 2001.
32. Husseini, G.A., et al., Ultrasonic release of doxorubicin from pluronic P105 micelles stabilized with an interpenetrating network of N, N-diethylacrylamide, *J. Contr. Rel.*, 83, 303, 2002.
33. Kurisawa, M. and Yui, N., Dual-stimuli-responsive drug release from interpenetrating polymer network-structured hydrogels of gelatin and dextran, *J. Contr. Rel.*, 54, 191, 1998.
34. Xuequan, L., et al., Radiation preparation and thermo-response swelling of interpenetrating polymer network hydrogel composed of PNIPAAm and PMMA, *Radiat. Phys. Chem.*, 57, 477, 2000.
35. Aso, Y., et al., Thermally controlled protein release from gelatin $\pm$ dextran hydrogels, *Radiat. Phys. Chem.*, 55, 179, 1999.
36. Grant, G.T., et al., Biological interactions between polysaccharides and divalent cations: the eggs-box model, *FEBS Lett.*, 32, 195, 1973.
37. Smidsrod, O. and Haug, A., Properties of poly(1, 4-hexuronates) in the gel state. II. Comparison of gels of different chemical composition, *Acta Chem. Scand.*, 26, 79, 1972.
38. Haug, A., Composition and properties of alginates, Report Norvegian Institute of Seaweed Research vol. 30, 1964, pp. 25–45.
39. Pasut, E., Analisi delle proprietà meccaniche e diffusive di gel omogenei costituiti da alginato di calcio, Thesis/Dissertation, University of Trieste (Italy), Dept. of Pharmaceutical Sciences, 2004.
40. Bühler, V., Kollidon: polyvinylpyrrolidone for the pharmaceutical industry, 4th, BASF Aktiengesellschaft, Fine Chemicals, D-67056 Ludwigshafen, 1998.
41. Magarotto, L., et al., Determinazione delle molecole reticolanti nel polivinilpirrolidone reticolato, in XVI Convegno Italiano di Scienza e Tecnologia delle Macromolecole, Pacini Editore, Pisa (Italy), 22–9–0003, 2003.
42. Ron, E.S. and Bromberg, L.E., Temperature-responsive gels and thermogelling polymer matrices for protein and peptide delivery, *Adv. Drug Deliv. Rev.*, 31, 197, 1998.
43. Brazel, C.S. and Peppas, N.A., Pulsatile and local delivery of thrombolytic and antithrombotic agents using poly (*N*-iso-propylacrylamide-co-methacrylic acid) hydrogels, *J. Contr. Rel.*, 39, 57, 1996.
44. Harsh, D.C. and Gehrke, S.H., Controlling the swelling characteristics of temperature-sensitive cellulose ether hydrogels, *J. Contr. Rel.*, 17, 175, 1991.
45. Schiller, M.E., Yu, X., Lupton, E.C., Roos, E.J., Holman, H.A., and Ron, E.S., Engineered Response hydrogels for medical application. *22* International Symposium on the Controlled Release of Bioactive Materials, 1995.
46. Omidian, H., Rocca, J.G., Park, K., Advances in superporous hydrogels, *J. Contr. Rel.*, 102, 3, 2005.
47. Grassi, M., et al., Preparation and evaluation of a melt pelletised paracetamol/stearic acid sustained release delivery system, *J. Contr. Rel.*, 88, 381, 2003.
48. Grassi, M., et al., Theoretical and experimental study on theophylline release from stearic acid cylindrical delivery systems, *J. Contr. Rel.*, 92, 275, 2003.

49. Charnay, C., et al., Inclusion of ibuprofen in mesoporous templated silica: drug loading and release property, *Eur. J. Pharm. Biopharm.*, 57, 533, 2004.

50. Unger, K., et al., The use of porous and surface modified silicas as drug delivery and stabilizing agents, *Drug Dev. Indus. Pharm.*, 9, 69, 1983.

51. Ahola, M., et al., Silica xerogel carrier material for controlled release of toremifene citrate, *Int. J. Pharm.*, 195, 219, 2000.

52. Kawanabe, K., et al., Treatment of osteomyelitis with antibiotic-soaked porous glass ceramic, *J. Bone Joint Surg. Br.*, 80, 527, 1998.

53. Ciesla, U. and Schuth, F., Ordered mesoporous materials, *Microporous Mesoporous Mater.*, 27, 131, 1999.

54. Kresge, C.T., et al., Ordered mesoporous molecular sieves synthesized by a liquid–crystal template mechanism, *Nature*, 359, 710, 1992.

55. Zhao, D., et al., Nonionic triblock and star diblock copolymer and oligomeric surfactant, *J. Am. Chem. Soc.*, 120, 6024, 1998.

56. Edler, K.J. and White, J.W., Further improvements in the long range order of MCM-41 materials, *Chem. Mater.*, 9, 1226, 1997.

57. Khushalani, D., et al., Mixed micelle templating of mesoporous silicas with sub-angstrom precision tuning of the pore sizes, *Chem. Mater.*, 8, 2188, 1996.

58. Alfredsson, V., et al., High-resolution transmission electron microscopy of mesoporous MCM-41 type materials, *J. Chem. Soc. Chem. Commun.*, 921, 1994.

59. Kruk, M., Jaroniec, M., Sayari, A., Adsorption study of surface and structural properties of MCM-41 materials of different pore sizes, *J. Phys. Chem. B*, 101, 583, 1997.

60. Adamson, A.W., *Physical Chemistry of Surfaces*, 4th ed., John Wiley & Sons, New York, Toronto, 1982.

61. Rathouský, J., et al., Adsorption on MCM-41 mesoporous molecular sieves. Part 2.—Cyclopentane isotherms and their temperature dependence, *J. Chem. Soc. Faraday Trans.*, 91, 937, 1995.

62. Beck, J.S., et al., A new family of mesoporous molecular sieves prepared with liquid crystal templates, *J. Am. Chem. Soc.*, 114, 10834, 1992.

63. Ekwall, P., *Advances in Liquid Crystals*, Brown, G.H., Ed., Academic Press, 1975.

64. Vallet-Regy, M., et al., Hexagonal ordered mesoporous material as a matrix for the controlled release of amoxicillin, *Solid State Ionics*, 172, 435, 2004.

65. Doadrio, A.L., et al., Mesoporous SBA-15 HPLC evaluation for controlled gentamicin drug delivery, *J. Contr. Rel.*, 97, 125, 2004.

66. Conte, U., et al., Swelling-activated drug delivery systems, *Biomaterials*, 9, 489, 1988.

67. Colombo, P., et al., Drug volume fraction profile in the gel phase and drug release kinetics in hydroxypropylmethyl cellulose matrices containing a soluble drug, *Eur. J. Pharm. Sci.*, 9, 33, 1999.

68. Lee, P.I., *Controlled Release Systems: Fabrication Technology*, Hsieh, D., Ed., CRC Press, 1987.

69. Ju, R.T., Nixon, P.R., Patel, M.V., Drug release from hydrophilic matrices. 1. New scaling laws for predicting polymer and drug release based on the polymer disentanglement concentration and the diffusion layer, *J. Pharm. Sci.*, 84, 1455, 1995.

70. Kararli, T.T., Hurlbut, J.B., Needham, T.E., Glass–rubber transitions of cellulosic polymers by dynamic mechanical analysis, *J. Pharm. Sci.*, 79, 845, 1990.
71. Linder, W.D., Mockel, J.D., Lippold, B.C., Controlled release of drugs from hydrocolloid embeddings, *Pharmazie*, 51, 263, 1996.
72. Grassi, M., Lapasin, R., Pricl, S., Modeling of drug release from a swellable matrix, *Chem. Eng. Commun.*, 169, 79, 1998.
73. Lee, P., Kinetics of drug release from hydrogel matrices, *J. Contr. Rel.*, 2, 277, 1985.
74. Vrentas, J.S., Jarzebski, C.M., Duda, J.L., A Deborah number for diffusion in polymer–penetrant systems, *AIChe. J.*, 21, 1, 1975.
75. Vrentas, J.S. and Duda, J.L., Molecular diffusion in polymer solutions, *AIChe. J.*, 25, 1, 1979.
76. Crank, J., *The Mathematics of Diffusion*, 2nd ed., Clarendon Press, Oxford, 1975.
77. Grassi, M., Diffusione in matrici di idrogel polimerici per sistemi farmaceutici a rilsacio controllato, Thesis/Dissertation, University of Trieste (Italy), Dept. of Chem. Eng., 1996.
78. Miller-Chou, A.A. and Koening, J.K., A review of polymer dissolution, *Progr. Polym. Sci.*, 28, 1223, 2003.
79. Siepmann, J. and Gopferich, A., Mathematical modeling of bioerodible, polymeric drug delivery systems, *Adv. Drug Deliv. Rev.*, 48, 229, 2001.
80. Mallapragada, S.K. and Peppas, N.A., Unfolding and chain disentanglement during semicrystalline polymer dissolution, *AIChe. J.*, 43, 870, 1997.
81. Carli, F., et al., Influence of polymer characteristics on drug loading into crospovidone, *Int. J. Pharm.*, 33, 115, 1986.
82. Nogami, H., Nagai, T., Suzuki, A., Studies on powdered preparation. XVII. Dissolution rate of sulfonamides by rotating disk method, *Chem. Pharm. Bull. (Tokyo)*, 14, 329, 1966.
83. Grassi, M., Colombo, I., Lapasin, R., Drug release from an ensemble of swellable crosslinked polymer particles, *J. Contr. Rel.*, 68, 97, 2000.
84. de Smidt, J.H., et al., Dissolution of theophylline monohydrate and anhydrous theophylline in buffer solutions, *J. Pharm. Sci.*, 75, 497, 1986.
85. Nogami, H., Nagai, T., Yotsuyanagi, T., Dissolution phenomena of organic medicinals involving simultaneous phase changes, *Chem. Pharm. Bull.*, 17, 499, 1969.
86. Higuchi, W.I., Bernardo, P.D., Mehta, S.C., Polymorphism and drug availability. II. Dissolution rate behavior of the polymorphic forms of sulfathiazole and methylprednisolone, *J. Pharm. Sci.*, 56, 200, 1967.
87. Singh, M., Lumpkin, J., Rosenblat, J., Mathematical modeling of drug release from hydrogel matrices via diffusion coupled with desorption mechanism, *J. Contr. Rel.*, 32, 17, 1994.
88. Singh, M., Lumpkin, J., Rosenblat, J., Effect of electrostatic interactions on polylysine release rates from collagen matrices and comparison with model predictions, *J. Contr. Rel.*, 35, 165, 1995.
89. Singh, M.P., et al., The effect of electrostatic charge interactions on release rates of gentamicin from collagen matrices, *Pharm. Res.*, 12, 1205, 1995.
90. Vozeh, S., Schmidlin, O., Taeschner, W., Pharmacokinetic drug data, *Clin. Pharmacokinet.*, 15, 254, 1988.

91. Hilt, J.Z., Nanotechnology and biomimetic methods in therapeutics: molecular scale control with some help from nature, *Adv. Drug Deliv. Rev.*, 56, 1533, 2004.

92. Byrne, M.E., Park, K., Peppas, N.A., Molecular imprinting within hydrogels, *Adv. Drug Deliv. Rev.*, 54, 149, 2002.

93. Peppas, N.A. and Langer, R., Origins and development of biomedical engineering within chemical engineering, *AIChe. J.*, 50, 536, 2004.

94. Sreenivasan, K., On the applicability of molecularly imprinted poly(HEMA) as a template responsive release system, *J. Appl. Polym. Sci.*, 71, 1819, 1999.

95. Hilt, J.Z. and Byrne, M.E., Configurational biomimesis in drug delivery: molecular imprinting of biologically significant molecules, *Adv. Drug Deliv. Rev.*, 56, 1599, 2004.

96. Percival, C.J., et al., Molecular imprinted polymer coated QCM for the detection of nandrolone, *Analyst*, 127, 1024, 2002.

97. Tang, A.X.J., et al., Immunosensor for okadaic acid using quartz crystal microbalance, *Anal. Chim. Acta*, 471, 33, 2002.

98. Kobayashi, T., et al., Molecular imprinting of caffeine and its recognition assay by quartz–crystal microbalance, *Anal. Chim. Acta*, 435, 141, 2001.

99. Tan, Y., et al., A study of a biomimetic recognition material for the BAW sensor by molecular imprinting and its application for the determination of paracetamol in the human serum and urine, *Talanta*, 55, 337, 2001.

100. Trotta, F., et al., A molecular imprinted membrane for molecular discrimination of tetracycline hydrochloride, *J. Membr. Sci.*, 254, 13, 2005.

101. Adrover, A., Giona, M., Grassi, M., Controlled release of theophylline from water-swollen scleroglucan matrices, *J. Membr. Sci.*, 113, 21, 1996.

102. Bunde, A. and Havlin, S., *Fractals and Disordered Systems*, Springer Verlag, Berlin, 1992.

103. Stauffer, D. and Aharony, A., *Introduction to Percolation Theory*, Taylor and Francis, London, 1992.

104. Havli, S. and Ben-Avraham, D., Diffusion in disordered media, *Adv. Phys.*, 695, 1987.

105. Giona, M., Statistical analysis of anomalous transport phenomena in complex media, *AIChe. J.*, 37, 1249, 1991.

106. Lee, P., Effect of non-uniform initial drug concentration distribution on the kinetics of drug release from glassy hydrogel matrices, *Polymer*, 25, 973, 1984.

107. Lee, P.I., Novel approach to zero-order drug delivery via immobilized nonuniform drug distribution in glassy hydrogels, *J. Pharm. Sci.*, 73, 1344, 1984.

108. Lee, P., Initial concentration distribution as a mechanism for regulating drug release from diffusion controlled and surface erosion controlled matrix systems, *J. Contr. Rel.*, 4, 1, 1986.

109. Colombo, P., Swelling-controlled release in hydrogel matrices for oral route, *Adv. Drug Deliv. Rev.*, 11, 37, 1993.

110. Huang, X. and Brazel, C.S., On the importance and mechanisms of burst release in controlled drug delivery—a review, *J. Contr. Rel.*, 73, 121, 2001.

111. Alhaique, F., et al., Studies on the release behaviour of polysaccharide matrix, *Pharmazie*, 48, 432, 2005.

112. Colombo, P., et al., Swellable matrices for controlled drug delivery: gel-layer behaviour, mechanisms and optimal performance, *Pharm. Sci. Technol. Today*, 3, 198, 2000.

113. Ritger, P.L. and Peppas, N.A., A simple equation for description of solute release II. Fickian and anomalous release from swellable devices, *J. Contr. Rel.*, 5, 37, 1987.

114. Siepmann, J. and Peppas, N.A., Modeling of drug release from delivery systems based on hydroxypropyl methylcellulose (HPMC), *Adv. Drug Deliv. Rev.*, 48, 139, 2001.

115. Grassi, M., et al., Apparent non-fickian release from a scleroglucan gel matrix, *Chem. Eng. Commun.*, 155, 89, 1996.

116. Grassi, M., et al., Analysis and modeling of release experiments, in *Proceedings of the 22nd International Symposium on Controlled Release of Biactive Materials*, Controlled Release Society, 364.

117. Peppas, N.A. and Sahlin, J.J., A simple equation for the description of solute release. III. Coupling of diffusion and relaxation, *Int. J. Pharm.*, 57, 169, 1989.

118. Moneghini, M., et al., Preparation of theophylline-hydroxypropylmethylcellulose matrices using supercritical antisolvent precipitation, *Drug Dev. Indus. Pharm.*, 2005.

119. Higuchi, W.I., Mir, N.A., Desai, S.J., Dissolution rates of polyphase mixtures, *J. Pharm. Sci.*, 54, 1405, 1965.

120. Banakar, U.V., *Pharmaceutical Dissolution Testing*, Marcell Dekker, New York, 1992, pp. 5–9.

121. Colombo, P., et al., Analysis of the swelling and release mechanisms from drug delivery systems with emphasis on drug solubility and water transport, *J. Contr. Rel.*, 39, 231, 1996.

122. Levich, V.G., *Physicochemical Hydrodynamics*, Prentice-Hall, 1962.

123. Wan, L.S.C., Heng, P.W.S., Wong, L.F., Matrix swelling: a simple model describing extent of swelling HPMC matrices, *Int. J. Pharm.*, 116, 159, 1995.

124. Cappello, B., et al., Water soluble drug delivery systems based on a non-biological bioadhesive polymeric system, *Farmaco*, 49, 809, 1994.

125. Grassi, M., Colombo, I., Lapasin, R., Experimental determination of the theophylline diffusion coefficient in swollen sodium-alginate membranes, *J. Contr. Rel.*, 76, 93, 2001.

126. Grassi, M., et al., Modeling of drug release from partially coated matrices made of a high viscosity HPMC, *Int. J. Pharm.*, 276, 107, 2004.

127. Fu, J.C., Hagemeir, C., Moyer, D.L., A unified mathematical model for diffusion from drug–polymer composite tablets, *J. Biomed. Mater. Res.*, 10, 743, 1976.

128. Coviello, T., et al., Scleroglucan/borax: characterization of a novel hydrogel system suitable for drug delivery, *Biomaterials*, 24, 2789, 2003.

129. Patankar, S.V., *Numerical Heat Transfer and Fluid Flow*, McGraw-Hill/Hemisphere Publishing Corporation, New York, 1990.

130. Coviello, T., et al., A new polysaccharidic gel matrix for drug delivery: preparation and mechanical properties, *J. Contr. Rel.*, 102, 643, 2005.

131. Desai, S.J., et al., Investigation of factors influencing release of solid drug dispersed in inert matrices. II. Quantitation of procedures, *J. Pharm. Sci.*, 55, 1224, 1966.

132. Higuchi, T., Mechanism of sustained-action medication. Theoretical analysis of rate of release of solid drugs dispersed in solid matrices, *J. Pharm. Sci.*, 52, 1145, 1963.

133. Muhr, A.H. and Blanshard, M.V., Diffusion in gels, *Polymer*, 23(Suppl.), 1012, 1982.

134. Grassi, M., Lapasin, R., Carli, F., Modeling of drug permeation through a drug interacting swollen membrane, in *Proceedings of the 27th International Symposium on Controlled Release of Bioactive Materials*, Controlled Release Society, 7–9–2000.

135. Kuijpers, A.J., et al., Controlled delivery of antibacterial proteins from biodegradable matrices, *J. Contr. Rel.*, 53, 235, 1998.

136. Brasseur, N., Brault, D., Couvreur, P., Adsorption of hematoporphyrin onto polyalkylcyanoacrylate nanoparticles: carrier capacity and drug release, *Int. J. Pharm.*, 70, 129, 1991.

137. Kwon, G.S., et al., Release of proteins via ion exchange from albumin–heparin microspheres, *Int. J. Pharm.*, 22, 83, 1992.

138. Frenning, G. and Stromme, M., Drug release modeled by dissolution, diffusion, and immobilization, *Int. J. Pharm.*, 250, 137, 2003.

139. Kim, H., Kaczmarski, K., Georges Guiochona, G., Mass transfer kinetics on the heterogeneous binding sites of molecularly imprinted polymers, *Chem. Eng. Sci.*, 60, 5425, 2005.

140. Elvira, C., et al., Covalent polymer–drug conjugates, *Molecules*, 10, 114, 2005.

141. Somasekharan, K.N. and Subramanian, R.V., Mechanism of release of organotin from thermoset polymers, in *Controlled Release of Bioactive Materials*, Baker, R., Ed., Academic Press, 1980.

142. Tani, N., Hydrophilic/hydrophobic controls of steroid release from a cortisol–polyglutamic acid sustained release system, in *Controlled Release of Pesticides and Pharmaceuticals*, Baker, R., Ed., Plenum Press, 1981.

143. Abdekhodaie, M.J. and Wu, X.Y., Modeling of a cationic glucose-sensitive membrane with consideration of oxygen limitation, *J. Membr. Sci.*, 254, 119, 2005.

144. Shah, S.S., Kulkarni, M.G., Mashelkar, R.A., Release kinetics of pendant substituted bioactive molecules from swellable hydrogels: role of chemical reaction and diffusive transport, *J. Membr. Sci.*, 51, 83, 1990.

145. Somasekharan, K.N. and Subramanian, R.V., *Structure, Mechanism and Reactivity of Organotin Carboxylate Polymers*, Carraher, E. and Tsuda M., Eds., American Chemical Society, 1980.

146. Peppas, N.A., Mathematical models for controlled release kinetics, in *Medical Applications of Controlled Release*, Langer, R. and Wise L.D., Eds., CRC Press, 1984.

147. Narasimhan, B., Accurate models in controlled drug delivery systems, in *Handbook of Pharmaceutical Controlled Release Technology*, Wise, L.D., Ed., Marcell Dekker, 2000.

148. Abdekhodaie, M.J. and Cheng, Y.L., Diffusional release of a dispersed solute from planar and spherical matrices into finite external volume, *J. Contr. Rel.*, 43, 175, 1997.

149. Grassi, M., Lapasin, R., Pricl, S., The effect of drug dissolution on drug release from swelling polymeric matrices: mathematical modeling, *Chem. Eng. Commun.*, 173, 147, 1999.

150. Byun, Y.R., et al., A model for diffusion and dissolution controlled release from dispersed polymeric matrix, *J. Korean Pharm. Sci.*, 20, 79, 1990.

151. Gurny, R., Doelker, E., Peppas, N.A., Modelling of sustained release of water-soluble drugs from porous, hydrophobic polymers, *Biomaterials*, 3, 27, 1982.

152. Peppas, N.A., A model of dissolution-controlled solute release from porous drug delivery polymeric systems, *J. Biomed. Mater. Res.*, 17, 1079, 1983.

153. Harland, R.S., et al., A model of dissolution-controlled, diffusional drug release from nonswellable polymeric microspheres, *J. Contr. Rel.*, 7, 207, 1988.

154. Bird, R.B., Stewart, W.E., Lightfoot, E.N., *Transport Phenomena*, John Wiley & Sons, New York, London, 1960.

155. Higuchi, T., Rate of release of medicaments from ointment bases containing drugs in suspension, *J. Pharm. Sci.*, 50, 874, 1961.

156. Zhou, Y., et al., Modeling of dispersed-drug release from two-dimensional matrix tablets, *Biomaterials*, 26, 945, 2005.

157. Mantovani, F., et al., A combination of vapor sorption and laser light scattering methods for the determination of the Flory parameter χ and the crosslink density of a polymeric microgel, *Fluid Phase Equilib.*, 167, 63, 2000.

158. Flory, P.J., *Principles of Polymer Chemistry*, Cornell University Press, Ithaca, NY, 1953.

159. Brannon-Peppas, L. and Peppas, N.A., Equilibrium swelling behavior of pH-sensitive hydrogels, *Chem. Eng. Sci.*, 46, 715, 1991.

160. Peppas, N.A., et al., Effect of degree of crosslinking on water transport in polymer microparticles, *J. Appl. Polym. Sci.*, 3, 301, 1985.

161. Baker, J.P., et al., Effect of initial total monomer concentration on the swelling behavior of cationic acrylamide-based hydrogels, *Macromolecules*, 27, 1446, 1994.

162. Prange, M.M., Hooper, H.H., Prausnitz, J.M., Thermodynamics of aqueous systems containing hydrophilic polymers or gels, *AIChe. J.*, 35, 803, 1989.

163. Hooper, H.H., et al., Swelling equilibria for positively ionized polyacrylamide hydrogels, *Macromolecules*, 23, 1096, 1990.

164. Huang, Y., Szleifer, I., Peppas, N.A., A molecular theory of polymer gels, *Macromolecules*, 35, 1373, 2002.

165. Vilgis, T.A. and Wilder, J., Polyelectrolyte networks: elasticity, swelling and the violation of the Flory–Rehener hypothesis, *Comput. Theor. Polym. Sci.*, 8, 61, 1998.

166. Tanaka, F., Thermodynamic theory of network-forming polymer, *Macromolecules*, 23, 3784, 1990.

167. Flory, P.J. and Rehner, J., Statistical mechanics of cross-linked polymer networks, *J. Chem. Phys.*, 11, 512, 1943.

168. Lele, A.K., et al., Thermodynamics of hydrogen-bonded polymer gel–solvent systems, *Chem. Eng. Sci.*, 50, 3535, 1995.

169. Erman, B. and Flory, P.J., Critical phenomena and transitions in swollen polymer networks and in linear macromolecules, *Macromolecules*, 19, 2342, 1986.

170. Iordanskii, A.L., et al., Diffusion and sorption of water in moderately hydrophilic polymers: from segmented polyetherurethanes to poly-3-hydroxybutyrate, *Desalinisation*, 104, 27, 1996.

171. Hasa, J., Ilavsky, M., Dusek, K., Deformational, swelling, and potentiometric behaviour of ionized poly(methacrylic acid), *J. Polym. Sci.*, 13, 253, 1975.

172. Oppermann, W., Swelling behavior and elastic properties of ionic hydrogels, in *Polyelectrolyte Gels, Properties, Preparations and Applications*, Harland, R.S. and Prud'homme R.K., Eds., American Chemical Society, 1992, Chapter 10.

173. Bird, R.B., Armstrong, R.C., Hassager, O., *Dynamics of Polymeric Liquids, Fluid Mechanics*, Wiley, New York, Santa Barbara, London, Sydney, Toronto, 1990, pp. 72–75.

174. Coviello, T., et al., A crosslinked system from scleroglucan derivative: preparation and characterization, *Biomaterials*, 22, 1899, 2001.

175. Breitkreutz, J., Prediction of intestinal drug absorption properties by three-dimensional solubility parameters, *Pharm. Res.*, 15, 1370, 1998.

176. Hancock, B.C. and Zografi, G., The use of solution theories for predicting water vapor absorption by amorphous pharmaceutical solids: a test of the Flory–Huggins and Vrentas models, *Pharm. Res.*, 10, 1262, 1993.

177. Van Campen, L., Amidon, G.L., Zografi, G., Moisture sorption kinetics for water-soluble substances. III: Theoretical and experimental studies in air, *J. Pharm. Sci.*, 72, 1394, 1983.

178. Van Campen, L., Amidon, G.L., Zografi, G., Moisture sorption kinetics for water-soluble substances. II: Experimental verification of heat transport control, *J. Pharm. Sci.*, 72, 1388, 1983.

179. Van Campen, L., Amidon, G.L., Zografi, G., Moisture sorption kinetics for water-soluble substances. I: Theoretical considerations of heat transport control, *J. Pharm. Sci.*, 72, 1381, 1983.

180. Nelson, J.R., Determination of molecular weight between crosslinks of coals from solvent-swelling studies, *Fuel*, 62, 112, 1983.

181. Sample Certificate of Analysis of EXPLOTAB, Mendell Penwest Data, 2981 Route 22, Patterson, NY 12563.

182. Delay, J.M. and Wissbrun, F., *Melt Rheology and its Role in Plastics Processing, Theory and Applications*, Kluwer Academic, Dordrecht, Boston, London, 1999.

183. Shellhammer, T.H., Rumsey, T.R., Krochta, J.M., Viscoelastic properties of edible lipids, *J. Food Eng.*, 33, 305, 1997.

184. Joshi, S. and Astarita, G., Diffusion–relaxation coupling in polymers which show two-stage sorption phenomena, *Polym.*, 20, 455, 1979.

185. Astarita, G. and Sarti, G.C., A class of mathematical models for sorption of swelling solvents in glassy polymers, *Polym. Eng. Sci.*, 18, 388, 1978.

186. Cohen, D.S., Theoretical models for diffusion in glassy polymers, *J. Polym. Sci.: Polym. Phys. Ed.*, 21, 2057, 1983.

187. Cohen, D.S., Sharp fronts due to diffusion and stress at the glass transition in polymers, *J. Polym. Sci.: Part B: Polym. Phys.*, 27, 1731, 1989.

188. Cox, R.W. and Cohen, D.S., A mathematical model for stress-driven diffusion in polymers, *J. Polym. Sci.: Part B: Polym. Phys.*, 27, 589, 1989.

189. Hayes, C.K. and Cohen, D.S., The evolution of steep fronts in non-Fickian polymer–penetrant systems, *J. Polym. Sci.: Part B: Polym. Phys.*, 30, 145, 1992.

190. Edwards, D.A. and Cohen, D.S., A mathematical model for a dissolving polymer, *AIChe. J.*, 41, 2345, 1995.

191. Frisch, H.L., Wang, T.T., Kwei, T.K., Diffusion in glassy polymers II, *J. Polym. Sci.*, 7, 879, 1969.

192. Peppas, N.A. and Sinclair, J.L., Anomalous transport of penetrants in glassy polymers, *Colloid Polym. Sci.*, 261, 404, 1993.

193. Hariharan, D. and Peppas, N.A., Modelling of water transport and solute release in physiologically sensitive gels, *J. Contr. Rel.*, 23, 123, 1993.

194. Adib, F. and Neogi, P., Sorption with oscillations in solid polymers, *AIChe. J.*, 33, 164, 1987.

195. Camera-Roda, G. and Sarti, G.C., Non-Fickian mass transport through polymers: a viscoelastic theory, *Transport Theory Stat. Phys.*, 15, 1023, 1986.

196. Singh, S.K. and Fan, L.T., A generalized model for swelling controlled release systems, *Biotech. Progr.*, 2, 145, 1986.

197. Lustig, S.R., Caruthers, J.M., Peppas, N.A., Continuum thermodynamics and transport theory for polymer–fluid mixtures, *Chem. Eng. Sci.*, 47, 3037, 1992.

198. Kim, D.J., Caruthers, J.M., Peppas, N.A., Experimental verification of predictive model of penetrant transport in glassy polymers, *Chem. Eng. Sci.*, 51, 4827, 1996.

199. Lustig, S.R., Solvent transport in polymers, in *Proceedings of the XII International Congress on Rheology*, 355.

200. Camera-Roda, G. and Sarti, G.C., Mass transport with relaxation in polymers, *AIChe. J.*, 36, 851, 1990.

201. Korsmeyer, S.W., Lustig, S.R., Peppas, N.A., Solute and penetrant diffusion in swellable polymers. I. Mathematical modeling, *J. Polym. Sci. Part B: Polym. Phys.*, 24, 395, 1985.

202. Korsmeyer, S.W., von Meerwall, E., Peppas, N.A., Solute and penetrant diffusion in swellable polymers. II. Verification of theoretical models, *J. Polym. Sci. Part B: Polym. Phys.*, 24, 409, 1986.

203. Li, Y. and Tanaka, T., Effects of shear modulus of polymer gels, in *Polymer Gels: Fundamental and Biomedical Applications*, DeRossi, D., et al., Eds., Plenum Press, 1991.

204. Press, W.H., et al., *Numerical Recepies in FORTRAN*, 2nd ed., Cambridge University Press, Cambridge, USA, 1992.

205. Durning, C.J. and Russel, W.R., A mathematical model for diffusion with induced crystallization: 2, *Polymer*, 26, 131, 1985.

206. Peppas, N.A. and Urdahi, K.G., Anomalous penetrant transport in glassy polymers VII. Overshoots in cyclohexane uptake in crosslinked polystyrene, *Polym. Bull.*, 16, 201, 1986.

207. Kim, D.K., Caruthers, J.M., Peppas, N.A., Penetrant transport in cross-linked polystyrene, *Macromolecules*, 26, 1841, 1993.

208. Grassi, M., Modeling of penetrant uptake in viscoelastic polymeric matrices, in *Proceedings of the Joint Conference of Italian, Austrian, Slovenian Rheologists*, RHEOtech, Trieste (Italy), 91.

209. Colombo, I., et al., Modeling phase transitions and sorption–desorption kinetics in thermo-sensitive gels for controlled drug delivery systems, *Fluid Phase Equilib.*, 116, 148, 1996.

210. Grassi, M., Yuk, S.H., Cho, S.H., Modelling of solute transport across a temperature-sensitive polymer membrane, *J. Membr. Sci.*, 152, 241, 1999.

211. Peppas, N.A. and Reinhart, C.T., Solute diffusion in swollen membranes. I. A new theory, *J. Membr. Sci.*, 15, 275, 1983.

212. Reinhart, C.T. and Peppas, N.A., Diffusion in swollen membranes. II. Influence of crosslinking on diffusive properties, *J. Membr. Sci.*, 18, 227, 1984.

213. Grassi, M., Drug activation by co-grinding: the effect of particle size, in *AAPS Conference on Pharmaceutics and Drug Delivery*, 6–7–0004, 35.

214. Manca, D., et al., High-energy mechanochemical activation of active principles, in Convegno GRICU 2004, Nuove Frontiere di Applicazione delle Metodologie dell'Ingegneria Chimica, CUES Srl, 9–12–0004, 123.

215. Joshi, A. and Himmelstein, K.J., Dynamics of controlled release from bioerodible matrices, *J. Contr. Rel.*, 15, 95, 1991.

216. Sparer, R.V., et al., Controlled release from erodible poly(ortho ester) drug delivery systems, *J. Contr. Rel.*, 1, 23, 1984.

217. Thombre, A.G. and Himmelstein, K.J., A simultaneous transport–reaction model for controlled drug delivery from catalyzed bioerodible polymer matrices, *AIChe. J.*, 31, 759, 1985.

218. Zygourakis, K., Discrete simulations and bioerodible controlled release systems, *Polym. Preparation ACS*, 30, 456, 1989.

219. Zygourakis, K., Development and temporal evolution of erosion fronts in bioerodible controlled release devices, *Chem. Eng. Sci.*, 45, 2359, 1990.
220. Zygourakis, K. and Markenscoff, P.A., Computer-aided design of bioerodible devices with optimal release characteristics: a cellular automata approach, *Biomaterials*, 17, 125, 1996.
221. Gopferich, A. and Langer, Modeling of polymer erosion, *Macromolecules*, 26, 4105, 1993.
222. Gopferich, A., Bioerodible implants with programmable drug release, *J. Contr. Rel.*, 44, 271, 1997.
223. Gopferich, A., Polymer bulk erosion, *Macromolecules*, 30, 2598, 1997.
224. Drake, A.W., *Fundamental of Applied Probability Theory*, McGraw Hill Publishing Company, New York, 1988.
225. Mallapragada, S.K., Peppas, N.A., Colombo, P., Crystal dissolution-controlled release systems. II. Metronidazole release from semicrystalline poly(vinyl alcohol) systems, *J. Biomed. Mater. Res.*, 36, 125, 1997.
226. Mallapragada, S.K. and Peppas, N.A., Crystal dissolution-controlled release systems: I. Physical characteristics and modeling analysis, *J. Contr. Rel.*, 45, 87, 1997.
227. Narasimhan, B. and Peppas, N.A., Molecular analysis of drug delivery systems controlled by dissolution of the polymer carrier, *J. Pharm. Sci.*, 86, 297, 1997.
228. Harland, R.S., et al., Drug/polymer matrix swelling and dissolution, *Pharm. Res.*, 5, 488, 1988.
229. De Gennes, P.G., Repetition of a polymer chain in the presence of fixed obstacles, *J. Chem. Phys.*, 55, 572, 1971.
230. Narasimhan, B., Mathematical models describing polymer dissolution: consequences for drug delivery, *Adv. Drug Deliv. Rev.*, 48, 195, 2001.
231. Ju, R.T., et al., Drug release from hydrophilic matrices. 2. A mathematical model based on the polymer disentanglement concentration and the diffusion layer, *J. Pharm. Sci.*, 84, 1464, 1995.
232. Ju, R.T., Nixon, P.R., Patel, M.V., Diffusion coefficients of polymer chains in the diffusion layer adjacent to a swollen hydrophilic matrix, *J. Pharm. Sci.*, 86, 1293, 1997.
233. Muray, W.D. and Landis, F., Numerical and machine solutions of transient heat-conduction problems involving melting or freezing, *J. Heat Transfer*, 81, 106, 1959.
234. Gao, P. and Fagerness, P.E., Diffusion in HPMC gels. I. Determination of drug and water diffusivity by pulsed-field-gradient spin-echo NMR, *Pharm. Res.*, 12, 955, 1995.
235. Ueberreiter, K., The solution process, in *Diffusion in Polymers*, Crank, J. and Park, G.S., Eds., Academic Press, 1968.
236. Siepmann, J., et al., HPMC-matrices for controlled drug delivery: a new model combining diffusion, swelling, and dissolution mechanisms and predicting the release kinetics, *Pharm. Res.*, 16, 1748, 1999.
237. Siepmann, J., et al., A new model describing the swelling and drug release kinetics from hydroxypropyl methylcellulose tablets, *J. Pharm. Sci.*, 88, 65, 1999.
238. Siepmann, J. and Peppas, N.A., Hydrophilic matrices for controlled drug delivery: an improved mathematical model to predict the resulting drug release kinetics (the "sequential layer" model), *Pharm. Res.*, 17, 1290, 2000.
239. Narasimhan, B. and Peppas, N.A., On the importance of chain repetition in models of dissolution of glassy polymers, *Macromolecules*, 29, 3283, 1996.

240. Fujita, H., Diffusion in polymer–diluent systems, *Fortschr. Hochpolym. Forsch*, 3, 1, 1961.

241. Tapia, C., Buckton, G., Newton, J., Factors influencing the mechanism of release from sustained release matrix pellets, produced by extrusion/spheronisation, *Int. J. Pharm.*, 92, 211, 1993.

242. Dappert, T. and Thies, C., Statistical model for controlled release microcapsules: rationale and theory, *J. Membr. Sci.*, 4, 99, 1978.

243. Gross, S.T., et al., Fundamentals of release mechanism interpretation in multiparticulate systems: the prediction of the commonly observed release equations from statistical population models for particle ensembles, *Int. J. Pharm.*, 29, 213, 1986.

244. Hoffman, A., et al., Fundamentals of release mechanism interpretation in multiparticulate systems: determination of substrate release from single microcapsules and relation between individual and ensemble release kinetics, *Int. J. Pharm.*, 29, 195, 1986.

245. Benita, S., et al., Relation between individual and ensemble release kinetics of indomethacin from microspheres, *Pharm. Res.*, 5, 178, 1988.

246. Chang, N.J. and Himmelstein, K.J., Dissolution–diffusion controlled constant-rate release from heterogeneously loaded drug-containing material, *J. Contr. Rel.*, 12, 210, 1990.

247. Wu, X.Y., Eshun, G., Zhou, Y., Effect of interparticulate interaction on release kinetics of microsphere ensembles, *J. Pharm. Sci.*, 87, 586, 1998.

248. Ritger, P.L. and Peppas, N.A., A simple equation for description of solute release I. Fickian and non-fickian release from non-swellable devices in the form of slabs, spheres, cylinders or discs, *J. Contr. Rel.*, 5, 23, 1987.

249. Zhou, Y., Chu, J.S., Wu, X.Y., Theoretical analysis of drug release into a finite medium from sphere ensembles with various size and concentration distributions, *Eur. J. Pharm. Sci.*, 22, 251, 2004.

250. Tenchov, B.G. and Yanev, T.K., Weibull distribution of particle sizes obtained by uniform random fragmentation, *J. Colloid Interface Sci.*, 111, 2, 1986.

251. Carli, F. and Motta, A., Particle size and surface area distributions of pharmaceutical powders by microcomputerized mercury porosimetry, *J. Pharm. Sci.*, 73, 197, 1984.

252. Mandelbrot, B.B., *The Fractal Geometry of Nature*, Freeman, San Francisco, 1982.

253. Orbach, R., Dynamics of fractal structures, *J. Stat. Phys.*, 36, 735, 1984.

254. Giona, M., First order reaction–diffusion kinetics in complex fractal media, *Chem. Eng. Sci.*, 47, 1503, 1992.

255. O'Shaughnessy, B. and Procaccia, I. Analytical solutions for diffusion on fractal objects, *Phys. Rev. Lett.*, 54, 455, 1985.

256. Vicksek, T., *Fractal Growth Phenomena*, World Scientific, Singapore, 1989.

257. Adrover, A., Giona, M., Grassi, M., Analysis of controlled release in disordered structures: a percolation model, *J. Membr. Sci.*, 113, 21, 1996.

8 Drug Release from Microemulsions

8.1 INTRODUCTION

Microemulsions are mixtures of two mutually insoluble liquids whose stability is essentially due to the presence of the surfactant, a substance characterized by both a hydrophilic part and a lipophilic part whose task is to dispose at the hydrophilic liquid/lipophilic liquid interface to reduce the interfacial tension. Thus, microemulsion becomes a thermodynamically stable and homogeneous system able to form spontaneously without requesting any additional external energy. Accordingly, microemulsion can be defined as *a system of water, oil, and amphiphile, which is a single optically isotropic and thermodynamically stable liquid solution* [1]. Three different approaches have been developed to explain microemulsions' formation mechanism and stability [2]: (1) interfacial or mixed film theories [3,4]; (2) solubilization theories [5]; (3) thermodynamic approaches [6]. In particular, according to a simplified thermodynamic view, the Gibbs free energy ΔG_F associated with microemulsion formation would result from the balance between the energy increase, due to surface area increase (ΔA) and oil–water surface tension (γ) decrease induced by the presence of surfactant, and the energy decrease imputable to system entropy (ΔS) increase:

$$\Delta G_F = \gamma \Delta A - T \Delta S, \tag{8.0}$$

where T is the temperature. Equation 8.0 clearly states that, for a spontaneous microemulsion formation, ΔG_F must be negative, so that $T\Delta S$ must exceed $\gamma \Delta A$. As microemulsion formation implies a huge increase of the oil–water interface area (ΔA), γ must be considerably lowered by surfactant to allow the prevalence of the entropic contribute. For this reason, very often, also for the purpose of reducing the amount of surfactant needed, the presence of a cosurfactant, usually an alcohol, is required. In this manner, the interfacial tension γ is generally lowered to less than 1 mN/m and the formation of lipophilic or hydrophilic domains ranging from a few to a hundred nanometers is allowed [7]. The entropic contribution ΔS arises from the mixing of one phase in the other in the form of a large number of small domains. Nevertheless, ΔS is also improved by other favorable entropic contributions arising from other dynamic processes such as the surfactant diffusion in the interfacial layer separating oil from water. In addition, surfactant exchange between solubilized surfactant molecules (monomer surfactant) and surfactant micelles (for sufficiently high concentration, surfactant molecules can organize as small micelles in equilibrium with monomer surfactant) improves the entropic contribute. Obviously, this energy balance, although strictly depending on the surfactant, cosurfactant, and oil kind, can also be heavily affected by temperature and a difference of just 1°C or less can considerably alter the equilibrium position in terms of component concentration.

Industrial development of microemulsions dates back to 1930s (30 years before the term microemulsion was adopted) when, for example, they were used to make floor brilliant (carnauba wax dispersion (droplet size ~140 nm) in water stabilized by denatured alcohol) and to work as lubricants/cooling fluids. Indeed, mineral oil (lubricant), soap, sulphonated petroleum (emulsifying and anticorrosion agent), ethylene glycol (coupling agent), and antifoaming were added to water (cooling fluid) to get the final microemulsion formulation. In the 1940s and 1950s, microemulsions' application extended to food industry (flavors), agriculture (pesticides), and painting industry (latex systems). Undoubtedly, however, one of the most important applications of microemulsions is that concerning tertiary oil recovery [8]. Indeed, after the primary and secondary oil recovery, a considerable oil amount is still present in the porous rocks constituting the wall of the natural oil tank. For this reason, an aqueous surfactant solution is injected to give origin to a microemulsion that can be, subsequently, recovered jointly with its oil load. As this operation is successful if oil–water surface tension is very small (10^{-2} mN/m), the use of microemulsions is very appropriate as, in this case, extremely low oil–water surface tension occurs (10^{-3} mN/m).

In virtue of microemulsion compartmental nature (they are constituted by lipophilic and hydrophilic domains), they are very interesting as microreactors. Indeed, heterogeneous reactions occurring between hydrophilic and lipophilic reactants can take a great advantage of the huge liquid/liquid surface available in the microemulsion environment.

Nowadays, microemulsions are largely used in industry for their spontaneous formation, transparency, thermodynamic stability, low-viscosity, high-solubilization attitude toward lipophilic and hydrophilic solutes, extremely high surface area, and very low surface tension. For example, applications in the pharmaceutical, alimentary, cosmetics, and agricultural industry can be cited. The aim of this chapter is to present and discuss microemulsions' structural and thermodynamic characteristics in the light of pharmaceutical applications. In particular, the problem of drug release from discrete microemulsions (the dispersed phase is represented by small droplets) will be matched.

8.2   MICROEMULSIONS: STRUCTURE

On the basis of temperature, pressure, surfactant, cosurfactant, and oil considered, microemulsions can assume a great variety of structures such as oil droplets in water and vice versa, random bicontinuous mixtures, ordered droplets, and lamellar mixtures with a wide range of phase equilibria among them and with excess oil or water phases [7–9].

Among these great varieties, however, the most important categories are represented by *discrete* and *bicontinuous* microemulsions. Discrete microemulsions are constituted by a continuous phase hosting a dispersed phase (in form of small droplets, for example) stabilized by at least one monolayer constituted by surfactant molecules (see Figure 8.1). This monolayer, forming

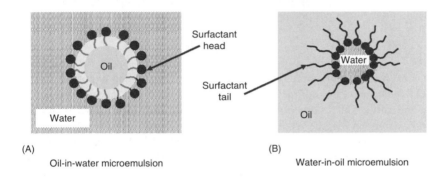

(A)

Oil-in-water microemulsion

(B)

Water-in-oil microemulsion

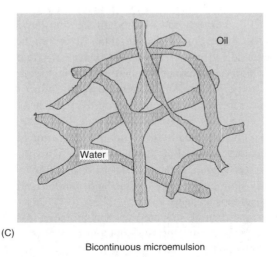

(C)

Bicontinuous microemulsion

FIGURE 8.1 Examples of microemulsion structures. (A) Oil in water discrete structure; (B) water in oil discrete structure; (C) bicontinuous structure.

a crown surrounding the droplet inner core, is, usually, 20 to 50 Å thick so that its volume is not negligible in comparison with that of the whole droplet. Indeed, as droplet size ranges between 100 and 1000 Å, crown volume represents the 14% of the whole volume if droplet diameter is 1000 Å and this percentage increases up to 88% when droplet radius is equal to 100 Å [7]. In virtue of droplets' reduced dimensions (i.e., lower than visible light wavelength), microemulsions are, typically, transparent. Droplets' shape depends on the surfactant properties that can be conveniently represented by the hydrophile–lipophile balance (HLB) [10] and the critical packing parameter (CPP) [11]. HLB takes into account the relative contribution of hydrophilic and hydrophobic fragments of the surfactant molecule. In general [2], low HLB (3–6) favors the formation of water-in-oil microemulsion whereas high HLB (8–18) is better suited for the realization of oil-in-water microemulsions. CPP, instead, highly influences the shape of the dispersed phase domains. This parameter is defined by

$$\text{CPP} = V/a_0 l, \tag{8.1}$$

where V is the partial molar volume of the hydrophobic portion of the surfactant, a_0 is the optimal surfactant (polar) head area and l is the length of the surfactant (hydrophobic) tail usually assumed to be 70%–80% of its fully extended length. When $\text{CPP} < 1/3$, oil-in-water globular structures are more probable, for $\text{CPP} \approx 1/2$, oil-in-water cylindrical structures form, for $\text{CPP} \approx 1$, planar structures are favored, for $\text{CPP} \approx 2$ water-in-oil cylindrical structures are more probable and for $\text{CPP} > 3$ oil-in-water droplets occur [2]. Obviously, changes in microemulsion composition or in ionic strength alter the surfactant microenvironment and this comports a variation of the surfactant apparent CPP. For example, although the introduction of small oil molecules, able to insert among the surfactant hydrophobic tails, would be probably responsible for an increase of the surfactant effective hydrophobic volume, this would not reasonably take place in presence of large oil molecules. Similarly, ionic strength increase could induce the reduction of a_0 due to the ions screening effect exerted on the polar heads that could stay closer to each other. This discussion makes clear that the great variety of structures that can be assumed by discrete microemulsions strongly depends on surfactant characteristics. For this purpose, it is worth mentioning that *nonionic* (polyoxyethylene, sugar ester), *zwitterionic* (phospholipids), *cationic* (quaternary ammonium alkyl salts), and *anionic* surfactants (sodium *bis*-2-ethylhexylsulphosuccinate) exist.

Bicontinuous microemulsions are more complicated to be visualized but their main characteristic relies on the contemporary existence of two intermeshing continuous phases represented by the lipophilic and the hydrophilic liquid. This means that it is possible to cross the entire microemulsion volume following a lipophilic or a hydrophilic path. In addition, these structures show high interfacial area characterized by a vanishing mean curvature [12]. On the basis of these peculiarities, an impressive representation of bicontinuous microemulsions is given by the *disordered open connected* model (Figure 8.1) assuming that the volume occupied by the surfactant molecules is negligible [7].

Microemulsions can also give origin to different phase equilibria implying the equilibrium with an aqueous phase, an oil phase, or with a differently structured microemulsion. As a matter of fact, the most-studied equilibria involving microemulsions are the so-called *Winsor phases* as their high-solubilization attitude make them suitable for the tertiary oil recovery process [13, 14]. Basically *Winsor* I, II, III, and IV equilibria exist. In *Winsor* I, a discrete oil-in-water microemulsion is in equilibrium with an excess oil phase, in *Winsor* II a discrete water-in-oil microemulsion is in equilibrium with an excess water phase, in *Winsor* III a bicontinuous microemulsion is in equilibrium with an excess oil and water phase whereas *Winsor* IV represent discrete microemulsions [15]. If *Winsor* III equilibrium is not so easy to be explained, *Winsor* I and II can be theorized assuming that the most stable microemulsion condition corresponds to a favorite value of droplet curvature (that, for a sphere, coincides with the inverse of sphere radius). Accordingly, the addition of oil to an oil/water microemulsion will reflect into droplet radius increase up to the favorite value above which the exceeding oil volume constitutes the excess oil phase. Analogous mechanism can be invoked for the *Winsor* II equilibrium. To make the scenario more complex, we must remember that the addition of surfactant concentration or the addition of salts makes possible a continuous transformation that starting from *Winsor* I ends in *Winsor* III passing through *Winsor* II. Figure 8.2 shows a schematic representation of a ternary phase diagram (water, oil, surfactant) showing the most important phases that can occur. In particular, we note that near the oil/water axis (low-surfactant concentration zone) a miscibility lacuna, corresponding to emulsion (two-phase region), occurs. When surfactant concentration is increased, oil-in-water discrete microemulsion or water-in-oil discrete microemulsion occurs if the low-oil concentration zone (near the water–surfactant axis) or the low-water concentration zone (near the oil–surfactant axis) is considered, respectively. In general, the addition of oil to a discrete oil-in-water microemulsion allows the transformation to the corresponding water-in-oil microemulsion passing through all or some of the *Winsor* I, II, and III equilibria. Obviously, an analogous behavior can occur starting from a discrete water-in-oil microemulsion by the addition of water. Moving on the water–surfactant axis, for increasing surfactant concentrations starting from zero, we find an aqueous surfactant solution, the formation of surfactant micelles in equilibrium with an aqueous surfactant solution and finally, micelles organization into higher order structures imparting the typical macroscopic gel-like behavior to the system. Similar behavior can be found on the oil–surfactant axis with the only difference that, obviously, reverse

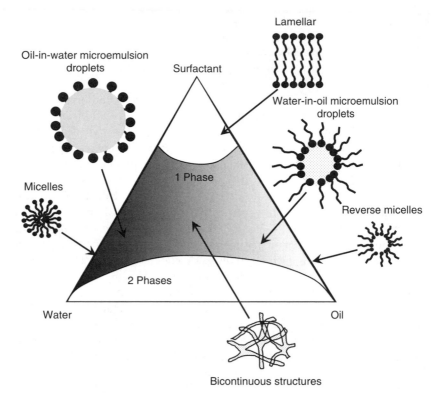

FIGURE 8.2 Schematic representation of a ternary phase diagram referring to a water, oil and surfactant system. The most important occurring phases are reported. (From Lawrence, M.J., and Rees, G.D., *Adv. Drug Deliv. Rev.*, 45, 86, 2000. With permission from Elsevier.)

micelles to occur. Finally, in the upper phase diagram corner (low water and oil content) lamellar structures can be formed.

8.3 MICROEMULSIONS: THERMODYNAMICS

As previously briefly discussed, surfactants and cosurfactants not only determine a liquid/liquid interfacial tension decrease, but they also affect interface curvature on the basis of their molecular structure. Indeed, the free energy F_c due to surface curvature and associated to an interfacial area A can be expressed by

$$F_c = \int\limits_A \left[\frac{K}{2}(c_1 + c_2 - 2c_0)^2 + \bar{K}c_1c_2\right]dA, \qquad (8.2)$$

where c_1 and c_2 are the surface first and second local curvatures, respectively, c_0 is the preferred or spontaneous curvature, K is the curvature elastic modulus, whereas $\bar{K}$ is Gauss modulus [7] (correctly, for a plane surface, $c_1 = c_2 = 0$ and, thus, $F_c = 0$). In the case of a single spherical droplet of radius R, Equation 8.2 becomes

$$F_c = 8\pi K(1 - c_0 R)^2 + 4\pi\bar{K} \tag{8.2'}$$

as $c_1 = c_2 = 1/R$. The term $4\pi\bar{K}$ can be seen as the free energy required to isolate a plane interfacial area and close it around a droplet. In *Winsor* I and II systems, the droplet interface tension σ can be calculated by summing up the interfacial tension γ associated to the macroscopic microemulsion/excess phase interface and the curvature energy per unit surface ($F_c/4\pi R^2$) assuming $c_0^2 \ll c_0$

$$\sigma(R) = \gamma - \frac{4Kc_0}{R} + \frac{2K + \bar{K}}{R^2}. \tag{8.3}$$

Assuming a positive curvature radius in the water-in-oil case and a negative curvature radius in the opposite situation, the condition of minimum energy requires that

$$\left(\frac{1}{R}\right)_{min} = \frac{2Kc_0}{2K + \bar{K}}. \tag{8.4}$$

Correctly, $1/R_{min}$ differs from c_0 as, according to Equation 8.2', an energy term $4\pi\bar{K}$ must be considered to form a droplet. In the case of polydisperse droplets characterized by a mean radius given by Equation 8.4, though the entropic contribute will be increased, the enthalpic one will be approximately constant. Accordingly, system equilibrium requires an increase of the enthalpic contribute that reflects into an equilibrium radius lower than R_{min}.

8.3.1 Discrete Microemulsions' Droplet Size Distribution

Assuming that these microemulsions can be viewed as the sum of two phases, the droplet and the continuous phase, Gibbs free energy G can be expressed by [7]

$$G = \sum_i N_i\mu_i = \sum_i N_{ci}\mu_{ci} + \sum_j N_{dj}\mu_{dj}, \tag{8.5}$$

where N_i and μ_i represent, respectively, the number of molecules and the chemical potential of the generic microemulsion component (ith), N_{ci} and μ_{ci} are, respectively, the number of molecules and the chemical potential of the *i*th component belonging to the continuous phase, *j*, referring to the number of surfactant molecules present at the droplet/continuous phase interface, identifies different droplet classes (bigger the droplet, higher the *j*), whereas N_{dj} and μ_{dj} are, respectively, the number of droplets belonging to the *j*th class and *j*th class droplets chemical potential. To make Equation 8.5 operative, μ_{ci} must be more precisely specified. Accordingly, assuming that surfactant and cosurfactant molecules lying at the droplet/continuous phase interface belong to the droplet and that we are dealing with spherical particles, the second addendum of Equation 8.5 right-hand side term can be expressed remembering that droplet phase Gibbs free energy is given by the sum of the chemical potential of all the present elements, of the interfacial energy and of the mixing free energy linked to the configurations the droplets can assume in the microemulsion. Consequently, Equation 8.5 becomes

$$G = \sum_i N_{im}\mu_{im} + \sum_j \sum_i N_{ij}\mu_{ij} - \sum_j \Delta P_j V_{dj} + \sum_j \gamma_j A_j - TS_{mix}, \quad (8.6)$$

where N_{ij} and μ_{ij} represent, respectively, the number of molecules and the chemical potential of the *i*th component belonging to the *j*th droplet class, ΔP_j is the pressure difference existing between the continuous phase and the *j*th droplet class (as later on discussed in more detail), V_{dj} is the volume occupied by all the droplets belonging to the *j*th class, γ_j and A_j are, respectively, the surface tension and the interface surface area competing to the *j*th droplet class, *T* is the temperature, whereas S_{mix} is the mixing entropy. Minimization of *G* leads to the following droplets' size distribution:

$$\phi_j = \frac{v_{dj}}{l^3} \exp\left[-\frac{(4/3)\pi R_j^3 \Delta\bar{\mu} + 4\pi R_j^2 \gamma_j}{kT}\right], \quad (8.7)$$

where ϕ_j is the class volume fraction of *j*th droplet, v_{dj} and R_j are, respectively, the volume and the radius of a single *j*th droplet class, *l* is a characteristic length (for monodisperse microemulsions, it coincides with the molecular volume of the droplet phase), *k* is Boltzmann constant whereas $\Delta\bar{\mu} = (\mu_i' - \mu_i)/v_i$, where v_i is the molar volume of *i*th element, whereas $\mu_i' - \mu_i$ is the difference between the *i*th component chemical potential in the droplet and the continuous phase evaluated at the same pressure.

8.3.2 LIQUID–LIQUID INTERFACIAL COMPOSITION

Although the experimental determination of the droplet/continuous phase composition in the case of microemulsions is not a trivial task, it can be useful to discuss a method that enables the determination of the liquid–liquid interfacial composition in the case of two immiscible liquid phases constituted by three components (hydrophilic liquid, hydrophobic liquid, and surfactant) as it can occur in the case of emulsions. Indeed, ultimately, microemulsion can be seen as the limiting case of an emulsion.

Assuming a negligible effect of surface curvature (this means that we are referring to a plane or to a spherical surface), the infinitesimal, reversible, variation of the internal energy U for a closed system made up by k components and three phases (α, β bulk phases; σ interfacial phase) is given by [16]

$$dU = dU^\alpha + dU^\beta + dU^\sigma, \tag{8.8}$$

$$dU^\alpha = T^\alpha \, dS^\alpha + \sum_{i=1}^{k} \mu_i^\alpha \, dn_i^\alpha - P^\alpha \, dV^\alpha, \tag{8.9}$$

$$dU^\beta = T^\beta dS^\beta + \sum_{i=1}^{k} \mu_i^\beta dn_i^\beta - P^\beta dV^\beta, \tag{8.10}$$

$$dU^\sigma = T^\sigma \, dS^\sigma + \sum_{i=1}^{k} \mu_i^\sigma \, dn_i^\sigma + \gamma_{\alpha\beta} dA, \tag{8.11}$$

where U^α, U^β, U^σ, represent the α, β, and σ phase internal energies, respectively, S^α, S^β, S^σ, T^α, T^β, and T^σ indicate entropy and temperature of α, β, and σ phases, respectively, V^α and V^β are the volumes of the α and β bulk phases, respectively, μ_i^α, μ_i^β, μ_i^σ, n_i^α, n_i^β, and n_i^σ represent, respectively, the chemical potential and the number of moles of the ith component belonging to α, β, and σ phases, A is the interface area and $\gamma_{\alpha\beta}$ is the interface tension between the α and β phases. The closed system hypothesis requires the system volume V, entropy S and moles' number n to be constant, so that the following relations hold:

$$dS = dS^\alpha + dS^\beta + dS^\sigma = 0, \tag{8.12}$$

$$dV = dV^\alpha + dV^\beta = 0, \tag{8.13}$$

$$dn = dn_i^\alpha + dn_i^\beta + dn_i^\sigma = 0. \tag{8.14}$$

Accordingly, the equilibrium condition ($dU = 0$) requires that

$$T^\alpha = T^\beta = T^\sigma = T \quad \text{(thermal equilibrium)}, \tag{8.15}$$

$$\mu_i^\alpha = \mu_i^\beta = \mu_i^\sigma = \mu_i \quad \text{(chemical equilibrium)}, \tag{8.16}$$

$$P^\alpha - P^\beta = \gamma_{\alpha\beta} \frac{dA}{dV^\alpha} = \gamma_{\alpha\beta} \left(\frac{1}{r_\alpha} + \frac{1}{r_\beta} \right) \quad \text{(mechanical equilibrium)}, \tag{8.17}$$

where r_α and r_β are the curvature radii of α and β phases, respectively. In the light of Equation 8.15 through Equation 8.17, Equation 8.9 through Equation 8.11 become

$$dU^\alpha = T\,dS^\alpha + \sum_{i=1}^{k} \mu_i\,dn_i^\alpha - P^\alpha\,dV^\alpha, \tag{8.18}$$

$$dU^\beta = T\,dS^\beta + \sum_{i=1}^{k} \mu_i\,dn_i^\beta - P^\beta\,dV^\beta, \tag{8.19}$$

$$dU^\sigma = T\,dS^\sigma + \sum_{i=1}^{k} \mu_i\,dn_i^\sigma + \gamma_{\alpha\beta}\,dA. \tag{8.20}$$

The integration of Equation 8.18 through Equation 8.20 and subsequent differentiation leads to

$$dU^\alpha = T\,dS^\alpha + S^\alpha dT + \sum_{i=1}^{k} \mu_i\,dn_i^\alpha + \sum_{i=1}^{k} n_i^\alpha\,d\mu_i - P^\alpha\,dV^\alpha - V^\alpha\,dP^\alpha, \tag{8.21}$$

$$dU^\beta = T\,dS^\beta + S^\beta\,dT + \sum_{i=1}^{k} \mu_i\,dn_i^\beta + \sum_{i=1}^{k} n_i^\beta\,d\mu_i - P^\beta\,dV^\beta - V^\beta\,dP^\beta, \tag{8.22}$$

$$dU^\sigma = T\,dS^\sigma + S^\sigma\,dT + \sum_{i=1}^{k} \mu_i\,dn_i^\sigma + \sum_{i=1}^{k} n_i^\sigma\,d\mu_i + \gamma_{\alpha\beta}\,dA + A\,d\gamma_{\alpha\beta}. \tag{8.23}$$

As Equation 8.18 through Equation 8.20 must be equal to Equation 8.21 through Equation 8.23, we have, at constant pressure and temperature:

$$\sum_{i=1}^{k} n_i^\alpha\,d\mu_i = 0, \tag{8.24}$$

$$\sum_{i=1}^{k} n_i^\beta\,d\mu_i = 0, \tag{8.25}$$

$$d\gamma_{\alpha\beta} = -\sum_{i=1}^{k} \frac{n_i^\sigma}{A}\,d\mu_i = -\sum_{i=1}^{k} \Gamma_i^\sigma\,d\mu_i. \tag{8.26}$$

Equation 8.24 and Equation 8.25 represent, respectively, the Gibbs–Duhem equation for the α and β bulk phases, whereas Equation 8.26 is the Gibbs equation for the interfacial phase σ. These equations represent the starting point for the interface composition determination resorting to interface tension and bulk composition determination. Indeed, for a three components system ($k = 3$) Equation 8.26 becomes

$$d\gamma_{\alpha\beta} = -\Gamma_1^\sigma \, d\mu_1 - \Gamma_2^\sigma \, d\mu_2 - \Gamma_3^\sigma \, d\mu_3, \qquad (8.27)$$

where $d\mu_1$ and $d\mu_2$ can be expressed as a function of $d\mu_3$ using Equation 8.24 and Equation 8.25:

$$d\mu_1 = d\mu_3 \left(\frac{n_3^\beta n_2^\alpha - n_2^\beta n_3^\alpha}{n_2^\beta n_1^\alpha - n_1^\beta n_2^\alpha} \right) = d\mu_3 C, \qquad (8.28)$$

$$d\mu_2 = d\mu_3 \left(\frac{n_1^\beta n_3^\alpha - n_3^\beta n_1^\alpha}{n_2^\beta n_1^\alpha - n_1^\beta n_2^\alpha} \right) = d\mu_3 B. \qquad (8.29)$$

Consequently, we have

$$d\gamma_{\alpha\beta} = -\Gamma_1^\sigma C \, d\mu_3 - \Gamma_2^\sigma B \, d\mu_3 - \Gamma_3^\sigma \, d\mu_3. \qquad (8.30)$$

Remembering the relation between the chemical potential and the activity, we finally have

$$\frac{\partial \gamma_{\alpha\beta}}{\partial \mu_3} = \frac{1}{RT} \frac{\partial \gamma_{\alpha\beta}}{\partial \ln a_3} = -\Gamma = -(\Gamma_1^\sigma C + \Gamma_2^\sigma B + \Gamma_3^\sigma), \qquad (8.31)$$

where R is the universal gas constant and a_3 is the activity of the third component. The interface composition determination requires measuring the dependence of $\gamma_{\alpha\beta}$, B, and C on a_3 to know how the left member of Equation 8.31 (*alias* $-\Gamma$) depends on B and C. Consequently, the knowledge of the experimental function $\Gamma(B,C)$ is achieved and the calculation of Γ_1^σ, Γ_2^σ, and Γ_3^σ directly descends from the fact that, locally ($\bar{B}, \bar{C}$), Γ can be approximated by a plane Γ^p:

$$\Gamma^p = \left(\left. \frac{\partial \Gamma}{\partial C} \right|_{\bar{C},\bar{B}} C + \left. \frac{\partial \Gamma}{\partial B} \right|_{\bar{C},\bar{B}} B + q \right) = \left(\Gamma_1^\sigma C + \Gamma_2^\sigma B + \Gamma_3^\sigma \right). \qquad (8.32)$$

Accordingly, we have

$$\Gamma_1^\sigma = \left. \frac{\partial \Gamma}{\partial C} \right|_{\bar{C},\bar{B}} \qquad (8.33)$$

$$\Gamma_2^\sigma = \left. \frac{\partial \Gamma}{\partial B} \right|_{\bar{C},\bar{B}}, \qquad (8.34)$$

$$\Gamma_3^{\sigma} = q = \Gamma^{\mathrm{p}}(\bar{C}, \bar{B}) - \frac{\partial \Gamma}{\partial C}\Big|_{\bar{C}, \bar{B}} \bar{C} - \frac{\partial \Gamma}{\partial B}\Big|_{\bar{C}, \bar{B}} \bar{B}., \qquad (8.35)$$

Finally, interface composition is given by

$$x_i^{\sigma} = \frac{\Gamma_i^{\sigma}}{\sum\limits_{i=1}^{i=3} \Gamma_i^{\sigma}} = \frac{n_i^{\sigma}}{\sum\limits_{i=1}^{i=3} n_i^{\sigma}}. \qquad (8.36)$$

Although the above mentioned approach shows the great advantage of yielding the interface composition regardless its position among the two bulk phases α and β (we recall here that, according to Gibbs' theory [16], the existence of a plane (interface) where a discontinuous variation of bulk properties, such as compounds concentration, takes place, is a pure idealization. Accordingly, also its exact positioning is not theoretically defined), the determination of plane Γ^{p} from experimental data is not a trivial task. For this reason, always in the light of Gibbs' theory, the shown general approach can be particularized for two convenient choices of the interface position, namely the planes where Γ_1^{σ} and Γ_2^{σ}, respectively, vanish. Accordingly, Equation 8.31 becomes

$$-\frac{\partial \gamma_{\alpha\beta}}{\partial \mu_3} = -\frac{1}{RT} \frac{\partial \gamma}{\partial \ln a_3} = (\Gamma_2^{\sigma_1} B + \Gamma_3^{\sigma_1}) = \Gamma = \frac{\partial \Gamma^{\mathrm{p}}}{\partial B}\Big|_{B} B + q_1, \quad \{\Gamma_1^{\sigma} = 0\}, \quad (8.37)$$

$$-\frac{\partial \gamma_{\alpha\beta}}{\partial \mu_3} = -\frac{1}{RT} \frac{\partial \gamma}{\partial \ln a_3} = (\Gamma_1^{\sigma_2} C + \Gamma_3^{\sigma_2}) = \Gamma = \frac{\partial \Gamma^{\mathrm{p}}}{\partial C}\Big|_{C} C + q_2, \quad \{\Gamma_2^{\sigma} = 0\}, \quad (8.38)$$

where $\Gamma_2^{\sigma_1}$, $\Gamma_3^{\sigma_1}$, q_1, $\Gamma_1^{\sigma_2}$, $\Gamma_3^{\sigma_2}$, and q_2 remind us that we are referring to the interface where Γ_1^{σ} or Γ_2^{σ} vanishes, respectively. The considerable advantage of this choice is that the plane Γ^{p} has now become a straight line whose slope and intercept has to be determined by means of experimental data.

Meriani and coworkers [17] apply the above discussed theoretical approach to the triacetin (hydrophobic liquid; $C_9H_{14}O_6$, m.w. 218.21; Fluka Chemika, Sigma–Aldrich, Italy), ethanol (C_2H_6O, m.w. 50; BDH, England), and distilled water system at 25°C. Starting point for this task is the determination of the water–triacetin–ethanol phase diagram. Figure 8.3 clearly shows that the limited water–triacetin mutual solubility increases with the added ethanol amount. Indeed, the biphasic region (denominating emulsion region in Figure 8.3) disappears when ethanol mass fraction in the ternary mixture exceeds approximately 0.2. In addition, the presence of tie-lines allows the determination of the composition of the triacetin-rich and water-rich phases into which the whole liquid mixture decomposes if its global composition falls in the immiscibility region. Finally, the interfacial tension between the water-rich and the triacetin-rich phases can be measured by means of the shape

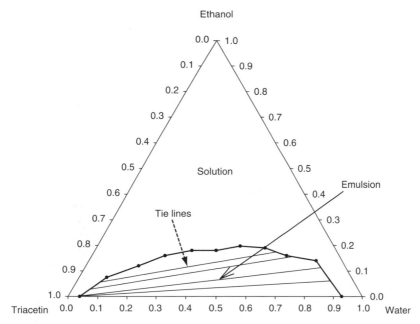

FIGURE 8.3 Water–triacetin–ethanol phase diagram (25°C). (From Meriani, F., et al., *J. Colloid Interface Sci.*, 263, 590, 2003. With permission from Elsevier.)

analysis of a liquid drop (triacetin-rich phase) hanging from a microsyringe into a continuous phase formed by the water-rich phase (G10 tensiometer; Krüss GmbH, Hamburg, D). Now, the determination of interface composition requires calculation of the activity of ethanol (component 3) regardless the particular interface chosen ($\Gamma_1^\sigma = 0$ or $\Gamma_2^\sigma = 0$). Chemical equilibrium conditions (Equation 8.16) require

$$a_{ETOH}^\alpha = X_{ETOH}^\alpha \gamma_{ETOH}^\alpha = X_{ETOH}^\beta \gamma_{ETOH}^\beta = a_{ETOH}^\beta, \qquad (8.39)$$

where a_{ETOH}^α, X_{ETOH}^α, γ_{ETOH}^α, X_{ETOH}^β, γ_{ETOH}^β, and a_{ETOH}^β are, respectively, the ethanol activity, molar fraction, and activity coefficient of phase α (water-rich phase) and β (triacetin-rich phase). Accordingly, the problem of activity determination moves on the activity coefficient calculation ($\gamma_{ETOH}^{\alpha,\beta}$) that can be undertaken by means of the following relation:

$$RT \ln(\gamma_{ETOH}^{\alpha,\beta}) = \frac{\partial n^{\alpha,\beta} G_{\alpha,\beta}^E}{\partial n_{ETOH}^{\alpha,\beta}} \Bigg|_{T,P,n_{TRIAC}^{\alpha,\beta}, n_{H_2O}^{\alpha,\beta}}, \qquad (8.40)$$

where T is the temperature, R is the universal gas constant, $G_{\alpha,\beta}^E$ is the excess molar Gibbs energy, $n^{\alpha,\beta}$ represents the total moles' number of phases α and

β, respectively. Remembering the simplest expression for $G_{\alpha,\beta}^{E}$ (Margules model [18]):

$$\frac{G_{\alpha,\beta}^{E}}{RT} = FX_{ETOH}^{\alpha,\beta}\, X_{TRIAC}^{\alpha,\beta}\, X_{H_2O}^{\alpha,\beta}, \tag{8.41}$$

where F is a model parameter, $X_{ETOH}^{\alpha,\beta}$, $X_{TRIAC}^{\alpha,\beta}$, and $X_{H_2O}^{\alpha,\beta}$ are the ethanol, triacetin, and water molar fraction of the α and β phases, respectively, we have

$$\ln(\gamma_{ETOH}^{\alpha,\beta}) = FX_{ETOH}^{\alpha,\beta}\, X_{TRIAC}^{\alpha,\beta}\, [1 - 2X_{H_2O}^{\alpha,\beta}]. \tag{8.42}$$

F determination (according to Margules model simplicity, F is not a constant) can be performed by inserting Equation 8.42 into Equation 8.39:

$$\ln\left(\frac{X_{ETOH}^{\alpha}}{X_{ETOH}^{\beta}}\right) = \ln\left(\frac{\gamma_{ETOH}^{\beta}}{\gamma_{ETOH}^{\alpha}}\right)$$

$$= F\,[X_{H_2O}^{\beta}X_{TRIAC}^{\beta}(1 - 2X_{H_2O}^{\beta}) - X_{H_2O}^{\alpha}X_{TRIAC}^{\alpha}(1 - 2X_{H_2O}^{\alpha})], \tag{8.43}$$

where the compositions of the two phases are experimentally determined. Table 8.1 shows the overall ethanol weight fraction W_{ETOH} (ethanol weight fraction referring to the α and β phases thought as a whole), and the corresponding values of the components mole (X) and mass (w) fraction in

TABLE 8.1
Bulk Composition (X, Molar Fraction; w, Mass Fraction), Surface Tension $\gamma_{\alpha\beta}$, and Margule Parameter F Variation with W_{ETOH} at $T = 25°C$

	α			β				
W_{ETOH}	X_{TRIAC}	X_{H_2O}	X_{ETOH}	X_{TRIAC}	X_{H_2O}	X_{ETOH}	$\gamma_{\alpha\beta}$ (mN/m)	$F(-)$
0	0.0063	0.9937	0	0.6678	0.3322	0	35.8	−31.6
0.034	0.0074	0.9657	0.0269	0.6619	0.3347	0.0034	27.7	26.3
0.064	0.0083	0.9398	0.0519	0.6596	0.3338	0.0066	20.8	25.8
0.108	0.0179	0.0901	0.892	0.5146	0.4104	0.075	12.6	1.5
0.135	0.0247	0.8783	0.097	0.4142	0.4586	0.1272	7.7	−8.4
	w_{TRIAC}	w_{H_2O}	w_{ETOH}	w_{TRIAC}	w_{H_2O}	w_{ETOH}		
0	0.071	0.928	0	0.968	0.039	0	35.8	−31.6
0.034	0.080	0.859	0.061	0.959	0.040	0.001	27.7	26.3
0.064	0.086	0.801	0.113	0.958	0.040	0.002	20.8	25.8
0.108	0.164	0.680	0.156	0.912	0.060	0.028	12.6	1.5
0.135	0.210	0.616	0.174	0.865	0.079	0.056	7.7	−8.4

Source: From Meriani, F., et al., *J. Colloid Interface Sci.*, 263, 590, 2003. With permission from Elsevier.

the α (water-rich) and β (triacetin-rich) phases jointly with the F and surface tension $\gamma_{\alpha,\beta}$ values. Consequently, it is possible to determine the $\gamma_{\alpha,\beta}$, B, and C dependences on $\ln(a_{ETOH})$. In particular, these dependences can be properly expressed by the following polynomial functions:

$$\gamma_{\alpha,\beta} = 35.3 + 4.4 \times 10^{-4}(\ln(a_{ETOH}) + 8.2)^2$$
$$- 2.1 \times 10^{-2}(\ln(a_{ETOH}) + 8.2)^4, \tag{8.44}$$

$$B = -18.8 \times 10^{-4} + 4.8 \times 10^{-4}(\ln(a_{ETOH}) + 6.63)^2 +$$
$$-2.5 \times 10^{-6}(\ln(a_{ETOH}) + 6.63)^4, \tag{8.45}$$

$$C = -5 \times 10^{-4} - 2.3 \times 10^{-4}(\ln(a_{ETOH}) + 8.03)^3$$
$$- 1.64 \times 10^{-6}(\ln(a_{ETOH}) + 8.03)^6. \tag{8.46}$$

Interestingly, the monotonic decrease of $\gamma_{\alpha,\beta}$ with $\ln(a_{ETOH})$ reveals a modification of the interface composition. Focussing the attention on $\Gamma^\sigma_{H_2O} = 0$ Equation 8.44 through Equation 8.46 permit to get the $\Gamma(= -\dfrac{1}{RT}\dfrac{\partial \gamma_{\alpha,\beta}}{\partial \ln a_{ETOH}})$ dependence on B as shown in Figure 8.4. Consequently, the interface triacetin and ethanol molar fractions, $X^{\sigma H_2O}_{TRIAC}$ and $X^{\sigma H_2O}_{ETOH}$, can be determined as the slope and the intercept of the generic straight line tangent to Γ (see Figure 8.5).

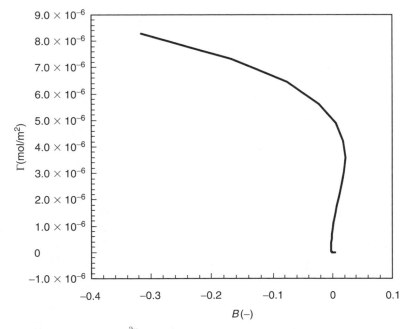

FIGURE 8.4 $\Gamma(= 1/RT\frac{\partial \gamma_{\alpha\beta}}{\partial \ln a_{ETOH}})$ dependence on parameter B(−). (From Meriani, F., et al., *J. Colloid Interface Sci.*, 263, 590, 2003. With permission from Elsevier.)

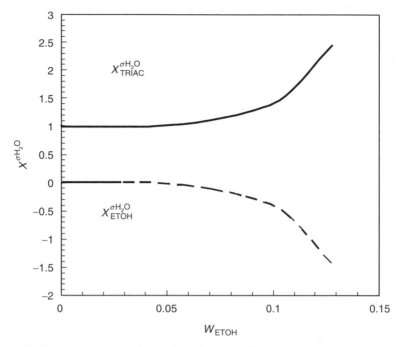

FIGURE 8.5 Dependence of the triacetin (solid line) and ethanol (dashed line) interfacial mole fraction vs. the overall ethanol mass fraction W_{ETOH}. The attention is focused on the interfacial surface where water concentration is zero. (From Meriani, F., et al., *J. Colloid Interface Sci.*, 263, 590, 2003. With permission from Elsevier.)

This analysis reveals that a W_{ETOH} increase (or a_{ETOH} increase) reflects in a triacetin interface enrichment and a corresponding ethanol impoverishment. The negative values assumed by $X_{\text{ETOH}}^{\sigma H_2O}$ are not meaningless recalling the meaning of the interface under consideration. Indeed, although the actual interface region has no sharply defined boundaries, Gibbs suggests simplifying this scenario by inventing a mathematical dividing surface so that the two phases continue uniformly up to the surface [16]. Accordingly, the extensive bulk properties do not show the real variation approaching the surface but they are always constant. Thus, their actual values for the system as a whole will then differ from the sum of the values for the two bulk phases by an excess or deficiency assigned to the surface region and this is the reason why a negative surface concentration can appear. Reasoning for the $\Gamma_{\text{TRIAC}}^{\sigma} = 0$ case, again we find that ethanol leaves the interface for increasing W_{ETOH}. At the same time, $X_{H_2O}^{\sigma \text{TRIAC}}$ shows a moderate increase. Although in this example a real surfactant was not involved, this approach can be developed if, for example, ethanol was replaced by a surfactant.

8.4 MICROEMULSIONS: EXPERIMENTAL CHARACTERIZATION

Due to the great variety of structures, microemulsions' characterization is a very hard task and this is the reason why many techniques have been developed for this purpose and, often, the combination of more than one of them is required for a reliable characterization. Although microemulsions can be characterized by different techniques (infrared beams absorption [19], surface tension measurements [8], ultrasounds absorption [20,13], fluoresce measurements [13], sound dispersion, and attenuation [21]), here we would like to present some of the most common and important ones.

The simplest characterization is the visual inspection that, sometimes, can allow the discrimination of different phases [22,23]. This approach can be improved by recurring to a turbidity analysis leading to absorbance or transmittance by means of a UV spectrophotometer [24]. Indeed, though the emulsion region is typically opaque, microemulsion is perfectly transparent and this transparency tends to decrease moving toward the high surfactant concentration zone.

8.4.1 SCATTERING TECHNIQUES

Scattering techniques, widely used in microemulsions' characterization, comprehend *small angle x-ray scattering* (SAXS) [25], *small angle neutron scattering* (SANS) [26], and *light scattering* (LS) [27]. In all these techniques, the intensity of the scattered radiation $I(q)$ is measured as a function of the scattering vector q defined by

$$q = \frac{4\pi}{\lambda} \sin\left(\frac{\theta}{2}\right),$$ (8.47)

where λ is the radiation wavelength and θ is the scattering angle. The general expression for I, in the case of monodisperse spherical particles, is

$$I(q) = n P(q) S(q),$$ (8.48)

where n is the particles' numerical density, $P(q)$ is a form factor related to particles' cross section, whereas $S(q)$ is a structural factor accounting for particle–particle interactions. Proper analytical expressions are available for P and S estimation [8]. Although SAXS can be used to get information about shape and dimension of microemulsion droplets, SANS can also provide information about the amphiphilic stratum dividing the oil phase from the water phase without big system perturbations. LS can be subdivided into bulk LS and surface LS. In bulk LS, widely used for the determination of droplets shape and dimension, the intensity of the scattered light is usually measured at different angles and for different droplets concentration in the microemulsion. Surface LS, also called *photon correlation spectroscopy* (PCS), is based on the reflection of a laser beam on the oil/water interface. Due to brownian motions, this surface is not perfectly plane so that the laser beam is reflected in all directions. The intensity variation of the reflected beam provides information about system dynamics. This technique can be used to estimate dispersed phase dimensions [28]. Scattering techniques allow to measure droplets diameter ranging between 2 and 100 nm (SANS and SAXS) or 2–3 μm (LS). The major problem in adopting scattering techniques relies on the necessity of sample dilution to reduce droplets interactions. Indeed, unfortunately, microemulsions' dilution can reflect into structure and composition modifications of the various phases [2,8].

8.4.2 NUCLEAR MAGNETIC RESONANCE TECHNIQUES

Nuclear magnetic resonance (NMR) is used to study microemulsions' structures and dynamics besides the phase behavior. Indeed, NMR self-diffusion and relaxation constitute selection tools for the determination of microstructural information [29,30]. Self-diffusion studies, coupled with microscopic analysis, probably represent the only direct way of proving the existence of microemulsion bicontinuous structures. On the other hand, relaxation is one of the most reliable methods for the determination of droplets shape and dimension, especially in concentrated systems as the rotational diffusion of droplets is evaluated. In addition, this technique is useful for the characterization of the surfactant film separating the oil phase from the water phase. However, we cannot forget that this technique can also provide useful information about microemulsions' phase diagrams. Indeed, we can get simultaneous information regarding water, oil, and surfactant regions. In the water-in-oil case, for example, though water molecules diffusion is slow (it corresponds to that of droplets), that competing to oil molecules is higher and surfactant molecules, belonging to the interfacial region, show diffusive properties similar to those of water. On the contrary, in the oil-in-water situation, water diffusion coefficient is higher than that of oil. In bicontinuous structures, both water and oil diffusion coefficients are high whereas that of surfactant is lower. Finally, it is worth mentioning that this technique can provide useful information about *critical micellar concentration* (CMC, this is the minimum concentration at which surfactant molecules organize into micelles (polar heads face to the external solvent in the case of polar fluids) giving origin to an equilibrium between micelles and solubilized surfactant molecules), micelles shape and dimension.

8.4.3 ELECTRONIC MICROSCOPY

Due to the intrinsic fragile nature of microemulsions, their characterization according to this technique poses serious problems. Indeed, sample preparation often comport microemulsion structure alteration or breakup [8]. However, the *freeze fracture electron microscopy* allows overcoming this problem as it consists in a very rapid microemulsion freezing followed by system fracture [31]. Accordingly, a microscopic analysis of the fractured surface can be done. The rapid freezing prevents structure modifications so that what we see is representative of the real microemulsion structure.

8.4.4 CONDUCTIVITY AND DIELECTRIC SPECTROSCOPY

These techniques are based on the natural observation that electrical properties modify in reason of microemulsion structure. Accordingly, discrete oil-in-water microemulsions should show a higher conductance than that associated with water-in-oil microemulsion [32]. However, in the passage from oil-in-water to water-in-oil structure, just before the formation of a bicontinuous structure, conductance should get the highest value as droplets exchange ions with each other. Analogously, static permittivity and dielectric relaxation frequency maximize in the presence of bicontinuous structures [33].

8.4.5 RHEOLOGY

Microemulsions are usually Newtonian fluids. As discussed in Chapter 3, this means that their viscosity does not depend on the shear rate (or on the shear stress). The reason for this behavior can be easily explained by remembering microemulsion structure. Indeed, its characteristic length (mean droplets diameter, in the case of discrete systems) is very small and velocity gradients employed in usual experimental conditions are not sufficiently high to determine a significant velocity variation at such a small scale [34]. In addition, variations of experimental velocity occur in a time range that is usually much bigger than that required by microemulsion structure to react to an external perturbation such as that induced by a velocity variation. Accordingly, microemulsions can exhibit non-Newtonian behavior only at very high-shear rates ($>10^4/s$) (this may not be completely true for some *Winsor* III microemulsions) and their viscoelastic character can be detected only at very high frequency (10^4–10^5 Hz) where it is possible to examine very fast relaxation processes [13].

Typically, the viscosity of oil-in-water microemulsions increases with the oil volume fraction. This behavior can be properly described by theoretical models (see Chapter 3) based on the assumption of a rigid spherical dispersed phase provided that droplets diameter considers also the presence of the surfactant layer surrounding the inner oil core. Sometimes, a correct description of the relative viscosity increase with oil volume fraction implies the introduction of a structural parameter accounting for possible particles clustering and for not spherically shaped particles. Indeed, in both cases, dispersed phase occupies a volume that is bigger than the real oil volume. The increase of the relative viscosity can be upto 30–40.

Water-in-oil microemulsions basically show the same behavior of the oil-in-water microemulsions. Nevertheless, for high concentrations and temperatures, attractive interactions among droplets favor a reversible droplets clustering that causes a viscosity increase. Consequently, though for low temperatures, microemulsion viscosity can be properly described by models based on the rigid sphere hypothesis, when temperature approaches a critical value, these models fail and viscosity increase with dispersed phase volume fraction is much more pronounced.

Typically, bicontinuous microemulsions also show a Newtonian behavior, although some rheological peculiarities distinguish them from discrete microemulsions. For this purpose it is useful to consider systems where the transition from the oil-in-water to the water-in-oil structure, passing through bicontinuous structures, is possible by simply changing one parameter such as temperature or salt concentration. This happens, for example, for proper compositions of the n-octanol/water/polyglycol–alkyl-ether systems in response to a temperature increase [35]. Interestingly, microemulsion viscosity increases with temperature and it reaches a maximum just before the transition from the oil-in-water to bicontinuous structure. Then, in the bicontinuous region, viscosity trend shows a decrease followed by another maximum in correspondence to the transition bicontinuous structure—water-in-oil. Accordingly, in the bicontinuous region, viscosity is characterized by a minimum that is, anyway, higher than that characterizing the original oil-in-water structure. Finally, these systems show a moderate shear thinning behavior in the shear rate range $0.1–10 \ s^{-1}$, where viscosity reduces to $1/3–1/5$ of the Newtonian plateau value measurable for shear rate lower than $0.1 \ s^{-1}$. This behavior is not only typical of the bicontinuous region but also of the water-in-oil and the oil-in-water regions. The reason for this lies on the fact that the dispersed phase is not only made up of spherical particles but also of droplet clusters or differently shaped droplets.

8.5   MICROEMULSIONS AND DRUG RELEASE

Microemulsions' thermodynamic stability, optical clarity, preparation simplicity, and the possibility of solubilizing both water-soluble and oil-soluble drugs make microemulsions very interesting as delivery system [36]. This versatility is also improved by the possibility of incorporating amphiphilic drugs that, sometimes, even lead to an increase in the extent of existence of the microemulsion region. Despite the great variety of microemulsion structures, for their ability of controlling release kinetics, discrete microemulsions (*Winsor* IV) represent the selection kind in the drug delivery field [2,37,38]. Indeed, as drug partitions between microemulsion droplets, continuous phase and surfactant micelle, a great variety of release kinetics can be obtained depending on the drug's preferred site of solubilization. For example, the preferred sites of incorporation of a lipophilic, water-insoluble drug into an oil-in-water microemulsion are the disperse oil phase and/or hydrophobic tail region of the surfactant molecules organizing into micelle. Interestingly, a hydrophobic drug must have a considerable high solubility in the oil phase such that microemulsion is convenient, in terms of drug solubilization, with respect to the simple water–surfactant mixture. In addition, oils exhibiting maximum drug solubility not always give origin to microemulsions with the highest solubilization capacity [2]. If oil phase is the preferred solubilizing environment, drug diffusion from the oil droplets to the living tissues can take place by crossing the surrounding aqueous medium which essentially acts as a barrier to drug transport owing to the very low drug solubility in water. Accordingly, microemulsion is used to retard drug delivery [39] and the oil phase works as a reservoir. On the contrary, it may also happen that microemulsion speeds up the drug uptake by the living tissues if oil droplets phagocytosis, led by particular biological structures [40], takes place. In this case, oil-in-water microemulsions can highly increase hydrophobic drugs bioavailability [36]. It is important to remember that, in the drug delivery field, the use of oil-in-water microemulsions is more straightforward than that of water-in-oil. Indeed, though the droplet structure of oil-in-water microemulsions is often retained on dilution by biological aqueous phase (this allows oral and parenteral administration), the same does not happen in the water-in-oil microemulsions. In this case, the aqueous phase volume fraction increase decreases the surfactant/water ratio which leads to droplets growth and, if the dilution continues, to phase separation or inversion so that original microemulsion properties can be lost. However, water-in-oil microemulsions can offer interesting advantages in the delivery of labile hydrophilic drugs such as peptides and oligonucleotides. Indeed, if, on one hand, they have little or no activity when delivered orally due to proteolysis in the gastrointestinal tract [41], their parenteral administration, especially for chronic conditions, can give rise to compliance problems. Accordingly, the possibility of incorporating them in the water droplets of a water-in-oil microemulsion prevents rapid enzymatic degradation [42]. Finally, drug delivery forms based on water-in-oil microemulsions can also be used when dilution by an aqueous phase is not so important as it happens after intramuscular injection [43].

All these considerations make clear that to have successful formulations, the knowledge of the phase behavior of water, surfactant, and oil is essential even if, unfortunately, the effects of *in vivo* conditions can considerably alter microemulsion characteristics. This is the reason why, for example, in the case of oral administration, it is unrealistic to predict the interactions of microemulsion components with the complex variety of food materials and digestive fluids present in the gastrointestinal tract [2].

8.5.1 MICROEMULSION ADMINISTRATION ROUTES

Typically, microemulsions can be administered according to oral, parenteral, pulmonary, ocular, topical, or transdermal strategy. Although for oral, parenteral, pulmonary, and ocular delivery low-viscosity systems are required, topical and transdermal administration requires that the system shows proper rheological properties allowing an easy application on the skin and this is the reason why gelling agents are usually added [44–46]. Typically, in oral administration, microemulsions are used as self-microemulsifying drug delivery system (SMEDDS). An oil–surfactant–drug solution is administered and the spontaneous formation of microemulsion takes place upon contact with the physiological water [2]. Probably, the most important example of SMEDDS is that related to the oral delivery of cyclosporine A (Neoral formulation), a hydrophobic undecapeptide used as immunosuppressant in transplantation surgery [47]. Neoral oil phase is constituted by medium chain length triglyceride whereas surfactant consists of an isotropic concentrated blend of medium chain length partial glycerides. Upon contact with water, this system gives rise to an initial water-in-oil microemulsion that, with further mixing, yields the phase inversion to an oil-in-water microemulsion. Neoral proved to be more efficient than the previous formulations [2].

Flurbiprofen delivery from an original ethyl oleate (oil phase)–Tween 20 (surfactant) mixture represents an example of parenterally administered SMEDDS [48]. Another example is that related to a microemulsion system composed of a medium-chain triglyceride (MCT), soybean phosphatidylcholine and poly(ethylene glycol) (660)-12-hydroxystearate (12-HSAEO15) as amphiphiles, and poly(ethylene glycol) 400 (PEG 400) and ethanol as cosolvents [49]. NMR analyses revealed that this system gives rise to bicontinuous structures even at high-oil concentrations but an oil-in-water microemulsion forms upon dilution in water. Felodipine was considered as drug.

Ocular delivery of pilocarpine via microemulsion proved to prolong the pharmacological effect with respect to a simple aqueous solution. Isopropyl myristate constitutes the oil phase, whereas lecithin, propylene glycol, and PEG 200 are cosurfactants [50].

As mentioned earlier, topical and transdermal administration of microemulsions poses the problem of their rheological properties that must guarantee a proper application on the skin since, usually, they are low-viscosity fluids. As for many other disperse systems, the typical way to solve the problem consists of adding a polymer suitable to impart particular rheological properties, without significantly modifying the other features (stability, high oil–water interface area). The selected polymer may be quite soluble in the continuous phase, thus forming a more or less concentrated ordinary solution, or display noncovalent intermolecular associations deriving from disparate forces such as Coulombic, van der Waals, dipole–dipole, hydrophobic, and hydrogen-bond interactions. Such physical interactions can cooperate leading to the formation of gel microdomains dispersed in a less-viscous sol phase or of a continuous gel network composed of polymer segments belonging to different chains and pervading the whole system [34]. In both cases, the polymer system generally displays weak gel properties and can be profitably used to infer marked elastic character as well as pronounced shear thinning behavior to the dispersion, with sufficiently high fluidity at high shear. Obviously, gelling polymer must be biocompatible and shows limited inter-actions with the surfactant. Although biocompatibility can be easily evaluated in advance, prediction of polymer–surfactant interactions represents a more difficult task, especially when we are dealing with dispersions of very fine droplets in weak gel polymeric matrices. It is well known that the formation of micellar aggregates is often facilitated by the presence of polymer leading to the decrease in critical micelle concentration, particularly when polymer chains contain hydrophobic groups [51]. Mixed hydrophobic clusters com-posed of polymer and surfactant molecules can be formed and affect the rheological behavior of polymer–surfactant dramatically. When significant polymer–surfactant interactions tend to occur, in the presence of disperse phase, they could compete with the stabilizing action of the surfactant at the oil/water interface and also with the polymer–polymer associations in the case of weak gel polymeric matrices so that the stability of the microemulsion could be substantially altered [44]. Examples of gelled microemulsions regard the iontophoretic transdermal release of sodium salicylate from an isopropyl myristate (oil phase)–Tween 85 (surfactant)–water microemulsion where gel-atine is added to increase viscosity and elastic properties [46]. On the basis of SANS measurements, this system should have the structure represented in Figure 8.6 [52]. Water droplets stay in an oil continuous phase that is pervaded by gelatine/water channels. Another example is about the addition of Carbopol 940 in a microemulsion composed of Labrafac Hydro (oil), Cremophor RH40 (surfactant), Transcutol (cosurfactant), and water [44]. Carbopol 940 confers an approximately plastic behavior to the system. The same polymer has been used for the gelation of a microemulsion composed of isopropyl myristate (oil), HTAB (surfactant), ethanol (cosurfactant), and water for the release of piroxicam, a sparingly water-soluble anti-inflammatory nonsteroidal drug [53].

8.5.2 MODELING OF DRUG RELEASE FROM MICROEMULSIONS

The study of drug release from a colloidal system is not a trivial task [38,54]. Indeed, only two ways exist for measuring drug release from microemulsions: membrane diffusion technique and *in situ* method [38]. The latter method is disadvantageous since the microemulsion should be put in the release envir-onment (usually an aqueous medium) which leads to its dilution. In this manner, as previously discussed, it is very difficult that microemulsion maintains its original structure and it will surely evolve toward other un-known structural conditions due to its changed location in the phase diagram. As a consequence, this technique does not allow a theoretical analysis of drug release. Conversely, the membrane diffusion technique prevents from vari-ations in the microemulsion structure and that is why a theoretical analysis of

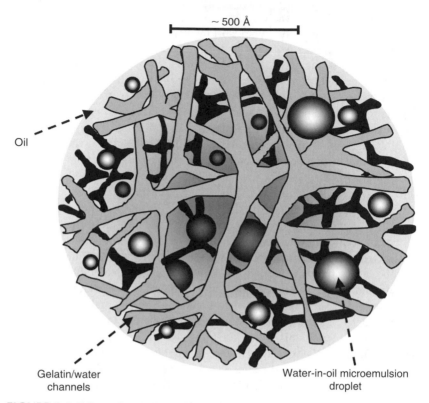

FIGURE 8.6 Schematic representation of a gelled microemulsion. This imagine is based on small angle neutron scattering data. (From Lawrence, M.J., and Rees, G.D., *Adv. Drug Deliv. Rev.*, 45, 86, 2000. With permission from Elsevier.)

drug release is allowed, provided that the membrane effect is properly accounted for. Although, in principle, this technique permits to study drug release regardless of microemulsion structure and structure-dependent mathematical models could be properly built on, in the light of their importance in pharmaceutical applications, only *Winsor* IV microemulsions are considered (i.e., discrete oil-in-water or water-in-oil microemulsions). In addition, as the theoretical analysis of drug release from oil-in-water or water-in-oil *Winsor* IV microemulsions is conceptually identical, the attention is focused on the oil-in-water system as they are suitable for hosting sparingly water-soluble drugs. Accordingly, the physical picture to be modeled is shown in Figure 8.7. At the beginning, the donor compartment contains the drug-loaded microemulsion whereas the receiver is filled by a drug-free aqueous medium containing a small percentage of surfactant to approximate the aqueous phase composition of the microemulsion [55]. In this way, it can be assumed

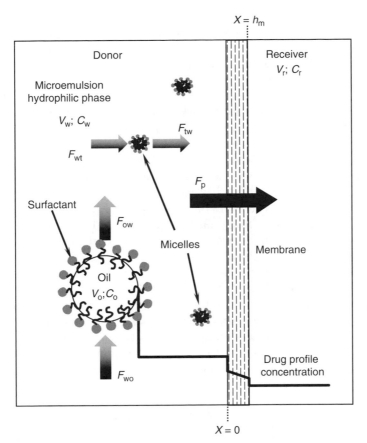

FIGURE 8.7 Schematic representation of the phenomena involved in drug release from a microemulsion coated by a drug membrane. Drug moves from the oil- and micellar-phases to the surrounding hydrophilic phase of the donor compartment and then it crosses the interposed membrane to reach the receiver compartment. (From Sirotti, C., et al., *J. Membr. Sci.*, 204, 401, 2002. With permission from Elsevier.)

that drug solubility in the receiver fluid is not the rate-determining step of the whole permeation process. Obviously, as it is virtually impossible to know the real composition of the microemulsion aqueous phase and in the light of the huge work developed by Trotta, Pattarino and Gasco [37,56], the least amounts of surfactant and cosurfactant ensuring a good drug solubility in the receiver environment should be considered. It is also evident that the membrane is considered impermeable to all substances present in the microemulsion except for the drug.

On the other hand, as the drug-loaded microemulsion can be retained under equilibrium conditions at the beginning of the permeation experiment (it is usually prepared in large advance with respect to the permeation experiment), drug distributes among the three phases: oil droplets, surfactant micelles, and aqueous continuum phase (containing part of the whole surfactant and cosurfactant amount). This means that the drug flux exiting from oil droplets (F_{ow}) is equal to the entering one (F_{wo}) and, similarly, the drug flux exiting from surfactant micelles (F_{sw}) equals the entering one (F_{ws}) (see Figure 8.7). The initial concentration gradient existing between the donor and receiver continuum aqueous phases induces drug diffusion through the interposed membrane. This process, though lowering the drug concentration in the microemulsion aqueous phase and altering the drug partition equilibrium among microemulsion phases, comports a consequent transfer of drug molecules from surfactant micelles and oil droplets to the surrounding aqueous phase in an attempt to restore new equilibrium conditions. As it is discussed hereafter and briefly mentioned earlier, this transfer process, in the presence of particular drugs, can alter the equilibrium existing between solubilized and micellized surfactant molecules. This can reflect into a modification of drug solubility in the continuum microemulsion aqueous phase.

Relying on this physical frame and neglecting the continuum phase–micelle drug partition (the only effect of the surfactant micelles is to increase drug solubility in the continuum aqueous phase), Yotsuyanagi and colleagues [57] modeled the mass transport from an external aqueous phase to a microemulsion. They resorted to relevant oil and aqueous mass balances assuming that drug fluxes (oil–water, F_{ow} and water–oil, F_{wo}) were linearly dependent on drug concentration

$$F_{ow} = k_{ow}C_o, \quad F_{wo} = k_{wo}C_w, \tag{8.49}$$

where C_o and C_w indicate, respectively, drug concentration in the oil and aqueous phases.

More recently, Guy and coworkers [58] solved the problem of drug release from an oil droplet phase into an outer aqueous environment not only by assuming the same hypotheses of Yotsuyanagi but also by considering that, at time zero, aqueous phase is drug free. This last condition clearly makes the model unsuitable to mimic the real conditions occurring in microemulsions. Boddé and Joosten [59] studied approximately the inverse situation studied by Yotsuyanagi, i.e., drug release from a two-phase system to perfect sink, always assuming linear expression for mass fluxes. Friedman and Benita [60] considered the existence of three phases: the continuous aqueous phase, the oil droplets, and the surfactant micelles. Moreover, in this case mass balances and linear dependences of mass fluxes on concentration were assumed. As it was discussed in Chapter 5, the description of drug transport between two liquid phases considering a linear flux–concentration relationship cannot be reliable for sparingly water-soluble (or oil-soluble) drugs.

8.5.2.1 Simple Case

Assuming with Friedman and Benita [60] that the continuum phase–micelle drug partition does not affect the whole release process, the only effect of the surfactant micelles is to modify (usually to increase) drug solubility in the continuum phase. Accordingly, key phenomena ruling drug release are drug flux from (F_{ow}) to (F_{wo}) oil droplets and drug permeation across the membrane separating microemulsion environment from the receiver chamber (see Figure 8.7). Although different ways exist for the mathematical expression of F_{ow} and F_{wo}, it is clear that F_{ow} must vanish when drug concentration in water, C_w, approaches solubility, C_{sw}, or when drug concentration in oil, C_o, is zero. Similarly, F_{wo} must vanish when C_o approaches solubility in oil, C_{so}, or when C_w is zero. Accordingly, a possible definition is that considered in Chapter 5:

$$F_{ow} = k'_{ow} C_o (C_{sw} - C_w), \tag{8.50}$$

$$F_{wo} = k'_{wo} C_w (C_{so} - C_o), \tag{8.51}$$

where k'_{ow} and k'_{wo} are the rate partition constants between the oil and the aqueous phases and vice versa. The superscript remembers that the following relation occur with the kinetics constants k_{ow} and k_{wo} defined in Equation 5.139 and Equation 8.49:

$$k'_{ow} = k_{ow}/C_{sw}, \quad k'_{wo} = k_{wo}/C_{so}. \tag{8.52}$$

In microemulsions, where, usually, the aim is to maximize drug loading and, consequently, drug concentration approaches the solubility limit in both the water and oil phases, the usual flux definition (Equation 8.49) can lead to unreliable models. It is also useful to emphasize that Equation 8.50 and Equation 8.51 implicitly suppose that the drug concentration in both the microemulsion water and oil phases is always uniform so that the two phases are assumed to be constantly well mixed.

Although a correct modeling of drug permeation through a membrane would require the solution of Fick's second law, we can assume to deal with a thin membrane where drug flux F_p can be given by [61–64]

$$F_p = \frac{D}{h_m} (k_{pd} C_w - k_{pr} C_r), \tag{8.53}$$

where h_m is the membrane thickness, D is the drug diffusion coefficient inside the membrane, C_r is the receiver drug concentration, k_{pd} is the drug partition coefficient between the membrane and the microemulsion aqueous phases, and k_{pr} is the drug partition coefficients between the membrane and the receiver medium. Equation 8.53 simply imposes that the drug transfer rate is proportional to gradient concentration across the membrane and that drug concentration in the receiver fluid is always uniform (receiver perfect mixing conditions).

In order to complete the model formulation, three more equations must be written down: two kinetic equations ruling the oil–water drug transfer and an overall mass balance guaranteeing the constancy of the drug mass in the whole system (microemulsion oil and water phase, membrane and receiver fluid)

$$\frac{dC_w}{dt} = A\frac{F_{ow}}{V_w} - A\frac{F_{wo}}{V_w} - S\frac{F_p}{V_w},\tag{8.54}$$

$$\frac{dC_o}{dt} = A\frac{F_{wo}}{V_o} - A\frac{F_{ow}}{V_o},\tag{8.55}$$

$$V_rC_r + V_oC_o + V_wC_w + \int_0^{h_m} S\left(k_{pd}C_w - \frac{k_{pd}C_w - k_{pr}C_r}{h_m}X\right)dX = M_0,\tag{8.56}$$

where t is the time, V_w is the microemulsion water phase volume, V_r is the receiver volume, X is the abscissa (see Figure 8.7), M_0 represents the whole drug amount present in the microemulsion at the beginning of the permeation experiment, S is the membrane surface, and A is the oil–water interface area defined as follows:

$$A = N_p 4\pi R_m^2; \quad N_p = \frac{3V_o}{4\pi R_m^3} \implies A = \frac{3V_o}{R_m},\tag{8.57}$$

where N_p and R_m are, respectively, the number and the average radius of the oil droplets. Equation 8.57 implicitly assumes that the oil droplets have a perfect spherical shape and that they are characterized by a monodisperse size distribution. Equation 8.54 indicates that the drug concentration in the micro-emulsion water phase is increased by the oil–water drug flux, but it is decreased by the water–oil drug flux and by the drug permeation through the membrane. On the contrary, drug concentration in the oil phase is only ruled by the water–oil and oil–water drug fluxes. Equation 8.56, allowing to close the balance between the unknowns (C_w, C_o, and C_r) and the solving equations, is the overall mass balance. It is interesting to note that the integral appearing in this balance, whose expression is due to the linearity assumption adopted for the drug profile concentration inside the membrane, represents the drug amount present in the membrane at the generic time t. The other three terms of the left-hand side of the same balance represent, respectively, the drug amount contained in the receiver and in the microemulsion oil and water phase at the generic time t.

In order to get model solution, it is convenient to express C_r as a function of C_w and C_o by means of Equation 8.56 and to insert Equation 8.50, Equation 8.51, and Equation 8.53 into Equation 8.54, and Equation 8.55 to finally get

$$C_r = \frac{M_0 - V_o C_o - C_w(V_w + 0.5Sk_{pd}h_m)}{V_r + 0.5Sk_{pr}h_m},$$ (8.58)

$$\frac{dC_w}{dt} = \frac{3V_o}{R_m V_w}k'_{ow}C_o(C_{sw} - C_w) - \frac{3V_o}{R_m V_w}k'_{wo}C_w(C_{so} - C_o)$$

$$- \frac{SD}{h_m V_w}(k_{pd}C_w - k_{pr}C_r),$$ (8.59)

$$\frac{dC_o}{dt} = \frac{3}{R_m}k'_{wo}C_w(C_{so} - C_o) - \frac{3}{R_m}k'_{ow}C_o(C_{sw} - C_w).$$ (8.60)

Model initial conditions descend from the hypothesis that drug equilibrium partitioning between the microemulsion oil and water is attained at the beginning ($t = 0$). Therefore, initial values of C_w (C_{w0}) and C_o (C_{o0}) can be derived, for example, by setting to zero the left-hand side of Equation 8.60 and by solving the following system of nonlinear algebraic equations:

$$k_{wo}C_{w0}(C_{so} - C_{o0}) = k_{ow}C_{o0}(C_{sw} - C_{w0}),$$ (8.61)

$$M_0 = V_o C_{o0} + C_{w0}V_w,$$ (8.62)

where the above equation is nothing more than the drug mass balance made up on the microemulsion oil and water phase and differs from Equation 8.56 only by the fact that, for $t = 0$, there is no drug neither in the membrane nor in the receiver. The solution of the above-mentioned equations lead to

$$C_{w0} = -0.5\left(\beta \mp \sqrt{\beta^2 - 4\gamma}\right), \quad C_{o0} = \frac{M_0 - C_{w0}V_w}{V_o},$$ (8.63)

where

$$\beta = \frac{M_0(R - 1) + C_{sw}V_wR + C_{so}V_o}{(1 - R)V_w}, \quad \gamma = \frac{M_0 C_{sw}R}{(R - 1)V_w}, \quad R = \frac{k'_{ow}}{k'_{wo}}.$$ (8.64)

Of course, the positive root in Equation 8.63 has to be chosen. Interestingly, as a direct consequence of the F_{ow} and F_{wo} definition, the drug partition coefficient k_p is defined by

$$k_p = \frac{C_{o0}}{C_{w0}},$$ (8.65)

is not only dependent on R as predicted by previous models [60]. Finally, model numerical solution can be obtained by means of the fifth order (adaptive stepsize) Runge–Kutta method [65].

It is now worth mentioning to show a particular situation leading to a model analytical solution. If the receiver volume is not very large in comparison with the donor one and if the water drug solubility is very low in comparison with the oil one, the F_{ow} and F_{wo} expressions can be properly approximated by

$$F_{ow} = k'_{ow} C_{o0}(C_{sw} - C_w), \tag{8.50'}$$

$$F_{wo} = k'_{wo} C_w(C_{so} - C_{o0}). \tag{8.51'}$$

Indeed, in the above-mentioned hypotheses, the drug concentration in the oil phase can be considered constant as the drug amount present in the microemulsion water phase and in the receiver volume is always small in comparison with that contained in the oil phase at the beginning of the permeation. Insertion of Equation 8.50' and Equation 8.51' in Equation 8.59 and Equation 8.60, and mass balance (Equation 8.56) consideration, where, on the contrary, C_o is considered time dependent, leads to model simplified form:

$$C_w(t) = W_1 + W_2 e^{A_1 t} + W_3 e^{A_2 t}, \tag{8.66}$$

$$C_o(t) = O_1 + O_2 e^{A_1 t} + O_3 e^{A_2 t}, \tag{8.67}$$

$$C_r = C_0^\nabla - E C_o(t) - F C_w(t), \tag{8.68}$$

where W_1, W_2, W_3, O_1, O_2, O_3, A_1, A_2, C_0^∇, E, and F are defined as follows:

$$F = \frac{V_w + 0.5 S k_{pd} h_m}{V_r + 0.5 S k_{pr} h_m}, \quad E = \frac{V_o}{V_r + 0.5 S k_{pr} h_m}, \quad C_0^\nabla = \frac{M_0}{V_r + 0.5 S k_{pr} h_m} \tag{8.69}$$

$$A_1 = 0.5\left(-\beta^* + \sqrt{(\beta^*)^2 - 4\alpha^* \gamma^*}\right), \quad A_2 = 0.5\left(-\beta^* - \sqrt{(\beta^*)^2 - 4\alpha^* \gamma^*}\right), \tag{8.70}$$

$$O_1 = \frac{\omega^* \gamma^* - \delta^* \beta^*}{\alpha^* \gamma^*}, \quad O_2 = -\frac{W_2}{\alpha^*}(\beta^* + A_1), \quad O_3 = -\frac{W_3}{\alpha^*}(\beta^* + A_2) \tag{8.71}$$

$$W_1 = \frac{\delta^*}{\gamma^*}, \quad W_2 = C_{w0} - W_3 - \frac{\delta^*}{\gamma^*}, \tag{8.72}$$

$$W_3 = \frac{1}{A_1 - A_2}\left[\alpha^* C_{o0} + \frac{\delta^* \beta^* - \omega^* \gamma^*}{\gamma^*} + (\beta^* + A_1)\left(C_{w0} - \frac{\delta^*}{\gamma^*}\right)\right], \tag{8.73}$$

where

$$\omega^* = \frac{3V_o}{R_m V_w} k_{ow} C_{oi} C_{sw} + \frac{SD}{h_m V_w} k_{pr} C_0^\nabla, \quad \alpha^* = \frac{SD}{h_m V_w} k_{pr} E, \qquad (8.74)$$

$$\beta^* = \frac{3V_o}{R_m V_w} (C_{o0} (k_{ow} - k_{wo}) + k_{wo} C_{so}) + \frac{SD}{h_m V_w} (k_{pd} + k_{pr} F), \quad (8.75)$$

$$\gamma^* = \frac{3k_{wo}}{R_m} (C_{so} - C_{o0}) + \frac{3k_{ow}}{R_m} C_{oi} \quad \delta^* = \frac{3k_{ow}}{R_m} C_{o0} C_{sw}. \qquad (8.76)$$

Grassi and coworkers [55] applied this model to study nimesulide (HEL-SINN, Pambio Noranco, CH; anti-inflammatory action) release from two microemulsions differing for the presence of the cosurfactant (Benzyl alcohol; Fluka Chemika, Sigma–Aldrich, MI, Italy). Other constituents are triacetin (Fluka Chemika, Sigma–Aldrich, MI, Italy) as oil phase, Tween 80 (Polyoxyethylene 20 sorbitan monooleate; Fluka Chemika, Sigma–Aldrich, MI, Italy) as surfactant, and distilled water. The first microemulsion, named MTB, is composed of 43% (v/v) surfactant, 24.5% (v/v) water, and 32.5% (v/v) oil phase, which is a (1:1) triacetin–benzilic alcohol mixture. The second microemulsion, named MT, is composed of 47% (v/v) surfactant, 37% (v/v) water, and 16% (v/v) oil phase, constituted only by triacetin. PCS revealed that average oil droplet's diameter is 29.0 nm for MTB and 26.9 nm for the MT. Nimesulide solubility (37°C) in triacetin is 65,180 $\mu g/cm^3$; whereas solubility (37°C) in triacetin–benzilic alcohol (1:1) is 68,400 $\mu g/cm^3$. The release experiments are performed resorting to the Franz cells apparatus consisting of two thermostatic (37°C) chambers (donor (microemulsion volume $= 4$ cm^3) and receiver ($V_r = 22$ cm^3) compartments) separated by an interposed silicon rubber membrane (surface $S = 3.46$ cm^2; SILASTIC, Dow Corning Corporation, Michigan, USA) 0.0149 cm thick permeable to nimesulide and impermeable to all other microemulsion components. Basically, this apparatus is the experimental representation of what is schematized in Figure 8.7. Receiver and the donor homogeneity are ensured by a magnetic stirrer and by a rotating impeller, respectively. The receiver is filled by an initial drug-free phosphate buffer saline pH 7.4 containing 1% Tween 80 to simulate the microemulsion water phase. This choice is done in an attempt to get the same drug solubility in both the receiver and in microemulsion aqueous phase. This is just a first approximation hypothesis, but it is supported by the fact that the model best fitting obtained using different solubility values, corresponding to different surfactant concentration R_{SC} ($R_{SC} = 0.0\%$, $C_{sw} = 2.8$ $\mu g/cm^3$; $R_{SC} = 2.0\%$, $C_{sw} = 20.4$ $\mu g/cm^3$), is never as good as that performed with the solubility corresponding to $R_{SC} = 1.0\%$ ($C_{sw} = 8.8$ $\mu g/cm^3$). Receiver nimesulide concentration is automatically measured and stored every 100 s by means of a computer managed

UV spectrophotometer (Lambda 6/PECSS System, Perkin–Elmer Corporation, Norwalk, CT, wavelength 391 nm).

A reliable determination of the rate partition constants k'_{ow} and k'_{wo} requires a preliminary characterization of both the membrane and the receiver medium adopted. Indeed, the nimesulide diffusion coefficient (D) in the Silastic membrane and the drug solubility in the microemulsion water phase (C_{sw}) have to be determined in advance. At this purpose, an *ad hoc* permeation experiment is performed. Although the donor volume ($V_d = 4$ cm^3) is filled by a water–surfactant mixture (surfactant amount $= 1\%$ w/w) saturated by the drug and containing an excess of solid drug, the receiver volume ($V_r = 22$ cm^3) is filled by an identical drug-free aqueous phase. The presence of the surfactant in the donor liquid is aimed to mimic the microemulsion water phase that, surely, contains a certain amount of surfactant. In this way, it is assumed that nimesulide solubility in the microemulsion water phase coincides with that of the fluid filling both the donor and the receiver of this permeation experiment. Moreover, the presence of an excess of solid drug guarantees the constancy of the drug concentration inside the donor liquid along the whole permeation time. Accordingly, the experimental permeation data are fitted by the following equation, suitable for describing drug permeation through a thin membrane [62, 64]:

$$C_r = \frac{k_{pd}}{k_{pr}} C_{sw} \left(1 - e^{(-\alpha k_{pd} t)} \right), \quad \alpha = \frac{SD}{V_r h_m}, \quad (8.77)$$

where k_{pd} and k_{pr} are set equal to 1. Knowing that $S = 3.46$ cm^2, model fitting leads to $D = (2.2 \pm 0.1) \times 10^{-6}$ cm^2/s and $C_{sw} = 8.8 \pm 0.2$ µg/cm^3.

On the basis of these parameters, it is now possible for the determination of the rate partition constants k'_{ow} and k'_{ow} of the two different microemulsions considered. In the case of the MT system, the donor is filled by the drug-loaded microemulsion whereas the receiver is filled by the water–surfactant mixture (surfactant amount $= 1\%$ w/w) already used. The following values of the parameters are adopted: $V_r = 22$ cm^3, $V_o = 0.604$ cm^3, $V_w = 3.36$ cm^3, $S = 3.46$ cm^2, $D = 2.21 \times 10^{-6}$ cm^2/s, $R_m = 1.345 \times 10^{-6}$ cm, $M_0 = 36680$ µg, $C_{so} = 65180$ µg/cm^3, $C_{sw} = 8.8$ µg/cm^3. Since the C_{sw}/C_{so} ratio is very low ($\leq 10^{-3}$), model analytical form can be used. The very good comparison between model best fitting (solid line in Figure 8.8) and experimental data (symbols; vertical bars indicate data standard error) confirms model reliability. Fitting parameters ($k'_{ow} = (1.00 \pm 0.02)10^{-2}$ cm^4/(µg s), $k'_{ow} = (0.121 \pm 0.002)$ cm^4/(µg s)) correctly indicate that water–oil drug transfer is favored with respect to the opposite one (k'_{ow} is one order of magnitude smaller than k'_{ow}).

In addition, in the MTB case, the donor is filled by the drug-loaded microemulsion and the receiver is filled by the water–surfactant mixture (surfactant amount $= 1\%$ w/w). The following values of the parameters are

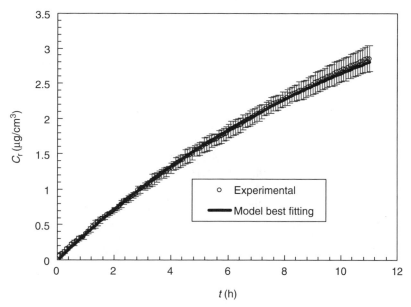

FIGURE 8.8 Comparison between model best fitting (solid curve; Equation 8.68) and the permeation data (open circles; vertical bars indicate data standard error) in the hypothesis of filling the donor compartment with the Nimesulide-loaded MT microemulsion. C_r indicates the receiver donor concentration whereas t is time. (From Grassi, M., Coceani, N., and Magarotto, L., *J. Colloid Interface Sci.*, 228, 141, 2000. With permission from Elsevier.)

adopted: $V_r = 22$ cm^3, $V_o = 1.3$ cm^3, $V_w = 2.7$ cm^3, $S = 3.46$ cm^2, $D = 2.21 \times 10^{-6}$ cm^2/s, $R_m = 1.45 \times 10^{-6}$ cm, $M_0 = 73280$ μg, $C_{so} = 68,400$ μg/cm^3, $C_{sw} = 8.8$ μg/cm^3. Again, as C_{sw}/C_{so} ratio is very low ($\leq 10^{-3}$), model analytical form can be used. Figure 8.9 shows that model best fitting (solid line) is good and fitting parameters read $k'_{ow} = (1.0612 \pm 0.0005)$ cm^4/(μg s) and $k'_{ow} = (2.200 \pm 0.004)$ cm^4/(μg s). Moreover, in this case water–oil drug transfer is favored with respect to the opposite one even if, now, the difference existing between the two rate constants is smaller than in the MT case (k'_{ow} is only two times k'_{ow}). This could be explained with the presence of benzilic alcohol (cosurfactant) enhancing the drug transfer from the oil to the water phase. It could be supposed that the benzilic alcohol, disposing inside the surfactant layer surrounding each droplet, could give rise to a sort of preferential channels connecting the inner oil phase with the outer aqueous phase.

8.5.2.2 More Complex Situation

In the previous section, we made the simplifying hypothesis that drug transfer from the various microemulsion domains (induced by drug permeation through the membrane separating the donor chamber from the receiver one)

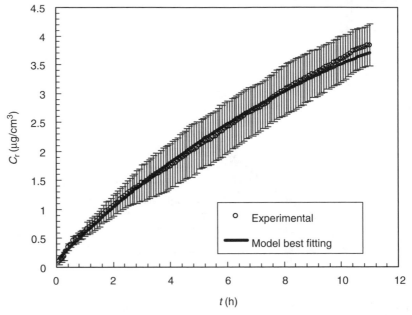

FIGURE 8.9 Comparison between model best fitting (solid curve; Equation 8.68) and permeation data (open circles; vertical bars indicate data standard error) in the hypothesis of filling the donor compartment with the Nimesulide-loaded MTB microemulsion. C_r is the receiver donor concentration and t is time. (From Grassi, M., Coceani, N., and Magarotto, L., *J. Colloid Interface Sci.*, 228, 141, 2000. With permission from Elsevier.)

did not alter microemulsion structure. In particular, we supposed that the equilibrium between micellized surfactant and solubilized surfactant in the aqueous phase was not altered by drug transfer. This was the reason why the effect of surfactant in the micellar and solubilized status was simply assumed to increase drug solubility in the aqueous phase. Consequently, drug exchange between surfactant micelles and aqueous phase was not considered. However, it may happen that the drug is able to alter the equilibrium existing between surfactant structures in the aqueous phase and, consequently, drug exchange between the micellar phase and the aqueous one cannot be neglected. This is exactly what Sirotti and coworkers [66] found working on nimesulide (HELSINN, Pambio Noranco, CH) release from a microemulsion composed of isopropyl myristate (oil phase; Fluka Chemika, Sigma–Aldrich, Italy; 11.75 wt.%), Tween 80 (surfactant; polyoxyethylene 20 sorbitan monooleate; Fluka Chemika, Sigma–Aldrich, Italy; 30.8 wt.%), benzyl alcohol (cosurfactant; Fluka Chemika, Sigma–Aldrich, Italy; 11.75 wt.%), and distilled water (45.7 wt.%). In particular, they find that drug transfer from the micellar phase may be combined with the disruption of some micelles and the consequent solubilization of surfactant molecules in the microemulsion aqueous phase. This, in turn, is responsible for an increase in the drug solubility c_w in the aqueous phase that, for the sake of simplicity, can be conveniently modeled by the following equation:

$$c_{sw}(t) = s_f + (s_0 - s_f)(1 - e^{-at}), \tag{8.78}$$

where s_0 and s_f are the initial and final solubilities, respectively, a is an adjustable parameter and, t is the time. On the basis of this fact, the model shown in the previous section needs to be improved. Accordingly, drug transfer among oil, aqueous solution, micelles, and receiver environment is governed by the following equations:

$$\frac{dC_o}{dt} = \frac{A_o}{V_o} F_{wo} - \frac{A_o}{V_o} F_{ow}, \tag{8.55}$$

$$\frac{dC_t}{dt} = \frac{A_t}{V_t} F_{wt} - \frac{A_t}{V_t} F_{tw}, \tag{8.79}$$

$$\frac{dC_w}{dt} = -\frac{V_o}{V_w} \frac{dC_o}{dt} - \frac{V_t}{V_w} \frac{dC_t}{dt} - \frac{S}{V_w} F_p, \tag{8.80}$$

$$V_r C_r + V_o C_o + V_w C_w + V_t C_t + S h_m \frac{k_{pd} C_w + k_{pr} C_r}{2} = M_0, \tag{8.81}$$

where A_t and V_t are the micelle–water interface area and volume, respectively, C_t is the drug concentration in the micelles whereas F_{wt} and F_{tw} are, respectively, the drug fluxes from the aqueous phase to micelles and vice versa. The subscripts o and w refer to analogous quantities pertaining, respectively, to the oil and water phases. The same logic leading to Equation 8.57 can be applied in this case to get oil–water interface area A_o and micelle–water interface area A_t supposing that both of them are monodisperse

$$A_o = \frac{3V_o}{R_o}, \quad A_t = \frac{3V_t}{R_t}, \tag{8.82}$$

where R_o and R_t are the mean radii of oil droplets and surfactant micelles, respectively. Finally, though F_p, F_{ow}, and F_{wo} are defined, respectively, by Equation 8.53, Equation 8.50, and Equation 8.51, F_{wt} and F_{tw} are defined by

$$F_{wt} = k'_{wt} C_w (C_{st} - C_t), \tag{8.83}$$

$$F_{tw} = k'_{tw} C_t (C_{sw}(t) - C_w), \tag{8.84}$$

where k'_{wt} and k'_{tw} are, respectively, the water–micelles and the micelles–water rate constants, whereas C_{st} represents drug solubility in the micelles and C_{sw} is the time-dependent drug solubility in the aqueous phase (see Equation 8.78). Basically, this model differs from the simpler one shown in previous section for the presence of a third phase represented by micelles. Accordingly, not only oil–water mass transfer occurs but also water–micelles drug exchange takes place. It is worth mentioning that in writing the mathematical equations relating to this model, it is assumed that micellar and water phase volumes V_t and V_w are time independent even if they should not be as micelles disruption leads to a simultaneous V_t decrease and V_w increase. Nevertheless, this simplification can be tolerated on condition that the breaking of few micelles is sufficient to cause a significant C_{sw} variation. Finally, also in this model, it is assumed that drug transport to the receiver chamber takes place from the aqueous phase and no direct drug transfer from oil droplets and micelle occurs.

As discussed earlier, the values of the initial drug concentration in the oil droplets, micelles, and aqueous phase can be determined assuming initial equilibrium conditions inside the microemulsion. This means zeroing Equation 8.55, Equation 8.79, and Equation 8.80 and properly modifying the mass balance (Equation 8.81)

$$\frac{dC_o}{dt} = \frac{A_o}{V_o}F_{wo} - \frac{A_o}{V_o}F_{ow} = 0, \tag{8.85}$$

$$\frac{dC_t}{dt} = \frac{A_t}{V_t}F_{wt} - \frac{A_t}{V_t}F_{tw} = 0, \tag{8.86}$$

$$V_oC_{o0} + V_wC_{w0} + V_tC_{t0} = M_0, \tag{8.87}$$

where C_{o0}, C_{w0}, and C_{t0} represent the initial values of C_o, C_w, and C_t, respectively. Rearrangement of Equation 8.85 through Equation 8.87 leads to a cubic expression in C_{o0}

$$\underline{A}c_{o0}^3 + \underline{B}c_{o0}^2 + \underline{C}c_{o0} + \underline{D} = 0, \tag{8.88}$$

where

$$\underline{A} = V_o(R_{ow} - 1)(R_{wt}R_{ow} - 1), \tag{8.89}$$

$$\underline{B} = R_{wt}R_{ow}(V_wR_{ow}C_{sw} + (R_{ow} - 1)(C_{st}V_t - M_0) + V_oC_{so}) + \\ + (R_{ow} - 1)(C_{so}V_o + M_0) - V_wR_{ow}C_{sw} - V_oC_{so}, \tag{8.90}$$

$$\underline{C} = C_{so}(R_{wt}R_{ow}(V_tC_{st} - M_0) + V_wR_{ow}C_{sw} + V_oC_{so} - M_0(R_{ow} - 2)) \tag{8.91}$$

$$\underline{D} = -C_{so}^2 M_0, \tag{8.92}$$

with $R_{ow} = k'_{ow}/k'_{wo}$ and $R_{wt} = k'_{wt}/k'_{tw}$. The only physically consistent root of Equation 8.88 represents the starting value of C_o. Consequently, the initial C_w and C_t values are given by

$$C_{w0} = \frac{R_{ow} C_{sw} C_{o0}}{C_{o0}(R_{ow} - 1) + C_{so}}, \tag{8.93}$$

$$C_{t0} = \frac{M_0 - V_o C_{o0} - V_w C_{w0}}{V_t}. \tag{8.94}$$

Due to the C_{sw} time dependence, only a numerical approach (fifth order Runge–Kutta method, with adaptive stepsize [65]) is possible to get model solution.

Similarly to the simpler case matched in Section 8.5.2.1, the Silastic membrane separating the donor environment from the receiver environment must be characterized in terms of nimesulide diffusion coefficient. Accordingly, the donor volume (37°C) is filled by a buffer Tween 80 (1 wt.%)–benzyl alcohol (2 wt.%) mixture (this is the same composition of the receiver fluid used in the release tests from microemulsion) saturated by the drug and containing an excess of solid drug, whereas the receiver volume is filled with an identical drug-free aqueous phase. Receiver and donor compartment homogeneities are ensured by a magnetic stirrer and a rotating impeller, respectively. Nimesulide concentration C_r is measured by a UV spectrophotometer (Lambda 6/PECSS System, Perkin-Elmer Corp., Norwalk, CT, wavelength 398 nm). Data fitting according to Equation 8.77 ($h_m = 149$ μm; $S = 3.46$ cm^2, $V_{donor} = 4$ cm^3, $V_r = 22$ cm^3, $k_{pd} = k_{pr} = 1$) leads to $D = (3.3 \pm 0.1) \times 10^{-6}$ cm^2/s and $C_{sw} = (9.4 \pm 0.2)$ μg/cm^3. Considering that this Silastic membrane belonged to a different lot with respect to that used in Section 8.5.2.1, and in the light of the different release environment (here we also have the cosurfactant), these results are perfectly in line with those determined in Section 8.5.2.1.

Permeation experiments from microemulsion are performed at 37°C by resorting to the Franz cells apparatus whose donor volume ($V_d = 4$ cm^3) contains the microemulsion loaded by nimesulide (3590 μg/cm^3), whereas the receiver compartment ($V_r = 22$ cm^3) is filled by a phosphate buffer (pH 7.4), which is initially drug free and contains Tween 80 (1 wt.%) and benzyl alcohol (2 wt.%) to simulate the microemulsion water phase. Receiver and donor compartment homogeneities are ensured by a magnetic stirrer and a rotating impeller, respectively. Nimesulide concentration C_r is measured by a UV spectrophotometer (Lambda 6/PECSS System, Perkin-Elmer Corp., Norwalk, CT, wavelength 398 nm).

Nimesulide solubility (37°C) in isopropyl myristate–benzyl alcohol (1:1) is 18 mg/cm^3 whereas its solubility in Tween 80 is 65 mg/cm^3. PCS reveals that mean oil droplets radius is $R_o = 27$ nm. Transmission electron microscope (Philips EM-208) images confirm this R_o measurement and indicate that surfactant micelles radius R_t is one order of magnitude smaller.

Figure 8.10 shows that the experimental C_r increase (symbols) is characterized by a sigmoidal trend, typical of drug permeation through thick

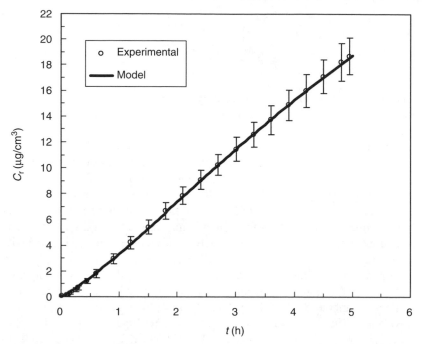

FIGURE 8.10 Comparison between the model best fitting (solid curve, Equation 8.85 through Equation 8.87) and permeation data (open circles). Vertical bars indicate data standard error. (From Sirotti, C., et al., *J. Membr. Sci.*, 204, 401, 2002. With permission from Elsevier.)

membranes [67] or through drug interacting membranes [68]. However, due to the high ratio D/h_m (2.2 μm/s) these two causes cannot be responsible for this peculiar behavior. Accordingly, microemulsion structural changes can be invoked. In particular, we can suppose that nimesulide exit from surfactant micelles (induced by drug permeation toward the receiver chamber) alters the equilibrium between micellized surfactant and solubilized surfactant. This could induce some surfactant micelles disruption causing the increase of surfactant molecules concentration in the aqueous phase. This, in turn, should increase nimesulide solubility in the aqueous phase. Due to permeation phenomenon, a new thermodynamic equilibrium between micelles, drug, and aqueous phase is not established for a long time. Accordingly, this frame could justify the sigmoidal experimental behavior revealed by Figure 8.10. To prove this hypothesis, we need to demonstrate, for example, that nimesulide can alter the CMC of surfactant and that the sigmoidal behavior tends to disappear when the whole surfactant and nimesulide amounts in the microemulsion formulation are decreased. Indeed, in this case, surfactant

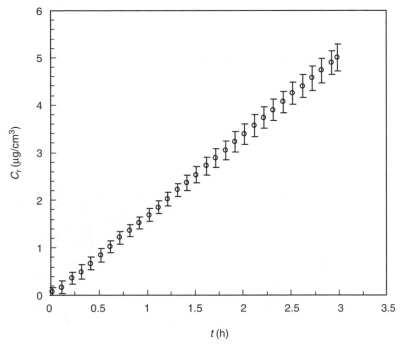

FIGURE 8.11 Nimesulide concentration (C_r) profile in the receiver phase during the release from a microemulsion containing low amounts of drug and surfactant. Vertical bars indicate data standard error. (From Sirotti, C., et al., *J. Membr. Sci.*, 204, 401, 2002. With permission from Elsevier.)

should be mainly at the oil–water interface. Accordingly, both nimesulide ($1000 \, \mu g/cm^3$) and surfactant amount (IPM + BA 23.5 wt.%; Tween 80 28%, water 48.5 wt.%) in the new microemulsion formulation are reduced. Figure 8.11 clearly shows that an almost linear C_r vs. t trend now occurs. This aspect can be more clearly appreciated by considering the initial C_r vs. t curves time derivatives pertaining to the two different situations (i.e., original and low nimesulide and surfactant concentration inside microemulsion). Figure 8.12 shows that in the case of low surfactant and nimesulide contents, slope variation vanishes after 6 min, whereas 1 h is needed for the microemulsion with original surfactant and nimesulide concentration to get a linear profile in the release curve. Such a time ratio (10) cannot be ascribed only to the different amounts of loaded drug (drug ratio equal to 3.5) so that we can conclude that surfactant micellar structures are responsible for the anomalous beginning of the drug release process.

Now we must verify that the presence of nimesulide affects the CMC of the water–Tween 80–benzyl alcohol system. Liquid–vapor surface tension

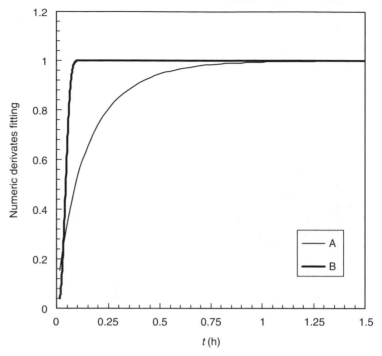

FIGURE 8.12 Smoothed derivates trends of the concentration curves (normalized on the 3 h values). A, release from a microemulsion containing original amounts of drug and surfactant (Figure 8.10 data); B, release from microemulsion containing low amounts of drug and surfactant (Figure 8.11 data). (From Sirotti, C., et al., *J. Membr. Sci.*, 204, 401, 2002. With permission from Elsevier.)

measurements are made on three systems (buffer–Tween 80, buffer + 2% benzyl alcohol–Tween 80, buffer + 2% benzyl alcohol + 14 μg/cm^3 Nimesulide–Tween 80) containing different amounts of surfactant. The buffer–Tween 80 system does not pose particular problems in the determination of the CMC value and it is considered as reference. On the contrary, the CMC determination is slightly more complex in the presence of benzyl alcohol (used as cosurfactant in our microemulsion) and nimesulide. The starting point for the CMC measurement is the Gibbs equation [69]:

$$\Gamma_2^1 = -\frac{1}{RT}\left(\frac{d\gamma_{LV}}{d\ln C_2}\right),\tag{8.95}$$

where C_2 is the surfactant (Tween 80) bulk concentration, T is the temperature, R is the universal gas constant, Γ_2^1 is the excess of surfactant molecules number per unit interface area, and γ_{LV} is the surface tension. In a binary system, Γ_2^1 is constant below and above the CMC, with different values in the two concentration ranges. Accordingly, the CMC value can be approximately individuated at the intersection between the tangents to the γ_{LV} vs. $\ln(C_2)$ curve before and after the curve slope variation [70]. Similar changes in surface tension occur in multicomponent systems with increasing surfactant concentration, even if the phenomenology appears more complicated by binary interactions. Indeed, though for the buffer–Tween 80 system the two tangents are easily located (see Figure 8.13), in the system containing benzyl alcohol, an initial increase in Tween 80 concentration does not generate a decrease in surface tension because of the above-mentioned Tween 80–benzyl alcohol binary interactions. Reasonably, the surfactant molecules are captured by benzyl alcohol until benzyl alcohol molecules are available and only afterward the γ_{LV} decrease with C_2 takes place. The CMC position can be located in correspondence with the highest γ_{LV} vs. $\ln(C_2)$ slope variation [71]. The presence of a small amount of nimesulide smoothes the γ_{LV} vs. $\ln(C_2)$ shape and lowers the γ_{LV} value along the whole concentration field examined. The CMC values achieved for the analyzed systems are as follows:

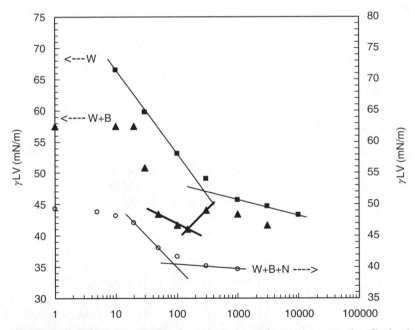

FIGURE 8.13 Surface tension γ_{LV} dependence on surfactant concentration C_2, for the three considered systems (W: water–Tween 80 (left ordinate axis), W + B: water–benzyl alcohol–Tween 80 (left ordinate axis), W + B + N: water–benzyl alcohol–Nimesulide–Tween 80 (right ordinate axis)). (From Sirotti, C., et al., *J. Membr. Sci.*, 204, 401, 2002. With permission from Elsevier.)

buffer–Tween 80, CMC = 276.2 $\mu g/cm^3$; buffer + 2% benzyl alcohol–Tween 80, CMC = 150 $\mu g/cm^3$; buffer + 2% benzyl alcohol + 14 $\mu g/cm^3$ nimesulide–Tween 80, CMC = 87.7 $\mu g/cm^3$. It is thus possible to observe how the presence of nimesulide reduces the CMC of the system. This confirms the assumption of micelle breakup occurring during the drug transfer to the receiver: when the drug concentration in the aqueous phase decreases, CMC increases; thus, causing the solubilization of surfactant molecules previously arranged in micellar structures.

Once experimentally demonstrated the reasonability of the hypothesis about nimesulide–surfactant interaction, it is possible evaluating model best fitting to experimental data. Assuming $V_r = 22$ cm^3, $V_d = 4$ cm^3, $V_w = 1.828$ cm^3, $V_o = 0.924$ cm^3, $V_t = 0.092$ cm^3 (as it is unknown, we assumed it to be equal to $1/10\ V_o$), $h_m = 0.0149$ cm, $S = 3.46$ cm^2, $D = 3.34 \times 10^{-6}$ cm^2/s, $R_o = 2.7 \times 10^{-6}$ cm, $R_t = 2.7 \times 10^{-7}$ cm, $M_0 = 14360$ μg, $C_{so} = 18,000$ $\mu g/cm^3$, $C_{st} = 65,000$ $\mu g/cm^3$, fitting procedure yields $A = 2.03 \times 10^{-4}$/s, $s_0 = 19.79$ $\mu g/cm^3$, $s_f = 44.06$ $\mu g/cm^3$, $k_{ow} = 4 \times 10^{-7}$ cm^4/(μg s), $k_{wo} = 1 \times 10^{-10}$ cm^4/(μg s), $k_{tw} = 1 \times 10^{-8}$ cm^4/(μg s), and $k_{wt} = 1 \times 10^{-12}$ cm^4/(μg μg s). Figure 8.10 shows the satisfactory agreement between the model best fitting (solid line) and the permeation data (symbols) also in the initial zone. It is important to remember that model intrinsic complexity (great number of parameters) and assumptions performed (i.e., V_t and r_t, values, for example) fitting parameter results are purely indicative. Nevertheless, the aim of this approach was not the exact calculation of k_{ow}, k_{wo}, k_{tw}, and k_{wt}, but to check whether the theoretical frame proposed is able to explain the unusual experimental findings.

8.6 CONCLUSIONS

Microemulsions are powerful drug delivery systems as they can be suitable for the release of both scarcely water-soluble drugs and enzyme-labile water-soluble drugs. In addition, ease of fabrication, versatility for what concerns administration route (topical, injection) and low production costs make them very attractive for the delivery field. Nevertheless, microemulsions are complex systems whose structure can be modified by external conditions jointly with the presence of the loaded drug. Accordingly, release kinetics designing is not an easy task and that is why a deep structural and thermodynamic characterization is needed. Obviously, this aspect arises in oral, injection, and topical administration as microemulsion is diluted in an external aqueous environment (oral, injection) or it is put in contact with an additional component (usually a polymer) aimed to impart the proper rheological properties.

This chapter was aimed to illustrate these aspects and to provide some possible interpretative tools clarifying the key mechanisms ruling drug release from microemulsions in the light of a proper release kinetics designing.

REFERENCES

1. Hoar, T.P. and Schulman, J.H., Transparent water-in-oil dispersions: the oleo-pathic hydro-micelle, *Nature*, 152, 102, 1943.
2. Lawrence, M.J. and Rees, G.D., Microemulsion-based media as novel drug delivery systems, *Adv. Drug Delivery Rev.*, 45, 86, 2000.
3. Schulman, J.H., Stoeckenius, W., and Prince, L.M., Mechanism of formation and structure of microemulsions by electron microscopy, *J. Phys. Chem.*, 63, 1677, 1959.
4. Prince, L.M., A theory of aqueous emulsion. I. Negative interfacial tension at the oil/water interface, *J. Colloid Interface Sci.*, 23, 165, 1967.
5. Shinoda, K. and Kunieda, H., Conditions to produce so-called microemulsions. Factors to increase the mutual solubility of oil and water by solubilizer, *J. Colloid Interface Sci.*, 42, 381, 1973.
6. Ruckenstein, E. and Chi, J.C., Stability of microemulsions, *J. Chem. Soc. Faraday Trans.*, 71, 1690, 1975.
7. Kegel, W.K., Overbeek, J.T.G., and Lekkerkerker, H.N.W., Thermodynamics of microemulsions I, in *Handbook of Microemulsion Science and Technology*, Kumar, P. and Mittal K.L., Eds., Dekker, 1999.
8. Solans, C., Pons, R., and Kunieda, H., Overview of basic aspects of microemulsions, in *Industrial Applications of Microemulsions, Surfactant Science Series*, Solans, C. and Kunieda H., Eds., Dekker, 1997.
9. Beekwilder, J.J., Thesis/Dissertation, Faculty of Applied Physics, University of Twente, 1990.
10. Carlfors, J., Blute, I., and Schmidt, V., Lidocaine in microemulsion—a dermal delivery system, *J. Disp. Sci. Technol.*, 12, 467, 1991.
11. Israelachvilli, J.N., Mitchell, D.J., and Ninham, B.W., Theory of self assembly of hydrocarbon amphiphiles into micelles and bilayers, *J. Chem. Soc. Faraday Trans. II*, 72, 1525, 1976.
12. Solans, C., Pons, R., and Kunieda, H., *Industrial Application of Microemulsions*, Marcel Dekker Inc, New York, 1997.
13. Quemada, D. and Langevin, D., Rheological modelling of microemulsions, *J. Theoret. Appl. Mech.*, numéro spécial, 201, 1985.
14. Winsor, P.A., Hydrotropy, solubilisation and related emulsification processes, *J. Chem. Soc. Faraday Trans.*, 44, 376, 1948.
15. Moulik, S.P. and Paul, B.K., Structure, dynamics and transport properties of microemulsions, *Adv. Colloid Interface Sci.*, 78, 99, 1998.
16. Adamson, A.W. and Gast, A.P., *Physical Chemistry of Surfaces*, Wiley, 1997, Chapter 3.
17. Meriani, F., et al., Characterization of a quaternary liquid system improving the bioavailabilty of poorly water soluble drugs, *J. Colloid Interface Sci.*, 263, 590, 2003.
18. Tassios, D.P., *Applied Chemical Engineering Thermodynamics*, Springer-Verlag, Berlin/New York, 1993.
19. Primorac, M., et al., The influence of water/oil phase ratio on the rheological behaviour of microemulsions, in The First World Meeting AGPI/APV, Budapest, 656.
20. Primorac, M., et al., Rheological properties of oil/water microemulsions, *Pharmazie*, 47, 645, 1992.

21. Blum, F.D., et al., Structure and dynamics in three-component microemulsions, *J. Phys. Chem.*, 89, 711, 1985.
22. Ghao, Z.G., et al., Physicochemical characterization and evaluation of a micro-emulsion for oral delivery of cyclosporin A, *Int. J. Pharm.*, 161, 75, 1998.
23. Arcoleo, V., et al., Study of AOT-stabilized microemulsions of formamide and *n*-methylformamide dispersed in *n*-heptane, *Mater. Sci. Eng.*, 5, 47, 1997.
24. Sirotti, C., Studio di microemulsioni per la formulazione di sistemi farmaceutici, Thesis/Dissertation, University of Trieste, Department of Chem. Eng., 2000.
25. Regev, O., et al., A study of microstructure of a four-component nonionic micro-emulsion by cryo-TEM, NMR, SAXS and SANS, *Langmuir*, 12, 668, 1996.
26. Bergenholtz, J., Romagnoli, A.A., and Wagner, N.J., Viscosity, microstructure and interparticle potential of $AOT/H_2O/n$-decane inverse microemulsions, *Langmuir*, 11, 1559, 1995.
27. Schurtenberger, P., et al., Structure and phase behaviour of lecithin-based micro-emulsions: a study of chain length dependence, *J. Colloid Interface Sci.*, 156, 43, 1993.
28. Gradzielski, M., Langevin, D., and Farago, B., Experimental investigation of the structure of nonionic microemulsions and their relation to the bending elasticity of the amphiphilic film, *Phys. Rev.*, E 53, 3900, 1996.
29. Giustini, M., et al., Microstructure and dynamics of the water-in-oil CTAB/*n*-pentanol/*n*-hexane/water microemulsion: spectroscopic and conductivity study, *J. Phys. Chem.*, 100, 3190, 1996.
30. Barnes, I.S., et al., The disordered open connected model of microemulsions, *Progr. Colloid Polym. Sci.*, 76, 90, 1988.
31. Bolzinger, M.A., et al., Characterisation of a sucrose ester microemulsion by freeze fracture electron micrograph and small angle neutron scattering experiments, *Langmuir*, 15, 2307, 1999.
32. D'Angelo, M., et al., Dynamics of water-containing sodium *bis*(2-ethylhexyl)sul-fosuccinate (AOT) reverse micelles: a high-frequency dielectric study, *Phys. Rev.*, E 54, 993, 1996.
33. Saidi, Z., et al., Percolation and critical exponents for the viscosity of microemul-sions, *Phys. Rev.*, A 42, 872, 1990.
34. Lapasin, R. and Pricl, S., *Rheology of Industrial Polysaccharides, Theory and Applications*, Chapman & Hall, London, 1995.
35. Mittal, L.K., *Handbook of Microemulsion Science and Technology*, Marcel Dek-ker, Inc, New York, 1999.
36. Constantinides, P.P., Lipid microemulsions for improving drug dissolution and oral absorption: physical and biopharmaceutical aspects, *Pharm. Res.*, 12, 1561, 1995.
37. Trotta, M., Gasco, M.R., and Morel, S., Release of drugs from oil–water micro-emulsions, *J. Contr. Rel.*, 10, 237, 1989.
38. Washington, C., Drug release from monodisperse systems: a critical review, *Int. J. Pharm.*, 58, 1, 1990.
39. Wallin, R., et al., Prolongation of lidocaine induced regional anaesthesia by a slow-release microemulsion formulation, in *Proc. Int. Symp. Control. Rel. Bioact. Mater.*, 555, 1997.
40. Thiele, L., et al., Particle-uptake by monocyte-derived dendritic cells *in vitro*: evaluation of particle size and surface characteristics, in *Proc. Int. Symp. Control. Rel. Bioact. Mater.*, 26, 163, 1999.

41. Lee, V.H.L., Enzymatic barriers to peptide and protein absorption, *Crit. Rev. Ther. Drug Carrier Syst.*, 5, 69, 1988.
42. Sarciaux, J.M., Acar, L., and Sado, P.A., Using microemulsion formulations for drug delivery of therapeutic peptides, *Int. J. Pharm.*, 120, 127, 1995.
43. Gasco, M.R., Pattarino, F., and Lattanzi, F., Long-acting delivery system for peptides: reduced plasma testosterone levels in male rats after a single injection, *Int. J. Pharm.*, 62, 119, 1990.
44. Lapasin, R., Grassi, M., and Coceani, N., Effects of polymer addition on the rheology of o/w microemulsions, *Rheol. Acta*, 40, 185, 2001.
45. Trotta, M., Morel, S., and Gasco, M.R., Effect of oil phase composition on the skin permeation of felodipine from (o/w) microemulsions, *Pharmazie*, 52, 50, 1997.
46. Kantaria, S., Rees, G.D., and Lawrence, M.J., Gelatin-stabilised microemulsion-based organogels: rheology and application in iontophoretic transdermal drug delivery, *J. Contr. Rel.*, 60, 355, 1999.
47. Holt, D.W., et al., The pharmacokinetics of Sandimmune Neoral: a new oral formulation of cyclosporine, *Transplant. Proc.*, 26, 2935, 1994.
48. Park, K.M., et al., Phospholipid-based microemulsions of flurbiprofen by the spontaneous emulsification process, *Int. J. Pharm.*, 183, 145, 1999.
49. von Corswant, C., Thoren, P., and Engstrom, S., Triglyceride-based microemulsion from intravenous administration of sparingly soluble substances, *J. Pharm. Sci.*, 87, 200, 1998.
50. Hasse, A. and Keiprt, S., Development and characterisation of microemulsions for ocular application, *Eur. J. Pharm. Biopharm.*, 43, 179, 1997.
51. Chew, C.H., et al., Interactions of polymerised surfactants and polymer, *Langmuir*, 11, 3312, 1995.
52. Atkinson, P.J., et al., Structure and stability of microemulsion-based organogels. *J. Chem. Soc. Faraday Trans.*, 87, 3389, 1991.
53. Dalmora, M.E., Dalmora, S.L., and Oliveira, A.G., Inclusion complex of piroxicam with b-cyclodexstrin and incorporation in cationic microemulsion. *In vitro* drug release and *in vivo* topical anti-inflammatory effect, *Int. J. Pharm.*, 222, 45, 2001.
54. Washington, C. and Evans, K., Release rate measurements of model hydrophobic solutes from submicron triglyceride emulsions, *J. Contr. Rel.*, 33, 383, 1995.
55. Grassi, M., Coceani, N., and Magarotto, L., Mathematical modeling of drug release from microemulsions: theory in comparison with experiments, *J. Colloid Interface Sci.*, 228, 141, 2000.
56. Pattarino, F., et al., Experimental design and partial least squares in the study of complex mixtures: microemulsions as drug carriers, *Int. J. Pharm.*, 91, 157, 1993.
57. Yotsuyanagi, T., Higuchi, W.I., and Ghanem, A.H., Theoretical treatment of diffusional transport into and through an oil–water emulsion with an interfacial barrier at the oil–water interface, *J. Pharm. Sci.*, 62, 40, 1973.
58. Guy, R.H., et al., Calculation of drug release rates from spherical particles, *Int. J. Pharm.*, 11, 199, 1982.
59. Boddé, H.E. and Joosten, J.G.H., A mathematical model for drug release from a two-phase system to perfect sink, *Int. J. Pharm.*, 26, 57, 1985.
60. Friedman, D. and Benita, S., A mathematical model for drug release from O/W emulsions: application to controlled release morphine emulsions, *Drug Dev. Ind. Pharm.*, 13, 2067, 1987.

61. Crank, J., *The Mathematics of Diffusion*, second edition, Clarendon Press, Oxford, 1975.

62. Grassi, M. and Colombo, I., Mathematical modelling of drug permeation through a swollen membrane, *J. Contr. Rel.*, 59, 343, 1999.

63. Flynn, G.L., Yalkowsky, S.H., and Roseman, T.J., Mass transport phenomena and models: theoretical concepts, *J. Pharm. Sci.*, 63, 479, 1974.

64. Coviello, T., et al., Novel hydrogel system from scleroglucan: synthesis and characterization, *J. Contr. Rel.*, 60, 367, 1999.

65. Press, W.H., et al., *Numerical Recipes in FORTRAN*, second edition, Cambridge University Press, Cambridge, UK, 1992.

66. Sirotti, C., et al., Modeling of drug release from microemulsions: a peculiar case, *J. Membrane Sci.*, 204, 401, 2002.

67. Grassi, M., Coceani, N., and Magarotto, L., Modeling of sparingly soluble drugs partitioning in a two-phase liquid system, *Int. J. Pharm.*, 239, 157, 2002.

68. Coviello, T., et al., A new polysaccharidic gel matrix for drug delivery: preparation and mechanical properties, *J. Contr. Rel.*, 102, 643, 2005.

69. Chattoraj, D.K. and Birdi, K.S., *Adsorption and the Gibbs Surface Excess*, Plenum Press, 1984.

70. Tadros, Th. F., Surfactants technologically important aspects of interface science, *Lecture*, 13, 1, 1983.

71. Goddard, E.D., Polymer/surfactant interaction, *J. Soc. Cosmet. Chem.*, 41, 23, 1990.

9 Drug Permeation through Membranes

9.1 INTRODUCTION

Membranes can be defined as a sheet of solid or semisolid material, insoluble in its surrounding medium, which separates phases that are usually (but not necessarily) fluid [1]. Based on mechanisms ruling mass transport, typically, membranes can be divided into two main classes: biological and synthetic membranes [2]. While in synthetic membranes, mass transport is usually due only to a chemical potential gradient, in biological membranes mass transport can be due also to other special mechanisms. These mechanisms can act cooperatively with chemical potential or they can act against it. In this last case, they need energy from the surrounding, and a typical energy source is that coming from the hydrolysis of high energy compounds such as ATP (adenosine triphosphate) that transforms into ADP (adenosine diphosphate) plus P (inorganic phosphate) + energy (ATP pump mechanism) [3]. While biological membranes permeability can be highly influenced by these non-chemical potential driven mechanisms, synthetic membranes permeability is essentially dependent on chemical–physical properties of the solute and membrane microscopic structure. As membranes are nothing more than special matrices where one-dimension is considerably smaller than the other two, they are characterized by the same structures of matrices discussed in Chapter 7. Nevertheless, it is important to recall some important structural aspects. In the most general case, they are composed of three different phases: (a) continuous, (b) shunt, and (c) dispersed [1]. In turn, these phases can be classified as primary, secondary, tertiary, etc., on the basis of their relative spatial relationships. A primary continuous phase is an uninterrupted phase between membrane surfaces as well as laterally or in the plane perpendicular to the flux vector. From the membrane permeability point of view, it can represent an uninterrupted diffusional path for a solute or it represents an inaccessible zone acting as a mere supporting structure. A primary shunt phase can be seen as an ensemble of pores or channel that completely cross membrane thickness (for example, in the direction perpendicular to membrane surfaces) but it is laterally discontinuous. Accordingly, a primary shunt phase cannot influence permeability or can affect it providing parallel diffusional pathways or can represent the sole diffusional pathway. Dispersed phases, embedded in continuous or shunt phases, are discontinuous along the flux vector and do not provide an uninterrupted pathway through the membrane or any of its subphases.

Very often, membranes adopted in the controlled release field are polymer based. Typically, they are constituted of a (primary) continuous phase (usually a liquid phase) trapped in a swollen solid phase (polymeric network), which can be seen as a continuous secondary phase in virtue of its low concentration with respect to that of the primary continuous one [4]. The presence of crosslinks between polymeric chains hinders polymer dissolution in the liquid phase that can only swell the network. In the case of strong crosslinks (typically chemical covalent bonds), the network does not modify with time while, on the contrary, if weak crosslinks prevail (typically physical interactions such as Coulombic, van der Waals, dipole–dipole, hydrophobic and hydrogen bonding interactions), polymeric chains are not so rigidly connected to each other. Consequently, polymer–polymer junction weakness makes these membranes easily erodable. This scenario can be more complex when the contemporary presence of two networks (interpenetrating structures) occurs. These structures are produced by an initial swelling of a monomer and reacting to form a second intermeshing network structure [5,6].

Although many different kinds of synthetic membranes exist (semipermeable, porous, and so on) [2], their tasks fall among drug release rate modulation (on a diffusive basis) [7], insulation of something from the external environment [8], making artificial implants biocompatible [9], allowing drug release (due to membrane dissolution) only when particular environmental conditions take place [10], and simulation of permeation properties of natural tissue properties [11–13].

The aim of this chapter is to study synthetic and natural membranes in the light of mass transport induced by a chemical potential gradient. Accordingly, drug diffusion and not convection will be considered as transport mechanism. In addition, neither osmosis nor electrotransport will be accounted for as drug transport mechanisms. The attention will be focused on mathematical models aimed to estimate membranes (both synthetic and natural) drug diffusion. In particular, for their importance in the pharmaceutical field, drug release from microencapsulated systems, drug permeation through gastrointestinal mucosa and skin will be matched.

9.2   SYNTHETIC MEMBRANES

As mentioned before, synthetic membranes are versatile tools that can be used, for example, in oral, transdermal, and implantable delivery systems [2]. Accordingly, the knowledge of their permeability is of paramount importance for the designing and optimization of modern controlled release systems. Consequently, proper experimental setup and related interpretative mathematical models are necessary for membrane permeability estimation.

9.2.1 SYNTHETIC MEMBRANES PERMEABILITY DETERMINATION

Membranes permeability P and drug diffusion coefficient inside membrane (D_m) are related by the following equation:

$$P = \frac{k_p D_m}{h_m},$$ (9.1)

where k_p is the drug partition coefficient between membrane and fluids (usually the same) facing membrane surfaces, while h_m represents membrane thickness. Accordingly, P measurement implies the D_m determination. Although several methods are available for the D_m experimental measure (for example, those based on nuclear magnetic resonance (NMR), dynamic light scattering (DLS), holographic relaxation spectroscopy, determination of drug concentration profile (sectioning and inverse sectioning methods), and drug concentration gradient under stationary and nonstationary gradient) [14–17], for its simplicity and versatility, the approach based on drug permeation will be discussed. The experimental setup consists of a "side-by-side apparatus" [18,19] composed of a membrane separating the donor compartment (see Figure 9.1), containing drug solution, from the receiver compartment, initially filled by pure solvent. Each compartment is equipped by a constant temperature jacket and a magnetic stirrer, housed at the bottom of the cell, to ensure constant temperature and fluid mixing, respectively. Due to the existing concentration gradient, a drug flux from the donor compartment to the receiver one arises. Drug concentration increase in the receiver compartment is measured and recorded by means of a personal computer managing a UV spectrophotometer as shown in Figure 9.1. In order to prevent bubbles formation in the whole system, it is desirable to eliminate CO_2 from the aqueous medium filling both compartments. For this purpose, pure solvent can be boiled or saturated by helium before usage. In addition, a surge chamber can be inserted between the UV and the peristaltic pump, responsible for solution circulation. The whole circulating system (peristaltic pump, surge chamber, and so on) can be substituted by an optical fiber apparatus allowing a direct measurement of drug concentration in the receiver environment [20]. While this last approach shows the considerable advantage of a simpler measurement and minor system perturbations, it cannot be considered when drug concentration via UV is not possible.

Practically, in this setup, D_m is deduced on the basis of the experimental permeation curve represented by drug concentration (C_r) increase in the receiver compartment. Indeed, knowing the side-by-side geometrical characteristics (donor and receiver volumes, membrane thickness, and surface available for diffusion), assuming that membrane thickness is constant during permeation, that no drug–matrix interactions occur, and that D is concentration independent, the permeation curve depends only on D. In an attempt to

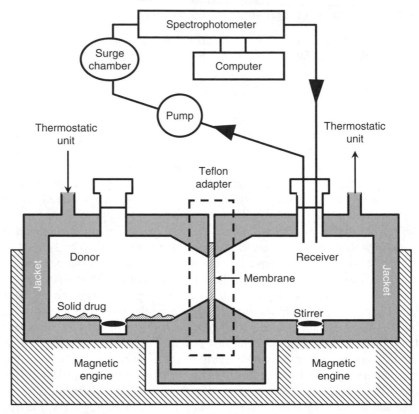

FIGURE 9.1 Schematic representation of the side-by-side apparatus. (From Grassi, M., Colombo, I., Lapasin, R., *J. Contr. Rel.*, 76, 93, 2001. With permission from Elsevier.)

make this approach as general as possible, we can assume that the donor compartment contains an excess amount of undissolved drug. This strategy relies on two practical reasons. The first concerns the possibility of getting a virtually constant drug concentration (C_d) in the donor compartment (this greatly simplifies the theoretical modeling of the permeation curve). The second is about the possibility of having the highest C_d values (theoretically, C_d coincides with drug solubility). Indeed, this turns out to be interesting for both soluble and poorly soluble drugs. If in the first case permeation should not be a long lasting test (of course this depends also on membrane permeability), in the second case it minimizes drug detection problems in the receiver compartment due to very low drug concentrations. Accordingly, in this experimental setup, drug permeation through the membrane induces the progressive dissolution of undissolved drug present in the donor compartment. If the receiver compartment volume is comparably larger than that of the donor one, the dissolution process will affect the permeation curve shape if its kinetics is smaller or comparable to that of drug permeation. Finally, a complete description of the physical frame realizing in the side-by-side apparatus must account for the presence of two stagnant layers (one on the donor compartment side and the other on the receiver compartment side) sandwiching the membrane. Indeed, despite compartments fluid stirring, their presence is unavoidable and increasing stirring can only reflect in stagnant layers thickness decrease. Obviously, stagnant layers presence is important only if they are thick enough in comparison to membrane thickness.

In order to determine D_m, a proper mathematical model is now needed. Assuming that membrane thickness is constant, that possible drug–matrix interactions can be properly accounted for by the partition coefficient k_p, that D_m is concentration independent, that sink conditions are attained in the receiver compartment (C_r can be retained always ~0 when low V_d/V_r ratio values apply), that donor drug concentration C_d can be retained constantly equal to C_{d0} (this happens when solid drug dissolution kinetics in the donor compartment is faster than permeation kinetics) and that the effect of stagnant layers is negligible, Flynn et al. [1] proposed:

$$C_r = \frac{C_{d0}k_p S}{V_r}\left(\frac{D_m}{h_m}t - \frac{h_m}{6} - \frac{2h_m}{\pi^2}\sum_{n=1}^{\infty}\frac{(-1)^n}{n^2}e^{-\left(\frac{n\pi}{h_m}\right)^2 D_m' t}\right), \qquad (9.2)$$

where t is time and S is membrane surface available for permeation. This model represents Fick's second law solution in the hypotheses above specified (steady-state hypotheses). For t approaching infinite, Equation 9.2 reduces to the following straight line:

$$C_r = \frac{C_{d0}PS}{V_r}(t - t_L); \quad P = \frac{K\,D_m}{h_m}; \quad t_L = \frac{h_m^2}{6D_m}, \qquad (9.3)$$

where t_L represents straight line intercept on the time axis. Interestingly, Barrie et al. [21] and Flynn et al. [1] demonstrate that Equation 9.3 works also in the presence of stagnant layers provided that P and t_L become

$$P = \frac{D_1 D_2 D_3 k_1 k_2 k_3}{h_1 D_2 D_3 k_2 k_3 + h_2 D_1 D_3 k_1 k_3 + h_3 D_1 D_2 k_1 k_2}, \qquad (9.4)$$

$$t_L = \left[\frac{h_1^2}{D_1}\left(\frac{h_1}{6D_1 k_1} + \frac{h_2}{2D_2 k_2} + \frac{h_3}{2D_3 K_3}\right) + \frac{h_2^2}{D_2}\left(\frac{h_1}{2D_1 k_1} + \frac{h_2}{6D_2 k_2} + \frac{h_3}{2D_3 k_3}\right)\right.$$
$$\left. + \frac{h_3^2}{D_3}\left(\frac{h_1}{2D_1 k_1} + \frac{h_2}{2D_2 k_2} + \frac{h_3}{6D_3 k_3}\right) + \frac{k_2 h_1 h_2 h_3}{D_1 D_3 k_1 k_3}\right] \Big/$$
$$\left[\frac{h_1}{D_1 k_1} + \frac{h_2}{D_2 k_2} + \frac{h_3}{D_3 k_3}\right], \qquad (9.5)$$

where $D_2 = D_m$, $h_2 = h_m$, h_1, and h_3 represent stagnant layers thickness (h_1 = donor side stagnant layer thickness; h_3 = receiver side stagnant layer thickness), and k_1, k_2, and k_3 are partition coefficients defined by

$$k_1 = k_{1d}; \quad k_2 = k_{21}k_{1d}; \quad k_3 = k_{3r}, \tag{9.6}$$

$$k_{1d} = \frac{C_1}{C_d}; \quad k_{21} = \frac{C_2}{C_1}; \quad k_{23} = \frac{C_2}{C_3}; \quad k_{3r} = \frac{C_3}{C_r}, \tag{9.7}$$

where $(C_d{:}C_1)$, $(C_1{:}C_2)$, $(C_2{:}C_3)$, and $(C_3{:}C_r)$ are drug concentrations at the (donor:first stagnant layer) interface, (first stagnant layer:membrane) interface, (membrane:second stagnant layer) interface, and (second stagnant layer:receiver) interface. In writing Equation 9.6 and Equation 9.7, it is assumed that k_1, k_2, and k_3 are constant and concentration independent. This means that equilibrium conditions are always attained at the four interfaces considered.

If steady-state conditions are not attained and only constant membrane thickness and diffusion coefficient are assumed, the physical frame pertaining to the side-by-side apparatus can be modeled according to the following equations:

$$\frac{\partial C_i}{\partial t} = \frac{\partial}{\partial X}\left(D_i \frac{\partial C_i}{\partial X}\right); \quad i = 1, 2, 3, \dots \tag{9.8}$$

where X is the abscissa, t is time, C_i and D_i represent, respectively, drug concentration and diffusion coefficient in the first stagnant layer ($i = 1$), the membrane ($i = 2$), and the second stagnant layer ($i = 3$). The above equations must be solved with the following boundary conditions:

$$V_d \frac{dC_d}{dt} = -\frac{dM}{dt} + D_1 S \frac{\partial C_1}{\partial X}\bigg|_{X=0}, \tag{9.9}$$

$$\frac{dM}{dt} = -V_d K_t (C_s - C_d), \tag{9.10}$$

$$D_1 \frac{\partial C_1}{\partial X}\bigg|_{X=h_1} = D_2 \frac{\partial C_2}{\partial X}\bigg|_{X=h_1}, \tag{9.11}$$

$$D_2 \frac{\partial C_2}{\partial X}\bigg|_{X=h_1+h_2} = D_3 \frac{\partial C_3}{\partial X}\bigg|_{X=h_1+h_2}, \tag{9.12}$$

$$V_r \frac{dC_r}{dt} = -D_3 S \frac{\partial C_3}{\partial X}\bigg|_{X=h_1+h_2+h_3}, \tag{9.13}$$

$$C_1(X = 0) = k_{1d} C_d, \tag{9.14}$$

$$C_2(X = h_1) = k_{21}C_1(X = h_1), \tag{9.15}$$

$$C_2(X = h_1 + h_2) = k_{23}C_3(X = h_1 + h_2), \tag{9.16}$$

$$C_3(X = h_1 + h_2 + h_3) = k_{3r}C_r, \tag{9.17}$$

and the following initial conditions:

$$C_d = C_1 = C_{d0}; \quad C_r = C_2 = C_3 = 0; \quad M = M_0, \tag{9.18}$$

where M is the time-dependent undissolved (solid) drug amount present in the donor compartment, K_t is drug dissolution constant, C_s is drug solubility, C_{d0} and M_0 are, respectively, the initial drug concentration and undissolved drug amount in donor compartment. While Equation 9.8 is Fick's second law, Equation 9.9 represents the drug mass balance made up on the donor compartment: the first right-hand side term takes into account the dissolution, whereas the second represents the matter flux leaving the donor through the first stagnant layer. Equation 9.10 describes the reduction of the solid drug M as the dissolution goes on. Obviously, when M is zeroed, the first term of the right-hand side of Equation 9.9 vanishes. Equation 9.11 and Equation 9.12 ensure, respectively, that no matter accumulation occurs at the first stagnant layer/membrane and membrane/second stagnant layer interfaces (matter fluxes are equal on both sides of the interfaces). Equation 9.13 is the drug mass balance made up on the receiver compartment: the right-hand side term is the entering drug flux coming from the second layer. Equation 9.14 through Equation 9.17 ensure partitioning conditions at the interface in $X = 0$, h_1, $(h_1 + h_2)$, and $(h_1 + h_2 + h_3)$. Finally, Equation 9.18 sets initial conditions: membrane, second stagnant layer, and receiver are drug free, while drug concentration in the donor compartment and in the first stagnant layer is equal to C_{d0}. By setting $M_0 = 0$, this model can describe the simpler situation represented by the absence of undissolved drug present in the donor compartment.

When drug dissolution takes place from a thin tablet, K_t is time independent as dissolution surface is, practically, constant. On the contrary, if drug dissolution occurs from an ensemble of spherical solid particles, this is no longer true as particle radius and, thus, particles surface (this is the dissolving surface) reduces as time goes on. Practically, this situation resembles that discussed in Chapter 7 about crystals drug dissolution in a polymeric matrix system. In that context, it was shown that the relation between K_t and the drug intrinsic dissolution constant k_d is given by (see Equation 7.81''):

$$K_t = k_d A_V, \tag{9.19}$$

where A_V is the solid surface per unit volume. In an analogy with what was done in Chapter 7, A_V expression for the present situation is

$$A_V = N_p \frac{4\pi R^2}{V_d}; \quad N_p = \frac{3M_0}{4\pi R_0^3 \rho_D}, \tag{9.20}$$

where R and R_0 indicate, respectively, the time-dependent particle radius (this approach implicitly assumes that all the solid particles are characterized by the same radius) and the initial particle radius, N_p is the number of particles constituting the whole solid drug amount M_0, and ρ_D is solid drug density. In this situation, Equation 9.19 and Equation 9.20 hold if the whole powder surface is available for dissolution. This is reasonably accomplished when the dissolution medium is highly stirred so that the particles cannot stick to each other or to the vessel walls. On the basis of Equation 9.20, it is possible measuring R_0 once the initial particles surface area per unit mass A_0 is experimentally determined. Indeed, we have $R_0 = 3/A_0 \rho_D$, where ρ_D is drug density. By this hypothesis, Equation 9.10 becomes

$$\frac{dM}{dt} = -N_p A_V k_d (C_s - C_d). \tag{9.10'}$$

The R time dependency derives from the following mass balance regarding the whole side-by-side system (donor and receiver volumes, stagnant layer volumes, and membrane):

$$M = M_0 + V_d(C_{d0} - C_d) - V_r C_r - \int_0^{h_1} C_1 S\, dX - \int_{h_1}^{h_1+h_2} C_2 S\, dX - \int_{h_1+h_2}^{h_1+h_2+h_3} C_3 S\, dX, \tag{9.21}$$

where, obviously, $M = N_p(4/3)\pi \rho R^3$ and, consequently, the well-known Hixon–Crowell equation [22],

$$R = \sqrt[3]{\frac{3M}{N_p 4\pi \rho}} = R_0 \sqrt[3]{\frac{M}{M_0}}$$

relates R and M.

The effects of the K_t reduction are more evident when the dissolution phenomenon implies a considerable decreasing of the particle radius. Indeed, in this case, the dissolution surface will be strongly reduced as well as dissolution kinetics. Although the above set of equations require a numerical solution (for example, control volume method [23]), it is interesting to discuss some analytical solutions applying in particularly favorable cases often encountered in practical applications. Indeed, if the drug concentration profile inside the two stagnant layers and the membrane is always linear (this holds for thin membranes) and K_t is time independent, the proposed numerical model leads to the following analytical solution:

$$C_d(t) = A_1 + A_2 e^{(m_1^* t)} + A_3 e^{(m_2^* t)}, \tag{9.22}$$

$$C_r(t) = B_1 + B_2 e^{(m_1^* t)} + B_3 e^{(m_2^* t)}, \tag{9.23}$$

$$M(t) = M_0 + E_1\left(e^{(m_1^* t)} - 1\right) + E_2\left(e^{(m_2^* t)} - 1\right), \tag{9.24}$$

where A_1, A_2, A_3, m_1, m_2, B_1, B_2, B_3, E_1, and E_2 are defined by the following equations:

$$m_1 = 0.5\Big(-(Tg - G + xb - Y) \\ + \sqrt{(Tg - G + xb - Y)^2 - 4((xb - Y)(Tg - G) - Tzxa)}\Big), \tag{9.25}$$

$$m_2 = 0.5\Big(-(Tg - G + xb - Y) \\ - \sqrt{(Tg - G + xb - Y)^2 - 4((xb - Y)(Tg - G) - Tzxa)}\Big), \tag{9.26}$$

$$A_1 = -\frac{(xb - Y)K_t C_s}{m_1 m_2}, \tag{9.27}$$

$$A_2 = \frac{K_t C_s - \dfrac{(xb - Y)K_t C_s}{m_1} - (m_2 - (Tg - G))C_s}{m_1 - m_2}, \tag{9.28}$$

$$A_3 = \frac{K_t C_s - \dfrac{(xb - Y)K_t C_s}{m_2} - (m_1 - (Tg - G))C_s}{m_2 - m_1}, \tag{9.29}$$

$$B_1 = -\frac{K_t C_s + (Tg - G)A_2}{Tz}, \tag{9.30}$$

$$B_2 = \frac{A_2}{Tz}(m_1 - (Tg - G)); \quad B_3 = \frac{A_3}{Tz}(m_1 - (Tg - G)), \tag{9.31}$$

$$E_1 = \frac{A_2}{m_1}; \quad E_2 = \frac{A_3}{m_2}, \tag{9.32}$$

$$G = K_t C_s + \frac{D_1 S K_{1d}}{V_d h_1}; \quad T = \frac{D_1 S}{V_d h_1 k_{21}}; \quad x = \frac{D_3 S}{h_3 V_r k_{23}}; \quad Y = \frac{D_3 S k_{3r}}{h_3 V_r}, \tag{9.33}$$

$$a = \frac{\gamma}{\alpha - \delta}; \quad b = \frac{\beta}{\alpha - \delta}; \quad g = \frac{\alpha\gamma}{\alpha - \delta}; \quad z = \frac{\delta\beta}{\alpha - \delta}, \tag{9.34}$$

$$\alpha = 1 + \frac{D_3 h_2}{D_2 h_3 k_{23}}; \quad \beta = -\frac{D_3 h_2 k_{3r}}{D_2 h_3};$$

$$\gamma = \frac{k_{1d}}{\dfrac{D_2 h_1}{D_1 h_2} + \dfrac{1}{k_{21}}}; \quad \delta = \frac{1}{\dfrac{D_1 h_2}{D_2 h_1 k_{21}} + 1}. \tag{9.35}$$

Interestingly, Grassi and Colombo [24] demonstrated that Equation 9.22 through Equation 9.24 can also be extended to the case of thick membranes. For this purpose, the lag time t_r, taking into account the time required to get a linear concentration profile in the two stagnant layers and the membrane, is introduced. Accordingly, it follows:

$$C_d(t) = A_1 + A_2 e^{(m_1(t-t_r))} + A_3 e^{(m_2(t-t_r))}, \qquad (9.36)$$

$$C_r(t) = B_1 + B_2 e^{(m_1(t-t_r))} + B_3 e^{(m_2(t-t_r))}. \qquad (9.37)$$

The above equations can describe the linear part of the permeation curve [24], where t_r is a fitting parameter. Model analytical solution further simplifies if $M_0 = 0$ (no dissolution phenomenon takes place in the donor environment) and stagnant layer presence can be neglected. In this case, its analytical solution reads [25]

$$C_r(t) = \frac{k_{pd} C_{d0} V_d}{k_{pd} V_r + k_{pr} V_d + S L_M k_{pr} k_{pd}} \left(1 - e^{\left(-S \frac{D}{h_m} \left(k_{pr} + k_{pd} \frac{V_r + 0.5 S h_m k_{pr}}{V_d + 0.5 S h_m k_{pd}} \right) t \right)} \right), \qquad (9.38)$$

where k_{pd} ($= k_{12}$) and k_{pr} ($= k_{23}$) are the partition coefficients on the donor and receiver side, respectively.

The above-discussed model, in its completely simplified form, has been considered to measure drug diffusion coefficient through different membranes. For example, it has been used to study monohydrated theophylline ($C_7H_8N_4$ $O_2 \cdot H_2O$; Carlo Erba, Milano) diffusion coefficient in crosslinked (Ca^{++}) calcium alginate membranes (Protanal LF 20/60, Na salt of alginic acid; Protan Biopolymer, Drammen, Norway) [26]. Permeation experiments ($T = 37°C$; $V_d = V_r = 100 \text{ cm}^3$; $S = 10.7 \text{ cm}^2$) are done by filling the donor compartment with saturated theophylline solution in the presence of undissolved theophylline ($M_0 = 10^6 \mu g$) while receiver compartment is initially filled with demineralized water. Theophylline concentration in the receiver compartment is measured by an UV spectrophotometer (271 nm, UV–Vis Spectrophotometer, Lambda 6, Perkin Elmer, USA). Finally, three different polymer concentrations (%P) (1, 2, and 4 w/w %) are considered. Once all model parameters are known ($C_s(37°C) = 12495 \pm 104 \mu g \text{ cm}^{-3}$; theophylline diffusion coefficient in water $D_{TheoW}(37°C) = (8.2 \pm 0.6) \times 10^{-6} \text{ cm}^2 \text{ sec}^{-1}$; theophylline powder dissolution constant $k_d(37°C) = 1.52 \times 10^{-3} \text{ cm sec}^{-1}$; $A_0 = 2941 \text{ cm}^2 \text{ g}^{-1}$; $\rho_{Theo} = (1.49 \pm 0.01) \text{ g cm}^{-3}$; $k_{1d} = k_{3r} = 1$, $k_{21} = k_{23} = k_p = 0.9$; $h_2 = h_m = 330 \mu m$ (%P = 1), 485 (%P = 2), 740 (%P = 4); $h_1 = h_3 = 61 \mu m$) the determination of theophylline diffusion coefficient D_{TheoM} through membranes can be performed by model fitting to experimental data. Table 9.1 shows that D_{TheoM}, in the range considered, is not substantially affected by polymer concentration. This can be explained

TABLE 9.1
Theophylline Diffusion Coefficient (D_{TheoM}; $T = 37°C$) in Different Crosslinked Polymeric Membranes

Polymer	%P (w/w)	D_{TheoM} (cm^2 sec^{-1})	References
ALF	1	4.3×10^{-6}	[26]
	2	4.4×10^{-6}	[26]
	4	4.2×10^{-6}	[26]
OS	1.6 ($r = 0.0$)	7.0×10^{-6}	[27]
	1.6 ($r = 0.5$)	4.8×10^{-6}	[27]
	1.6 ($r = 2.0$)	7.1×10^{-7}	[27]

ALF: alginate LF 20/60; OS: oxidized scleroglucan; %P indicates polymer concentration in the membrane, while r is proportional to membrane crosslink density. In the OS case, the partition coefficient is assumed equal to 1.

supposing that %P increase, in the 1%–4% range, does not substantially reflect in a reduction of network mesh size, but it reflects in higher membrane thickness. Nevertheless, we cannot exclude that, due to reduced theophylline dimension (≈ 3.8 Å [27]), polymeric network meshes are always very large even in the case of %$P = 4$. It is also interesting to notice that regardless of the model fitted to experimental data (numerical model (Equation 9.8 through Equation 9.18); Equation 9.3 with P given by Equation 9.4; Equation 9.37 with t_r and D_{TheoM} as fitting parameters), D_{TheoM} is always the same. This shows the equivalence of these approaches, provided that all other model parameters are known on the basis of independent experiments. Obviously, Equation 9.4 and Equation 9.37 must be fitted only on the permeation curve linear part.

Coviello et al. [27] study theophylline permeation through different oxidized scleroglucan membranes crosslinked by 1,6-hexanedibromide. Scleroglucan, a linear chain of 1,3 β linked D-glucopyranose units with single D-glucopyranose residues 1,6 β linked to every third unit of the chain [28], is oxidized using periodate to obtain the aldehydic derivative and then chlorite for the further oxidation to carboxylic groups [29]. Sclerox solution is then eluted through an H$^+$ exchange column (Dowex 50×8, Fluka, Switzerland) that was previously obtained in its tetrabutylammonium (TBA) form by treatment with tetrabutylammonium hydroxide (Fluka). The crosslinking reaction between sclerox (in its TBA form) and 1,6-hexanedibromide is carried out in DMSO, at 37°C, in the dark, for 24 h. Fixing polymer concentration to %$P = 1.6$ w/w, three different membranes are produced by varying the ratio (r) between the equivalents of reagent and the polymer ($r = 0.0, 0.5,$ and 2.0). Table 9.1 shows that, now, D_{TheoM} depends on crosslink density (that increases with r) as one order of magnitude differentiates the $r = 0$ from the $r = 2$ case.

9.2.2 Microencapsulated Systems

The knowledge of drug diffusion coefficient inside a membrane finds a direct application, for example, in microencapsulated delivery systems. Indeed, drug permeation plays a key role in affecting release kinetics from these particular delivery systems constituted by an inner solid core (drug) coated by an external membrane. Microencapsulation is widely used in the pharmaceutical, chemical, and agricultural field to protect various substances from environmental impact, as well as for extending their action [30]. In the pharmaceutical field, in particular, microencapsulation is widely used to control drug release kinetics, minimize side effects, reduce gastric irritations and mask the unpleasant taste of the contained drug [30–35]. In addition, it is a very versatile tool suitable for parenteral, pulmonary, oral, and nasal administration routes [36]. Another important feature of microencapsulation consists in providing sustained and controlled release of the encapsulated drug, while the nonreleased portion may be protected from degradation and physiological clearance. For vaccines, microspheres may provide additional adjuvancy [37,38] and allow for direct targeting to professional antigen-presenting cells [39]. Moreover, microparticles surface modification may allow targeted delivery to specific cells [40] and tissues [41]. By virtue of excellent biocompatibility, the biodegradable polyesters poly(lactic acid) (PLA) and poly(lactic-co-glycolic acid) (PLGA) are the most frequently used biomaterials for the microencapsulation of therapeutics and antigens [42]. Although less frequently, proteins [41], polymer blends [43], polysaccharides such as chitosan [9,44], and lipids [45] have also been considered.

Many different active compounds are microencapsulated: analgesics, antibiotics, antihistamine, cardiovascular agents, iron salts, antipsychotics [46], vitamins, peptides [47], proteins [48], antiasthma [33,49], broncodilators, diuretics, anticarcinogens, tranquilizers, antihypertensives [30,36], DNA [50], viruses [51], bacteria derived compounds [52], and microbial cells [53].

Basically, microencapsulated delivery systems can be produced by spray drying, phase separation (coacervation), and solvent extraction/evaporation [36]. Spray drying is relatively simple and of high throughput but it is not recommended for highly temperature-sensitive compounds. Additionally, control of the particle size is difficult and yields for small batches are moderate [54]. Coacervation consists of enclosing a water-soluble material within a polymeric coating by using an organic solvent where the addition of a second polymeric material induces phase separation [30]. In particular, phase separation is achieved by cooling the heated organic solution containing the dispersed drug to get polymeric coating deposition on drug particles. This technique is not suitable for the production of small microparticles (diameter <100 μm) and can show problems for what concerns residual solvents and coacervating agents in the microspheres [55]. The use of supercritical fluids as phase separating agents was intensively studied to minimize the amount of potentially harmful residues in the microparticles. Accordingly, processes such as precipitation with compressed antisolvent (PCA) [56], gas or supercritical fluid antisolvent (GAS or SAS), and aerosol solvent extraction system (ASES) [57] have been studied. Solvent extraction/evaporation neither requires elevated temperatures nor phase separation-inducing agents. It requires the co-dissolution or the dispersion or the emulsification of the drug in a solution of the coating material to get liquid phase 1 (drug/matrix dispersion). Then, liquid phase 1 is added, under stirring, to a proper solvent (liquid phase 2) to get a fine dispersion of liquid 1 into liquid 2 (this step is crucial for the determination of particles size distribution and particles mean diameter) (droplet formation). Then, liquid 2 removal follows (extraction agents or evaporation) (solvent removal). Microparticles harvest and drying completes the whole production cycle. This technique allows controlled particle sizes in the nano- to micrometer range, high encapsulation efficiencies, and low residual solvent contents provided that a careful selection of encapsulation conditions and materials is performed [36].

To achieve maximum efficiency from using microencapsulated particles, it is necessary to know the time needed to extract the solid phase out of the polymeric capsule. The extraction rate of active components through a polymeric coating depends on many factors such as the nature of the polymeric coating, the conditions under which the coating was applied, its structure, thickness, and porosity [58]. As a consequence, a proper designing requires a deep knowledge of the release mechanism and the drug-coating physical properties [59,60].

9.2.2.1 Physical Frame

The physical frame to model is represented by an ensemble of differently shaped particles each one constituted by an inner solid drug core coated by an external membrane. As soon as particles are put in contact with the release environment, the external fluid crosses the coating, progressively dissolves the drug core and fills the inner void space generated by core dissolution. Consequently, the drug present in the inner solution diffuses through the coating determining its concentration increase in the release environment. Thus, drug is distributed among outer solution (release environment), coating, inner solution, and solid core. Obviously, it is possible that coating thickness is not constant due to (1) a swelling process induced by the release environment fluid income (thickness increases); (2) an erosion process determined by release fluid hydrodynamic conditions or other reasons discussed in detail in Chapter 7, Section 7.2.2 (thickness decreases). This picture applies for non-semipermeable membrane where osmotic effects play key role in determining release kinetics [61]. A little variation in this scenario implies that the inner solid core is replaced by a drug solution.

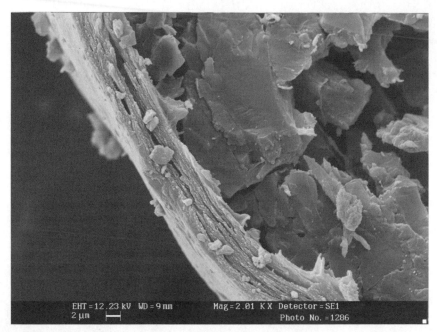

FIGURE 9.2 SEM picture of microcapsule coating (ethylcellulose, Hercules, Salford, USA). The existence of channels, shunt, and holes confers an inhomogeneous nature to the membrane. (From Sirotti, C., Colombo, I., Grassi, M., *J. Microencapsul.*, 19, 603, 2002. With permission.) http://www.tandf.co.uk/journals/titles/10799893.asp

It is more than evident that the complexity of this picture would make any reasonable attempt of mathematical modeling meaningless (at least with usual computer). This is the reason why, at least at our knowledge, all the models developed till now share the common hypotheses of spherical particles, constant coating thickness, and existence of an effective drug diffusion coefficient (D_{ec}) inside the coating [62,63]. The necessity of defining D_{ec} arises from the complexity of coating structure and topology. Indeed, while only rarely coating is represented by a homogeneous membrane, more often it is an inhomogeneous structure characterized by very tortuous irregular channels connecting the inner core with the external environment (see Figure 9.2). In this light, the release process can be reviewed and the scheme depicted in Figure 9.3 can be chosen as a useful reference to enucleate release kinetics mechanism: (1) external fluid income across the coating; (2) solid drug dissolution; (3) drug spreading into the inner solution; (4) drug partitioning between the inner solution and the coating; (5) drug diffusion through the membrane; (6) mass transfer into the release environment.

Accordingly, fundamental parameters ruling release kinetics from micro-encapsulated particles are external fluid and drug (D_{ec}) effective diffusion

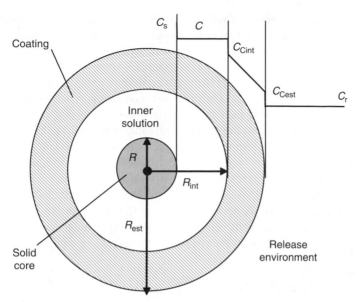

FIGURE 9.3 Schematic representation of microcapsule during drug release. (From Sirotti, C., Colombo, I., Grassi, M., *J. Microencapsul.*, 19, 603, 2002. With permission.) http://www.tandf.co.uk/journals/titles/10799893.asp

coefficient in the coating, membrane thickness (h_m), drug partition coefficients between coating/inner (k_{pd}) and coating/outer (k_{pr}) liquids (these liquids can be different), mass transfer coefficients relative to steps (3) and (6), and, finally, particle size distribution.

9.2.2.2 Modeling

Starting from the above-mentioned physical frame, different mathematical models have been proposed in the last 20 years and, among them, in a rigorous chronological order, we can mention, for example, the works of Liu et al. [34], Lu and Chen [59], Demchuk et al. [58], Lavasanifar et al. [33], Borgquist et al. [64], Sirotti et al. [65], Manca and Rovaglio [66], Ito et al. [67], Muro-Suòé et al. [68]. These models can be subdivided into two main categories: those who account for polydisperse particle size distribution and those who do not (monodisperse system). Clearly, while older models fall in the first category, more recent ones belong to the second category. Indeed, while, originally, the basic target was to understand the mechanisms determining release kinetics from microencapsulated systems, nowadays, the necessity to be as close as possible to real delivery systems obliges to consider polydispersion. Accordingly, this paragraph is aimed to discuss the delivery mechanisms' mathematical modeling first and, then, the problem of polydispersion will be matched.

If step (1) is not very fast, release kinetics is characterized by a "lag time" effect consisting of a delayed drug concentration appearance in the release environment. Indeed, drug release can occur only after the external fluid has completely crossed the coating. The practical importance of this lag time clearly arises only when step (1) kinetics is small in comparison with those of other steps. On the contrary, a burst effect, consisting in a rapid initial drug concentration increase in the release environment, takes place if the drug is already present in the coating at the beginning of the release experiment. If this aspect is, *de facto*, absolutely negligible in the presence of a solid inner core, it can be important when a liquid drug solution fills the inner core. In this case, indeed, it is probable that, at test time, the drug is completely distributed between the inner solution and the coating. Step (1) is always assumed instantaneous, as water diffusion coefficient in the coating is usually neatly higher than that of the drug [64]. Assuming homogeneous solid core, its dissolution (step 2) is accounted for by the following equation:

$$\frac{dM_D}{dt} = \frac{d}{dt}\left(\frac{4}{3}\pi R^3 \rho_D\right) = 4\pi R^2 \rho_D \frac{dR}{dt} = -4\pi R^2 k(C_s - C), \qquad (9.39)$$

where M_D and R are, respectively, the solid core mass and radius at time t, ρ_D is solid core density, k is the mass transfer coefficient, C_s and C are, respectively, drug solubility and concentration in the inner fluid. If drug concentration C is expressed as mass/volume, due to the inner solution volume increase with time (solid core volume decreases with time due to dissolution), it is better to rewrite Equation 9.39 as follows:

$$\frac{dR}{dt} = -\frac{k}{\rho_D}\left(C_s - \frac{M}{\frac{4}{3}\pi(R_{int}^3 - R^3)}\right), \qquad (9.40)$$

where R_{int} is the coating inner radius (see Figure 9.3), M and $(4/3)\pi(R_{int}^3 - R^3)$ are, respectively, the drug amount present in the inner solution and inner solution volume at time t. Manca and Rovaglio [66] and Borgquist et al. [64] suggest evaluating k according to the following correlation [63]:

$$k = Sh\frac{D_{H_2O}}{R} = \left(1 + 0.3Re^{0.5}Sc^{0.33}\right)\frac{D_{H_2O}}{R}, \qquad (9.41)$$

$$Re = \frac{v2R\rho_{H_2O}}{\mu_{H_2O}}; \quad Sc = \frac{\mu_{H_2O}}{\rho_{H_2O}D_{H_2O}}, \qquad (9.42)$$

where Sh, Re, and Sc are, respectively, Sherwood, Reynolds, and Schmidt number, μ_{H_2O} and ρ_{H_2O} are, respectively, water viscosity and density, D_{H_2O}

is drug diffusion coefficient in the inner solution (in so doing it is assumed that inner solution properties coincide with that of water), while ν is the flow velocity outside the core surface created by the agitation of inner core solution. In particular, on the basis of a rigorous demonstration, Manca and Rovaglio [66] assume that the product $Re \times Sc$ is very small such that $k = D_{H_2O}/R$. Finally, Sirotti et al. [65] assume that k is constant and equal to drug intrinsic dissolution constant k_d (see Chapter 5) (this is true when drug solubility in the inner fluid is low <50 mg cm^{-3}). It is evident that Equation 9.40 disappears as soon as R becomes zero.

As drug spreading inside the inner solution is always assumed to be instantaneous (step 3), drug concentration in the inner solution can be assumed coinciding with drug solubility C_s until the solid core exists. Alternatively, the following mass balance is usually adopted [64–66]:

$$\frac{dM}{dt} = 4\pi R^2 k \left(C_s - \frac{M}{\frac{4}{3}\pi(R_{int}^3 - R^3)} \right) - k_T 4\pi R_{est}^2 (C_{Cint} - C_{Cest}), \qquad (9.43)$$

$$C_{Cint} = k_{pd}C = k_{pd}\frac{M}{\frac{4}{3}\pi(R_{int}^3 - R^3)}; \quad k_{pr} = \frac{C_{Cest}}{C_r}, \qquad (9.43')$$

where k_T represents the overall mass transport coefficient accounting for mass transfer resistance due to the inner solution/coating interface (k_{ic}), coating thickness (k_c), and coating/external solution interface (k_{ce}),

$$\left(\frac{1}{k_T} = \frac{1}{k_{ic}} + \frac{1}{k_c} + \frac{1}{k_{ce}} \right),$$

R_{est} is particle external radius, C_{Cint}, C_{Cest}, and k_{pd}, k_{pr} are, respectively, drug concentration in the coating and partition coefficients at R_{int} and R_{est}, while C_r is drug concentration in the release environment. Muro-Suòé et al. [69] propose an interesting approach for k_{pd} and k_{pr} determination based on activity coefficients that are, in turn, estimated with a group-contribution model for polymers (step 4). Equation 9.43 states that the drug amount M increases due to solid core dissolution (first right-hand side term), whereas it decreases because of matter flux across the coating (second right-hand side term). As, in general, coating thickness is small, R_{est} could be substituted by R_{int} in the second right-hand side of Equation 9.43. Alternatively, as suggested by Sirotti et al. [65], the term $4\pi R_{est}^2(C_{Cint} - C_{Cest})$ can be replaced by $4\pi(R_{int}^2 C_{Cint} - R_{est}^2 C_{Cest})$. While coating resistance to diffusion (k_c) is set equal to $k_c = D_{ec}/(R_{est} - R_{int})$ and k_{ic} is evaluated by Equation 9.41, different suggestions are given for the estimation of k_{ce}. Indeed, while Borgquist et al. [64] again suggest Equation 9.41, Manca and Rovaglio [66] suggest the Armenante equation [70]

$$k_{ce} = \frac{D_{H_2O}}{2R_{est}}\left[2 + 0.52\left(\frac{P_s^{1/3}D_{imp}^{4/3}}{\nu_{H_2O}}\right)\left(\frac{\nu_{H_2O}}{D_{H_2O}}\right)^{1/3}\right], \qquad (9.44)$$

where ν_{H_2O} is water kinematic viscosity, P_s is the specific power dissipated into the release medium by impeller blade (power/mass) while D_{imp} is impeller diameter. It is evident that the first right-hand side term of Equation 9.43 disappears when R becomes zero.

A theoretically more correct expression of the drug flux leaving the inner solution (second right-hand side term; step 5) should be given by

$$D_c\frac{\partial C_C}{\partial R}\bigg|_{R=R_{int}},$$

where C_C is drug concentration evaluated in R ($R_{int} < R < R_{est}$). This implies solving Fick's second equation inside coating thickness

$$\frac{\partial C_C}{\partial t} = \frac{1}{R^2}\frac{\partial}{\partial R}\left(D_{ec}\frac{\partial C_C}{\partial R}R^2\right). \qquad (9.45)$$

However, Borgquist et al. [64,71] and Ito et al. [67] clearly demonstrate that this approach is unnecessary as a linear drug concentration profile rapidly arises and holds inside coating thickness.

Finally, assuming that the volume of the release environment fluid going inside the particle is negligible, the release environment drug concentration C_r (step 6) can be determined according to the following equation [64,66,67,71]:

$$\frac{dC_r}{dt} = -\frac{4\pi R_{est}^2}{V_r}D_{ec}\frac{\partial C_C}{\partial R}\bigg|_{R=R_{est}} = \frac{4\pi R_{est}^2}{V_r}k_{ce}(C_{Cest} - C_r); \; C_r = \frac{C_{Cest}}{k_{pr}}. \qquad (9.46)$$

If $k_{pr} = 1$ ($k_{ce} \longrightarrow \infty$, no external mass transfer resistance), Equation 9.46 becomes

$$\frac{dC_r}{dt} = -\frac{4\pi R_{est}^2}{V_r}D_{ec}\frac{\partial C_C}{\partial R}\bigg|_{R=R_{est}} = \frac{4\pi R_{est}^2}{V_r}\frac{D_{ec}}{R_{est} - R_{int}}(C_{Cint} - C_r). \qquad (9.46')$$

This equation simply states that the speed of C_r increase is determined by the drug flow-crossing particle coating surface. Alternatively, to improve the accuracy of model numerical solution (this is particularly important in the case of polydisperse particles ensemble), Equation 9.46 can be replaced by the following drug balance extended to the whole system (release volume, coating, inner solution, and solid core) [65]:

$$V_r C_r + \int_{R_{int}}^{R_{est}} C_C(R) 4\pi R^2 \, dR + CV_{int} + \frac{4}{3}\pi R^3 \rho_D = M_0, \qquad (9.47)$$

where V_{int} is the inner solution volume while M_0 is the initial solid core mass (if a linear C_C trend is assumed, the integral appearing in Equation 9.47 yields an analytical solution). Equation 9.47 allows expressing the C_r dependence on C.

Before extending the above-mentioned model to polydisperse particles ensemble, it is worth discussing the approach proposed by Muro-Suòé et al. [68] applying to a liquid drug solution filling the inner space. Indeed, this approach, with little modifications, turns out to be useful also to model release kinetics in the presence of an inner solid core as discussed in the following. Assuming that C_0 is the initial drug concentration in the inner solution, its time reduction can be described according to the usual kinetics equation:

$$\frac{dC}{dt} = -\frac{k_T 4\pi R_{est}^2}{V_{int}}(C_{Cint} - C_{Cest}). \qquad (9.48)$$

Assuming that the drug amount contained in the coating is negligible, the following mass balance holds at anytime:

$$V_r C_r + CV_{int} = C_0 V_{int}. \qquad (9.49)$$

Expressing C_r as function of C by means of this last equation, and remembering the partitioning conditions

$$\left(k_{pd} = \frac{C_{Cint}}{C}; \; k_{pr} = \frac{C_{Cest}}{C_r} \right),$$

it is possible to get Equation 9.48's analytical solution:

$$C(t) = C_0 \left(\frac{k_{pr}V_{int}}{k_{pd}V_r + k_{pr}V_{int}} + \frac{k_{pd}V_r}{k_{pd}V_r + k_{pr}V_{int}} e^{-\frac{k_T 4\pi R_{est}^2 k_{pd}}{V_r}\left(\frac{V_r}{V_{int}} + \frac{k_{pr}}{k_{pd}}\right)t} \right). \qquad (9.50)$$

Assuming that V_{int} is negligible in comparison to V_r (this is usually more than reasonable), Equation 9.50 becomes

$$C(t) = C_0 e^{-\frac{k_T 4\pi R_{est}^2 k_{pd}}{V_r}\left(\frac{V_r}{V_{int}} + \frac{k_{pr}}{k_{pd}}\right)t}. \qquad (9.50')$$

On the basis of this last equation and assuming sink conditions ($C_r \approx 0$), Muro-Suòé et al. [68] could derive C_r time variation speed:

$$\frac{dC_r}{dt} = \frac{k_T 4\pi R_{est}^2}{V_r}(C(t) - 0). \quad (9.51)$$

In the case of solid core presence in the inner space, drug flow from micro-particle can be approximately given by

$$\frac{dC_r}{dt} = \frac{k_T 4\pi R_{est}^2}{V_r} k_{pd} C_s. \quad (9.52)$$

This equation holds until solid core radius $R > 0$. Indeed, only in this condition inner solution drug concentration can be approximated by drug solubility C_s. Obviously, in this case, V_{int} is not a constant as solid core shrinks. However, it can be assumed to be constantly equal to particle inner cavity without making a substantial error. After solid core disappears, Equation 9.52 must be replaced by the following one:

$$\frac{dC_r}{dt} = \frac{k_T 4\pi R_{est}^2}{V_r} k_{pd} C(t) = \frac{k_T 4\pi R_{est}^2}{V_r} k_{pd} C_s e^{-\frac{k_T 4\pi R_{est}^2 k_{pd}}{V_r}\left(\frac{V_r}{V_{int}} + \frac{k_{pr}}{k_{pd}}\right)(t - t_{dis})} \quad (9.53)$$

that immediately descends from the Muro-Suòé et al. [68] approach. Accordingly, Equation 9.52 and Equation 9.53 can be merged to give

$$\frac{dC_r}{dt} = \frac{k_T 4\pi R_{est}^2}{V_r} k_{pd} C_s \left(I + (1 - I)e^{-\frac{k_T 4\pi R_{est}^2 k_{pd}}{V_r}\left(\frac{V_r}{V_{int}} + \frac{k_{pr}}{k_{pd}}\right)(t - t_{dis})} \right), \quad (9.54)$$

where $I = 1$ until $R > 0$, $I = 0$ for $R \geq 0$, although t_{dis} represents solid core disappearing time. A rough estimation of t_{dis} can be given by Equation 9.40 assuming zero inner solution concentration (this is not a reasonable hypothesis but it is the only one allowing t_{dis} estimation by this simplified view). In this hypothesis, R reduction follows a linear trend:

$$R = R_0 - \frac{kC_s}{\rho_D}t. \quad (9.55)$$

Accordingly, R zeroing will take place for $t_{dis} = R_0 \rho_D / kC_s$.

Now, the theoretical frame presented (Equation 9.39 through Equation 9.55) can be extended to polydisperse particles ensemble. The key point is determining drug flow contribution pertaining to each particles class (j) into which the continuous particle size distribution can be ideally subdivided (N_c particles class). Accordingly, the first issue to match is the determination of the particles number N_j constituting class j. This can be done remembering that the differential particle size distribution $f(R_j)$ expresses the fraction of the whole particles volume (V_0) competing to particles characterized by mean radius R_j and constituting class j [64–66]. Consequently, it follows:

$$N_j = \frac{f(R_j)V_0}{\frac{4}{3}\pi R_j^3}. \tag{9.56}$$

It is evident that R_j is the mean radius calculated on the basis of particles class j minimum ($R_{j\min}$) and maximum ($R_{j\max}$) radius.

On this basis, the release model referring to an ensemble of polydisperse particles reads

$$\frac{dR_j}{dt} = -\frac{k_j}{\rho_D}\left(C_s - \frac{M_j}{\frac{4}{3}\pi(R_{\text{int }j}^3 - R_j^3)}\right), \tag{9.57}$$

$$\frac{dM_j}{dt} = 4\pi R_j^2 k_j\left(C_s - \frac{M_j}{\frac{4}{3}\pi(R_{\text{int }j}^3 - R_j^3)}\right) - k_{Tj}4\pi R_{\text{est }j}^2(C_{\text{Cint }j} - C_{\text{Cest }j}), \tag{9.58}$$

$$C_{\text{Cint }j} = k_{\text{pd}}C_j = k_{\text{pd}}\frac{M_j}{\frac{4}{3}\pi(R_{\text{int }j}^3 - R_j^3)}; \quad k_{\text{pr}} = \frac{C_{\text{Cest }j}}{C_r}, \tag{9.58'}$$

$$V_r C_r + \sum_{j=1}^{N_c} N_j\left[\int_{R_{\text{int }j}}^{R_{\text{est }j}} C_{Cj}(R)4\pi R^2 dR + C_j\frac{4}{3}\pi(R_{\text{int }j}^3 - R_j^3) + \frac{4}{3}\pi R_j^3\rho_D\right] = M_0, \tag{9.59}$$

where M_j, R_j, $R_{\text{int }j}$, $R_{\text{est }j}$, $C_{\text{Cint }j}$, $C_{\text{Cest }j}$, C_{Cj}, k_j, and k_{Tj} are, respectively, drug mass inside the inner solution, solid core radius, internal and external coating radius, drug concentration at the inner solution/coating interface, drug concentration at the coating/external release environment interface, drug concentration inside the coating and transfer mass coefficients referring to the generic particles class j. Obviously, k_j is evaluated according to Equation 9.41 and Equation 9.42, where R is replaced by R_j while k_{Tj} is evaluated on the basis of $k_{\text{ic}j}$, $k_{\text{c}j}$, and $k_{\text{ce}j}$ also R, R_{int}, R_{est} are replaced by R_j, $R_{\text{int }j}$, $R_{\text{est }j}$ in Equation 9.41 or Equation 9.44. Equation 9.57 through Equation 9.59 allow to calculate the time variation of the ($4 \times N_c + 1$) unknowns (R_j, M_j, $C_{\text{Cint }j}$, $C_{\text{Cest }j} + C_r$) (we have N_c Equation 9.57, N_c Equation 9.58, $2 \times N_c$ Equation 9.58' plus the whole mass balance Equation 9.59). It is evident that a numerical method is required to solve these equations (for example, a fifth-order adaptive stepsize Runge-Kutta method [72]) and that the greater N_c, the longer the computing time.

Interestingly, the approach of Muro-Suòé et al. [68] (Equation 9.54) can be easily generalized to the polydisperse case. Indeed, Equation 9.54 becomes

$$\frac{dC_r}{dt} = \frac{4\pi k_{pd}C_s}{V_r} \sum_{j=1}^{N_c} k_{Tj} R_{est\,j}^2 \left(I_j + (1 - I_j)e^{-\frac{k_{Tj}4\pi R_{est\,j}^2 k_{pd}}{V_r}\left(\frac{V_r}{V_{int\,j}} + \frac{k_{pr}}{k_{pd}}\right)(t - t_{dis\,j})} \right), \quad (9.54')$$

where $V_{int\,j} = (4/3)\pi R_{int\,j}^3$, $I_j = 1$ until $R_j > 0$, $I_j = 0$ for $R_j \geq 0$, while $t_{dis\,j}$ represents solid core disappearing time and, as before, it can be roughly estimated by means of the following equation:

$$R_j = R_{0j} - \frac{k_j C_s}{\rho_D} t; \quad t_{dis\,j} = \frac{R_{0j}\rho_D}{k_j C_s}. \quad (9.55')$$

Undoubtedly, Equation 9.54', although approximated, allows to reduce a lot of computational heaviness.

It is now interesting to discuss model application (Equation 9.57 through Equation 9.59) to a real situation. For example, Sirotti et al. [65] apply it to study release kinetics from coacervation produced microcapsules (ethylcellulose, Hercules, Salford, USA) containing theophylline (Carlo Erba, Milano, Italy). Figure 9.4 shows a cross-section of a typical microcapsule. They assume that $k_j = k_d \, \forall \, j$, $k_{Tj} = D_{ec}/(R_{est\,j} - R_{int\,j})$ and $k_{pd} = k_{pr} = 1$. Accordingly, model fitting parameters are k_d and D_{ec}. Microparticles size distribution is determined by means of a static laser light diffraction technique (Coulter LS200, Instrumentation Laboratory, Italy) and analyzing the scattering data with a Fraunhofer method. On the basis of both this distribution and microparticles drug content T_r (0.925 g/g), it is possible to calculate the internal radius (and thus, polymeric coating thickness) size distribution. Indeed, remembering that T_r is defined as the ratio between the drug mass and the total microparticles mass, we have

$$T_r = \frac{\frac{4}{3}\pi R_{int\,j}^3 \rho_D}{\frac{4}{3}\pi R_{int\,j}^3 \rho_d + \frac{4}{3}\pi [R_{est\,j}^3 - R_{int\,j}^3]\rho_C}, \quad (9.60)$$

where ρ_D (= 1.49 g cm^{-3}) and ρ_C (= 1.13 g cm^{-3}) are drug and coating densities, respectively (helium picnometry (micrometrics multivolume pycnometer 1305)). Accordingly, the internal microparticle radius of class j is given by

$$R_{int\,j} = R_{est\,j} \sqrt[3]{\frac{\rho_C T_r}{\rho_D(1 - T_r) + \rho_C T_r}}. \quad (9.61)$$

Table 9.2 and Figure 9.4 report, respectively, coating thickness for each class and particle size distribution ($V\%$). Release experiments are performed by dispersing 280 mg of microcapsules in water (37°C) (receiver volume $V_r = 500$ cm^3). Every 15 min, for 4 h, 2 cm^3 are sampled from the receiver phase and they are

TABLE 9.2
Particle Size Distribution and Coating Thickness ΔR,
R_{est}, and R_{int} Represent, Respectively, Particles
External and Internal Radius

Class	ΔR (μm)	R_{int} (μm)	R_{est} (μm)
1	1.8	65.1	66.9
2	2.0	71.4	73.4
3	2.1	78.5	80.6
4	2.4	86.0	88.4
5	2.6	94.5	97.1
6	2.8	103.8	106.6
7	3.1	113.9	117.1
8	3.4	125.0	128.4
9	3.8	137.3	141.1
10	4.1	150.7	154.8
11	4.5	165.4	169.9
12	5.0	181.6	186.6
13	5.5	199.3	204.8
14	6.0	218.9	224.9
15	6.6	240.2	246.8
16	7.2	263.7	271.0
17	7.9	289.5	297.5
18	8.7	317.8	326.5
19	9.6	348.9	358.5
20	10.5	383.0	393.5
21	11.5	420.4	432.0
22	12.6	461.5	474.1

Source: From Sirotti, C., Colombo, I., Grassi, M., *J. Microencapsul.*, 19, 603, 2002. With permission. http://www.tandf.co.uk/journals/titles/10799893.asp

immediately replaced by an equal amount of water. Consequently, the concentration data collected are properly corrected for dilution. Drug concentration in each sample is measured by means of an UV spectrophotometer (Lambda 6/PECSS System, Perkin-Elmer Corp., Norwalk, CT, wavelength 271 nm). Figure 9.5 clearly shows a very good agreement between experimental data (open circles) and model best fitting (solid line). Fitting parameters are $k_d = 2.7 \times 10^{-4}$ cm sec^{-1} and $D_{ec} = 2 \times 10^{-8}$ cm^2 sec^{-1}. On the basis of these D_{ec} values and coating thicknesses (see Table 9.2), microparticles permeability ($= D_{ec}/h_m$) spans between 2×10^{-5} and 2×10^{-4} cm sec^{-1}. This means that theophylline release is mainly ruled by diffusion through the coating as k_d is 2–10 times higher than permeability. Interestingly, the reliability of this approach has also been proved in other experimental situations [64,66–68].

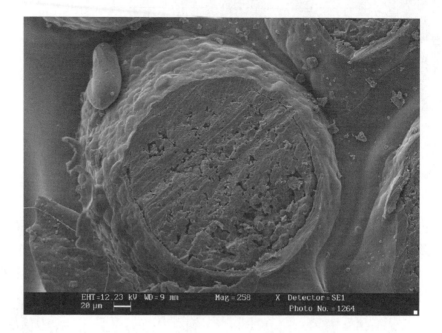

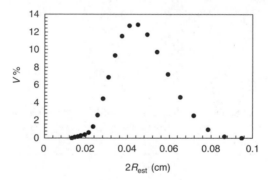

FIGURE 9.4 Cross-section and size distribution of microencapsules made up of a theophylline solid core coated by an ethylcellulose (Hercules, Salford, USA) membrane. (From Sirotti, C., Colombo, I., Grassi, M., *J. Microencapsul.*, 19, 603, 2002. With permission.) http://www.tandf.co.uk/journals/titles/10799893.asp

Figure 9.6 and Figure 9.7 report the effect of different particle size distributions on drug release kinetics assuming $k_d = 2.7 \times 10^{-4}$ cm sec^{-1}, $D_{ec} = 2 \times 10^{-8}$ cm^2 sec^{-1}, and the other conditions pertaining to Figure 9.5. In particular, Figure 9.6 shows model prediction supposing to shift upward (bigger particles) and downward (smaller particles) the experimental distribution depicted in Figure 9.4. The entity of this shifting is represented by

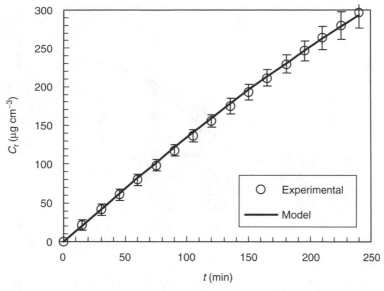

FIGURE 9.5 Comparison between experimental data (circles, experimental theophyl-line concentration, vertical bars indicate datum standard error, and model best fitting, solid line). (From Sirotti, C., Colombo, I., Grassi, M., *J. Microencapsul.*, 19, 603, 2002. With permission.) http://www.tandf.co.uk/journals/titles/10799893.asp

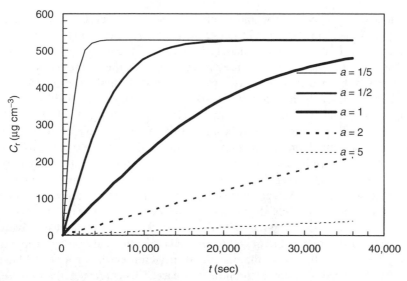

FIGURE 9.6 Model prediction about the effect of reference particle size distribution (see Figure 9.4) shifting toward higher (shifting factor $a > 1$) and lower (shifting factor $a < 1$) diameters. (From Sirotti, C., Colombo, I., Grassi, M., *J. Microencapsul.*, 19, 603, 2002. With permission.) http://www.tandf.co.uk/journals/titles/10799893.asp

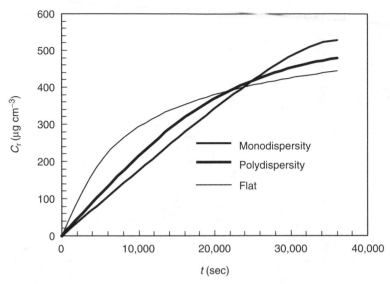

FIGURE 9.7 Model prediction about the effect of different particle size distributions (thickest line refers to the original size distribution depicted in Figure 9.4). (From Sirotti, C., Colombo, I., Grassi, M., *J. Microencapsul.*, 19, 603, 2002. With permission.) http://www.tandf.co.uk/journals/titles/10799893.asp

parameter a indicating the ratio between the shifted class radius and its original value. This figure evidences how the bigger the average distribution radius ($a = 1$ corresponds to model best fitting of Figure 9.5), the lower the release kinetics, as a direct consequence of the smaller total surface area. Figure 9.7 shows different release profiles in case of monodispersed ($N_c = 1$ and $R_{est1} = 450$ μm), polydispersed (the experimental distribution of Figure 9.4), and flat (134 μm $< R_{est\ j} < 948$ μm and $f(R_{est\ j}) \times 100 = 4.5\%$ for $j = 1$ to 22) distributions. It is interesting to notice that flat distribution yields to an initial improved release kinetics (thinnest line), monodisperse distribution determines the slowest release kinetics (thickest line) while polydisperse case lies between the former two situations. This is connected with the specific surface area S_A (powder surface/powder volume) competing to the three different cases considered. Indeed, we have $S_A = 199.6$ cm^{-1}, 147.3 cm^{-1}, 133.4 cm^{-1}, for the flat, polydisperse, and monodisperse, respectively. A further inspection of Figure 9.7 reveals that the three curves meet approximately after 25,000 sec so that the flat distribution ensures a sort of "burst effect" followed by a slower, prolonged release. This characteristic decreases in the polydisperse case and it completely disappears in the monodisperse case. Therefore, a possible way of designing the microencapsulated system by selecting the best distribution in relation to the therapeutic aim arises.

9.3 BIOLOGICAL MEMBRANES

Among the existing different biological membranes, for their importance in drug administration, the attention of this paragraph will be devoted to intestinal mucosa and skin. While the reader can refer to Chapter 2 for what concerns anatomy and physiology of these biological membranes, this paragraph deals with the mathematical modeling of drug transport through them. Indeed, main target is to provide proper theoretical tools to determine their permeability from laboratory experimental tests.

9.3.1 INTESTINAL MUCOSA

As discussed in the introduction of Chapter 5, drug solubility in aqueous media and drug permeability through intestinal mucosa are key parameters for what concerns bioavailability of orally administered drugs. In particular, Amidon's biopharmaceutical classification [73] subdivides drugs on the basis of their solubility and permeability so that four classes arise (low and high permeability/solubility). Obviously, high-solubility and high-permeability drugs constitute the desirable class. While various technologies exist to improve drug solubility [74], nothing similar exists for what concerns intestinal permeability enhancement, at least when drug transport across the mucosa occurs according to passive diffusion (that is induced by the presence of a chemical potential gradient across the membrane). This aspect assumes an important relevance if we remember that although drug transport through intestinal mucosa can take place according to different and complex mechanisms (see Chapter 2), it has been demonstrated [75] that, usually, passive diffusion is the most important one. In particular, both in the rat and the human being, while lipophilic drugs follow a transcellular pathway, hydrophilic ones can undertake both paracellular (passage through the water filled tight junctions) and transcellular pathways [76]. However, hydrophilic drugs paracellular crossing is thought to be relevant only for small molecules (molecular weight <200) [77].

This frame is induced to develop several approaches for the evaluation of permeability from both an experimental and theoretical viewpoint and Figure 9.8 tries to summarize them [78].

9.3.1.1 *In Silico* Methods

The so-called *in silico* methods [78] predict intestinal permeability on the basis of drug characteristics or descriptors, such as lipophilicity, H-bonding capacity, molecular size, polar surface area, and quantum properties. One of the most famous is the *Rule of five* by Lipinski et al. [79]. The model is based on an analysis of 2287 compounds possessing United States Adopted Name (USAN) or International Nonproprietary Name (INN) designations. USAN/INN designations are usually acquired before Phase II clinical trials,

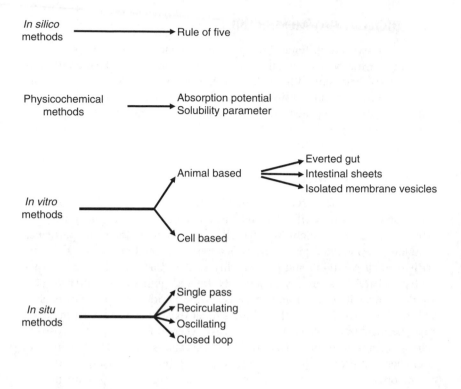

FIGURE 9.8 Experimental and theoretical techniques used for the evaluation of intestinal mucosa permeability.

meaning that these compounds had probably completed Phase I clinical trials evaluating safe dosage levels. This approach predicts low permeability or solubility when the following cut-offs are exceeded:

1. There are more than five H-bond donors (expressed as the sum of OHs and NHs).
2. The molecular weight MW is over 500.
3. The log P is over 5 (P is the drug 1-octanol/water partition coefficient).
4. There are more than 10 H-bond acceptors (expressed as the sum of Ns and Os).
5. Compound classes that are substrates for biological transporters are exceptions to the rule.

The Rule of five has the advantages of being simple, easy to interpret, and fast to compute even if, as other similar approaches [80], it does not take into account the interactions between drug properties (descriptors). Accordingly, it can yield to erroneous predictions [81] and that is why other approaches have been performed [82,83].

9.3.1.2 Physicochemical Methods

Dressman et al. [84] propose to incorporate various basic drug physicochemical properties into a unique parameter, the absorption potential (AP), defined by

$$AP = \log\left(\frac{P F_{un}}{D_o}\right); \quad D_o = \frac{X_o}{C_s V_L}, \quad (9.62)$$

where P is the drug 1-octanol/water partition coefficient, F_{un} is the fraction of drug not ionized at pH 6.5, C_s is the aqueous solubility of not ionized species at 37°C, V_L is the volume of water taken with the dose (it is usually set = 250 mL), X_o is the drug dose, and D_o is dose number. The incorporation of F_{un} into AP implies the acceptance of the pH-partition hypothesis [85] according to which ionizable compounds diffuse through biological membranes mainly in their nonionic forms [86]. It is evident that AP accounts for both drug solubility and permeability and that is why it is an estimation of drug bioavailability. However, if the drug does not show solubility problems, AP can be considered an estimation of permeability (it accounts only for passive diffusion transport mechanism). A relatively good correlation was demonstrated between AP and fraction absorbed in human subjects. Macheras and Symillides [87] established a quantitative relation between AP and the dose fraction F_a absorbed

$$F_a = \frac{(10^{AP})^2}{(10^{AP})^2 + F_{un}(1 - F_{un})}. \quad (9.63)$$

This equation undergoes the following conditions: $AP = 10^3$ if $AP > 10^3$ and $D_o = 1$ if $D_o > 1$.

Another simple and interesting physicochemical approach aimed to estimate gastrointestinal absorption site is that based on the Hildebrand's solubility parameter δ [88,89] and developed by Breitkreutz [90]. While, originally, δ was defined only for nonpolar substances

$$\delta = \sqrt{\frac{E_{coh}}{V_M}}, \quad (9.64)$$

where E_{coh} and V_M are, respectively, the cohesive energy (J mol^{-1}) and the molar volume (cm^3 mol^{-1}), Hansen [91] extended the concept of solubility parameters also to polar systems. Accordingly, three partial solubility parameters were defined: δ_d, accounting for intermolecular dispersion or Van der Waals' forces; δ_p, accounting for intermolecular polar forces; and δ_h, accounting for intermolecular hydrogen bonding. Consequently, the total solubility parameter δ_{tot} was defined as

$$\delta_{tot} = \sqrt{\delta_d^2 + \delta_p^2 + \delta_h^2}. \tag{9.65}$$

The partial solubility parameters describe the ability of a molecule to interact with another one of the same or a different type via intermolecular forces. Assuming the concept of group contribution (each molecule is constituted by different physicochemical groups) for molecule molar volume [92] and molecular forces [93], δ_d, δ_p, and δ_h can be calculated as follows:

$$\delta_d = \frac{\sum_i F_{d_i}}{\sum_i V_i}, \tag{9.66}$$

$$\delta_p = \frac{\sqrt{\sum_i F_{p_i}^2}}{\sum_i V_i}, \tag{9.67}$$

$$\delta_h = \frac{\sqrt{\sum_i E_{h_i}}}{\sqrt{\sum_i V_i}}, \tag{9.68}$$

where i is the group within the molecule; F_d is the group contribution to dispersion forces; F_p is the group contribution to polar forces; E_h is the group contribution to hydrogen bond energy; and V_i is the group contribution to molar volume. Breitkreutz developed a program (SPWin, version 2.1) for the calculation of δ_d, δ_p, and δ_h (the program is available upon request to the author: breitkr@uni-muenster.de) based on Fedors [92] and Van Krevelen and Hoftyzer [93] group-contribution values. Introducing the combined solubility parameters δ_v and δ_a:

$$\delta_v = \sqrt{\delta_d^2 + \delta_p^2}, \tag{9.69}$$

$$\delta_a = \sqrt{\delta_d^2 + \delta_h^2}, \tag{9.70}$$

Bagley et al. [94] made possible the projection of δ_d, δ_p, and δ_h into a two-dimensional space called Bagley diagram. Interestingly, a study done on several drugs reveals that drugs belonging to the region delimited by $\delta_v = 20 \pm 2.5$ $(\text{J cm}^{-3})^{0.5}$ and $\delta_h = 11 \pm 3$ $(\text{J cm}^{-3})^{0.5}$ are absorbed along the whole gastrointestinal tract. Conversely, drugs characterized by $\delta_h > 17$ $(\text{J cm}^{-3})^{0.5}$ are absorbed in the upper part of intestine rather than the cecum and colon.

9.3.1.3 *In Vitro* Methods

While *in vitro* techniques represent a simpler approach compared to *in vivo* studies, they do not account for the effect of physiological factors such as gastric emptying rate, gastrointestinal transit rate, gastrointestinal pH, etc. Among the various *in vitro* strategies that can be considered, we can mention (1) *animal tissue* and (2) *cell-based* methods [78].

Animal tissue methods, in turn, comprehend the *everted gut* technique [95], *intestinal sheets* [96], and *isolated membrane vesicles* [97]. Everted gut technique is aimed to determine intestinal wall permeability on the basis of drug transport across the membrane from the donor drug solution to the receiver environment. This technique implies fixing the intestinal tract on the left side of a "U" glass capillary connected to a cylindrical glass vessel. In this disposition, intestinal mucosa faces the donor environment while the serosal side faces the receiver environment (this is exactly the opposite of the anatomic configuration) (see Figure 9.9). Oxygen suppliers ensure tissue viability. The advantage of this configuration consists in a wide donor volume and a small receiver volume so that drugs accumulate faster. This model is ideal for studying the absorption mechanism of drugs since both the passive and active transport can be studied. The drawbacks of this technique are the lack of active blood and nerve supply that can lead to a rapid loss of viability and tissue damage due to intestine everting. Intestinal sheets technique involves the isolation of the intestinal tissues, cutting it into strips of appropriate size and clamping it on a suitable device so that the donor and receiver environments are separated by a flat intestinal membrane. Tissue permeability is evaluated resorting to drug concentration increase in the receiver environment (that is in contact with the intestine serosal side). This setup, also known as Ussing chamber [96], allows the measurement of membrane electric resistance and short-circuit current during experiment. Both of them are routinely used as an indicator of the viability of the intestinal tissue during the transport study. Lack of blood and nerve supply, rapid loss of tissues viability, changes in morphology and functionality represent main drawbacks of this approach. Isolated membrane vesicles were originally used in transport studies by Hopfer et al. [97] and only later brush border membrane vesicles have been isolated from numerous species (including humans) and widely used in transport studies. Membrane vesicles can be prepared from either intestinal scrapings or isolated enterocytes according to the methodologies reported by Hillgren et al. [98]. When prepared from intestinal tissues, they

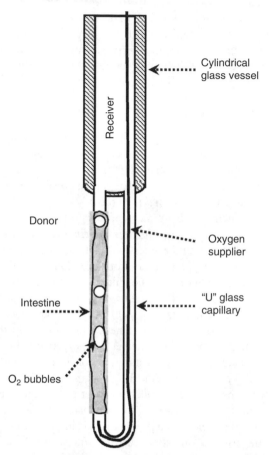

FIGURE 9.9 Schematic representation of the intestine holder used in the everted gut techniques.

allow the examination of drugs interaction with a specific membrane (brush border membrane vs. basolateral membrane of enterocytes, for example). Additionally, as they allow studying drug and nutrient transport at the cellular level (brush border as well as basolateral side), they are ideal for mechanistic absorption studies and for the isolation and identification of transporter proteins (specifically expressed either on the brush border or the basolateral side of the membranes). One of the most important advantages of vesicles over the everted gut and intestinal sheets approaches is the very small amount of drug required. In addition, the possibility of vesicles cryopreservation permits storing them for a long time. Conversely, it is practically impossible to isolate pure brush border membrane vesicles (or basolateral vesicles) without the contamination with the other type of vesicles. Finally, vesicles isolation process often implies transporter proteins and enzymes getting damaging.

Cell-based methods make use of different kinds of cell monolayer models mimicking human intestinal epithelium. Contrary to enterocytes, human immortalized (tumor) cells grow rapidly, organize into monolayers and, spontaneously, differentiate providing an ideal system for transport studies. Just to cite some of the most commonly used cell models we can mention Caco-2 (human/colon; epithelial), HT-29 (human/colon; epithelial), T-84 (human/colon; epithelial), MDCK (canine/kidney; epithelial), and LLC-PK1 (porcine/kidney; epithelial) [78]. Caco-2, undoubtedly the most extensively characterized and useful cell model in the field of drug permeability and absorption [99], is human colon adenocarcinoma undergoing spontaneous enterocytic differentiation leading to a monolayer (on a semi-permeable porous filter) where cell polarity and tight junctions are well established. While Caco-2 cells are an ideal tool to rank drugs based on the AP so that well-absorbed drugs are easily distinguished from poorly absorbed ones, they cannot be used to estimate absolute permeability values, especially when drugs mainly follow the paracellular route. Indeed, cell monolayer characteristics are extremely variable [100]. In addition, as they do not produce mucus, permeability evaluation is altered by the absence of the diffusive resistance offered by mucus [101]. Finally, it is worth mentioning that Caco-2 cell model accounts only for passive drug transport (both transcellular and paracellular) as they underexpress pharmaceutically important transporters such as peptide transporters, OCT, and OAT.

9.3.1.4 *In Situ* Methods

These methods involve perfusion of drug solution, prepared in physiological buffer, through isolated cannulated intestinal segments that are in anatomical position. Absorption is evaluated on the basis of drug disappearance from the intestinal lumen. Typically, intestinal segments of rodents (rats or rabbits) are used even if Lennernas [102] extended the perfusion studies to humans also. The most important advantage of *in situ* methods compared to the *in vitro* techniques consists in intact blood and nerve supply. Accordingly, this methodology is highly accurate for predicting the permeability of passively transported compounds, while the use of a scaling factor is recommended for predicting permeability of carrier-mediated compounds [102]. Different variants of the perfusion techniques exist [103]: single pass perfusion [104,105], recirculating perfusion [106], oscillating perfusion [107], and the closed loop method [108].

9.3.1.5 *In Vivo* Methods

These methods are aimed to evaluate drug absorption (and not only intestinal permeability) on the basis of drug plasma concentration following oral

TABLE 9.3
Human Intestinal Permeability P_H (cm sec^{-1}) Referring to Different Compounds

No	Compound	MW	pKa	P_H (cm sec^{-1})	References
1	Amiloride	230	5.26 (b)	1.6×10^{-4}	[112]
2	Antipyrine	188	—	5.3×10^{-4}	[113]
3	Atenolol	266	10.08 (b)	5.0×10^{-5}	[73]
4	Carbamazepine	236	—	4.3×10^{-4}	[112]
5	Cimetidine	241	6.71 (b)	3.0×10^{-5}	[112]
6	Creatinine	113	—	3.0×10^{-5}	[112]
7	Digoxin	781	—	1.0×10^{-3}	[114]
8	Desipramine	266	10.63 (b)	4.4×10^{-4}	[112]
9	Fluvastatine	411	4.32 (a)	2.4×10^{-4}	[112]
10	Furosemide	331	4.06 (a)	5.0×10^{-6}	[112]
11	Griseofulvin	353	—	1.6×10^{-3}	[114]
12	Hydrochlorothiazide	298	—	4.0×10^{-6}	[112]
13	Ketoprofen	254	3.49 (a)	8.3×10^{-4}	[112]
14	Metoprolol	267	10.08 (b)	2.0×10^{-4}	[73]
15	Naproxen	230	4.06 (a)	8.3×10^{-4}	[73]
16	PEG 400	400	—	8.0×10^{-5}	[115]
17	Piroxicam	331	4.66 (b)	1.6×10^{-4}	[115]
18	Propanolol	259	10.08 (b)	3.5×10^{-4}	[115]
19	Ranitidine	314	9.04 (b)	2.7×10^{-5}	[112]
20	Terbutaline	225	12.01 (b)	3.0×10^{-5}	[112]

Note: a and b indicate, respectively, acid or base; MW, molecular weight.

administration. These methods pose the very important question about the possible transfer of permeability/absorption data from animals to humans. Despite several similarities existing among mammals for what concerns the composition of epithelial cell membranes, different pH conditions, intestinal motility, transit time, and enzymes/transporters distribution lead to species variability [78,109–111]. Table 9.3 reports human permeability P_H data referring to different compounds [73,112–115].

9.3.1.6   Modeling: Everted Gut

In vitro animal-based techniques need to be coupled to appropriate mathematical models to evaluate membrane permeability from experimental data. Assuming that drug transport occurs on a pure diffusive basis, the intestinal sheets and the everted gut approaches can be studied recurring to Fick's law. While the analysis of the intestinal sheets case can be performed according to models discussed in Section 9.2, everted gut approach needs some more comments due to its cylindrical nature. In order to greatly simplify the

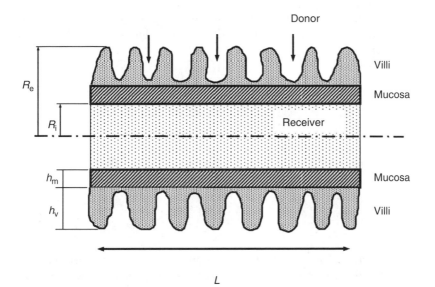

FIGURE 9.10 Schematic representation of everted intestine. (From Meriani, F., et al., *J. Pharm. Sci.*, 93, 540, 2004. With permission from John Wiley & Sons.)

analysis, intestine can be considered as a perfect cylinder (see Figure 9.10) characterized only by its length L, internal and external radii R_i and R_e, respectively. Accordingly, the presence of villi and microvilli, greatly increasing permeation area, is neglected. As the geometrical cylinder surface (cylinder external surface as we are dealing with everted gut configuration) is much smaller (one or two orders of magnitude) than that competing to real intestinal mucosa, we are obliged to define an apparent drug diffusion coefficient D_A accounting for surface reduction. A possible D_A definition is given by

$$D_A S \nabla C = D_{REAL} S_{REAL} \nabla C, \tag{9.71}$$

where D_{REAL} and S_{REAL} are, respectively, the real drug diffusion coefficient and luminal surface, S is the geometrical cylinder surface, while ∇C is the concentration gradient across the membrane. Accordingly, as the ratio D_{REAL}/D_A equals S/S_{REAL} (it is much smaller than 1), D_{REAL} knowledge can be achieved once D_A is known. Nevertheless, as we are interested in estimating the drug flow crossing the intestinal membrane ($D \times S \times \nabla C$), the exact D_{REAL} knowledge is unnecessary. In these hypotheses and assuming intestine as a uniform membrane, drug permeation can be described by Fick's second law:

$$\frac{\partial C(r)}{\partial t} = \frac{1}{r} \frac{\partial}{\partial r} \left(D_A \frac{\partial C(r)}{\partial r} r \right), \tag{9.72}$$

where r is the radial coordinate, $C(r)$ indicates drug concentration in r, and t is time. This equation must accomplish the following initial and boundary conditions:

initial conditions:

$$C(r) = 0; \quad R_i < r < R_e; \quad C(R_e) = C_{d0}k_{pd}, \tag{9.73}$$

boundary conditions:

$$D_A \frac{\partial C(r)}{\partial r}\bigg|_{r=R_e} 2\pi R_e L = V_d \frac{dC_d}{dt}; \quad C_d = \frac{C(R_e)}{k_{pd}}, \tag{9.74}$$

$$-D_A \frac{\partial C(r)}{\partial r}\bigg|_{r=R_i} 2\pi R_i L = V_r \frac{dC_r}{dt}; \quad C_r = \frac{C(R_i)}{k_{pr}}, \tag{9.75}$$

where V_d and V_r are, respectively, donor and receiver environment volumes, C_d and C_r are, respectively, donor and receiver environment drug concentrations, C_{d0} is initial C_d value, while k_{pd} and k_{pr} are, respectively, drug partition coefficient at the liquid/intestine interfaces located in R_e and R_i. Equation 9.73 imposes that intestinal wall is drug free at the beginning of the experiment, whereas initial drug concentration in the donor environment is C_{d0}. Equation 9.74 states that C_d time variation depends on the drug flow crossing the intestinal wall and, similarly, Equation 9.75 states that the C_r time increase is due to the drug flow coming out from the intestinal membrane. Both equations are coupled with the obvious drug partitioning conditions in R_e and R_i. Equation 9.75 could be replaced by the usual overall drug mass balance

$$M_0 = V_d C_d + V_r C_r + \int_{R_i}^{R_e} C(r) 2\pi L r \, dr, \tag{9.76}$$

where M_0 is total drug amount present in the system ($= V_d \times C_{d0}$). Numerical solution of Equation 9.72 through Equation 9.75 (or Equation 9.76) allows the determination of C_r time increase. Additionally, drug permeability P through the intestinal membrane is given by

$$P = \frac{D_A k_p}{h_m + h_v}, \tag{9.77}$$

where $k_p = k_{pd} = k_{pr}$, h_m is the real wall thickness and h_v the villi layer thickness (see Figure 9.10). Considering villi layer morphology, it is acceptable to consider $h_m + h_v$ as the effective intestine wall thickness. Meriani et al.

[74] apply this model to evaluate nimesulide permeability through rat intestine. In order to assure intestine cells homeostasis, donor environment ($V_d = 1000$ cm^3) is filled by a Krebs Ringer modified buffer (pH = 7.4; 37°C) containing nimesulide ($C_{d0} = 100$ μg cm^{-3}) while the receiver ($V_r = 4 \times 12$ cm$^3 = 48$ cm^3) is filled by initially pure Krebs Ringer modified buffer. In addition, both the donor and the receiver environments are continuously oxygenated (95% O_2, 5% CO_2) to maintain cell viability (oxygenation also ensures mixing in the receiver environment). Male Wistar rats (Centro servizi di Ateneo, Settore Stabulario e Sperimentazione animale, Trieste University, Italy) weighing approximately 250 g were fasted for 12 h (water *ad limitum*), then sacrificed by CO_2. Small intestine (duodenum, jejunum, and ileum) is removed, separated from the mesentery, rinsed with the buffer using a 10 mL syringe, then cut into four different sections. Each section is then everted on a teflon rod and fixed on its own holder by means of surgical thread (accordingly, the receiver environment is the sum of four receiver environments each one pertaining to each intestine portion: $L = 9 \times 4 = 36$ cm, $R_e = 0.27$ cm, and $R_i = 0.21$ cm). In order to avoid preferential fluxes inside the donor due to stirring, a symmetrical intestine holder disposition on the holding plate is required. Assuming $k_p = 1$, $h_v + h_m = 0.06$ cm, model best fitting yields $D_A = 1.6 \times 10^{-6}$ cm^2 sec^{-1} and $P = 2.7 \times 10^{-5}$ cm sec^{-1}. Figure 9.11 shows the good comparison between experimental data (open circles) and model best fitting (solid line). Following the same approach, Grassi et al. [116] study acyclovir permeation through rat intestine getting $P = 3 \times 10^{-5}$ cm sec^{-1} ($V_d = 140$ cm^3 (Krebs Ringer modified buffer pH = 7.4; 37°C), $V_r = 1 \times 12$ cm^3; $C_{d0} = 1$ mg cm^{-3}; $L = 1 \times 6$ cm, $k_p = 1$, $R_e = 0.27$ cm, and $R_i = 0.21$ cm). Again, Figure 9.11 reports the good agreement between experimental data (filled circles) and model best fitting (dashed line).

9.3.1.7 Modeling: Single Pass and Recirculating Perfusion

In the above discussion about the everted gut modeling, the complexity of intestinal mucosa, reach in villi and microvilli, has been simply accounted for by an effective diffusion coefficient. This assumption is absolutely reasonable also in the light of the physical frame related to the everted gut setup. This assumption becomes less obvious when dealing with the single pass and the recirculating perfusion techniques. Indeed, in both these cases we must account for the flow of a drug solution in contact with the highly heterogeneous intestinal mucosa. This means that drug uptake in the gastrointestinal tract is a heterogeneous process since it takes place at a complex and heterogeneous interface under variable stirring conditions [117,118]. A possible way to account for this fact is to consider surface geometrical complexity by means of fractal geometry. Indeed, in Section 7.3.8 we noticed that diffusion is highly influenced by medium fractal nature and this aspect can be

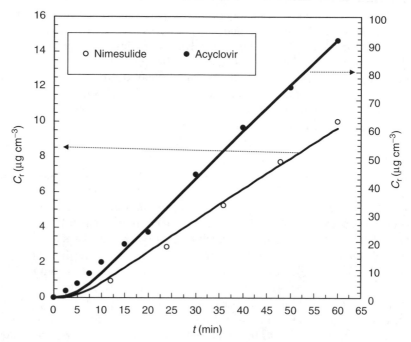

FIGURE 9.11 Comparison between model best fitting (solid lines) and experimental data relative to nimesulide (open cricles) and acyclovir (filled circles).

represented by the relation connecting the mean squared displacement $\langle r(t) \rangle$ with time

$$\langle r^2(t) \rangle \propto t^\beta \quad \text{for } t \to \infty, \quad \beta = \frac{d_s}{d}, \quad (7.224)$$

where d_s is the fracton dimension, while d is the surface fractal dimension. If d represents medium fractal properties, d_s refers to the diffusion process fractal characteristics. In Euclidean media $\beta = 1$ (Fick's law holds) while in fractal environments $\beta < 1$ (typically we have $\beta = 0.714$ in two-dimensions and $\beta = 0.857$ in three-dimensions [119]). Practically speaking, Macheras and Argyrakis [117] verified that the fractal character of gastrointestinal tract plays an important role for low-solubility and low-permeability drugs while it is not important for high-solubility and high-permeability drugs.

With this warning in mind, we can now consider the nonfractal (heterogeneous) modeling for the single pass and recirculating perfusion techniques. Basically, drug absorption can be studied according to four models: laminar flow, plug flow, complete radial mixing, and mixing tank [120]. In the laminar flow model, the velocity profile is that of Newtonian fluid flowing in a tube. This means that it does not depend on axial position (z) but it depends, in a parabolic manner, on the distance r from velocity tube axis (velocity maximum occurs on tube axis and velocity zeroes at the inner tube surface $r = R_i$). In this model, drug concentration C_d depends on both radial (r) and axial positions (z). Plug flow model is a simplification of the laminar flow model as velocity profile is assumed flat, this meaning that velocity is constant and does not depend on both the axial and radial positions. On the contrary, C_d is still assumed to be dependent on both r and z. In the complete radial mixing, again velocity profile is flat but C_d becomes dependent only on the axial position. In other words, it is assumed that radial mixing is perfect. Finally, mixing tank model assumes perfect axial and radial mixing so that C_d and the velocity profile do not depend on either the radial or axial position. Experimental data fitting provides satisfactory results for all the four models presented even if different membrane permeability is achieved. In particular, wall permeabilities fall in the order-mixing tank < complete radial mixing < plug flow < laminar flow, the laminar flow approach is the most reliable one [120]. Accordingly, the laminar flow approach should be preferred. Nevertheless, the real physical frame to be modeled is more complicated than that supposed by the four models above discussed. Indeed, several other aspects such as the presence of a boundary layer arising at the intestinal wall surface [121,122], the perfusion rate entity [105], the existence of a carrier-mediated transport [123], gut metabolism [124], and water exchange between solution and intestinal wall [125,126] must be accounted for. Assuming that the effect of the boundary layer can be neglected [75], that drug transport occurs only on a passive diffusion basis (no carrier-mediated transport and gut metabolism occur), we can focus the attention on a model accounting for the simultaneous drug absorption and water exchange. For this purpose, the complete radial mixing model provides a reasonable and simple starting point for further implementation.

The *single pass perfusion* implies a solution, of known drug concentration C_{di}, flows at a constant volumetric flow rate q_i, through an intestinal segment of length L at the end of which drug concentration C_{do} is detected (Figure 9.12). Due to water exchange, C_{do} value depends on both the drug intestinal absorption and the water volumetric flow rate (q_w) exchanged between the flowing solution and the intestinal wall. Accordingly, a correction accounting for q_w is necessary. Starting point is the stationary drug differential mass balance (see Figure 9.12):

$$q(x)C_d(x) = (q(x) + dq(x))(C_d(x) + dC_d(x)) + E_p \, PC_d(x) \, dx, \qquad (9.78)$$

where $q(x)$ and $C_d(x)$ are the volumetric flow rate and the drug concentration, respectively, at position x, R_i is intestine internal radius, E_p ($= 2\pi R_i$) is the internal perimeter, and P is the drug permeability through the intestinal wall. Dividing Equation 9.78 for dx and neglecting the second-order differential $dq(x)dC_d(x)$, we get

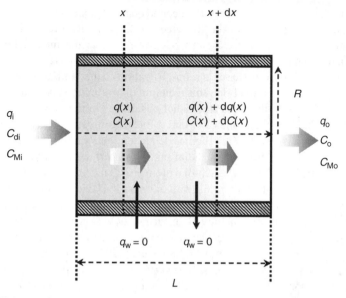

FIGURE 9.12 Mass balance scheme referring to the single pass perfusion technique. (From Grassi, M., Cadelli, G., *Int. J. Pharm.*, 229, 95. With permission from Elsevier.)

$$q(x)\frac{dC_d(x)}{dx} + C_d(x)\frac{dq(x)}{dx} + E_pPC_d(x) = 0. \qquad (9.79)$$

Supposing that $q(x)$ has a linear trend through the intestinal length L (this is the simplest assumption to adopt)

$$q(x) = q_i + \frac{q_w}{L}x; \quad \frac{dq(x)}{dx} = \frac{q_w}{L}, \qquad (9.80)$$

we finally get

$$\frac{dC_d(x)}{dx} = -\frac{\left(\frac{q_w}{L} + E_pP\right)C_d(x)}{\left(\frac{q_w}{L}x + q_i\right)}. \qquad (9.81)$$

The integration of Equation 9.81, in the interval $x = 0$ to $x = L$, yields

$$C_{do} = C_{di}\left(\frac{q_i + q_w}{q_i}\right)^{-\left(1 + \frac{E_pPL}{q_w}\right)}. \qquad (9.82)$$

Consequently, it follows:

$$P = \frac{q_w}{LE_p} \ln\left(\frac{C_{di}}{C_{do}} \times \frac{q_i}{q_i + q_w}\right) \Big/ \ln\left(\frac{q_i + q_w}{q_i}\right). \tag{9.83}$$

This equation allows the P determination once q_w is known. It is important to mention that q_w is positive if the intestinal wall supplies water to perfusate, while it is negative in the opposite case. In the limiting case of a vanishing q_w, it is easy to demonstrate (L'Hôpital–Bernoulli theorem [127]) that Equation 9.83 becomes

$$\lim_{q_w \to 0} (P) = \frac{q_i}{LE_p} \ln\left(\frac{C_{di}}{C_{do}}\right). \tag{9.84}$$

Although different strategies exist for the q_w evaluation [128], one of the most common consists in adding a marker in the flowing drug solution. This second solute should not be absorbable by the intestinal wall. Thus, a difference between the marker inlet (C_{Mi}) and outlet (C_{Mo}) concentration indicates the presence of water exchange. Accordingly, considering Equation 9.83 form for vanishing E_p, we get

$$q_w = q_i \left(\frac{C_{Mi}}{C_{Mo}} - 1\right). \tag{9.85}$$

In the light of Equation 9.85, Equation 9.83 becomes

$$P = \frac{q_i}{LE_p} \ln\left(\frac{C_{di}C_{Mo}}{C_{do}C_{Mi}}\right) \times \left(\left(\frac{C_{Mi}}{C_{Mo}} - 1\right) \Big/ \ln\left(\frac{C_{Mi}}{C_{Mo}}\right)\right). \tag{9.83'}$$

It is also easy to verify that, in the limit $C_{Mo}/CM_i = 1$, Equation 9.83 degenerates into Equation 9.84. Usually, the second term of Equation 9.83′

$$\left(\left(\frac{C_{Mi}}{C_{Mo}} - 1\right) \Big/ \ln\left(\frac{C_{Mi}}{C_{Mo}}\right)\right)$$

is neglected but it is easy to verify that its contribution cannot be negligible. Indeed, it varies almost linearly with the ratio C_{Mi}/C_{Mo}, which is equal to 0.72 for $C_{Mi}/C_{Mo} = 0.5$ and 1.23 for $C_{Mi}/C_{Mo} = 1.5$.

Recirculating perfusion can be seen as a particularization of the single pass approach for poorly permeable drugs. Indeed, for poorly absorbable drugs, the difference between C_{di} and C_{do} may be negligible. Recirculating perfusion consists of connecting an intestinal tract of length L to a closed loop containing a reservoir characterized by an initial drug concentration C_{r0} (see Figure 9.13). While time-dependent drug concentration C_r is measured inside the reservoir, a pump guarantees solution circulation. The fact that drug

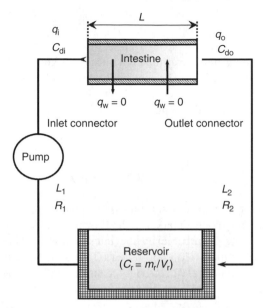

FIGURE 9.13 Schematic representation of the recirculating perfusion set up. (From Grassi, M., Cadelli, G., *Int. J. Pharm.*, 229, 95. With permission from Elsevier.)

solution is obliged to pass many a times through the intestinal tract, allows the measurement of drug disappearance for poorly absorbable drugs also. Obviously, in these conditions, the use of a marker for the q_w evaluation is no longer reliable as the hypothesis of $E_p = 0$ fails. Nook et al. [129] suggest to measure initial (V_{si}) and final (V_{sf}) solution volume so that q_w can be calculated from solution volume variation

$$q_w = \frac{V_{sf} - V_{si}}{t_{exp}},$$ (9.86)

where t_{exp} is the whole experimental time. On this basis, it is possible matching the modeling of this technique resorting to the drug differential mass balance in transitory conditions (always referred to a complete radial mixing model):

$$\frac{\partial C_d}{\partial t} \pi R_i^2 dx = qC_d - (q + dq)(C_d + dC_d) - E_p dx P C_d,$$ (9.87)

where C_d and q depend on both x and t. Assuming a linear trend for q (see Equation 9.80), neglecting the second-order differential ($dq \times dC_d$) and dividing Equation 9.87 for $\pi R_i^2 dx$ leads to

$$\frac{\partial C_d}{\partial x}\left(\frac{q_i}{\pi R_i^2}+\frac{q_w}{\pi R_i^2 L}x\right)+\frac{\partial C_d}{\partial t}=-C_d\left(\frac{q_w}{\pi R_i^2 L}+\frac{2P}{R_i}\right). \tag{9.88}$$

The characteristic method [130] provides Equation 9.83 solution

$$C_{do}=C_{di}\left(\frac{q_i+q_w}{q_i}\right)^{-\left(1+\frac{E_p PL}{q_w}\right)}. \tag{9.89}$$

This equation differs from Equation 9.82 (single pass case) as, now, C_{di} is time dependent. In particular, it is drug concentration in the reservoir fluid at time $t-t_1$, where t_1 is the time required by the solution to pass from the reservoir to the intestinal beginning (see Figure 9.13) and it can be defined by

$$t_1=\frac{L_1}{v_1}=\frac{S_1 L_1}{q_i}, \tag{9.90}$$

where L_1 and S_1 are, respectively, the inlet connector length and internal section. It is convenient to set the beginning of the experiment ($t=0$) when solution has completed the first loop (see Figure 9.13). Indeed, before this time, drug concentration in the reservoir fluid is always constant (C_{r0}). Accordingly, for $t=0$, reservoir solution volume V_{ri} is

$$V_{ri}=V_{si}-V_1-V_{int}-q_w(t_{int}+t_2)-V_2, \tag{9.91}$$

where V_1 and V_2 represent the internal volume of the inlet and outlet connectors, respectively, V_{int} is the internal intestine volume, while t_{int} and t_2 are, respectively, the time required by the solution to go through the intestine and the outlet connectors. Analogous to t_1, t_2 can be evaluated by

$$t_2=\frac{L_2}{v_2}=\frac{S_2 L_2}{q_o};\quad q_o=q_i+q_w, \tag{9.92}$$

where L_2 and S_2 are, respectively, the outlet connector length and internal section, whereas q_o is the flow rate entering the outlet connector. The t_{int} determination requires considering the space dependence of the flow rate $q(x)$:

$$q(x)=\pi R_i^2 v_x=\pi R_i^2\frac{dx}{dt}=q_i+\frac{q_w}{L}x. \tag{9.93}$$

Integration of Equation 9.93 in the intervals $0<x<L$ and $0<t<t_{int}$, leads to t_{int} determination

$$t_{int}=\frac{\pi R_i^2}{q_i}\ln\left(\frac{q_i+q_w}{q_i}\right). \tag{9.94}$$

It can be easily verified [127] that if q_w vanishes, reasonably, $t_{int} = \pi R_i^2 L/q_i$. According to Equation 9.86, the reservoir fluid volume V_r variations follow a linear trend:

$$V_r = V_{ri} + q_w t. \qquad (9.95)$$

Now, it is possible to get model final form remembering that it is nothing more than a transitory drug mass balance made on the reservoir. For this purpose, it is worth mentioning that it assumes different expressions for $t < t_1 + t_2 + t_{int}$ and for $t > t_1 + t_2 + t_{int}$. Indeed, if $t < t_1 + t_2 + t_{int}$, C_{di} in Equation 9.89 is always constantly equal to C_{r0} (a single pass perfusion takes place) while it becomes time dependent for $t > t_1 + t_2 + t_{int}$. Accordingly, it follows:
$t < t_1 + t_2 + t_{int}$:

$$\frac{dm_r}{dt} = -q_i \frac{m_r}{V_r} + (q_i + q_w)C_{r0}\left(\frac{q_i + q_w}{q_i}\right)^{-\left(\frac{E_P PL}{q_w}+1\right)}, \qquad (9.96)$$

$t > t_1 + t_2 + t_{int}$:

$$\frac{dm_r(t)}{dt} = -q_i \frac{m_r(t)}{V_r(t)} + (q_i + q_w)\frac{m_r(t - t_{int} - t_1 - t_2)}{V_r(t - t_{int} - t_1 - t_2)}\left(\frac{q_i + q_w}{q_i}\right)^{-\left(\frac{E_P PL}{q_w}+1\right)}, \qquad (9.97)$$

where m_r is the drug amount present in the reservoir at time t. Both equations impose that m_r variation depends on the exiting flow (first right-hand side of Equation 9.96 and Equation 9.97) and on the incoming fluid (second right-hand side of Equation 9.96 and Equation 9.97). Both equations need a numerical solution (fourth-order Runge-Kutta method [72]).

Grassi and Cadelli [131] test both the single pass and the recirculating perfusion models on antipyrine absorption from rat intestine. Male Wistar rats (Centro servizi di Ateneo, Settore Stabulario e Sperimentazione animale, Trieste University, Trieste, Italy) weighing approximately 300–350 g are fasted for 16 h (water *ad limitum*) before each experiment. Anaesthesia is induced with i.m. injection of ethyl urethane (Sigma Chemical, Steinheim, Germany) dissolved in physiological solution (1.5 g kg^{-1}). The abdomen is opened with a midline longitudinal incision and about 20 cm jejunal segment is cannulated (plastic tubing; outlet and inlet tubes diameter equal to 0.4 cm) inserting the outlet tube at 30 cm far from the cecal intestine. The rat was placed on a heated slide (37°C), intestinal tract is positioned in its anatomical position so that the inlet and outlet tubes are out of the abdomen. The surgical area is covered with a cotton sheet constantly wetted by means of a 37°C physiological solution. Subsequently, jejunal is cleaned by a saline solution (flow rate 2.5 cm^3 min^{-1}) until a clear perfusate appears (approximately 10 min). In the single pass perfusion, $q_i = 0.45 \pm 0.14$ cm^3 min^{-1}, inlet antypirine

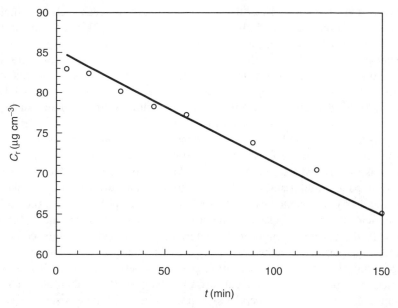

FIGURE 9.14 Comparison between experimental data (open circles) and model best fitting (solid line) relative to the recirculating perfusion technique.

concentration $C_{di} = 105$ µg cm^{-3} while marker (phenol red) inlet concentration $C_{Mi} = 20$ µg cm^{-3}. Solution is allowed to flow for 105 min while C_{Mo} and C_{do} are evaluated after 60, 75, 90, and 105 min. Antipyrine concentration is detected by an UV spectrophotometer (wavelength 254 nm). In the recirculating perfusion, after intestinal cleaning, q_i is set to 0.62 cm^3 min^{-1} while the outlet tube and the pump are connected to a thermostatic (37°C) reservoir filled by 30 cm^3 antipyrine solution ($C_{r0} = 110$ µg cm^{-3}). Remembering that $R_i = 0.173$ cm and $L = 22 \pm 9$ cm, the single pass case yields $q_w = (-0.022 \pm 0.01)$ cm^3 min^{-1}, $C_{di}/C_{do} = (1.1 \pm 0.08)$ and, consequently, $P = (2.5 \pm 1) \times 10^3$ cm min^{-1}. Figure 9.14 reports an example of the comparison between experimental data (open circles, recirculating perfusion) and model (Equation 9.91 and Equation 9.92) best fitting (solid line). It is clear that the agreement is satisfactory and this proves both model and experimental data reliability. Interestingly, this data fitting yields $P = (2.8 \pm 0.4) \times 10^{-3}$ cm min^{-1} that is statistically equal to the P value found in the single pass perfusion. Finally, it is worth mentioning that, in the recirculating case, $q_w = -0.028 \pm 0.007$ cm^3 min^{-1} that, again, is statistically equal to the value determined in the single pass perfusion.

9.3.2 SKIN

Skin drug delivery can be, substantially, subdivided into topical and transdermal. As in topical administration, the drug is intended to act at skin level, this strategy is indicated for the treatment of skin diseases such as follicle-related disorders (acne and alopecias, for example) [132]. On the contrary, as transdermal administration is aimed to get a systemic release, skin represents a barrier and not a target [133]. Obviously, topical administration presents far less problems than transdermal one for what concerns the delivery system designing. However, transdermal administration shows interesting advantages over other systemic administration routes such as (1) the reduction of first-pass drug degradation as the liver is initially bypassed, (2) the reduction of overdosage peaks that can occur in oral administration, (3) the existence of variable delivery conditions typical of the gastrointestinal tract [134,135]. As skin offers a considerable resistance to drug penetration, transdermal administration can take advantage of chemical and physical strategies aimed to improve skin permeability. Basically, chemical strategies consist in the use of enhancers, compounds able to modify the barrier function of the outermost part of skin (stratum corneum (SC)). In particular, inducing SC reversible changes, they increase skin permeability and, usually, also drug partition coefficient [136]. While this strategy works for small molecules (molecular weight less than 500), larger compounds (proteins and peptides) require physical strategies such as electrically based techniques (iontophoresis, electroporation, ultrasound, photo-mechanical wave), structure-based techniques (microneedles), and velocity-based techniques (jet propulsion) [137]. Regardless of the necessity or not of physical/chemical enhancing, fundamental prerequisite for a reliable and effective designing of transdermal delivery systems is the knowledge of skin structure and properties. Although this topic has been widely discussed in Chapter 2, for the purpose of this section, skin can be seen as a tri-layer membrane composed by the SC, the epidermis, and the dermis (see Figure 9.15). SC (thickness 10–20 μm) is composed by a lipid-rich matrix where flattened, interdigitated, partially desiccated, keratinized dead epidermal cells are organized into a layered, close-packed array [138]. It is by far the most important barrier to drug permeation through the skin. Epidermis, a viable tissue devoid of blood vessels 50–100 μm thick, is just below the SC. The innermost skin layer is represented by the dermis (1–2 mm thick) containing capillary loops able to take up administered drugs for systemic distribution [134]. Skin structure is completed by the presence of sweat ducts and hair follicles, vertical holes, characterized by an external orifice of 50–100 μm diameter, crossing the three mentioned layers [132]. In principle, these structures favor drug permeation as they represent channels connecting the external environment with the dermis capillary bed. Traditionally, both sweat glands and follicular drug transport in humans have been neglected due to the low skin surface competing to the outer part of these structures (1%) [138]. Nowadays, on the contrary, the importance of follicular transport has been reconsidered also in the light of the fact that the potential area available for penetration also includes the internal follicular

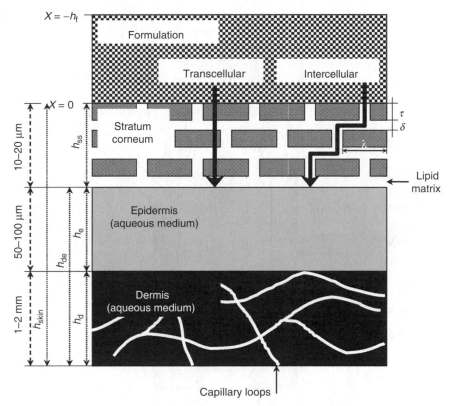

FIGURE 9.15 Schematic representation of human skin and most important geometric characteristics.

surface [132]. Of course, the problem of transdermal drug delivery is made more complex by the fact that follicles number and skin permeability are not constant all over human surface but vary from site to site [139].

Typically, *in vitro* techniques aimed to measure skin permeability are based on the experimental apparatus proposed in mid-1970s by Franz [140,141]. Basically, Franz cell is a double-walled beaker characterized by a top donor chamber separated from the receiver chamber by skin. While thermostatic fluid flows inside the beaker double wall, receiver sampling takes place from the sampling port (see Figure 9.16). Skin permeability is evaluated on the basis of drug concentration increase in the receiver environment. *In vivo* tests are usually conducted on human, rodent, and pig skin, which is the best approximation of human skin. Sometimes monkey skin is also used [139].

Historically speaking, the first attempt to mathematical model of drug permeation phenomenon through skin is that of Higuchi [142]. Then, many other authors developed excellent works on this topic and they generated

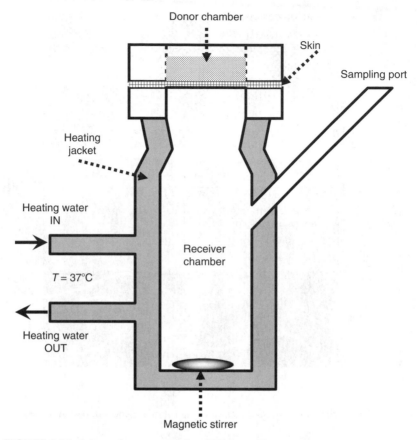

FIGURE 9.16 Schematic representation of Franz cell.

models that can be subdivided into two main classes: empirical and mechanistic models. Although mechanistic models should be preferred as they can provide very useful information about the relative importance of the various phenomena concurring in the determination of skin permeability, empirical ones can overcome inaccuracies of mechanistic models (due to uncertainties) providing better predictions from the practical point of view [143]. The following sections are devoted to present and discuss some of the most important mechanistic and empirical models.

9.3.2.1 Mechanistic Models

Drug transport through the skin is a complex phenomenon comprehending physical, chemical, and biological interactions [143]. In particular, it implies (a) drug dissolution within and release from the formulation, (b) drug partitioning into the SC, (c) diffusion through the SC, (d) partitioning from the SC into the aqueous viable epidermis, (e) diffusion through the viable epidermis and into the upper dermis, and (f) uptake into the local capillary network and eventually the systemic circulation [144]. In addition, skin metabolic activity toward entering drug should be considered [143]. It is worth mentioning that although drug diffusion across SC can be transcellular or intercellular (see Figure 9.15), it is widely affirmed that the intercellular lipidic pathway represents the most important route for drug permeation [143,144]. Therefore, an ideal drug candidate would have sufficient lipophilicity to partition into the SC, but also sufficient hydrophilicity to enable the second partitioning step into the viable epidermis and eventually the systemic circulation. While for the majority of the drugs the importance of transfollicular contribution to drug permeation is under debate [137], it seems sure that this mechanism plays an important role for ions and polar compounds as demonstrated by experiments carried out on afollicular animals [145].

As discussed in Chapter 1, once the physical frame has been defined, mathematical modeling can follow. Obviously, the above-discussed physical frame must be further detailed as nothing precise has been said about drug transport inside our formulation. In the simplest situation, the formulation coincides with a drug solution, but it can be drug dispersion in a liquid phase, a polymeric matrix containing the drug in dissolved or undissolved form, or it can be a reservoir containing a drug solution separated from the skin by a polymeric membrane [134]. Although other, more complex situations can occur [144], this section is focused on the above-cited four conditions for what concerns the release properties of the formulation.

Assuming that the formulation consists in a well-stirred drug solution and considering the skin as a unique barrier characterized by proper and constant diffusion (D_{skin}) and partition (k_{skin}) coefficients, drug transport can be approached according to Fick's second law

$$\frac{\partial C_{skin}}{\partial t} = \frac{\partial}{\partial X}\left(D_{skin}\frac{\partial C_{skin}}{\partial X}\right), \tag{9.98}$$

where t is time, X is the abscissa, and $C_{skin}(t,X)$ is drug concentration inside the skin. Assuming that initial drug concentration in the solution C_{d0} is constant and that drug concentration in $X = h_{skin}$ (skin thickness) is always zero, Equation 9.98 must be solved with the following initial and boundary conditions:
initial:

$$C_{skin}(X) = 0, \quad 0 < X < h_{skin}, \tag{9.99}$$

boundary:

$$C_{skin} = k_{skin}C_{d0}, \quad X = 0; \quad C_{skin} = 0, \quad X = h_{skin}, \tag{9.100}$$

where Equation 9.99 states that the skin is initially drug free, while Equation 9.100 states the partitioning conditions at the formulation/skin interface (left equation) and sink conditions in the receiver environment (right equation). Equation 9.99 holds in the presence of a wide solution volume, while Equation 9.100 holds in the presence of a wide receiver volume. Equation 9.98 analytical solution reads [143]

$$M_{\mathrm{t}} = S h_{\mathrm{skin}} k_{\mathrm{skin}} C_{\mathrm{d0}} \left(\frac{D_{\mathrm{skin}}}{h_{\mathrm{skin}}^2} t - \frac{1}{6} - \frac{2}{\pi^2} \sum_{i=1}^{\infty} \frac{(-1)^i}{i^2} e^{\left(\frac{D_{\mathrm{skin}} i^2 \pi^2 t}{h_{\mathrm{skin}}^2} \right)} \right), \quad (9.101)$$

where M_{t} is the drug amount permeated at time t and S is the permeation area (obviously, Equation 9.2 is newly found). This equation can be used to model both *in vitro* and *in vivo* experiments even if the assumption of zero drug concentration in $X = h_{\mathrm{skin}}$ is more reasonable for *in vivo* conditions as systemic drug concentration is lowered by drug elimination or metabolization. On the contrary, the assumption of pure one-dimension diffusion is much more reasonable in *in vitro* setup than in *in vivo* experiments. It is worth noticing that D_{skin} is the skin effective diffusion coefficient (see Chapter 7) accounting for skin heterogeneity (follicles, sweat gland, SC, epidermis, dermis). Interestingly, Equation 9.101, for very long times, degenerates into

$$M_{\mathrm{t}} = S h_{\mathrm{skin}} k_{\mathrm{skin}} C_{\mathrm{d0}} \left(\frac{D_{\mathrm{skin}}}{h_{\mathrm{skin}}^2} t - \frac{1}{6} \right). \quad (9.102)$$

Consequently, drug flux J (mass/(surface × time)) is given by

$$J = k_{\mathrm{skin}} C_{\mathrm{d0}} \frac{D_{\mathrm{skin}}}{h_{\mathrm{skin}}} = C_{\mathrm{d0}} P_{\mathrm{skin}}, \quad (9.103)$$

where P_{skin} is skin permeability. Michaels et al. [138], starting from Equation 9.103, assume that skin resistance is substantially due to SC, develop a very interesting simple model that can also account for the permeation of dissociating compounds. Practically, they suppose that SC is made up of by parallel array of thin plates (constituted by proteins) regularly dispersed in a lipid homogeneous continuous matrix (see Figure 9.15). Consequently, drug diffusion occurs according to two different mechanisms: (a) solely through the continuous, tortuous lipid matrix, or (b) alternatively crossing the continuous and dispersed (plates) phase. Thus, J will be the sum of two contributes (J_{a} and J_{b}) and the problem shifts on their evaluation which, in turn, implies the determination of the corresponding permeability P_{a} and P_{b}. Assuming that plates are rectangles characterized by length λ and thickness τ and that they are equidistantly separated by nearest neighbors by an interstitial channel of thickness δ (see Figure 9.15), $h_{skin} = n(\delta + \tau)$ ($n =$ layers number) while lipidic and plate phases thickness will be $h_L = n\delta$ and $h_P = n\tau$, respectively. Thus, P_b can be evaluated remembering that the diffusive resistance of a layered membrane is given by the sum of the resistances pertaining to each layer [1]

$$\frac{1}{P_b} = R_L + R_P = \frac{n\delta}{k_{pL}D_L} + \frac{n\tau}{k_{pP}D_P} = \frac{n(\delta + \tau)\left(\dfrac{\beta}{\sigma}\dfrac{D_P}{D_L} + 1\right)}{(\beta + 1)k_{pP}D_P},$$ (9.104)

$$\beta = \frac{\delta}{\tau}; \quad \sigma = \frac{k_{pL}}{k_{pP}}$$

P_a is estimated considering that the surface fraction (normal to the diffusion direction) pertaining to the lipid phase is $2\beta/\alpha$ and that the effective path length through the lipid channels is $1 + (\alpha + \beta) / 2(1 + \beta)$. Accordingly, we have

$$P_a = \frac{k_{pL}D_L \dfrac{2\beta}{\alpha}}{n(\delta + \tau)\left(1 + \dfrac{\alpha + \beta}{2(1 + \beta)}\right)}; \quad \alpha = \frac{\lambda}{\tau}.$$ (9.105)

Finally, model expression is

$$M_t = S(J_a + J_b)t = SC_{d0}(P_a + P_b)t.$$ (9.106)

On the basis of experimental evidences, the authors suggest $\alpha = 20$, $\beta = 0.16$, and $n(\delta + \tau)$ is approximately 40 μm. This approach can also be extended to compounds undergoing dissociation. Let us focus the attention on an acid that dissociates according to the following equilibrium:

$$AH \leftrightarrow A^- + H^+; \quad K_A = \frac{A^- H^+}{AH},$$ (9.107)

where AH, A^-, and H^+ represent, respectively, molar concentration of undissociated acid, dissociated acid, and cation, while K_A is the acid dissociation constant. In the light of Equation 9.107 and knowing the whole acid molar concentration in solution ($AH_0 = A^- + AH$), it is possible deducing A^- and AH:

$$AH = \frac{AH_0}{1 + K_A/H^+}; \quad A^- = \frac{AH_0}{1 + H^+/K_A}.$$ (9.108)

Thus, Equation 9.101 can be generalized as follows:

$$M_t = S(J_{AH} + J_{A^-})t = S(C_{AH}(P_a + P_b)|_{AH} + C_{A^-}(P_a + P_b)|_{A^-})t,$$ (9.109)

where C_{AH} and C_{A^-} are, respectively, undissociated and dissociated acid concentration (mass/volume) while $(P_a + P_b)|_{AH}$ and $(P_a + P_b)|_{A^-}$ indicate, respectively, AH and A^- permeability where P_a and P_b are expressed by Equation 9.104 and Equation 9.105. Obviously, it is possible that, for instance, A^- permeability $((P_a + P_b)|_{A^-})$ is negligible and only AH can permeate. The application of this model on experimental data reveals the following human skin permeability (30°C): atropine (2.31×10^{-9} cm sec^{-1}), ephedrine (1.67×10^{-6} cm sec^{-1}), estradiol (1.48×10^{-6} cm sec^{-1}), nitroglycerin (2.78×10^{-6} cm sec^{-1}), and scopolamine (1.41×10^{-8} cm sec^{-1}).

Coceani et al. [146] study drug permeation in presence of a formulation consisting in a liquid phase containing dispersed drug in form of all equal spherical particles of initial radius R_0. This configuration, suitable for *in vitro* test more than for *in vivo* applications, allows keeping almost constant drug concentration in the liquid donor phase and as close as possible to drug solubility. Basically, this approach focuses the attention on the solid drug dissolution process inside the liquid phase and on the possibility that a stagnant layer exists between skin and receiver environment. In addition, the skin is considered as a two-layer membrane composed by the SC and the sum of the epidermis–dermis layer. One-dimension Fick's second law is solved on the SC, dermis–epidermis, and stagnant layer reads

$$\frac{\partial C_{sc}}{\partial t} = \frac{\partial}{\partial X}\left(D_{sc}\frac{\partial C_{sc}}{\partial X}\right), \tag{9.110}$$

$$\frac{\partial n C_{de}}{\partial nt} = \frac{\partial}{\partial nX}\left(D_{de}\frac{\partial C_{de}}{\partial nX}\right), \tag{9.111}$$

$$\frac{\partial n C_{ss}}{\partial nt} = \frac{\partial}{\partial nX}\left(D_{ss}\frac{\partial C_{ss}}{\partial nX}\right), \tag{9.112}$$

where D_{sc}, D_{de}, D_{ss}, C_{sc}, C_{de}, and C_{ss} are, respectively, the drug effective diffusion coefficient and concentration in the SC, in the dermis–epidermis and in the stagnant layer. Equation 9.110 through Equation 9.112 must accomplish to the following boundary conditions (see Figure 9.15):

$$V_d \frac{dn C_d}{dnt} = V_d K_t (C_s - C_d) + D_{sc} S \frac{\partial C_{sc}}{\partial nX}\bigg|_{X=0}, \tag{9.113}$$

$$D_{sc} \frac{\partial C_{sc}}{\partial X}\bigg|_{X=h_{sc}} = D_{de} \frac{\partial C_{de}}{\partial X}\bigg|_{X=h_{sc}}, \tag{9.114}$$

$$D_{de} \frac{\partial C_{de}}{\partial X}\bigg|_{X=h_{sc}+h_{de}} = D_{ss} \frac{\partial C_{ss}}{\partial X}\bigg|_{X=h_{sc}+h_{de}}, \tag{9.115}$$

$$V_r \frac{dC_r}{dt} = -D_{ss} S \frac{\partial C_{ss}}{\partial X}\bigg|_{X=h_{sc}+h_{de}+h_{ss}}, \qquad (9.116)$$

$$k_{p1} = \frac{C_{sc}}{C_d}; \quad k_{p2} = \frac{C_{de}}{C_{sc}}; \quad k_{p2} = \frac{C_{de}}{C_{ss}}; \quad k_{p4} = \frac{C_{ss}}{C_r} = 1, \qquad (9.117)$$

and the following initial conditions:

$$C_d = C_{d0}; \quad C_r = C_{sc} = C_{de} = C_{ss} = 0; \quad M = M_0, \qquad (9.118)$$

where V_d and V_r are the donor and receiver environment volumes, respectively, K_t is the dissolution rate constant, C_d and C_r are drug concentration in the donor and receiver compartment, respectively, C_{d0} and M_0 are the starting drug concentration in the donor compartment and the starting amount of undissolved drug in the donor compartment, C_s is drug solubility in the fluid filling the donor compartment, S is the permeation area, and K_{p1}, K_{p2}, K_{p3}, and K_{p4} are the partition coefficients. Equation 9.113 is the drug mass balance made up on the donor compartment: the first right-hand side term takes into account powder dissolution, whereas the second accounts for matter flux leaving the donor through the SC. Equation 9.114 and Equation 9.115 impose, respectively, that no matter accumulation occurs at the SC/dermis–epidermis interface ($X = h_{sc}$) and at the dermis–epidermis/stagnant layer interface ($X = h_{sc} + h_{de}$). Equation 9.16 states that the increase in concentration of the receiver compartment depends on drug flux exiting from the stagnant layer. Finally, Equation 9.117 expresses interfaces partitioning conditions while Equation 9.118 imposes that, in the beginning, skin and receiver are drug free, while donor compartment contains a drug solution (initial concentration C_{d0}) and an amount M_0 of undissolved drug. As discussed in Chapter 5, Section 5.2.4, Chapter 7, Section 7.3.4, and Chapter 9, Section 9.2.1, particles dissolution constant K_t is time dependent as particle radius decreases upon dissolution ($K_t = 4\pi R^2 k_d/V_d$). Accordingly, once k_d (intrinsic drug dissolution constant) is known, K_t evaluation implies the knowledge of particle radius R reduction. For this purpose, as shown in Section 9.2.1, recourse may be done to a drug mass balance made up on the skin, the donor and receiver compartments:

$$M = M_0 + V_d(C_{d0} - C_d) - V_r C_r - \int_0^{h_{sc}} C_{cs} S\, dX - \int_{h_{sc}}^{h_{sc}+h_{de}} C_{de} S\, dX - \int_{h_{sc}+h_{de}}^{h_{sc}+h_{de}+h_{ss}} C_{ss} S\, dX, \quad (9.119)$$

where M is the drug amount not yet dissolved at time t. Bearing in mind the Hixon–Crowell equation [22]:

$$R = R_0 \sqrt[3]{\frac{M}{M_0}}, \qquad (9.120)$$

it is possible to calculate R reduction and, thus, K_t reduction. Obviously, the effects of K_t decrease are more evident when the dissolution phenomenon implies a considerable variation of the particle radius. Indeed, in this case, the solid surface will be strongly reduced and, as a consequence, the drug flux feeding the donor compartment will be decreased. Thus, the importance of K_t reduction should be more relevant for few big particles than for many small particles, the total particles mass remains the same. Coceani et al. apply this model to study acyclovir permeation through male hairless rat skin (Rnu eutimic, Charles River, Milan, Iyaly) at 37°C. In particular, they studied (Franz cell apparatus) acyclovir permeation through the full skin and the dermis–epidermis layer after removing SC. Interestingly, they found that the effect of the stagnant layer is not very important as its permeability P_{ss} ($P_{ss} = 7.1 \times 10^{-4}$ cm sec^{-1}; $D_{ss} = 7.8 \times 10^{-4}$ cm^2 sec^{-1}) is neatly larger than that of the dermis–epidermis layer ($P_{de} = 9.5 \times 10^{-6}$ cm sec^{-1}; $D_{de} = 10^{-6}$ cm^2 sec^{-1}, $K_{p3} = 0.95$) and that of the SC ($P_{sc} = 10^{-6}$ cm sec^{-1}; $D_{sc} = 10^{-9}$ cm^2 sec^{-1}, $K_{p1} = 0.5$). Based on these data they concluded that skin permeability is $P_{skin} = 9.05 \times 10^{-7}$ cm sec^{-1} ($P_{skin}^{-1} = P_{sc}^{-1} + P_{de}^{-1}$). Finally, they noted that skin permeability tends to decrease with age; this is worth remembering in transdermal drug delivery systems designing.

Fernandes et al. [147] propose a very interesting analytical solution to the problem of skin (thought as a unique membrane) permeation when formulation is represented by a matrix containing dissolved drug. Practically, their model consists in the following equations:

$$\frac{\partial C_f}{\partial t} = \frac{\partial}{\partial X}\left(D_f \frac{\partial C_f}{\partial X}\right), \tag{9.121}$$

$$\frac{\partial C_{skin}}{\partial t} = \frac{\partial}{\partial X}\left(D_{skin} \frac{\partial C_{skin}}{\partial X}\right), \tag{9.122}$$

where $C_f(X)$ and $C_{skin}(X)$ represent, respectively, drug concentrations in the formulation and the skin, while D_f and D_{skin} are drug diffusion coefficients in the formulation and the skin, respectively. These equations are solved under the following initial and boundary conditions (see Figure 9.15):
initial conditions:

$$C_f = C_{f0} \quad -h_f < X < 0; \quad C_{skin} = 0 \quad 0 < X < h_{skin}, \tag{9.123}$$

boundary conditions:

$$\left. \frac{\partial C_f}{\partial X} \right|_{X=-h_f} = 0, \tag{9.124}$$

$$\left. D_f \frac{\partial C_f}{\partial X} \right|_{X=0} = \left. D_{skin} \frac{\partial C_{skin}}{\partial X} \right|_{X=0}, \tag{9.125}$$

$$C_{\text{skin}}(X = 0) = k_{\text{p}}C_{\text{f}}(X = 0), \tag{9.126}$$

$$C_{\text{skin}}(X = h_{\text{skin}}) = 0. \tag{9.127}$$

Although, initially, matrix (formulation) is uniformly loaded by the drug (concentration C_{f0}) (Equation 9.123), skin is drug free. Equation 9.124 imposes the existence of a drug impermeable wall on the matrix back while Equation 9.125 excludes drug accumulation at the matrix/skin interface. Finally, Equation 9.126 imposes partitioning conditions at the matrix/skin interface and Equation 9.127 fixes to zero drug concentration in the blood circulation.

Model analytical solution reads

$$\frac{M_{\text{t}}}{M_{\infty}} = 1 + \beta p \sum_{i=1}^{\infty} \left\{ \frac{\text{sen}(\lambda_i/\sqrt{p})}{\text{Den}_i} e^{\left(-\lambda_i^2 \frac{D_{\text{skin}}}{h_{\text{skin}}^2} t\right)} \right\}, \tag{9.128}$$

$$\text{Den}_i = \text{Den}_{i1} + \text{Den}_{i2}, \tag{9.129}$$

$$\text{Den}_{i1} = -\frac{1}{2}\lambda_i \text{sen}(\lambda_i) \left[\sqrt{\frac{1}{p}}(k_{\text{p}} + \beta p)\lambda_i \cos\left(\frac{\lambda_i}{\sqrt{p}}\right) + 3k_{\text{p}}\text{sen}\left(\frac{\lambda_i}{\sqrt{p}}\right) \right], \tag{9.130}$$

$$\text{Den}_{i2} = \frac{1}{2}\lambda_i \cos(\lambda_i) \left[3\beta\sqrt{p} \cos\left(\frac{\lambda_i}{\sqrt{p}}\right) - \lambda_i(k_{\text{p}} + \beta)\text{sen}\left(\frac{\lambda_i}{\sqrt{p}}\right) \right], \tag{9.131}$$

$$\beta = \frac{D_{\text{skin}}h_{\text{f}}}{D_{\text{f}}h_{\text{skin}}}; \quad p = \frac{D_{\text{f}}h_{\text{skin}}^2}{D_{\text{skin}}h_{\text{f}}^2}, \tag{9.132}$$

eigen-values λ_i are given by the solution of

$$\tan(\lambda_i)\tan\left(\frac{\lambda_i}{\sqrt{p}}\right) + \frac{\beta\sqrt{p}}{k} = 0. \tag{9.133}$$

Manitz et al. [148], extending the above-shown model to the two-dimensional case, consider the contemporary diffusion of drug and enhancer. For this purpose, they solve one Fick's second law for the drug and another for the enhancer, where the drug ($D_{\text{sc}}^{\text{drug}}$) and the enhancer ($D_{\text{sc}}^{\text{enh}}$) diffusion coefficients in the SC are defined as follows:

$$D_{\text{sc}}^{\text{drug}} = D_{\text{sc0}}^{\text{drug}} + \beta_1 \frac{C_{\text{enh}}(X,Y)}{k_1 + C_{\text{enh}}(X,Y)}, \tag{9.134}$$

$$D_{\text{sc}}^{\text{enh}} = D_{\text{sc0}}^{\text{enh}} + \beta_2 \frac{C_{\text{enh}}(X,Y)}{k_2 + C_{\text{enh}}(X,Y)}, \tag{9.135}$$

where D_{sc0}^{drug} and D_{sc0}^{enh} represent, respectively, the drug and enhancer diffusion coefficients in the SC when local enhancer concentration C_{enh} is zero, while β_1, β_2, k_1, and k_2 are parameters whose values are >0 (β_1, β_2 would be negative in presence of a reducer). On the contrary, diffusion coefficients in the formulation, epidermis, and dermis are considered constant. A further improvement of this model is about the solution of the mass transport equations. Indeed, the authors correctly note that the solutions of the parabolic equations like those constituting their model are characterized by an infinite propagation speed. In other words, this means that both $C_{drug}(t, X, Y)$ and $C_{enh}(t, X, Y)$ assume positive values in the whole diffusion domain for each time $t > 0$. It is obvious that this is a contradiction since diffusants can move into and through human skin only at a finite penetration velocity. Accordingly, they suppose that both $C_{drug}(t, X, Y)$ and $C_{enh}(t, X, Y)$ are positive only in a subdomain of the whole diffusion domain until the whole domain boundaries are met. The definition of subdomain boundaries is based on the fact that the whole diffusant mass must be inside the subdomain until the whole domain boundaries are met. Obviously, formulation belongs to this subdomain. This idea translates in the following condition:

$$C(t, X, Y)\phi(t)\eta = -D(C(t, X, Y))\frac{\partial C}{\partial \eta}\bigg|_{t,X,Y}, \qquad (9.136)$$

where C and D refer to drug or enhancer, $\phi(t)$ is the velocity of the moving boundary, and $\eta(t, X, Y)$ is its outer normal. Equation 9.136 affirms that the penetration velocity of the diffusant is proportional to the concentration gradient on the boundary. The complete boundary definition finally requires defining the concentration value on it

$$C(t, X, Y)|_{(X,Y)\in boundary} = \lambda. \qquad (9.137)$$

The authors suggest $\lambda = 0.05C_0$, where C_0 is the initial drug (or enhancer)-concentration in the formulation. This approach is used to evaluate the differences in model prediction when Equation 9.136 and Equation 9.137 are considered (moving boundary) and when they are neglected. Hydrocortisone is chosen as model drug ($D_{formulation}^{hydro} = 1\times10^{-10}$ cm^2 sec^{-1}, $D_{sc0}^{hydro} = 3.85\times10^{-11}$ cm^2 sec^{-1}, $C_0 = 1$ mg cm^{-1}, $\beta_1 = 19$, $k_1 = 0.5\times C_0$, $D_{epidermis}^{hydro} = D_{dermis}^{hydro} = 4.4\times10^{-8}$ cm^2 sec^{-1}, $k_p^{form/sc^1} = 3.8$, $k_p^{sc/ep^1} = 1.55$, $k_p^{ep/de^1} = 1$) while the enhancer is characterized by $D_{formulation}^{enh} = 1\times10^{-10}$ cm^2 sec^{-1}, $D_{sc0}^{enh} = 10^{-10}$cm^2 sec^{-1}, $C_0 = 1$ mg cm^{-3}, $\beta_2 = 0$, $k_2 = 0$, $D_{epidermis}^{hydro} = D_{dermis}^{hydro} = 4.4\times10^{-8}$ cm^2 sec^{-1}, $k_p^{form/sc^1} = k_p^{sc/ep^1} = k_p^{ep/de^1} = 1$). Formulation radius is set to 1 cm while skin diffusion area is characterized by 2 cm radius. Formulation, SC, epidermis, and dermis thicknesses are, respectively, 55, 25, 160, and 1600 μm. The comparison is done based on $t_{1/2}$, which is the time needed for the permeation of the 50% of the initial drug load in the formulation. The authors find that moving boundary approach yields to $t_{1/2}$ values that are, respectively, 1.0447, 1.131, and 1.365 times that competing to the not moving boundary approach when $\lambda = 0.001 \times C_0$, $0.01 \times C_0$, and $0.05 \times C_0$.

The last configuration considered implies that the formulation is composed by a drug solution reservoir separated from the skin by a matrix. Again, assuming the skin as a one-layer membrane, mass transport in the matrix (our old formulation) and in the skin is ruled by Equation 9.121 and Equation 9.122, respectively. Obviously, different initial and boundary conditions apply: initial conditions:

$$C_{res} = C_{res0},$$
(9.138)

$$C_f = k_p^{fr} C_{res0} \quad -h_f < X < 0; \quad C_{skin} = 0 \quad 0 < X < h_{skin},$$
(9.139)

boundary conditions:

$$V_{res} \frac{dC_{res}}{dt} = S_f D_f \frac{\partial C_f}{\partial X}\bigg|_{X=-h_f},$$
(9.140)

$$D_f \frac{\partial C_f}{\partial X}\bigg|_{X=0} = D_{skin} \frac{\partial C_{skin}}{\partial X}\bigg|_{X=0},$$
(9.141)

$$C_{skin}(X = 0) = k_p^{sf} C_f(X = 0),$$
(9.142)

$$C_{skin}(X = h_{skin}) = 0,$$
(9.143)

where S is the surface area available for permeation, C_{res} is the drug concentration in the reservoir, C_{res0} is its initial value, k_p^{fr} and k_p^{sf} are matrix/reservoir and skin/matrix partition coefficients while V_{res} is the reservoir solution volume. Equation 9.139 implicitly assumes that, at test time, a drug equilibrium partitioning is fully established between the solution and the matrix whereas Equation 9.140 states that C_{res} reduction speed is equal to the drug flow crossing matrix surface in $X = 0$.

In conclusion of this paragraph, it is worth mentioning that, to render skin permeation mathematical modeling closer to *in vivo* conditions, the boundary condition set in $X = h_{skin}$ should be modified. Indeed, if at an initial stage drug concentration can be set equal to zero, a more sophisticated approach is required to account also for drug accumulation in the blood and drug elimination from the blood [135]. Accordingly, the new boundary condition reads

$$V_b \frac{dC_b}{dt} = -S_{skin} D_{skin} \frac{\partial C_{skin}}{\partial X}\bigg|_{X=h_{skin}} - k_e C_b$$

$$\left. \frac{C_{\text{skin}}}{C_b} \right|_{X=h_{\text{skin}}} = k_p^{\text{sb}}, \tag{9.144}$$

where S_{skin} is the area available for permeation, V_b and C_b are, respectively, blood volume and drug concentration in it, k_e is the drug elimination constant while k_p^{sb} is the skin/blood drug partition coefficient. Equation 9.144 simply affirms that C_b variation depends on drug flow crossing the skin and on drug elimination (here supposed, for the sake of simplicity, a first-order kinetics process).

9.3.2.2 Empirical Models

Among the different types of existing empirical models [143], quantitative structure–activity relationship (QSAR) and neural network-based models merit to be mentioned. Basically, QSAR attempts to statistically relate skin permeability (P_{skin}) to drug structural descriptors or physicochemical parameters. For example, Potts and Guy [149], on the basis of many P_{skin} values, propose the following equation:

$$\log(P_{\text{skin}}) = -6.3 + 0.71 \log(P_{\text{o/w}}) - 0.0061 \, \text{MW}, \tag{9.145}$$

where $P_{\text{o/w}}$ and MW are, respectively, drug n-octanol/water partition coefficient and molecular weight (this equation yields to a correlation coefficient $r^2 = 0.67$ on 93 data. P_{skin} is measured in cm sec^{-1}). In this particular case, drug descriptors are identified with $P_{\text{o/w}}$ and MW. Of course, other, more sophisticated models falling in this category exist [143].

Artificial neural networks (ANNs) are theoretical–mathematical tools mimicking the learning processes of the human brain. Indeed, they are constituted by elaboration units (ANN neurons or nodes) that are interconnected with each other [150]. This means that, as real neurons do, the elaboration unit receives information from and sends information to the other elaboration units. On the basis of the interneurons connections, ANN assumes different architectures and different ANN typologies arise [151]. A feed-forward ANN (FF-ANN) is the most popular arrangement for the simulation of causal and effect relationships. Structurally, FF-ANN consists of an input layer, an output layer, and any number of intermediate layers called hidden layers (see Figure 9.17). The units in the input layer are neurons that receive input information, elaborate them, and send them to the hidden layer(s). Hidden layer(s), in turn, further elaborates the information to finally send them to output layer units that will provide FF-ANN output. Without going into the mathematical details of how neurons exchange information each other [151], it is interesting underlying that FF-ANN becomes operative (i.e., can correctly predict output values on the basis of determined set of input data information) after a proper training called learning step. Briefly,

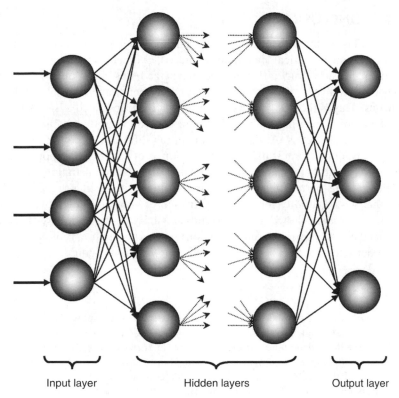

Input layer   Hidden layers   Output layer

FIGURE 9.17 Typical neural network structure.

FF-ANN is presented by many input/output sets represented by, for example, drug properties (MW and $P_{o/w}$; input data) and the corresponding P_{skin} (output data). In so doing, FF-ANN learns the relation between input and output data. Accordingly, when a new drug, characterized by new properties, is presented to FF-ANN, it should be able to yield a reasonable P_{skin} value. Obviously, prediction reliability is strongly dependent on the quality and reliability of the learning step. One of the most important advantages of ANN consists in establishing a relation between input and output data without the necessity of knowing the exact mechanisms of the process leading to input data transformation into output data. For example, Agatonovic-Kustrin et al. [152] use ANN to establish a quantitative drug structure–permeability relationship across polydimethylsiloxane membranes used as human skin surrogates. Lim et al. [153] use ANN (4–4-1 FF structure) to predict human P_{skin} starting from three-dimensional drug molecular structure properties (dipole moment, polarizability, sum (N_2O), sum(H)). Interestingly, in this case, FF-ANN proves to be much superior to the conventional multiple linear regression approach in terms of root mean squares errors.

9.4 CONCLUSIONS

Drug diffusion through membranes is an important phenomenon both for synthetic and biological membranes. Indeed, synthetic membranes can be used to measure drug diffusion coefficient and this turns out to be very important in all the delivery systems based on membrane controlled release kinetics. This is the case, for example, of microencapsulated systems where synthetic membrane permeability is one of the most important parameters for ruling release kinetics. At the same time, drug diffusion in biological membranes, specifically intestinal mucosa and skin, is of paramount importance for the wide practical application of oral and transdermal administration routes. In addition, drug permeability is one of the most important drug properties together with solubility for the determination of drug bioavailability. In this frame, mathematical modeling collocates as a unifying methodology aimed to extract the basic principles ruling drug diffusion. This, in turn, is of paramount importance for a correct and reliable delivery systems designing.

REFERENCES

1. Flynn, G.L., Yalkowsky, S.H., Roseman, T.J., Mass transport phenomena and models: theoretical concepts, *J. Pharm. Sci.*, 63, 479, 1974.
2. Grassi, M., Membranes in drug delivery, in *Handbook of Membrane Separations: Chemical, Pharmaceutical, and Biotechnological Applications*, Sastre, A.M., Pabby, A.K., Rizvi, S.S.H., Eds., Marcell Dekker, 2007.
3. Rubino, A., Field, M., Shwachman, H., Intestinal transport of amino acid residues of dipeptides, *J. Biol. Chem.*, 246, 3542, 1971.
4. Grassi, M., Grassi, G., Mathematical modelling and controlled drug delivery: matrix systems, *Curr. Drug Deliv.*, 2, 97, 2005.
5. Husseini, G.A., et al., Ultrasonic release of doxorubicin from Pluronic P105 micelles stabilized with an interpenetrating network of N,N-diethylacrylamide, *J. Contr. Rel.*, 83, 303, 2002.
6. Kurisawa, M., Yui, N., Dual-stimuli-responsive drug release from interpenetrating polymer network-structured hydrogels of gelatin and dextran, *J. Contr. Rel.*, 54, 191, 1998.
7. Robinson, J.R., Lee, V.H.L., *Controlled Drug Delivery*, Marcel Dekker, New York, Basel, 1987.
8. Desai, T.A., et al., Microfabricated immunoisolating biocapsules, *Biotechnol. Bioeng.*, 57, 118, 1998.
9. Leoni, L., Desai, T.A., Micromachined biocapsules for cell-based sensing and delivery, *Adv. Drug Deliv.*, 56, 211, 2004.
10. Järvinen, K., et al., Drug release from pH and ionic strength responsive poly(acrylic acid) grafted poly(vinylidenefluoride) membrane bags *in vitro*, *Pharm. Res.*, 15, 802, 1998.
11. Iyer, M., et al., Prediction blood–brain barrier partitioning of organic molecules using membrane-interaction QSAR analysis, *Pharm. Res.*, 19, 1611, 2002.

12. Geinoz, S., et al., Quantitative structure–permeation relationship for solute transport across silicone membranes, *Pharm. Res.*, 19, 1622, 2002.
13. Taillardt-Bertschinger, A., et al., Molecular factors influencing retention on immobilised artificial membranes (IAM) compared to partitioning in liposomes and *n*-octanol, *Pharm. Res.*, 19, 729, 2002.
14. Westrin, B.A., Axelsson, A., Zacchi, G., Diffusion measurement in gels, *J. Contr. Rel.*, 30, 189, 1994.
15. Johansson, L., Löfroth, J.E., Diffusion and interactions in gels and solution. I. Method. *J. Colloid Int. Sci.*, 142, 116, 1991.
16. Inoue, S.K., Guenther, R.B., Hoag, S.W., Algorithm to determine diffusion and mass transfer coefficients, in *Proceedings of the Conference on Advances in Controlled Delivery*, 145, 1996.
17. Colombo, I., et al., Determination of the drug diffusion coefficient in swollen hydrogel polymeric matrices by means of the inverse sectioning method, *J. Contr. Rel.*, 47, 305, 1997.
18. Harland, R.S., Peppas, N.A., On the accurate experimental determination of drug diffusion coefficients in polymers, *S.T.P. Pharm. Sci.*, 3, 357, 1993.
19. Tojo, K., et al., Characterization of a membrane permeation system for controlled delivery studies, *AIChE J.*, 31, 741, 1985.
20. Grassi, M., et al., Effect of milling time on release kinetics from co-ground drug polymer systems, in *Proceedings of the 2003 AAPS Annual Meeting and Exposition,* # M1201, Grassi, M., Magarotto, N.L., Ceschia, D., Eds., 2003.
21. Barrie, J.A., et al., Diffusion and solution of gases in composite rubber membranes, *Trans. Faraday Soc.*, 59, 869, 1962.
22. Banakar, U.V., *Pharmaceutical Dissolution Testing*, Marcel Dekker, 1992.
23. Patankar, S.V., *Numerical Heat Transfer and Fluid Flow*, Hemisphere Publishing, 1990.
24. Grassi, M., Colombo, I., Mathematical modelling of drug permeation through a swollen membrane, *J. Contr. Rel.*, 59, 343, 1999.
25. Grassi, G., et al., Propaedeutic study for the delivery of nucleic acid based molecules from PLGA microparticles and stearic acid nanoparticles, *Int. J. Nanomed.*, 2006, in press.
26. Grassi, M., Colombo, I., Lapasin, R., Experimental determination of the theophylline diffusion coefficient in swollen sodium-alginate membranes, *J. Contr. Rel.*, 76, 93, 2001.
27. Coviello, T., et al., Novel hydrogel system from scleroglucan: synthesis and characterization, *J. Contr. Rel.*, 60, 367, 1999.
28. Lapasin, R., Pricl, S., *Rheology of Industrial Polysaccharides, Theory and Applications*, Chapman & Hall, London, 1995.
29. Crescenzi, V., Gamini, A., Paradossi, G., Solution properties of a new polyelectrolyte derived from the polysaccharide scleroglucan, *Carbohydr. Polym.*, 3, 273, 1983.
30. Kas, H.S., Öner, L., Microencapsulation using coacervation/phase separation, in *Handbook of Pharmaceutical Controlled Release Technology*, Wise, D.L., Ed., Marcel Dekker, Inc., New York, Basel, 2000.
31. Laghoueg, N., et al., Oral polymer–drug devices with a core and an erodible shell for constant drug delivery, *Int. J. Pharm.*, 50, 133, 1998.

32. Gazzaniga, A., et al., On the release mechanism from coated swellable minimatrices, *Int. J. Pharm.*, 91, 167, 1993.
33. Lavasanifar, A., et al., Microencapsulation of theophylline using ethilcellulose: *in vitro* drug release and kinetic modeling, *J. Microencapsul.*, 14, 91, 1997.
34. Liu, H., et al., Spherical dosage form with a core and shell. Experiments and modeling, *Int. J. Pharm.*, 45, 217, 1998.
35. Lorenzo-Lamosa, M.L., et al., Design of microencapsulated chitosan microspheres for colonic drug delivery, *J. Contr. Rel.*, 52, 109, 1998.
36. Freitas, S., Merkle, H.P., Gander, B., Microencapsulation by solvent extraction/evaporation: reviewing the state of the art of microsphere preparation process technology, *J. Contr. Rel.*, 102, 313, 2005.
37. Johansen, P., et al., Revisiting PLA/PLGA microspheres: an analysis of their potential in parenteral vaccination, *Eur. J. Pharm. Biopharm.*, 50, 129, 2000.
38. Hanes, J., Cleland, J.L., Langer, R., New advances in microsphere-based single-dose vaccines, *Adv. Drug Deliv. Rev.*, 28, 97, 1997.
39. Walter, E., et al., Hydrophilic poly(DL-lactide-co-glycolide) microspheres for the delivery of DNA to human-derived macrophages and dendritic cells, *J. Contr. Rel.*, 76, 149, 2001.
40. Faraasen, S., et al., Ligand-specific targeting of microspheres to phagocytes by surface modification with poly(L-lysine)-grafted poly(ethylene glycol) conjugate, *Pharm. Res.*, 20, 237, 2003.
41. Lu, B., Zhang, J.Q., Yang, H., Lung-targeting microspheres of carboplatin, *Int. J. Pharm.*, 265, 1, 2003.
42. Shive, M.S., Anderson, J.M., Biodegradation and biocompatibility of PLA and PLGA microspheres, *Adv. Drug Deliv. Rev.*, 28, 5, 1997.
43. Cleek, R.L., et al., Microparticles of poly(DL-lactic-co-glycolic acid)/poly (ethylene glycol) blends for controlled drug delivery, *J. Contr. Rel.*, 48, 259, 1997.
44. Kato, Y., Onishi, H., Machida, Y., Application of chitin and chitosan derivatives in the pharmaceutical field, *Curr. Pharm. Biotechnol.*, 4, 303, 2003.
45. Reithmeier, H., Herrmann, J., Gopferich, A., Lipid microparticles as a parenteral controlled release device for peptides, *J. Contr. Rel.*, 73, 339, 2001.
46. Borgquist, P., Zackrisson, G., Axelsson, A., Simulation and parametric study of single pellet release, in *Proceedings of the International Symposium on Controlled Release of Bioactive Materials*, 736, 2001.
47. Cuña, M., et al., Microencapsulated lipid cores for peptide colonic delivery, in *Proceedings of the International Symposium on Controlled Release of Bioactive Materials*, 500, 2000.
48. Ransone, C.M., et al., A novel microencapsulation process for sustained delivery of proteins, in *Proceedings of the International Symposium on Controlled Release of Bioactive Materials*, 603, 1999.
49. Lee, S.J., Rosenberg, M., Preparation and properties of glutaraldehyde cross-linked whey protein-based microcapsules containing theophylline, *J. Contr. Rel.*, 61, 123, 1999.
50. Capan, Y., et al., Influence of formulation parameters on the characteristics of poly(D,L-lactide-co-glycolide) microspheres containing poly(L-lysine) complexed plasmid DNA, *J. Contr. Rel.*, 60, 279, 1999.

51. Sturesson, C., et al., Encapsulation of rotavirus into poly(lactide-co-glycolide) microspheres, *J. Contr. Rel.*, 59, 377, 1999.

52. Rafati, H., et al., Protein-loaded poly(DL-lactide-co-glycolide) microparticles for oral administration: formulation, structural and release characteristics, *J. Contr. Rel*, 43, 89, 1997.

53. Wang, G., et al., A porous microcapsule membrane with straight pores for the immobilization of microbial cells, *J. Membr. Sci.*, 252, 279, 2005.

54. Johansen, P., Merkle, H.P., Gander, B., Technological considerations related to the up-scaling of protein microencapsulation by spray-drying, *Eur. J. Pharm. Biopharm.*, 50, 413, 2000.

55. Thomasin, C., et al., A contribution to overcoming the problem of residual solvents in biodegradable microspheres prepared by coacervation, *Eur. J. Pharm. Biopharm.*, 42, 16, 1996.

56. Falk, R.F., Randolph, T.W., Process variable implications for residual solvent removal and polymer morphology in the formation of gentamycin-loaded poly (L-lactide) microparticles, *Pharm. Res.*, 15, 1233, 1998.

57. Jung, J., Perrut, M., Particle design using supercritical fluids: literature and patent survey, *J. Supercrit. Fluids*, 20, 179, 2001.

58. Demchuk, I.A., Nagurskii, O.A., Gumnitskii, Ya., M., Mass transfer from a solid spherical particle coated with an insoluble polymeric capsule, *Theor. Found. Chem. Eng.*, 31, 339, 1997.

59. Lu, S.M., Chen, S.R., Mathematical analysis of drug release from a coated particle, *J. Contr. Rel.*, 23, 105, 1993.

60. Narasimhan, B., Langer, R., Zero-order release of micro- and macromolecules from polymeric devices: the role of the burst effect, *J. Contr. Rel.*, 47, 13, 1997.

61. Wang, C.Y., et al., Asymmetric membrane capsules for delivery of poorly water-soluble drugs by osmotic effects, *Int. J. Pharm.*, 297, 89, 2005.

62. Narasimhan, B., Accurate models in controlled drug delivery systems, in *Handbook of Pharmaceutical Controlled Release Technology*, Wise, D.L., Ed., Marcel Dekker, 2000.

63. Bird, B.R., Stewart, W.E., Lightfoot, E.N., *Transport Phenomena*, John Wiley & Sons Inc., Press, New York–London, 1960.

64. Borgquist, P., et al., Simulation and parametric study of a film-coated controlled-release pharmaceutical, *J. Contr. Rel.*, 80, 229, 2002.

65. Sirotti, C., Colombo, I., Grassi, M., Modelling of drug-release from poly-disperse microencapsulated spherical particles, *J. Microencapsul.*, 19, 603, 2002.

66. Manca, D., Rovaglio, M., Modeling the controlled release of microencapsulated drugs: theory and experimental validation, *Chem. Eng. Sci.*, 58, 1337, 2003.

67. Ito, R., Golman, B., Shinohara, K., Design of multi-layer coated particles with sigmoidal release pattern, *Chem. Eng. Sci.*, 60, 5415, 2005.

68. Muro-Suòé, N., et al., Model-based computer-aided design for controlled release of pesticides, *Comput. Chem. Eng.*, 30, 28, 2005.

69. Muro-Suòé, N., et al., Predictive property models for use in design of controlled release of pesticides, *Fluid Phase Equilibria*, 228–229, 127, 2005.

70. Armenante, P., Kirwan, D., Mass transfer to microparticles in agitated systems, *Chem. Eng. Sci.*, 44, 797, 1989.

71. Borgquist, P., et al., Simulation of the release from a multiparticulate system validated by single pellet and dose release experiments, *J. Contr. Rel.*, 97, 453, 2004.

72. Press, W.H., et al., *Numerical Recepies in FORTRAN*, second ed., Cambridge University Press, Cambridge, USA, 1992.
73. Amidon, G.L., et al., A theoretical basis for a biopharmaceutic drug classification: the correlation of *in vitro* drug product dissolution and *in vivo* bioavailability, *Pharm. Res.*, 12, 413, 1995.
74. Meriani, F., et al., *In vitro* nimesulide absorption from different formulations, *J. Pharm. Sci.*, 93, 540, 2004.
75. Fagerholm, U., Lennernäs, H., Experimental estimation of the effective unstirred water layer thickness in the human jejunum, and its importance in oral drug absorption, *Eur. J. Pharm. Sci.*, 3, 247, 1995.
76. Fagerholm, U., Johansson, M., Lennernas, H., Comparison between permeability coefficients in rat and human jejunum, *Pharm. Res.*, 13, 1336, 1996.
77. Lennernas, H., Human jejunal effective permeability and its correlation with preclinical drug absorption models, *J. Pharm. Pharmacol.*, 49, 627, 1997.
78. Balimane, P.V., Chong, S., Morrison, R.A., Current methodologies used for evaluation of intestinal permeability and absorption, *J. Pharmacol. Toxicol. Methods*, 44, 301, 2000.
79. Lipinski, C.A., et al., Experimental and computational approaches to estimate solubilità and permeabilità in drug discovery and development settings, *Adv. Drug Deliv. Rev.*, 46, 3, 2001.
80. van de Waterbeemd, H., et al., Property-based design: optimization of drug absorption and pharmacokinetics, *J. Med. Chem.*, 44, 1313, 2001.
81. Egan, W.J., Merz, K.M., Baldwin, J.J., Prediction of drug absorption using multivariate statistics, *J. Med. Chem.*, 43, 3867, 2000.
82. Palm, K., et al., Polar molecular surface properties predict the intestinal absorption of drugs in humans, *Pharm. Res.*, 14, 568, 1997.
83. Stenberg, P., et al., Prediction of the intestinal absorption of endothelin receptor antagonists using three theoretical methods of increasing complexity, *Pharm. Res.*, 16, 1520, 1999.
84. Dressman, J., Amidon, G., Fleisher, D., Absorption potential: estimating the fraction absorbed for orally administered compounds, *J. Pharm. Sci.*, 74, 588, 1985.
85. Jacobs, M.H., Some aspects of cell permeability to weak electrolytes, in Cold Spring Harbour Symp. *Quant. Biol.*, 8, 30, 1940.
86. Lawrence, X.Y., et al., Transport approaches to the biopharmaceutical design of oral drug delivery systems: prediction of intestinal absorption, *Adv. Drug Deliv. Rev.*, 19, 359, 1996.
87. Macheras, P., Symillides, M.Y., Towards a quantitative approach for the prediction of the fraction of dose absorbed using the absorption potential concept, *Biopharm. Drug Dispos.*, 10, 43, 1989.
88. Hildebrand, J., Scott, R.L., *The Solubility of Nonelectrolytes*, third ed., Reinhold, New York, 1950.
89. Hildebrand, J., Scott, R.L., *Regular Solutions*, Prentice-Hall, Englewood Cliffs, N.J, 1962.
90. Breitkreutz, J., Prediction of intestinal drug absorption properties by three-dimensional solubility parameters, *Pharm. Res.*, 15, 1370, 1998.
91. Hansen, C.M., The universalità of the solubilità parameter, *Ind. Eng. Chem. Res.*, 8, 2, 1969.

92. Fedors, R.F., A method for estimating both the solubility parameters and molar volumes of liquids, *Polym. Eng. Sci.*, 14, 147, 1974.
93. Van Krevelen, D.W., Hoftyzer, P.J., *Properties of Polymers. Their Estimation and Correlation with Chemical Structures*, Elsevier, Amsterdam, 1976.
94. Bagley, E.B., Nelson, T.P., Scagliano, J.M., Three-dimensional solubilità parameters and their relationship to internal pressure measurements in polar and hydrogen bonding solvents, *J. Paint Technol.*, 43, 35, 1971.
95. Wilson, T.H., Wiseman, G., The use of sacs of everted small intestine for the study of the transference of substances from the mucosal to the serosal surface, *J. Physiol.*, 123, 116, 1954.
96. Ussing, H., Zerahn, K., Active transport of sodium as a source of electric current in the short-circuited isolated frog skin, *Acta Physiol. Scandinavica.*, 23, 110, 1951.
97. Hopfer, U., et al., Glucose transport in isolated brush border membrane from rat small intestine, *J. Biol. Chem.*, 248, 25, 1973.
98. Hillgren, K.M., Kato, A., Borchardt, R.T., *In vitro* systems for studying intestinal drug absorption, *Med. Res. Rev.*, 15, 83, 1995.
99. Artursson, P., Palm, K., Luthman, K., Caco-2 monolayers in experimental and theoretical predictions of drug transport, *Adv. Drug Deliv. Rev.*, 22, 67, 1996.
100. Walter, E., Kissel, T., Heterogeneity in the human intestinal cell line Caco-2 leads to differences in transepithelial transport, *Eur. J. Pharm. Sci.*, 3, 215, 1995.
101. Khanvilkar, K., Donovan, M.D., Flanagan, D.R., Drug transfer through mucus, *Adv. Drug Deliv. Rev.*, 48, 173, 2001.
102. Lennernas, H., Human intestinal permeability, *J. Pharm. Sci.*, 87, 403, 1998.
103. Schurgers, N., et al., Comparison of four experimental techniques for studying drug absorption kinetics in the anesthetized rat *in situ*, *J. Pharm. Sci.*, 75, 117, 1986.
104. Amidon, G.E., et al., Predicted absorption rates with simultaneous bulk fluid flow in the intestinal tract, *J. Theor. Biol.*, 89, 195, 1981.
105. Komiya, I., et al., Quantitative mechanistic studies in simultaneous fluid flow and intestinal absorption using steroids as model solutes, *Int. J. Pharm.*, 4, 249, 1980.
106. Tsuji, A., et al., GI absorption of beta-lactam antibiotics II: Deviation from pH-partition hypothesis in penicillin absorption through *in situ* and *in vitro* lipoidal barriers, *J. Pharm. Sci.*, 67, 1705, 1978.
107. Schurgers, N., DeBlaey, C., Effect of pH, buffer concentration and buffer composition on the absorption of theophylline from the small intestine of the rat, *Int. J. Pharm.*, 19, 283, 1984.
108. Doluisio, J.T., et al., Drug absorption. I. An *in situ* rat gut technique yielding realistic absorption rates, *J. Pharm. Sci.*, 58, 1196, 1969.
109. Dressman, J., Comparison of canine and human gastrointestinal physiology, *Pharm. Res.*, 3, 123, 1986.
110. Kararli, T.T., Comparison of the gastrointestinal anatomy, physiology, and biochemistry of humans and commonly used laboratory animals, *Biopharm. Drug Dispos.*, 16, 351, 1995.
111. Lin, J.H., Species similarities and differences in pharmacokinetics, *Drug Metab. Dispos.*, 23, 1008, 1995.

112. Obata, K., et al., Prediction of oral drug absorption in humans by theoretical passive absorption model, *Int. J. Pharm.*, 293, 183, 2005.
113. Lennernäs, H., et al., Regional jejunal perfusion, a new *in vivo* approach to study oral drug absorption in man, *Pharm. Res.*, 9, 1243, 1992.
114. Yu, L.X., An integrated model for determining causes of poor oral drug absorption, *Pharm. Res.*, 16, 1883, 1999.
115. Takamatsu, N., et al., Human intestinal permeability of piroxicam, propanolol, phenylalanine, and PEG 400 determined by jejunal perfusion, *Pharm. Res.*, 14, 1127, 1997.
116. Grassi, M., et al., Permeation of acyclovir through the intestinal mucosa of the rat: experiments with an *in vitro* mathematical model, *Boll. Chim. Farm.*, 138, 121, 1999.
117. Macheras, P., Argyrakis, P., Gastrointestinal drug absorption: is it time to consider heterogeneity as well as homogeneity? *Pharm. Res.*, 14, 842, 1997.
118. Kalampokis, A., Argyrakis, P., Macheras, P., Heterogeneous tube model for the study of small intestinal transit flow, *Pharm. Res.*, 16, 87, 1999.
119. Havlin, S., Molecular diffusion and reactions, in *The Fractal Approach to Heterogeneous Chemistry*, Avnir, D., Ed., John Wiley, 1989.
120. Amidon, G.L., et al., Analysis of models for determining intestinal wall permeabilities, *J. Pharm. Sci.*, 69, 1369, 1980.
121. Johnson, D.A., Amidon, G.L., Determination of intrinsic membrane transport parameters from perfused intestine experiments: a boundary layer approach to estimating the aqueous and unbiased membrane permeabilities, *J. Theor. Biol.*, 131, 93, 1988.
122. Kou, J.H., Fleisher, D., Amidon, G.L., Calculation of the aqueous diffusion layer resistance for absorption in a tube: application to intestinal membrane permeability determination, *Pharm. Res.*, 8, 290, 1991.
123. Tsuji, A., Tamai, I., Carrier-mediated intestinal transport of drugs, *Pharm. Res.*, 13, 963, 1996.
124. Ito, K., Kusuhara, H., Sugiyama, Y., Effects of intestinal CYP3A4 and P-glycoprotein on oral drug absorption—theoretical approach, *Pharm. Res.*, 16, 225, 1999.
125. Lu, H.H., et al., Intestinal water and solute absorption studies: comparison of *in situ* perfusion with chronic isolated loops in rats, *Pharm. Res.*, 9, 894, 1992.
126. Yuasa, H., Matsuda, K., Watanabe, J., Influence of anesthetic regimens on intestinal absorption in rats, *Pharm. Res.*, 10, 884, 1993.
127. Demidovic, B., in *Esercizi e Problemi di Analisi Matematica*, Editori Riuniti, 1975, Chapter II.
128. Sutton, S.C., Rinaldi, M.T.S., Comparison of the gravimetric, phenol red and 14C-PEG-3350 methods to determine water absorption in the rat single-pass intestinal perfusion model, *AAPS Pharm. Sci.*, 3, 1, 2001.
129. Nook, T., Doelker, E., Burri, P., Intestinal absorption kinetics of various model drugs in relation to partition coefficients, *Int. J. Pharm.*, 43, 119, 1988.
130. Lapidus, L., Pinder, G.F., *Numerical Solution of Partial Differential Equations in Science Engineering*, Wiley, New York, 1981.
131. Grassi, M., Cadelli, G., Theoretical considerations on the *in vivo* intestinal permeability determination by means of the single pass and recirculating techniques, *Int. J. Pharm.*, 229, 95, 2001.

132. Meidan, V.M., Bonner, M.C., Michniak, B.B., Transfollicular drug delivery—is it a reality? *Int. J. Pharm.*, 306, 1, 2005.
133. Langer, R., Transdermal drug delivery: past progress, current status, and future prospects, *Adv. Drug Deliv. Rev.*, 56, 557, 2004.
134. Prausnitz, M.R., Mitragotri, S., Langer, R., Current status and future potential of transdermal drug delivery, *Nat. Rev. Drug Discov.*, 3, 115, 2004.
135. Ouriemchi, E.M., Vergnaud, J.M., Processes of drug transfer with three different polymeric systems with transdermal drug delivery, *Comput. Theor. Polym. Sci.*, 10, 391, 2000.
136. He, N., et al., Model analysis of flux enhancement across hairless mouse skin induced by chemical permeation enhancers, *Int. J. Pharm.*, 297, 9, 2005.
137. Cross, S.E., Roberts, M.S., Physical enhancement of transdermal drug application: is delivery technology keeping up with pharmaceutical development? *Curr. Drug Deliv.*, 1, 81, 2004.
138. Michaels, A.S., Chandrasekaran, S.K., Shaw, J.E., Drug permeation through human skin, theory and *in vitro* experimental measurement, *AlChe. J.*, 21, 985, 1975.
139. Hadgraft, J., Lane, M.E., Skin permeation: the years of enlightenment, *Int. J. Pharm.*, 305, 2, 2005.
140. Franz, T.J., Percutaneous absorption on the relevance of *in vitro* data, *J. Invest. Dermatol.*, 64, 190, 1975.
141. Franz, T.J., The finite dose technique as a valid *in vitro* model for the study of percutaneous absorption in man, *Curr. Probl. Dermatol.*, 7, 58, 1978.
142. Higuchi, T., Rate of release of medicaments from ointment bases containing drugs in suspension, *J. Pharm. Sci.*, 50, 874, 1961.
143. Yamashita, F., Hashida, M., Mechanistic and empirical modeling of skin permeation of drugs, *Adv. Drug Deliv. Rev.*, 55, 1185, 2003.
144. Kalia, Y.N., Guy, R.H., Modeling transdermal drug release, *Adv. Drug Deliv. Rev.*, 48, 159, 2001.
145. Illel, B., et al., Follicles play an important role in percutaneous absorption, *J. Pharm. Sci.*, 80, 424, 1991.
146. Coceani, N., Colombo, I., Grassi, M., Acyclovir permeation through rat skin: mathematical modelling and *in vitro* experiments, *Int. J. Pharm.*, 254, 197, 2003.
147. Fernandes, M., Simon, L., Loney, N.W., Mathematical modeling of transdermal drug-delivery systems: analysis and applications, *J. Membr. Sci.*, 256, 184, 2005.
148. Manitz, R., et al., On mathematical modeling of dermal and transdermal drug delivery, *J. Pharm. Sci.*, 87, 873, 1998.
149. Potts, R.O., Guy, R.H., Predicting skin permeability, *Pharm. Res.*, 9, 663, 1992.
150. Yamashita, F., Takayama, K., Artificial neural network modeling for pharmaceutical research, *Adv. Drug Deliv. Rev.*, 55, 1117, 2003.
151. Baughman, D.R., Liu, Y.A., *Neural Networks in Bioprocessing and Chemical Engineering*, Academic Press, San Diego, New York, 1995.
152. Agatonovic-Kustrin, S., Beresford, R., Yusof, A.P., ANN modeling of the penetration across a polydimethylsiloxane membrane from theoretically derived molecular descriptors, *J. Pharm. Biomed. Anal.*, 26, 241, 2001.
153. Lim, C.W., et al., Prediction of human skin permeability using a combination of molecular orbital calculations and artificial neural network, *Biol. Pharm. Bull.*, 25, 361, 2002.

Index